World Atlas of Coral Reefs

The publisher gratefully acknowledges the generous contribution to this book provided by the Moore Family Foundation.

Published in association with UNEP-WCMC by The University of California Press
University of California Press
Berkeley and Los Angeles, California
University of California Press, Ltd.
London, England

UNEP-WCMC
219 Huntingdon Road
Cambridge, CB3 0DL, UK
Tel: +44 (0) 1223 277314
Fax: +44 (0) 1223 277136
E-mail: info@unep-wcmc.org
Website: www.unep-wcmc.org

Cloth edition ISBN:
0-520-23255-0

Cataloging-in-Publication data is on file with the Library of Congress

Citation: Spalding M.D., Ravilious C. and Green E.P. (2001). *World Atlas of Coral Reefs*. Prepared at the UNEP World Conservation Monitoring Centre. University of California Press, Berkeley, USA.

Printed in Hong Kong

09 08 07 06 05 04 03 02
9 8 7 6 5 4 3

The paper used in this publication meets the minimum requirements of ANSI/NISO Z39.48-1992 (R 1997) (*Permanence of Paper*). ♾

World Atlas of Coral Reefs

Mark D. Spalding, Corinna Ravilious, and Edmund P. Green

UNIVERSITY OF CALIFORNIA PRESS
BERKELEY LOS ANGELES LONDON

World Atlas of Coral Reefs

Prepared at
The UNEP World Conservation
Monitoring Centre
219 Huntingdon Road
Cambridge CB3 0DL, UK
Website: www.unep-wcmc.org
Director: Mark Collins

Authors
Mark D. Spalding
Corinna Ravilious
Edmund P. Green

Research assistants
Sarah Carpenter
Rachel Donnelly

Space Shuttle photographic research
Julie A. Robinson
Marco Nordeloos

Photographs
Edmund P. Green and Mark D. Spalding
unless otherwise stated

Cartography
Corinna Ravilious

Layout
John Dunne

Color separations
Swaingrove

Printed in Hong Kong

A Banson Production
3 Turville Street
London, E2 7HX, UK
banson@ourplanet.com

The UNEP World Conservation Monitoring Centre provides objective, scientifically rigorous products and services that include ecosystem assessments, support for implementation of environmental agreements, regional and global biodiversity information, research on threats and impacts, and development of future scenarios for the living world.

The Centre became the biodiversity information and assessment arm of the United Nations Environment Programme in June 2000. It was founded in 1979 by IUCN and in 1988 was transformed into a joint activity of IUCN, WWF and UNEP. The financial support and guidance of these organizations in the Centre's formative years is gratefully acknowledged.

Supporting institutions

The United Nations Environment Programme is the principal United Nations body in the field of the environment. Its role is to be the leading global environmental authority that sets the global environmental agenda, that promotes the coherent implementation of the environmental dimension of sustainable development within the United Nations system and that serves as an authoritative advocate for the global environment. Its objectives include analysis of the state of the global environment and assessment of global and regional environmental trends, provision of policy advice and early warning information on environmental threats, and to catalyze and promote international cooperation and action, based on the best scientific and technical capabilities available. Website: www.unep.org

ICLARM–The World Fish Center is an international, non-profit research center working to alleviate poverty and promote food security through the sustainable development and use of aquatic resources based on environmentally sound management. The focus of ICLARM's work is in developing countries and coral reefs are one of the key resources systems studied. A major coral reef project led by ICLARM is ReefBase: A Global Database on Coral Reefs. ReefBase aims to provide data and information on coral reefs and associated shallow tropical habitats, in order to facilitate better understanding of the relationship between human activities and the status and dynamics of these environments. Over 110 institutions and individuals have contributed information and expert advice to ReefBase. Websites: www.iclarm.org and www.reefbase.org

Scientists in the Earth Sciences and Image Analysis Laboratory at the Johnson Space Center work closely with astronaut crews and manage the Earth photography by astronauts on space missions. They also facilitate public access to the imagery, with an emphasis on using astronaut photographs for scientific studies. The cataloged data and imagery is located at http://eol.jsc.nasa.gov

Aventis foundation

The Aventis Foundation, based in Strasbourg, France, was formed in 2000. The Aventis Foundation promotes projects at the interface of culture, science, business and society. The Foundation aims to select projects that are international, interdisciplinary, and looking towards the future. One of its prime aims is to identify the people who will shape tomorrow and to enable them to contribute to sustainable development through their activities in science, politics and society. Website: www.aventis-foundation.org

PADI (Professional Association of Diving Instructors) Project AWARE seeks to increase both the diving and non-diving communities' environmental awareness, to encourage responsible interactions between humans and the aquatic environment and to emphasize the diver's role in preserving the aquatic realm. Aquatic World Awareness, Responsibility and Education at www.projectaware.org

Marine Aquarium Council

The non-profit Marine Aquarium Council is an international network that brings together environment organizations, the aquarium industry, aquarium keepers (hobbyists), public aquariums, government agencies and others to ensure quality and sustainability in the collection, culture and commerce of marine ornamentals. MAC is doing this by developing an international system of certification and labeling that will: establish standards for quality products and practices; document compliance with these standards and label the results; and create consumer demand and confidence for labeled "products" from certified industry operators. Paul Holthus, Executive Director, 923 Nu'uanu Ave, Honolulu, Hawai'i 96817, USA. Tel: (1 808) 550 8217; Fax: (1 808) 550 8317; E-mail: paul.holthus@aquariumcouncil.org; Website: www.aquariumcouncil.org

The International Coral Reef Initiative is a voluntary partnership that allows representatives of over 80 countries with coral reefs to work with major donor countries and development banks, international environmental and development agencies, scientific associations, the private sector and NGOs to decide on the best strategies to conserve the world's coral reef resources. ICRI is not a permanent structure or organization, but an informal network linked by a global Secretariat. Website: http://icriforum.org/

Dulverton Trust

The Dulverton Trust is a UK grant-making charitable trust, with an interest in the field of conservation. It was founded by Lord Dulverton in 1949.

Acknowledgments

The authors would like to thank the many organizations that have lent financial or other support at various stages in the preparation of this work. These include the United Nations Environment Programme Division of Early Warning and Assessment, the Dulverton Trust, ICLARM, NASA, the Aventis Foundation, the Marine Aquarium Council and PADI. We would also like to express our gratitude to the Moore Foundation for providing support for the production costs of this book.

Both Sarah Carpenter and Rachel Donnelly provided excellent support as research assistants during the preparation of this work. The information behind these maps has been compiled over seven years and many thanks are due to all those who have helped, including Mary Edwards, Simon Blyth, Jonathan Rhind and the "placement students" Annabel Lee, Ivor Wheeldon, Alastair Grenfell, Susannah Hirsh, Joanna Hugues and Chantal Hagen. Also in Cambridge, many thanks are owed to the staff of the University Library Map Room for all their help over the years. In UNEP we are also very grateful to Dan Claasen, Salif Diop, Agneta Nilsson and Arthur Dahl for their constant support.

Colin Watkins has been an incredible support on this project, not only with fund-raising but also with his persistence, vision and optimism; thank you Colin. Thanks also to Heather Cross, Mary Cordiner and Will Rogowski at UNEP-WCMC. In the evenings and weekends, Mania Spalding and Stephen Grady have shown great patience and tolerance of the crazy hours we put in to prepare this book – many thanks indeed to both of them.

We are extremely grateful to Charlie Veron, who not only reviewed part of the text and supplied some photographs, but also provided his newly prepared coral distribution data and thoroughly reviewed our resulting species lists. Thanks to Clive Wilkinson, Bernard Salvat and Lauretta Burke for supplying data and general encouragement. Thanks to the many others in the International Coral Reef Initiative for their kind support. Jerry Kemp, Doug Perrine, Giotto Castelli and Colin Fairhurst also provided some valuable additions to the photographs.

Julie A Robinson has worked long and hard on the images from Space Shuttle and Mir, while, at an earlier stage, Marco Nordeloos spent countless hours sorting through thousands of images to select those clearly showing coral reefs. In addition to this, without the many efforts of Kamlesh P Lulla it would not have been possible for the Earth Sciences and Image Analysis Laboratory at Johnson Space Center to contribute to this book. Thanks are due to NASA astronauts for their continuing efforts to photograph coral reef areas from orbit, and to all the members of the Earth Sciences and Image Analysis Laboratory who have over the years supported astronaut photography of Earth. Particular thanks are due to the support staff and those who helped in checking the annotations. The Digital Imaging Laboratory at Johnson Space Center gave special attention to making the high-quality digital copies of film products that were the starting point for the images that appear in the book.

Thanks must also go to Lonely Planet/Pisces Books who provided a number of copies of their Diving and Snorkelling Guides.

In addition, considerable thanks are owed to the many reviewers, listed below, who have checked over large parts of the text. These have greatly improved the final quality of the texts. However, any errors which remain are solely the responsibility of the authors.

In Part I: Stephen Grady, Lucy Conway and Sarah Carpenter (Chapters 1-3), David Woodruff (primarily Chapter 1) and Paul Holthus (aquarium trade and certification).

In Part II: Jeremy Woodley (Chapters 4 and 5), Hector Reyes Bonilla (Mexico), Juan Manuel Díaz (Colombia), Hector Guzman (Honduras, Nicaragua, Costa Rica and Panama), Sheila Marques Pauls (Venezuela) and Clive Petrovic (British Virgin Islands).

In Part III: David Obura (East Africa), Nyawira Muthiga (Kenya), Chris Horrill, Martin Guard, Matthew Richmond and Jason Reubens (Tanzania), Jean Pascal Quod (Eastern Africa), Arjan Rajasuriya (South Asia), Charles Anderson (Maldives), Charles Sheppard (Chagos Archipelago), Alain de Grissac (northern Red Sea), Lyndon DeVantier (Middle Eastern reefs), Jeremy Kemp (Red Sea, southern Arabia), Rupert Ormond (northern Red Sea), Hansa Chansang (Thailand), HM Ibrahim (Malaysia), Laura David (Philippines), Vo Si Tuan (Vietnam), CF Dai (Taiwan) and Andre Jon Uychiaoco (Southeast Asia).

In Part IV: JEN Veron (Australia), Robin South (Melanesia and Polynesia), Aaron Jenkins (Papua New Guinea), Duncan Vaughan (Fiji), John Gourley (Mariana Islands), Darrin Drumm (Cook Islands) and Flinn Curren (American Samoa).

Thanks are also due to James Nybakken for his overall appraisal of the text.

Contents

Introduction

Coral reefs are one of the world's most spectacular ecosystems. They straddle the tropics and cut a broad swathe around the globe. They are clearly visible, even from space, as patterns of dazzling colors tracing the edges of coastlines and scattering far out into the oceans. Up close, the magic of coral reefs is magnified. These ecosystems are packed with the highest densities of animals to be found anywhere on the planet. Thronging with life, they rival even the tropical rainforests in terms of diversity.

From a human perspective coral reefs are not only a source of wonder and fascination. They also represent a critical resource for millions of people. For millennia coastal peoples have relied on coral reefs as a source of food. The wide strips of coral reefs lining their shores have also provided protection from the worst onslaughts of tropical storms. Over the centuries, these same reefs have actually provided the sand for the beaches and even the rocks which make up the islands on which so many people live. In more recent times coral reefs have become the treasured destination for millions who have sought peace and rest on tropical shores, or adventure, diving into the world of the coral reef. These same travelers are providing a new source of income and employment for some of the world's most impoverished nations. Into the future, reefs have the capability to provide new resources for the world's burgeoning populations, most notably with the development of new pharmaceuticals.

How little we know

It is possible, even today, to pick up the best navigational charts for certain areas and find quite shocking gaps in our knowledge. For some parts of the world, the best information about the location and dimensions of coral reefs was gathered by Captain James Cook and others in the 18th century. On some of these "modern" charts there remain dotted lines showing "possible" locations of reefs, or notes describing reefs as "position unconfirmed". While sea monsters no longer populate our maps, many of the gaps where they once sat still remain.

This lack of knowledge is not simply confined to knowing where the reefs are. Efforts to quantify the total numbers of species which are found on reefs remain largely restricted to wild extrapolations and educated guesses. As many as 100 000 species may have been named and described from coral reefs, but the total number inhabiting the world's reefs may be anything between half and 2 million, perhaps more.

In some ways this lack of knowledge is not surprising. Many reefs are remote and, as they are far from regular shipping traffic, efforts to map these areas have not been prioritized. Without good charts other navigators remain cautious about sailing in such areas. From an ecological perspective our knowledge has been further hampered by the fact that humans are terrestrial, air-breathing creatures. Early scientists could only peer down with fascination through the intervening waters which separated them from the reefs, or haul up dead or dying samples for inspection. Only in the 1950s did scuba-diving become a popular and relatively safe activity, and our scientific knowledge of the ecology of reefs has almost entirely been amassed over the last 50 years.

The World Atlas of Coral Reefs

This atlas presents a unique compendium of information. It provides a summary of what we know about the geographic distribution and status of coral reefs at the start of the third millennium. Unfortunately, even as we have begun to gather this information, the reefs themselves have been changing. The atlas also provides information on the changes which have already occurred, and on the human impacts on the coral reefs in every country.

This atlas is primarily an information resource. Putting such information together at the global level is more than a summary, however, and provides us with an entirely new perspective.

The first three chapters provide a global review of the coral reefs, firstly taking an ecological and geological perspective, then a human perspective, and finally looking more specifically at the task of mapping coral reefs. The main bulk of the book is then focussed towards a region-by-region review of coral reefs.

The most important resource in any atlas is the maps themselves. The UNEP World Conservation Monitoring Centre first commenced its global coral reef mapping work in 1994 and has now developed the most detailed global maps of coral reefs in existence. These maps show the distribution of the vast majority of the world's shallow coral reefs. Equally important with the maps in this atlas has been to place the location of coral reefs in a wider context. The maps in this book thus show major natural features (forests, rivers, topography and bathymetry), but also significant human factors, including settlements, dive centers and marine protected areas.

The texts and tables provide information which enables a more detailed interpretation of the information provided on the maps, including information which cannot directly be shown on the maps themselves. For all countries and territories where there are reefs, basic information is provided describing the distribution of the reefs and some ecological features. Human uses and impacts on coral reefs are further considered, including efforts to control such impacts or protect coral reefs. Data tables list all the protected areas with coral reefs, but also provide directly comparable information describing the countries, their reefs, and the human impacts on these.

Users of this book can learn about the location and status of coral reefs around the world. Those traveling regularly to coral reef areas, for leisure or for work, can use the *World Atlas of Coral Reefs* to learn about new areas before they visit, to get a basic grounding in the ecology of coral reefs, and to consider the issues and challenges facing reefs in particular areas. Experts from particular locations, or in particular subjects, can learn about other areas, and gain useful information about different parts of the world.

A considerable amount of information held within the pages of this atlas has never been published before.

■ The work includes a new, revised global estimate of the total area of coral reefs worldwide. In Chapter 1 it is estimated that shallow coral reefs worldwide occupy some 284 300 square kilometers, an area about half the size of Madagascar. This is less than 1.2 percent of the world's continental shelf area, and only 0.09 percent of the total area of the world's oceans. Coral reefs are a scarce, but critically important resource.

■ An assessment of the area of coral reefs in individual countries provides an important perspective on the ownership and responsibilities associated with this critical heritage. Indonesia is the largest coral reef country in the world, followed by Australia and the Philippines. Also high up the list are many small nations in terms of land area: Papua New Guinea, Fiji, the Maldives, the Marshall Islands, Solomon Islands, Bahamas and Cuba.

■ The same statistics also point to the important role which a number of the world's very wealthy nations could play in protecting the world's coral reefs. Australia, France, the UK, the USA and even New Zealand hold jurisdiction or significant influence over coral reefs in their own waters and in the waters of their overseas territories and associated states. Together these cover over one quarter of the world's coral reefs.

■ Using information from the new taxonomic work *Corals of the World*, JEN Veron has provided the very latest information on coral biodiversity around the world. National statistics have been calculated for all countries and clearly illustrate the critical heritage which is currently being threatened by human activities. The most diverse region of the world for coral reefs is centered on the Philippines, Indonesia, Malaysia and Papua New Guinea, with between 500 and 600 species of coral in each of these countries. Unfortunately these are also some of the most threatened coral reefs in the world.

■ Reef tourism is now a major global industry. Visitors to the Great Barrier Reef increased from 1.1 million in 1985 to over 10 million in 1995. Scuba diving is probably the most popular adventure sport in the world, and vast numbers of scuba divers visit coral reefs every year. A new database has been gathered which gives the location of dive centers around the world. This contains information on over 2 000 dive centers, marked on the maps throughout this work. They show, quite clearly, that diving tourism is now ubiquitous, and is located in 91 countries and states.

■ Marine protected areas are becoming a critical tool for the protection of coral reefs worldwide. They are being widely used, not only for conservation, but also to enhance fish catches, by protecting small stocks of fish which are able to resupply adjacent areas. There are now some 660 marine protected areas worldwide which incorporate coral reefs. These include two of the world's largest protected areas, Australia's Great Barrier Reef and the northwestern Hawaiian Islands, covering entire large ecosystems.

■ Unfortunately, many protected areas exist on paper only – they are poorly managed and have little or no support or enforcement. Equally worrying is that in almost every single case, protected areas are aimed solely at controlling the direct impacts of humans on coral reefs. Fishing and tourist activities may be controlled, but the more remote sources of threats to reefs, notably pollution and sedimentation from the adjacent land, continue unabated. Without a more concerted effort to control all of the impacts of humans on coral reefs even the best managed marine protected areas may be managed in vain.

■ There are other stories, however, which provide valuable examples of success. Fisheries reserves in a few areas are now revitalizing the food supplies and economies of local villages, while tourist income is paying for the wise management of a number of important areas. It is vital that the messages from these sites are carried as swiftly as possible to all countries and communities who depend on coral reefs.

Aside from such clear statistics, the pages of this atlas reveal a startling, recurring tale of degradation and loss.

■ Corals are extremely sensitive to increases in temperature, exhibiting a stress response known as coral bleaching. Records of such bleaching have increased considerably in recent years, and in 1998 a global mass bleaching event occurred, with devastating mass mortalities of corals in many areas. Recovery is now underway, but there are very real concerns about the recurrence of such events with global climate change.

■ In the Caribbean apparently natural damage from disease and hurricanes has been exacerbated by the impacts of human activities, and reefs have lost coral cover and diversity in almost every country, even in many apparently remote and protected locations.

■ In Southeast Asia burgeoning populations and rising living standards are placing untenable pressures on the coral reefs, and many are succumbing, no longer able to provide the fish and other resources which have supported coastal populations for generations.

■ Even the more remote reefs worldwide are not secure. In the past, remote atolls in the Pacific have been used for testing nuclear weapons and for dumping waste, and even today a number are still used for military target practice. More widespread has been the impact of fisheries. In many places traditional management and restraint has enabled sustainable use of fish resources, but such traditional systems are breaking down in some areas, while better transport and high prices are driving stocks of some target species towards complete disappearance, even in quite remote locations.

The problems facing the world's reefs

Natural changes are a part of any ecosystem, and we are still at the early stages of understanding the natural dynamics of coral reefs. However, the 20th century saw the near exponential growth of human populations, combined with even more rapid increases in consumer demands being placed on the planet's limited resources, and such trends are set to continue through the 21st century. Humans are thus bringing new pressures to bear on the world's coral reefs and driving more profound changes, more rapidly, than any natural impact has ever done. Overfishing has become so widespread that there are few, if any, reefs in the world which are not threatened. This, combined with such destructive practices as blast fishing, is shifting the patterns and balances of life in many reef ecosystems. From onshore a much greater suite of damaging activities is taking place. Often remote from reefs, deforestation, urban development and intensive agriculture are now producing vast quantities of sediments and pollutants which are pouring into the sea and rapidly degrading coral reefs in close proximity to many shores.

The impacts of these activities affect not only the reefs, but also the many millions of coastal peoples who depend upon them for sustenance and income. In many areas these changes are so rapid that we are unable to document the existence of reefs before they are degraded. We have no idea how much has already gone.

A further specter overshadowing the world of coral reefs is that of global climate change. It is now universally accepted that the global climate is changing at an accelerated rate as a result of human activities. Coral reefs, it would appear, are among the most vulnerable ecosystems to rising sea surface temperatures. Coming on top of the other threats already mentioned, it seems highly probable that the predicted rises in sea surface temperatures over the next century may well cause the total demise of at least some of these critical, valuable and beautiful ecosystems.

Faint glimmers of hope

As our knowledge and our concern about coral reefs is increasing, so are the efforts to redress the problems. Overfishing is a worldwide problem, and its most damaging impact is on the fishing communities themselves. Thankfully, examples are now cropping up around the globe of successful management efforts which can remedy the problem. By setting aside small areas as "no-take" zones, local communities are finding that there are enormous benefits. Fish stocks build up in these zones and spill over into the surrounding area such that the overall yield of fish from the wider area is increased. Everyone benefits.

Tourism has caused considerable damage, through the unplanned coastal development and pollution which are so often linked to it. The sewage systems of many hotels empty directly into the waters where the guests swim, and the damage to reefs can be considerable. Increasing awareness, however, is leading to better controls on development and major efforts to improve sewage treatment. As such measures develop, tourism can become a force for good, giving an added value to reefs in the eyes of the local communities, and often providing a direct income, through park fees, for the management of marine protected areas.

Most importantly, our increasing understanding of the interactions between humans and reefs, and between terrestrial activities and their downstream impacts in the coastal zone, are allowing for the development of integrated planning. We are aware of the problems, and have the solutions. The challenge is to apply them.

Essential information

Key to all maps in Chapters 4 to 14, labeled a-j

Coral reef

Mangrove

Dive center

Population center

International boundary

River

Water body

Land

Forest

National marine protected area

National marine protected area (boundary unknown)

International protected area

International protected area (boundary unknown)

Bathymetry

0-200 meters

200-2 000 meters

> 2 000 meters

Space Shuttle photographs

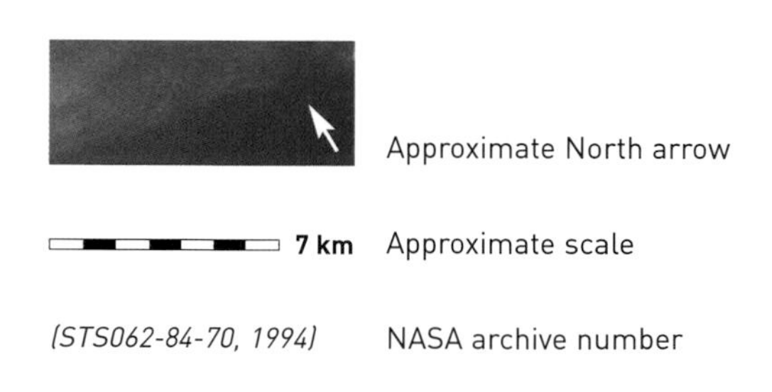

Approximate North arrow

Approximate scale

NASA archive number

Throughout this publication the use of na *indicates that no relevant information is available.*

For technical notes regarding the text, maps and data tables, see page 401.

Part I

Understanding Coral Reefs

Chapter 1
The World of Coral Reefs

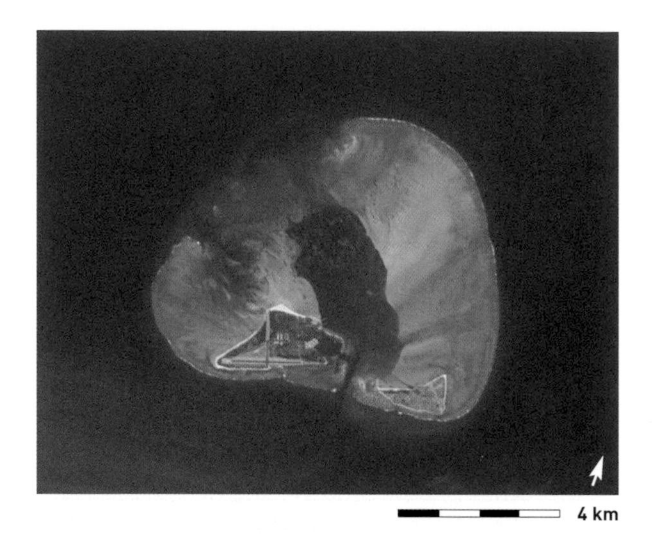

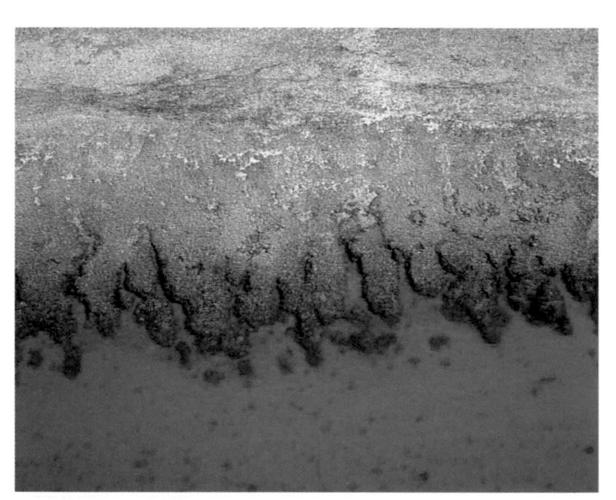

Coral reefs are among the most diverse and complex of all ecosystems; they are among the most heavily utilized and economically valuable to humankind; and they are also among the most beautiful and fascinating. In order to understand what lies behind such accolades it is important to appreciate exactly what coral reefs are, how they are formed and where they are found. Building on such a foundation it is also valuable to develop a basic understanding of some of the organisms that make up the complexity of life on coral reefs, and what role they play in maintaining these ecosystems. Such knowledge provides the basis for a wider understanding of the interactions of humans and reefs. It is also critical for understanding the changes that are now occurring on coral reefs, and for responding to such changes.

This chapter offers a simple definition and description of coral reefs. It goes on to provide an overview of their distribution, and of the organisms that make them up. It considers the elements determining these distribution patterns, from factors of geological history to present day limiting processes and the very important role of ocean currents. The chapter also looks briefly into some of the patterns of biodiversity which are observed at finer resolutions, patterns which are observed between neighboring reefs, and zonation patterns across individual reefs. Finally, the chapter provides an overview of the main organisms which make up the patterns of life on coral reefs.

Above, left: Midway Islands (STS055-82-63, 1993). Above, right: The edge of the reef, with spur and groove formations, Great Barrier Reef. Below, left: Shallow waters of an atoll lagoon. Below, right: The intricate branches of an Acropora *coral.*

Defining coral reefs

For all those who have seen one, a coral reef is relatively simple to describe. From land or from the air, reefs are usually clearly visible, marked by a complex patterning of bright colors. These arrays of blues, turquoises and greens delimit a diverse and complex physical structure coming close to the ocean surface. The shallowest points are frequently shown by the brilliant white of breaking surf, and may even briefly become dry land during the lowest tides. From underwater the complexity is still more clearly shown – reefs are typified by the presence of large stony corals growing in profusion and by an often bewildering array of species growing or moving among them. Moving across a reef, patterns or zones become apparent, each dominated by different organisms, depending on factors such as depth, shelter and water movements.

Although simplistic, such descriptions incorporate the key elements of a more thorough scientific definition of a coral reef. Coral reefs are shallow marine habitats, defined both by a physical structure and by the organisms found on them.

Corals themselves are very simple organisms. They are found in all the world's oceans, at all depths. Although described in more detail later in this chapter, typically they have a very small cylindrical body, topped with a ring of tentacles which are used to capture food from the surrounding waters. A large number of corals have developed the ability to live in colonies and to build up a communal skeleton. Among these are many species which lay down a stony skeleton of calcium carbonate. These corals are known as hermatypic or reef-building corals. They are almost entirely confined to areas of warm, shallow water, and it is their skeletons, essentially built of limestone, which are critical to the formation of coral reefs.

Even in ideal conditions, these hermatypic corals are slow growing. Some massive corals, which typically grow as large dome-shaped structures, may build up a skeleton at rates of just a few millimeters per year. The faster growing tips of branching corals may extend at rates of 150 millimeters per year or more.

Over centuries or millennia the active growth of these corals (alongside other organisms such as coralline algae, which also lay down calcium carbonate skeletons) leads to the building up of vast carbonate structures. The process is not simple, and numerous additional factors come into play. Storms frequent many areas of tropical coastlines and the waves they produce can, quite literally, pound a reef to rubble in a few hours. Over longer time scales, corals are eroded by countless organisms. Some fish bite large chunks out of them, digesting the coral tissues and algae on their surface. Unseen but equally important is a great diversity of bio-eroding organisms that burrow into or chemically dissolve the coral rock, weakening and destroying its structure. Sand and rubble from these apparently destructive activities often fill the interstices of the reef, while certain algae and other corals may then bind or overgrow such loose materials, cementing them together with more calcium carbonate to form a yet more solid structure.

In this way a coral reef is built. Only a tiny fraction of the growth of individual corals is converted into upwards development of a reef structure, and so their formation takes place over geological time scales. The most rapid periods of reef "growth" have shown upwards accumulation of reef structures reaching 9-15 meters in 1 000 years in some areas, but much lower figures are probably

Above: Individual polyps of the great star coral Montastrea cavernosa, *clearly showing the cylindrical body, with a ring of tentacles. Below: The growth of numerous corals builds up the massive physical structure of an Indian Ocean reef.*

more normal. In fact the majority of reef structures that exist today are not the result of continuous growth, but of pulses of growth interspersed with quiescent periods, or even periods of erosion, when the reefs might be defined as fossil or non-living reefs. Sea levels in the oceans have varied dramatically, particularly during the recent ice ages, and many reefs have intermittently become dry land, or have been flooded by waters too deep to allow corals to grow. Between these extremes, however, some of these fossil structures become recolonized by corals and reef development recommences.

Over shorter time scales, the division between an actively growing coral reef and a fossil reef is, in many areas, unclear. No reef is in a constant state of growth. During major tropical storm events, all reefs undergo losses in coral cover and often considerable erosion of their physical structure. Over years or decades, the extent of actively growing coral cover also varies considerably. Recently observed events, including coral disease, coral bleaching, outbreaks of the coral-feeding crown-of-thorns starfish, or the die-off of important grazers such as the long-spined sea urchin (see page 61), have all produced considerable losses of live coral cover to some reefs. Recovery from such events points to a natural resilience, but also shows that any understanding of a "reef" measured from only one particular moment in time will be limited.

Taking such points into consideration, a coral reef can thus be more rigorously defined as a physical structure which has been built up, and continues to grow over decadal time scales, as a result of the accumulation of calcium carbonate laid down by hermatypic corals and other organisms. The manner in which such structures develop has led to the recognition of a number of types of reef, while there are also many other communities which, while not as obviously covered by these definitions, are clearly related and equally important.

Types of reef

Corals can only grow in warm, well lit waters and require a solid surface on which to settle. These factors restrict the initial appearance of hermatypic corals to shallow rocky substrates in the tropics. As corals proliferate, their skeletons provide a solid substrate for the appearance and settlement of more corals and other organisms. The upward growth of a physical reef structure can also allow corals to continue to grow in shallow well lit waters, even if the basement on which they are growing subsides or sea levels rise.

Fringing reefs are perhaps the simplest structures to understand. These develop from the simple upward growth of a calcium carbonate platform from a shelving coastline. Because growth is most rapid and prolific in shallow water the corals quickly grow to the surface and produce a shallow platform which is usually around the level of the lowest tides. Further offshore growth is slower, but the typical structure of a mature fringing reef includes a shallow platform out to a sharply defined edge, the reef crest, beyond which there is a steeply shelving reef front dropping down to the sea floor.

Barrier reefs are usually older structures rising up from a deeper base at some distance from the shore, with a lagoon separating them from the coast. Some have their origins as fringing reefs on shelving coastlines, but develop when the coastline on which they are growing subsides or is flooded by rising sea levels. Under these conditions the fringing reef continues to grow upwards, but deeper waters fill in a lagoon between this structure and the coastline. In other cases barrier reefs may have simply developed in offshore locations, but still remain separated from the coast by a lagoon.

Atolls are unique reef formations, broadly circular, and enclosing a wide lagoon. They are typically found in oceanic locations, away from the continental shelf.

Figure 1.1: The main types of coral reef structure

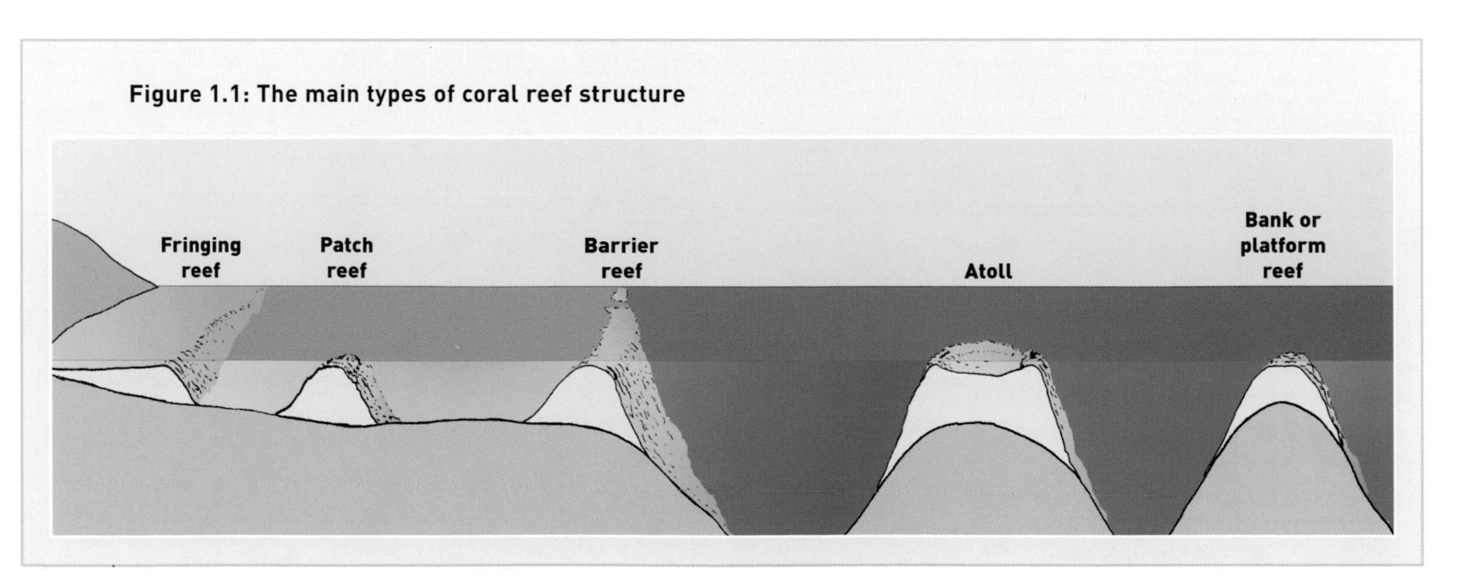

Darwin was the first to correctly understand their origin. They initially form as fringing reefs around isolated, usually volcanic, islands. Such islands then subside, but the reefs continue to grow, first forming a barrier around the sinking island, but then, as the island disappears beneath the surface, forming a single ring of coral. The depths of coral limestone which may accrue on these structures are considerable – drilling in the Marshall Islands has revealed reef deposits up to 1.4 kilometers in depth, dating back over 50 million years.

Bank or platform reefs are simple physical structures with a variety of origins. They are essentially reefs with no obvious link to a coastline, but without the clear structure of a barrier reef or atoll. In some cases they may have similar origins to either of the latter, but do not hold back or encircle a lagoon, in other cases they may have simply grown up over natural rises in the coastal shelf. Larger or slightly submerged reef structures of this type are also sometimes referred to as shoals.

Other types of reef and coral communities

These reef types can be clearly illustrated (Figure 1.1). However, the reality often reveals many other structures which do not conform quite so easily to strict definitions. Near-atolls are described in a few areas where there is a tiny remnant of the original high island in the center of an atoll ring. There are also a considerable number of atoll-like platform reefs which may not have the true geological origin of an atoll (around a subsiding volcanic island), but where the surface structure is almost exactly that of an atoll. There are also a number of structures which lie offshore in the location of a true barrier reef, but which may not quite conform to the definition or geological origin of a barrier reef. Bank barriers are commonly described in parts of the Caribbean where small banks lie at some distance offshore and sometimes do not rise all

Figure 1.2: The development of an atoll, based on Darwin's original theory

A volcanic island is colonized by corals and becomes surrounded by a fringing reef.

The island itself subsides, the corals continue to grow and a barrier reef is formed.

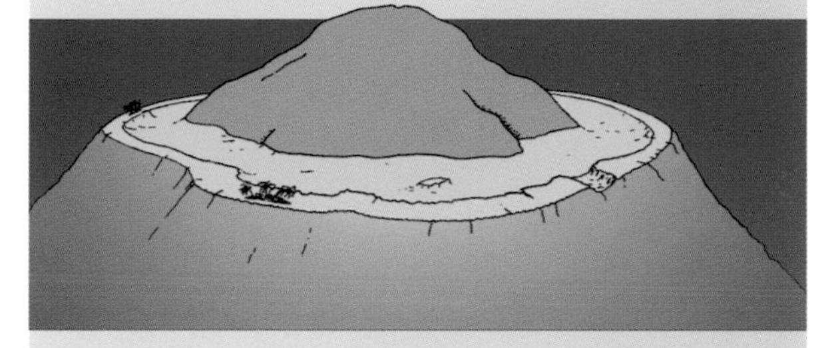

The island is lost, but coral maintains upward growth and a ring-shaped atoll is formed.

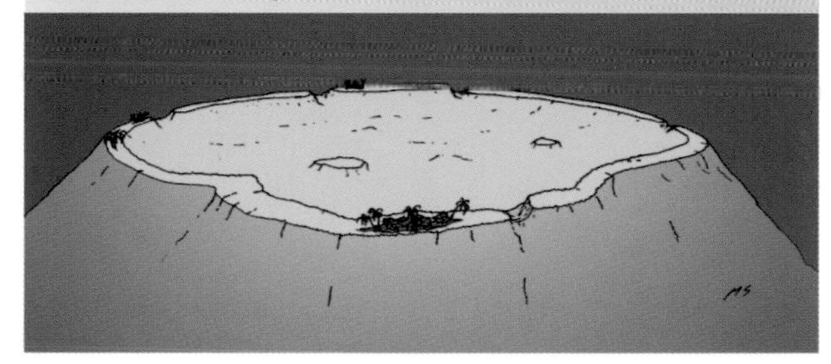

Table 1.1: Estimates of global reef area calculated from the reef maps

Region	Area (km²)	% of world total
Atlantic and Caribbean	**21 600**	**7.6**
Caribbean	20 000	7.0
Atlantic	1 600	0.6
Indo-Pacific	**261 200**	**91.9**
Red Sea and Gulf of Aden	17 400	6.1
Arabian Gulf and Arabian Sea	4 200	1.5
Indian Ocean	32 000	11.3
Southeast Asia	91 700	32.3
Pacific	115 900	40.8
Eastern Pacific	**1 600**	**0.6**
Total	**284 300**	

Figures are rounded to the nearest 100 square kilometers, and percentage figures to one decimal place. National level statistics are provided in the regional accounts later in this book. In order to avoid the problems associated with using maps prepared at multiple scales, such calculations are made by first simplifying the global coverage down to a 1 kilometer grid, each grid cell being simply marked as reef or non-reef. Reef area is then calculated as the total of 1 square kilometer cells with reef. Although this method exaggerates the total area from that actually shown on the maps, this can be justified on the grounds that the maps only show reef flat to reef crest areas, while the true reefs extend beyond these areas (see also Chapter 3).

the way to the sea surface. The long offshore reef tracts of Florida, Cuba and elsewhere rival many true barrier reefs in length, but are frequently not regarded as true barrier reefs because they are only separated from the mainland by a very shallow lagoon, or because they are not located on the edge of the continental shelf. Small physical structures, often lying within the wider formation of a barrier or atoll lagoon, are often referred to as patch reefs.

Perhaps more importantly, there are significant areas around the world where there are coral communities which perform the same ecological function as coral reefs, but lack a clear physical structure. These include recent formations where there may be a thin veneer of live coral, or they may be physical reefs, but not yet mature or clearly visible. For clarity such structures are frequently referred to as coral communities, submerged reefs, or sub-surface reefs.

Global distribution

Charles Darwin was probably the first person to prepare a global map of coral reefs. His and other efforts are described in Chapter 3. Coral reefs are restricted to a broad swathe, roughly confined to the tropics, and circling most of the globe (Map 1.1). Within this range they are far from evenly distributed, with large areas confined to remote island regions and offshore areas far from major land masses. Further investigation shows that coral reefs are largely absent from the Central Atlantic and the shores of West Africa, they are highly restricted along the western (Pacific) shores of the Americas, and are also restricted along the coastline of South Asia from Pakistan to Bangladesh.

Using the maps shown in this publication it is possible to estimate the total area of coral reefs in the world. Although there are clear limitations to such estimates, these figures are clearly valuable for getting an overall perspective on the area of coral reefs in the world, and in allowing for regional comparisons. There are an estimated 284 300 square kilometers of coral reefs worldwide[1]. This figure represents only 0.089 percent of the world's oceans and less than 1.2 percent of the world's continental shelf area. Thus, at the global scale, coral reefs are a rare habitat. Further analysis clearly shows that the great majority of coral reefs are found in the region known as the Indo-Pacific, which stretches from the Red Sea to the Central Pacific. Less than 8 percent of the world's reefs are found in the Caribbean and Atlantic.

Zooming in to these maps, new patterns emerge at finer resolutions. Reefs are often limited in their development in the nearshore waters of large continental land masses, although barrier structures are widespread in such places. They are poorly developed close to large river mouths. In contrast, they are particularly well developed around islands and along the coastlines of drier continental areas.

In order to understand these patterns of reef distribution it is necessary to look at the organisms which make up the coral reef ecosystem. The factors impinging on their evolution, dispersion, and survival are the same factors which have created the patterns in coral reef distribution that we see today.

1. The reef area figures used throughout this work are based on a new calculation, and replace the early estimate provided by Spalding and Grenfell (1997) of 255 000 square kilometers. It is likely, as mapping work continues, that such figures will continue to be refined and improved. This may lead to further upwards adjustment of the global total, although in some areas there is also likely to be some reduction of figures as maps are improved. Thus it seems unlikely that a "final" figure would exceed 300 000 square kilometers.

Map 1.1: The coral reefs of the world

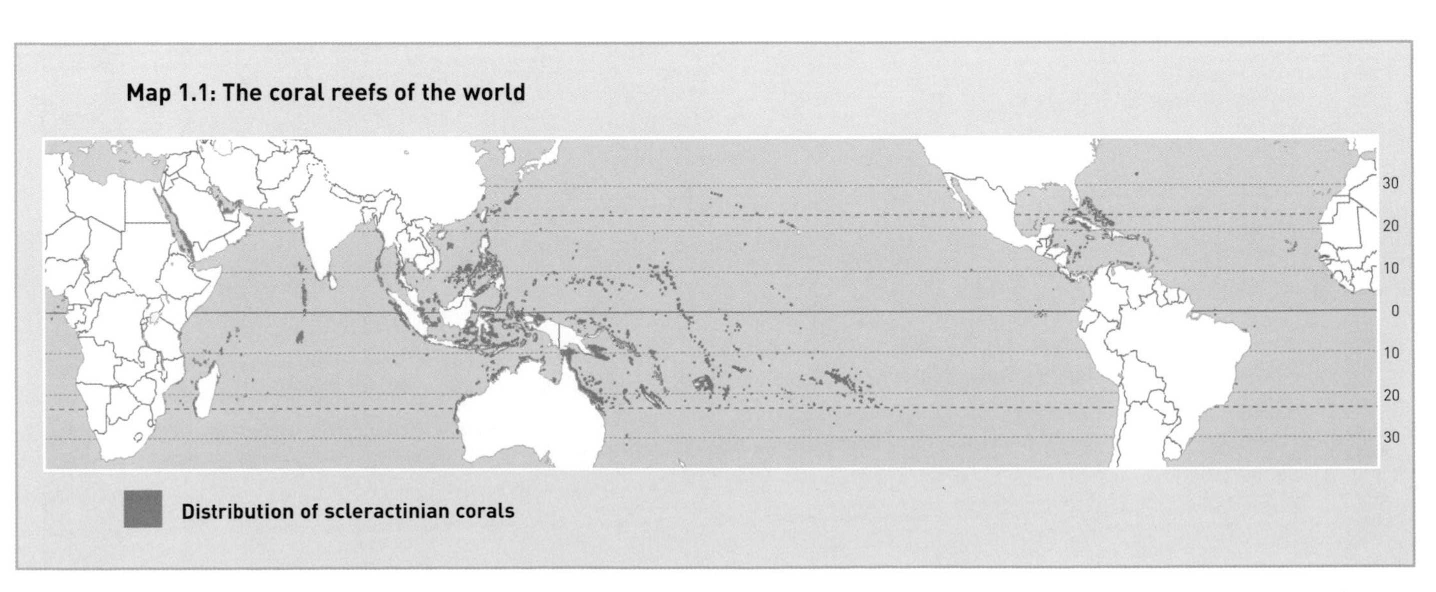

Patterns of diversity

Observations of life on coral reefs reveal a number of striking patterns in the distribution of species. At the global level, few species are ubiquitous. Some may be widespread across one or even two ocean basins, but many others are restricted to certain oceans or particular seas.

As a larger picture is built up through looking at many species, certain patterns emerge. Some regions are highly distinctive with large numbers of endemic species, found nowhere else. The total diversity of species is also uneven, with centers of particularly high diversity, and with clear gradients in diversity mirroring environmental gradients.

When looking at finer resolutions, new patterns emerge. Certain species appear to predominate in near continental reefs, while others are found on oceanic reefs. Closer still, and the position on the continental shelf, or that relative to the prevailing wind or currents, appears to hold sway in determining the species composition. At the scale of tens or hundreds of meters, patterns of zonation are observed across individual reefs, with species adapted to different depths, exposure, water circulation and so on.

Finally, at the scale of individual points or quadrats, the pattern of which species are found where seems to disappear in a random noise. Even here, however, the factors driving the settlement and survival of individuals may be far from chaotic, but driven by highly complex interactions, both in the immediate sense and over the life history of the individual.

Patterns at the global scale

Corals are clearly the most important organisms when it comes to understanding the factors that drive the distribution of coral reefs. The majority of reef-building corals fall within the group known as Scleractinia. They have been the subject of continuing studies by biologists and taxonomists for many years, and a considerable amount is now known about their distribution and about the factors which influence it. Some 794 species of scleractinian coral are considered to be reef builders, and Map 1.2 shows a plot of their distribution, highlighting the patterns of varying diversity. A number of points can be observed:

- Corals, like the reefs they build, are restricted to a narrow band of low latitudes, with diversity diminishing fairly rapidly along latitudinal clines.
- There are two distinctive regions of coral distribution, one centered around the Wider Caribbean (the Atlantic), the other reaching from East Africa and the Red Sea to the Central Pacific (the Indo-Pacific).
- Diversity is far lower in the Atlantic than in the Indo-Pacific.
- Coral diversity is at its highest around insular Southeast Asia.
- Coral diversity and reef development are very restricted along the western shores of the Americas and West Africa.

Although only relating to corals, these patterns are reflected in most other groups found in tropical coastal

Map 1.2: Patterns of diversity in reef-building scleractinian corals

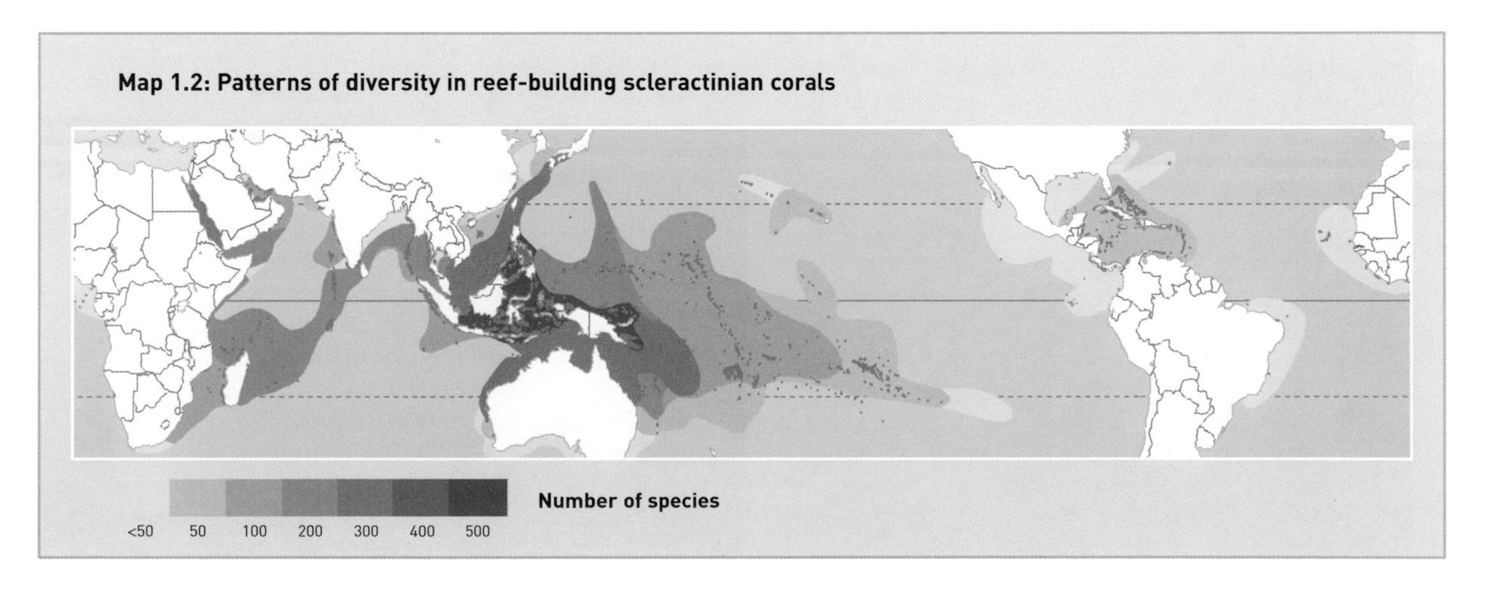

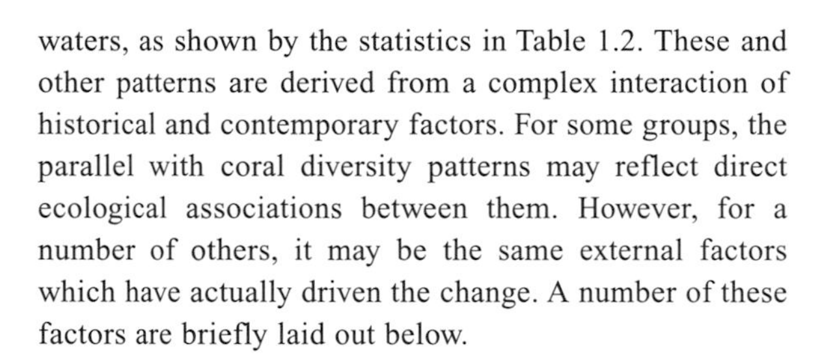

waters, as shown by the statistics in Table 1.2. These and other patterns are derived from a complex interaction of historical and contemporary factors. For some groups, the parallel with coral diversity patterns may reflect direct ecological associations between them. However, for a number of others, it may be the same external factors which have actually driven the change. A number of these factors are briefly laid out below.

The influence of temperature

To a large extent, both scleractinian corals and the reefs they build are restricted to a latitudinal band between 30°N and 30°S. This general observation is entirely related to the decreasing temperatures which generally follow increasing latitude. Most reef corals cannot survive in temperatures much below 16-18°C for even a few weeks. In conditions of extreme cold, corals can die within a matter of hours or days, while under slightly less extreme conditions, their growth rates are reduced. There is some evidence that overgrowth by algae rather than the direct influence of cold water may restrict coral development in some high latitude areas.

High temperatures are also inimical to coral growth. Extreme high temperatures drive the phenomenon known as "coral bleaching", during which the corals expel their symbiotic algae (see Chapter 2). Aside from human-induced climate change, it has been suggested that occasional high temperatures associated with El Niño events may be at least a partial explanation for the highly limited reef development which is observed, for example, along the western shores of the Americas.

The role of currents

While temperature influences can be broadly equated with latitude, ocean currents can disrupt these simple patterns. In a few areas of the world, major warm currents flow all year round from the tropics into higher latitudes. These have allowed the development of reefs quite beyond their normal limits. Notable are the Leeuwin Current in Western Australia; the East Australian Current; the Kuroshio Current in Japan; and the Gulf Stream warming the isolated oceanic reefs of Bermuda. In a similar way, cold waters prevent reef growth. Cold water upwellings along the coastlines of northeastern Somalia and southern Arabia are perhaps the clearest example, while the extremely limited development of reefs and coral communities along the western coastlines of the Americas and West Africa may also be influenced by cold water upwellings.

Another role of currents is in the transport of larvae to areas of reef. The establishment of corals in new areas is dependent on the transport of coral larvae in ocean

Left, above: A small coral cay on the reef flat of an atoll, Salomon Atoll, Chagos Archipelago. Left, below: The Beqa Barrier Reef in Fiji. Right: Fringing reefs, near Suva, Fiji.

Table 1.2: Regional patterns of species diversity in coral reefs and related ecosystems: the clear pattern of maximum diversity in the Indo-Pacific region is shown in all species groups

Taxonomic group	Indo-West Pacific	Eastern Pacific	Western Atlantic	Eastern Atlantic
Scleractinian corals[1]	719	34	62	
Alcyonarian corals	690+	0	6	
Sponges (genera)	244		117	
Gastropods:				
Cypraeidae	178	24	6	9
Conus	316	30	57	22
Bivalves	2 000	564	378	427
Crustaceans:				
Stomatopods	249	50	77	30
Caridean shrimps	91	28	41*	
Echinoderms	1 200	208	148	
Fish	4 000	650	1 400	450
Butterflyfish and angelfish[2]	175	8	15	7
Seagrasses[3]	34	7	9	2
Mangroves[4]	59	13	11	7

* All Atlantic

Source: Paulay (1997) except:
1. Veron (2000).
2. Allen et al (1999).
3. WCMC database – figures include species with warm temperate distributions.
4. Spalding et al (1997).

currents. Unfavorable currents may prevent the colonization of areas by new species, notably in the remote reef regions of Brazil and the Eastern Pacific. The mechanisms, and the importance of this transport, are further considered on page 23.

Changing patterns over geological time scales

Separate faunas – Atlantic and Indo-Pacific. Many of the global patterns in reef and coral development can be explained by looking at the tectonic and climatic history of reefs. Scleractinian corals evolved during the Triassic (205-250 million years ago) and quickly developed a circum-global distribution, only restricted by areas of suitably shallow substrate. As the continents broke up and shifted, the global connection of tropical oceans became more restricted. With the closure of the Tethys Sea, the waters of the Indian Ocean and Western Pacific were separated from those of the Atlantic and far Eastern Pacific, and the coral reef communities in each began to develop distinctive characteristics.

Low diversity in the Atlantic. The closure of the isthmus of Panama divided the "western" fauna into two. This entire region was then subjected to massive extinctions during the Pliocene/Pleistocene glaciations, removing many of the species which were once commonly found on all coral reefs. The Atlantic corals now share only seven genera with the Indo-Pacific. Even as environmental conditions improved, continued eustatic disruption may have prevented subsequent re-expansion and diversification of the coral reef fauna, and there has been little time since the end of these glaciations for any further species radiation. The result today is clearly shown in the far lower species diversity in the Atlantic reefs. For scleractinian corals the Atlantic only holds about one tenth the number of species as the Indo-Pacific, while similar patterns hold for almost all other species groups, with between one third and one tenth of the diversity on Atlantic reefs as compared to Indo-Pacific reefs.

High diversity in the Indo-Pacific. The same period of extinctions was not so extreme in the "eastern fauna", the area now known as the Indo-Pacific. Right across the region there are large areas of shallow coastal shelf spanning considerable latitudinal ranges. Over these areas there were more locations or refuges offering opportunities for survival of species during periods of environmental adversity. Species diversity remains high across much of this region, although there is a clearer decline in diversity moving east across the Central Pacific.

The Southeast Asian center of diversity. Quite apart from the generally high diversity recorded across the Indo-Pacific, there is an area of outstanding diversity centered on a triangle encompassing the Philippines and central and

eastern Indonesia. Species numbers here outstrip any region of the world, and species counts from individual bays or islands typically outstrip total species counts from the entire Caribbean. Some of this great diversity may in fact be linked to the same period of glaciations which caused destruction elsewhere. This region is believed to have maintained somewhat benign conditions during this time, allowing the survival of many species. Additionally, during certain periods, species may only have survived in relict populations restricted to small refugia. Their isolation, exacerbated by changing sea levels, may have allowed the independent evolution of populations and the formation of new species. These would have repopulated the wider region as conditions ameliorated. Further species may have accumulated here from outside the region, driven by patterns of ocean currents flowing westwards from the islands of the Pacific Ocean.

A number of other historical and contemporary factors are responsible for driving regional patterns in biodiversity, notably the low diversity observed in the Eastern Pacific, Brazilian and West African faunas, and the sustained high diversity in the Red Sea and low diversity in the Arabian Gulf. These are considered more fully in the regional chapters.

Patterns at finer scales

Moving in to study reef distribution at finer resolutions, the discontinuous nature of coral reefs within countries or along particular coastlines is highly apparent. Corals, and the reefs they form, are highly sensitive to factors such as salinity, sediments and nutrients. Where conditions are inappropriate they do not occur. More importantly in recent times, where conditions change, the corals, and the reefs themselves, may die.

Sediments and sedimentation

The initial growth of a coral is dependent on a larval animal finding the right substrate on which to settle. Corals cannot grow on fine muds or shifting sediments, and such sediments are a common feature along many of the world's coastlines. Where corals cannot settle and grow, reefs do not form. This is at least part of the explanation for the absence of reefs close to large river mouths and along other stretches of sediment-laden coastlines. Another influence of sediments is that of turbidity – in areas with large amounts of suspended sediments in the water column, the loss of light further reduces or prevents coral growth.

Once established, corals can cope with limited amounts of sediments settling upon them from the waters above, actively removing sediments which smother their tissues and block out the light. Similarly, once a reef is formed it is often able to maintain a presence in areas of otherwise shifting sediments. The reef structure lifts itself above the sediments, and provides the hard substrate on which new corals can grow. Reefs can also reduce the influence of currents and waves which, in some areas, are responsible for resuspending sediments that might otherwise smother corals.

Where conditions of sediments and turbidity change considerably this can lead to the rapid demise of corals and the death of reefs. There is an energetic cost to a coral in removing sediments which settle upon it, while the loss of light associated with increasing turbidity greatly reduces a coral's chances of survival.

Salinity

Corals are wholly marine organisms, unable to grow in freshwater. It is sometimes quite hard to distinguish the effects of freshwater from the influence of the sediments, typically also carried by streams and rivers. However, the absence of corals from wide areas associated with major rivers is at least in part related to the low salinities in these areas.

Nutrients

The considerable biomass and wealth of diversity observed on coral reefs around the world has led to a common misconception among non-specialists that reefs may be dependent on considerable inputs of nutrients. In fact reefs are highly efficient at nutrient recycling, and are widespread in some of the most nutrient-poor parts of the oceans. Where nutrient levels are higher, often close to coastlines or areas of upwelling, reefs still survive, but in very high nutrient situations other opportunistic species

Damselfish and butterflyfish around a blue coral Heliopora coerulea.

Movements between reefs

One critical issue when it comes to understanding the establishment of patterns in species distribution is the movement of individuals between localities. Reefs in general are ecological islands, typically surrounded by non-reef areas and often separated from one another by tens or hundreds of kilometers. Many reef organisms are sessile, and do not move at all. Even for the most mobile groups, movements of adult animals between reefs would be so hazardous as to be almost impossible, and such journeys are rarely undertaken. From the largest to the smallest, almost all coral reef species have a larval life history which survives for some time in the plankton. It is these tiny animals which move, or are swept, from place to place within a reef, and from reef to reef.

Typically, corals and other reef species produce vast numbers of eggs – many coral reef fish produce between 10 000 and a million eggs. These may be fertilized internally or in the waters above the reef. Either way, larvae are formed and enter the plankton where they may remain for weeks or even months – larval survival in the plankton has been recorded to over 120 days in some reef fish.

Whilst in the plankton, eggs and larvae may be carried distances ranging from meters to hundreds of kilometers. Many larvae have quite considerable swimming ability, but sea surface currents, more than any other factor, determine the long-distance transport of most organisms. Studies on reef fish distribution have shown that the species with the shortest larval phases tend to be geographically restricted while those with long larval phases are often geographically widespread. The great majority do not survive, or may be carried to areas where they are unable to settle, but it is this same movement which allows genetic flow between widely separated reefs. It also enables the establishment of new species and new reef communities in areas where they may not currently occur, or the recovery of populations which have been lost for any reason. A number of reef communities surviving at the edge of their natural ranges, such as those on the latitudinal limits of reef development in Western Australia, or those periodically impacted by extreme El Niño conditions in the Eastern Pacific, may be entirely dependent on larval recruitment from other, distant, reefs. This also has important implications for management, particularly for the recovery of reefs that are destroyed by pollution or blast fishing, or when overfishing removes all adult fish from an area.

There is still a great deal that remains unknown about this critical dispersive phase of reef organisms. The mass spawning event of reef corals on the Great Barrier Reef was first discovered only in the early 1980s – here it was observed that the great majority of corals released their eggs and sperm during a few nights associated with a particular full moon. Such synchronous spawning events flood the nearby waters, reducing the ability of predators to consume all the eggs and larvae and so increasing the chances of individual survival. Such mass spawning events are being discovered in other areas too, and in other groups. Certain reef fish, such as the larger groupers, have been observed to travel many kilometers to congregate at spawning grounds.

At the same time as these mass spawning events are being discovered, recent genetic studies have shown that patterns of connections between reefs are not a simple reflection of surface currents, but may also reflect other factors, both contemporary and historic. Some work suggests that species may not always travel vast distances or be as "interconnected" as previously thought. Certain "species" are now being broken down into geographically distinct sibling species groups which are sufficiently different from one another in genetic terms to suggest that there may be no gene flow between them, and that they may at the present time be ecologically isolated.

An Acropora *coral releasing clouds of egg and sperm bundles, Western Australia (photo: Bette Willis).*

may out-compete them. These typically include algae and sponges which may compete for space and light and overgrow corals. It can also include algae living within the plankton which can literally block out the light and increase the turbidity to levels which the corals cannot survive.

Patterns across the reef

Where conditions allow, coral reefs form and continue to thrive, marking out a colorful barrier along many coastlines and far out across the Pacific and Indian Oceans. As individual coral reefs are examined more closely, new and distinctive patterns emerge, formed by the species which make up the reef community.

Moving across a reef from the beach to the open sea, environmental conditions vary considerably. Close to the shore there may be freshwater runoff, loose sediments of sand or mud, and regular exposure to the air and sunlight with the shifting tides. Further out conditions are shallow and bright, but there may be little circulation of the water. At the outer edge of the shallow reef, the waters change dramatically. Waves may break on the reef top. Lower down, the light diminishes rapidly with depth. Light and depth, tides, water circulation, wave action, sediments, nutrients, temperature variation and salinity all have a part to play in determining which species are found where on a reef, and clear zones have been recognized. A number of these are illustrated in Figure 1.3.

Beach and intertidal communities

Although considered beautiful by millions of tourists, beaches and other intertidal areas are among the harshest communities for many species. Daily exposure to drying air and hot bright sunshine is inimical to most marine species, while regular or occasional soaking by saltwater is equally difficult for most terrestrial species. Beaches themselves are places of constantly shifting sediments, offering no solid substrate on which to establish, and only tiny interstices between the sediment particles for refuge. The coastline is also the point at which terrestrial inputs are at their most concentrated, with runoff, pollutants and sediments greatly influencing life in some areas. Life on sandy beaches is not abundant. There are many microscopic life forms within the sand, and a range of species, notably crabs, patrol the shores for food. In rocky areas a greater diversity of life occurs, notably molluscs, algae and bryozoans, and a complex pattern of communities may be found associated with tide pools and their position relative to the tidal range. Mangroves are a group of highly adapted plants which thrive in intertidal waters. Although frequently associated with reefs they are somewhat restricted in where they can grow, and only build extensive communities in areas where there are fine silts and muds, particularly where there is some freshwater input.

Above: Mangroves are important intertidal communities in many reef areas. Center: The reef crest, the shallowest part of the reef, northern Red Sea. Below: Coral diversity is highest on the reef slope, typically reaching a peak below the areas of highest exposure to waves, but still in shallow areas where loss of light is not a limiting factor.

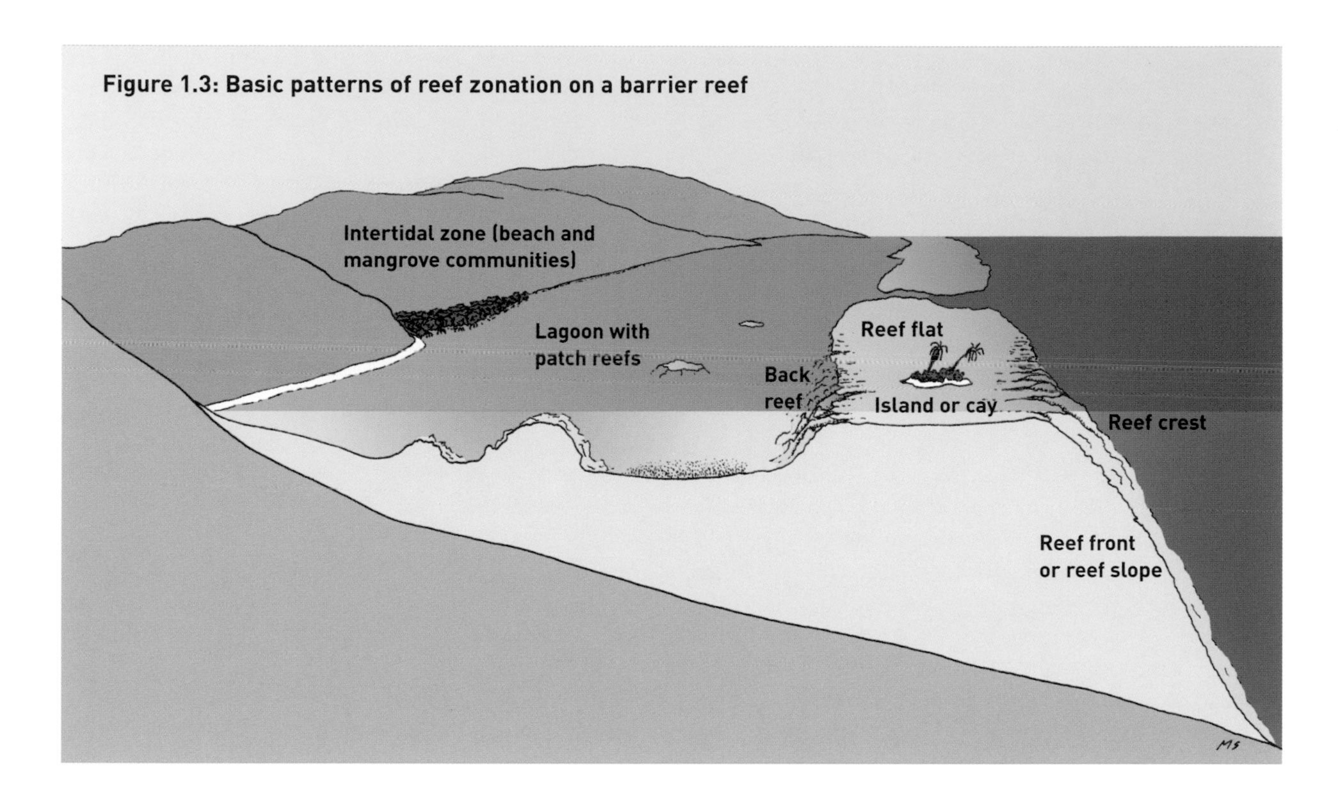

Figure 1.3: Basic patterns of reef zonation on a barrier reef

Lagoons

On barrier reefs the shoreline drops relatively rapidly towards depths of a few meters, sometimes a few tens of meters, before rising again to the shallow waters of the main reef structure. This area of deeper water is known as the lagoon. It is paralleled by a similar area at the center of most atolls. Although true fringing reefs do not have such deep water, in fact the division between fringing reefs and barrier reefs is sometimes hazy, and there may be shallow lagoons even on fringing reefs. Conditions in lagoons vary considerably. In some cases the lagoon is enclosed, and the flow of water is restricted by the high rim of the atoll, or by the shallow waters of the barrier reef. The degree of enclosure greatly influences conditions within the lagoon. Relatively shallow, enclosed lagoons may be areas of considerable temperature extremes as the waters cool at night or become rapidly heated during the day. They may also be areas where nutrients and sediments build up. At the same time, the bright, calm waters of the lagoon can provide ideal conditions for many species.

Seagrass communities are a common feature of many reef lagoons, but bare sandy sediments are perhaps even more widespread. Corals also thrive in many lagoon areas. In a few cases they are widespread across the lagoon floor, but more commonly they build up large structures, often known as bommies or patch reefs, which may be a few meters to many tens of meters across. Active coral growth can lead to the development of even more complex structures, such as the faros of the Maldives which have a circular structure very similar to a tiny atoll, but growing within the lagoon of a true atoll.

Back reef

At their seaward edge, most lagoons rise up quite sharply towards the shallow waters of the reef flat (see below). If there is good water circulation in the lagoon itself this area can be ideal for coral growth, with bright conditions, undisturbed by wave action. This area is known as the back reef, and may consist of a simple slope with a surface cover of corals, or may be an area of intricate gardens of coral rising and falling, interspersed with sandy patches.

Reef flat

In a mature reef, the active, upward growth of corals and coralline algae is eventually inhibited by the water surface. Upward growth can no longer occur, although there may be some consolidation and infilling of the reef rock. Outward growth of the reef into deeper water continues, and gradually a wide shallow platform is produced, the reef flat. In fringing reefs the reef flat extends outwards directly from the shore, but atolls and barrier reefs are also topped with reef flats. Small sandy islands or cays may form on the reef flat from the accumulation of sand and coral rubble during storms. Typically, reef flats range between a few tens of centimeters and 1 or 2 meters in depth, but they may reach many hundreds of meters wide. Physical conditions in the reef flat are quite harsh. Water temperature may fluctuate considerably through a 24 hour period, and some parts may be exposed to the air at low tides. Water circulation is also quite limited, and oxygen levels are often rapidly depleted.

The base of this zone is usually coral rock, but it is often covered by a mix of sand and rubble patches, over which there may be algae or seagrass. Corals continue to grow in deeper depressions, and small coral communities develop in larger areas of deep water (such areas are sometimes known as moats). A considerable number of small invertebrates live permanently on the reef flat. Large numbers of organisms feed on the reef flat at high tide, while foraging birds visit during the lowest tides.

Reef crest

The edge of the reef flat facing the ocean is an area of high energy, with almost constant wave action, and occasional exposure to the air. There is a rapid and constant circulation of the water, and water temperatures are generally more constant than those of the reef flat. Conditions for coral growth are not ideal, but a few species, dominated by branching forms, have adapted to them. On some reefs coralline algae are even better adapted and may predominate. Their combined growth builds up to make this the shallowest zone of the reef, often drying out at low tides. In more exposed reefs deep surge channels may be gouged into this reef crest and serve to dissipate the wave energy.

Reef front or reef slope

Beyond the reef crest is the zone with the greatest diversity and abundance of life. Typically this reef front or reef slope falls quite steeply towards the seabed. In this zone conditions change quite rapidly with depth and exposure. The shallowest waters, particularly on exposed reefs, may still be subject to considerable wave action and the growth of corals may be restricted. In such places branching corals predominate, and in the most exposed areas their growth forms are typically low and compact. Wave action often leads to the development of deep channels and high ridges known as spur and groove formations.

Below the influence of wave action diversity is unparalleled. Reefs are rarely dominated by single species, and both the corals and most other species groups form highly complex mixed communities. As depths increase, light is rapidly filtered out by the overlying water. Certain species can only grow in bright waters, and so are limited to depths of only 10 or 20 meters. A smaller number of species have adapted to darker conditions and may begin to dominate. The depth limits to coral growth are variable, as the water clarity determines the degree of light penetration. Reefs on more turbid continental margins typically have no active coral growth below about 50 meters and active reef accumulation may stop at 20 meters or less. In the clear waters of oceanic atolls extensive coral growth has been observed as deep as 100 meters, although this is probably unusual.

There is clearly enormous variation across the world of coral reefs, but these broad zones are widespread. Even among the less developed reefs, the natural propensity towards these patterns of zonation is often visible.

It is also important to remember that most of the reefs visible today are in fact ancient structures and much of their present-day shape has been developed over millennia, under quite varied conditions. In some cases there are vestigial structures marking former sea levels. Terraces are often observed on reef slopes, indicating patterns of growth towards an earlier low sea level, while entire submerged reefs may show many of the structures described above, but with reef flats and lagoons now at considerable depths below the present sea surface. Similarly, raised reefs are quite common, with atolls or platform reefs raised up to form modern islands, and the subsequent development of fringing reefs around their margins.

Patterns of diversity on a coral reef are the subject of a great range of influences, from the patterns produced by history, including the massive perturbations of recent ice ages to the present day patterns of temperature, sediments and nutrients. On particular reefs new patterns emerge, the result of a great complexity of local influences, including light, exposure and water circulation. The final section of this chapter examines some of the great wealth of diversity which makes up life on the reef. It provides an overview of all of the major groups which are so critical, not only to the development and functioning of the reef, but also to the great value of coral reefs as a natural resource for humanity.

A juvenile lemon shark Negaprion brevirostris *crosses a Caribbean lagoon at high tide. Lagoon areas are often dominated by wide expanses of bare sand.*

Quantifying diversity

Coral reef diversity is directly comparable to that of the most diverse terrestrial habitats, the lowland tropical rainforests. At levels of higher taxa (the more generalized "groups" of species), reefs greatly outstrip these other mega-diversity ecosystems. Densities of species per unit-area are also staggering. Species are often regarded as the building blocks of biodiversity, and, although reefs occupy only a small area of the planet, there are probably more species per unit-area of coral reefs than in any other ecosystem. There are an estimated 4 000 coral reef fish species worldwide, almost a quarter of all marine fish species. Nearly 1 500 fish species have been recorded at the Great Barrier Reef in Australia, and up to 200 species have been recorded from single samples on individual dives.

Fish represent the dominant vertebrates on coral reefs, perhaps comparable with the birds of a rainforest, but their numbers pale into insignificance when compared with total species composition of reefs. A 5 square meter reef microcosm sampled in the Caribbean yielded 534 species from 27 phyla, with a further 30 percent of species not fully identified. One sample of "boring cryptofauna" (animals which burrow holes and live within the coral rock) from a single dead coral colony yielded 8 265 individuals from some 220 species. We are only just beginning to comprehend the scale and depth of this diversity. Further parallels with tropical rainforests and other high diversity ecosystems abound.

With this wealth of species a great diversity of interactions has evolved between species. No organism lives in isolation, but on reefs the ecological processes which so often drive evolution have pushed the coexistence of species to extremes. Through pressures such as predation and competition, many species have become highly specialized to live in tight niches, with highly specific diets, cryptic habits, or highly evolved defense mechanisms. Others have become masters of stealth and capture or camouflage and escape. Co-evolution has also led to complex two-way interactions between species, including mutualistic partnerships where both organisms benefit. The relationship between corals and their algal partners is perhaps the most important example of such a partnership, having led to the proliferation and success of the reef-building corals.

Like forests, coral reefs also show a considerable structural diversity. Across the reef zones described above, but particularly in the areas of most active coral growth, a coral reef represents a highly complex three dimensional

A dense school of blue-lined snapper Lutjanus kasmira, *Seychelles.*

environment. Wave action creates deep grooves in the shallow reef front, while the corals themselves, with their complexity of forms, create a highly convoluted surface. Even the limestone at the base of the living reef surface is a complex mass of holes formed by the older patterns of coral growth, together with processes of erosion. This not only provides a large area for the settlement of other reef organisms, but also a complex background for the drama of life on the reef, providing passages and holes at all scales for the movement of animals, and for their concealment, shelter, ambush and escape.

One final comparison with rainforests is that the diversity of life in both ecosystems remains remarkably poorly known. It has been estimated that less than 10 percent of the organisms found on reefs have been described by scientists. But not all experts agree on species identification and definition and there is no central record, even of the species which have been described. It is thus impossible, at present, to estimate accurately the total numbers of species occurring on coral reefs. Using a number of broad assumptions, one recent attempt has suggested that there may be some 93 000 described coral reef species. The global total, including the vast number of undescribed species, could thus be closer to 1 million. Others have estimated that there may be over 3 million reef species. Perhaps the greatest problem hindering a more detailed assessment of coral reef biodiversity is the lack of basic taxonomic research and inventory, combined with the lack of sufficiently qualified taxonomists to undertake the work. Defining and describing species is a complex task, and detailed observation and description of external morphological characteristics of animals and plants have traditionally been key tools. A number of recent studies, however, have suggested that many of these morphologically similar "species" may in fact be species complexes, groups of sibling species, each highly distinct in genetic terms. If such examples prove to be commonplace, the final analysis of species diversity may lead to massive increases in the total species numbers.

Left: Expansive beds of branching Acropora *with damselfish above, at the Great Barrier Reef. Right: A barrel sponge, encrusting red algae and corals in the Philippines.*

Organisms of the coral reef

In order to better understand the ecology of the coral reef environment it is important to have an overview of the main species groups which occur there. This final section of the present chapter provides a background to some of the main groups of organisms on coral reefs, focussing on the larger or more conspicuous life forms. A number of major groups are taken in turn, each being briefly described, with their role in the reef ecosystem receiving particular attention. Although the major headings refer to broad taxonomic groupings (such as phyla), a strict taxonomic hierarchy has not been followed. Particular groups have rather been selected based on their importance in the reef environment. For more detailed taxonomic information readers are referred to the sources at the end of this chapter.

Algae and higher plants

As with other ecosystems, sunlight provides the primary energy source for life on the coral reef, and photosynthetic organisms capture this light and convert it to the organic molecules which are the building blocks of life. Higher plants (the more complex life forms, which dominate on land) have an insignificant role to play in most reefs. In contrast, algae are present throughout the reef and are critical, not only as the basis of the complex trophic pathways, but also as a structural component in the building of the reefs themselves. Despite this, algae are not highly conspicuous on the reef, either when compared to terrestrial ecosystems, or even to the marine ecosystems of temperate waters. Four main groups of algae are recognized.

Blue-green algae (Cyanophyta or cyanobacteria)

These are the simplest forms, being prokaryotic (with a simple cell structure and lacking a central nucleus) and related to bacteria. They can be unicellular or filamentous (with cells arranged in long chains) and are widespread throughout the reef, although their role and importance remain little known.

Red algae (Rhodophyta)

These include a great variety of forms and species, ranging from unicellular to filamentous to complex forms. A number of species secrete calcareous skeletons and are referred to as coralline algae. The encrusting coralline algae, such as *Porolithon*, are among the most important plants on the reef, playing a critical role in binding

A shallow scene with branching Acropora *corals and various damselfish, Seychelles.*

sediments, particularly in the shallowest waters. In some places, including many reefs in the western Indo-Pacific, these are the dominant benthic organisms in the shallower parts of coral reefs and may play a more important role in reef building than the corals themselves.

Brown algae (Phyaeophyta)

These are more familiar in temperate rocky shore areas, where they often form the major plant communities. There are no unicellular species and many form quite complex "plants". Although not dominant on reefs, a number of species are widespread, including *Lobophora*, *Padina*, and *Sargassum* in the Indo-Pacific and *Dictyota* in the Caribbean.

Green algae (Chlorophyta)

This is a large and diverse group, including unicellular and complex forms. As with the red algae, some produce secondary calcification. Among these, *Halimeda* is widespread and the calcified remains of its disc-shaped segments are often a major component of reef sand. *Caulerpa* is another common genus in both Caribbean and Indo-Pacific reefs, forming complex and intricate plant structures. There are some 75 species, the majority of which are found in coral reef areas.

In addition to these main groups, there are several other algal groups, such as the diatoms (Bacillariophyta) which, although not important components of the benthos, form a dominant part of the marine phytoplankton. Another group, the dinoflagellates (Dinophyceae) are sometimes considered alongside the algae, but here are considered separately, below.

Higher plants

Two groups of higher plant are often discussed in association with coral reefs, although in reality they form distinctive habitat types which may, or may not, be found in close proximity to reefs. In contrast to coral reefs, the habitats associated with these species have low species diversity.

Seagrasses

Seagrasses are actually a polyphyletic group of marine angiosperms (flowering plants) which are broadly distributed from the tropics to the Arctic, although there is a peak in their diversity in the tropics. All species belong to the monocotyledon families Potomogetonaceae and Hydrocharitaceae. Only one genus, *Thalassodendron*, is able to grow on rocky substrates and is found in very close association with corals, although many species are frequently associated with the soft sediments of reef flat and lagoon areas.

Mangroves

Mangroves are a similarly varied group, and are typically defined as trees or shrubs which normally grow in, or adjacent to, the intertidal zone and which have developed special adaptations in order to survive in this environment. Interpretations of this definition vary, and hence there is no fully agreed list of what does and does not constitute a mangrove. The association between mangroves and coral reefs is somewhat opportunistic: although they are sometimes observed growing on coral rock, mangroves usually require soft sediments and sheltered environments. In many areas the calm waters behind fringing and barrier reef systems provide such areas. The ability of mangrove communities to bind silts and muds may reduce levels of siltation in offshore areas and enable reefs to flourish. There is also a considerable movement of fish species between the two habitats, but again this would appear to be opportunistic rather than essential. Globally the distribution of mangroves and reefs is quite distinct. While both are largely restricted to the tropics and near-tropics (with the exception of mangroves in southern Australia and New Zealand), mangroves flourish in many

Left: Encrusting red algae can be a major structural component of the reef crest. Right: Seagrasses are a common component of deeper reef flat and lagoon areas.

areas where reefs are absent, notably the coasts of West Africa and the Bay of Bengal. Unlike reefs, they are absent over most of the Central and Western Pacific and are very sparsely distributed in the arid regions of the northern Red Sea and the Arabian Gulf, and on many oceanic atolls.

Dinoflagellates (Dinoflagellata)

This is a common group of microscopic organisms generally found in the plankton. Most are heterotrophic, but a few photosynthesize. They are characterized by the possession of two flagella, and are sometimes considered to be algae (Dinophyceae), but more commonly grouped with the Protozoa. The dinoflagellates are particularly important in the coral reef ecosystem, as it is this group which contains the zooxanthellae.

Zooxanthellae are capable of living freely in the plankton, although they are regularly associated with a broad range of coral reef organisms, living as endosymbionts within the tissues of these organisms. As photosynthetic organisms, they are able to supply a considerable amount of the nutrition required by their hosts, but also benefit both from the waste products of their hosts and from the shelter provided by their tissues. The vast majority of reef-building corals are dependent on these organisms. It was long considered that the zooxanthellae inhabiting corals were from only one or two species, but this view is now strongly challenged and the full diversity of this group is in need of further investigation.

Another important dinoflagellate, at least from a human perspective, is *Gambierdiscus toxicus*, which grows on benthic algae and dead coral rock. This species produces a toxin known as ciguatera which is not broken down by the organisms which unwittingly ingest it. This toxin can build up through the food chain reaching concentrations, in some larger predatory fish, that are highly toxic to humans who eat them. Outbreaks of ciguatera have, in some cases, been linked to extensive coral reef disturbance, the dead and bare surfaces perhaps providing a greater surface area for this species to inhabit.

Sponges

Sponges are among the most primitive multicellular organisms (with ancestral-like organisms detected from pre-Cambrian deposits some 650-700 million years old), and yet they have a high diversity and are widespread across the globe. Although they do not form true bodies with differentiated organs, most sponges grow into well structured forms, with a network of internal canals through which sea water is passed, aided by the movement of flagella and microvillae. Water is drawn into the sponge through specialized cells, and wastewater is then flushed through exhalent pores, which are usually clearly marked on the surface of the sponge. The majority of sponges are filter feeders and are able to process considerable volumes of water every day, filtering out nutrients. Other sponges, including many which live in the nutrient-poor waters of the reef, rely on associations with blue-green algae (cyanobacteria) or zooxanthellae and are effectively autotrophic. A number of sponges are capable of chemically dissolving (etching) into corals in a process which is a major part of bioerosion on coral reefs. Sponges have a great variety of physical structures, and indeed many show considerable plasticity in their growth forms. Within their cellular matrix, certain specialized cells lay down skeletal tissues. Skeletons are formed from numerous smaller elements called spicules made from silica or calcite, while in others they are formed from spicules or longer fibres made from collagen. With these strengthening skeletons sponges produce large structures, which may be encrusting, lattice-, ball-, vase-, or barrel-shaped, or longer rope-like or branching forms.

Unlike many other groups it would appear that the

Left: Bright clumps of the green alga Chlorodesmis. *Right: A conspicuous tubular grey sponge, Indonesia.*

sponge faunas of the Caribbean are at least equal to those of the Indo-Pacific in terms of diversity (per unit-area), while sponge biomass is considerably greater on many Caribbean reefs. One further difference is that Caribbean sponges are more strongly heterotrophic, which could reflect the higher amounts of nutrients available on these reefs. In the Indo-Pacific autotrophic sponges are rather more common.

Despite having high diversity, much of which remains undescribed, sponges are often not highly visible or dominant in the reef benthos. In the tropical island regions of Oceania some 1 000 species have currently been described. For many countries the known species may number no more than 30-40. However, an estimated 500 species have been recorded at Chuuk Atoll (Federated States of Micronesia, in the Pacific) alone. Alongside this genetic diversity, many sponges produce complex chemical compounds, often as a form of defense against predators. The investigation of these chemicals for pharmaceutical products is proving increasingly interesting.

Cnidarians

This is a large group of relatively simple organisms. They are characterized by a basic body structure, with two primary cell layers, an epidermis and an endodermis, separated in most species by a simple, supportive, jelly-like matrix, the mesoglea. A rudimentary nervous system has developed in this group, with a nerve net but no centralized nervous system. Carnivory is common, although some species have developed close associations with endosymbiont algae (see above). One feature of this group is possession of specially adapted cells known as cnidocytes, which incorporate a highly complex capsule or "nematocyst" which, when triggered, is inverted to release a long, whip-like thread with a barbed or pointed tip and often releasing highly potent toxins. These may be used to capture prey or for defense. There are two basic body forms: the medusa is disc shaped, solitary and pelagic, while the polyp is typically sessile, and consists of an upright body with a fringe of tentacles encircling a single opening which acts as both mouth and anus. Colonial living has arisen in a number of members of this group. There are four classes.

The Hydrozoa are a fairly mixed group, and include some complex colonial planktonic members such as the Portuguese man-o'-war *Physalia* spp. There are also a number of sessile groups which are common on reefs worldwide, including colonial hydroids, but also a number of species which lay down a calcareous skeleton. These include the members of the orders Milleporina and Stylasterina. The former are the fire corals, *Millepora* spp. which are widespread in all coral areas and can form an important part of the substrate on the reef crest and reef slope. Growth forms are typically branching or encrusting. The stylasterinids are also known as lace corals and typically form fairly small and fragile branching colonies in darker areas and overhangs. In both milleporids and stylasterinids there is some specialization of polyps, with numbers of specialized stinging polyps surrounding a single feeding polyp.

A sea whip Junceella. *Unlike stony corals, sea whips have flexible skeletons made predominantly of protein.*

The Scyphozoa, or jellyfish, is a large group, though not dominant on reefs. One genus, the upside-down jellyfish *Cassiopea* is often found resting in reef flat areas. Like many corals these have algal endosymbionts living within their body tissues. Another class, the Cubozoa, are like the jellyfish, but with a clearly four-sided body wall and tentacles concentrated at the corners. Also known as box jellyfish or sea wasps, these include some highly toxic species, including the box jellyfish *Chironex fleckeri* in the waters off Australia and the sea wasp *Carybdea alata* from Caribbean waters.

The most important class on the coral reefs of the world is the anthozoans, and these are considered separately below.

Anthozoans

These are a very large group of cnidarians which lack any medusoid form and have polyps with a central gastrovascular cavity divided into partitions by septae. They are divided into two main groups, the Octocorallia (or Alcyonaria) which have eight tentacles and body partitions and are all colonial, and the Zoantharia (or

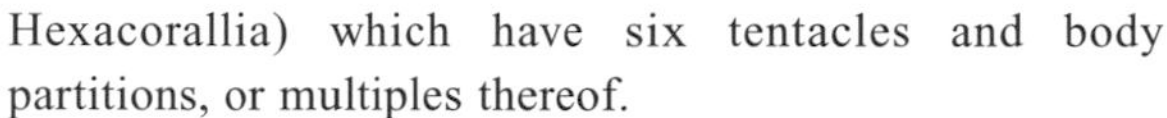

Hexacorallia) which have six tentacles and body partitions, or multiples thereof.

Octocorals are predominantly a tropical group of photic (sunlit) waters, although some species are found in cooler and deeper waters. Many of the reef species contain symbiotic zooxanthellae within their tissues. Perhaps the best known are the gorgonians (Gorgonacea), which are widespread on coral reefs globally. These include the sea whips and sea fans that are often dominant in deeper parts of reefs. Their colonies are strengthened by a central scleroprotein skeleton. Another spectacular group on coral reefs are the soft corals (Alcyonacea), which are common on many Indo-Pacific reefs, but less significant in the Caribbean. These do not have a clear skeletal structure, and body structure is maintained through hydrostatic pressure. Most species secrete spicules of calcium carbonate. Well known and widespread genera include the high, branching colonies of *Dendronephthya* and the spreading, lobed or branching forms of *Lobophyton*, *Sarcophyton* and *Sinularia*.

A number of smaller groups are also regularly found on coral reefs. The organ-pipe corals (Stolonifera) lay down parallel calcareous tubes connected with cross-plates to form massive hemispherical domes, and have a distinctive red skeleton. The blue corals (Helioporacea) are a true contributor to reef development, laying down strong calcareous skeletons, and forming large branching colonies in shallow areas. Both of these latter groups are restricted to the Indo-Pacific. Two other groups, the telestaceans (Telestacea) and sea pens (Pennatulacea), are more widespread but not of major importance on most reefs.

Zoantharians are a diverse group of solitary and colonial species. Many live in close association with symbiotic zooxanthellae. The most important of the zoantharians are the Scleractinia, which include the majority of reef-building corals and are treated separately below. The remainder of this group can best be described at the level of the orders within the group.

The Actinaria are the familiar sea anemones, which are simple non-colonial zoantharians, some of which can reach considerable sizes. Although primarily carnivores a number of reef species have developed a dependence on symbiotic zooxanthellae, while many have also developed tight symbiotic relationships with anemonefish (Pomacentridae). The Actinaria are a diverse group, with over 1 000 species worldwide, although they are not especially diverse on coral reefs.

Three other smaller orders are also commonly found on reefs, but remain poorly known. The Ceriantharia or tube anemones are another non-colonial group of about 50 species worldwide, which construct a tube buried into soft substrates. The Coralliomorpharia are the disc or coral anemones, with an internal body morphology quite similar to that of corals. The Zoanthidea are a fairly important group within the tropics, and may be abundant in shallow areas such as reef flats and shallow lagoon floors. They are solitary or colonial anemone-like actinarians, which do not secrete a skeleton, but often

Left: The highly colorful soft corals of the genus Dendronephthya *are common in the Indo-Pacific. Right: The central "mouth" of a giant sea anemone* Heteractis.

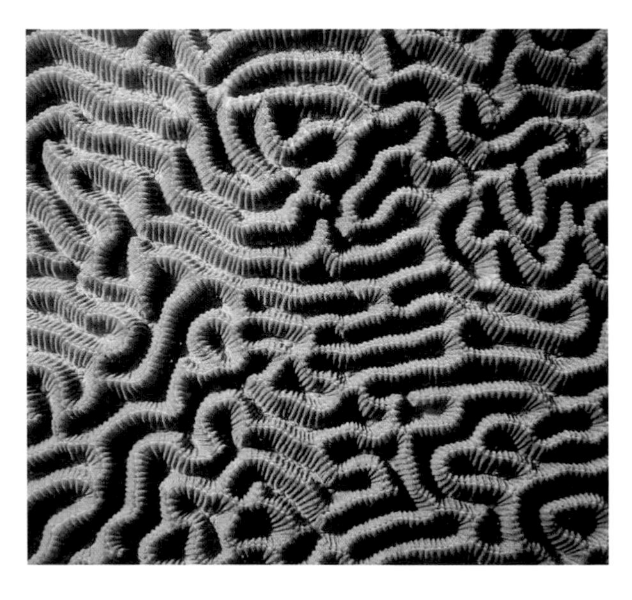

incorporate sediments into their mesoglea for support or protection.

The Antipatharia are commonly known as the black corals. They are all colonial, and secrete a horny proteinaceous skeleton. Although well known and economically important, they are not a major component of most reef communities and are not common in depths of less than 20 meters, with the majority of species being found below 100 meters.

Scleractinia

The Scleractinia, or stony corals, are a very large order within the zoantharians, all of which secrete a calcium carbonate skeleton. Although widespread throughout the world they reach their greatest extent and abundance in shallow tropical waters where the majority of species are colonial and lay down large skeletal structures, the basic building blocks of reefs. Some 794 species of hermatypic Scleractinia have now been described and the great center of scleractinian diversity lies in insular Southeast Asia, the center of the Indo-Pacific region.

The Scleractinia have an ancient lineage, and leave a good fossil record which can be traced back to at least the mid-Triassic over 200 million years ago. There is no clear evidence that they evolved from a single ancestor, however, and many of the features of this group may in fact have arisen independently.

The skeleton of the individual coral polyp is called a corallite, with a base-plate from which a number of divisions known as septa rise up, radiating in towards the center. The outer edge of the polyp is often defined by a wall forming a tube-like structure enclosing the septae. New polyps are formed in colonial species by budding from existing polyps, or by growth upwards from the connecting tissues between existing polyps. Gradually new skeletal material is laid down over existing material. The skeletal structure of individual polyps forms the basis for species identification, and in many cases full identification can only be completed with dried skeletal material.

The larger structures built by colonies can become highly complex, with massive corals producing domes or towers, encrusting corals, and a vast range of branching (ramose), columnar, foliacious (sheet or leaf-like) and tabular (plate-like) structures. Many ecological studies utilize this coral morphology as a means of describing a reef. The dominance of different growth forms is often indicative of environmental conditions such as wave exposure and varies across the reef profile. It also provides a partial measure of structural complexity. While morphology can appear highly distinctive, it can also be highly varied within a species, influenced by these same external environmental parameters, and hence it is often of limited value in species identification.

Above: The elkhorn coral Acropora palmata, *once a dominant coral on many Caribbean reefs, has been decimated by disease in most areas. Center: The laminar or foliaceous coral* Echinopora lamellosa. *Below: The complex surface of a brain coral* Platygyra.

Most species are involved in a tight symbiosis with zooxanthellae and derive the majority of their nutrients from these algae. They are all equipped with tentacles and capable of feeding independently to some degree, typically on plankton or minute organic particles. However, the dependence on their algal partners is considerable, and many species can be considered virtually autotrophic.

Aside from asexual reproduction during colony growth, corals undergo sexual reproduction. Some species are hermaphroditic, while others have separate sexes. The majority of species release eggs and sperm during a spawning event – such events can be tightly harmonized within and between species leading to spectacular mass spawning events. In a few species the fertilized egg is kept within the polyp and free-swimming larvae or planulae are released some days or weeks later. Both eggs and planulae spend some days or weeks living in the plankton prior to settling and this is critical to the genetic flow between reefs and the establishment of corals in new areas.

Scleractinian corals are one of the few groups on reefs which have been sufficiently well studied to provide a global picture of their distribution and abundance (see Map 1.2).

Worm-like groups

There are several large, unrelated groups in the animal kingdom which have soft, elongated bodies and a general worm-like appearance. Many of these, while inconspicuous, are important residents of the reef.

Bristle worms (Polychaeta)

These are segmented worms with a pair of paddle-like legs on each segment. The head bears a number of sensory organs, which may be highly adapted in different species. They include almost every feeding habit: carnivores, herbivores, omnivores, detritus feeders and filter feeders. Many burrow inside coral or rock, chemically dissolving or physically grinding their way in and then remaining to filter feed or to gather and digest sheets of mucus secreted by the coral. Perhaps the most familiar are the sabellid worms, sessile burrowing forms that extend a feathery net of tentacles to filter the passing water. Such conspicuous species are just the tip of the iceberg, however, and in one study over 1 400 individual polychaete worms representing 103 species were extracted from a single 4.7 kilo lump of branching coral.

Ribbon worms (Nemertea)

These are typically highly elongated and flattened worms, free-living carnivores often feeding on polychaetes. They have very soft bodies, and some produce complex protective chemicals to deter predators.

Peanut worms (Sipuncula)

This is a group of unsegmented worms which typically burrow into sand or bore into rocks and corals and are detrital or algal feeders.

Flatworms (Platyhelminthes)

The phylum Platyhelminthes is a large group of small, elongated animals with highly flattened bodies. Many species are parasitic, however there is one highly active carnivorous group, the polyclad flatworms (Polycladida) which are relatively widespread on reefs. Their bodies are covered in cilia and some are capable of swimming. A number of coral reef species are highly colorful and can be confused with nudibranchs (see Molluscs, page 37). Few detailed inventories have been produced and identification to species is usually very difficult.

Crustaceans

The Crustacea, one of the largest groups of organisms on the reefs, are not the most conspicuous. They are defined by having two pairs of antennae, and typically have a chitinous exoskeleton and jointed biramous limbs. Beyond this definition, the group includes a vast array of species with highly different body forms. The class

Above: A flatworm on a reef in Pulau Redang, Malaysia. Below: The spiny lobsters Panulirus *spp. are of considerable commercial importance on reefs around the world.*

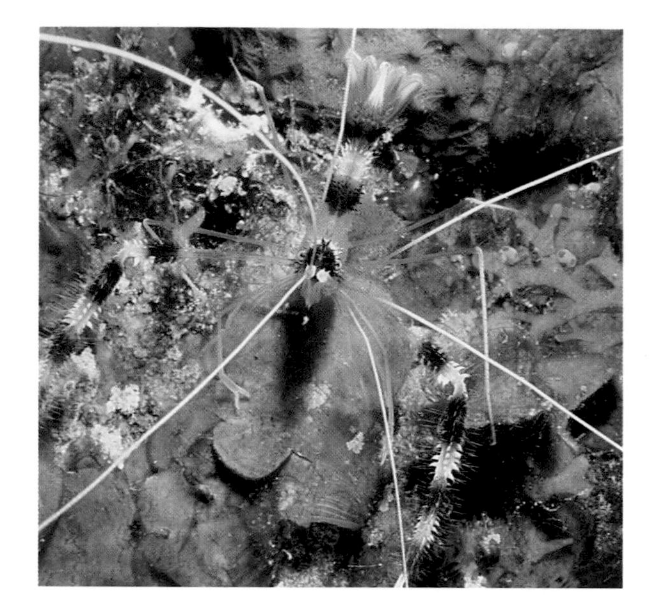

Maxillopoda contains the abundant copepods which are usually very small, and includes many planktonic species which are found in the coral reef environment. It also includes the barnacles which are commonly observed on reefs and intertidal areas. The class Ostracoda is another highly diverse group, often showing a bivalved appearance. These again are very small, mostly filter feeders or detritivores. The most important and widely recognized group on coral reefs is the class Malacostraca, and in particular, within this, the Decapoda and the Stomatopoda.

Decapoda

These are the shrimps, lobsters and crabs, with some 10 000 species worldwide, including numerous species found on the reef. A brief list of the major groups is provided below.

Penaeidea – these are the commercially important prawns, often associated with inshore lagoon and mangrove areas, but not well represented on the reef.

Stenopodidea (boxing or coral shrimps) – a small but well known group usually found in pairs and living in "cleaning stations" where they regularly remove parasites from fish or other crustaceans. They have a highly enlarged third pair of thoracic legs, with pincers on the tips.

Caridea – a large group of shrimps with a number of sub-groups:

Palaemonidae – on reefs, the palaemonid shrimps are well represented by commensal species which live in facultative or obligate partnerships with corals, anemones, molluscs and echinoderms. The genus *Periclimenes* is particularly widespread. Many species have striking colors which they are capable of adapting to suit their hosts.

Alpheidae (snapping shrimps) – also known as pistol shrimps, these are perhaps the commonest crustacean family on reefs. They are able to make a cracking sound with their pincers and are largely responsible for the almost constant background snapping noises heard on many reefs. Most are detrital feeders. Some of the best known snapping shrimps are those which share their burrows with certain species of goby; the former maintain a burrow in which they both live, while the latter provide warning when predators approach.

Other caridean shrimps are the hump-backed or cleaner shrimps (Hippolytidae) which include more colorful commensal and cleaner species, and the harlequin shrimps (Gnathophyllidae), also very colorful, which include some species that prey on starfish such as crown-of-thorns.

Palinura (spiny lobsters) – although not a diverse group this includes the familiar crayfish, which are a large, colorful and commercially important group of species found on reefs around the world. The group also includes the less commonly observed slipper lobsters.

Anomura (hermit crabs, squat lobsters and porcelain crabs) – the hermit crabs are widespread on reefs and nearby intertidal areas. They are well known for their habit of utilizing discarded mollusc shells as a form of protection. They have an extended and soft abdomen which fits well into the coiled whorls of these shells, and they regularly exchange shells as they grow. Most are scavengers or detrital feeders. Porcelain crabs are less diverse and less obvious on the reef, but are often found living in association with anemones. They resemble true crabs, but only have three pairs of walking legs, and have elongated antennae.

Brachyura (true crabs) – one of the most diverse crustacean groups on coral reefs, with more than 2 000 species described from the tropical and sub-tropical waters of the Indo-Pacific. The true crabs are recognizable by their strong and usually broad thoracic carapace and their greatly reduced abdomen which remains tucked up

Left: A banded coral shrimp Stenopus hispidus. *These play an important role as "cleaners" on the reef. Right: The peacock mantis shrimp* Odontodactylus scyllarus, *a powerful predator in the Indo-Pacific.*

on the underside of their thorax. All have four pairs of walking legs and often a well developed pair of pincers.

Stomatopoda

Also known as mantis shrimps, these are an ancient group which are thought to have diverged among the Crustacea around 400 million years ago. Over 400 species have been described, the majority of which are to be found in shallow tropical seas. All are active predators, with highly developed visual acuity and a specially adapted second pair of thoracic legs. In one major group, the smashing mantis shrimps (Gonodactylidae), these legs are strengthened into club-like appendages, while in the other group, the spearing mantis shrimps (Lysiosquillidae), they are adapted into barbed spears. Both groups are able to unfold these appendages at remarkable speeds to hit their prey. Smashing mantis shrimps are capable of breaking open the shells of molluscs and crabs, while spearing mantis shrimps are able to impale softer bodied shrimps and fishes.

Molluscs

Molluscs are another highly diverse group found on the reefs, with one estimate of more than 10 000 described species from coral reefs. Members of this phylum all have a body which can be broadly divided into a head, a central visceral mass and a strong muscular foot. Most also have a mantle which to varying degrees folds around the body. A rasping tongue, or radula, is common and most species secrete a calcareous shell. Four groups predominate, and all are present on reefs.

Chitons (Polyplacophora)

These are regarded as the most ancient of the molluscs, recognizable by their low, oval shape dominated by the presence of eight transverse and overlapping shell plates. They are grazers, and are most commonly found in shallow and intertidal areas.

Snails (Gastropoda)

The largest and most diverse group, the Gastropoda usually have a single coiled shell. The simplest forms (Archaeogastropoda) include the limpets, abalones, trochus, turbans and nerites, all of which are algal grazers. Another major group is the Mesogastropoda, which encompasses many reef species, including the cowries, periwinkles and conches. Many are algal grazers, although some have developed specialized diets – the helmet shells, tritons and tun shells feed on echinoderms. The Neogastropoda are a more advanced group. Many have an elongated siphonal canal and highly developed proboscis which can be used for capturing prey. In this way, the murex shells are capable of boring through the shells of other molluscs and injecting them with venom, while the cone shells have developed a highly specialized radula tooth attached to a poison sac. They are able to fire this, rather like a harpoon, and rapidly kill even highly mobile prey such as fish.

Opisthobranch gastropods are another well known sub-group, with some shelled forms such as the bubble shells, but also a large number of shell-less forms including the algal-grazing sea hares and the highly diverse and colorful nudibranchs. The latter are all carnivorous and many have relatively specialized diets. Some are able to maintain nematocysts from their prey and use them in defense, while others utilize the toxic chemicals their prey have developed, again for their own protection.

Bivalves (Bivalvia)

These are a large group of bilaterally symmetrical molluscs with a shell completely split into two matching halves and

Left: A cowrie Cypraea, *clearly showing its muscular foot and the thin mantle of tissue partly covering its shell. Right: A nudibranch* Nembrotha cristata *amidst tunicates and coral.*

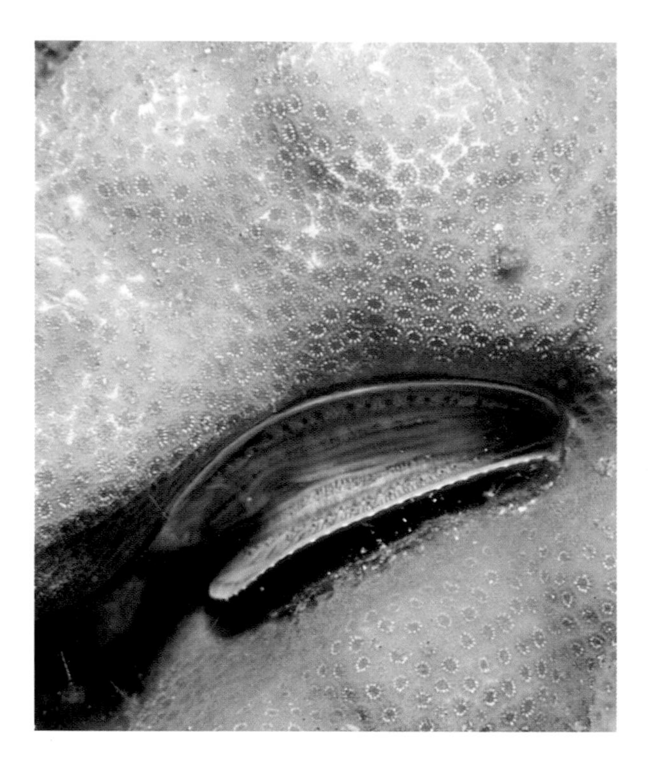

joined with a hinge ligament. Many reef species are found to burrow into soft substrates, or become incorporated into the reef matrix as coral or calcareous algae grow around them. The majority of bivalves are filter feeders. Groups include oysters, thorny oysters, scallops, mussels and the giant clams. This latter group (family Tridacnidae) is restricted to the Indo-Pacific and all species live in a close association with zooxanthellae. The giant clam *Tridacna gigas* can reach over 1.3 meters in length and weigh over 300 kilos.

Cephalopods (Cephalopoda)

These are the most highly modified molluscs in which the head, and the eyes in particular, are highly developed, while the foot has been modified into a number of tentacles or arms. One major group, the Nautiloidea or nautili, is largely restricted to deep water and not found on reefs. The other group (Coleoidea) includes the cuttlefish, squids and octopuses. All are active predators, with horny "beaks" developed around their mouth and specialized suction cups on their arms or tentacles for holding prey and other objects. All have chromatophores in their skin and are capable of extremely rapid color changes which they utilize for camouflage, but also as a form of communication between individuals.

Most cuttlefish maintain a significant calcareous "shell" which is internalized, while some squid also contain vestigial traces of a chitinous shell. Both of these groups are highly active free-swimming predators, but neither are numerous or diverse in reef environments. Octopuses are more widespread, although many remain hidden during the day.

Bryozoans

Individual bryozoans are tiny animals with a highly characteristic feeding device known as a lophophore, a ring of ciliated tentacles to capture and direct food into a central mouth. Most lay down a horny or calcareous skeleton, and are capable of withdrawing into this, and sometimes closing it with a hard operculum. They are sessile and colonial. Individual "zooids" in a colony may show particular specializations. Many are encrusting, but a number of species form erect plate or plant-like structures and are known as lace corals. Bryozoans, although inconspicuous, are numerous on all reefs around the world, and are often among the first organisms to colonize newly exposed surfaces. They can play an important role in cementing fragments and consolidating the reef structure.

Echinoderms

The echinoderms are a diverse and highly conspicuous phylum. They are divided into five groups, which appear highly differentiated, but have a few traits in common. Unlike most organisms, which can be divided symmetrically into two halves or are simply radial (corals and some worm-like groups), the echinoderms exhibit penta-radial symmetry – their bodies radiate into five symmetrical parts. All echinoderms also lay down a calcareous skeleton. Extending from their body surface they have small tube feet which are important in respiration and in most groups also serve a role in locomotion.

Left: A scallop Pedum spondyloideum *– this species does not bore into the coral, but the coral has grown up around it.*
Right: A cuttlefish Sepia *sp. hovers over a solitary mushroom coral* Fungia.

Feather stars (Crinoidea)

These have a very simple, small body or calyx from which five arms radiate. These branch almost immediately and hence most feather stars appear to have numerous arms. Each is equipped with many short pinnules and the arms are used to sweep the water for plankton. Below the calyx are numbers of short, dextrous cirri which are used for locomotion and to grip the substrate. Most feather stars are nocturnal.

Starfish (Asteroidea)

The starfish, or sea stars, are a well known group. Most have five arms, and in many cases the body organs are housed, or extend into, these arms. They have a mouth on their underside and anus facing upwards. Many species are capable of extruding their stomachs through this opening in order to facilitate digestion. They move using the large numbers of tube feet on their underside. Starfish include detritus feeders, omnivores and predators. One of the better known is the crown-of-thorns starfish *Acanthaster planci*, a large and unusual looking starfish with a large central body, numerous legs and a covering of sharp spines. It is a regular predator of scleractinian corals (see page 60).

Brittle stars (Ophiuroidea)

These are somewhat similar to starfish in general body plan. They have a distinct central disc containing the body organs, with a ventral mouth, but they have a very simple digestive system and no anus. Most have only five arms,

which are slender, highly mobile and typically covered in spines. They use these arms, rather than their simple tube feet, for locomotion. Most are detritus feeders, grazing the substrate beneath them. Others utilize their arms, sometimes with a mucous web, to sweep the water and capture prey, while some are more active predators. The basket stars are a sub-group with highly branched arms used to filter the water at night.

Urchins (Echinoidea)

Sea urchins are a highly distinctive group. They have no arms, and the small plates of the skeleton have fused to form a "test" which acts like a shell to protect the internal organs, but is in turn covered by a thin layer of living tissue. The body is typically further protected by a large number of spines. All urchins are grazers or detrital feeders, and have a powerful set of scraping jaws on their underside. Among the most familiar species on the reef are the long-spined species of the family Diadematidae, bearing highly elongated (typically 20 centimeters long) dark spines. These perform a critical function as grazers in many reef ecosystems and their loss on some Caribbean reefs has been linked to rapid declines in coral cover as algal growth predominates (see page 61). A number of urchins have developed secondary bilateral symmetry and have adapted to a burrowing lifestyle. These include the heart urchins and the highly flattened sand dollars.

Sea cucumbers (Holothurians)

These are elongated, sometimes even worm-like, creatures in which the calcareous skeleton is highly reduced to a mass

Left: A group of colorful feather stars on a reef in the Philippines. Right: The central disc and highly mobile arms of a brittle star in the Caribbean.

of tiny spicules in the body wall. They have a mouth at one end which is typically ringed by tentacles, while the anus lies at the other end. Tube feet are concentrated along the bottom of the body and used in locomotion in some groups. Some are detrital feeders, and many of the more conspicuous forms on reefs feed by ingesting sand and digesting the microfauna it contains. Others, mainly burrowing species, have long and highly branched tentacles which they use to collect plankton. As a form of defense, many sea cucumbers eject a number of sticky tubules from their anus when threatened. If these do not sufficiently deter their predators they may also eviscerate, expelling most of their internal organs through their anus. These are highly edible and the predator may feed on these while the animal escapes and begins to regenerate its internal organs.

Tunicates

This group includes a number of planktonic salps, but most important on the reefs are the ascideans or sea squirts. These are sessile animals and typically have a tube-like structure with a large opening, the inhalent siphon into which water is drawn, passing through narrow pores or gill slits before being exhaled through a slightly smaller exhalent siphon. Food is captured from this water onto mucus. This group is well represented and quite diverse on most reefs. Some are solitary, although even these are often found in aggregations, while others are colonial and the individual zooids may be more difficult to distinguish. Quite a number of species on the reef have developed a tight association with blue-green algal symbionts.

Fishes

Fish are one of the most conspicuous elements of reef life, being diverse, highly active, and often among the most colorful elements of reef communities. Over 4 000 species of fish inhabit coral reefs, representing over 25 percent of all marine fish species. Although not restricted to reef environments, quite a number of groups are distinctively associated with the reefs and a number of these are briefly described below.

Groupers (Serranidae)

A large group of highly active carnivorous fishes, typically with large mouths and more than one row of teeth. One highly distinctive sub-group are the anthiases, small and colorful zooplankton feeders often forming dense schools over coral heads. Most of the remainder are large stocky fishes which may be active or ambush-based predators, mostly feeding on fishes and crustaceans. The giant grouper *Epinephelus lanceolatus* is the largest true reef fish, recorded to over 270 centimeters long and more than 400 kilos in weight.

Above: A pineapple sea cucumber or prickly redfish Thelenota ananas. *Center: A small group of ascideans or sea squirts* Rhopalaea. *Below: Goatfish (Mullidae) and surgeonfish (Acanthuridae) in the Central Indian Ocean.*

Snappers (Lutjanidae)

This is a family of about 100 medium to large, elongate fishes, all of which are predatory. The majority feed on fishes, with some crustaceans and other invertebrates, while a small number feed on plankton. They are popular food fish in many countries. The majority are found on reefs, although a few commercially important species are found in depths between 100 and 500 meters.

A related family, the fusiliers (Caesionidae) are restricted to the Indo-Pacific. Most of the 20 species are also reef-associated, however they roam more widely, often in large schools, feeding on zooplankton during the day.

Grunts and sweetlips (Haemulidae)

In many ways these are very similar to the snappers, being elongate, but heavy bodied. They are generally nocturnal feeders and largely feed on invertebrates, with some plankton feeders. They are called grunts because of a common habit of grinding their pharyngeal teeth which, amplified by their gas-filled swim bladder, produces a grunting sound. The term sweetlips comes from the highly thickened lips of the Indo-Pacific genus *Plectorhinchus*.

Butterflyfish (Chaetodontidae)

Among the best known of the reef fishes, these are small disc-shaped fish, highly colorful with distinctive, almost flag-like patterns. Most of the 121 recorded species are reef-associated, and only eight are recorded outside the tropics. They have small mouths, and many pick at the substrate, feeding on a mixed diet of invertebrates and algae. Some are more specialist and feed largely or exclusively on live coral polyps, while a few feed on plankton.

Angelfish (Pomacanthidae)

Closely related to the butterflyfish, these also have relatively flattened bodies, though more rectangular in profile. Again this is a highly reef-associated family, with the vast majority of the 83 known species restricted to shallow tropical seas. Most are also highly colorful, but certain smaller species are relatively cryptic. Some species feed on detritus and algae, others specialize on sponges, and a few feed on plankton.

Damselfish (Pomacentridae)

These are an abundant and diverse group found on the coral reefs of the world, with over 320 species. All are small, and often highly colorful. Many are schooling species, and feed on plankton. Some are grazers and a number are known as farmer fish as they actively guard a patch of algal turf from other grazers. The anemonefish live in a close symbiotic association with large sea anemones.

Above: A giant grouper Epinephelus lanceolatus, *the largest true reef fish. Center: A school of snapper* Lutjanus ehrenbergii *and* Gnathodentex aurolineatus. *Below: Oriental sweetlips* Plectorhinchus orientalis, *with a small cleaner wrasse* Labroides dimidiatus.

Wrasses (Labridae)

It is difficult to generalize about this group, which is not only one of the largest groups, but also the most diverse in terms of appearance and lifestyle. All wrasses are carnivorous, but their diet varies considerably. The humphead or Napoleon wrasse *Cheilinus undulatus* is the largest member of the family. Reaching 229 centimeters and over 190 kilos, it feeds primarily on molluscs and crustaceans. Many of the smaller members of the group feed quite generally on benthic invertebrates including some large and quite diverse genera such as *Thalassoma* and *Halichoeres*. A number of species feed on zooplankton, including the genera *Cirrhilabrus* and *Paracheilinus*. In the Indo-Pacific the cleaner wrasses *Labroides* spp. feed on diseased or damaged tissues or external parasites of other fishes – they establish cleaner stations and solicit the attention of other fish, or may be approached by particular fish requiring their services. This role is of considerable importance, and many would-be predators allow these wrasses to perform this service and even to enter the mouth and gill areas without attempting to eat them.

Parrotfish (Scaridae)

Closely related to the wrasses, parrotfish are morphologically all relatively similar: elongate robust fishes, with a powerful beak formed from the fusion of their teeth. The majority are extremely colorful, although these color patterns are also observed to change dramatically over the course of the fish's lifetime. They are a predominantly herbivorous group, and feed by scraping or excavating the rock surface, often ingesting significant amounts of rock with the benthic algae they eat. A few of the larger species also feed in part on live coral. The largest, the bumphead parrotfish *Bolbometopon muricatum,* reach 120 centimeters, and have been estimated to remove between 2.5 and 5 tons of reef rock per year, converting it to sand and thus acting as a major erosive force on some reefs.

Surgeonfish (Acanthuridae)

This group is named for the sharp spines carried towards the base of the tail and used in defense. They are another highly reef-associated group, with relatively compressed, oval-shaped bodies. Of the 72 species described only six are recorded in the Atlantic. The majority of surgeonfish are algal grazers, but a number, including the unicornfish *Naso* spp., feed on plankton.

In addition to these large and conspicuous groups there are very many others. Some, such as the highly diverse blennies and gobies, and also the moray eels, soldierfish, cardinalfish and scorpionfish, include many reef-associated species, but may be less conspicuous on the reefs. Others are regular visitors, including the sharks

Above: A long-nosed butterflyfish Forcipiger flavissimus, *its fine mouth parts enabling it to forage for invertebrate food in the fine structure of the reef. Center: The Indian dascyllus* Dascyllus carneus, *a small damsel which gains shelter amidst branching corals. Below: A queen parrotfish* Scarus vetula *resting at night in a mucus bubble. Note the powerful "beak".*

and rays, jacks or trevallies, and barracuda. Reef fish play a vitally important role in the wider functioning of the reef ecosystem, as has been borne out by the observed impacts of overfishing in many areas (see Chapter 2). They are also among the best studied of all species found on reefs, and many are widely regarded as indicators in the study of wider patterns of biodiversity on reefs.

Reptiles

The overall diversity of reptiles in the oceans is very low. Most modern reptiles have kidneys which are unable to tolerate high salinities, so the two main groups which are found on or near reefs are the ancient group of marine turtles and the modern group of sea snakes. There are seven species of marine turtle, all of which are found in tropical and sub-tropical waters. None are strictly reef species, but several regularly make use of reefs as a source of food, notably hawksbill and loggerhead turtles, which feed on invertebrates. Green turtles feed on marine plants and algae and are often seen feeding in seagrass areas near reefs. All marine turtles regularly nest on tropical coastlines, often close to reefs.

There are some 55 species of sea snake belonging to the family Elapidae, only found in the Indo-Pacific. The largest group (sub-family Hydrophiinae) are the most highly adapted, many never leave the water and all give birth to live young. Another group, the sea kraits (Laticaudinae) still leave the water to lay their eggs. All are highly adapted to their aquatic environment, with flattened tails to aid swimming, and considerable breath-holding capabilities. Most eat fish, and have developed highly toxic venom to ensure that their prey die quickly before they have time to swim off.

Seabirds

Although not exhibiting spectacular diversity, a number of seabirds are found regularly in coral reef environments. These include predominantly pelagic seabirds which nest on tropical oceanic islands, notably boobies (Sulidae), tropicbirds (Phaethontidae), terns and noddies (Sternidae), frigatebirds (Fregatidae) and shearwaters (Procellariidae). These often breed in spectacular numbers on small coral islands, especially where there is little human disturbance, and no predation from introduced species such as rats. Although they primarily feed on offshore pelagic species they may take some nearshore species.

Smaller numbers of waders and other seabirds are also found on or near reefs. These include sandpipers, oystercatchers, turnstones and plovers. Egrets and herons are also widespread, often feeding across the reef flat at low tide. Pelicans are quite common on reefs in the Caribbean region, and in a few places flamingos have

Above: A school of jacks, the silver pompano Trachinotus blochii. *Center: A banded sea snake* Laticauda *coming ashore to a small coral cay. Below: Hawksbill turtle* Eretmochelys imbricata *on a Caribbean reef.*

been recorded on coral reefs. Birds of prey, including ospreys and sea eagles, are likewise occasional visitors to the reef.

Marine mammals

With the exception of humans, mammals are not widespread on coral reefs. One important group, the sirenians, is often found in close proximity to reefs. These animals, the manatees of the Caribbean and the dugongs of the Indo-Pacific, are large herbivores that feed in seagrass areas, rarely venturing over the reefs themselves.

Another group is seals and sea lions (pinnipeds). Historically, monk seals were distributed in the Caribbean and Hawai'i (with a third Mediterranean species). The Caribbean monk seal is now extinct, and the Hawaiian monk seal is still declining despite extensive measures for its protection. Two other species, the Galapagos fur seal and the Galapagos sea lion, are found in the Galapagos, where there are coral communities but no true reefs.

Perhaps the best known and most diverse group is that of the whales and dolphins (cetaceans). A number of species are found in tropical waters and may be observed near reefs. Dolphins in particular regularly shelter in bays and lagoons near reefs, and occasionally feed on reef organisms. Humpback whales return annually to breed in tropical waters, and have a number of regular breeding grounds close to coral reefs, including locations in Hawai'i, the Great Barrier Reef and the Caribbean. Despite these associations, cetaceans are typically only visitors to coral reefs, and rarely dependent on them.

The human presence

Life on the coral reef is complex and diverse. Our understanding of the diversity of life, of the complexity of interactions, and of the structures and patterns that occur on coral reefs around the world is still extremely limited. Humans have lived in very close proximity to reefs for millennia, and in many areas can certainly be considered to be a part of these ecosystems. At the same time, however, changes almost entirely related to human activities are being imposed on coral reefs around the world. In many areas structures are being degraded, diversity is diminishing, and the complex interactions of the reef are being reduced and undermined. These issues, together with the efforts which are being made to redress them, are considered in the next chapter.

Above: A grey heron stalking prey on the reef flat. Center: A dugong Dugong dugon *swims along the edge of a coral reef. The lines on its back are probably scars from boat propellers* (photo: Doug Perrine/Seapics.com). *Below: A healthy reef off Nusa Penida, Indonesia.*

Selected bibliography

Allen GR, Steene R (1999). *Indo-Pacific Coral Reef Field Guide.* Tropical Reef Research, Singapore.

Allen GR, Steene RC, Allen M (1999). *A Guide to Angelfishes and Butterflyfishes.* Odyssey Publishing/Tropical Reef Research, Singapore.

Benzie JAH (1999). Genetic structure of coral reef organisms: ghosts of dispersal past. *Amer Zool* 39: 131-145.

Birkeland C (ed) (1997). *Life and Death of Coral Reefs.* Chapman and Hall, New York, USA.

Connell JH (1978). Diversity in tropical rain forests and coral reefs. *Science* 199: 1302-1310.

Crossland CJ (1988). Latitudinal comparisons of coral reef structure and function. *Proc 6th Int Coral Reef Symp* 1: 221-226.

Done TJ, Ogden JC, Wiebe WJ, Rosen BR (1996). Biodiversity and ecosystem function of coral reefs. In: Mooney HA, Cushman JH, Medina E, Sala OE, Schulze E-D (eds). *Functional Roles of Biodiversity: A Global Perspective.* John Wiley and Sons Ltd, Chichester, UK.

Hubbell SP (1997). A unified theory of biogeography and relative species abundance and its application to tropical rain forests and coral reefs. *Coral Reefs* 16 (Supplement): S9-S21.

Huston MA (1985). Patterns of species diversity on coral reefs. *Ann Rev Ecol Syst* 16: 149-177.

Huston MA (1994). *Biological Diversity: The Coexistence of Species on Changing Landscapes.* Cambridge University Press, Cambridge, UK.

Jackson JBC (1991). Adaption and diversity of reef corals. *BioScience* 41(7): 475-482.

Knowlton N, Jackson JBC (1994). New taxonomy and niche partitioning on coral reefs: jack of all trades or master of some? *Trends in Ecology and Evolution* 9(1): 7-9.

Lieske E, Myers R (1994). *Collins Pocket Guide: Coral Reef Fishes: Indo-Pacific and Caribbean.* Harper Collins Publishers, London, UK.

McAllister DE, Schueler FW, Roberts CM, Hawkins JP (1994). Mapping and GIS analysis of the global distribution of coral reef fishes on an equal area grid. In: Miller RI (ed). *Mapping the Diversity of Nature.* Chapman and Hall, London, UK.

Mather P, Bennett I (eds) (1993). *A Coral Reef Handbook: A Guide to the Geology, Flora and Fauna of the Great Barrier Reef,* 3rd edn. Surrey Beatty and Sons Pty Ltd, Chipping Norton, NSW, Australia.

Ogden JC (1997). Ecosystem interactions in the tropical coastal seascape. In: Birkeland C (ed). *Life and Death of Coral Reefs.* Chapman and Hall, New York, USA.

Ormond RFG, Roberts CM (1997). The biodiversity of coral reef fishes. In: Ormond RFG, Gage JD, Angel MV (eds). *Marine Biodiversity Patterns and Processes.* Cambridge University Press, Cambridge, UK.

Paulay G (1997). Diversity and distribution of reef organisms. In: Birkeland C (ed). *Life and Death of Coral Reefs.* Chapman and Hall, New York, USA.

Polunin NVC, Roberts CM (eds) (1996). *Reef Fisheries.* Chapman and Hall, London, UK.

Pyle RL (1996). Exploring deep coral reefs: how much biodiversity are we missing? *Global Biodiversity* 6: 3-7.

Reaka-Kudla ML (1997). The global biodiversity of coral reefs: a comparison with rain forests. In: Reaka-Kudla ML, Wilson DE, Wilson EO (eds). *Biodiversity II: Understanding and Protecting our Biological Resources.* Joseph Henry Press, Washington DC, USA.

Roberts CM (1997). Connectivity and management of Caribbean coral reefs. *Science* 278: 1454-1457.

Roberts CM, Ormond RFG (1987). Habitat complexity and coral reef fish diversity and abundance on Red Sea fringing reefs. *Mar Ecol Prog Ser* 41: 1-8.

Sale PF (ed) (1991). *The Ecology of Fishes on Coral Reefs.* Academic Press, San Diego, USA.

Spalding MD, Blasco F, Field CD (1997). *World Mangrove Atlas.* The International Society for Mangrove Ecosystems, Okinawa, Japan.

Spalding MD, Grenfell AM (1997). New estimates of global and regional coral reef areas. *Coral Reefs* 16: 225-230.

Veron JEN (1995). *Corals in Space and Time: The Biogeography and Evolution of the Scleractinia.* UNSW Press, Sydney, Australia.

Veron JEN (2000). *Corals of the World.* 3 vols. Australian Institute of Marine Science, Townsville, Australia.

Chapter 2
Signs of Change

Coral reefs are a rare but critically important resource. For the most part they are located far from major towns and cities, and many of the largest expanses of reefs in the world are located in some of the most remote areas of the planet. Despite this apparent scarcity and remoteness, coral reefs are probably the most familiar marine habitat to peoples across the globe. In the developed nations they feature in ecology lessons, holiday brochures, advertising campaigns, tropical aquaria and numerous wildlife documentaries. In many of the countries in which they occur they are a critical part of everyday existence, providing coastal defense, a source of food, recreation and income. These same reefs are increasingly being seen at the global level as a rare resource of incalculable value for current and future generations, a global heritage of incredible beauty, immense productivity and unparalleled diversity. In this chapter we explore the different ways in which humans have used reefs over the millennia and particularly in the present day. We go on to consider the problems which many of these uses are now creating for the survival of coral reefs. Finally we consider some of the positive actions that are being taken to prevent further damage or to reverse the decline; actions which benefit not only the reefs themselves, but also the populations who rely on them.

Left: Dar es Salaam, Tanzania. Sprawling coastal towns and cities are bringing countless pressures to bear on adjacent reefs. Right: Tourism to coral reefs has boomed in recent years, and is bringing critical income to many coral reef nations.

The importance of reefs

Around the world, reefs are an enormously valuable resource. In some cases, efforts have been made to calculate simple economic statistics to capture this value, but it is equally important to look beyond dollar values. Reefs are a source of employment, a source of food, a source of protection and a source of leisure. In many places where there are reefs there are few other natural resources which can replace such functions, and in this sense a simple economic value pales into insignificance against social, nutritional, cultural and other measures. Perhaps the earliest services to humans ever provided by reefs were those of protecting coastlines and even creating new land. Former reefs underlie many coastal lands, while at the present time many islands, and indeed entire nations, are built on the coral rock and sand which is an integral part of a living reef ecosystem. Coastal populations have similarly relied upon reefs for food since pre-history. Today many coastal communities rely on reefs in a similar way, catching fish for their own use, but fisheries have also diversified considerably, and reef fish, molluscs and crustaceans are now a major part of several commercial fisheries. Recent times have witnessed other uses of reefs, most notably in their explosive popularity as places for recreation and tourism in the last few decades.

Fisheries

Reefs, and particularly reef flats, have been used as an abundant and productive source of food for millennia. The earliest dated use comes from middens (waste dumps) containing shells and other remains in localities as far apart as Australia and the Americas, showing active gathering of reef flat organisms for food. The oldest site currently known is Matenkupum in Papua New Guinea, where fish bones and shells gathered by humans from reef flat and mangrove areas have been dated back some 32 000 years.

The initial use of coral reefs for food was undoubtedly restricted to the reef flats. Fish and shellfish were probably captured directly by hand, or by nets, spears, poison or traps. There is little or no evidence of outer reef-slope species in these early middens, and no evidence of fish hooks. However, offshore fisheries probably developed very early. The travels of Polynesians, ancient Egyptians and others are clearly dated back at least 3 000-4 000 years and show considerable seafaring abilities. It seems likely that coastal navigation pre-dates this by several millennia, and would undoubtedly have been linked to the utilization of the rich resources which lay beyond the reef flat.

Today, reef fisheries are globally significant. In many nations, particularly those of the Pacific islands, coral reefs provide one of the major sources of animal protein. In a number of countries, including the Maldives, Kiribati, Tokelau and Tuvalu, an average of over 100 kilos of fish are consumed per person every year. In each of these cases the majority of these fish are nearshore species taken by artisanal fisheries. At the same time, commercial fisheries have been developing rapidly in many areas, with sales not only locally but also to export markets.

Fishing methods

Coral reef fisheries are typically multi-species fisheries and, because of this and the complex nature of the reef environment, they are also typically multi-gear fisheries. Only large industrial fishing gears are excluded (long lines, drift nets and most trawls) but, apart from these, almost every technique has been utilized at some point or other.

Gleaning, or harvesting by hand, is still one of the most widely used and effective methods for collecting certain species. This is especially common on the reef flat, where crustaceans, molluscs, sea cucumbers, sea urchins and a host of other species may be collected by hand. In many societies such work is traditionally undertaken by women and children, while men may fish from boats

Fisherman with a cast net, Antigua.

offshore. Away from artisanal communities, harvesting by hand is still the predominant method for collecting many species, including the commercial fisheries for trochus, conch, clams, lobster and sea cucumbers.

The use of nets on reefs is widespread. Cast nets are often used in shallow water. These are circular nets with a weighted perimeter. Typically the fisher stands in the water or walks towards a school of fish, then casts the nets over them. The weighted edges fall first, encircling and trapping the fish. Fixed nets are also placed on the reef flat and reef slope. Although they are a highly effective means of fishing, nets often get snagged on coral and cause breakage, and are sometimes abandoned.

Fish traps range in size from small portable structures carried on boats through to sizeable structures built of stone, wood or bamboo directly on the reef flat. Typically the former use baits to attract fish, while the latter rely on fish becoming trapped at low tide, or may involve actively driving fish into the traps.

The use of spears is still widespread in most areas. Prior to the availability of masks to see underwater, most spearing was done from the surface, and this is still the case in some places. Single or multi-pronged spears are stabbed or thrown at the targets. Underwater spearing was originally dominated by long spears which were lunged or stabbed at the target species. The use of spear guns is now a widespread and highly effective means of fishing. Spears are typically propelled by a rubber cord, and are attached to the gun by a narrow line which enables the fisher to hold on to the spear and fish after firing.

Hook and line fishing is also widely used on reefs. Historical records of hooks go back many centuries, and there are examples made from mother of pearl, turtle shell, wood and many other materials. Today they are almost exclusively made from metal and attached to modern monofilament lines (metal traces are also commonly used, particularly for sharks which can bite through polypropylene line). Hook and line is used both on reef flats and over the reef slope. Techniques involve stationary baited lines and also trolling, in which lines are towed behind a boat and usually attract fish with a lure rather than bait.

Driving fish into traps or nets is still undertaken in a few areas. Traditional methods involved the setting of a trap (fixed structures of the type described above, or an equivalent structure made up of fixed nets) and then chasing or frightening the fish into the trap using several people, sometimes carrying leaves or other items to help corral the fish. Two further destructive methods of driving fish into nets on the reef slope have been noted in Southeast Asia. *Muro-ami* fishing involves a line of breath-holding divers swimming down below a large suspended net armed with poles or rocks on lines. The divers literally smash the reef with the rocks or poles to chase the fish upward into the nets. (The divers are often children, and there is little or no concern for their welfare or safety.) Now illegal everywhere, some *muro-ami* fishing has been replaced by a technique known as *paaling* in which the divers hold hoses with compressed air and drive the fish towards the nets with walls of bubbles.

Poisons or stupefactants have been used as a means of catching fish for many years. In traditional societies particular species of plant have been found to stun or kill fish. In more recent times detergents and bleaches have been found to have the same effect, while sodium cyanide is now the most widely used chemical. These materials are placed in tide pools, poured into the water or more directly squirted into the nooks and crannies of the reef slope. The fish are then simply collected by hand.

Blast fishing is the most destructive fishing method on reefs. Explosives are usually home-made, often using fertilizers, although dynamite is sometimes used. They are typically thrown by hand towards the reef and explode on the water surface. The shock wave from the blast kills all the fish which have gas-filled swim bladders (the majority of species). It is non-selective, and many of the individuals are lost as they sink to the bottom or become caught among the coral.

The targets

Within the great diversity of life on the reef there is a similar diversity in the species that are targeted for food, and tastes and traditions vary across the globe. Some groups, such as groupers and lobsters, are taken just about everywhere. Rabbitfish, parrotfish, snappers and emperors are also widely taken. In areas where tourism is dominant, certain species are often particularly popular and highly valuable, including grouper and snapper,

Sharks have increasingly become a target, and are becoming rare in almost all areas.

lobster and conch. Among other groups, taste varies between localities and certain species are not popular, either a result of long-standing traditions or more recent reputations. The taste for shark flesh, for example, is highly localized. While many cultures do not like to eat shark there are others where it is very popular. Certain Pacific islanders have traditionally hunted sharks for many years, luring them alongside small fishing boats and capturing them with nooses in the days before strong hooks and steel traces. Marine turtles and their eggs were once popular worldwide, but the last century has seen their decimation in almost all areas and most countries now forbid or greatly restrict their hunting. The presence of ciguatera, a natural toxin (see page 31), in certain fish species appears to have some regional variation, but in many areas where it is more prevalent this has led to the avoidance of certain groups, such as jacks, barracuda and large snappers, which are more likely to carry the toxins.

Many coral reef animals have complex breeding cycles which involve regular spawning at the same locality. Such occurrences are often linked to the phases of the moon and may be monthly or annual, and in some cases they involve very large aggregations of fish, which may have swum many kilometers to reach the spawning site. Scientific knowledge of these patterns is very recent, but in many cases they have been known to fisherfolk, particularly in the Pacific islands, for many years. Such aggregations make highly productive fishing sites, but they are also extremely susceptible to overfishing and it is possible in the course of only one or two spawning events to decimate an entire population of a species over a wide area of reef. Traditional societies have often recognized the importance of these sites and imposed restrictions to prevent overharvesting. One rather unusual species targeted by traditional fishers in a number of Pacific islands is the palolo worm, which is the breeding phase of a rock-dwelling polychaete worm *Palola siciliensis*. These swarm to the surface typically for two nights in succession every 12 to 13 lunar months, when they become the center of a great celebration, collected in great numbers and eaten either raw or cooked.

The demands of non-traditional societies combined

The live fish trade

Recent years have seen a spectacular growth in the demand for live fish in a number of Chinese restaurants in East Asia, notably in Hong Kong. Particular species are targeted, predominantly groupers and the humphead wrasse *Cheilinus undulatus*. In 1997 it was estimated that Hong Kong imported some 32 000 tons of live food fish. Typical wholesale prices for these were US$40-100 per kilo, such that the overall annual value of this industry was estimated at US$500 million for Hong Kong alone. Market prices are considerably higher.

While smaller fish are favored by many, another aspect to this trade has been the purchase of the largest possible fishes, which have become a status symbol in major celebrations and business dinners. There are records of individual fish measuring more than 2 meters in length retailing for more than US$10 000. The impact on the natural stocks of these species has been considerable. With such high values, many of these species have been all but decimated from the reefs around Southeast Asia, and vessels are collecting live fish from locations in the Western Indian Ocean and into the Pacific to meet demands in Asia. Unfortunately it is these largest individuals which have the greatest reproductive potential and so recovery from such drastic overfishing is likely to be slow.

Live fish in a holding pen awaiting shipment to the restaurants of East Asia.

The aquarium trade

An estimated 1.5-2.0 million people worldwide keep marine aquaria, approximately half in the USA and a quarter in Europe. For the most part, these are hobbyists who maintain tropical fish stocks in home aquaria. Dedicated enthusiasts are able to propagate many species of coral and fish, but most aquaria are stocked from wild-caught species.

In recent years the aquarium industry has attracted some controversy. Opponents to the trade draw attention to the damaging techniques sometimes used to collect fish and invertebrates, and to high levels of mortality associated with insensitive shipping and poor husbandry along the supply chain. Aquarium species are typically gathered by local fishers using live capture techniques (such as slurp guns or barrier and hand nets) or chemical stupefactants such as sodium cyanide. The latter is non-selective, and adversely affects the overall health of the specimens as well as killing non-target organisms.

Supporters of the aquarium industry maintain that it is potentially highly sustainable, that proper collection techniques have minimal impact on the coral reef, and that the industry is relatively low volume but very high value. There is little disagreement about the latter – a kilo of aquarium fish from one island country was valued at almost US$500 in 2000, whereas reef fish harvested for food were worth only US$6. Aquarium species are a high value source of income in many coastal communities with limited resources, with the actual value to the fishers determined largely by market access. In Fiji many collectors pay an access fee to the villages to collect on their reefs, but by selling directly to exporters they can have incomes many times the national average. By contrast, in the Philippines there are many middlemen, and collectors themselves typically earn only around US$50 per month.

The controversy over the benefits and costs, in terms of environmental impact, of the trade persists largely because of a lack of quantitative data. For some species, trade data are actually available under the provisions of the Convention on International Trade in Endangered Species of Wild Fauna and Flora (CITES). Under this convention regulated trade is permitted in species listed in Appendix II, which are vulnerable to exploitation but not yet at risk of extinction. All species of hard coral and giant clams are listed under Appendix II. Shipments of corals and clams involving Parties to the Convention must be accompanied by a CITES permit issued by the national CITES management authority. Parties to CITES are then obliged to

with improvements in transportation and the availability of refrigeration have greatly affected reef fisheries in many areas. These new markets are typically highly focussed towards single species. A number of species which were only eaten at the local level have hence become highly valuable, and are now largely taken by export fisheries. Eastern Asia, particularly China and Japan, is a large market for a number of species, including sea cucumbers, clams and sharks. The live fish trade is also extremely popular in a few markets in East Asia (see page 49). Western markets have tended to focus on lobster, conch and particular finfish.

Productivity

Considerable effort has been expended in trying to determine the size of fishery a reef can sustain. Obviously any such figure will depend on the degree of impact considered acceptable; the number of species which are likely to be utilized; and on numerous external factors such as light, temperature and nutrients. In fact almost any level of fishing will have some impact, not only on the population of the target species, but on the reef community. Every species exists in relationship with other species, whether as predator, competitor or prey, so that removal of individuals will alter some of these balances. At low levels, such changes are impossible to detect and may be indistinguishable from natural variation. But as fishing pressure increases it is possible to detect impacts through changes in the size, density and biomass of individuals and the age structure of the population. From the fishers' perspective there can also be changes in the catch per unit of effort as fishing levels increase, and in extreme conditions particular species may disappear completely from the reef.

Quite a number of studies have attempted to look at the actual levels of yield on selected reefs and have shown figures ranging from about 0.2 to 40 tons per square kilometer per year. Such figures do not indicate sustainability, only utilization, and they are highly dependent on the calculations of area fished. One of the best studied and most heavily utilized reefs is Bolinao in the Philippines, where some 17 000 people are "employed" in the utilization of about 68 square kilometers of coral reef.

produce annual reports specifying the quantity of trade that has taken place in each listed species.

In 1997 a total of 1 200 tons of coral was traded internationally, with 56 percent imported by the USA and 15 percent by the European Union. Approximately half of this was live coral for aquaria, a tenfold increase on the amount of live coral traded in the late 1980s.

No marine aquarium fish or invertebrates other than clams or corals are listed under CITES. Only very rough estimates of trade figures are available, suggesting that 15-20 million fish from approximately 1 000 species may be traded per year. The trade in individual fish species is unknown, and no data exist on the extent of the trade in invertebrates other than corals and clams.

Thus, while the current impacts of the aquarium trade remain poorly known, the potential for this industry is considerable. It is quite possible to manage aquarium fisheries at low and sustainable levels. It is further possible to capture live species using non-destructive techniques. With well managed shipping and husbandry, mortality can also be kept very low, as has already been shown by some operators in the industry. Targeting mostly non-food species, aquarium fisheries can, in theory at least, provide an alternative economic activity for low income coastal populations and an important source of foreign exchange for national economies, as well as a stronger economic incentive for the sustainable management of reefs. The application of international certification schemes may provide an important tool for achieving this.

A branching coral ready for shipment in a trade which is now monitored by CITES.

It seems likely that sustainable yields from reefs will actually vary considerably depending on local ecological conditions. There may also be substantial regional differences. Estimates from the Caribbean suggest that 4-5 tons of fish per square kilometer per year may be sustainable, while figures may be higher in Southeast Asia. Such figures are calculated on the basis of a multi-species fishery. Where only a small sub-set of fish species is taken, much lower figures would be expected.

Aquaculture

In addition to the fisheries already mentioned, aquaculture is of increasing importance in many coral reef countries. In the broader coral reef environment, shrimp farming is probably the most widespread and high value form of aquaculture. It is also, typically, one of the most damaging to the environment, and often to the human communities nearby. By contrast there are several other aquaculture activities which have little or no impact on the surrounding ecosystems and appear to provide a sustainable and potentially high value use of coral reef habitats. Such aquaculture includes pearl oysters, clams and algae, as well as a range of corals and other species for the marine aquarium trade. The coral reef environment not only provides both the space and the ideal conditions for cultivating these species, but in many of these industries remains critical to the supply of nutrients as well as wild stock with which to initiate and maintain the industries.

Shrimp farming is more typically associated with mangrove areas than with coral reefs, and large areas of mangrove in the tropics have been converted to shrimp farms. Conversion itself often involves near complete destruction of forests, with the digging of wide pools for the rearing of the prawns. The process of developing and running such an operation can lead to the production of considerable sediments and the release of large amounts of nutrients and other chemicals utilized in the intensive production process, and these can be damaging to nearby reef ecosystems. All too often poor management means that these farms have a highly limited lifespan, although the profits may be very large indeed during the few (often 5-10) years of operation. Closure of the farms is rarely tied to any ecological restoration and, all too often, wastelands are left with unproductive and still highly polluted pools in the place of former mangroves.

Pearl collection can be traced back several millennia to the pearl divers of the Middle East. However, the farming of pearl oysters is more recent, occurring only since the techniques for pearl culture were developed. It is now widespread in a number of reef regions, but most notably with the culture of blacklip pearl oysters *Pinctada margaritifera*. One of the largest producers is French Polynesia, where extensive pearl farms have been developed in the lagoons of a number of atolls. The oysters, which are filter feeders, are suspended in the water column. Thus far no adverse long-term effects from these farms have been noted on the surrounding reefs, although there have been mass mortalities, with associated temporary eutrophication of parts of the lagoons on a few occasions.

The culture of particular organisms on the reef flat is also increasing. Aquaculture of giant clams (several species of the family Tridacnidae) has been developed in a number of Pacific islands and in Australia. Farmed clams are used both in the aquarium trade and as food, both as a high value export and for local consumption. Another perhaps more widespread activity in reef flat and shallow lagoon areas is the cultivation of algae. Seaweeds are used both as a source of food and as a raw material from which a group of natural gums are extracted for use as thickening or gelling agents, including sodium alginate, carrageenin and agar. Various species are exploited, including *Eucheuma* and *Sargassum*. Algal farming is widespread in countries of the Pacific and Indian Oceans, with the largest tropical exporters being the Philippines, Indonesia and Tanzania, and it is also being developed in some parts of the Caribbean. These and other activities clearly impact the natural ecosystem function, both through the smothering effect by the cultured organism over existing benthic species, but also from the trampling of the reef flat. At the same time, however, their impact is limited to a small area and may be more than countered by the reduced impact on other parts of the reef as local communities turn their attention to aquaculture.

Other types of aquaculture in reef regions include the rearing of trochus (a mollusc used for food but also an important species in the mother-of-pearl industry), and a small but growing interest in sponge farming. Efforts are also underway to develop techniques for culturing sea cucumbers (holothurians). The culture of corals and fish for the marine ornamental trade is a more challenging activity, and is often undertaken under more controlled conditions in large-volume aquaria adjacent to coral reefs, or in the importing countries of North America and Europe.

Other reef products

Harvesting of reef resources also goes beyond food values, and mention has already been made of the use of aquaculture for pearl production. The aquarium trade is a relatively new use of reefs with significant economic importance in some areas (see page 50). Reefs are also widely used for the collection of materials for jewellery and other ornaments, while the bare materials of the reef have often been used as a base for construction.

Mother-of-pearl ornamentation can be traced back to Thebes in Egypt in 3200 BC, and pearls themselves have been found in China dating back to 2500 BC. While their

Left: Giant clam Tridacna gigas *amidst lines of cages of juvenile clams. Mariculture of clams is now taking place in a number of countries in the Pacific. Right: Shells for sale on the streets of the Seychelles. At low levels such activities may be sustainable.*

exact origins remain unclear it is highly likely that reefs were being used as one source of such products. The use of cowrie shells as currency was another ancient and widespread utilization, and at least some of them would have come from reefs. Examples of shell currency have been found across Africa and South Asia to China. There is evidence of a trade in these shells from India to China from records back to 400 AD, while the Maldives became known as the Money Islands because of the preponderance of shells in those areas. Cowries were widely used in some cultures into the mid-19th century.

The jewellery and curio industry

Pearls and coral are still widely harvested today for international trade. For pearls, the development of aquaculture techniques has greatly influenced the value and geographic spread of pearl culture. Today the pearl industry is dominated by Australia (especially Western Australia), Japan and French Polynesia, but pearl aquaculture is widespread across island locations in the Pacific and there is a continuing small-scale natural pearl industry, notably in India.

Corals, particularly red, pink and black, are widely used in the jewellery industry, and in many cases this use can be traced back to antiquity. While these corals are not restricted to coral reef regions, reefs are a major source, particularly for black corals. This industry was unsustainable in many countries and, like the trade in scleractinian corals (see box on the aquarium trade on page 50), is now strictly regulated under CITES. The USA is again the major consumer of these products, while Taiwan and the Philippines have tended to dominate the export statistics. Within the USA – in the Hawaiian islands – there is also a substantial black coral harvest. This is relatively well managed, with a minimum size restriction preventing overutilization.

The use of turtle shell was once very popular in jewellery and other decorative ornaments, but the scarcity of marine turtles combined with strict controls on the trade in turtle products has now greatly reduced the size of this trade.

Building materials

The earliest structures to have been built from coral and reef rock were probably walls and pens associated with the capture of fish, but there is also a long tradition of removing materials from coral reefs and nearby lagoons and sand flats, particularly for building purposes. The use of coral rock in the construction of stone buildings also goes back many centuries, notably in houses along the Red Sea and in the Maldives.

Today reef rock is still widely used in the Maldives, and in other coralline island nations where there is no other natural building material. Similarly, sand and reef rock are often dredged from reef flats and lagoons for the construction industry, despite the immediate consequences that such actions have on the reefs and nearby beaches. Apart from high levels of siltation which often smother and kill the adjacent reefs, the extraction of sand and rock frequently leads to coastal erosion and to the massive costs associated with trying to maintain or stabilize the coast, or to losses associated with the collapse of buildings and roads into the sea. In many areas the reef flat is also used as a location for land reclamation. Such activity frequently leads to the partial or complete loss of live corals on the adjacent fringing reef. On the Egyptian coast of Hurghada the entire fringing reef has been killed as hotels have encroached it. This reef was one of the main attractions of this coastline, but tourists must now travel by boat to see coral reefs which were once thriving just a few meters from their hotel beds.

Genetic treasure house

The genetic diversity to be found on reefs is unparalleled. Furthermore, within the ecological complexity of the reef there are countless examples of interactive evolutionary processes that have driven genetic diversification down interesting and potentially valuable paths from a human perspective. In such a world, the evolution of complex secondary metabolites is common, particularly the development of toxins as a means of defense or attack. Such compounds may be mirrored by the evolutionary development of other chemicals to counter their impacts. Toxins are of particular interest in pharmaceutical research, and the coral reef abounds in such substances. Stonefish, sea snakes, box jellyfish, cone shells and pufferfish contain some of the most toxic compounds

A wall of coral rock, Maldives.

known to date, but these are just the tip of the iceberg. Huge numbers of species, but particularly the sessile or slow-moving invertebrates such as sponges, bryozoans, ascideans and nudibranchs, contain a vast panoply of complex and potent chemical compounds.

Traditional societies have used compounds from the reefs in traditional medicine and, while perhaps the majority of examples are lost and forgotten, there are some, such as the use of terebellid worms in Hawai'i and of rabbitfish gall bladders in Palau, which indicate ongoing perceived benefits from the active properties of particular species.

Bioprospecting for potential new pharmaceuticals has been ongoing in terrestrial environments for many decades but is still relatively new in the marine environment. Now, however, considerable efforts are being put into gathering and screening new genetic materials from coral reefs. Anti-cancer properties are being investigated in a number of products derived from marine life, including cryptophycins from blue-green algae, and other metabolites from planktonic dinoflagellates. Pseudopterosins derived from Caribbean sea-whip corals are being investigated for skincare and anti-inflammatory properties. The compound manoalide, derived from a Pacific sponge, has now spawned more than 300 chemical analogues being tested for anti-inflammatory properties. Initial tests on a neurotoxin derived from a Pacific marine snail are showing very powerful painkilling properties. These are only published examples; there are likely to be many others under development which remain undisclosed to the wider public.

Such activities are not without controversy. There is concern that host countries may lose ownership and a share of any profits from new pharmaceuticals or other compounds developed by the drug companies. There is also concern that collection of more unusual and rare species may actually impact the global populations of these species. In both cases these problems can be solved with detailed management and cooperation.

The value of coral reefs as a standing stock of genetic diversity is considerable, but in many ways this value is potential, just beginning to be utilized. For this reason it is almost impossible to value in an economic sense, but the benefits could be vast.

Tourism and recreation

Coral reefs are among the most visually impressive habitats on the planet, burgeoning with life and dazzling with color. Despite this, the visual appreciation of these habitats is a recent phenomenon. The first use of modern mask, fins and snorkel can be traced back to the Mediterranean in the 1920s and 1930s, while the development of the first fully automatic aqualung and regulator did not occur until 1942. Diving as a popular activity began to spread in the 1950s, and probably the first diving club in the tropics was established in Jamaica in 1957. In less than 50 years diving has grown from being an obscure (and dangerous) sport to being one of the world's most popular adventure activities. PADI, the world's largest diver certification organization, provided certification for over 800 000 divers in 1999, and has provided certification for over 8 million divers since its establishment. Worldwide there are now over 15 million recreational divers, certified under the various existing dive training organizations.

Just as the sport of diving has developed, the concept of rapid and relatively cheap international travel and the economic boom and development of many countries around the world have led to tourism becoming one of the world's most important industries. These two factors have combined, and coral reefs are the favored destination of amateur divers around the world. In addition to certified divers, many millions more undertake single dives, enjoy

Divers pay premium prices for close encounters with charismatic creatures such as this loggerhead turtle, Caretta caretta.

snorkelling over reefs, or view the reefs from glass-bottomed boats. As a part of the present work an entirely new dataset has been compiled, listing dive centers which offer certified dive training in all the coral reef areas of the world. These centers are shown on the detailed maps in the second section of the book, but are broadly illustrated here, showing that there are now few areas of the world where it is not possible to dive on coral reefs.

Tourist numbers to particular reefs can be very large indeed. In 1985 there were around 1.1 million visitors per year going to the Great Barrier Reef in Australia, but by 1995 this figure was over 10 million. In 1997, the value of this tourism to the Great Barrier Reef was estimated at over US$700 million. In a much smaller area, the boom of tourism in the southern parts of the Egyptian Sinai Peninsula is vast, with numbers of tourist rooms increasing from under 600 in 1988 to over 6 000 by 1995, and 16 000 by 1999. Visitor numbers in many parts of the Caribbean are often even greater. Calculations of the total value of tourists to reef areas are difficult. While direct spending may include hotels, diving and park fees, the indirect spending on associated travel, subsequent travel to other areas within a country, or other activities undertaken as part of a tourist package, may more than double the total value of this tourism to a country or region. It also provides considerable employment in a range of sectors (hotels, diving, fishing), as well as presenting a strong incentive for the sound management of reef resources.

For the most part, scuba diving is a recreational activity only enjoyed by the relatively wealthy, and there exists, in many places, a considerable dichotomy between those who enjoy coral reefs as a place for recreation and those who may be resident in coral reef areas, but are unable to do likewise. In a few places considerable efforts are being made to develop a greater appreciation and understanding of coral reefs among those resident people who are unable to afford expensive holidays or diving. As such an appreciation is developed so local recreation on coral reefs is likely to increase, and with this a desire to protect reefs not only because of their economic value, but also their great beauty and high recreational potential.

Coastline protection

Many of the world's reefs can be traced as narrow strips running like colorful barriers close to tropical coastlines. These barriers have a critical role to play in defending those same coastlines from the daily pounding of waves and the scouring of currents and, even more important, the worst ravages of storms. Despite the apparently fragile nature of the individual corals, reefs are able to grow in highly wave-swept environments and gradually build up the vast structures which buffer and defend the coastline. In fact the reefs not only protect coastlines from the worst excesses of storm damage, but are also the source of sand which builds up or replenishes beaches.

During the biggest storms many corals may be broken up, but the coral rubble and sand are often forced up during these same storms into islands or onto beaches, creating new land. The presence of vegetation on these small islands produces organic deposits, while the roots, together with various chemical and mechanical processes, may further help to bind the substrate, creating habitable islands, permanently raised above sea level. Some nations, consisting only of these small coral islands, are wholly dependent on these processes for their very existence.

This function of reefs is again difficult to quantify in terms of its value. Furthermore, in the short run, the ecological degradation of a reef may not greatly impair this function, although there is good evidence that the powers of erosion are relatively rapid and that a dead reef will begin to lose even its physical structure within years or decades.

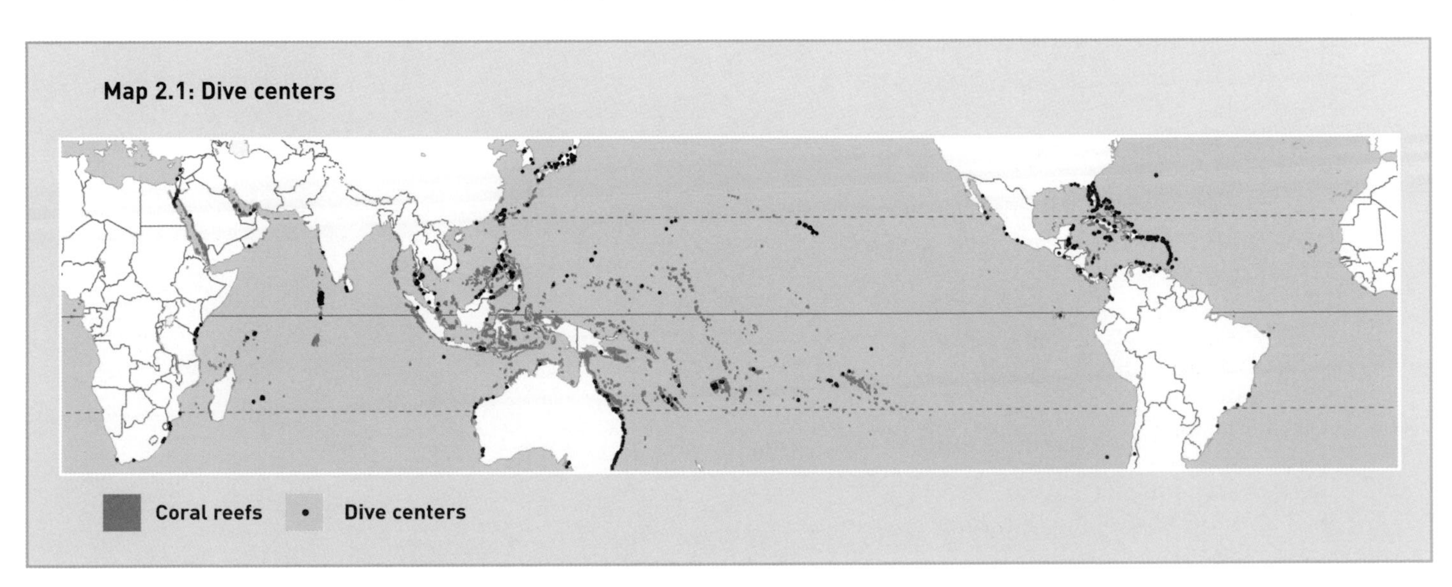

Threats to reefs

Coral reefs in their natural state have often been considered as highly stable ecosystems with their tight ecological complexity protecting a finely balanced and highly sensitive ecosystem. Even minor changes to the species balance, it is suggested, will have repercussions across the ecosystem. Other ecologists have challenged this view and pointed to the many apparently catastrophic events which are regular occurrences on some of the world's reefs. There is evidence that certain levels of disturbance, such as physical damage from major storms, or even the impacts of diseases, may actually be of benefit to a reef ecosystem over the longer term. Without the constant shifts and changes of conditions which these natural events produce, conditions might favor particular species to the detriment of others, and diversity on the reefs might be reduced. Arguments about the true stability and complexity of interrelationships on coral reefs will continue for many years, but it remains clear that reefs have survived and even thrived in areas where disturbance can be a regular occurrence. Unfortunately, it is equally true that recent decades have created levels of stress to coral reefs that are unprecedented in the recent history of the planet. These human disturbances are not only intense, but are widespread in every part of every ocean. In many areas they also compound one another, or are further exacerbated by natural disturbances.

The true resilience of coral reefs is now being tested to extreme levels over large areas of the planet, and the result has been considerable degradation, with loss of ecosystem function in many areas. In a few places reefs seem to have disappeared completely, and have been replaced by other ecosystems of lower diversity and productivity. As such changes occur, the many important goods and services which the reef formerly provided become diminished or are lost.

Most human-induced impacts fall into four broad categories: pollution, sedimentation, overfishing and climate change. Direct physical damage presents a fifth factor, more limited in spatial extent, but in some cases more damaging and irreversible than any of the others. The actual effects of these impacts on particular reefs may vary. They may be complicated by their interactions with one another, or be countered by natural mitigating factors or further human intervention. In a few cases, changes or apparent degradation on the reefs cannot be directly traced to single, or even multiple stressors, and yet there may still be a link to human impacts.

In the following account, the effects of these broad categories of stress are first considered. More complex phenomena are then addressed independently: the crown-of-thorns starfish, the *Diadema* die-off and the global patterns of coral diseases. As threats combine so the resulting impacts may be greatly exacerbated. A final section looks at how this may work in practice, and further summarizes a recent global study which attempted to map the global patterns of stress to the world's coral reefs.

Left: Shark fins are being gathered from around the world, predominantly for export to East Asia in a trade which is far from sustainable. Right: Coral bleaching has become a widespread phenomenon. In extreme events other species, such as this anemone, have also been observed to bleach and die.

Pollution

The major form of pollution on coral reefs is nutrient enrichment, primarily linked to human waste, but also to agricultural runoff. Recent decades have seen burgeoning human populations in coastal areas close to reefs, and rapid urbanization of many societies. Many countries, particularly the poorer nations, have failed to develop sewage treatment systems able to cope with growing urban areas, and vast amounts of sewage enter coastal zones through the drainage system or are directly piped into nearshore waters. Even away from urban centers, coastal development has occurred at a considerable pace. Tourism in particular has led to this kind of development. While it is smaller scale, it actively seeks out sites immediately adjacent to the coastline, and in many cases there is again little or no sewage treatment. In addition to sewage pollution, agricultural development in many countries has entailed the enrichment of the land with nutrients in the form of artificial fertilizers, many of which enter the drainage system and are carried to coastal waters.

The impacts of eutrophication on reefs are complex. Typically there are changes in the community structure. Algae flourish in situations of high nutrient loads, and can overgrow and kill corals or prevent new settlement of coral larvae. Algal blooms are also sometimes observed in the water column, with the further effect of reducing light levels for the corals below. Among the rest of the reef community, particle-feeding organisms such as sponges and zoanthids may become more common, and plankton-eating and herbivorous fish may increase. More subtle changes, such as a reduction in the reproductive and growth capacity of particular species may also result, but may not be immediately apparent. These changes may reduce the reef's resilience to further changes or its ability to recover from events such as hurricanes. They may also affect the sustainable yield of reef fisheries. In extreme conditions, nutrient pollution alone is sufficient to kill a reef and lead to its replacement by algal communities.

Toxic pollutants receive less attention and have been less well studied. Oil is perhaps the most common pollutant in many remote areas. Spills have occurred widely in coral reef regions. Continuous chronic oil pollution results from minor spills or deliberate discharges of oil. These occur in oil drilling areas such as the Arabian Gulf, and in certain localities such as the Straits of Malacca, the Straits of Hormuz, the Gulf of Aden and the approaches to the Panama Canal, where vessels often discharge ballast water or clean out their tanks. Other toxic chemicals are released by a large range of industries, including tailings from mining. The effects of oil pollution on corals are varied. In some cases polluted reefs have shown slightly higher mortality rates than unaffected reefs, and the reproductive potential of surviving corals is significantly reduced. A major spill in Panama led to a decrease in total coral cover of 50-75 percent in shallow water reefs.

Sedimentation

Reef corals are highly sensitive to the impacts of sediments. As sediments hang suspended in the water column they block light from the reef below, preventing the growth or even survival of corals, particularly in deeper waters. As the sediments settle they may smother corals. While corals are able to remove sediments, secreting mucus and then sloughing this off into the water, such an activity uses energy and nutrients. In the long term the smothering effects of sediments weakens many corals, reducing growth or reproductive potential and leaving them less able to compete with other benthic organisms such as algae or filter feeders. Even as sediments settle they create a surface which is inimical to new coral settlement and growth, while light sediments are often resuspended by wave action and hence may remain in the system for considerable lengths of time.

Increasing levels of sediment in coastal environments in recent times can be linked to coastal development, dredging, land reclamation, and also distant terrestrial activities, including deforestation and poor agricultural practices. In many cases the influences of sedimentation and pollution are combined, and the differential impact of each is somewhat difficult to determine. Mine tailings in Papua New Guinea are reported to have degraded or destroyed significant areas of reef around Bougainville, an area of little or no nutrient impacts. However, even here these impacts may have been exacerbated by toxic compounds present in the tailings themselves.

Unsustainable fishing

Reefs are highly productive ecosystems. It is possible to harvest some of this productivity for human consumption, without compromising the overall functioning of the reef ecosystem, and humans have utilized and depended on reefs in this way for millennia. For many traditional cultures the relationship between local populations and their adjacent coral reefs can best be seen as an ecological one, where humans play a role as a part of the reef ecosystem.

Just as with other fisheries, there are limits to this utilization. Beyond these limits fish stocks decline and catch per unit of effort begins to fall. Fisheries scientists regularly talk in terms of maximum sustainable yields, although exact figures are difficult to ascertain and may be highly variable over space and time. A more cautious statistic, which is probably more appropriate for the management of poorly understood coral reef fisheries, is that of maximum economic yield. This is defined as that yield which gives the highest possible economic return for the effort expended. This represents a lower total catch for the simple reason that catch per unit of effort begins to decline as fish stocks begin to be impacted and before the maximum sustainable yield is reached.

Worldwide, reef fish stocks are regularly harvested

Table 2.1: Target species being decimated for specialist markets worldwide

Target species	Notes
Lobster	High value, popular in tourist centers and for export, particularly in the Caribbean
Queen conch	High value, popular for tourist and local consumption, Caribbean
Trochus and green snail	Collected in Pacific islands for meat and mother-of-pearl Have become very rare in some islands
Marine turtles	Popular food (meat and eggs), often for local consumption Use is now highly restricted in most areas
Sharks	Dried fins have very high value and are exported to the Far East Sharks are being decimated worldwide, including in coral reef areas
Sea cucumbers or bêche-de-mer	Although there is some local consumption, high value in the Far East has led to widespread exploitation across the Indo-Pacific
Giant clam	Popular for local consumption and export has led to local extinction in many Pacific islands
Sea urchins	Certain species again popular in Far East export markets
Seahorses	Popularity for Chinese medicine and the aquarium trade has led to widespread losses in some areas
Large groupers and Napoleon wrasse	These and other very large reef fish have been widely removed from many reef areas to supply the live fish trade

beyond sustainable limits. This may not be an entirely modern phenomenon. Changes in fish catch from the midden records in the Pacific suggest that even millennia-old cultures may have overexploited particular stocks. The impact of European culture also led to rapid over-exploitation of some stocks in the Caribbean as long ago as the 17th and 18th centuries, when the Caribbean monk seal began its decline towards eventual extinction, and major turtle rookeries were decimated.

Modern society, particularly in the last few decades, has greatly increased fishing pressure on reefs. Human populations have boomed in almost all coral reef areas. Fishing methods have become more effective, and now include cheap gears such as mass-produced metal hooks and mono-filament line and nets. The means of access to reefs have also greatly increased, with outboard motors and modern building materials for boats. Refrigeration has also enabled the storage of catches taken at remote reefs and the export of catches to overseas markets.

A number of different types of overfishing have been recognized. Economic overfishing occurs when fishing effort is estimated to be in excess of the maximum economic yield and yields per unit of effort are no longer maximized. Growth overfishing is recognized as occurring when the average size of the target fish is reduced towards that of immature individuals. Recruitment overfishing is

Map 2.2: The 1998 coral bleaching event

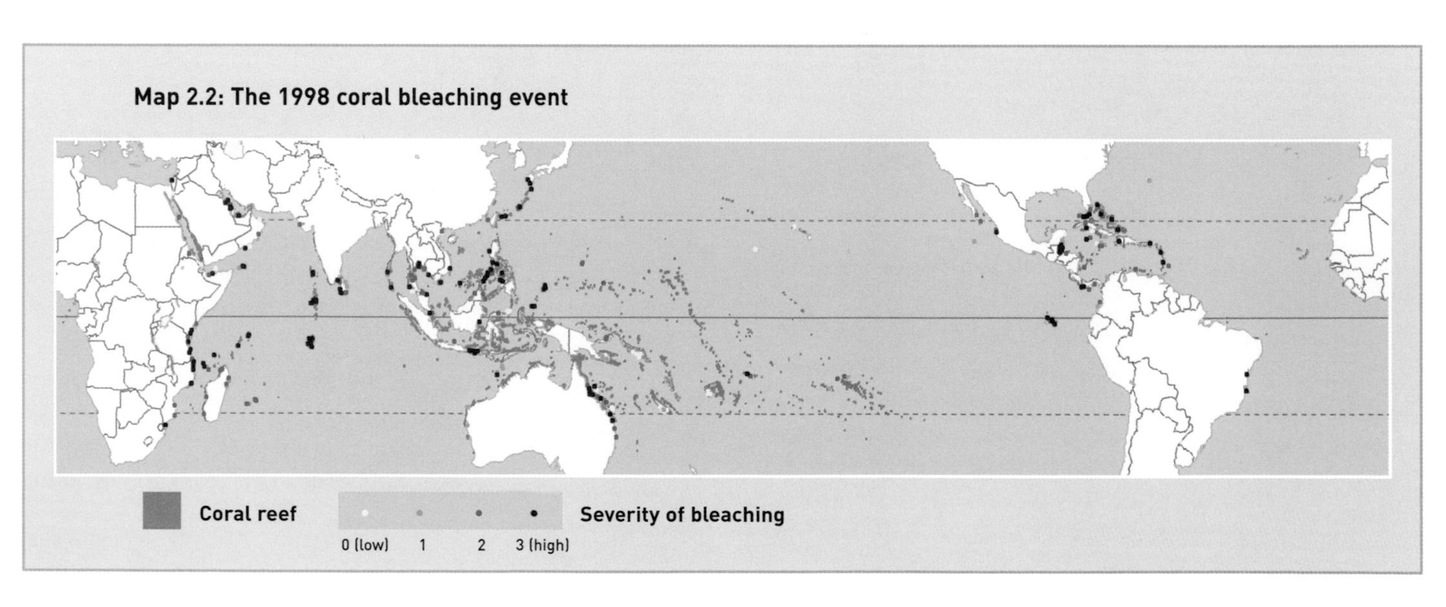

defined as occurring when the size of the adult stock is reduced sufficiently to impair recruitment (the important role of *ex situ* recruitment in the reef environment makes this stage of overfishing difficult to identify). Ecosystem overfishing has been reported where fishing has impacts on the wider community structure. In the multi-species environment of reef fisheries the concept of "fishing down the food chain" is relatively common – when stocks of the more popular piscivorous species have been depleted fishing effort moves down to planktivorous or herbivorous species. One final term has been coined for the most heavily fished reefs of all, that of Malthusian overfishing. This is seen to occur where there are too many fishers for any sustainable form of fishery, but fishing continues, often for reasons of poverty or economics, to the detriment and sometimes the complete destruction of reef communities.

Even far from human populations particular species may be targets for fishing, and the term "target species overfishing" has been coined to describe the focussed removal of particular species for special markets. Some examples of target species are listed in Table 2.1. The key force driving many of these fisheries is economic. Extremely high values commanded by particular products are supporting the often illegal harvesting of even some of the more remote coral reefs around the world. One of the major markets is the Far East. There are records of a single bowl of shark fin soup costing over US$100, and Hong Kong was recorded as importing some 6 400 tons of shark fin in 1999, equating to more than 28 million sharks.

The impacts of fishing are not simply those of over-harvesting. Destructive fishing practices have reduced the productivity of coral reefs in many areas. These include blast fishing and the fish-driving methods of *muro-ami* and *paaling* already mentioned. Trawling is another practice which can affect reefs. Although trawls are not dragged over large reef structures it seems likely that many smaller coral communities on continental shelves have been completely destroyed by large trawl gear in recent years. As most of these smaller structures were never documented the scale of this loss may never be determined. Fishing with poisons may also damage reefs. There is some evidence to suggest, for example, that sodium cyanide used in the capture of live fish may have a detrimental impact on corals.

Most of these destructive fishing methods lead to the flattening or pulverization of the reef substrate, which is of critical importance, providing food and shelter for countless organisms. By limiting the surface area for coral and algal growth and reducing topographic complexity, many species are denied the shelter on which they depend. Recovery of the reef structure from a single blast may take years or decades, and in some of the worst affected areas several blasts per hour are being recorded. Although blast fishing has been recorded in many countries, including parts of the Caribbean and East Africa, it is at its worst, and remains widespread, throughout Southeast Asia.

The polyps of the boulder star coral Montastrea annularis, *each just a few millimeters across. Those to the lower right are bleached, while the remainder are mostly their original color, although even the tips of these polyps are beginning to lose their color.*

Climate change and bleaching

Coral reefs are highly sensitive to climatic influences and appear to be among the most sensitive of all ecosystems to temperature changes, exhibiting the phenomenon known as coral bleaching when stressed by higher than normal sea temperatures.

Coral bleaching is the term used for a loss of color in reef-building corals and the subsequent visibility of the underlying (white) skeleton. Reef-building corals are highly dependent on a symbiotic relationship with microscopic algae (zooxanthellae, see Chapter 1) which live within the coral tissues. The bleaching results from the ejection of the zooxanthellae by the coral polyps and/or by the loss of chlorophyll by the zooxanthellae themselves. This reaction of corals has been widely observed for many years: corals usually recover from bleaching but they can die in extreme cases.

Bleaching is caused by various types of stress, including temperature extremes, pollution and exposure to air. It is temperature-related stresses, however, which have been most widely reported, and are of particular concern in relation to climate change. On any given reef slope, the normal range of sea temperatures throughout the year is narrow – usually about 4°C – though the range of temperatures tolerated by reef-building corals worldwide is much wider (16-36°C). It would appear that corals in individual regions and localities have become highly adapted to these quite narrow temperature regimes. Studies have shown that temperatures of only 1-2°C above the normal maximum (threshold temperatures) for a few weeks are enough to drive a "mass bleaching" event (where high proportions of corals across the reef are bleached).

Crown-of-thorns starfish

A crown-of-thorns starfish consuming a branching coral.

The crown-of-thorns starfish *Acanthaster planci* is a large and distinctive starfish which is found across the Indo-Pacific from the Red Sea to the Eastern Pacific. Adults can grow up to about 60 centimeters in diameter, and have up to 21 short arms around the edge of a wide central disc. The entire dorsal surface is covered in short, poisonous spines. They feed exclusively on live coral.

For many years crown-of-thorns starfish were considered rare, but in the early 1960s the same species was observed reaching plague proportions on a number of reefs in the Great Barrier Reef in Australia. The initial plague recorded on Green Island numbered hundreds of thousands and killed 80 percent of the coral during the period 1962-64. Throughout the 1960s and 1970s new plagues were observed in numerous locations throughout the Pacific, including Guam, Japan, Hawai'i and Micronesia. With each plague up to 95 percent of living coral cover was lost.

A number of theories have been advanced as to the cause of these plagues, and as to whether they are natural or human-induced. There is some evidence of outbreaks occurring prior to the 1960s, although it seems unlikely that these were as frequent or widespread as the plagues which have occurred since. Three general theories have gained particular prominence:

- Most plagues have been observed close to high islands, and have followed severe storms. It has been suggested that the lower salinities and/or higher nutrients associated with terrestrial runoff during these storms may favor larval survival and massive juvenile recruitment.
- Aggregation of adults has been observed following destruction of coral areas by storms or other events and the concentrated plagues may be a behavioral response to these events.
- The starfish have a small number of natural predators, including certain pufferfish, triggerfish and emperors, but also the giant triton (a large mollusc). Low populations of these predators have been found on some infested reefs and it has been suggested that reduced predation might allow population explosions of the starfish.

In reality, it seems likely that any or all of these factors could be combined, with high levels of recruitment, behavioral aggregation and reduced predation all allowing the build-up of massive populations capable of destroying a reef.

Considerable debate has also taken place as to whether these plagues are natural or human-induced, and again the answer is controversial. Forest clearance and agriculture on many island and mainland coasts has undoubtedly led to higher rates of runoff, creating pulses of low salinity and higher nutrient inputs. Observations of the Great Barrier Reef appeared to tally closely with significant increases in fishing efforts, including the popular targeting of at least one of the key predators. In reality it seems likely that crown-of-thorns outbreaks could occur under entirely natural conditions, but may now be occurring more frequently as a result of human activities.

The long-spined sea urchin

In 1983 a mass mortality of the long-spined (or black-spined) sea urchin *Diadema antillarum* was first observed in Panama. The cause of the deaths remains unclear, although there is evidence that a bacterial pathogen may have been involved. The disease spread rapidly to almost every reef in the Caribbean during 1983 and 1984, typically leading to the loss of at least 93 percent of the urchins. One year later a second wave of the disease swept through and removed many of the surviving urchins.

The impact of these events has been severe across the Caribbean. *Diadema* was a critical grazing herbivore on many reefs. Its loss led to massive increases in filamentous and fleshy algae. On reefs already impacted by coral disease (see page 62) or hurricanes, this proliferation of algae greatly slowed or prevented coral recruitment and recovery, while even on healthy reefs, the growth of algae has led to the deterioration of many coral colonies through shading by the rapidly growing algae. It has been suggested that overfishing of herbivorous fish species in many areas has further encouraged algal growth. In most areas there has been little or no recovery of *Diadema* populations.

Humans have not been directly implicated in the cause of this die-off. The sudden appearance of a pathogen may be the result of a natural mutation, although it could also be linked to the arrival of an existing pathogen from another region. In the latter case, the first appearance of the disease close to the Panama Canal suggests that the pathogen may have entered the Caribbean either directly through this canal, or in ballast water from a ship. Whether its ultimate origins are natural or human-induced, the impacts of this die-off have been exacerbated by human activities. High levels of fishing had probably already increased the ecological dependence on this single species of herbivore prior to this event, while continued overfishing is slowing or preventing recovery of many reefs as there are few other herbivores to take its place. In many areas, pollution and sedimentation may have also raised nutrient levels, favoring conditions for rapid algal growth, and allowing for the establishment of new, algal-dominated communities in many areas.

The long-spined sea urchin Diadema antillarum.

Reports of coral bleaching have increased greatly since 1979, with all records of mass bleaching occurring after this date. The number of coral reef provinces (geographic divisions) in which mass bleaching has been reported varies widely between years, but shows a close correlation with El Niño Southern Oscillation (ENSO) events. The most significant mass bleaching event to date was associated with the 1997-98 ENSO event, when there were reports of bleaching from all over the world (see Map 2.2). In certain areas, most notably the Central Indian Ocean, this event was followed by mass mortality, where up to 90 percent of all corals died over thousands of square kilometers, including virtually all reefs in the Maldives, the Chagos Archipelago and the Seychelles. Although new coral growth has been observed in these areas, full recovery from such an event will take many years or decades. There is some concern that mortality on such a massive scale could lead to local disappearance of certain species, driving a loss in diversity and changes in community structure.

Although there are no clear records of mass bleaching events prior to 1979, it is possible that such events could be rare but recurrent phenomena that reefs have recovered from in the past. However, the extent of coral bleaching observed during recent ENSO events provides a clear indication of the wider long-term impacts of rising sea surface temperatures. Although mass bleaching is largely driven by ENSO events at the present time, most climate models predict that the threshold temperatures which currently drive these events will be reached on an annual basis in 30-50 years.

At both the regional and local level, certain corals have adapted to warmer or more variable temperature regimes. These include some of the same species that have been observed to be highly sensitive to temperature variations in other areas. Such adaptation is clearly seen in the reefs of the Arabian Gulf (Chapter 9). Some observations of the 1998 event showed local-scale survival of corals in a few areas of reef flats and lagoons even in the most hard-hit regions. These areas are likely to be subject to more extreme temperatures on a regular basis, from the reduced water circulation and exposure to solar radiation and/or cold conditions in these areas. It remains to be seen whether larvae from these corals can recolonize reefs where more sensitive corals have died, or whether there is indeed sufficient genetic resilience within these species to adapt to the continuing increases in temperatures predicted under current models of climate change.

Further concerns compound the problems from rising sea surface temperatures. Corals may be placed under additional stress by the projected increases in concentrations of atmospheric carbon dioxide. It is believed that the presence of aragonite in surface waters will be reduced by such increases, and aragonite is an important mineral component of the coral skeleton. Lower concentrations will reduce calcification rates and skeletal strength, which in turn may lead to reduced rates of reef growth and weaker skeletal structures.

All reef development is the result of coral growth out-pacing natural processes of erosion, from bioeroding organisms and also physical processes such as storms. Slower coral growth rates and weaker skeletal structures may shift the balance of many reefs from that of gradually accreting structures to that of gradually eroding structures, and this change will be further compounded by increasing rates of sea-level rise. The impacts of this will affect not only the coral reef ecosystems, but also the long-term survival of island peoples and even of entire nations.

Direct physical impacts

The threats described thus far are predominantly indirect in their action, their impacts working by degrees from subtle shifts to more extreme or longer-term impacts. Humans are occasionally far more destructive to coral reefs. Dynamite fishing has already been mentioned, and is an activity which leads to the instantaneous destruction of small patches of reef. Similar damage is wrought at various scales through a number of activities, as listed in Table 2.2.

Many of these activities are highly localized. But, depending on their location, even small impacts may be important. This is particularly true of popular tourist areas. There have been a number of cases now where insurance companies have been forced to compensate extremely high sums per square meter of reef damaged where ships have grounded in popular recreational areas. Likewise the

Coral disease

The first reports of disease affecting scleractinian corals did not appear until the early 1970s when it was quite shocking to read of rapid tissue degradation occurring in reef-building corals. In all cases there was loss of living tissue and, in many, complete mortality of colonies occurred. Since then increasingly frequent observations of coral diseases have been reported. Diseases have been observed on 106 species of coral (including some soft corals) on reefs in 54 countries around the world. Coral diseases appear to be particularly prevalent in the Caribbean.

Two patterns of disease were described at first, both being characterized by a distinctive band, a few millimeters to a few centimeters wide, of diseased tissue separating healthy tissue from exposed dead coral skeleton. Black band disease was observed affecting a variety of massive Caribbean corals, principally species of *Montastrea* and *Diploria*, whereas white band disease was noted on branching acroporid corals.

These bands of infected tissue move across and progressively destroy coral tissue at rates of several millimeters per day.

Since the 1970s a plethora of conditions have been described as new coral diseases, and many claims have been made for the decline in coral reefs due to disease. The exact causes of most of these conditions remain unclear. Only two have been unquestionably linked to a pathogen – aspergillosis (a disease of gorgonians in the Caribbean) and white plague type II. Part of the problem is undoubtedly the difficulty of finding and identifying pathogens in marine organisms, combined with the difficulty of linking any such pathogens to specific conditions.

Despite the uncertainty of the causal agent, it is commonly accepted that white band disease has been a major contributor to the massive decline in Caribbean acroporid corals. For example, palaeological studies in Belize documented a disappearance of acroporid corals from 70 percent of the coral canopy, and replacement by species of *Agaricia*, between 1985 and 1996. While the reasons for the impacts of direct diver and anchor damage at popular dive sites may reduce the real economic value of those sites in the longer term. Land reclamation probably has the most significant impact in terms of the total area of reef destroyed and is one of the few activities that genuinely leads to a change in absolute reef area. Building on coral reefs has been widespread in many countries, including Egypt, the Seychelles, the Maldives, Singapore and a number of the smaller Pacific atolls.

Aside from urban or industrial activities, military activities have had a major impact on many coral reefs. Military installations exist in a number of coral reef areas, including several remote coral islands and atolls. Weapons testing grounds and military dumps have also been located in such areas. These facilities can have significant impacts on the surrounding coral reefs. The reefs of Vieques off Puerto Rico and Fallaron de Medinilla in the Northern Mariana Islands are still regularly used by the US military for target practice, with significant impacts on the surrounding reefs. Many other facilities have reclaimed areas of reef flat for construction, or have dredged or blasted channels for boats. Johnston Atoll has been used as a dump for large quantities of hazardous waste, while a number of other remote Pacific atolls have been used, historically, as testing grounds for nuclear weapons, and in a few, high levels of radiation remain.

Compounded problems

The different human-induced stresses that impinge on coral reefs rarely act in isolation. Climate change appears set to produce changes over wide areas of the planet. Sedimentation and pollution are often found together, resulting from coastal development and changes in landuse. The risk of overfishing is equally widespread in all areas where human populations are growing. Even a single activity, such as the development of a new tourist resort, can create such a combination of threats. Sedimentation may arise during the clearance of vegetation and building works; toxic and nutrient pollution may result from wastewaters; direct physical damage from the construction of jetties or the impacts of boat anchors; and fish populations may be impacted in the provision of food for guests, staff and dependents.

Natural impacts, particularly storms, often exacerbate these human-induced problems. In Jamaica a number of coral reefs have lost so much coral cover that it may no longer be accurate to call them coral reefs. Although there are records suggesting that chronic overfishing had been occurring on these reefs since at least the 1960s (and some authors have suggested since the previous century), the reefs still maintained high coral cover until Hurricane Allen swept over the island in 1980, reducing much of the coral to rubble. There was some recovery over the next

sudden susceptibility of acroporids to white band disease is unknown, it was dramatic. Such a shift had not occurred in the previous 4 000 years.

A variety of environmental factors appear to influence the distribution and virulence of coral diseases, as well as the susceptibility of their hosts. Diseases are more prevalent in the Caribbean than elsewhere, and there also appears to be some sub-regional variation. A comparison between the *Reefs at Risk* data from the Caribbean (described on page 65) and the distribution of coral diseases reveals that less than 3 percent of the locations at which disease has been observed were on reefs under low threat from environmentally damaging human activities. The possibility therefore exists that the incidence of disease may be a suitable bio-indicator of disturbance to coral reefs at regional levels.

Map 2.3: Coral disease

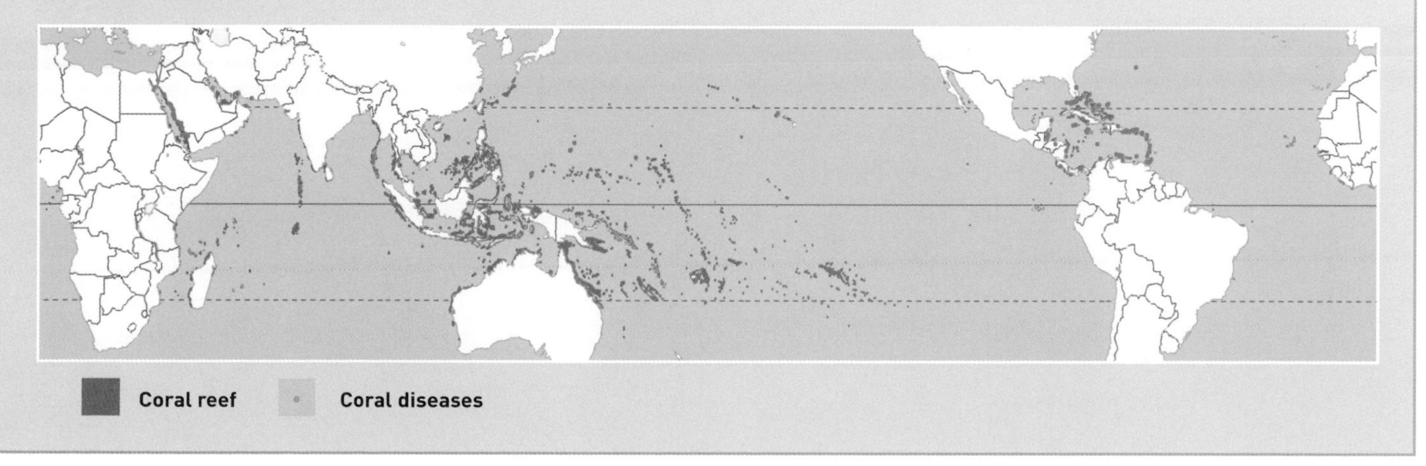

Coral reef **Coral diseases**

Table 2.2: Major types of direct physical damage to coral reefs

Activity	Notes
Ship grounding	Direct impact on relatively small areas of shallow reef
Blast fishing	Individual blasts may only impact a few square meters, but can be frequent
Weapons testing	Remote atolls were used in the early testing of nuclear weapons, some smaller-scale damage continues in military testing grounds to the present day
Reef walking	Tourists and local people walking on reefs can kill coral and flatten the reef matrix
Diver damage	Coral breakage or death from frequent handling, only a major problem on very popular dive sites
Direct smothering	Solid waste, and thicker elements of spilled oil can kill corals through direct contact and smothering
Anchor damage	Apart from initial impact, anchors may drag and anchor chains sweep over wide arcs, smashing corals over large areas
Reef mining	Includes direct removal of coral, sand and aggregate for construction purposes
Dredging/construction	Dredging of channels and lagoons for ships' passage is common in many reef areas, also construction of deep water channels with the blasting of reefs, and the construction of jetties or roads over the reef flat
Land reclamation	Perhaps the most complete and irreversible destruction of wide areas of reef, as reef flats and shallow lagoons are infilled and converted to land

two years, but then in 1983 the *Diadema* die-off (see page 61) killed the majority of these sea urchins in the waters all around Jamaica and much of the rest of the Caribbean. Jamaica was particularly impacted by the loss of these herbivores, as the other large herbivores – fish – had already been drastically reduced. Algal cover on the reefs increased dramatically, smothering and killing some of the remaining coral cover. It appeared that a new stable state had been established. Hurricane Gilbert impacted the islands in 1988 and further reduced coral cover, while the algae regenerated rapidly. Fleshy macroalgae covered 92 percent of surveyed "reefs" in the early 1990s. It has been suggested that a return to a coral-dominated community would require a major increase in herbivory and may be impossible in the short to medium term.

Reports of degradation of the world's reefs are widespread from all parts of the ocean, but a detailed assessment of the scale and distribution of these losses remains impossible. Many reports are anecdotal, while vast areas of the world's reefs are beyond the reach of scientists to produce even irregular assessments. Various efforts are now underway to improve the levels of available information. ReefCheck is a global coral reef monitoring scheme which has been using volunteer divers

Map 2.4: Reefs at risk

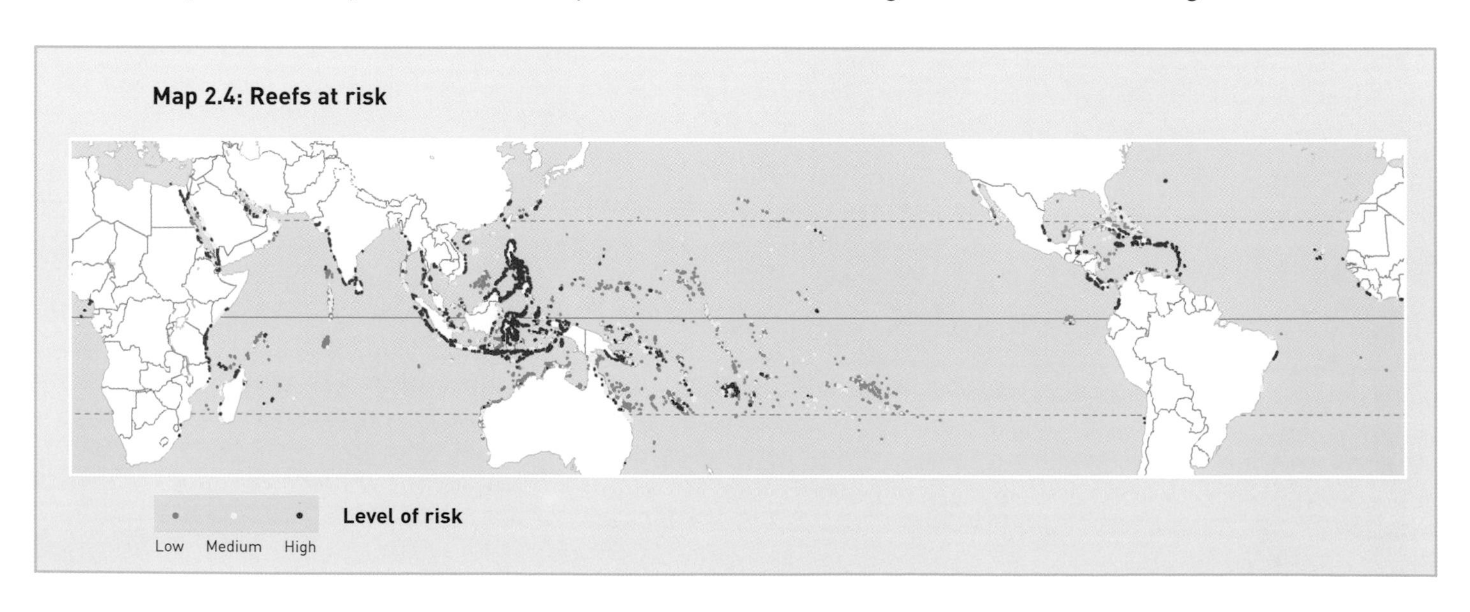

to assess reef health using standard procedures across hundreds of coral reefs worldwide. At the same time, the efforts of ReefCheck's partner organization, the Global Coral Reef Monitoring Network, have led to regular reporting by national experts enabling a parallel overview based on expert opinion. Even with these systems, however, the possibilities of looking at reefs and assessing stress are limited – the majority are only visited by experts every few years, and many have never been studied.

Reefs at risk

One alternative approach to mapping reef stress is to model the threats to reefs using existing datasets, combined with expert knowledge on the sensitivities of reef ecosystems. In an attempt to undertake such an objective global assessment, in 1998 the World Resources Institute led a team of organizations and experts to construct a global model of the different threats to reefs. Data were not available to analyze all the major potential human impacts, so the model ignored both climate change and direct physical destruction and concentrated on what are currently the most widespread and potent threats – pollution, sedimentation and unsustainable fishing practices. For these, a number of proxy indicators were utilized, enabling the development of four main threat layers.

Coastal development (representing the primary source of nutrient pollution, but also a source of sedimentation): reefs were considered to be threatened by a simple proximity measure for population centers of varying size, airports, military bases, mines and tourist centers. An estimate of sewage treatment levels was also factored into this layer.

Marine pollution (representing a secondary source of mostly toxic pollutants): reefs were considered to be threatened according to proximity to major ports, oil tanks and wells, and major shipping routes.

Overexploitation and destructive fishing: the degree to which reefs were considered threatened was based on their proximity to differing levels of population density. Known incidents of destructive fishing were further buffered out as a potential threat to the wider reef areas nearby.

Inland pollution and erosion (representing the primary source of sediment impacts, but also the important pollutants associated with inland and agricultural areas): a detailed surface model of "relative erosion potential" was developed based on land cover, slope and rainfall. This was modelled through the drainage basins to give a measure of threat at river mouths, which was then buffered to nearby reefs, the distance depending on intensity of input.

Using an earlier version of the reef maps presented in this atlas, these threat layers were combined. After a number of expert consultations, the refined model produced a global map (see Map 2.4). It was estimated that, overall, some 58 percent of the world's reefs were under medium to high threat. The global figure hides a number of important regional patterns clearly illustrated in the map.

For the Pacific, which harbors the largest proportion of the world's coral reefs, the majority are still largely unthreatened. But in Southeast Asia, the center of coral reef diversity, with many coastal populations heavily reliant on the reefs for sustenance, over 80 percent of the reefs are considered threatened. The Caribbean, with its very high dependence on the coastal tourism industry, has over 60 percent under threat. In some cases it can be assumed that these reefs are already degraded, but in other areas this is a measure of the potential for degradation. In reality there are many factors which may compound, or may mitigate, the outcome of these threats. One important factor is the wise management and utilization of reefs.

Table 2.3: Reefs at risk – statistics generated from the global analysis

Region	Proportions of reef area at different levels of risk (%)		
	Low	**Medium**	**High**
Middle East	39	46	15
Caribbean	39	32	29
Atlantic (excl. Caribbean)	13	32	55
Indian Ocean	46	29	25
Southeast Asia	18	26	56
Pacific	59	31	10
Global total	**42**	**31**	**27**

Source: Bryant et al (1998).

Responses

Humans need coral reefs. They are of critical value to communities and countries around the world, not only in economic terms, but also in terms of protein supplied to millions of people, in terms of jobs, of coastal protection and of recreational value.

The value of reefs is dependent on their continued functioning as ecosystems. It is this alone that maintains their worth for coastal tourism and the protection of coastlines, and as a store of vast genetic diversity. It is also this continued functioning which gives reefs their productivity and enables the harvesting of reef resources by humans.

In terrestrial ecosystems our concept of "harvesting" or otherwise utilizing "natural" ecosystems is typically associated with massive modification. But the sea presents a major contrast, as our harvesting of almost all the ocean's resources is dependent on the maintenance of natural ecosystems as they are. If this harvesting is turned to mining, sustainability is lost, and with it food, jobs and entire economies.

The array of threats to reefs has already been outlined. The ultimate causes of these threats, or the failure to react to them, can in most cases be related back to two broad problems: a lack of understanding about reefs and a lack of ownership or responsibility for them. The lack of understanding works at all levels, from a poor scientific knowledge, to the inadequate transmission of existing knowledge (scientific or traditional) to decision makers and a broader public. Detailed knowledge about reefs is actually rapidly improving. We now know a great deal about how coral reefs function, and about the consequences of actions such as blast fishing or pumping raw sewage into coastal waters. A large number of monitoring programs have been established, using both scientists and trained amateurs to amass large amounts of information about the status of coral reefs, and how this changes over time. Social and economic studies on human communities adjacent to reefs are also widespread, and we have a clear, though still rapidly evolving, understanding of how to manage and sustainably utilize reefs into the future.

A sail-powered dhow, a common fishing vessel in East Africa.

With this existing knowledge, powerful facts have emerged. A broad swathe of "experts", including social scientists, ecologists, lawyers and economists, hold various parts of the same message. Coral reefs are an incredibly valuable resource, they can be utilized sustainably, and such utilization can bring immediate economic and social benefits. There are no examples whatsoever which show the contrary. Non-sustainable use of coral reefs, or blind degradation by remote actions such as deforestation, poor agriculture and pollution, never pays. The social and economic consequences are bad in the short term, and terrible in the longer term.

If these facts could be clearly and accurately transmitted to a global public many of the problems would be immediately diminished, but the transmission of knowledge is slow, and too often the managers and decision makers, let alone the wider public, are unaware of these valuable facts. There remains an urgent need for education in all quarters to ensure that the lessons being learned by the experts are rapidly passed on to all people who have an interest in, or an impact upon, the future of coral reefs – from fishers to schoolchildren to government members and city dwellers.

The problem of a lack of ownership and responsibility for coral reefs is a greater challenge. Lying offshore, reefs in almost every country of the world are seen as a common resource, available to anyone. Unfortunately, as populations have boomed and traditional understanding of coral reef utilization has been rapidly lost in many areas, so this common access has led to a "tragedy of the commons": too many people trying to exploit a shared resource. There is no incentive for fishers to reduce their catch if all others will not do the same. As the number of fishers increases and fishing methods become more efficient, so the pressures on reefs become unsustainable. Open access and common ownership becomes a free-for-all, as people try to grab as much as possible before anyone else does.

The lack of ownership at a community level also typically means that there are few concerted efforts by those using the reefs to control external factors impinging on "their" resources. In terrestrial ecosystems it is unthinkable that another individual, company or community should be allowed to dump raw sewage or industrial waste onto a neighbor's grazing land, or to smother it with sediments, preventing the growth of new grass. This is precisely what is happening to coral reefs, but because they are a common resource, and because they lie below the water surface, invisible to the wider public, little or nothing is done.

There are thus a number of key responses to address the problems facing reefs. Study and research into ecological functions, human interactions and management techniques must continue. At the same time the transference of existing knowledge must be a priority. Finally, to deal directly with the real problems facing reefs on the ground there must be management interventions to reconcile the issues of protecting a common resource. The remainder of this chapter looks at a number of the management interventions now being widely put into practice.

In many countries active reef management is not new. Many of the lessons now being learned have, in fact, been known and put into practice in traditional societies adjacent to reefs for centuries or millennia, and some of these continue to be the most effective means of protecting coral reefs to the present day. For most countries, however, new management regimes have to be developed from scratch. Complex negotiations with a broad multiplicity of reef users and adjacent communities are a major part of designing systems to protect coral reefs, and to enable their continued existence and continued contribution to human society. Measures include controls on fishing activities or fishing methods, wider controls on human activities, and the partitioning of the reefs themselves, establishing protected areas or more complex systems across entire coastal zones. Such measures, effectively applied and with community support, are already being seen to have spectacular results, reversing declines and providing a considerable support to coastal communities. In most places, however, the challenges are still enormous, and it remains to be seen how much more will be brought back from the brink amidst the spiralling challenges of population growth and climate change.

Traditional management

Humans have used reefs as a source of food for millennia. As this has become more intensive so systems of control have developed, including cultural controls, taboos, patterns of reef ownership, closed seasons and restrictions on types of fishing gear. The best examples, some of which still exist, are to be found in the Pacific islands.

Perhaps the most widespread and often highly effective form of reef management is customary tenure, where reefs, or the reef fish and other resources, are the property of particular communities. The consequences of such ownership are considerable, especially in cohesive, traditional communities. Under restricted ownership it is entirely in the interest of the reef's owners not to overfish.

Within these areas of customary tenure many additional rules, traditions and customs have arisen which provide considerable further controls. In many cases certain areas are closed to fishing for given periods, or even permanently. The placing of taboos on certain reefs is still common in some countries, closing them to use for periods of months or even years. Other traditions have prevented permanent use of certain resources or eating of certain species.

Table 2.4: Restrictions in regular use worldwide as a means of controlling levels of fishing on coral reefs

Legal measure	Notes and examples
Licensing	Placing some degree of control over the numbers of fishers or vessels entering a fishery by requiring possession of a license, and perhaps further restricting the granting of such licenses, for example, to local community members
Gear restrictions	Limiting the types or amounts of fishing gear which may be used; restricting the numbers or design of fish traps; restricting or prohibiting the use of particular lines, nets or trace; prohibiting the use of scuba equipment in lobster fisheries
Species protection	Complete prohibition on the capture of particular species or species groups, often rare species such as clams, lobsters, butterflyfish, sharks
Catch restrictions	Total numbers caught may be restricted on a daily basis, or per fisher. Minimum size limits are often set for species such as lobster to ensure they reach breeding size before capture
Seasonal restrictions	Closing particular fisheries for certain periods of the year
Area restrictions	The closure of particular designated areas to certain activities, or to all fishing

An understanding of the reef environment is as important for traditional management as it is for any other form of management, and such knowledge in traditional societies can be considerable. For example, the spawning cycles and locations of many species of fish are well known. While such knowledge is useful for the exploitation of species it has also often led to restrictions. In certain villages on Palau it was forbidden to catch certain species of fish at their spawning sites, while other species were protected on their first day of spawning, but could be caught subsequently. Turtles were often only allowed to be caught after they had already laid one or more batches of eggs, and similarly only a certain proportion of the eggs could be harvested. Certain easily caught species could not be harvested during fair weather conditions, ensuring a supply when it was not possible to use boats during storms, or when fishing was poor.

In many societies further restrictions have provided different rights to particular social groups. In Yap in the Caroline Islands, Micronesia, traditional societies once operated on a complex hierarchy. Women and children, together with certain members of "lower" classes were only allowed to use simple fishing gears, and could only fish in rivers and tide pools; a wider group of individuals could use various techniques including hook and line and traps; certain techniques required involvement of the wider community; while there were also methods, such as net fishing from canoes and trolling behind the reef, which were restricted to prestigious members of society.

The influence of Western culture has radically altered many traditional societies. New fishing gears, including metal hooks, monofilament lines and lightweight nets, were among the first introductions, rapidly changing the effectiveness of fishing techniques. Subsequently there has been a more gradual erosion of cultural and traditional values, in many places supported by colonization or by efforts to embrace Western lifestyles and government. In some areas, even in the remote Pacific islands, all methods of traditional management have been lost, however elsewhere they remain, and in a number of countries there are now increasing efforts to recognize customary marine tenure within national constitutions. Where this is taking place there is, to varying degrees, a reawakening of the potential for reef management at village level. Under such systems the implementation of more Western styles of reef management, including legally gazetted protected areas or coastal zone management systems, may be inappropriate or even impossible.

Legal controls

The use of the legal system as a means to control individual actions which are known to damage reefs is widespread. To be effective, legal measures require enforcement. While this can be done through intensive policing, many of the world's reefs are too remote, and many countries too poor to carry out such expensive activities. Increasing awareness of this problem has led some countries to legislate more generally, while developing more detailed control measures at the local level in collaboration with local communities. Such approaches enable important education of these communities and often lead to a sense of ownership, both of the common resources and of the regulations, which are perceived more clearly to be for their own benefit.

Fishing activities are among those most commonly controlled by legal measures. As an extreme example, blast fishing is now illegal in every country where it is known to occur. Other fishing controls are also widely used in some countries, and a number of these are listed in Table 2.4.

As tourism grows and diving and snorkelling become widespread so a number of measures may be taken to restrict activities, with regulations prohibiting such actions as spearfishing or anchoring boats in coral reef areas. Pollution controls are also of increasing importance, particularly in areas under development for tourism. Many new developments now require the undertaking of environmental impact assessments prior to getting permission to build, and there are growing numbers of laws governing new buildings, including measures such as proximity to the sea and sewage treatment.

Most of the legislation designed directly to deal with coral reef protection is focussed toward the immediate or adjacent threats, but many of the problems facing coral reefs are actually derived from quite remote activities. Here too, however, legislation can be utilized which may be directly linked to coral reef protection, but may also have wider applications. Policy or legislation to control sewage and other pollution is one such example, while another may be various efforts to control agricultural or forestry practices. The prevention of clear felling of forests on steep slopes or in buffer zones close to rivers is a clear example, often enacted to prevent soil erosion, but also having positive consequences for coral reefs at some distance away. One of the most widespread legal mechanisms for protecting coral reefs is the designation of protected areas, and this is considered separately in the following section.

Marine protected areas and no-take zones

The earliest examples of setting aside areas for conservation are predominantly terrestrial. Sacred forests and royal hunting grounds dating back many centuries are scattered across Europe and Asia. In the marine realm, the earliest protected areas were probably some of the reefs of the Pacific, where local communities or community chiefs placed restrictions or total bans on fishing. The growth of legally declared marine protected areas outside such traditional systems is, in comparison, a more recent phenomenon, with only a few sites declared by the end of the 19th century.

Strict definitions of marine protected areas vary. One of the most widely used, and one of the broadest, is provided by IUCN–The World Conservation Union, which states that a marine protected area is "any area of intertidal or subtidal terrain, together with its overlying water and associated flora, fauna, historical and cultural features, which has been reserved by law or other effective means to protect part or all of the enclosed environment". Such a definition includes sites such as mangrove forests, even if they do not incorporate open sea, but it also leads to the inclusion of sites which are predominantly terrestrial, simply because they include small areas of intertidal land. The maps throughout this atlas show the locations of all marine protected areas, but the associated data tables in the text list only those with coral reefs.

These sites have been established for a number of

Market research has shown that divers, many of whom spend thousands of dollars on dive vacations as well as on diving and photographic equipment, are enthusiastic supporters of entrance fees which are used for the maintenance of marine protected areas.

purposes, and cover a broad range of management regimes, from strict protection for the total preservation of natural ecosystem processes, to the measured and interactive management of a seascape with multiple human uses, through to a fisheries protection role within a wider system of fisheries management.

Overall, the majority of existing coral reef protected areas have been established with nature conservation as their primary motive. Scientists and non-governmental organizations (NGOs) have been key in driving for the designation of such sites, while commitments at national and international levels have strongly encouraged governments to act to set aside areas for biodiversity conservation.

In reality, sites which are established without broader consideration of local communities, and without clearly accounting for the various costs and benefits involved in their establishment, are invariably met with opposition, or simply ignored. A great many of the sites listed as protected around the world are poorly managed or ineffective. "Paper parks" are a very real problem worldwide. These are sites with a legal status, but are unmarked and often forgotten on the ground.

Without adequate management many of the threats facing parks will continue or increase. With the exception of remote sites and some privately protected areas, any such management requires endorsement and support from adjacent communities, while this in turn usually requires recognition of clear economic benefits. Thus, although biodiversity protection may be the driving force behind the protection of sites, the most successful attempts at developing it have considered other issues, most notably recreational and fisheries values.

Protected areas and fisheries

Overfishing is a problem on reefs worldwide. In the areas where it occurs, the dependence of local communities on reefs is often higher than anywhere else and so, while control of fishing activities is urgently needed, it is also particularly difficult to apply. Many protected areas have failed and been ignored as poverty and the basic need for food or income have driven people to continue utilizing these sites for fishing.

Against this background, recent initiatives have focussed very heavily on community involvement. A number of small areas within fishing grounds have been set aside where strict rules of no-take are applied, with full community cooperation. The results have been quite remarkable. In sites such as Apo Island in the Philippines and the Hol Chan Marine Reserve in Belize, fish stocks in reserves have grown rapidly in number, while individuals of particular species have also reached considerable sizes. This abundance leads to a significant net export of fish from the reserve area, and yields immediately adjacent to the reserve (where the fishers, of course, now choose to fish) have boomed. The social and economic arguments are incontrovertible. Within only a year or two of the closure of an area to fishing, the total yields from the wider fishing grounds around these and many other sites around the world have risen, and continue towards a plateau of higher and more sustainable yields after five or ten years. Further economic benefits have been gained from tourism in a number of these no-take zones. The high fish abundance makes for very popular dive sites, and carefully managed tourism has little or no impact on their continuing function as fish reserves.

Protected areas and tourism

In many areas around the world the economic and social values associated with tourist arrivals are beginning to compete with, or even outweigh, the value of reefs for fisheries. Reef-based tourism is attracting millions of divers per year, and these tourists will often select their

Map 2.5: The global distribution of marine protected areas containing coral reefs

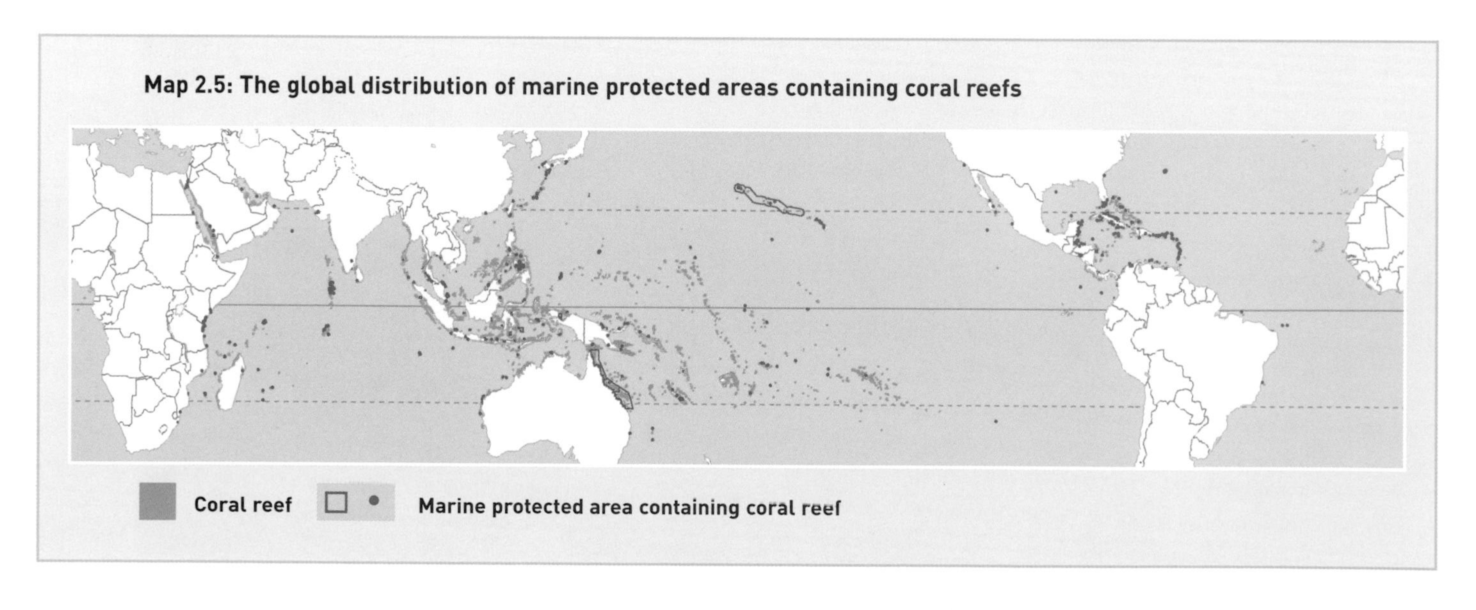

location and pay more to observe undamaged reefs. The value of tourism has been critical for some sites in providing direct income for management and enforcement activities. For example, user fees in the marine parks on Saba and Bonaire in the Netherlands Antilles provided 60-70 percent of the annual park running costs in 1999, with much of the remainder being provided by sales of souvenirs and yacht fees. Even where such direct benefits cannot be calculated, however, the income provided to individual hotels, dive companies and national economies from dive tourism is clearly enhanced in many countries by the presence of protected areas.

Multiple-use protected areas

Small protected areas are relatively simple to designate and manage, but in many areas fail to address the complex problems facing coral reefs. Most reefs are utilized by a broad range of "stakeholders", often with widely differing or conflicting requirements. Another approach to coral reef management has been the designation of typically very large areas, within which zones or sectors are marked out for different uses.

The Great Barrier Reef Marine Park is the largest coral reef protected area in the world, and is the best known multiple-use protected area. Designated in 1979 it covers some 344 800 square kilometers. In fact the majority of the park area is open to a considerable range of activities, including trawling and most other fishing methods. About 21 percent of the park is closed to trawling and, included in this, about 5 percent (but 12 percent of the reefs) is closed to all fishing. In fact the system of zoning also has restrictions on all access in "preservation zones" and "scientific research zones", and other fisheries restrictions including areas of periodic or seasonal fisheries closures. The size of this park ensures the integrated management of the largest interconnected reef system in the world. It provides a clear example of a holistic approach to reef management, with a clearly planned subdivision of reef zones. It has benefited from having a powerful and independent management authority, but has also made considerable progress in recent years through the encouragement of wide public participation in the planning and management process.

At a much smaller scale, the Soufriere Marine Management Area in St. Lucia provides an example of how the same principles of zoning and multiple use may be applied in developing countries. This area covers 11 kilometers of the western coastline of the island. It was developed following a long period of public consultation, and legally established in 1994. Zones set aside particular areas for recreation, yacht anchoring, marine reserves (with no fishing, but diving and/or diving permitted with a ticket) and also fishing priority areas (typically adjacent to the marine reserves). Since 1999 the annual fees paid by the 6 300 divers and 3 600 visiting yachts have made the management authority self-financing. Fish biomass has tripled in the marine reserve areas, and fishers are reporting increased yields from the adjacent fishing priority areas.

International designations

While the vast majority of protected areas are established at the national or even local level, a large number of important sites around the world are receiving international recognition through a number of global and regional agreements and conventions. The best known and most widely applied of these are the Convention on Wetlands of International Importance especially as Waterfowl Habitat (Ramsar Convention), the Convention Concerning the Protection of World Cultural and Natural Heritage (World Heritage Convention), and the UNESCO Man and the Biosphere (MAB) Programme.

Originally focussed towards waterfowl protection, the Ramsar Convention has been highly successful, with thousands of sites designated worldwide, and active encouragement of the inclusion of marine sites. Member states are required to identify and conserve sites considered to be of international importance, which according to the convention includes sites to a depth of 6 meters below sea level. Some 20 sites have thus far been designated which include coral reefs.

The World Heritage Convention focusses towards the identification and protection of areas of "outstanding and universal value", including both cultural and natural sites. Acceptance on the World Heritage List is only awarded to sites after a rigorous selection process. Thus far some 18 sites have been declared which include coral reefs.

UNESCO's Man and the Biosphere Programme is not a strict convention, but a scientific program under the auspices of the United Nations Educational, Scientific and Cultural Organization. Under this program biosphere reserves are designated to encourage a broad range of objectives tying together humans and their environment. Sites therefore typically encompass management towards sustainable utilization, research, monitoring and biodiversity conservation. They also serve an important role as demonstrations of human interaction with the environment. To date some 17 sites have been declared which include coral reefs.

In the majority of cases these international designations are applied to sites which are already protected under national legislation. International recognition remains very important for a number of reasons. It provides an additional "layer" of legal protection, which can further restrict damaging activities, or attempts by national governments to allow, or ignore, damaging activities. It can provide support for the management and maintenance of sites, allowing the networking of managers, the sharing of ideas, and often

Figure 2.1: The development of a global network of coral reef protected areas since 1930

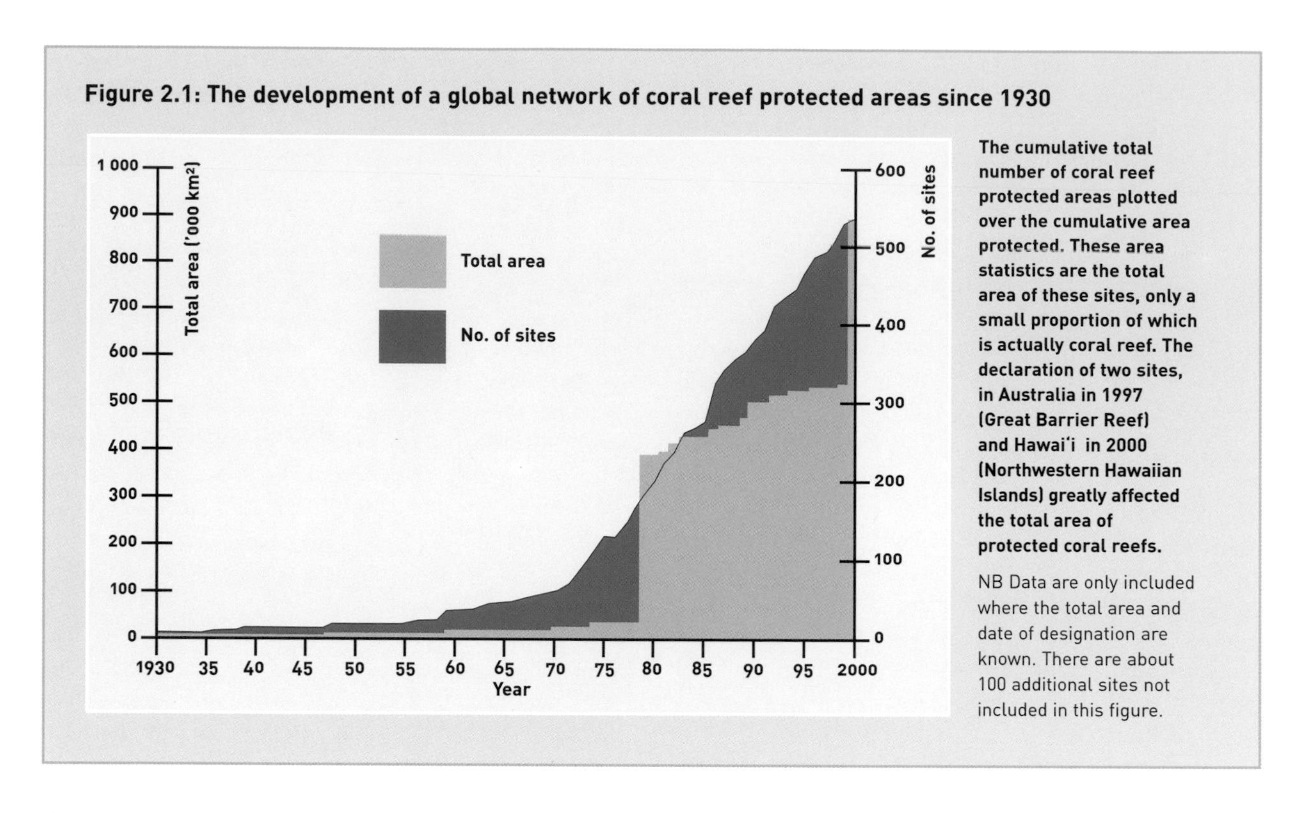

The cumulative total number of coral reef protected areas plotted over the cumulative area protected. These area statistics are the total area of these sites, only a small proportion of which is actually coral reef. The declaration of two sites, in Australia in 1997 (Great Barrier Reef) and Hawai'i in 2000 (Northwestern Hawaiian Islands) greatly affected the total area of protected coral reefs.

NB Data are only included where the total area and date of designation are known. There are about 100 additional sites not included in this figure.

further logistical or financial support for specific activities. It also provides recognition and prestige, often raising the profile of important sites and habitats in particular countries, but also serving to draw international interest.

Worldwide protection

Until the 1960s there was only a very small number of protected coral reefs, but from about this time onwards there has been a dramatic increase in numbers, as illustrated in Figure 2.1. At the close of 2000 there were over 660 marine protected areas which included coral reefs. These range in size from tiny marine reserves to two of the world's largest protected areas. The total area of these sites is more than 900 000 square kilometers, but three quarters of this area lies in only two sites, the Great Barrier Reef Marine Park and the Northwestern Hawaiian Islands Coral Reef Ecosystem Reserve. Unfortunately, these area figures cannot be equated with the reef area statistics presented in the previous chapter as the great majority of these sites include large areas of non-coral. Without boundary details for many sites in any global database, it is not yet possible to accurately calculate the proportion of the world's coral reefs which are protected. However, this total is likely to be quite high if Australia and the USA are included, but significantly lower otherwise.

Looking at a map of the global distribution of these sites (Map 2.5) it is clear that, while there are protected areas in all regions, a number of regions stand out as having relatively few. This would include much of the Middle East, with the exception of the northern Red Sea, but also the Pacific islands. In the latter, the urgency of establishing more protected areas is lower because of the existence of traditional management systems, as well as the overall lack of intense human pressures on many of these reefs.

A partial solution

These statistics and maps show that there is now a considerable network of protected areas containing coral reefs, and the number of sites within this network is growing fast. Coral reefs thus appear to be well protected in comparison with other ecosystems, although there are problems with such an assessment. First and foremost, many of these parks are ineffective. In countless cases worldwide, perhaps even the majority of sites, these protected areas are weakly enforced or completely ignored. Many others are weak in legal terms. These include a number of sites which are essentially terrestrial parks, but with a marine component. While their boundary includes marine areas, the legislation contains few or no provisions for protection of marine resources, and fishing and other activities continue unabated.

While direct comparisons between terrestrial protected areas and their marine counterparts are easily made, there are important differences. Most important is that many terrestrial sites are effectively fenced off from their surroundings, and are to some degree self-sustaining. By contrast the fluid environment of the coral reef cannot be ring-fenced. Most coral reefs are part of a large and tightly interconnected coral reef realm. Two points arise: firstly that individual reefs may actually be dependent on other reefs "upstream" for the provision of larval recruits and the maintenance of diversity; and secondly that legal protection of particular reefs provides little protection from external threats such as pollution and sedimentation originating elsewhere.

In a few cases marine protected areas are of sufficient size to have some potential to be self-sustaining, while a similarly small number of sites include both coral reefs and sufficient areas of the adjacent land to provide significant protection from sedimentation and pollution. Even Australia's Great Barrier Reef is affected by terrestrial activities occurring beyond the park boundaries. Protected areas, particularly in the marine environment, cannot be considered sufficient in themselves. They are part of the solution, but are most effective when placed within the context of a broader suite of coastal policy and planning measures, including wider legal measures (for example on land-based activities), but also programs on awareness raising and education. The concepts of integrated coastal zone management are increasingly being embraced in countries around the world, but before considering these, a brief overview of other systems or regimes which provide protection to coral reefs is laid out.

Other approaches

In addition to the conventional designation of protected areas with some legal status, a number of other coral reefs around the world receive some degree of protection from other regimes or factors.

Private ownership: private protected areas are rare in the marine environment, as few countries allow for private ownership of marine resources, but there are a number of privately owned coastal sites, including entire islands, which provide some degree of *de facto* protection of their adjacent marine resources.

Private and NGO initiatives: these have led to the adoption of codes of practice, or even the recognition of voluntary reserves in a number of areas. Recognizing the importance of the coral reef resources to their own business, it is now relatively common to see dive organizations setting restrictions on their own customers' activities (such as banning fishing, or fish feeding, and ensuring adequate buoyancy control so that divers do not touch the reef). A number of dive schools and NGOs have become involved in monitoring coral reefs, with the most notable scheme being ReefCheck, an international initiative which is using volunteer divers to monitor hundreds of reefs worldwide. While this does not provide protection *per se* it is a very powerful tool for reef assessment, raises awareness of conservation issues, and gives a clear message to other national agencies about the public concern for coral reefs. Another increasingly common activity, often supported by the dive industry, has been the undertaking of reef clean-up activities, where teams of volunteer divers go out to remove solid waste such as fishing lines and nets from coral reefs.

Military use: despite the damage caused to coral reefs by a number of military facilities mentioned above, there is also evidence, from a few areas, of more positive impacts. Many of these areas are closed to other activities on land, and to fishing in nearby waters. The presence of a military force further acts as a considerable deterrent to illegal fishing activities which abound in most areas. The large US base on Diego Garcia in the Central Indian Ocean, combined with British personnel on this island, undoubtedly has some deterrent effect for the large areas of coral reefs in the Chagos Archipelago.

The role of consumers

Such approaches by the dive industry may be motivated by direct conservation concern, or by consumer demand. "Consumers", the paying users of the reef, may become an increasingly powerful tool, providing an economic incentive for coral reef conservation. Hotels, as well as dive

A fairy tern Gygis alba *returning to roost on Cousine Island, a privately owned reserve in the Seychelles.*

Certification schemes

One tool which is increasingly being used as a means to promote sustainable utilization of the world's resources is that of certification, or "ecolabeling". Utilizing reliable and independently derived international standards in a certification process, industries or accredited organizations have the right to place a certificate or label on their products informing consumers that the product was produced, gathered or harvested in a sustainable fashion, with little or no impact on the environment. Consumers are equally encouraged to select these products, confident that they come from sustainable and well managed sources. In many cases consumer concern about the environment and sustainability is high, and many consumers are prepared, if necessary, to pay higher costs for these products.

These ideas have been applied to a limited number of fisheries, with perhaps the most notable being the certified "dolphin-friendly" tuna being marketed in many Western nations. One leading organization in this field is the Marine Stewardship Council (MSC), established in 1996 by WWF, the conservation organization, in partnership with Unilever, a major multinational corporation with an interest in fisheries. The focus of the MSC has been towards food fisheries, but no coral reef food fishery has yet been certified to MSC standards.

One other major fishery in the coral reef environment is that of the aquarium trade, which has a very high value, but relatively low volume of trade, predominantly operating between developing coral reef nations and aquarium hobbyists in North America and Europe. To date this trade remains relatively poorly documented, and largely uncontrolled (with the exception of hard coral species). In 1998, the Marine Aquarium Council (MAC) was established as an international organization to achieve market-driven quality and sustainability for the collection of marine ornamental species, most of which come from coral reefs. To achieve these aims, MAC has been developing standards for products and practices (ecosystem management, collection, handling and husbandry), but also establishing a system to certify and label compliance with these standards, and creating consumer demand for certification and labeling. Pilot schemes were to be tested in early 2001, linking collection and export operations with importers and retailers. Parallel programs were to be run raising awareness amongst hobbyists, industry and the public, prior to release of a full certification system later the same year.

Certification systems established by independent international multi-stakeholder organizations such as the MSC and MAC have considerable potential for helping to ensure that reef fisheries are sustainable. The critical factors that will determine the success or failure of these schemes are consumer interest and acceptance, combined with rigorous and reliable standards to ensure sustainability. If consumers do not select certified fish or fish products, and/or if the certification does not actually contribute significantly to sustainability, the ecolabeling program will not achieve its goals.

schools, recognize tourists' interest in a clean environment and healthy reefs and may adjust their environmental practices accordingly. At another level, some communities or individual fishers in Southeast Asia have found more lucrative and sustainable incomes by taking tourist boats to coral reefs, and in a few areas local communities have set certain areas aside for tourist use without the need for legal designations.

Future directions with this may include certification schemes or other systems which allow tourists to select their destinations and hotels in advance, and to avoid those areas or hotels which are major polluters or are making no efforts towards conservation. Schemes have been adopted in some parts of the world where tourist beaches are provided awards for environmental quality, and have been seen to have a significant influence on consumer choice. Other initiatives for marking hotels according to their environmental impact are being considered in some regions.

Even away from the reefs themselves, consumers can have an impact. The aquarium trade is a major industry bringing coral reef species from coral reefs around the world to consumers, mostly in the USA and Europe. At the present time this trade is poorly monitored, but efforts to develop a certification scheme as a means of supporting the sustainable development of this industry are now underway (see box).

Mariculture and fisheries enhancement

The use of mariculture (the farming, or aquaculture, of marine organisms) is growing in many reef areas. Clearly, by providing an alternative source of income and employment in coastal areas, mariculture can reduce the numbers who fish on the reefs themselves, and replace some of the demand for reef-caught protein. Mariculture is also occasionally used in replenishment schemes where species that have been greatly diminished in abundance are restored to certain areas. This has been successful for returning species such as giant clams and trochus to reefs where they have been largely or completely exterminated by overfishing.

Another form of reducing fishing pressure has been the establishment of artificial reefs. These are artificial structures which are placed onto the seabed and serve as a complex habitat attracting fish, and hence being popular with fishers. They are often purpose-built structures, utilizing materials from car tyres, to boulders, to moulded concrete shapes, but alternatively may be pre-existing structures, typically boats, which are taken out and sunk in relatively shallow waters. The deployment of these has been highly controversial in many coral reef areas. There have been failures in the design of some artificial reefs (car tyres have come adrift, or boats have not been properly cleaned and leak oil). Also, some concerns have been expressed that these structures may not be increasing overall fish stocks, just attracting species away from the reefs to areas where they are more readily caught. Despite this controversy, artificial reef technology is quite widely used in many non-reef areas, and may prove to be valuable and sustainable in some environments.

Reef recovery and restoration

As reefs are degraded so it is critical to establish management regimes which may promote their recovery. Increasingly it is possible to pinpoint the causes of a reef's degradation or loss, and management measures can be found to reduce or remove these causal factors and to reverse the conditions. These may include changes to the fishing regime, or to adjacent activities on agricultural land or in industrial or urban areas.

Such restoration techniques may be enhanced by more active processes of restocking, although the value of such approaches, when compared to allowing natural recovery processes, is sometimes questionable. A number of suggestions have been made regarding "planting out" of artificially reared corals, transplanting coral fragments from other areas, or enhancing coral growth rates with various processes. These may well prove successful in a few cases, but the high cost of such activities means that they are rarely, if ever, a viable process for restoring wide areas of reef ecosystems.

Enjoying a dive on a healthy reef in the Indian Ocean.

Integrating measures

A considerable raft of measures are thus available for the protection of coral reefs, including fisheries controls, protected areas, and other schemes ranging from diver clean-ups, to consumer or market driven controls on reef utilization. All of these measures are heavily dependent on awareness raising and on education, and the establishment of training programs for managers and other authorities.

Education needs to be aimed at all levels, including politicians and senior managers, artisanal and commercial fishers, recreational users of the reefs, tourists and aquarium hobbyists, but also the vast numbers of people whose lifestyles or businesses may affect reefs through pollution or sedimentation. The problems of global climate change represent an even larger challenge, which needs to be met with global education in order to make people aware of the massive changes required to reverse greenhouse gas emissions.

There remains, however, a considerable weakness in any of these or other measures, when they are taken in isolation. Protected areas, if they have the support of the local community, can be highly effective in controlling overfishing problems and preventing direct damage to reefs. They will fail to serve their purpose, however, if tourists are allowed to destroy the same reefs through anchor damage, or pollution from hotels, or if a forestry or mining permit inland allows massive sedimentation to pour onto the reefs downstream.

The concept of integrated coastal management has been widely accepted and promoted in many countries. In essence this involves developing a policy, not for particular locations, but for the entire coastal zone, including inland watersheds, and also offshore waters. Such policies, if developed in full consultation, and with active participation of local stakeholders, can be a highly effective means to protect not only coral reefs, but the livelihoods of all those living in the coastal zone. Typically the drafting of legal measures to implement policy is required, but often there is also devolution of controls to local agencies, further engendering the sense of ownership which can overcome the problems of common access. The development of such integrated measures is critical, but also challenging, requiring considerable coordination of disparate groups of people, and involving complex negotiation and processes of conflict resolution.

Provision of information and ongoing research remain a further priority. This includes establishing or expanding systems to monitor coral reefs around the world, in order to establish a clearer base line, and provide a warning of change. Equally important are studies on different management systems and fisheries techniques, and further research into aquaculture and reef restoration techniques. Such studies have already enabled the development of new and innovative management regimes, greatly improving the lives of many reef users around the world. More detailed studies into ecology, genetics and oceanography may further our understanding of natural processes and the connections and interactions between reefs, which may prove critical in the design of nature reserves and wider protection and management systems.

Perhaps the most important message, which is still not widely appreciated, is that active, sustainable management of coral reefs is always the most sensible approach. There is still a perception that reefs are of low economic or social value, and there is little or no connection between land-based activities and the impacts on the coral reefs.

Coral reefs are extremely valuable in social and economic terms. They are also highly sensitive ecosystems. Examples from the establishment of effective no-take zones, to well managed tourism development, to the establishment of small-scale aquaculture, show, again and again, that reefs can be well managed and that good management pays. Even in the relatively short term the benefits from wise management are rapidly observed, in both economic and social terms. With the application of such measures the value of reefs is a permanent one, sustainable across generations.

Left: A snorkeller holds up nets which have become tangled in shallow coral and abandoned. Right: A dried sea cucumber. Demand for these has led to their over-harvesting from reefs across the Indo-Pacific.

Selected bibliography

Barber CV, Pratt VR (1997). *Sullied Seas: Strategies for Combating Cyanide Fishing in Southeast Asia and Beyond.* World Resources Institute and International Marinelife Alliance, Washington DC, USA.

Birkeland C (ed) (1997). *Life and Death of Coral Reefs.* Chapman and Hall, New York, USA.

Brown BE (1997). *Integrated Coastal Management: South Asia.* University of Newcastle, Newcastle upon Tyne, UK.

Bryant D, Burke L, McManus J, Spalding M (1998). *Reefs at Risk: A Map-based Indicator of Threats to the World's Coral Reefs.* World Resources Institute, International Center for Living Aquatic Resources Management, World Conservation Monitoring Centre and United Nations Environment Programme, Washington DC, USA.

Cesar HSJ (ed) (2000). *Collected Essays on the Economics of Coral Reefs.* CORDIO, Kalmar University, Kalmar, Sweden.

Chadwick-Furman NE (1996). Reef coral diversity and global change. *Global Change Biology* 2: 559-568.

Done TJ (1992). Phase-shifts in coral reef communities and their ecological significance. *Hydrobiologia* 247: 121-132.

Ginsburg RN (ed) (1994). *Proceedings of the Colloquium on Global Aspects of Coral Reefs: Health, Hazards and History, 1993.* Rosenstiel School of Marine and Atmospheric Sciences, University of Miami, Miami, USA.

Green EP, Hendry H (1999). Is CITES an effective tool for monitoring trade in corals? *Coral Reefs* 18: 403-407.

Green EP, Bruckner AW (2000). The significance of coral disease epizootiology for coral reef conservation. *Biological Conservation* 96(3): 347-361.

Hatziolos ME, Hooten AJ, Fodor M (eds) (1998). *Coral Reefs: Challenges and Opportunities for Sustainable Management.* The World Bank, Washington DC, USA

Hawkins JP, Roberts CM, Clark V (2000). The threatened status of restricted range coral reef fish species. *Animal Cons* 3: 81-88.

Hoegh-Guldberg O (1999). Climate change, coral bleaching and the future of the world's coral reefs. *Mar Freshwater Res* 50: 839-866.

Hughes TP (1994). Catastrophes, phase-shifts, and large-scale degradation of a Caribbean coral reef. *Science* 265: 1547-1551.

Jackson JBC (1997). Reefs since Columbus. *Coral Reefs* 16 (Supplement): S23-S32.

Jennings S, Kaiser MJ (1998). The effects of fishing on marine ecosystems. *Adv Mar Biol* 34: 201-352.

Jennings S, Polunin NVC (1996). Impacts of fishing on tropical reef ecosystems. *Ambio* 25: 44-49.

Kleypas JA, Buddemeier RW, Archer D, Gattuso J-P, Langdon C, Opdyke BN (1999). Geochemical consequences of increased atmospheric carbon dioxide on coral reefs. *Science* 284: 118-120.

McManus JW (1997). Tropical marine fisheries and the future of coral reefs: a brief review with emphasis on Southeast Asia. *Proc 8th Int Coral Reef Symp* 1: 129-134.

Polunin NVC, Roberts CM (eds) (1996). *Reef Fisheries.* Chapman and Hall, London, UK.

Roberts CM (1997). Connectivity and management of Caribbean coral reefs. *Science* 278: 1454-1457.

Russ GR, Alcala AC (1996). Do marine reserves export adult fish biomass? Evidence from Apo Island, central Philippines. *Mar Ecol Prog Ser* 132: 1-9.

Salm RV, Clark JR, Siirila E (eds) (2000). *Marine and Coastal Protected Areas: A Guide for Planners and Managers.* IUCN–The World Conservation Union, Washington DC, USA.

Salvat B (ed) (1987). *Human Impacts on Coral Reefs: Facts and Recommendations.* Antenne Museum EPHE, French Polynesia.

Sapp J (1999). *What is Natural? Coral Reef Crisis.* Oxford University Press, New York, USA.

Silvestre GT, Pauly D (1997). *ICLARM Conference Proceedings, 53: Status and Management of Tropical Coastal Fisheries in Asia.* International Center for Living Aquatic Resources Management, Manila, Philippines.

Wilkinson CR (ed) (2000). *Status of Coral Reefs of the World: 2000.* Australian Institute of Marine Science, Cape Ferguson, Australia.

Chapter 3
Reef Mapping

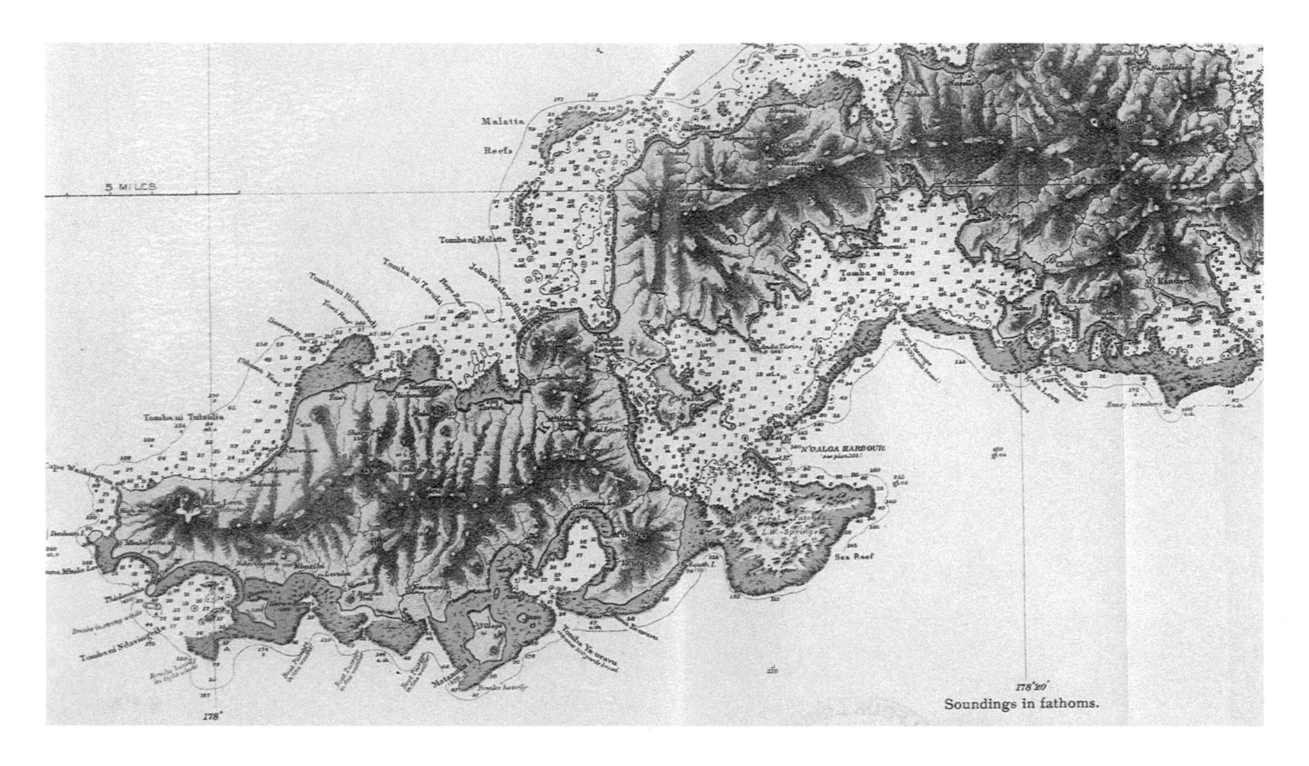

As long as humans and coral reefs have existed in close proximity, a knowledge of reef distribution has been important. The earliest navigators in coastal environments were in particular need of such information: reefs were a terrible hazard to be avoided, but also a source of food, while shelter in their calm lagoons was important during rough weather. This chapter traces the early development of knowledge on reef distribution, and subsequently of reef mapping, including the development of the first global maps of coral reefs. It then reviews the contemporary methods of reef mapping, including both hydrographic techniques and remote sensing. A final section looks at the current state of global reef mapping, centered on the present work.

Historical background

Navigation among coral reefs may be almost as old as navigation itself. Many early cultures achieved considerable feats of navigation in coral reef regions. In Egypt, friezes on the walls of the mortuary temple of Queen Hatshepsut clearly describe a voyage of considerable distance by boat along the Red Sea. This famous expedition to the Land of Punt took place in 1496 BC. Navigation in these areas was not solely undertaken by the Egyptians. The Babylonian and Sumerian empires traded across the Arabian Gulf and with many island kingdoms such as Dilmun in present-day Bahrain. Herodotus reports that Phoenician sailors circumnavigated the entire coast of Africa under orders from Pharaoh Necho II around 600 BC.

The production of maps in the ancient world, incorporating the coastlines of the Arabian Gulf and the Red Sea, can be traced back over 4 000 years, to Babylonian stone tablets and Egyptian papyrus maps. Following on from this, the Greeks and Romans developed more detailed maps, including of the seas around the Arabian Peninsula.

In the Pacific, the timing of movements and the directions taken by Pacific islanders are still under debate, although it has been suggested that Polynesian navigators may have reached Hawai'i from the Marquesas Islands by 400 AD. Almost every tropical Pacific island had thus been settled by this date. Such movements indicate phenomenal navigational skills. Little is known about the existence of physical tools which might have been utilized to aid their navigation, but in the 19th century a number of stick charts were discovered in the Marshall Islands. These consisted of

Kandavu Island, Fiji, from a British Admiralty Chart annotated by Agassiz (1899). Fringing and barrier reefs are clearly marked, demonstrating the considerable attention to detail in many hydrographic charts.

bound frameworks of sticks, with their intersections corresponding to the location of islands (often marked with cowrie shells). Other sticks showed the direction of waves or currents, or an indication of distance or direction to other islands. Although the stick-maps themselves are not old, it seems quite likely that such maps could have played a key role in the navigational feats of the Polynesian peoples for many centuries.

Detailed mapping of the location of coral reefs began in the ages of discovery in the West. Since classical times maps or charts had been widely developed in the Mediterranean and were often kept by navigators in coastal pilot books. Travels outside this region began in the 15th and 16th centuries. Columbus made his first historic visits to the Caribbean islands in 1492-93, while Vasco da Gama rounded the Cape of Good Hope and traveled along the coast of East Africa before crossing to India just a few years later.

Charts were originally produced by individuals and later by the large trading companies. Being of high political and economic value, many of these were kept as closely guarded secrets. From the early 18th century onwards, the same charts became the primary responsibility of national hydrographic offices. The representation of coral reefs on these charts was as a navigational feature rather than a biological phenomenon, although in many cases the charts were drawn with at least some knowledge of the biological and geological setting of coral reefs.

Much of the early scientific knowledge of coral reefs, from the 18th and 19th centuries, was actually gathered on the exploratory and hydrographic expeditions setting out from Western Europe. The voyages of Captain James Cook in the Pacific and later ones such as those of the Beagle with Charles Darwin and the Wilkes Exploring Expedition in the Pacific from 1838 to 1842 led to a great expansion of knowledge about reefs. Such expeditions were concerned not only with charting and the expansion of empires, but with many wider issues of research and discovery, including natural history and geology.

The first global maps of reefs

The first major global treatise specifically on coral reefs, which considered both the biological and geological origins of these, and which included a map showing global coral reef distribution, was that produced by Charles Darwin in 1842. While much of what is contained in this work was the result of Darwin's own observations during five years on the Beagle, an even larger proportion is based on his readings of the reports of other such expeditions and on his discussions with the various ships' captains and others who took part in these. In his own words, Darwin's work "is the result of many months' labour. [He] consulted, as far as [he] was able, every voyage and map" (Darwin, 1842).

In 1912, the French scientist Joubin published a much larger-scale global map of coral reefs, covering the world in five large sheets at a scale of 1:10 000 000. Joubin's work was based not only on a survey of existing maps and charts, but also on the results of a much larger data-gathering exercise, which included correspondence with many interested people throughout the world. (He includes in his acknowledgement thanks to the *abbés* throughout the world who passed his information requests on to missions in quite a number of countries.)

Until very recently, few other works have made such systematic attempts to map coral reefs globally. In general, following on from the early global reviews, there was a trend towards more detailed studies at the local

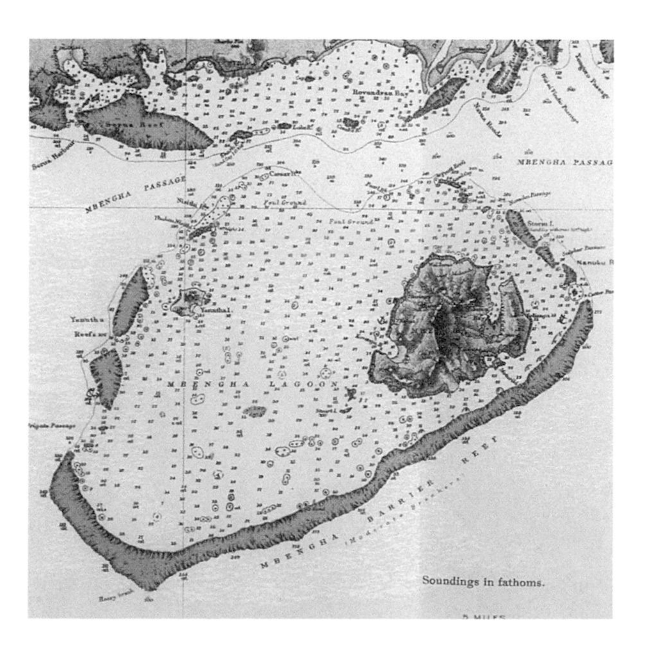

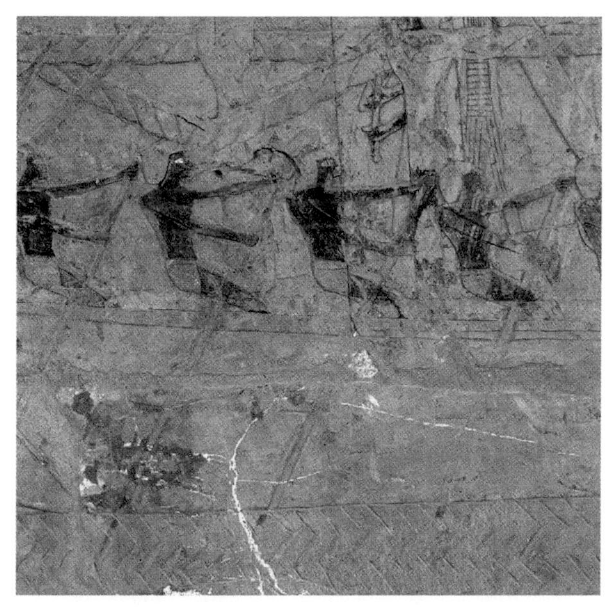

Left: Beqa (Mbenga) Island and Barrier, Fiji, from a British Admiralty Chart annotated by Agassiz (1899). Right: The walls of the mortuary temple of Queen Hatshepsut in Egypt illustrating the expedition to the Land of Punt (Sudan or Eritrea) in 1496 BC. The carvings show an intimate knowledge of these waters, including several identifiable fish species (photo: Giotto Castelli).

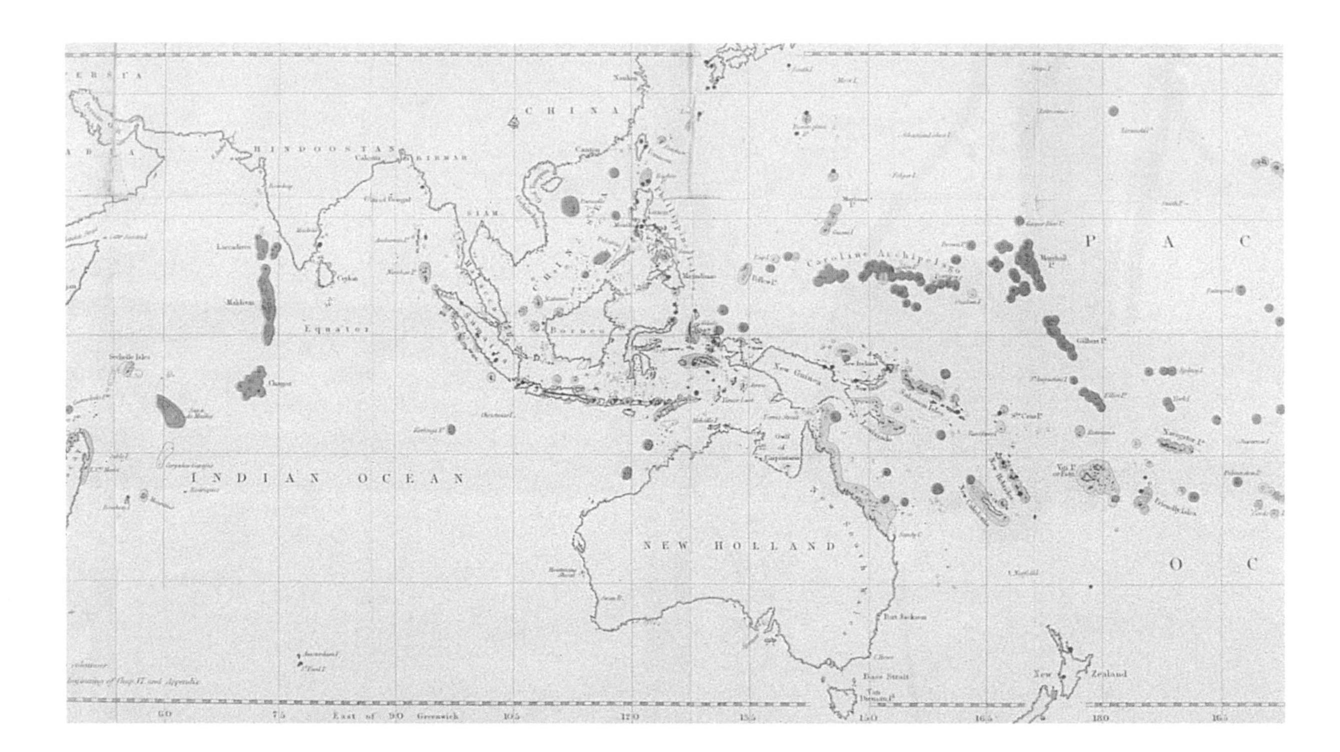

level. This is exemplified by the early work of Agassiz and the later studies such as those carried out on the Great Barrier Reef Expedition of 1928-29 and the works supported by the Coral Atoll Program of the Pacific Science Board. More recently, the mapping of individual reefs and those of particular areas and countries at higher resolution has developed considerably, though at widely differing rates, throughout the world. In addition to charts, reefs are increasingly shown on topographic maps, as well as on other more specialized maps of natural resources. The availability of remote sensing technology has greatly improved the information base for reef mapping and this, together with other techniques used in the preparation of detailed reef maps, is considered in the next section.

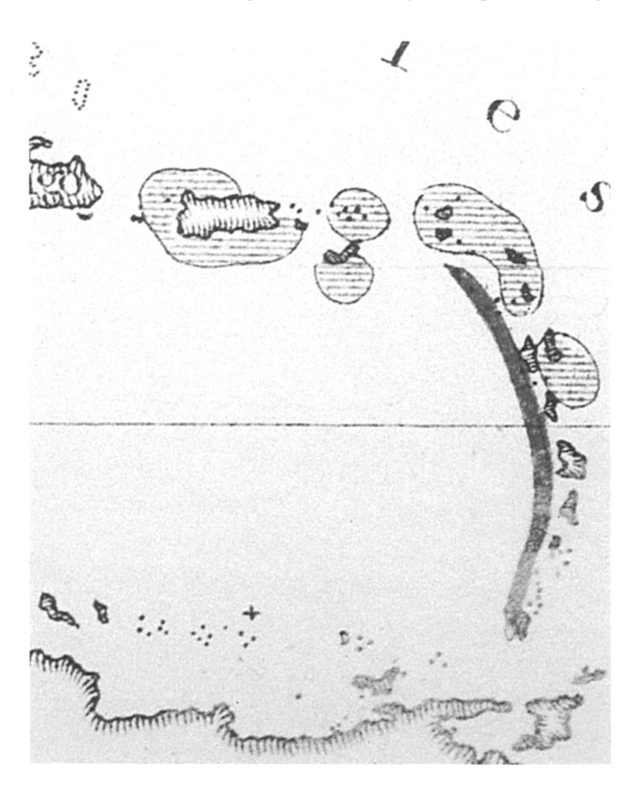

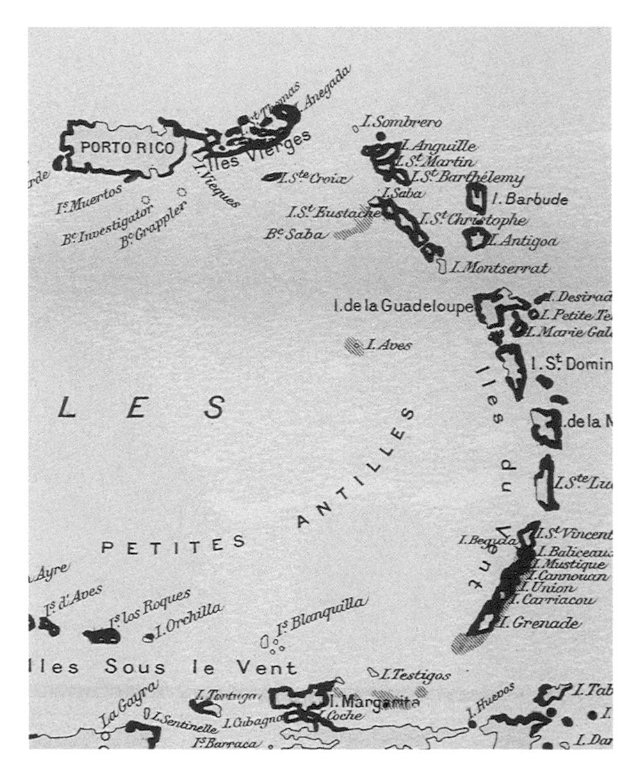

Above: Darwin's world map of coral reefs. This was prepared in 1842 from a study of multiple charts and voyage reports. Below, left: Further detail of Darwin's world map of 1842, the West Indies. Below, right: The considerably greater detail provided in Joubin's world map of coral reefs of 1912.

Reef mapping techniques

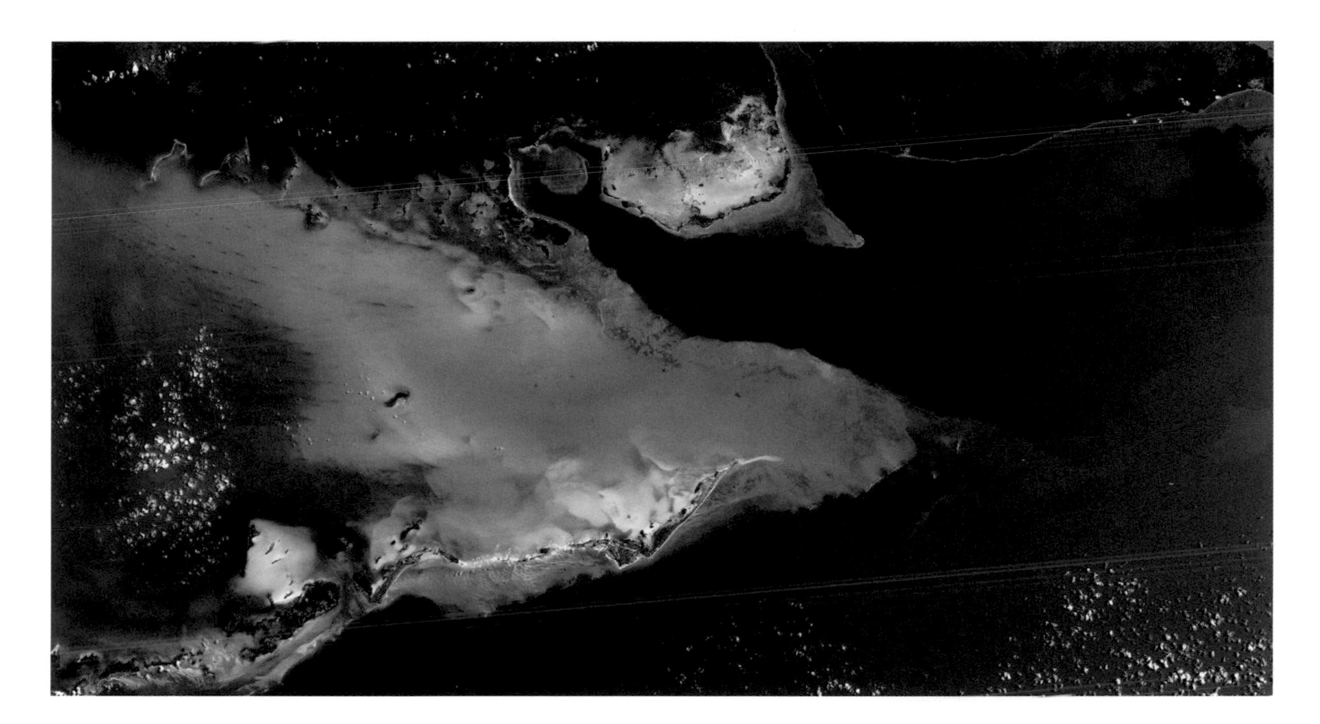

Over the years a wide array of mapping techniques have been utilized to chart coral reefs. The choice of techniques is clearly influenced by the primary purpose for which a map is required, the scale of the work and the availability of resources. This latter point is partly a question of funds and personnel, but the availability of resources and their costs have also changed considerably over time. Many older maps were severely constrained by the mapping techniques available, but for large areas of the world these still remain the best available maps.

Ground surveying and mapping

On land, a variety of techniques have been widely applied for ground surveying, from the very simple preparation of hand sketches and chain surveying to more sophisticated techniques of plane table surveying, geodetic surveying and theodolite mapping. These techniques cannot easily be applied to the coral reef environment, although it is usually possible to gain information on distances from shore of particular features, and also to carry out some ground-based surveys on areas of very shallow reef flat which may be exposed at low tide.

Historically, most marine mapping has relied on boat-based surveys. One of the great challenges to such surveys in the past has been the fixing of the boat's position. It was of course possible to utilize existing maps to establish position relative to known terrestrial features. Far offshore, or in completely uncharted areas, however, detailed astronomical methods were required to determine absolute location, alongside careful measurement of physical movement over shorter distances. Highly accurate positioning is now routinely available through the use of Global Positioning Systems (GPS), which determine position to within a few meters by linking position on the ground to the known positions of a number of satellites.

The calculation of depth is another important element in reef mapping. Prior to the development of sonar, this was done using soundings. Long cables were lowered off the edge of the boat until they hit the sea bed, at which point the depth was measured. In many cases tallow was placed in a small indentation in the bottom of the weight at the base of the sounding line or cable and inspected on return to the surface. This provided some indication of the benthos: sand or mud, and occasionally pieces of coral would adhere to this substance.

Although many of these techniques have now been surpassed they cannot be ignored. For vast areas of the world the only maps in existence were prepared using these methods, and the great skills developed during the

A Landsat image of southwest Cuba, showing a wide area of shallow banks, edged with coral reefs (image provided by National Remote Sensing Centre, UK).

Scale and resolution

The scale of a map is a measure of the reduction of the size of features in their representation on that map. Resolution is a related term, but is a more direct measure of the detail shown on a map and generally refers to the minimum size of an object visible on a map, or the minimum distance by which two objects must be separated for them to appear as such. The characteristics portrayed in the spatial representation of any natural system are strongly related to scale and resolution, and these need to be seriously considered both by those responsible for the preparation of maps and by the end-users. A number of problems can arise, especially in the generation of statistics and the comparison of maps drawn at different scales.

This is illustrated by attempts to measure coastline length: the complexity (and hence the length) of any coastline shown on a map is a function of map resolution. In the Caribbean, for example, maps of individual islands frequently show great detail, including tiny offshore rocks and islets. Regional maps, by contrast, may summarize complex coastlines into a few simple lines, while world maps often leave out many Caribbean islands altogether. Measuring the length of coastline of any island from each of these maps would clearly yield three very different results.

With traditional photography the resolution or level of detail can be altered by varying the lens and photographic material, or by altering the height at which images are taken, although a balance must be struck between the area covered by the image and the detail or resolution to be found in the resulting image. Optical scanners gather light from a continuous grid of defined blocks known as pixels. The sensor records, for each pixel, the reflectance levels for those wavebands for which it is tuned. The resolution is largely a function of pixel size which is, in turn, dependent on the optics of the sensor and the height at which the images are taken (although this cannot easily be varied for satellites).

The problems posed by scale are not simply those of statistics. Entire natural phenomena will be hidden or exposed depending on the scales of study. This is of particular relevance in the coastal zone where many narrow linear features are likely to be entirely lost at low resolutions.

This atlas provides examples of some "problems" of scale. Many of the maps have base-scales of 1:1 000 000 or lower. At this scale 1 millimeter on the map represents 1 kilometer on the ground, and it is very difficult to see features with a diameter of less than 400-500 meters. Many of the world's coral reefs are fringing or patch structures which are narrower than this. Drawn strictly to scale, such features would not be visible in many maps. In an effort to address this problem the boundaries or lines marking the locations of reefs have been slightly exaggerated to ensure that they are clearly visible on the maps, although this also has the effect of exaggerating apparent reef area. This same problem can of course be carried back to the source materials (each of which is listed for the particular maps) – source materials prepared from remote imagery at scales of 1:1 000 000 will be unable to pick out the smaller reef features.

detailed surveying expeditions led to levels of accuracy which are quite extraordinary. Many of the sources used in this atlas can be traced back, at least in part, to surveys over 150 years old. In quite a number of cases such maps have now been verified against air photographs and found still to be highly accurate.

Remote sensing

Coral reefs largely occur in clear, shallow water. These two factors make many reefs highly visible when observed from above the ocean surface, and thus make them highly amenable to being mapped using remote sensing. Remote sensing is the term widely applied for the "acquisition of information about the land, sea and atmosphere by sensors located at some distance from the target of study". Typically this means the gathering of images from aircraft and satellites, although it also includes such techniques as radar and, in the marine environment, sonar.

It was not long after aircraft became widely available in the early 20th century that reefs began to be explored and mapped from above. Vertical aerial photographs of the Great Barrier Reef were first taken in 1925 and these same images were used as a baseline for the survey by the Great Barrier Reef Expedition of 1928-29. Great advances were made in sonar technology during the Second World War

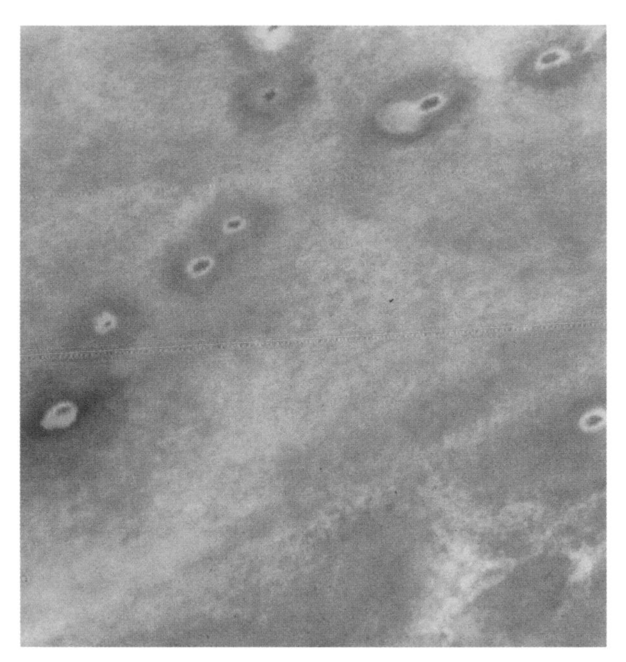

and this remained the only alternative to aerial photography for remote reef mapping until the advent of satellites in the 1970s. At the beginning of the 21st century there is an impressive, and technologically bewildering, array of optical devices mounted on satellites and aircraft available for coral reef mapping, as well as highly sophisticated sonar devices on boats.

The common principle behind the use of remote sensing is that coral reefs modify light or sound in a different manner to their surroundings: coral reefs "look different" to adjacent open water, seagrass beds, sand or other substrates when viewed from above. In photographs, as with satellites, it is the sun's light (usually the visible portion of the electromagnetic spectrum) which is modified by the coral reef, and enables it to be distinguished from other habitats. This particular range of wavelengths of light reflected by any subject is known as its spectral signature.

A raw image prepared by any remote sensing tool is known as an unclassified image. To convert any such image into a map it must be interpreted or classified. Using digital images, one very simple approach is to produce an unsupervised classification, in which no prior knowledge is used, but the features on the map are grouped together based on similar characteristics (such as a range of shades of the same color). More useful, particularly for the subtle but important variances in reflectance characteristics on a coral reef, is to undertake a supervised classification. This requires some prior knowledge of the features to be classified, which may be undertaken through field study (ground-truthing) or utilizing existing independent knowledge, including maps or photographs. Using digital techniques, mapping software can be trained on particular localities within an image and then extrapolated to the entire image, to classify all areas with similar spectral signatures.

In attempting to identify the features on an image, the interpreter is able to use considerably more than colors reflected off the surface. A whole series of characteristics including texture, relationships to other features, patterns and scale can aid identification of habitats or other features in a remotely sensed image. Image interpretation can be a highly skilled process which, if undertaken with care and consideration, may greatly enhance the effective resolution of an image. It is possible, for example, that a coral patch may be too small to be picked up by a remote sensor, but that patterns of grazing in surrounding seagrass beds may produce a wide halo effect of bare sand around such reef patches. Such halos may be clearly visible and give an indication of the presence of a central reef patch. Such interpretation clearly requires some knowledge of coral reef ecology.

Remote sensing, particularly using optical sensors, provides the opportunity of rapidly advancing beyond the simple mapping of geographical location. It is now possible to map different zones and ecological communities within a coral reef, and to look for changes in these patterns over time. Remote sensing can thus be used to look more closely at patterns of human impacts, and of management measures on reef environments, and to monitor changes over time.

Satellite sensors

Several different sensors are currently available, but the most widely used for reef mapping are Landsat Thematic Mapper (Landsat TM) and the Système pour l'observation de la Terre (SPOT). These have similar spatial resolutions, 10-30 meters depending on the mode of data collection,

Left: A scientist surveying a Caribbean reef. The quadrat is used to estimate percentage cover of different species. Right: A high resolution CASI image of patch reefs. Each is 20-50 meters in diameter and surrounded by clearly visible halos.

Astronaut photography of coral reefs

Astronauts have been photographing our planet through spacecraft windows ever since the beginning of human spaceflight. To date, nearly 400 000 photographs have been taken by astronauts on NASA missions using hand-held cameras. Most are in natural color and, due to selective photography by astronauts, tend to have relatively low cloud cover. They are taken from a variety of angles out of the spacecraft, including near vertical views down at Earth, low oblique views at an angle, and high oblique views that include the horizon. Once converted to digital form, these images typically have a high resolution, with pixel sizes of 20-80 meters in most images.

Earth observation training for astronauts includes the study of phenomena in the realms of ecology, geology, geograpy, oceanograpy, meteorology and the environment. Not surprisingly, the photographs they bring back to Earth are used by scientists of many different disciplines. Near-vertical or low-oblique angle photographs can be digitized at high resolution and used as three-band (red, green, blue) remote sensing images in the same way a scientist would use Landsat or SPOT data. Image processing techniques can be applied to determine landuse, cover or change over time.

Nearly 30 000 photographs of coral reef areas have been taken by astronauts on board the Space Shuttle providing a valuable but underutilized data source for coral reef scientists and managers. To facilitate the use of these public domain images, NASA's Earth Sciences and Image Analysis Laboratory has been collaborating with the International Center for Living Aquatic Resources Management (ICLARM) to include astronaut-acquired photographs in the global coral reef database, ReefBase.

Many astronaut photographs clearly show shallow reefs, and can show submerged features up to depths of about 15 meters in clear waters. It is possible to distinguish major geomorphological features within reef systems, including reef crests and patch reefs. They can be combined with traditional satellite data to help distinguish between clouds and lagoon features such as pinnacles. Furthermore, astronaut photographs may provide reef scientists and managers with information on the location and extent of river plumes and sediment runoff, or facilitate identification of land cover types, including mangroves.

Photographs taken by astronauts are used to illustrate coral reefs throughout this book. They have been selected to show the range of coral reefs found in every region, and to further illustrate places described in the text and features of particular interest, including human developments, the plumes of sediments in river mouths, shallow banks and remote atolls. Selected photographs were scanned, and color and contrast were hand corrected to give an approximation of natural color. None of the photographs shown here has been georeferenced, and a number were clearly taken at

and both have sensors which detect light in discrete portions of the visible spectrum (bands). This is important because infrared radiation will not penetrate water and, generally speaking, more bands enable more subtle changes in the light returning from the seabed to be detected. Satellite images can be geometrically corrected (assigned map properties) and used in a geographical information system (GIS) quite easily.

Cloud cover is an important constraint in some areas. It is often cloudy over reefs in the humid tropics and this limits the availability of useful satellite images. There are also some very real problems of accuracy. The technical specifications of the sensors mean that, while coral reefs may be mapped accurately and separately (in about 70 percent of cases) from seagrasses, algal beds and sand, more detailed mapping of different reef habitats is only possible to an accuracy of 20-40 percent. A new generation of high resolution satellites will become available over the next few years, though their utility for reef mapping will probably depend on commercial factors. Initial data collection from new sensors is usually concentrated outside the tropics as a greater economic return can be generated from temperate regions. At the present time, therefore, satellite sensors are well suited for broad-scale mapping of reefs over large areas, but other techniques are required for more detailed interpretation.

In addition to satellites, manned spacecraft provide an important opportunity for obtaining images of the Earth from space (see box). Examples of such images taken from the Space Shuttle are found throughout the present work.

oblique angles through spacecraft windows. When feasible, near vertical photographs have been rotated so that north is toward the top of the page. An approximate scale bar and north arrow have been added based on reference to a 1:1 000 000 scale navigation chart.

Astronaut photographs provide a unique source of moderate resolution reef remote sensing data. They provide global coverage, with free and immediate availability in the public domain. The database of photographs can be searched and browsed online, and high resolution digital copies of photographs in this atlas can be accessed via the website of Earth Science and Image Analysis at NASA's Johnson Space Center: http://eol.jsc.nasa.gov

European Space Agency astronaut Gerhard PJ Thiele photographs Earth from the Space Shuttle Endeavour *in February 2000 (STS099-305-12).*

Aerial photography

No other remote technique has produced as many maps of coral reefs as the simple use of photography from the air, typically from airplanes. Both conventional films and digital cameras are used. Cameras can be mounted on a wide variety of aircraft, and the techniques for analyzing the images are well proven. High resolution mapping can be carried out if the aircraft is operated at low altitude (sub-meter resolution is achieved by flying at 1 000 meters or less) although the area covered by a single flight path is much reduced as a result. The main constraints to this technique are the time required to process, geometrically correct and combine overlapping aerial photographs. In addition, many nations are highly sensitive about aerial surveys of their coastlines. Without a doubt the military archives of many countries represent a valuable, but generally unavailable, repository of coral reef images.

Airborne multispectral imagers

A more complex system of aerial imaging involves sensors similar to those used on satellites, picking up radiation on a number of specified spectral wavelengths. These multispectral imagers are of a size that may be operated from aircraft, enabling similar spatial resolution to aerial photography. The use of aircraft permits data to be collected specifically from areas of interest, at times when conditions are most favorable. Another major benefit of airborne multispectral sensors is that in many cases the optical configuration of the sensor can be set up as required. The two systems which have been used most for reef mapping are the NASA Airborne Visible Infrared Imaging Spectrometer (AVIRIS) and the Compact Airborne Spectrographic Imager (CASI). The costs and the complex processing requirements associated with these systems are still the main constraints to their general use for coral reef mapping. The minimum capital required just to obtain the imagery is typically many tens, if not hundreds, of thousands of dollars. At the close of the 1990s, however, these represented the most accurate available systems for the preparation of fine-scale coral reef maps.

Active sensors

Satellites and cameras are sometimes termed passive sensors – they gather information from existing light as it is reflected from objects on the ground. By contrast, a number of active sensors are used in the marine environment which direct their own source of light or sound towards their subject and measure the reflection. Acoustic signals, or sonar, are perhaps the best known. Sonar systems are carried by ship-borne sensors. Using higher frequencies, good spatial resolution can be achieved (1-4 meters), while in shallow water they are unaffected by water turbidity or depth. Sonar is used in most bathymetric mapping, and thus is an important element of many reef maps, although it does not provide a direct measure of ecological features.

Light detection and ranging (LIDAR) is a light-based form of remote sensing which involves emitting pulses from an airborne laser and receiving energy which has been reflected from both the water surface and submerged features. The time difference between the two types of return provides a highly accurate (±15 centimeters) measurement of depth. The result is an extremely high resolution bathymetric chart. LIDAR is much less affected by water clarity than normal optical sensors and in clear conditions can operate to depths of about 50 meters. Like sonar, LIDAR only maps the topography of the seabed, not ecological features. Furthermore, because of the vast amounts of data processing involved and the requirement for specialist aircraft, it is yet to be used routinely in coral reef mapping.

Ground-truthing

The process of producing a supervised classification of an image is highly reliant on the correct interpretation. Ultimately this is linked to a detailed sampling on the ground, either directly by the cartographers, or using existing information gathered by others in photographs, maps or ecological surveys. Ground-truthing is an expensive but critical element in preparing maps from remote sensors. Natural variation in reefs between locations and over time is often significant and too much extrapolation from previous work, or from work undertaken in other areas, can lead to substantial errors. At the same time, ground-truthing can present considerable opportunities for the refinement of individual maps, greatly increasing the accuracy and allowing for the discernment of particular features or habitats which might not have been visible in other areas.

Remote sensing now dominates mapping in almost every field, and is a critical component of reef mapping around the world. Apart from specialist maps covering relatively small areas, however, most existing maps which show reefs are composite productions which may have been prepared using satellite or aerial imagery in combination with bathymetric data from sonar surveys and even with much older data from early charts. A more widespread and rapid updating of reef maps worldwide is limited by the high cost of remote sensing and the detailed technical skills required for image interpretation and map production.

Despite the power of remote sensing as a mapping tool, there are also several practical constraints to its use. Many of the world's reefs are located in the humid tropics where cloud cover is frequent, which greatly restricts the acquisition of images. The nature of mapping a submarine feature also creates its own set of problems, not least of which is the inability to map deeper reefs. Although the depth limits vary it is rarely possible to map features more than 20-30 meters below the ocean surface with conventional satellite imagery. Even above these depths, the water column greatly affects the light returning to the sensor, changing the spectral signatures of particular seabed characteristics depending on depth and water clarity. Although these effects can be partially compensated during image processing they cannot be totally removed or corrected. Moreover, the nature of the water column above a reef, especially turbidity and depth, is highly variable. Reef geometry, too, does not lend itself easily to being mapped by remote sensing – few parts are flat and most coral tends to be concentrated on steeply sloping edges.

One further weakness is that the remote sensing tools used for seabed mapping are typically different from those used to draw bathymetric maps. Many reef maps prepared using remote sensing do not give detailed bathymetric data, although these are clearly an important feature of many reef maps.

The only alternative to remote sensing for mapping coral reefs is the use of boat-based surveys to map surface features such as reef crests, plotting bathymetry, or even undertaking detailed sampling of the seabed. Unlike remote sensing, which samples the entire seascape, errors arise in this method because of the possibility of overlooking some habitats between adjacent sampling points. Conversely, one advantage of ground survey methods is that they allow for the mapping of additional resources such as different habitats or benthic species, which cannot be distinguished by remote sensors because of similar reflectance patterns or sparse distribution. Ground methods may also allow greater

Habitat map of Cockburn Harbour, South Caicos, based on a supervised classification of CASI multispectral imaging data, and showing the very good resolution and habitat differentiation which can be achieved from these sensors (reproduced by permission of UNESCO, from Green et al (2000)).

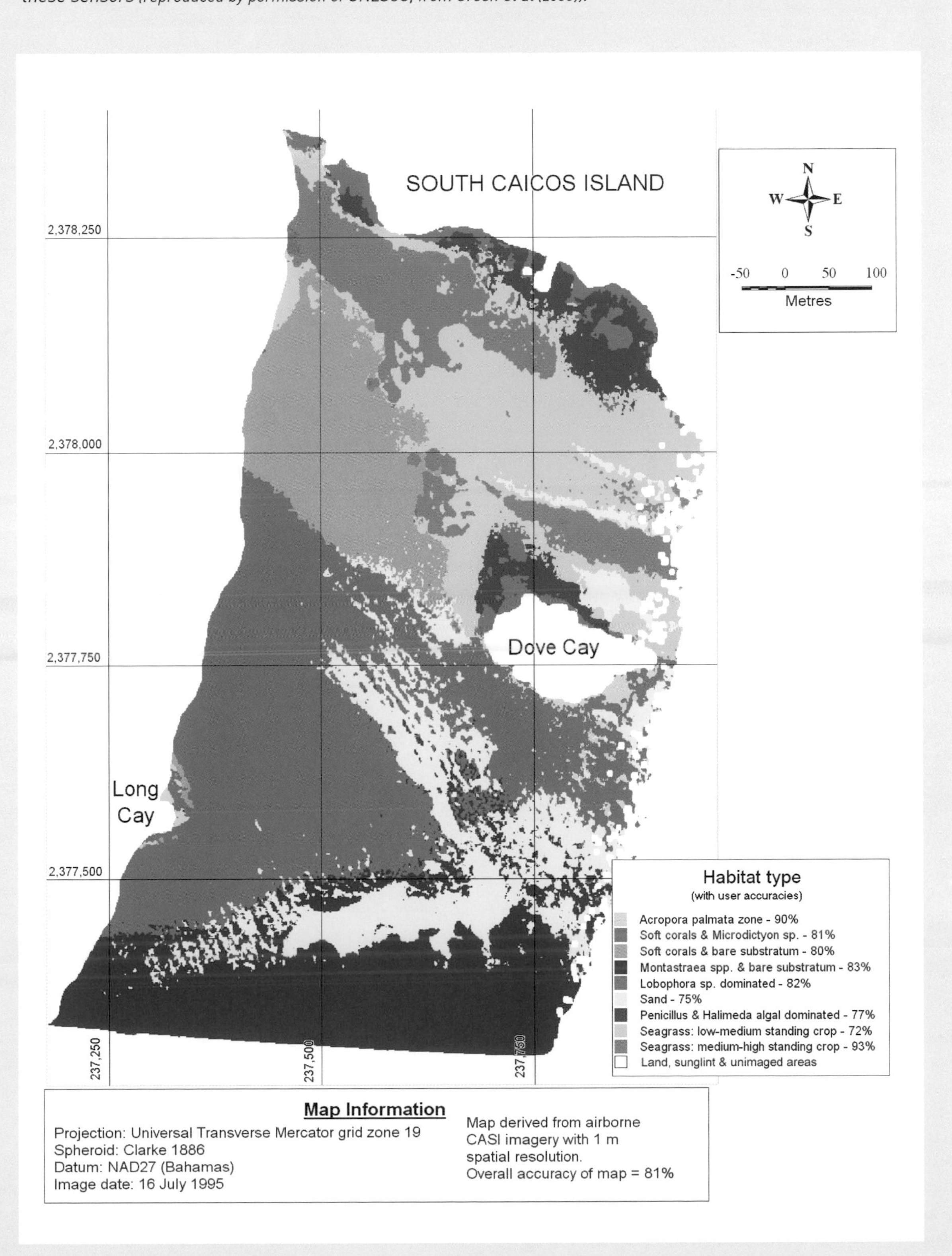

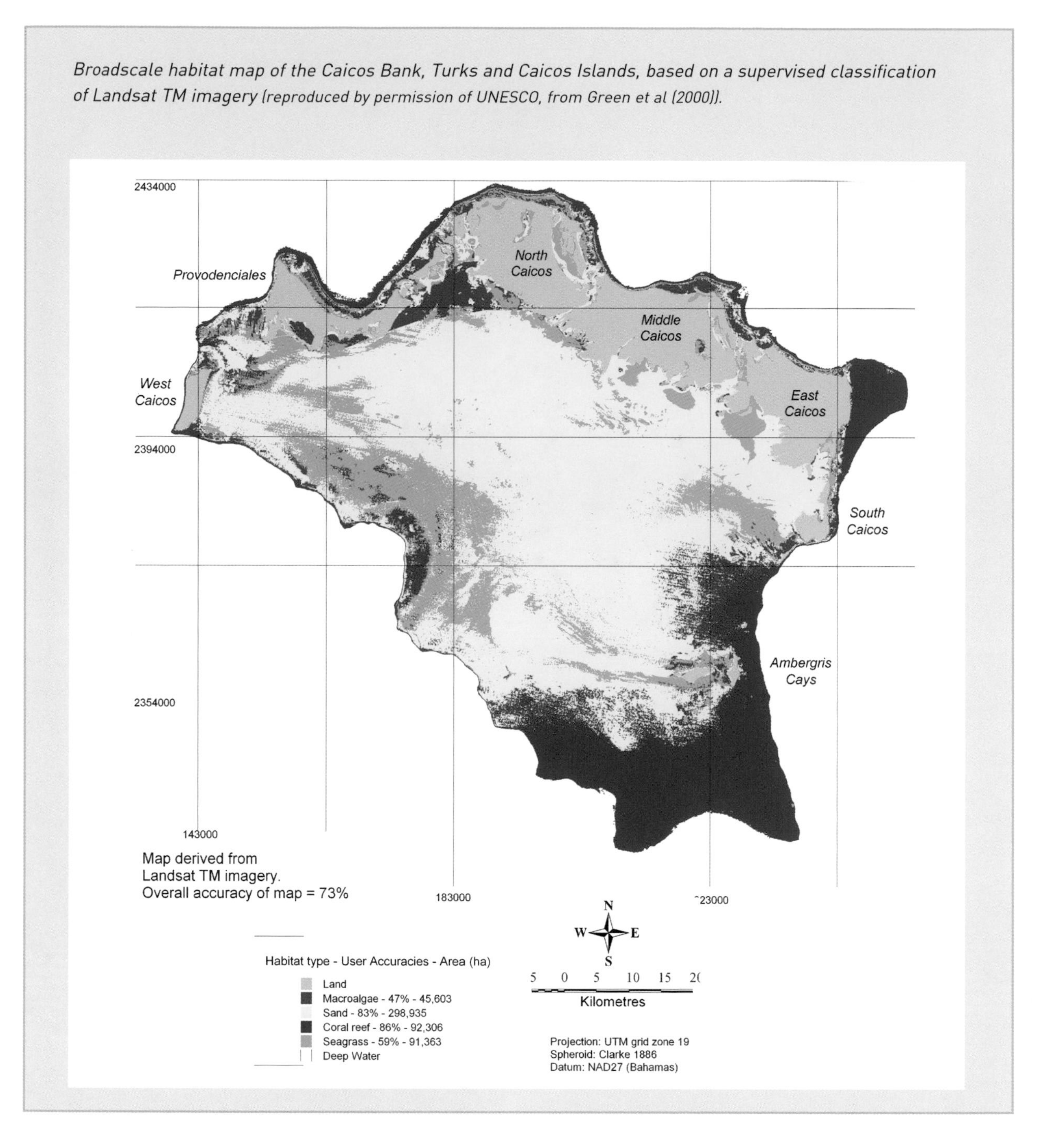

Broadscale habitat map of the Caicos Bank, Turks and Caicos Islands, based on a supervised classification of Landsat TM imagery (reproduced by permission of UNESCO, from Green et al (2000)).

visual depth penetration than many remote sensing methods. Furthermore, the inability of most sensors to detect small but important navigational hazards (such as rocks 1 or 2 meters in size) mean that many new charts still utilize old hydrographic techniques, often in combination with remotely gathered data.

Generally speaking, even for small areas, boat-based surveys are even more costly than using remote sensing, and the latter will continue to provide the dominant reef mapping techniques in coming years. New priorities will undoubtedly be to increase resolution and the differentiation of zones and communities on the reefs. One important aspect of this will be the development of comprehensive "spectral libraries", providing a constantly expanding reference library of spectral signatures associated with different reef communities at different depths and in different conditions. Such libraries will provide critical reference material for the truthing of new images without the need for extensive field surveys. Another priority will be to begin to utilize these images to make comparisons over time, and to use remote sensing in coral reef monitoring.

Global reef mapping

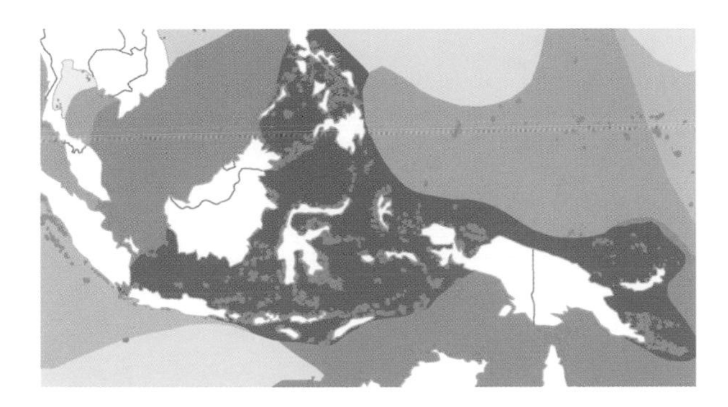

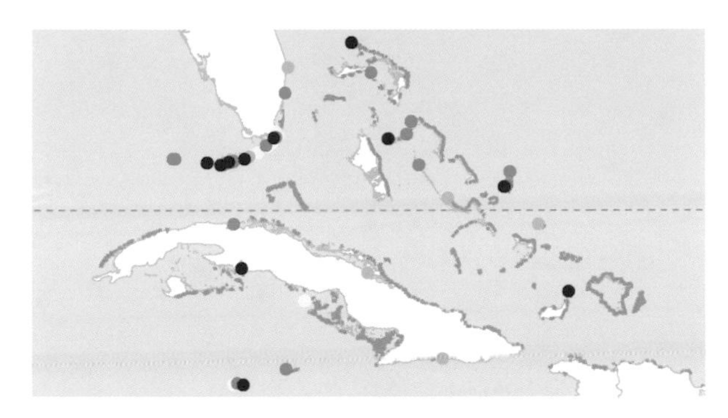

Recent years have seen a burgeoning of interest in obtaining global information about the natural environment. The values of having a global perspective on coral reef distribution are manifold, supporting an increased understanding of ecological and geological processes which, in turn, can fuel more locally based research into reef processes, or more widely based studies of global change. Linking this same information to knowledge of human uses and demographic change can have considerable value for development and resource management issues, and can again help in the understanding of these at finer scales.

The increasing role of regional and global organizations in decision-making processes, and the growing awareness of large-scale patterns within marine ecosystems across national boundaries, further increase the need for understanding at the global level. Improved communications and data availability have strengthened the potential for developing and utilizing existing global-level datasets.

Global maps of coral reefs have been developed and are maintained by a number of organizations. As significant topographic and navigational features, reefs remain important on global map and chart coverages, such as those produced by the UK and US Hydrographic Offices, and also the air navigational series, such as the US Defense Mapping Agency Operational Navigational Charts. Reefs on such charts are generally not given a clear geological or biological definition, as discussed below, but they remain valuable tools in developing a global understanding of reef distribution. Increasingly, such datasets are being prepared in digital formats.

One other recent global synthesis of maps with a clearer ecological and geological focus was the three-volume *Coral Reefs of the World*, prepared at the IUCN Conservation Monitoring Centre (now UNEP-World Conservation Monitoring Centre, UNEP-WCMC). The volumes contain regional maps and national maps which were prepared from numerous sources, including scientific papers, navigational charts and personal communications. Since 1994, UNEP-WCMC has been developing a global map of coral reefs on its GIS. These maps have been published and widely distributed in the global coral reef database, ReefBase, and utilized in a variety of publications and analyses. They are now widely used over the Internet.

The global coral reef map in 2001

The only practical means for the preparation of a global map of coral reefs are through the compilation of existing

Left, above: Detail of Map 1.2, plotting scleractinian coral diversity. Left, below: Detail of Map 2.2, showing coral bleaching events in 1998. Right: Part of a detailed map showing the location of coral reefs in relation to mangrove forests, protected areas and dive centers.

material prepared at finer resolutions for specific areas. Although it is theoretically possible to prepare a more standardized global map using remote sensing, the realization of this goal remains some way off.

The approach adopted by UNEP-WCMC in this work has been hierarchical. As a starting point, a commercially available global base map of digital coral reef data was used plotting the world's reefs at a scale of 1:1 000 000, based on the US Defense Mapping Agency Operational Navigational Chart series. Although broadly accurate, this information was at too low a resolution to show many reefs. Further data were then added from the *Coral Reefs of the World* volumes, focussing on those countries where the 1:1 000 000 source was deemed particularly inadequate, notably the Caribbean and parts of the Western Indo-Pacific. In parallel with this work a data-gathering phase was initiated. Funds were insufficient for a systematic outreach, but improvement of national data has continued on an opportunistic basis, with particular focus directed towards those countries where existing data were considered particularly poor.

The result of this mapping work, by 2001, was a comprehensive and detailed GIS dataset, outputs from which are presented in the maps contained in this volume. These maps represent a summary of the best global map of coral reefs available. A list of the source materials is presented for each map. In all, around 70 percent of countries include source material from new sources at scales finer than 1:1 000 000. Many of these were at scales of 1:250 000 or finer, including navigational charts, topographic map series, processed satellite images and, more occasionally, specialist coral reef or shallow substrate maps.

With a world map derived from multiple sources there will be variation in the definition of coral reefs which is used, although remarkably few maps, other than detailed habitat surveys, provide these definitions in their keys or in accompanying documentation. Despite this weakness, it is usually possible to determine what has been mapped from an understanding of reef geomorphology, combined with some independent knowledge of the reefs of each region. Coral reefs portrayed on navigational maps and charts typically include shallow reef flat and reef crest areas only. High resolution resource inventories may give less attention to areas of reef flat where bottom cover is primarily bare rock or sand, but typically extend beyond the reef crest. The outer limits of coral reef areas as shown on maps may be set by the depth limitations of the survey techniques rather than by any clear ecological boundary.

With this information it is possible to distil out a more generalized definition common to all sources. Such a definition, in fact, conforms quite closely to the broad definition provided in Chapter 1. As they are mapped in this volume, coral reefs are shallow structures built by corals and other hermatypic organisms, and in every case they are associated with an important living component, including hermatypic corals.

These shallow reef areas are among the most important areas in terms of reef growth, productivity, coastal protection and diversity, but it is important to realize that there are considerable additional areas, including sub-surface structures and coral communities which lack clear physical structures and are not shown in these maps. Other authors, using broader definitions of reef, have calculated significantly higher reef areas than those presented in Chapter 1, although the lack of available data means that such calculations are inclined to be more predictive, and are of little value at regional or national resolutions.

Work on coral reef mapping at UNEP-WCMC is an ongoing activity. The contribution of remotely sensed data is likely to expand in coming years, with increasing collaboration with partners around the world. It will be equally important to try to capture some of the sub-surface reefs, particularly in those parts of the world where these are more common, and may harbor important reservoirs of biodiversity. New priorities include the improved mapping of biodiversity patterns to overlay the coral reef distribution information, and the mapping of related habitats, notably seagrasses. It will also be important to develop a better understanding of human interactions with coral reefs, through continued mapping of the threats to reefs, and also improving the available information on marine protected area boundaries. This latter information will enable a more complete assessment of the global distribution and coverage of marine protected areas with coral reefs and will help to highlight the gaps in this network.

Corinna Ravilious incorporating coral reef data into UNEP-WCMC's global GIS.

Selected bibliography

Agassiz A (1899). The islands and coral reefs of Fiji. *Bull Mus Comp Zool* 33: 1-167 (and 120 plates).

Dana JD (1872). *Corals and Coral Islands.* Sampson Low, Marston, Low and Searle, London, UK.

Darwin C (1842). *The Structure and Distribution of Coral Reefs.* Smith, Elder and Co., London, UK.

Green EP, Mumby PJ, Edwards AJ, Clark CD (1996). A review of remote sensing for the assessment and management of tropical coastal resources. *Coast Man* 24: 1-40.

Green EP, Mumby PJ, Ellis AC, Edwards AJ, Clark CD (ed Edwards AJ) (2000). *Remote Sensing Handbook for Tropical Coastal Management.* Coastal Management Sourcebooks 3, UNESCO, Paris, France.

Joubin ML (1912). Carte des bancs et récifs de coraux (Madrépores). *Annales de l'Institut Océanographique* IV: 7 (with 5 maps in separate volume).

LeDrew E, Holden H, Peddle D, Morrow J, Murphy R, Bour W (1995). Towards a procedure for mapping coral stress from SPOT imagery with *in situ* optical correction. Third Thematic Conference on Remote Sensing for Marine and Coastal Environments.

Lulla KP et al (1996). The Space Shuttle Earth Observations Photography Database: an underutilized resource for global environmental sciences. *Environmental Geosciences* 3: 40-44.

McManus JW, Vergara SG (eds) (1998). ReefBase: A Global Database on Coral Reefs and their Resources, Version 3.0, CD-ROM. International Center for Living Aquatic Resources Management, Manila, Philippines.

Peddle DR, LeDrew FF, Holden HM (1995). Spectral mixture analysis of coral reef abundance from satellite imagery and *in situ* ocean spectra, Savusavu Bay, Fiji. Third Thematic Conference on Remote Sensing for Marine and Coastal Environments.

Petroconsultants SA (1990). MUNDOCART/CD. Version 2.0. 1:1 000 000 world map prepared from the Operational Navigational Charts of the United States Defense Mapping Agency. Petroconsultants (CES) Ltd, London, UK.

ReefBase (2000). ReefBase 2000: Improving Policies for Sustainable Management of Coral Reefs. Version 2000. CD-ROM. ICLARM, Philippines. (See also http://www.reefbase.org)

Robinson JA, Feldman GC, Kuring N, Franz B, Green E, Noordeloos M, Stumpf RP (2000). Data fusion in coral reef mapping: working at multiple scales with SeaWiFS and astronaut photography. *Proceedings of the 6th International Conference on Remote Sensing for Marine and Coastal Environments* 2: 473-483.

Smith SV (1978). Coral-reef area and the contributions of reef to processes and resources of the world's oceans. *Nature* 273: 225-226.

Spalding MD (1997). Mapping global coral reef distribution. *Proc 8th Int Coral Reef Symp* 2: 1555-1560.

Stanley Gardiner J (1931). *Coral Reefs and Atolls.* Macmillan and Co. Ltd, London, UK.

UNEP/IUCN (1988a). *Coral Reefs of the World. Volume 1: Atlantic and Eastern Pacific.* UNEP Regional Seas Directories and Bibliographies. UNEP and IUCN, Nairobi, Kenya, Gland, Switzerland and Cambridge, UK.

UNEP/IUCN (1988b). *Coral Reefs of the World. Volume 2: Indian Ocean.* UNEP Regional Seas Directories and Bibliographies. UNEP and IUCN, Nairobi, Kenya, Gland, Switzerland and Cambridge, UK.

UNEP/IUCN (1988c). *Coral Reefs of the World. Volume 3: Pacific.* UNEP Regional Seas Directories and Bibliographies. UNEP and IUCN, Nairobi, Kenya, Gland, Switzerland and Cambridge, UK.

Wells JW (1954). Recent corals of the Marshall Islands. Bikini and nearby atolls, part 2, oceanography (biologic). *US Geol Survey Prof Pap* 260: 385-486.

Winkler (1901). On sea charts formerly used in the Marshall Islands, with notices on the navigation of the islanders in general. *Smithsonian Institution Annual Report for 1898.* 487-508.

Part II

The Atlantic and Eastern Pacific

The Atlantic Ocean covers one fifth of the surface area of the planet, second only in size to the Pacific. It is the world's youngest ocean, only beginning to form around the time of the break-up of the supercontinent of Pangea about 180 million years ago. It also has the biggest drainage basin of any ocean, and the large amounts of sediment entering it from the great rivers such as the Amazon, Orinoco, Mississippi, Niger and Congo certainly have a role to play in inhibiting coral reef development along much of its perimeter.

In the eastern tropical areas of the Atlantic the coastline is relatively simple, following the continental coast of West Africa, and with only a few significant island groups, notably the Cape Verde Islands and São Tomé and Principé. The Central Atlantic has few shallow water features and, although there is some volcanic activity associated with the mid-ocean ridge, there are only a few oceanic islands in the tropics. The western coastline of the Atlantic is quite different: between the eastern coast of Venezuela and the southern tip of Florida there is a chain of islands separating the Atlantic proper from the Caribbean Sea in the south and from the Gulf of Mexico in the north. These semi-enclosed seas actually have an older geological history than the Atlantic itself. As the Central American Sea they would have been directly connected to the Mediterranean until the break-up of Pangea and the appearance of the Atlantic. There is volcanic activity in a few areas on the eastern and western edges of the Caribbean Basin.

The only major coral reef development in this entire region is centered around the Caribbean Sea and to the north of Cuba in an area bounded by the Bahamas and Florida. There are smaller reef developments in a few locations in the Gulf of Mexico and around Bermuda in the Northern Atlantic. Brazil too has some reef structures, although in general these are small and intermittent. There are no true reefs on the other oceanic islands or in the Eastern Atlantic. Even in the Caribbean Sea, reef development rarely reaches the extent of reefs in the Indo-Pacific, and there appear to be real physical differences in reef structures between these regions. Many Caribbean reefs are deeper submerged features, while many of those with a clear reef crest often only have a narrow reef flat. Although there are a small number of barrier reefs and atolls these are clearly not as prolific as in many parts of the Indo-Pacific. In all, less than 8 percent of the world's coral reefs are found in this region.

In terms of biodiversity the coral reefs of this region are depauperate, but they are also unique. The Atlantic corals now share only seven genera with the Indo-Pacific. There are three clear regions in the Atlantic reef province, with the highest diversity focussed in the Caribbean (from Bermuda to Trinidad), but with other small centers of coral diversity in Brazil and the Eastern Atlantic. Current coral faunas are uniform across the Caribbean sub-region, with few geographically restricted species. The Brazilian sub-region is isolated from the Caribbean by a substantial barrier posed by the long, sediment-rich coastal areas of the Guyanas and the Brazilian Amazon. Although these reefs have a very low diversity of species, many are endemic. Some appear to be relict species with evidence of a wider distribution shown in the fossil record, while others are probably the result of allopatric speciation, and appear to have sister species in the Caribbean. The Eastern Atlantic has even lower levels of species diversity and there only one is endemic. The remainder show affinities to both

Caribbean and Brazilian sub-regions, suggesting immigration from both areas.

In many areas of the Caribbean, including remote localities, there have been wide declines in coral cover together with dramatic increases in algal cover. While these can sometimes be directly related to human impacts in specific localities, there also appears to have been a regional decline. Much can be linked to the widespread die-off of the long-spined sea urchin *Diadema antillarum*, which occurred in 1983-84 (see page 61). This species was a major algal grazer on the reefs and, in places where overfishing was particularly high, it was often the major herbivore on reefs. The die-off appears to have been pathogenic – caused by a bacterial infection – and may have been natural or may have been carried to the region in ballast water from ships. At the same time wide areas have been afflicted by coral diseases. White band disease has infected and killed many populations of staghorn coral *Acropora cervicornis* and elkhorn coral *Acropora palmata*, which were once among the main structural components of reefs. Dead corals are rapidly overgrown with algae, especially because of the loss of grazing *Diadema*, and large areas of reef have now become algae-dominated and are showing little sign of a return to previous conditions.

The degree to which these problems are the result of entirely natural processes remains unclear, but this region is also widely affected by many direct human impacts, including sedimentation, nutrient pollution and overfishing. Tourism is probably the most important industry across the wider Caribbean, with most tourists seeking beach-based holidays. In many places, coastal development associated with the growing tourism industry has greatly exacerbated the problems facing the reefs of the region. In the 1998 *Reefs at Risk* analysis, 71 percent of the reefs in the Wider Caribbean were described as threatened by human activities. In areas where these problems are most extreme, the rates of recovery of reef systems are certainly being slowed. Efforts to improve reef status and recovery through the implementation of marine protected areas and other coastal management regimes are varied, but include important success stories. At the same time, the indirect nature of many threats means that, even with legal protection, many reefs are still in decline.

The Eastern Pacific

The western shores of the Americas are completely separated from the Caribbean Sea and have very different coral reef communities. Although part of the Pacific, this region is quite distinct from the rest of this ocean in terms of its reefs. Much of the continental coastline plunges into relatively deep water as the oceanic plates are subducted under the continental plates of North and South America. There are only a few offshore islands lying beyond the continental shelf. Water conditions on the continental shelf fluctuate considerably, with cool water upwellings in most years, occasionally reversed to warm water upwellings during El Niño Southern Oscillation (ENSO) events.

The isthmus of Panama closed between 3 and 3.5 million years ago and, apart from possible very minor breaches of this gap, there has been no marine connection between the coral reefs on either side of this land bridge since that time. Dramatic changes have been wrought on the marine communities since the closure of the isthmus. Initially the two separated communities were probably very diverse. However, the Pliocene/Pleistocene glaciations wiped out large numbers of species, and almost completely removed coral species from the Pacific shores of the Americas. There has been some recolonization of these shores by corals, but they have come from the Pacific. Recolonization has been slow and sporadic due to the great physical distance between these shores and the nearest reefs around the Central Pacific islands. This "East Pacific Barrier" is enhanced by unfavorable patterns of ocean currents, further reducing the chances of larval transport across the Pacific.

The coral reefs of the Eastern Pacific are thus highly distinctive communities. Their closest affinities are with the reefs of the Pacific, but they have a much lower diversity of species and many are endemic. The reefs themselves are rarely well developed as physical structures – most are simply coral communities. The few structural reefs are mostly small in overall extent, and consist of only a few meters depth of carbonate deposits. The one exception to this generalization is Clipperton Atoll, a true atoll administered by French Polynesia (and described in Chapter 14).

Fluctuating oceanographic conditions have a considerable impact on coral reefs. The first region-wide mass mortalities associated with coral bleaching were linked to a major ENSO event in 1983. The 1997-98 ENSO event also caused high levels of bleaching, although mortality appears to have been lower. Localized centers of upwelling appeared to suffer least and may be important refuges for the East Pacific fauna during such events.

Human impacts on this region are generally low. Most of the reefs are associated with offshore islands and therefore not heavily impacted by terrigenous influences. Overfishing is certainly a problem in some areas. There is virtually no tourism, with the exception of the highly controlled visits to the Galapagos.

MAP 4

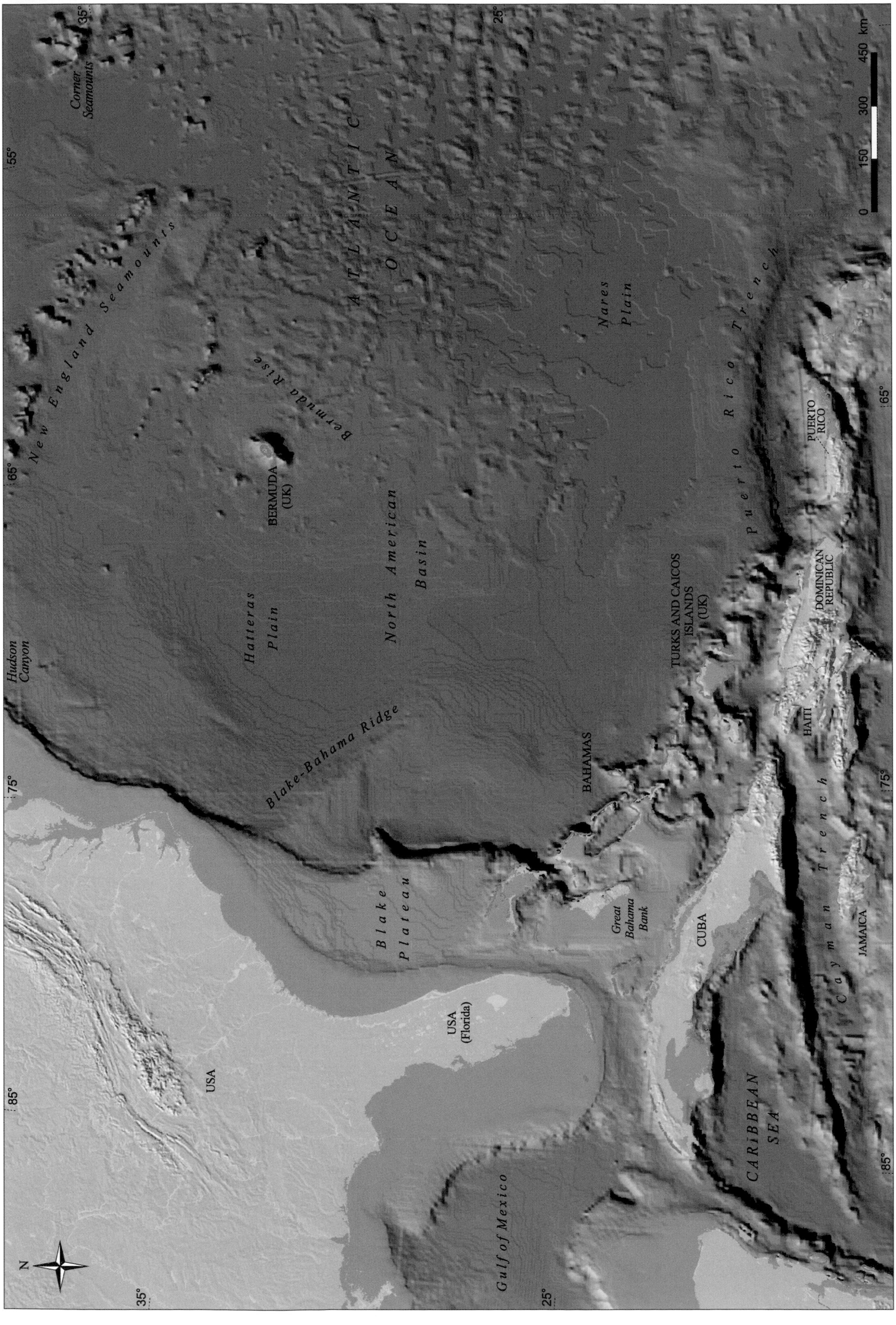

New England Seamounts
Corner Seamounts
ATLANTIC OCEAN
Bermuda Rise
BERMUDA (UK)
Hudson Canyon
Hatteras Plain
North American Basin
Nares Plain
Puerto Rico Trench
Blake-Bahama Ridge
BAHAMAS
TURKS AND CAICOS ISLANDS (UK)
PUERTO RICO
DOMINICAN REPUBLIC
HAITI
Blake Plateau
Great Bahama Bank
CUBA
Cayman Trench
JAMAICA
USA (Florida)
USA
CARIBBEAN SEA
Gulf of Mexico
N
35° 25° 55° 65° 75° 85°
0 150 300 450 km

Chapter 4
Northern Caribbean

The northernmost reefs of the Wider Caribbean region lie outside the true Caribbean Basin, stretching in a broad sweep from the Turks and Caicos Islands in the south to Florida and the northern Bahamas. Far out into the Atlantic, the island of Bermuda forms an outlier to this group, connected to the region by the warm waters of the Gulf Stream.

This is an area of great biological interest. Because of its northerly location it encompasses the outer limits of the distribution range of many coral reef species. There are clear decreases in biodiversity with latitude and oceanographic processes probably play an important role in these biodiversity patterns. The general northward flow of currents maintains a supply of warm waters to latitudes even quite far outside the tropics, and supports the active growth of the most northerly reefs in the world, in Bermuda. These same currents may further maintain biodiversity on some reefs by transporting new larvae of coral reef organisms from reefs "upstream". Reef development in the region has built up complex reef systems around older carbonate structures and islands, and the overall extent of reefal shelf is really very large indeed.

In human terms, this is a region of great contrasts. The reefs of the Florida Reef Tract are among the most intensively studied in the world, but they are also among the most heavily utilized and have become highly degraded. By contrast, the vast extent of reefs of the Bahamas are, for the most part, poorly described, and human impacts (other than fishing) are concentrated in a few locations, leaving much of the rest relatively undisturbed.

The legal protection of reefs, through the designation of protected areas or the implementation of other strict management controls, is well established in the region. However, the condition of Florida's reefs, where human-induced stresses have been apparent for many decades, suggest that such protection may not be enough. Though they have been "protected" for some years, there remain considerable conflicts in the "user" demands of those living beside and visiting the reefs.

In contrast to the Florida reefs, the Flower Garden Banks off Texas (Map 5a), as well as Bermuda and the Turks and Caicos Islands, provide examples of relatively well protected reefs. In the case of Bermuda, this protection comes in spite of high population densities and fairly heavy use of reef resources.

Left: Broad view of the Florida peninsula, with some of the reefs of the northern Bahamas visible to the right (STS095-743-33, 1998). Right: Blue chromis Chromis cyanea *hovering above the branches of the increasingly rare staghorn coral* Acropora cervicornis.

MAP 4a

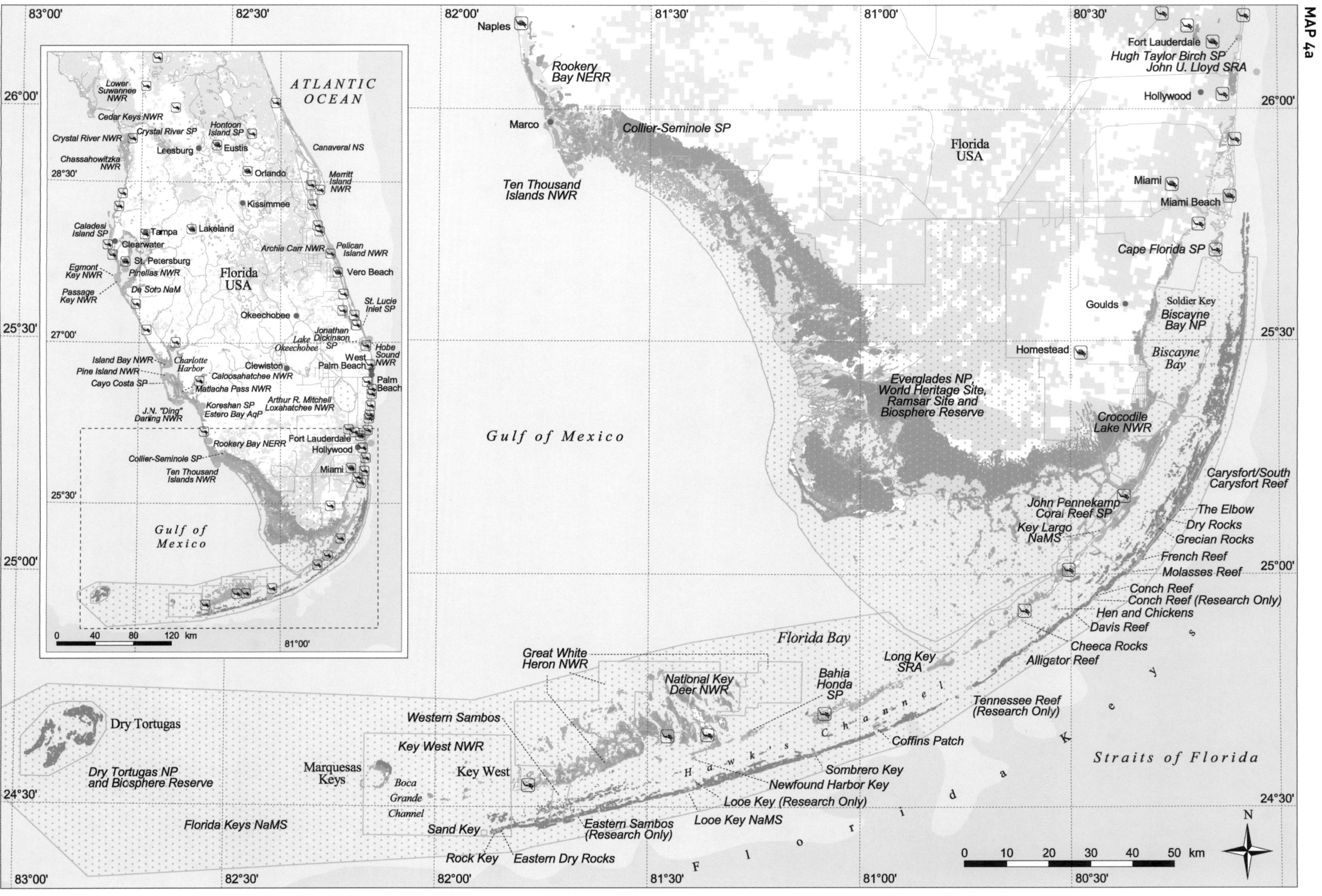

Naples
Rookery Bay NERR
Marco
Collier-Seminole SP
Ten Thousand Islands NWR
Florida USA
Fort Lauderdale
Hugh Taylor Birch SP
John U. Lloyd SRA
Hollywood
Miami
Miami Beach
Cape Florida SP
Goulds
Soldier Key
Biscayne Bay NP
Biscayne Bay
Homestead
Everglades NP, World Heritage Site, Ramsar Site and Biosphere Reserve
Crocodile Lake NWR
Gulf of Mexico
Carysfort/South Carysfort Reef
John Pennekamp Coral Reef SP
The Elbow
Key Largo NaMS
Dry Rocks
Grecian Rocks
French Reef
Molasses Reef
Conch Reef
Conch Reef (Research Only)
Hen and Chickens
Davis Reef
Cheeca Rocks
Alligator Reef
Florida Bay
Long Key SRA
Great White Heron NWR
National Key Deer NWR
Bahia Honda SP
Tennessee Reef (Research Only)
Coffins Patch
Dry Tortugas
Western Sambos
Key West NWR
Marquesas Keys
Key West
Boca Grande Channel
Dry Tortugas NP and Biosphere Reserve
Sombrero Key
Newfound Harbor Key
Looe Key (Research Only)
Looe Key NaMS
Hawk's Channel
Florida Keys
Straits of Florida
Florida Keys NaMS
Sand Key
Eastern Sambos (Research Only)
Rock Key
Eastern Dry Rocks
N
0 10 20 30 40 50 km
83°00'
82°30'
82°00'
81°30'
81°00'
80°30'
26°00'
25°30'
25°00'
24°30'
ATLANTIC OCEAN
Lower Suwannee NWR
Cedar Keys NWR
Crystal River NWR
Crystal River SP
Hontoon Island SP
Leesburg
Eustis
Canaveral NS
Chassahowitzka NWR
28°30'
Orlando
Merritt Island NWR
Kissimmee
Caladesi Island SP
Tampa
Lakeland
Clearwater
Archie Carr NWR
Pelican Island NWR
St. Petersburg
Egmont Key NWR
Pinellas NWR
Florida USA
Vero Beach
Passage Key NWR
De Soto NaM
St. Lucie Inlet SP
Okeechobee
Jonathan Dickinson SP
27°00'
Lake Okeechobee
Hobe Sound NWR
West Palm Beach
Island Bay NWR
Charlotte Harbor
Clewiston
Pine Island NWR
Caloosahatchee NWR
Palm Beach
Cayo Costa SP
Matlacha Pass NWR
Arthur R. Mitchell Loxahatchee NWR
Koreshan SP
Estero Bay AqP
J.N. "Ding" Darling NWR
Rookery Bay NERR
Fort Lauderdale
Hollywood
Collier-Seminole SP
Ten Thousand Islands NWR
Miami
25°30'
Gulf of Mexico
0 40 80 120 km
81°00'

Florida and the US Gulf of Mexico

MAP 4a

The coral reefs of mainland USA are largely restricted to two areas: the coastline of southern Florida and a few small but important reef patches in the Gulf of Mexico.

Florida

The Florida Reef Tract is one of the most extensive reef systems in the region. Starting directly offshore from Miami Beach there is a near continuous offshore reef structure which stretches in a barrier-like formation for some 260 kilometers. Further west the shallow platform continues, and there are isolated reef patches, including the Dry Tortugas and a number of submerged structures. The reef front has some well developed spur and groove structures, with coral mounds rising from the bottom below this. Overall coral cover in the region was typically 14 percent in the late 1990s, reaching 30-40 percent on some patch reefs behind the reef crest. A slightly deeper channel lies behind the reef (Hawk's Channel). Behind this there is a chain of islands, the Florida Keys. These are low-lying (less than 2 meters elevation), and are composed of Pleistocene limestone, stretching from Soldier Key near Miami to Key West. Behind the Florida Reef Tract and into the area of Florida Bay the waters are generally very shallow and harbor some of the most extensive seagrass areas in the entire region. Mangroves dominate the shoreline, both of the Florida Everglades and around many areas of Biscayne Bay and the Florida Keys. North of Miami there is some reef development as far as Vero Beach. Generally these are not major reef structures and coral cover is low, although on deeper reefs down to 22 meters staghorn corals are reported to be increasing.

Human impacts on the reefs of Florida have been apparent for many years. The Florida Keys were first joined to the mainland by a railway in 1912 and then by road in 1938. They have now developed into one of the most popular tourist destinations of continental USA. Over 4 million people visit the Keys annually, joining some 100 000 permanent residents. The majority are attracted to the area by the marine environment, and sailing, diving and fishing are critical to the local economy.

The reefs of Florida, afflicted by an enormous range of impacts, include some of the most degraded in the region. Initial changes were noted following the construction of the railway causeway out to the Keys in the early 1900s. Patterns of water flow from Florida Bay became severely disrupted, and although channels have subsequently been re-opened, other impacts have continued and now include ship groundings, anchor damage and the scouring of sea-grass by propellers. Between 1980 and 1993 approximately 500 vessels were reported to have grounded in the Looe Key and Key Largo Sanctuaries alone, while some 500 ship groundings are now reported annually in the Florida

Detailed view of the western Florida Keys, clearly showing the intense human development, including airstrips and roads, as well as sediments in the surrounding waters (STS038-85-103, 1990).

Keys National Marine Sanctuary. Another major problem is the eutrophication and pollution of nearshore waters associated with the extensive agricultural areas that drain into the bay, and from sewage. There are some 200 sewage treatment plants, 22 000 septic tanks, 5 000 cesspools and 139 marinas harboring over 15 000 boats in the Florida Keys. Fishing pressure is considerable throughout the area, and many fish stocks are considered overfished.

Although some declines in ecological conditions and coral growth have been linked back to the construction of the causeway, far more rapid declines have been observed since the 1980s. In 1981, *Acropora* corals covered up to 96 percent of the reef substrate in places, but by 1986 this cover had fallen to about 3 percent, linked to the impacts of white band disease. Even since 1996, there have been declines in both remaining hard coral cover and diversity at a majority of permanent monitoring sites. More than ten coral diseases have been observed. Coral bleaching events have occurred with increasing frequency since the 1980s, typically when warm weather is further exacerbated by very calm doldrum-like conditions. The most recent bleaching was associated with similar calm conditions in 1997, which continued to a second major warm period with intense bleaching in 1998, followed by the impacts of Hurricane Georges and Tropical Storm Mitch. In combination, these led to considerable losses of coral in shallow areas.

Reefs in Florida are possibly the most intensely monitored in the world, with 16 programs active in 2000. All coral reefs in Florida are protected either at the federal or state level. The Florida Keys National Marine Sanctuary, designated in 1990, extends over nearly 10 000 square kilometers of critical marine habitat, including coral reef, hard bottom, seagrass meadows, mangrove communities and sand flats. The sanctuary only provides partial protection and many unsustainable activities continue. However in 1997 a system of 23 no-take marine reserves was established within the area – while these only cover 1 percent of the total sanctuary area, they include some 65 percent of the shallow reef zones. Within three years there already appeared to be signs of some recovery.

In addition to the marine zoning program, key sanctuary initiatives include a water quality protection program, extensive education and volunteer programs, channel marking initiatives, and installing and maintaining mooring buoys to prevent anchor damage to the reef.

The remaining reefs in Florida are around the Dry Tortugas and lie within the Dry Tortugas National Park.

USA, Atlantic

General Data	
Population (thousands)	275 563
GDP (million US$)	6 392 711
Land area (km²)	9 451 035
Florida	*152 000*
Marine area (thousand km²)	na
Per capita fish consumption (kg/year)	21
Status and Threats	
Reefs at risk (%)	91
Recorded coral diseases	16
Biodiversity	
Reef area (km²)	1 250
Coral diversity	na / 58
Mangrove area (km²)	na
No. of mangrove species	na
No. of seagrass species	na

US Gulf of Mexico

In the US waters of the Gulf of Mexico there are a number of banks rising up from the continental shelf which are derived from salt domes. Although corals are found on a number of these they generally lack diversity and cannot be called true reefs, with the exception of the East and West Flower Garden Banks. The banks themselves cover less than 90 square kilometers, with the reef areas occupying only a small proportion (about 1.4 square kilometers). They are located 200 kilometers south of Galveston, Texas, and are among the most isolated reefs in the Wider Caribbean. Little known and studied until the advent of oil exploration during the 1970s, these reefs only appeared on nautical charts in the 1930s and fewer than ten papers were published on the Flower Gardens before 1969. They are colonized by 20 species of hard corals. Dominated by species of *Diploria*, *Montastrea* and *Porites*, they have a live coral cover of around 47 percent at a depth of 15-30 meters. Shallow water species of soft corals are not found here. These reefs are naturally protected by their distance from shore and their depth: the bank crests are 15-20 meters deep and so even hurricanes inflict relatively little damage. Environmental conditions are generally more stable than they are in Florida – the surrounding water is oceanic and exceptionally clear all the year round. Temperatures range from 19 to 30°C.

Live coral cover has remained relatively high since monitoring first began in 1972, and the incidence of coral disease is relatively low (2 percent of colonies). Algal cover increased rapidly from negligible amounts to a maximum of 14 percent after the die-off of the urchin *Diadema*, but this was reversed within a year after the populations of large, herbivorous parrotfishes increased.

The Flower Garden Banks are the setting for some spectacular seasonal events. Each year, at the last quarter of the moon in August, many coral species reproduce in a syncronchized mass spawning. Each winter the reefs

Protected areas with coral reefs

Site name	Designation	Abbreviation	IUCN cat.	Size (km²)	Year
Florida and US Gulf of Mexico					
Biscayne Bay	National Park	NP	II	729.00	1980
Dry Tortugas	National Park	NP	II	262.03	1992
Everglades	National Park	NP	II	6 066.88	1947
Florida Keys	National Marine Sanctuary	NaMS	IV	9 603.73	1990
John Pennekamp Coral Reef	State Park	SP	V	226.84	1959
John U Lloyd	State Recreation Area	SRA	V	1.02	1973
Key Largo	National Marine Sanctuary	NaMS	V	323.88	1975
Key West	National Wildlife Refuge	NWR	IV	979.43	1908
Looe Key	National Marine Sanctuary	NaMS	V	15.54	1981
Flower Garden Banks	National Marine Sanctuary	NaMS	V	145.04	1992
Everglades and Dry Tortugas National Parks	**UNESCO Biosphere Reserve**			**8 716.59**	**1976**

witness gatherings of large schools of hammerhead sharks. Manta rays can be seen throughout the year – juveniles in the summer; adults in the winter. Whale sharks are periodically abundant.

The Flower Gardens were declared a US National Marine Sanctuary in 1992. An additional smaller bank, Stetson Bank, which lies further north in an area of greater temperature variation and higher turbidity, was added to this designation in 1996. While there is some coral growth on Stetson Bank there is no active reef accretion. The marine sanctuary has ensured that there has been remarkably little impact from the petroleum industry, although there are about 4 000 hydrocarbon production facilities and over 35 000 kilometers of pipeline in the northwestern Gulf of Mexico. Harvesting of reef organisms is restricted to hook and line fishing, anchoring by commercial vessels is prohibited, and mooring buoys for dive boats have been installed. Some 2 000 divers visit each year.

A stoplight parrotfish Sparisoma viride *amidst massive corals and a gorgonian coral or sea fan.*

MAP 4b

N
ATLANTIC OCEAN
65°00'
64°56'
64°52'
64°48'
64°44'
64°40'
64°36'
32°28'
32°24'
32°20'
32°16'
Cristobal Colon PA
North East Breaker PA
North Rock PA
Aristo PA
Caraquet PA
Madiana PA
Hog Breaker PA
Snake Pit PA
North Shore
Coral Reef Pr
Xing Da
Area PA
Mills Breaker PA
Murray's Anchorage
Eastern Blue Cut PA
St. George's I.
Lovers Lake NR
and Ramsar Site
St. George's
Harbour
St. David's I.
Lartington PA
Montana PA
Constellation Area PA
Pelinaion & Rita Zovetto PA
Harrington
Sound PrivR
Walsingham
PrivR
Castle Harbour
Castle Harbour Islands NR
Harrington
Sound
Shelly Bay Marsh NR
The Cathedral PA
Commissioner's Point Ar PA
Ireland I. North
Ireland I. South
Kate PA
L'Herminie PA
South Shore
Coral Reef Pr
Daniel's Island NR
Great
Sound PrivR
HAMILTON
BERMUDA
(UK)
Spittal Pond NR
and Ramsar Site
Somerset I.
Darlington PA
Great Sound
Hungry Bay NR
Hungry Bay Mangrove
Swamp Ramsar Site
Blanche King PA
North Carolina PA
Hamilton Harbour PrivR
Little Sound
Tarpon Hole PA
Godet's
Rock NR
Evans Bay PrivR
Airplane PA
Godet's
Island NR
Hermes & Minnie Breslauer PA
Marie Celeste PA
South West Breaker Area PA
0 2 4 6 8 10 km

Bermuda

MAP 4b

Bermuda is an isolated group of 150 limestone islands in the Sargasso Sea area of the Western North Atlantic Ocean more than 1 000 kilometers from continental USA. Most of the land area is represented by five islands which are joined together by causeways. These islands are the high points of the Bermuda Platform – the nearby Plantagenet and Challenger Banks rise to 50 meters below sea level. Together the three banks crown the Bermuda Rise, a mid-plate hotspot of similar origin to Hawai'i, although geologically older. Water in excess of 8 000 meters depth occurs just 6 kilometers to the northwest of Bermuda, as the sides of the Bermuda Rise fall steeply to the ocean floor.

The Bermuda Platform is 1 400 kilometers from the nearest hermatypic coral reefs, in Florida and the Bahamas, yet supports the northernmost coral reefs of the Atlantic. A sub-tropical climate maintains shallow water temperatures above 19°C in winter, with a summer maximum of 27°C. The northerly occurrence of such warm temperatures, at a similar latitude to the Canary Islands, is due to the Gulf Stream which passes to the north and west of Bermuda.

The reef flora and fauna of Bermuda are much less diverse than in the Caribbean. Only one third of Caribbean corals occur here, with the most notable absence being the genus *Acropora*. Approximately 120 species of reef fish have been recorded. Fringing, bank barrier and lagoonal patch reefs are found on the Bermuda Platform and the health of these small reefs is good overall. Coral cover averaged 30-35 percent in 2000, reaching 50 percent on the outer terrace. Grazing by parrotfishes and surgeonfishes was sufficient to prevent even a temporary increase in the cover of fleshy algae after most of the *Diadema* died in 1983.

Bermuda has a very high human population density. It also enjoys one of the highest per-capita incomes in the world, through both the provision of financial services and the development of luxury tourist facilities for 600 000 annual visitors. The tourist industry accounts for an estimated 28 percent of gross domestic product and attracts 84 percent of its business from North America. The industrial sector is small, and agriculture is severely limited by a lack of suitable land. Domestic waste is the main source of terrestrial pollution. The occasional grounding of large vessels is also a problem for the reefs. Queen conch are reported to be commercially extinct, but other reef fisheries are at a low level and appear to be

Bermuda

General Data	
Population (thousands)	63
GDP (million US$)	1 797
Land area (km^2)	39
Marine area (thousand km^2)	450
Per capita fish consumption (kg/year)	44
Status and Threats	
Reefs at risk (%)	100
Recorded coral diseases	2
Biodiversity	
Reef area (km^2)	370
Coral diversity	26 / na
Mangrove area (km^2)	0.16
No. of mangrove species	3
No. of seagrass species	4

Above: A large hogfish Lachnolaimus maximus. *These fish are popular food items and increasingly rare in many areas.*
Below: Massive corals Montastrea *and gorgonians or sea fans.*

sustainable. In general, marine conservation enjoys priority status, with a high level of protection afforded to about a quarter of the Bermudan coral reefs through two coral reef preserves, three seasonally protected no-take fishing areas, nine large protected dive sites and a further 20 smaller ones. The latter are mostly wrecks on the reef substrate, and many are now important reef habitats in their own right. All fishing is prohibited in these areas. Recreational fishers elsewhere have bag limits, and commercial trap fishing for finfish was totally banned in 1990. Recently the number of convictions for use of illegal fish traps has increased in Bermuda.

Protected areas with coral reefs

Site name	Designation	Abbreviation	IUCN cat.	Size (km²)	Year
Bermuda					
Airplane	Protected Area	PA	III	0.28	na
Aristo	Protected Area	PA	III	0.28	na
Blanche King	Protected Area	PA	III	0.28	na
Caraquet	Protected Area	PA	III	0.28	na
Commissioner's Point Area	Protected Area	PA	III	0.12	na
Constellation Area	Protected Area	PA	III	0.79	na
Cristobal Colon	Protected Area	PA	III	0.28	na
Darlington	Protected Area	PA	III	0.28	na
Eastern Blue Cut	Protected Area	PA	III	1.13	na
Hermes and Minnie Breslauer	Protected Area	PA	III	0.79	na
Hog Breaker	Protected Area	PA	III	0.28	na
Kate	Protected Area	PA	III	0.28	na
Lartington	Protected Area	PA	III	0.28	na
L'Herminie	Protected Area	PA	III	0.28	na
Madiana	Protected Area	PA	III	0.28	na
Marie Celeste	Protected Area	PA	III	0.28	na
Mills Breaker	Protected Area	PA	III	0.28	na
Montana	Protected Area	PA	III	0.28	na
North Carolina	Protected Area	PA	III	0.28	na
North East Breaker	Protected Area	PA	III	0.28	na
North Rock	Protected Area	PA	III	3.14	na
North Shore Coral Reef	Preserve	Pr	IV	130.50	1966
Pelinaion and Rita Zovetto	Protected Area	PA	III	0.79	na
Snake Pit	Protected Area	PA	III	0.28	na
South Shore Coral Reef	Preserve	Pr	IV	4.50	na
South West Breaker Area	Protected Area	PA	III	1.13	na
Tarpon Hole	Protected Area	PA	III	0.28	na
Taunton	Protected Area	PA	III	0.28	na
The Cathedral	Protected Area	PA	III	0.28	na
Vixen	Protected Area	PA	III	0.03	na
Xing Da Area	Protected Area	PA	III	0.12	na

Bahamas

MAP 4c

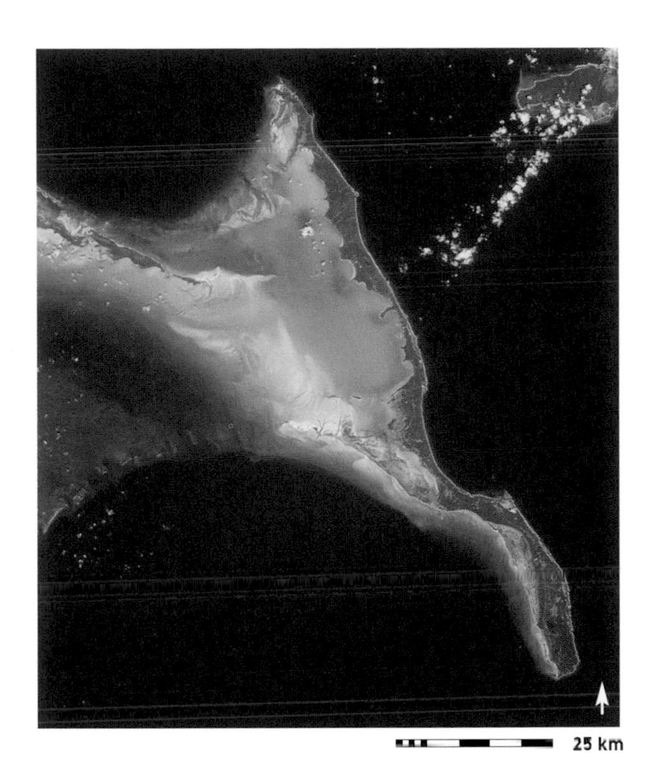

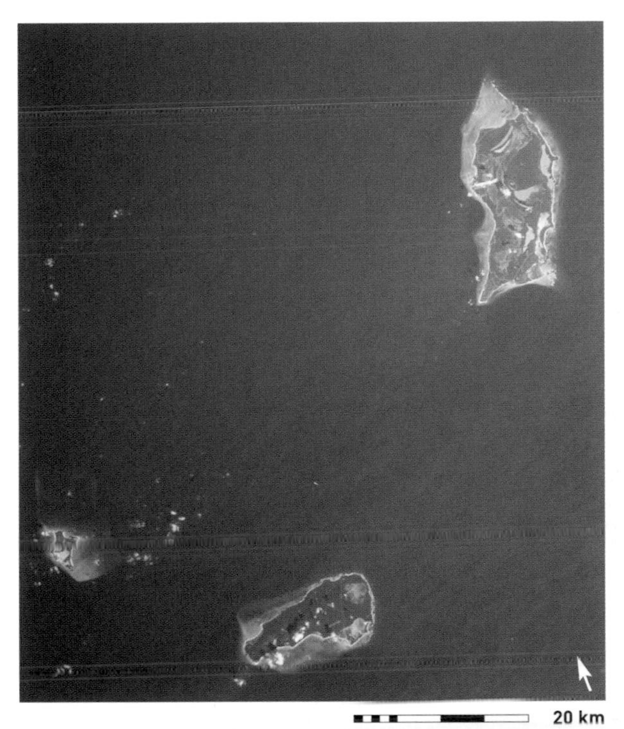

The Bahamas, an archipelago of some 700 islands and numerous reefs, stretch over 1 225 kilometers from north to south. Most of these islands are scattered over two shallow banks, the Little Bahama Bank and the Great Bahama Bank, with depths of 10 meters or less bounded by extremely deep water of up to 4 000 meters. The Bahamas are named after these banks: *baja mar* is Spanish for "shallow sea". The other islands occur on smaller, more isolated, banks to the southeast (principally the Crooked, Mayaguana and Inagua Banks) and the west (Cay Sal Bank). To the south, Hogsty Reef is one of the few atoll-type structures in the Caribbean. All the Bahamian islands have low relief and are formed from carbonate material, laid down by corals and calcareous algae, or by physical deposition from saturated water. Successive ice ages exposed these carbonate platforms, and wind-blown sand dunes created at much the same time subsequently lithified, further raising the elevation in some areas.

Two major currents affect the Bahamas. The North Equatorial Current, part of the North Atlantic Gyre, flows up from the southeast where it diverges: part passes along the east of the archipelago, while the remainder passes through the Old Bahama Channel which separates the country from Cuba. The Gulf Stream flows through the Straits of Florida from the west, before flowing north between Florida and the Bahamas. One effect of its powerful flow is that most of the land-based runoff from Florida is diluted and dispersed without reaching the Bahamas.

Reef development in much of the Bahamas is naturally limited by the exposure to hurricanes of the windward sites, by unusually cold winters in the northern islands and by turbid, high salinity waters on many leeward bank margins. However there are thousands of small patch reefs, dozens of narrow fringing reefs and some bank barrier reefs, such as the Andros Barrier Reef which is one of the longest reef systems in the Western Atlantic. Many Bahamian reefs are in fairly good condition, which is probably due to limited anthropogenic disturbance associated with their remoteness and the country's low population density.

White band and other diseases have affected corals from San Salvador in the east to Andros in the west. Macroalgal cover is usually low to moderate and the abundance of both herbivorous and commercially important fish is high. Corals of the central Bahamas showed extensive bleaching in August 1998, with over 60 percent of all hard corals bleached to a depth of 20 meters

Left: Long Island, Bahamas (STS055-73-38, 1993). Right: San Salvador, Rum Cay and Conception Island (STS095-705-61, 1998).

MAP 4c

No.	Protected Area Name
1	A rock in the Exuma Cays WBR
2	Betty Cay WBR
3	Big Darby Island WBR
4	Big Galliot Cay WBR
5	Black Sound Cay NP
6	Channel Cays and Flat Cay WBR
7	Conception Island NP
8	Exuma Land and Sea Park NP
9	Goat Cay WBR
10	Grassy Creek Cays and Rocks WBR
11	Guana Cay WBR
12	High Cay WBR
13	Inagua NP
14	Lake Cunningham WBR
15	Lightbourn Creek (Waterloo) WBR
16	Little Derby Island WBR
17	Little San Salvador (Little Island) WBR
18	Lucayan NP
19	Mammy Rhoda Cay WBR
20	Paradise Island WBR
21	Pelican Cays Land and Sea NP
22	Peterson Cay NP
23	Union Creek MNR
24	Washerwomans Cut Cays WBR
25	Water Cay WBR
26	Wood Cay WBR

Protected areas with coral reefs

Site name	Designation	Abbreviation	IUCN cat.	Size (km2)	Year
Bahamas					
Conception Island	National Park	NP	II	8.09	1973
Exuma Land and Sea Park	National Park	NP	II	455.84	1958
Inagua	National Park	NP	II	743.33	1965
Little San Salvador (Little Island)	Wild Bird Reserve	WBR	IV	1.82	1961
Pelican Cays Land and Sea	National Park	NP	II	8.50	1981
Peterson Cay	National Park	NP	II	0.01	1971

around New Providence Island. Near complete bleaching of all the hard corals and some gorgonians was seen at Little Inagua, Sweetings Cay, Chubb Cay, Little San Salvador, San Salvador and Egg Islands. Samana Cay was much less affected. There was also extensive bleaching at Walker's Cay in the northern Bahamas, with many types of coral affected. Some subsequent mortality was noted, particularly at the Exuma Cays.

Edible reef animals are still common on many Bahamian reefs, and fish stocks are generally abundant. There is a well developed commercial and export fishery, with total landings in 1999 close to 5 000 tons, valued at over US$70 million. This figure includes over 2 700 tons of the very high value spiny lobster tails. There is local overexploitation of certain stocks, including whelk *Cittarum pica*, queen conch, spiny lobster and several species of grouper. Concern has been expressed that spawning aggregations of groupers have become the target for spearfishers. A number of illegal fishing activities occur which include the use of toxic chemicals, the harvesting of hawksbill turtles, the taking of undersized or juvenile queen conch, and the collection of spiny lobster out of season or with prohibited diving gear. Artificial shelters are often positioned close to reefs to attract spiny lobsters, although there is concern that these may simply aggregate existing spiny lobsters rather than enhancing natural stocks. There is a limited legal harvest of adult green turtles during an open season (April-July). Sand is still being mined from a few reef sites on a fairly small scale. Over half of the commercial dive sites have mooring buoys. Declines in coral cover have been recorded in some locations. On New Providence dredging, landfill, sedimentation and the construction of a cruiseship port have led to the loss of 60 percent of the coral reef habitat.

The Bahamas is a stable, developing nation with an economy heavily dependent on tourism and offshore banking. Tourism alone accounts for more than 60 percent of gross domestic product and directly or indirectly employs 40 percent of the archipelago's labor force. Moderate growth in tourism receipts and a boom in the construction of new hotels, resorts and residences has led to localized pressures on coral reefs, but the total area is so great that the majority of reefs are probably little affected.

Shark feeding, a new speciality of the Bahamian dive industry, is attracting increasing numbers of tourists. Several protected areas have been established, though in 2000 these were poorly funded and had only one paid warden. Overall prospects for the conservation of the marine environment in the Bahamas will depend heavily on the fortunes of the tourism sector and continued income growth in the USA, which accounts for the majority of tourist visitors.

Bahamas

General Data	
Population (thousands)	295
GDP (million US$)	3 712
Land area (km2)	12 869
Marine area (thousand km2)	652
Per capita fish consumption (kg/year)	22
Status and Threats	
Reefs at risk (%)	49
Recorded coral diseases	8
Biodiversity	
Reef area (km2)	3 150
Coral diversity	32 / 58
Mangrove area (km2)	2 332
No. of mangrove species	4
No. of seagrass species	2

Turks and Caicos Islands

MAP 4d

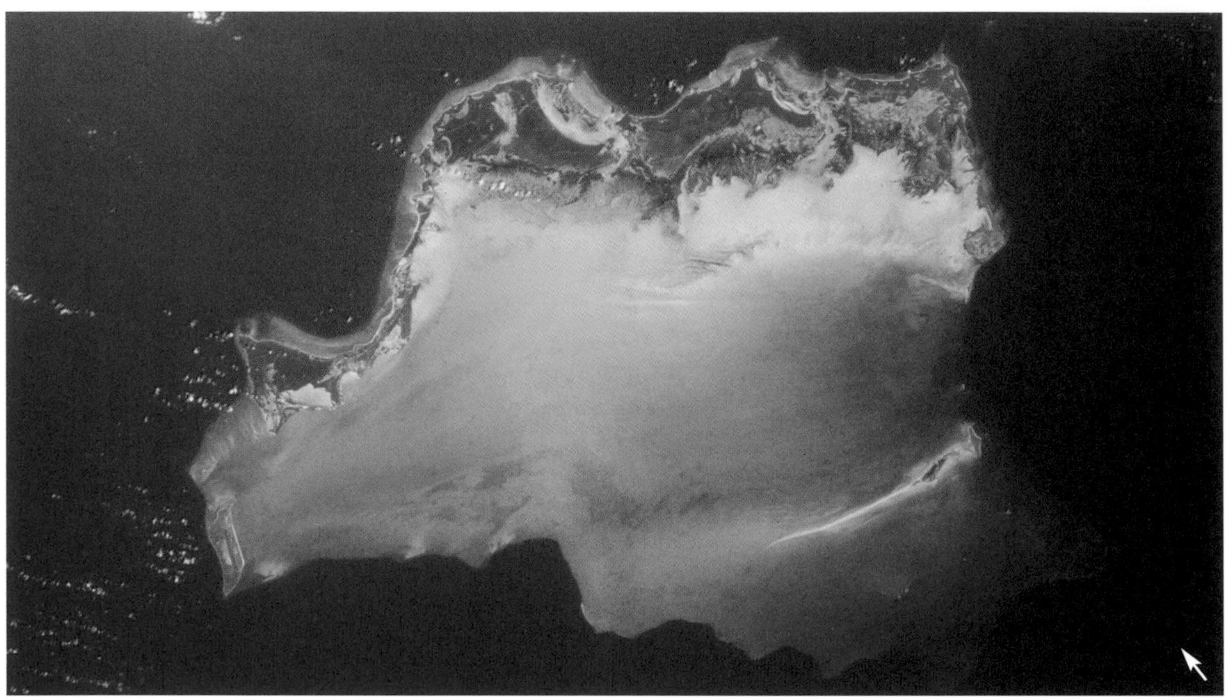

The Turks and Caicos Islands consist of two archipelagos of limestone islands distributed across the relatively small Turks Bank and the much larger Caicos Bank, with an area of some 8 000 square kilometers. The margins of these are defined by sharp "drop-offs" to deep oceanic water. Geologically the Turks and Caicos Islands are similar to the Bahamas, consisting of oolitic limestone sediments, eolianite hills and karst limestone cliffs on the windward shores. There are three submerged banks to the southeast – the Mouchoir, Silver (La Plata) and Navidad Banks, the last two of which are claimed by the Dominican Republic.

The edges of the main banks are dominated by coral, algae and gorgonian communities. Hard coral cover averaged 18 percent in 1999, with 30 percent of the substrate being covered by the alga *Microdictyon marinum*, and a low density (five per square meter) of soft corals. There is true fringing reef along the southern tip of Long Cay, the southern coast of South Caicos and the northern coasts of Middle and North Caicos, and in many locations this protects a lagoon with dense beds of seagrasses, mainly *Thalassia testudinum* and *Syringodium filiforme*. Diversity is high, with some 37 coral species and more than 400 fish species recorded. Large patch reefs, sometimes several hundred meters in diameter, occur across the Caicos Bank, which is mainly covered by sparse seagrass and calcareous green algae. A number of coral diseases

Turks and Caicos Islands

General Data	
Population (thousands)	18
GDP (million US$)	na
Land area (km²)	491
Marine area (thousand km²)	153
Per capita fish consumption (kg/year)	na
Status and Threats	
Reefs at risk (%)	47
Recorded coral diseases	1
Biodiversity	
Reef area (km²)	730
Coral diversity	29 / 57
Mangrove area (km²)	111
No. of mangrove species	5
No. of seagrass species	na

The Caicos Bank. Much of the central bank is dominated by sand, but there are also important seagrass and mangrove communities (STS050-82-98, 1992).

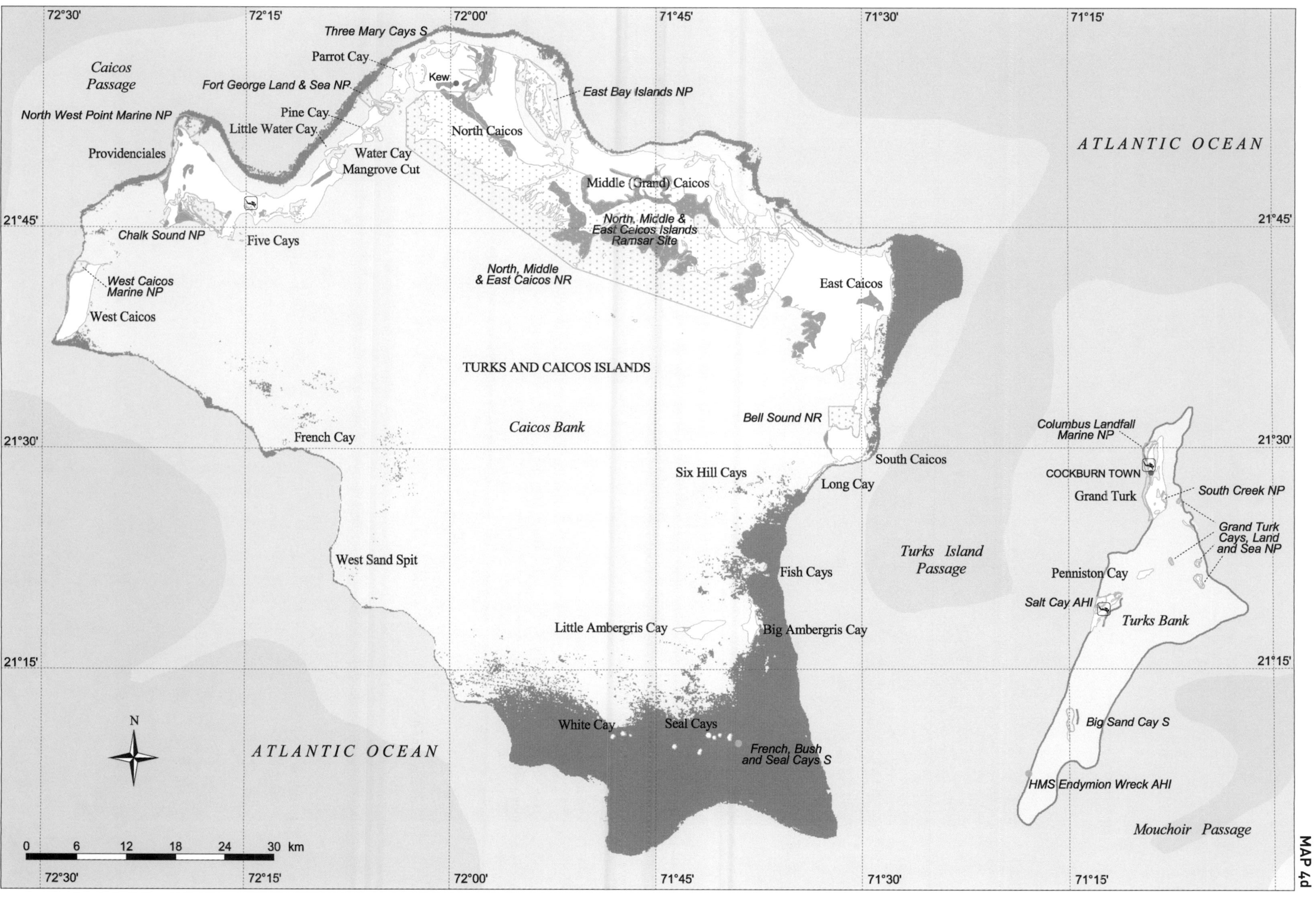
Three Mary Cays S
Caicos Passage
Parrot Cay
Kew
Fort George Land & Sea NP
East Bay Islands NP
North West Point Marine NP
Pine Cay
Little Water Cay
North Caicos
Providenciales
Water Cay
Mangrove Cut
ATLANTIC OCEAN
Middle (Grand) Caicos
North, Middle & East Caicos Islands Ramsar Site
Chalk Sound NP
Five Cays
North, Middle & East Caicos NR
West Caicos Marine NP
East Caicos
West Caicos
TURKS AND CAICOS ISLANDS
Bell Sound NR
Caicos Bank
Columbus Landfall Marine NP
French Cay
South Caicos
COCKBURN TOWN
Six Hill Cays
Long Cay
South Creek NP
Grand Turk
Grand Turk Cays, Land and Sea NP
Turks Island Passage
West Sand Spit
Fish Cays
Penniston Cay
Salt Cay AHI
Turks Bank
Little Ambergris Cay
Big Ambergris Cay
N
White Cay
Seal Cays
Big Sand Cay S
ATLANTIC OCEAN
French, Bush and Seal Cays S
HMS Endymion Wreck AHI
Mouchoir Passage
0
6
12
18
24
30 km
72°30'
72°15'
72°00'
71°45'
71°30'
71°15'
21°45'
21°30'
21°15'

Protected areas with coral reefs

Site name	Designation	Abbreviation	IUCN cat.	Size (km²)	Year
Turks and Caicos					
Fort George Land and Sea	National Park	NP	IV	4.94	1987
French, Bush and Seal Cays	Sanctuary	S	IV	0.20	1987
Grand Turk Cays, Land and Sea	National Park	NP	IV	1.56	1987
West Caicos Marine	National Park	NP	IV	3.97	1992
North, Middle and East Caicos Islands	**Ramsar Site**			**544.00**	**1990**

have been reported and are thought to have contributed to extensive losses of *Acropora*, though considerable stands of *Acropora palmata* still remain.

The Turks and Caicos economy is based on tourism, fishing and offshore financial services. Nearly all capital goods and food for domestic consumption are imported, and there is minimal manufacturing and agriculture. With no rivers, terrestrial runoff is very low. In 1999, over 120 000 visitors were recorded, with the USA being the primary source of tourists. The primary export fisheries are conch and lobster, with exports of 646 tons and 314 tons respectively in 1998, mainly to the USA. Reef fish are largely taken for internal markets, generally sustainably.

Nutrient discharge is a localized problem, notably on Providenciales, resulting from coastal development, including hotels and marinas, but also conch aquaculture and fish processing plants. Direct damage to the reefs by divers may be a localized problem at some dive sites. Generally the reefs of the Turks and Caicos Islands show little signs of being adversely affected by human activity, however development in the tourism sector may pose a significant threat, notably newly proposed harbor and tourist developments on East and South Caicos. A considerable number of marine protected areas have been designated, although active management of these is limited away from Providenciales.

Eagle rays Aetobatis narinari *rising up from the deep waters on the edge of the Caicos Bank.*

Selected bibliography

FLORIDA AND THE US GULF OF MEXICO

Chiappone M, Sullivan KM (1996). Distribution, abundance and species composition of juvenile scleractinian corals in the Florida Reef Tract. *Bull Mar Sci* 58(2): 555-569.

Gittings SR, Hickerson EL (eds) (1998). Dedicated Issue – Flower Garden Banks National Marine Sanctuary. *Gulf of Mexico Science* 16(2): 127-237.

Jaap WC, Hallock P (1990). Coral reefs. In: Myers RL, Ewel JS (eds). *Ecosystems of Florida.* 574-618.

Lee TN, Rooth C et al (1992). Influence of Florida current, gyres and wind-driven circulation on transport of larvae and recruitment in the Florida Keys coral reefs. *Continental Shelf Res* 12(7-8): 971-1002.

Murdoch TJT, Aronson RB (1999). Scale-dependent spatial variability of coral assemblages along the Florida Reef Tract. *Coral Reefs* 18: 341-351.

Ogden JC, Porter JW et al (1994). A long-term interdisciplinary study of the Florida-Keys seascape. *Bull Mar Sci* 54(3): 1059-1071.

Porter JW, Meier OW (1992). Quantification of loss and change in Floridian reef coral populations. *Amer Zool* 32(6): 625-640.

Suman DO (1997). The Florida Keys National Marine Sanctuary: a case study of an innovative federal-state partnership in marine resource management. *Coast Man* 25(3): 293-324.

BERMUDA

Cook CB, Logan A et al (1990). Elevated temperatures and bleaching on a high latitude coral reef – the 1988 Bermuda event. *Coral Reefs* 9(1): 45-49.

Ministry of the Environment (2000). *Marine Resources and the Fishing Industry in Bermuda: A Discussion Paper.* Ministry of the Environment, Government of Bermuda.

Schultz ET, Cowen RK (1994). Recruitment of coral reef fishes to Bermuda – local retention or long-distance transport. *Mar Ecol Prog Ser* 109(1): 15-28.

Smith SR (1992). Patterns of coral recruitment and post-settlement mortality on Bermuda's reefs – comparisons to Caribbean and Pacific reefs. *Amer Zool* 32(6): 663-673.

Thomas MLH, Logan A (1992). *A Guide to the Ecology of Shoreline and Shallow-Water Marine Communities of Bermuda.* Bermuda Biological Station for Research, Special Publication Number 30.

BAHAMAS

Anthony SL, Langg JC, Maguire B (1997). Causes of stony coral mortality of a central Bahamian reef: 1991-95. *Proc 8th Int Coral Reef Symp 2*: 1789-1794.

Hearty PJ, Kindler P (1997). The stratigraphy and surficial geology of New Providence and surrounding islands, Bahamas. *J Coast Res* 13(3): 798-812.

Liddell WD, Avery WE, Ohlhorst SL (1996). Patterns of benthic community structure, 10-250 m, the Bahamas. *Proc 8th Int Coral Reef Symp* 1: 437-442.

Smith GW (1994). Effects of temperature and UV-B on different components of coral reef communities from the Bahamas. In: Ginsburg RN (ed). *Proceedings of the Colloquium on Global Aspects of Coral Reefs: Health, Hazards and History, 1993.* Rosenstiel School of Marine and Atmospheric Science, University of Miami, USA. 126-131.

Steneck RS, Macintyre IG, Reid RP (1997). A unique algal ridge system in the Exuma Cays, Bahamas. *Coral Reefs* 16(1): 29-37.

TURKS AND CAICOS ISLANDS

Gaudian G, Medley P (1995). Evaluation of diver carrying capacity and implications for reef management in the Turks and Caicos Islands. *Bahamas J Sci* 3(1): 9-14.

Mitchell BA, Barborak JR (1991). Developing coastal park systems in the tropics – planning in the Turks and Caicos Islands. *Coastal Man* 19(1): 113-134.

Mumby PJ, Green EP, Clark CD, Edwards AJ (1998). Digital analysis of multispectral airborne imagery of coral reefs. *Coral Reefs* 17(1): 59-69.

Map sources

Map 4a

Coral reef and mangrove data were obtained from the Florida Marine Research Institute in digital format. Highly detailed files from numerous sources (from 1980), including aerial photography and ground surveys. Location of the Texas Flower Garden Banks has been extracted from hydrographic charts.

Map 4b

Coral reef data are taken from Hydrographic Office (1984). Last major edits to this chart were in 1959.

Hydrographic Office (1984). Bermuda Island. *British Admiralty Chart No. 344.* 1:60 000. Taunton, UK.

Map 4c

Coral data have been digitized from UNEP/IUCN (1988a)* which presents coral reefs as arcs at a scale of approximately 1:2 600 000. Mangrove data are based on B&B (c.1995a and c.1995b) and Sealey and Burrows (1992).

B&B (c.1995a). *Bahamas North 1:500 000 Road Map.* Berndtson and Berndtson Publications, Fürstenfeldbruck, Germany. (Used for: Bimini Island – 1:100 000.)

B&B (c.1995b). *Bahamas South 1:500 000 Road Map.* Berndtson and Berndtson Publications, Fürstenfeldbruck, Germany. (Used for Aklins Island only – 1:500 000; Mayaguana – 1:500 000; Great Inagua – 1:500 000; Exuma Islands – 1:500 000; Cat Island – 1:500 000; San Salvador – 1:250 000; Long Island – 1:500 000.)

Sealey N, Burrows EJ (eds) (1992). *School Atlas for the Commonwealth of the Bahamas.* Longman Group UK Ltd, Harlow, UK. (Used for Grand Bahama – 1:600 000; Abaco – 1:650 000; New Providence – 1:110 000; Andros – 1:730 000.)

Map 4d

For the Caicos Bank, reef and coastline are taken from a broad-scale habitat map based on Landsat TM imagery (path, row 45, 22 Nov 1990). Supervised classification of marine areas produced four categories of marine habitat, of which one was defined as "coral reef", shown here. Mangrove information, together with all information for the Turks Bank, is taken from DOS (1984).

DOS (1984). *Turks and Caicos Islands.* 1:200 000. Series DOS 609 2nd edn. Department of Overseas Surveys. London, UK.

* See Technical notes, page 401

CHAPTER 5
Western Caribbean

This region incorporates some of the largest islands in the Caribbean, including Cuba, as well as the mainland of Central America from Mexico south to Colombia. It also harbors considerable areas of coral reef. The reefs of Discovery Bay in Jamaica have been the subject of intensive studies since the 1950s, and together with the extensive barrier and fringing reef systems of Belize and the Caribbean coastline of Mexico are among the best known reefs of the region.

Reef development across the continental shelf of Nicaragua is poorly documented but believed to be extensive. Cuba has considerable though little-known reefs on all sides, particularly associated with the long-shore archipelagos which border more than half of its coastline.

Human impacts on the reefs in this region are highly varied. Some, such as those in Jamaica, have been heavily utilized by humans for many decades. In recent years these have been severely degraded by a combination of the *Diadema* die-off and disease, together with direct human impacts, notably overfishing, but also sedimentation and nutrient pollution. Others have been protected by their distance from the coast along with relatively low human population densities.

The East Pacific coastline of the Americas holds a number of unique tropical coral assemblages. For the most part these do not form true reefs, although there are some reef structures in a few locations. Biodiversity is low on the reefs, though they contain a number of unique and important species. The diversity of reef corals in this region was much reduced during the Pliocene/Pleistocene glaciations and has not recovered.

This region is regularly impacted by extreme environmental conditions. Coastal waters are typically cool and rich in nutrients, associated with upwellings, but there is also occasional extreme warming linked to El Niño Southern Oscillation (ENSO) events. This can cause widespread coral bleaching and subsequent mortality, driving localized extinctions of particular species. The frequent occurrence of such events further helps to explain the lack of more extensive reef development. Clipperton Atoll is the only major reef structure on the western edge of this region. Because it is administered from French Polynesia and shares some features with the Indo-Pacific reefs, it is covered in Chapter 14.

Left: Grand Cayman. The shallow lagoon is surrounded by extensive and important mangrove areas (STS062-84-70, 1994).
Right: The newly described coral Pocillopora effusus *(top of picture) growing on a rock outcrop exposed to strong wave action on Clarion Island, Mexico. Thus far this species is only known from Clarion Island and Mexico (photo: JEN Veron).*

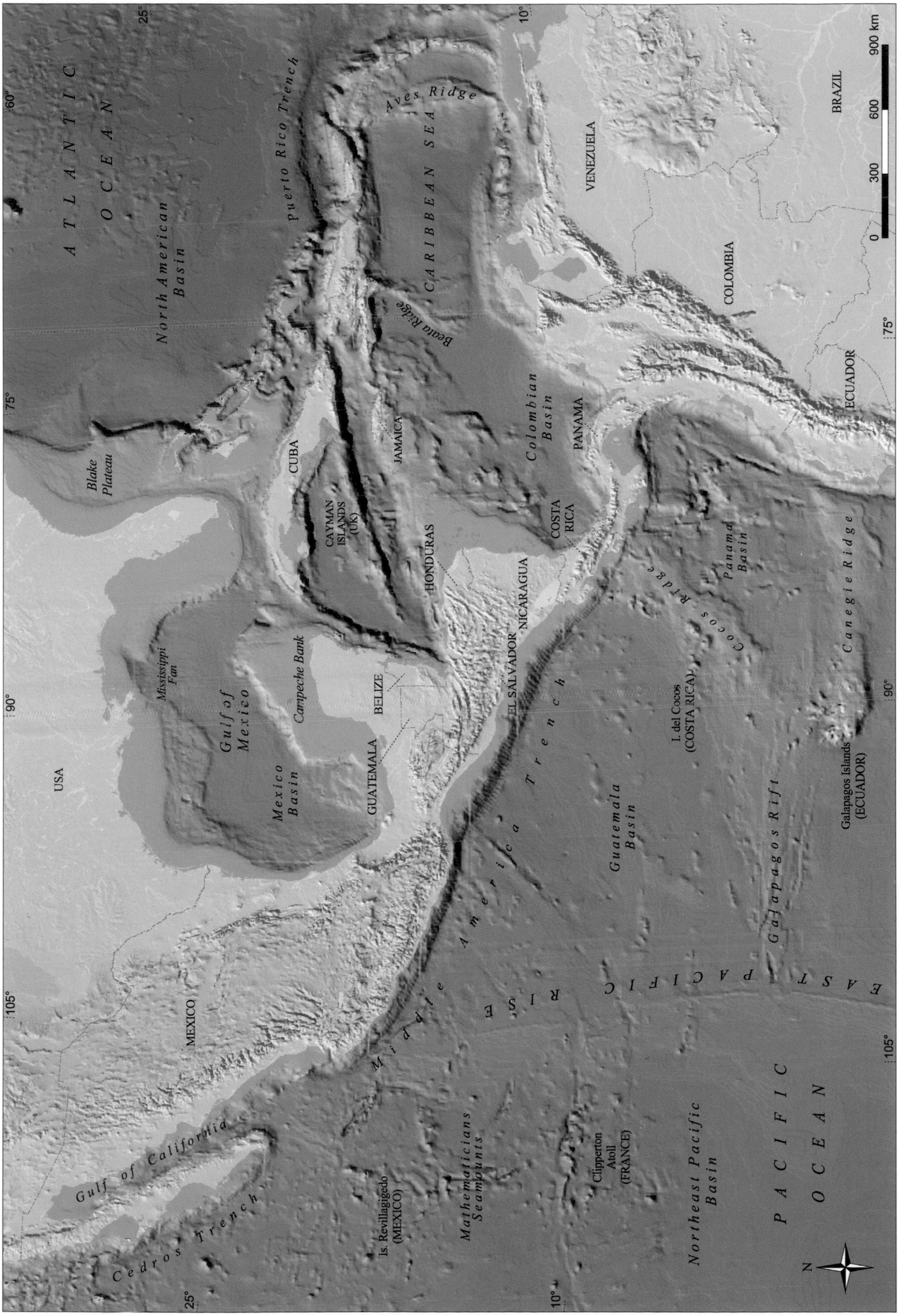

ATLANTIC OCEAN
North American Basin
Blake Plateau
Puerto Rico Trench
USA
MEXICO
Gulf of California
Cedros Trench
Mississippi Fan
Gulf of Mexico
Mexico Basin
Campeche Bank
CUBA
CAYMAN ISLANDS (UK)
JAMAICA
Beata Ridge
CARIBBEAN SEA
Aves Ridge
GUATEMALA
BELIZE
HONDURAS
EL SALVADOR
NICARAGUA
COSTA RICA
PANAMA
Colombian Basin
VENEZUELA
COLOMBIA
ECUADOR
BRAZIL
Is. Revillagigedo (MEXICO)
Mathematicians Seamounts
Middle America Trench
EAST PACIFIC RISE
Clipperton Atoll (FRANCE)
Guatemala Basin
I. del Cocos (COSTA RICA)
Cocos Ridge
Panama Basin
Northeast Pacific Basin
Galapagos Rift
PACIFIC OCEAN
Galapagos Islands (ECUADOR)
Canegie Ridge
N
0 300 600 900 km
105°
90°
75°
60°
25°
10°

MAP 5a

No.	Protected Area Name
	Mexico
1	Archipiélago de Revillagigedo BR(N)
2	Arrecife Alacranes NMP
3	Arrecifes de Cozumel NP
4	Arrecifes de Puerto Morelos NP
5	Arrecifes de Sian Ka'an BR(N)
6	Bahía de Loreto NMP
7	Banco Chinchorro BR(N)
8	Cabo Pulmo NMP
9	Cajón del Diablo ETC
10	Cajón del Diablo HR
11	Chamela BS
12	Costa Occidental de Isla Cozumel APFFS
13	Costa Occidentel de Isla Mujeres APFFS
14	Dzilam SR
15	El Pinacate y Gran Desierto de Altar BR(N)
16	El Veladero NP
17	El Vizcaíno BR(N)
18	Fondo Cabo San Lucas APFFS
19	Isla Contoy NP
20	Isla Isabel NP
21	Isla Mujeres, Punta Cancún y Punta Nizuc NP
22	Isla Rasa ETC
23	Isla Tiburón ETC
24	Isla de Guadalupe ETC
25	Islas del Golfo de California ETC
26	La Blanquilla ETC
27	La Encrucijada BR(N)
28	Laguna de Chankanaab PNat
29	Lagunas de Chacahua NP
30	Los Arcos ETC
31	Los Tuxtlas BS
32	Pantanos de Centla ER(N)
33	Playa Ceuta RZSTP
34	Playa Cuitzmala RZSTP
35	Playa El Tecuán RZSTP
36	Playa El Verde Camacho RZSTP
37	Playa Mexiquillo RZSTP
38	Playa Mismaloya RZSTP
39	Playa Piedra de Tlacoyunque RZSTP
40	Playa Rancho Nuevo RZSTP
41	Playa Teopa RZSTP
42	Playa Tierra Colorada RZSTP
43	Playa adyacente a Río Lagartos RZSTP
44	Playa de Escobilla RZSTP
45	Playa de Isla Contoy RZSTP
46	Playa de Maruata y Cololoa RZSTP
47	Playa de Puerto Arista RZSTP
48	Playa de la Bahía de Chacahua RZSTP
49	Ría Celestún ETC
50	Ría Lagartos ETC
51	Sian Ka'an BR(N)
52	Sierra de Santa Marta (Los Tuxtlas) NarPA
53	Sistema Arrecifal Veracruzano NMP
54	Tulum NP
55	Volcán de San Martín (Los Tuxtlas) NarPA
56	Xcalak NMP
57	Yum-Balam FFPA

Alto Golfo de California Biosphere Reserve
USA
El Paso
Ensenada
Isla Angel de la Guardia
Islas del Golfo de California Biosphere Reserve
Whale Sanctuary of El Vizcaino World Heritage Site
El Vizcaíno Biosphere Reserve
Bahía Concepción
Gulf of California
Rocas Alijos
Baja California
Isla Espíritu Santo
Cabo Pulmo
Punto Chileno
Mazatlan
PACIFIC OCEAN
Marismas Nacionales Ramsar Site
Islas Marías
Isla Jaltemba
Bahía Banderas
Guadalajara
Isla San Benedicto
Isla Roca Partida
Isla Socorro
Isla Clarión
Islas Revillagigedo
MEXICO
MEXICO CITY
Puébla
Veracruz
Alvarado
Oaxaca
Acapulco
Carrizalillo
Huatulco Bays
Puerto Angel
Puerto Angelito
Pueblo Nuevo
GUATEMALA
HONDURAS
BELIZE
Dallas
Alvarado
San Antonio
Tampico
Isla Lobos
Bon Secour NWR
Grand Bay NWR
St. Marks NWR
Breton NWR
Bayou Sauvage NWR
Shell Keys NWR
Gulf Islands (Florida) NS
Sea Rim SP
St. Joseph Peninsula SP
McFaddin NWR
New Orleans
Anahuac NWR
St. Vincent NWR
Sabine NWR
Apalachicola NERR
Brazoria NWR
Galveston Island SP
Delta NWR
Big Boggy NWR
Flower Garden Banks NaMS
Aransas NWR
Padre Island NS
Gulf of Mexico
Laguna Atascosa NWR
Campeche Bank
Bajos del Norte
Arrecife Alacranes
Bajo Granville
Bajo Madagascar
Roca Ifigenia
Bajo Serpiente
Cayos Arenas
Banco y Cayo Nuevo
Banco Ingles
Bajo Sisal
Arrecife Triángulos
Merida
Bancos Pera y Perlas
Bancos Obispo
Banco Nuevo
Cayos Arcas
Yucatan Peninsula
Bahía de Campeche
Pantanos de Centla Ramsar Site

N

0 100 200 300 400 500 km

115° 110° 105° 100° 95° 90° 85°
30° 25° 20° 15°

Mexico

MAPS 5a and b

Coral reefs and communities occur throughout Mexico but are concentrated in four main areas: the Gulf of California and Pacific Coast; the nearshore reefs between Tampico and Veracruz in the western Bahía de Campeche; the more distant offshore reefs of the Campeche Bank; and the fringing reef and atolls of the Caribbean Sea.

Hermatypic corals were originally considered to be rare in the Mexican Pacific, but recent research has described abundant coral populations in these reefs, despite their small size (mostly a few hectares or less) and their discontinuous occurrence. True reefs with an elevated structure occur at Cabo Pulmo, Ensenada Grande on Isla Espíritu Santo, Punto Chileno, Islas Marías and at scattered locations along the southern coast of Oaxaca. Coral communities, sometimes with abundant coral growth but little net accretion, are present in the central Gulf of California from Isla Angel de la Guardia to Bahía Concepción. They consist of just two species – *Porites panamensis* and *P. sverdrupi* – which are tolerant of the low temperatures of the upper gulf. The latter is an endemic species believed to be a relict from the Pliocene undergoing a natural process of extinction. Other communities occur all along the Pacific coast, occupying rocky areas at 0-15 meters in depth. The communities at Isla Jaltemba, Huatulco Bays, east of Puerto Angel, Puerto Angelito and Carrizalillo are particularly well developed but are composed of only a few species, mainly *Pocillopora* spp., *Porites* spp., *Pavona* spp., *Psammocora* spp. and *Fungia* spp. The latest El Niño and post-El Niño events in 1997 and 1998 caused considerable bleaching and mortality around Bahía Banderas and Huatulco, but had considerably less impact at some other localities.

Some 200 kilometers south of Baja California and 600 kilometers west of mainland Mexico, lies a small but important group of four volcanic islands, the Islas Revillagigedo. They lie in deep oceanic water and are broadly impacted by a westward flowing North Equatorial Current, which is fed by the cold California Current and the warmer Costa Rica Coastal Current. These relatively harsh conditions are exacerbated by regular tropical storms. Despite this, the islands harbor the most diverse fish and coral communities in the Mexican Pacific. Reef development is limited, but there are some true reef structures, notably in the more sheltered bays. Twenty hermatypic coral species have been recorded around these islands, dominated by *Pocillopora* spp., as well as *Porites lobata* and *P. lichen*. Many gorgonian species have also

Left: Bahia La Paz, in the south of Baja California. Although there are important marine communities and some corals, true reef development is highly limited (STS030-71-9, 1989). Right: French angelfish Pomacanthus paru *(photo: Colin Fairhurst).*

been recorded. Biogeographically the islands appear to be more closely linked with Clipperton Atoll than the Mexican mainland and up to three of the hermatypic coral species found on the islands may be endemic to these two areas. Additionally, six molluscs and 12 reef fish have been found to be endemic to the islands.

In the Gulf of Mexico reefs occur in the south, and are mostly located along the edge of the continental shelf, both around Veracruz and on the Campeche Bank, which follows the western and northern edges of the Yucatan Peninsula. The majority of the Veracruz reefs are platform reefs, some with emergent parts, as in Isla Lobos, although patch reefs do exist in El Giote off Anton Lizardo and at Punta Gorda, Punta Majagua, Hornos and Punta Mocambo. Sedimentation is very high on reefs close to the port of Veracruz where the rivers Antigua, Papaloapan and Alvarado limit coral growth, and diversity in these areas is low.

The Campeche reefs have ecological and morphological characteristics that distinguish them from the Caribbean reefs of Mexico, although their fauna is similar. Both emergent (eg Cayos Arcas, Cayos Arenas and Arrecife Triángulos) and submerged (eg Banco Nuevo, Banco Ingles, Bajo Serpiente, Madagascar and Sisal) reefs are present. All are platforms rising from a pre-Holocene base located at a depth of 50-60 meters. Arrecife Alacranes is an atoll.

The most extensive reef development in the country is in the state of Quintana Roo on the east coast of the Yucatan Peninsula. Here the continental shelf is very narrow, in many places less than 2 kilometers wide. There are partly submerged fringing reefs along much of this coastline, while from Xcalak southwards there is a fully developed fringing reef which continues to Ambergris Key in Belize, and then extends into the Belize Barrier Reef. Extensive spur and groove systems have developed in the center and south. This coastline is noted for its lack of rivers but numerous limestone sink holes result in an outflow of freshwater at various points. Offshore are two further important features: Cozumel Island, a relatively large island in the north with a number of reefs on both windward and leeward shores; and close to the Belize border the large atoll of Banco Chinchorro, which is separated from the mainland by a 1 000 meter deep channel. The reefs are well developed on the eastern (windward) side of this atoll: coral cover is lower in the shallow waters, and a spur and groove system has developed. The lagoon is generally sandy, with extensive seagrass cover and some patch reefs. Both Banco Chinchorro and Cozumel modify the northerly flow of water in the Caribbean Current. South of Cozumel, part of the current is funnelled into the channel and accelerates up to 4 knots to form the Yucatan Current. Its speed is believed to influence sedimentation rates and possibly coral larvae settlement, particularly in the Playa del Carmen area. Considerable declines in coral cover at Puerto Morelos and nearby reefs have been related to the impacts of Hurricane Gilbert in 1988 and a large but unquantified bleaching event in 1995. However, unlike in Belize to the south, the combined impacts of the bleaching and Hurricane Mitch – the Atlantic's fourth strongest hurricane on record – did not lead to widespread coral mortality along this coastline.

Mexico

General Data	
Population (thousands)	100 350
GDP (million US$)	264 715
Land area (km²)	1 962 948
Marine area (thousand km²)	3 289
Per capita fish consumption (kg/year)	11
Status and Threats	
Reefs at risk (%)	39
Recorded coral diseases	1
Biodiversity	
Reef area (km²)	1 780
Coral diversity	78 / 81
Mangrove area (km²)	5 315
No. of mangrove species	5
No. of seagrass species	6

Little information exists on the anthropogenic impacts on coral reefs and communities in the Mexican Pacific. Most of the coral and reef communities occur in places subject to recent intensive development for tourism, and sedimentation arising from deforestation in adjacent watersheds is increasing. Even on the remote Islas Revillagigedo sedimentation – arising from overgrazing by goats and pigs – may be a problem. A small aquarium fishery in the Gulf of California has recently been expanded, with three operators and combined permits for the collection of nearly 90 000 individual fish (from 20 species), 1 000 corals and 80 000 other invertebrates per year. These figures are of some concern because the coral communities from which these collections are derived are small and scattered, and their natural vulnerability is further exacerbated by the extreme environmental conditions of the region.

In the Gulf of Mexico, the Veracruz reefs have probably suffered the greatest damage from human impacts due to their proximity to the coast and their location near important ports such as Veracruz and Tuxpan. Campeche Bank reefs have suffered from oil-related activities over the last 25 years. A deep water oil port was installed by Cayos Arcas and the chronic effects

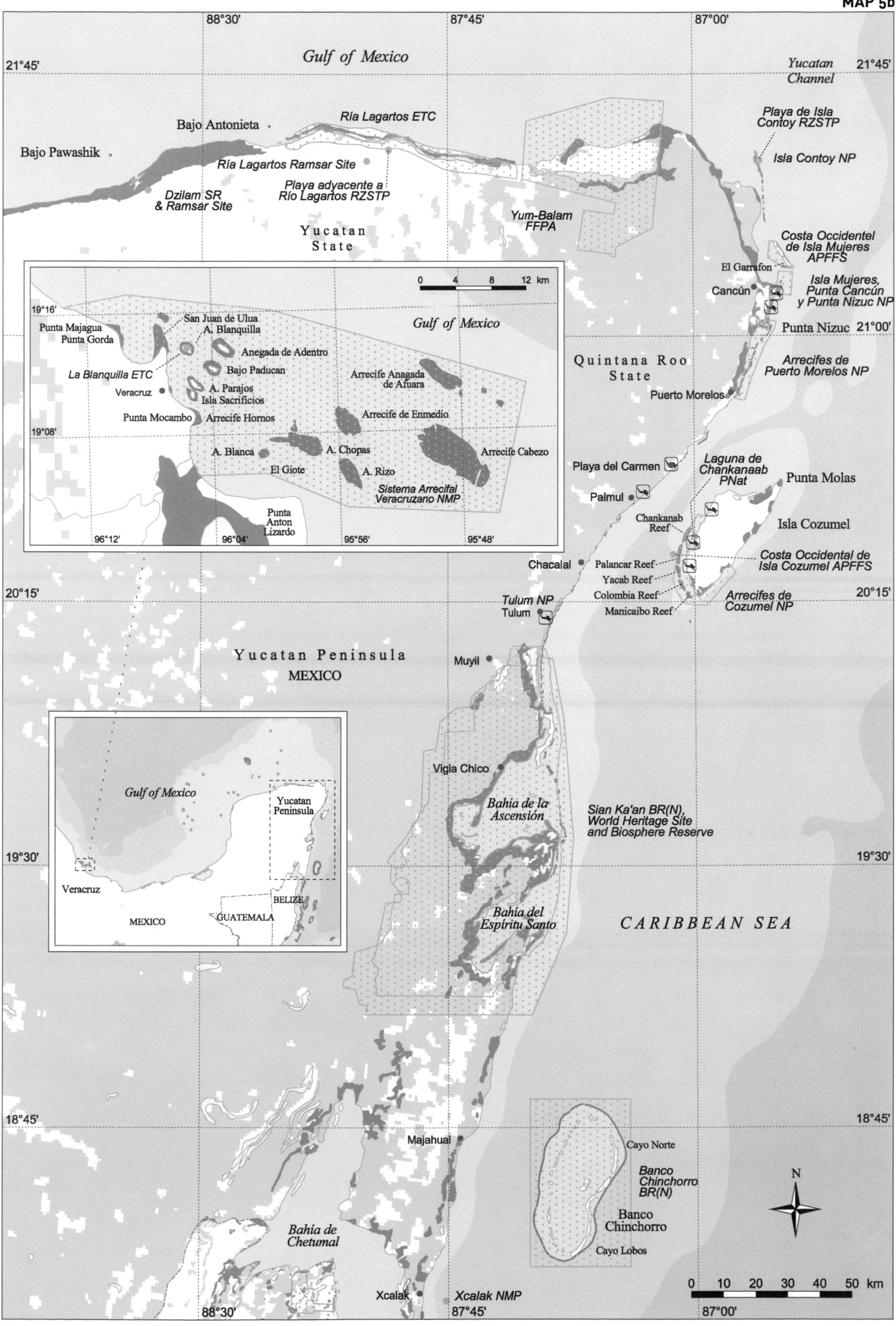

88°30'
87°45'
87°00'
21°45'
Gulf of Mexico
Yucatan Channel
Bajo Antonieta
Ría Lagartos ETC
Bajo Pawashik
Ría Lagartos Ramsar Site
Playa de Isla Contoy RZSTP
Isla Contoy NP
Dzilam SR & Ramsar Site
Playa adyacente a Río Lagartos RZSTP
Yum-Balam FFPA
Yucatan State
Costa Occidentel de Isla Mujeres APFFS
El Garrafon
Cancún
Isla Mujeres, Punta Cancún y Punta Nizuc NP
Punta Nizuc
21°00'
Quintana Roo State
Arrecifes de Puerto Morelos NP
Puerto Morelos
0 4 8 12 km
19°16'
San Juan de Ulua
Punta Majagua
Punta Gorda
A. Blanquilla
Gulf of Mexico
Anegada de Adentro
La Blanquilla ETC
Bajo Paducan
Arrecife Anagada de Afuara
Veracruz
A. Parajos
Isla Sacrificios
Punta Mocambo
Arrecife Hornos
Arrecife de Enmedio
19°08'
A. Blanca
A. Chopas
Arrecife Cabezo
El Giote
A. Rizo
Sistema Arrecifal Veracruzano NMP
Punta Anton Lizardo
96°12'
96°04'
95°56'
95°48'
Playa del Carmen
Laguna de Chankanaab PNat
Punta Molas
Palmul
Chankanab Reef
Isla Cozumel
Chacalal
Palancar Reef
Costa Occidental de Isla Cozumel APFFS
Yacab Reef
20°15'
Colombia Reef
Arrecifes de Cozumel NP
Tulum NP
Tulum
Manicaibo Reef
Yucatan Peninsula
MEXICO
Muyil
Vigia Chico
Gulf of Mexico
Yucatan Peninsula
Bahía de la Ascensión
Sian Ka'an BR(N), World Heritage Site and Biosphere Reserve
19°30'
Veracruz
BELIZE
MEXICO
GUATEMALA
Bahía del Espíritu Santo
CARIBBEAN SEA
18°45'
Majahual
Cayo Norte
N
Banco Chinchorro BR(N)
Banco Chinchorro
Bahía de Chetumal
Cayo Lobos
0 10 20 30 40 50 km
Xcalak
Xcalak NMP

of numerous small oil spills in addition to occasional large spills are believed to have adversely affected the coral reefs. Fishing is at least partially regulated in the reefs near Veracruz, while Campeche Bank is heavily exploited.

The Caribbean reefs have been subject to intense artisanal fishing since the 1960s, and tourism has developed enormously since the mid-1970s. Small reef patches, such as El Garrafon at Isla Mujeres and Punta Nizuc at Cancun, have been completely destroyed by tourism and impacts are becoming more evident elsewhere along the Cancun-Tulum touristic corridor in places such as Akumal and Puerto Morelos, as well as the offshore island of Cozumel. The impacts of construction and inadequate sewage systems in porous karst limestone, combined with the direct impacts of anchor damage and diver damage, are a cause for concern throughout this area.

A large number of marine protected areas have been declared which include coral reefs. Active management in some of these is supporting increased protection of coral reef resources in these areas.

Protected areas with coral reefs

Site name	Designation	Abbreviation	IUCN cat.	Size (km²)	Year
Mexico					
Archipiélago de Revillagigedo	Biosphere Reserve (National)	BR(N)	VI	6 366.85	1994
Arrecife Alacranes	National Marine Park	NMP	II	3 337.69	1994
Arrecifes de Cozumel	National Park	NP	II	119.88	1996
Arrecifes de Puerto Morelos	National Park	NP	II	108.28	1998
Arrecifes de Sian Ka'an	Biosphere Reserve (National)	BR(N)	VI	349.27	1998
Bahía de Loreto	National Marine Park	NMP	II	2 065.81	1996
Banco Chinchorro	Biosphere Reserve (National)	BR(N)	VI	1 443.60	1996
Cabo Pulmo	National Marine Park	NMP	II	71.11	1995
Costa Occidental de Isla Cozumel	Area de Protección de Flora y Fauna	APFFS	IV	na	1980
Costa Occidental de Isla Mujeres	Area de Protección de Flora y Fauna	APFFS	IV	6.64	1973
Fondo Cabo San Lucas	Area de Protección de Flora y Fauna	APFFS	na	na	1973
Isla Contoy	National Park	NP	II	51.26	1998
Isla Mujeres, Punta Cancún y Punta Nizuc	National Park	NP	V	86.73	1996
La Blanquilla	Other Area	ETC	IV	668.68	1975
Laguna de Chankanaab	Parque Natural	PNat	Unassigned	na	1983
Los Arcos	Other Area	ETC	V	na	1975
Sian Ka'an	Biosphere Reserve (National)	BR(N)	VI	5 281.47	1986
Sistema Arrecifal Veracruzano	National Marine Park	NMP	II	522.39	1992
Xcalak	National Marine Park	NMP	II	na	2000
Islas del Golfo de California	**UNESCO Biosphere Reserve**			**3 603.60**	**1995**
Sian Ka'an	**UNESCO Biosphere Reserve**			**5 281.48**	**1986**
Sian Ka'an	**World Heritage Site**			**5 280.00**	**1987**

Belize

MAP 5c

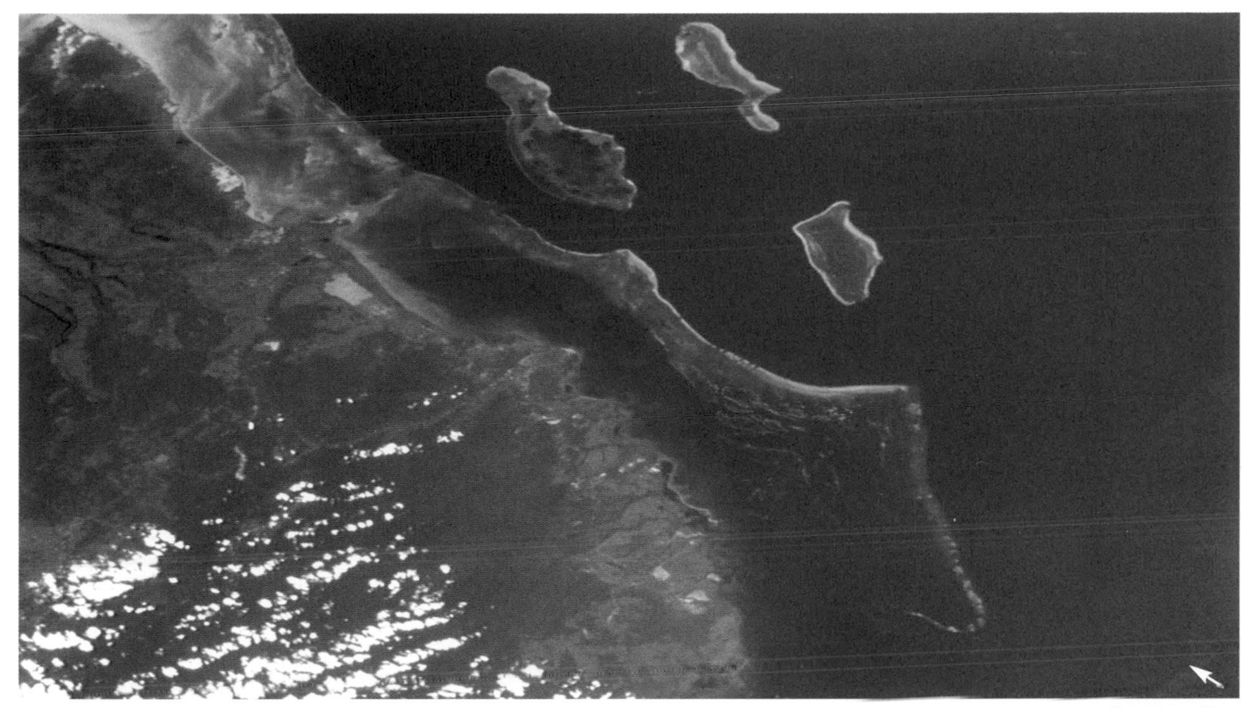

Although a relatively small country, Belize has some of the most extensive coral reef resources in the region. The coastline is fringed by a shallow shelf with a barrier reef running along its outer edge. The Belize Barrier Reef is the longest in the Caribbean, 230 kilometers in length, though there are barrier-like reef tracts in Florida and Cuba which are considerably longer. To the north the barrier reef becomes joined to the mainland at Ambergris Cay, a southerly extension of the Yucatan Peninsula. At this point the reef system becomes fringing, and continues north along the coastline of Mexico. These reefs, together with others to the south in Honduras, are sometimes known as the Meso-American Reef, in recognition of the inter-connected nature of their ecosystems.

The mainland coast is dominated by narrow sandy beaches or mangrove forests, often associated with river deltas. The development of reefs along the mainland is extremely limited by fluctuations in turbidity and sediments. Some reefs do occur in the south between Placencia and Punta Ycacos, but have low species richness and are dominated by sediment-resistant genera such as *Siderastrea* and *Porites*. The lagoon is 20-40 kilometers wide, typically only a few meters deep in the north, but reaching 50 meters in the south. It supports some of the most extensive seagrass beds in the Caribbean. Patch reefs occur across the whole shelf, though they are much more abundant in the south. These patch reefs vary considerably in size from small collections of corals to large reefs many tens of meters in diameter, as their form and species composition are determined largely by the location on the shelf, wave and current energy, and depth. Rhomboidal atoll-like structures called faros are very unusual features associated with the southern shelf. They are believed to be formed by corals growing on top of submerged sand or rubble cays. The lagoon also houses regionally important populations of the Caribbean manatee, although there are concerns that illegal hunting may be reducing its numbers, particularly in the south of the country. The barrier reef itself typically consists of a rubble strewn reef flat with numerous mangrove cays on its central and landward side, fronted by a reef crest. The outer slope is best developed (and studied) in the central section, where the reefs are typically long and unbroken with a deep spur and groove system which in some areas becomes a double ridge separated by a rubble-filled channel. The reef is split by a series of channels, and in the south it breaks up and becomes partially submerged.

The other striking feature of Belizean reefs is three

The Belize Barrier reef and the three offshore atolls (STS060-85-W, 1994).

MAP 5c

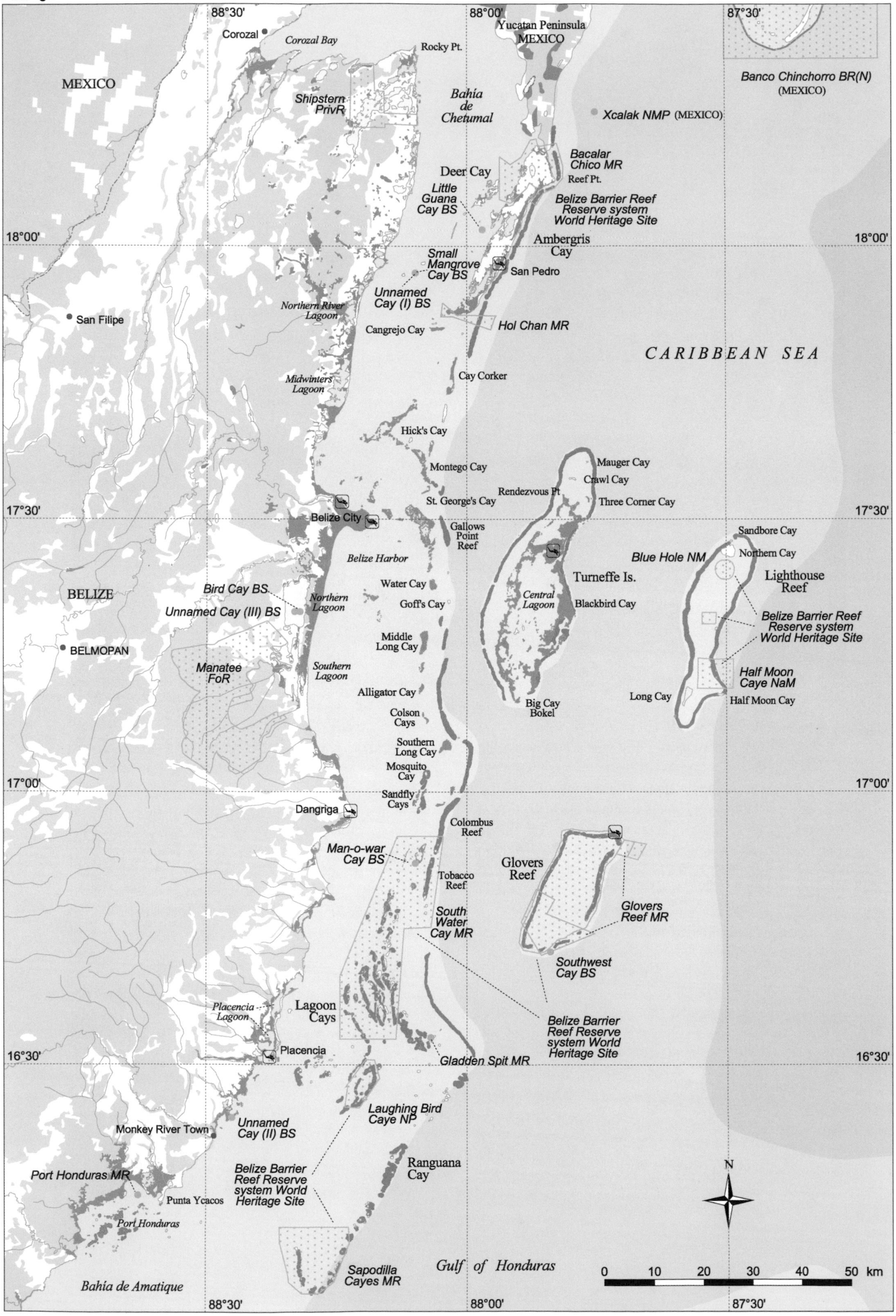

88°30'
88°00'
87°30'
Corozal
Corozal Bay
Rocky Pt.
Yucatan Peninsula
MEXICO
MEXICO
Shipstern PrivR
Bahía de Chetumal
Banco Chinchorro BR(N) (MEXICO)
Xcalak NMP (MEXICO)
Bacalar Chico MR
Reef Pt.
Deer Cay
Little Guana Cay BS
Belize Barrier Reef Reserve system World Heritage Site
18°00'
Ambergris Cay
18°00'
Small Mangrove Cay BS
San Pedro
Unnamed Cay (I) BS
Northern River Lagoon
San Filipe
Cangrejo Cay
Hol Chan MR
CARIBBEAN SEA
Midwinters Lagoon
Cay Corker
Hick's Cay
Montego Cay
Mauger Cay
Crawl Cay
Rendezvous Pt
Three Corner Cay
St. George's Cay
17°30'
Belize City
Gallows Point Reef
17°30'
Sandbore Cay
Northern Cay
Blue Hole NM
Belize Harbor
Turneffe Is.
Lighthouse Reef
Water Cay
Bird Cay BS
BELIZE
Northern Lagoon
Goff's Cay
Central Lagoon
Blackbird Cay
Unnamed Cay (III) BS
Belize Barrier Reef Reserve system World Heritage Site
Middle Long Cay
BELMOPAN
Manatee FoR
Southern Lagoon
Half Moon Caye NaM
Alligator Cay
Big Cay Bokel
Long Cay
Half Moon Cay
Colson Cays
Southern Long Cay
Mosquito Cay
17°00'
17°00'
Sandfly Cays
Dangriga
Colombus Reef
Man-o-war Cay BS
Glovers Reef
Tobacco Reef
Glovers Reef MR
South Water Cay MR
Southwest Cay BS
Placencia Lagoon
Lagoon Cays
Belize Barrier Reef Reserve system World Heritage Site
16°30'
Placencia
Gladden Spit MR
16°30'
Laughing Bird Caye NP
Unnamed Cay (II) BS
Monkey River Town
Ranguana Cay
N
Port Honduras MR
Belize Barrier Reef Reserve system World Heritage Site
Punta Ycacos
Port Honduras
Gulf of Honduras
Sapodilla Cayes MR
Bahía de Amatique
0 10 20 30 40 50 km
88°30'
88°00'
87°30'

large atolls further offshore: the Turneffe Islands, Lighthouse Reef and Glovers Reef. All three show distinct differences between the leeward and windward slopes, with the development of spur and groove formations on the windward (eastern) sides, but also some of the most highly developed reef structures. Lighthouse and Glovers Reefs are exposed to higher wave energy on these eastern slopes and as a result they have a higher coverage of *Acropora palmata* and *Lithothamnion* than Turneffe. Both of these atolls also have deep lagoons with numerous patch reefs and very little land cover. Turneffe, by contrast, has a land area of 22 percent of the atoll and a shallow lagoon with only a few patch reefs in the north.

Hurricanes have regularly impacted Belize's reefs. Hurricane Hattie in 1961 was reported to have reduced live coral cover by 80 percent in some places, although the reefs subsequently made a good recovery. As elsewhere in the region, *Acropora* cover has fallen dramatically since the late 1970s, linked to the impact of white band disease. In 1998, the El Niño-related coral bleaching event, followed by Hurricane Mitch, hit the reefs of Belize particularly hard. Corals remained bleached for a considerable period, but by early 1999 mortality was high, with a 62 percent loss of live coral cover in the south, 55 percent in the north, 45 percent in the atolls and 36 percent in the central reefs.

Belize has a long history of human activities in the coastal zone, which can be traced back to 300 BC. The Mayan Indians used cays in the lagoons as stations for fishing conch, finfish, turtle eggs and manatees, as well as ceremonial centers and burial sites. Nowadays the major threats to the reefs of Belize are fishing, sedimentation, agrochemicals, sewage, solid wastes and dredging. Fishing occurs on a relatively small scale given the reef area, but in 1998 employed 2 000 fisherfolk with 350 boats. The dominant fisheries are lobster (mainly *Panulirus argus*), which was considered to be near to its maximum sustainable yield in the early 1980s, and conch (mainly *Strombus gigas*). The latter produces catches averaging 180 tons per year. The adults aggregate in the shallow back reef and seagrass areas and although there are signs that the populations may be overexploited, catches have remained consistent. A deeper and unfished reproductive population could be responsible for maintaining the catch. Nearly two thirds of lobster and conch are exported to the USA. By contrast 80 percent of

Belize

General Data	
Population (thousands)	249
GDP (million US$)	504
Land area (km²)	22 169
Marine area (thousand km²)	31
Per capita fish consumption (kg/year)	8
Status and Threats	
Reefs at risk (%)	63
Recorded coral diseases	4
Biodiversity	
Reef area (km²)	1 330
Coral diversity	46 / 57
Mangrove area (km²)	719
No. of mangrove species	5
No. of seagrass species	na

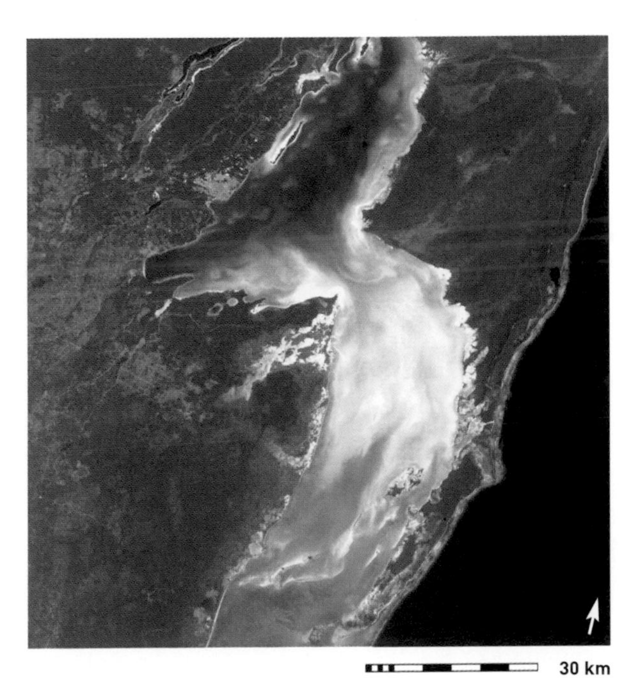

Left: Soft corals and a tubular sponge. Right: Chetumal Bay, on the border between Belize and Mexico. To the right is Ambergris Cay where the Belize Barrier Reef connects to the Yucatan Peninsula and becomes a fringing reef (ISS001-ESC-5317, 2001).

Protected areas with coral reefs

Site name	Designation	Abbreviation	IUCN cat.	Size (km²)	Year
Belize					
Bacalar Chico	Marine Reserve	MR	IV	107.00	1996
Blue Hole	Natural Monument	NM	III	41.00	1986
Gladden Spit	Marine Reserve	MR	IV	na	2000
Glovers Reef	Marine Reserve	MR	IV	308.00	1993
Half Moon Caye	National Monument	NaM	III	39.25	1982
Hol Chan	Marine Reserve	MR	IV	4.11	1987
Man-o-war Cay	Bird Sanctuary	BS	IV	0.01	1977
Port Honduras	Marine Reserve	MR	IV	na	2000
Sapodilla Cayes	Marine Reserve	MR	IV	127.00	1996
South Water Cay	Marine Reserve	MR	IV	298.00	1996
BELIZE BARRIER REEF RESERVE SYSTEM	WORLD HERITAGE SITE			**963.00**	**1996**

finfish, especially higher quality species such as groupers (Serranidae) and snappers (Lutjanidae), are caught for local consumption. Shrimp mariculture is now an important industry in Belize. There are considerable concerns about the impacts this industry may already be having on coastal fisheries and further expansion is likely to impact mangrove areas. A number of major fish spawning aggregations are known in Belize, and many are considered to be overfished. One of the largest of these, Gladden Spit, has recently been declared a protected area.

The Belize economy is heavily dependent on agriculture. Sugar, the chief crop, accounts for nearly half of Belize's exports, while the banana industry is the country's largest employer. Increased sedimentation from forest or savannah clearances and eutrophication from fertilizers are suspected, although few effects on the marine environment can be directly attributed to agricultural landuse practices. Adjacent to the banana and citrus growing regions of Belize, shallow marine habitats such as patch reefs and seagrass beds are located much further offshore than elsewhere, and there is a higher growth of algae under certain conditions. This is an area of intense ongoing research. Belize's trade deficit grew throughout the 1990s, mostly as a result of low export prices for sugar and bananas, and so tourism has been assuming an increasingly important role. This may also bring a greater range of threats to the reef systems, although there have been particular efforts to develop Belize as an environmentally sustainable "ecotourism" destination.

Considerable efforts have also been directed towards the development of a system of marine protected areas. The Hol Chan Marine Reserve in the north of the country is widely cited as an example of an effective no-take zone, implemented with the support and collaboration of the local population. This site has significantly higher fish numbers and biomass than surrounding areas, but, more importantly, its protection has demonstrably increased fish yields from the surrounding areas. For many of the other marine protected areas the legislation and infrastructure are largely in place for full and effective reef management, although further enforcement is still required.

A tourist on the Belize Barrier Reef. Walking on reef flats can become a problem when there are large numbers of visitors to a small area of reef.

Honduras, Nicaragua, Guatemala and El Salvador

MAP 5d

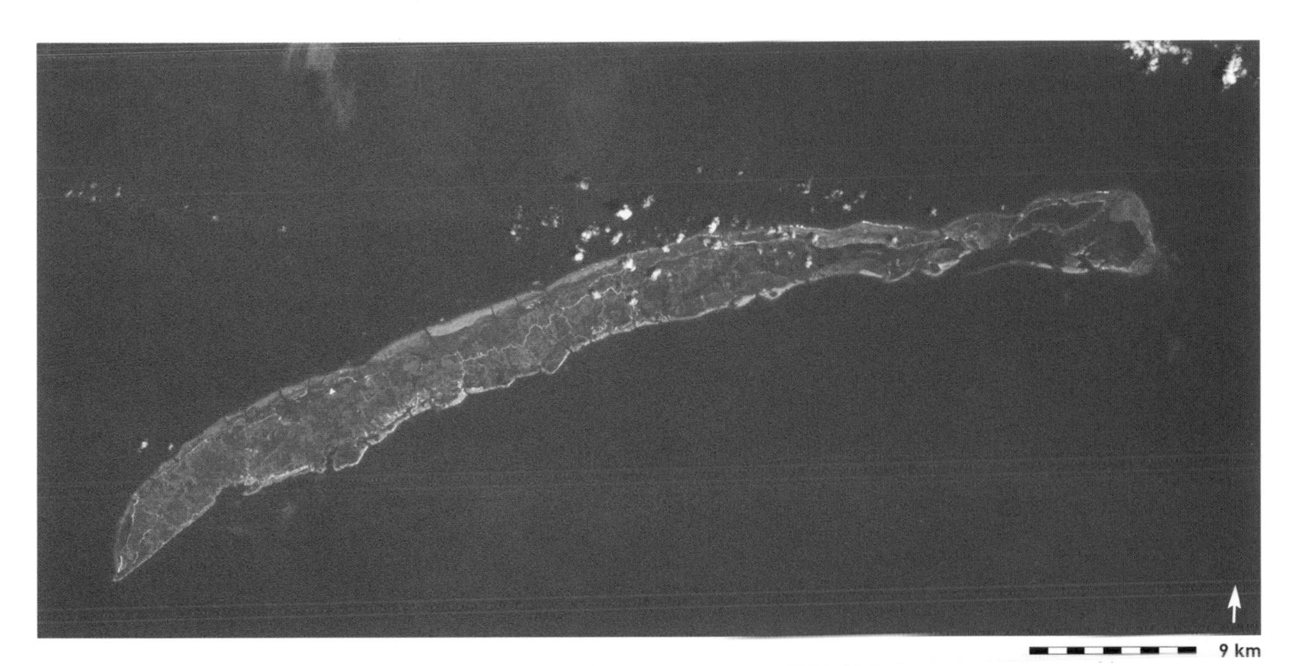

Honduras has a long mainland coast facing the Caribbean Sea but dominated by heavy riverine inputs and extensive mangrove communities. There are no recorded coastal coral reefs, although small, poorly developed coral communities are recorded from Puerto Cortes, La Ceiba and Trujillo. Important coral reefs occur around the Bay Islands (Utila, Roatán, Guanaja) and also the Cayos Cochinos which lie between Roatán and the mainland. Fringing and patch reefs also occur to the east associated with the Misquitia Cays and Banks, which are a continuation of the reef systems on the Nicaraguan shelf to the south. There are also reefs associated with the remote Swan Islands (Islas del Cisne) some 150 kilometers northeast of the mainland. These are three raised coralline islands which lie close to the edge of the Cayman Trench. They are surrounded by fringing reefs, with the most extensive reef development on their northern shores.

The Bay Islands lie relatively close to the shore, but are also near to the deep water of the Cayman Trench, and are surrounded by well developed fringing reefs. The typical seaward profile of Roatán's reefs is a gradation from terrestrial muds to coarse calcareous sand and seagrass beds (mainly *Thalassia testudinum*). Sparse corals and algae such as *Turbinaria* and *Sargassum* occur on a limestone pavement about 100-200 meters from shore, eventually merging into a spur and groove zone. *Agaricia tenuifolia* is the principal reef-building species in these shallow waters, though in higher energy areas *Acropora palmata* is more common. On the fore reef, at 10-15 meters, *Montastrea annularis*, *Colpophyllia natans* and *Diploria* spp. are very abundant, with live coral cover averaging 28 percent on the deep fore reef, but ranging between 24 and 53 percent in the Sandy Bay/West End Marine Reserve. The shelf edge is nearly vertical in many places and also has high coral cover, with species of *Agaricia* and colonies of *Eusimilia fastigiata* growing to an unusually large size. In total, 44 species of coral have been recorded here, but a complete inventory of marine biodiversity is planned as part of a five-year natural resources management project for the Bay Islands. Relatively healthy until 1998, the Bay Islands reefs experienced extensive bleaching during the El Niño event and were damaged during Hurricane Mitch. The most pressing threats to reefs in Honduras are a projected increase in diving-related tourism and associated migration from the mainland.

The Cayos Cochinos consist of two larger and 12 very small volcanic islands. The northern coasts of the larger

Roatán, in the Bay Islands, Honduras (STS050-80-52, 1992).

Honduras

General Data	
Population (thousands)	6 250
GDP (million US$)	3 725
Land area (km²)	112 851
Marine area (thousand km²)	238
Per capita fish consumption (kg/year)	4
Status and Threats	
Reefs at risk (%)	57
Recorded coral diseases	1
Biodiversity	
Reef area (km²)	810
Coral diversity	31 / 57
Mangrove area (km²)	1 458
No. of mangrove species	5
No. of seagrass species	1

islands are subject to high wave energies and are dominated by massive corals, while the southern shores, and the more protected shores of the smaller islands, have a greater diversity of corals, dominated by *Agaricia*. There are extensive seagrass beds.

A number of fish stocks are considered to be overexploited, and there is a large prawn trawling industry. Tourism is a major industry in the Bay Islands, and present to a lesser extent in the Cayos Cochinos.

Efforts to protect the marine resources of the Bay Islands have been spearheaded by the local community and an unofficial marine reserve has been set up around the West End and Sandy Bay. There are several other marine protected areas, notably the Cayos Cochinos Biological Reserve, which covers the entire island and reef system of this area and is actively managed with support from the private sector.

Nicaragua

The Caribbean coast of Nicaragua runs for over 350 kilometers along a north-south axis. Offshore, the coastal shelf drops quickly to 20-40 meters then maintains this depth to the abrupt shelf edge, which lies some 250 kilometers from shore in the north but as close as 20 kilometers in the south. This is significant because 90 percent of Nicaragua's watersheds drain into the Caribbean, and this coast receives more than 3 meters of rainfall per year in the north and more than 7 meters per year in the south, among the highest precipitation rates in the world. The coastal ecology of this country is generally quite poorly studied, but the marine resources are believed to be extremely important in a regional context. Large areas are covered by coral reefs, mangroves and seagrass beds, while human impacts are thought to be minimal because of the low population density.

Reefs occur along the entire coastline, but especially around the offshore islands, notably the Miskito Cays in the north and Corn Cays towards the center of the country. These, together with other shelf edge reefs, may be a true barrier reef system. Reefs also occur around a group of inshore cays: Man O'War Cays, Crawl, Set Net and Taira Cays and the Pearl Cays. Seagrass beds, predominately *Thalassia testudinum*, cover huge areas in between the mainland, these cays and the shelf edge. Although these beds have never been mapped they are believed to be some of the most extensive in the Caribbean, if not the world, and may provide food and refuge for more than half the remaining green turtles *Chelonia mydas* in the Caribbean. Undoubtedly they also play an important role as nursery habitat and feeding grounds for coral reef fish and invertebrates, and buffer coral reefs from much of the low salinity water and sediment flowing from the coastal rivers.

Increased sedimentation from the clearing of forests is believed to be responsible for the low coverage of live coral on reefs within 25 kilometers of the mainland. The reefs of the Miskito Coast Marine Reserve have not been extensively surveyed. A total of 27 species of scleractinian corals and 12 gorgonians have been recorded. However the diversity of habitat available and healthy conditions suggest that there may be many more species present. The Pearl Cays reefs lie close to shore on the edge of a turbid coastal boundary current which sometimes enters into this archipelago. In 1998 these cays supported a thriving community of *Acropora palmata* colonies on their windward eastern sides. The reefs of Great Corn Island are better known, largely due

Nicaragua

General Data	
Population (thousands)	4 813
GDP (million US$)	2 534
Land area (km²)	129 047
Marine area (thousand km²)	127
Per capita fish consumption (kg/year)	2
Status and Threats	
Reefs at risk (%)	58
Recorded coral diseases	1
Biodiversity	
Reef area (km²)	710
Coral diversity	22 / 57
Mangrove area (km²)	1 718
No. of mangrove species	9
No. of seagrass species	1

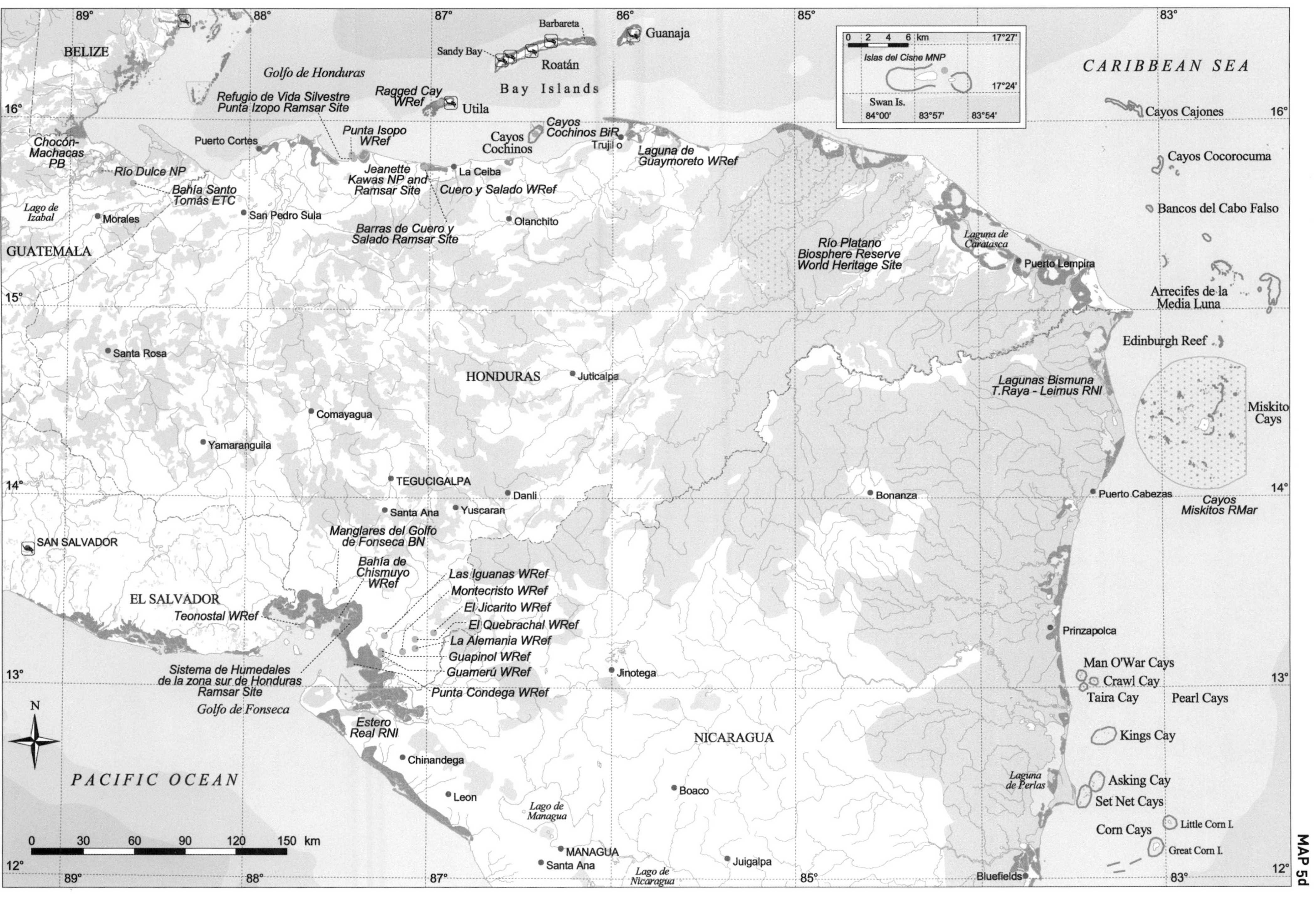

BELIZE
Golfo de Honduras
Refugio de Vida Silvestre Punta Izopo Ramsar Site
Ragged Cay WRef
Utila
Sandy Bay
Barbareta
Roatán
Bay Islands
Guanaja
Islas del Cisne MNP
0 2 4 6 km
17°27'
17°24'
Swan Is.
84°00'
83°57'
83°54'
CARIBBEAN SEA
Cayos Cajones
Cayos Cocorocuma
Bancos del Cabo Falso
Arrecifes de la Media Luna
Edinburgh Reef
Miskito Cays
Cayos Miskitos RMar
Chocón-Machacas PB
Río Dulce NP
Bahía Santo Tomás ETC
Lago de Izabal
Morales
GUATEMALA
Puerto Cortes
San Pedro Sula
Punta Isopo WRef
Jeanette Kawas NP and Ramsar Site
Barras de Cuero y Salado Ramsar Site
La Ceiba
Cuero y Salado WRef
Olanchito
Cayos Cochinos
Cayos Cochinos BiR
Trujillo
Laguna de Guaymoreto WRef
Río Platano Biosphere Reserve World Heritage Site
Laguna de Caratasca
Puerto Lempira
Lagunas Bismuna T.Raya - Leimus RNI
Puerto Cabezas
Santa Rosa
HONDURAS
Juticalpa
Comayagua
Yamaranguila
TEGUCIGALPA
Danli
Santa Ana
Yuscaran
Bonanza
SAN SALVADOR
Manglares del Golfo de Fonseca BN
Bahía de Chismuyo WRef
Las Iguanas WRef
Montecristo WRef
El Jicarito WRef
El Quebrachal WRef
La Alemania WRef
Guapinol WRef
Guamerú WRef
Punta Condega WRef
EL SALVADOR
Teonostal WRef
Sistema de Humedales de la zona sur de Honduras Ramsar Site
Golfo de Fonseca
Estero Real RNI
Jinotega
Prinzapolca
Man O'War Cays
Crawl Cay
Taira Cay
Pearl Cays
Kings Cay
NICARAGUA
Chinandega
Leon
Boaco
Lago de Managua
MANAGUA
Santa Ana
Juigalpa
Lago de Nicaragua
Laguna de Perlas
Asking Cay
Set Net Cays
Corn Cays
Little Corn I.
Great Corn I.
Bluefields
N
PACIFIC OCEAN
0 30 60 90 120 150 km
89°
88°
87°
86°
85°
83°
16°
15°
14°
13°
12°

to the presence of a monitoring site. Most of the shallow nearshore reefs have declined over the last decade so that live coral cover is now less than 10 percent. An increase in sewage is the probable cause, especially because the island is quite densely populated (500 people per square kilometer) in comparison to the rest of the coast (5 people per square kilometer). Waste from two fish processing plants, which produce 40 percent of the country's seafood exports, may be polluting nearshore waters. Deeper reefs have remained more stable, with an average cover of 38 percent algae, 22 percent hard coral, 1 percent sponges and 1 percent soft corals. No bleaching has been observed. The effects of Hurricane Mitch on the coral reefs of Nicaragua in late 1998 are unknown.

Indigenous Miskito Indians, together with some other communities, use the reefs in the north of the country for artisanal fisheries. For the most part this would appear sustainable, however the green turtle harvest is very high (14 000 per year) and in urgent need of control. Illegal fishing from neighboring countries may be reducing fish stocks in some places.

Guatemala and El Salvador

There are no true reefs in either Guatemala or El Salvador. Guatemala has a few small coral communities in the Gulf of Honduras, but none are known from its longer Pacific coast. El Salvador is reported to have small coral communities at Los Cobanos although there is no available information describing them.

Protected areas with coral reefs

Site name	Designation	Abbreviation	IUCN cat.	Size (km²)	Year
Honduras					
Bahía de Chismuyo	Wildlife Refuge	WRef	IV	290.00	1992
Cayos Cochinos	Biological Reserve	BiR	V	460.00	1993
El Jicarito	Wildlife Refuge	WRef	IV	15.41	1992
El Quebrachal	Wildlife Refuge	WRef	IV	1.98	1992
Guamerú	Wildlife Refuge	WRef	IV	na	1992
Guapinol	Wildlife Refuge	WRef	IV	na	1992
Islas del Cisne	Marine National Park	MNP	II	na	1991
Jeanette Kawas	National Park	NP	II	781.62	1988
La Alemania	Wildlife Refuge	WRef	IV	na	1992
Laguna de Guaymoreto	Wildlife Refuge	WRef	IV	50.00	1992
Las Iguanas	Wildlife Refuge	WRef	IV	14.26	1992
Montecristo	Wildlife Refuge	WRef	IV	na	1992
Punta Isopo	Wildlife Refuge	WRef	IV	112.00	1992
Ragged Cay	Wildlife Refuge	WRef	IV	na	na
Teonostal	Wildlife Refuge	WRef	IV	na	1992
PARQUE NACIONAL JEANETTE KAWAS	RAMSAR SITE			**781.50**	**1995**
REFUGIO DE VIDA SILVESTRE PUNTA IZOPO	RAMSAR SITE			**112.00**	**1996**
Nicaragua					
Cayos Miskitos	Marine Reserve	RMar	Ia	500.00	1991

Queen angelfish Holacanthus ciliaris.

Costa Rica and Panama

MAP 5e

The Caribbean coast of Costa Rica is dominated by wide areas of alluvial sediments and there are considerable riverine inputs. These conditions greatly inhibit the development of coral reefs, although there are fringing communities at Limón (northwest of Isla Uvita) and Punta Cahuita towards the south. Less developed coral communities are also found from Puerto Viejo to Punta Mona.

By contrast, Panama has a more complex coastline, including rocky shores and two areas of extensive offshore islands, and there is some important reef development, notably at Bocas del Toro in the west and from Cristóbal eastwards. The eastern third of Panama's coastline lies in a province known as San Blas or the Kuna Yala. From Punta San Blas, the 175 kilometer long San Blas Archipelago runs parallel to the coast with several hundred small nearshore sand cays stretching to the Colombian coastline.

These Caribbean coastlines lie well south of the main westward flow of the Caribbean Current. This current sets up two counter-clockwise eddies, the first producing eastward currents flowing from southern Costa Rica and around the Golfo de los Mosquitos, and the second sweeping east along the San Blas islands. This area also lies to the south of the main Caribbean hurricane belt and there has only been one record of a hurricane along the Panama coast in the last 120 years.

In the past three decades the Costa Rican coral reefs have suffered a spectacular decline, undoubtedly exacerbated by an increase in sedimentation caused by deforestation on the mainland. This is particularly apparent at Cahuita, where live coral cover was 40 percent in the late 1970s but had decreased to 11 percent by 1993, while the cover of rubble and algae increased from 60 percent to 90 percent. By 1999, coral cover had declined to about 3 percent. In the same period sediment load increased significantly. Branching corals suffered particularly badly. *Acropora cervicornis* largely disappeared and the abundance of *Agaricia agaricites* decreased by 15 percent, although it remained the dominant species. Massive species were much less affected. For example, while the abundance of *Porites porites* fell by 9 percent, *Siderastrea siderea* colonies increased in abundance by 16 percent. Part of this decline can be linked to a severe earthquake in 1991 which affected the whole of the Costa Rican Caribbean coast, in some places causing the reef slope to slump. In Limón the reefs largely recovered from this event, but they have not done so in Cahuita, possibly

Left: Densely packed houses in a Kuna village in the San Blas Islands, northeast Panama. The mainland coast remains largely undeveloped. Right: The scattered reefs and occasional islands of the San Blas Archipelago are some of the region's best developed reefs.

MAP 5e

CARIBBEAN SEA

PACIFIC OCEAN

NICARAGUA

COSTA RICA

PANAMA

COLOMBIA

Corn Cays (NICARAGUA)

Cayos de Albuquerque (COLOMBIA)

Bluefields

Juigalpa

Granada

Lago de Nicaragua

Golfo de Santa Elena

Bahía Culebra

Liberia

Area de Conservación Guanacaste World Heritage Site

Santa Rosa NP

Tamarindo Ramsar Site

Palo Verde NP

Ostional NWR

Curú NWR

Golfo de Nicoya

Cabo Blanco SNR

Isla Pájaros BiR

Islas Guayabo BiR

Quesada

Tortuguero NP

Santiago

SAN JOSE

Barra del Colorado NWR

Matina RFor

Limón

Isla Uvita

Cahuita NP

Punta Cahuita

Puerto Viejo

Punta Mona

Manuel Antonio NP

Marino Ballena NP

Térraba-Sierpe Ramsar Site

Golfo Dulce FoR

Isla del Caño BiR

Corcovado NP

Puerto Cortes

Golfo Dulce

Coto Brus AR

La Amistad International Park World Heritage Site

Gandoca-Manzanillo NWR and Ramsar Site

San San - Pond Sak Ramsar Site

Bocas del Toro

Isla Bastimentos NP

Laguna de Chiriquí

Golfo de los Mosquitos

David

Punta Entrada

Ensenada de Muertos

Golfo de Chiriquí

Isla Coiba NP

Bahía Honda

Golfo de Montijo Ramsar Site

Sarigua NP

Peñón de la Onda WRef

Cenegón del Mangle WRef

Isla Iguana WRef

Portobelo NP

Punta Galeta

Colón

Cristóbal

Panama Canal

PANAMA CITY

Taboga WRef

Las Perlas Archipelago

Golfo de Panama

Isla de Cañas WRef

San Blas Archipelago

Punta San Blas

Golfo de San Blas

Punta Patiño Ramsar Site

Chepigana FoR

Canglón FoR

Darien National Park World Heritage Site

Isla del Coco NP World Heritage Site and Ramsar Site

Isla del Coco

5°30'

5°20'

87°10'

87°00'

12°

10°

8°

85°

83°

81°

79°

N

0 40 80 120 160 200 km

due to intense sedimentation smothering coral recruits as they settle onto the substrate.

In Panama, some 64 hard coral species have been recorded off the Caribbean coastline. The inshore reefs at Bocas del Toro receive a high sediment load, rich in pesticides and fertilizers, from mainland banana plantations. Reef development is less heavily impacted further offshore and live coral cover in these areas remains at about 25 percent, reaching as high as 70 percent in some places. The Laguna de Chiriqui is lined with extensive mangroves and the beaches of the area are still important rookeries for hawksbill and loggerhead turtles. Further east the reefs near Cristóbal and at Punta Galeta have experienced little disturbance as they were within a US military reservation until the return of the Panama Canal to the Panamanian government in 1999. Punta Galeta has an emergent reef with a shallow fore reef pavement and a substantial coral boulder reef flat which is dominated by foliose and crustose algae. During calm weather and low tides it is exposed for long periods, up to 30-40 times a year, from 1 to 14 hours at a stretch. In the central region, proximity to the Panama Canal presents a real and continuing threat of oil pollution. A major spill in 1987 entered into mangrove communities and has continued to leach out into the surrounding waters where studies have shown significant declines in coral cover and diversity.

Some of the best developed reefs occur along the San Blas coastline. A number of islands and reefs lie on the outer edge of the continental shelf as patch reefs or fringing reefs around coral cays in a barrier-type structure. Further east, the reefs and islands are mostly located closer to the shore. Fifty-seven species of scleractinian coral have been recorded. This is an area which has better protection than many other reefs in the Caribbean, being an autonomous region run by the Kuna Indians. The mainland coast remains heavily forested and so there is little sediment runoff. The Kuna have tightly packed villages on 41 of the offshore islands, and while all waste from these villages passes straight into the surrounding waters, there appears to be sufficient dilution in most areas. Growing population numbers have led to expansion of some islands by reclamation, destroying the reefs in the immediate vicinity. Most fishing is subsistence level, but there is some export of spiny lobster, conch, spider crab and octopus, and there is evidence of significant target species overfishing for

Costa Rica

General Data	
Population (thousands)	3 711
GDP (million US$)	7 130
Land area (km²)	51 608
Marine area (thousand km²)	566
Per capita fish consumption (kg/year)	7
Status and Threats	
Reefs at risk (%)	93
Recorded coral diseases	2
Biodiversity	
Reef area (km²)	970
Coral diversity	25 / 83
Mangrove area (km²)	370
No. of mangrove species	9
No. of seagrass species	na

A schoolmaster Lutjanus apodus, *with tubular sponges and soft corals. Along with other snappers, this species is often heavily fished.*

some of these. Despite the apparently healthy picture overall, there have been dramatic reductions in coral cover, with equally dramatic increases in algae. In decline since the early 1970s, coral cover at various monitoring stations averaged 40 percent in 1983, and by 1997 had dropped to below 15 percent. Wide areas of *Agaricia* spp., which dominated many patch reefs, have died, and *Acropora cervicornis* and *A. palmata* have more or less vanished everywhere. The shallow inner reefs of the Golfo de San Blas, which were once formed by extensive mounds of *Porites porites*, have also shown considerable declines. Although there is some tourism to these islands, diving is not common.

Eastern Pacific

The Pacific coastline of both Costa Rica and Panama is strongly affected by extremes of water temperature associated with warm El Niño (~33°C) events and more frequent cool upwelling episodes (~15°C). These restrict offshore reef development in many areas, while terrestrial runoff greatly restricts reef development on mainland coasts. In general, reef development is sporadic and mostly at point locations around offshore islands. Most reefs in this region consist of shallow (less than 10 meters) sub-tidal *Pocillopora* banks bound together with calcareous algae, while *Porites lobata* is also a major reef builder in Costa Rica. Species diversity is low, but 23 species of hermatypic corals have been recorded on the Pacific side of Panama, and 18 in Costa Rica. Despite their simple community structure and low diversity, coral cover on these small reefs can be very high, reaching over 90 percent on healthy reefs. Cores through these reefs have shown carbonate accretions up to 10-12 meters thick, suggesting vertical accretion rates similar to many reefs in the Indo-Pacific. The Pacific reefs were severely impacted by the 1982-83 El Niño event, which drove mass bleaching and mortality in all areas. In Costa Rica recovery has generally been good and, despite repeated bleaching in 1992 and 1997-98, coral cover remains high in most areas. By contrast, recovery on many reefs in Panama has not been great.

Panama

General Data	
Population (thousands)	2 808
GDP (million US$)	7 114
Land area (km²)	74 697
Marine area (thousand km²)	332
Per capita fish consumption (kg/year)	14
Status and Threats	
Reefs at risk (%)	65
Recorded coral diseases	2
Biodiversity	
Reef area (km²)	720
Coral diversity	52 / 84
Mangrove area (km²)	1 814
No. of mangrove species	12
No. of seagrass species	3

In Costa Rica the main areas with coral communities or partial reef development are near Santa Elena, Bahía Culebra, Isla del Caño and Golfo Dulce. Golfo Dulce, in eastern Costa Rica, has largely escaped the impacts of El Niño events, although the reefs here have been severely affected by sedimentation from deforestation, mining and road construction. In 1993 live coral cover was less than 2 percent. From the surface to a depth of 1 meter the substrate consisted almost entirely of a mesh of dead *Pocillopora damicornis* and *Psammocora stellata*. Severely bio-eroded colonies of *Porites lobata*, a species which is especially resistant to sedimentation, covered the fore reef slope to a depth of about 12 meters. Bahía Culebra has the most diverse coral reefs on mainland Costa Rica, and here coral cover is much higher – some 20-50 percent – with *Pocillopora elegans*, *Pavona clavus* and *Leptoseris payracea* being the dominant species. Reefs on Isla del Caño are reported to be in a phase of recovery, with high levels of new coral recruitment. Recreational diving and unplanned tourism are the main threats in Costa Rica, although damage from commercial fishing nets and the collection of aquarium organisms has also been reported.

Boats waiting to enter the Panama canal. There is a continuous threat of oil spills, and the ballast water on these vessels can contain numerous marine organisms which are sometimes released in other parts of the world, threatening native species.

In Panama, Pacific reefs are best developed around offshore islands in the Gulf of Chiriqui in the west and the Gulf of Panama in the east. At the latter, the largest areas of reef are around Las Perlas Archipelago, a group of 53 basaltic islands. Reefs are best developed on northern and eastern sides, facing away from the upwelling currents. These islands were severely impacted by the 1982-83 bleaching, and live coral cover on some reefs remains below 2 percent, although Isla Iguana has over 30 percent. In the Gulf of Chiriqui the reefs are less affected by upwellings and El Niño events, and tend to be larger and more diverse – these are probably the best developed reefs on the continental shelf of the Eastern Pacific. There are also some pocilloporid reefs associated with the mainland around Ensenada de Muertos, Bahía Honda and Punta Entrada. On the offshore islands fringing reefs have clear patterns of zonation. Most of these reefs remain inaccessible to recreational divers.

In addition to these reefs, Costa Rica also incorporates the remote Isla del Coco which lies half-way between mainland Costa Rica and the Galapagos. This island is reported to have coral cover over much of its offshore slopes, dominated by *Porites lobata*.

Protected areas with coral reefs

Site name	Designation	Abbreviation	IUCN cat.	Size (km²)	Year
Costa Rica					
Cabo Blanco	Strict Nature Reserve	SNR	Ia	11.72	1963
Cahuita	National Park	NP	II	140.22	1970
Gandoca-Manzanillo	National Wildlife Refuge	NWR	IV	94.49	1985
Isla del Caño	Biological Reserve	BiR	Ia	2.00	1978
Isla del Coco	National Park	NP	II	23.64	1978
Manuel Antonio	National Park	NP	II	6.82	1972
Marino Ballena	National Park	NP	II	42.00	1990
Area de Conservación Guanacaste	**World Heritage Site**			**1 310.00**	**1999**
Cocos Island National Park	**World Heritage Site**			**997.00**	**1997**
Gandoca-Manzanillo	**Ramsar Site**			**94.45**	**1995**
Isla del Coco	**Ramsar Site**			**996.23**	**1998**
Panama					
Comarca Kuna Yala (San Blas)	Indigenous Commarc	IndCo	na	3 200.00	1938
Isla Bastimentos	National Park	NP	II	132.26	1988
Portobelo	National Park	NP	II	359.29	1976
Punta Patiño	**Ramsar Site**			**138.05**	**1993**

A Kuna Indian in Panama with a catch of spiny lobster Panulirus argus.

Colombia and Ecuador

MAP 5f

Colombia enjoys 1 700 kilometers of Caribbean coastline, but coral reefs are restricted to less than 150 kilometers, located away from major estuaries and sediment plumes. The Caribbean Current forms a gyre in the Colombian Basin, moving water in a north to northeasterly direction off the Colombian coast. This creates localized upwellings, bringing cold water to the surface and further curtailing the distribution of coral reefs. Reefs occur off Acandi in the far west and Punto Lopez in the east, but the most extensive structures are those off Santa Marta (at Punta Betin, Isla Morro Grande, Bahía Granate, Bahía Chengue and Bahía Gayraca) and Cartagena (at Islas de San Bernardo and Islas del Rosario). Several hundred kilometers northwest of Colombia there are a number of islands and reefs which also form part of Colombia (although they actually lie closer to Nicaragua) on the Nicaraguan Rise. These include the larger populated islands of San Andrés and Providencia, but also a number of shallow reefs including those on the banks of Quitasueño, Serrana and Roncador, and the atolls of Courtown and Albuquerque.

All the reefs off Santa Marta and Cartagena have experienced great changes in the last 20 years. The live coral cover at the Islas del Rosario declined from 41 percent in 1983 to 21 percent in 1990, concomitant with a threefold rise in algae cover. Similar though less severe changes have occurred in the Islas de San Bernardo. The most affected species were the acroporids which lost about 80 percent of their live cover, then *Agaricia tenuifolia* and *Porites porites* which suffered 30-40 percent mortality. Between 6 and 12 percent of *Diploria strigosa*, *Montastrea annularis* and *Siderastrea siderea* also died over the same period. This change has been attributed to a combination of bleaching, coral disease and pollution from the area's major cities and ports. There appears to be a gradient in reef state moving eastwards from Santa Marta, with 19 percent live coral cover at Punta Betin, 37 percent at Bahía Granate and 49 percent at Bahía Gayraca in 1993.

The offshore reefs, coral banks and atolls on the Nicaraguan Rise are well developed and diverse, with 44 scleractinian corals recorded off San Andrés. But unfortunately these complexes, which represent about 75 percent of the total coral reef area in Colombia, also appear to be in decline. San Andrés is a densely populated island, with 80 000 people living on less than 25 square kilometers, and a major tourist destination. Here coral cover has declined by about half – mortality in excess of 50 percent from

Broad view of the Galapagos Islands, with Fernandina and Isabela in the fore. Coral reef development in these islands is restricted to very small structures (STS068-168-28, 1994).

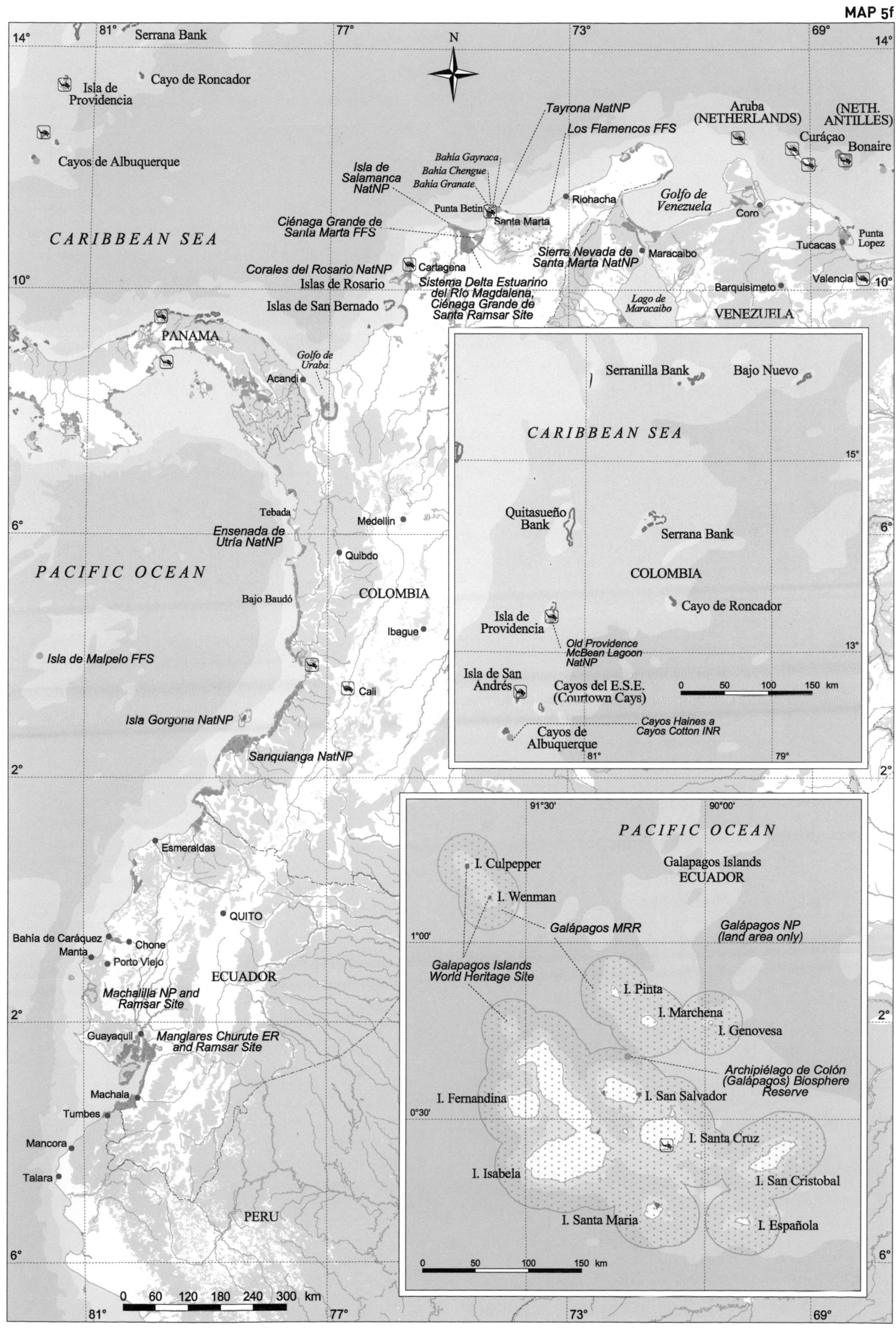
Serrana Bank
Isla de Providencia
Cayo de Roncador
Cayos de Albuquerque
CARIBBEAN SEA
Tayrona NatNP
Los Flamencos FFS
Aruba (NETHERLANDS)
(NETH. ANTILLES)
Curáçao
Bonaire
Isla de Salamanca NatNP
Bahía Gayraca
Bahía Chengue
Bahía Granate
Punta Betín
Santa Marta
Riohacha
Golfo de Venezuela
Coro
Ciénaga Grande de Santa Marta FFS
Punta Lopez
Tucacas
Sierra Nevada de Santa Marta NatNP
Maracaibo
Corales del Rosario NatNP
Cartagena
Islas de Rosario
Sistema Delta Estuarino del Río Magdalena, Ciénaga Grande de Santa Ramsar Site
Valencia
Barquisimeto
Lago de Maracaibo
VENEZUELA
Islas de San Bernado
PANAMA
Golfo de Uraba
Acandi
Tebada
Medellin
Ensenada de Utría NatNP
Quibdo
PACIFIC OCEAN
COLOMBIA
Bajo Baudó
Ibague
Isla de Malpelo FFS
Cali
Isla Gorgona NatNP
Sanquianga NatNP
Esmeraldas
QUITO
Bahía de Caráquez
Chone
Manta
Porto Viejo
ECUADOR
Machalilla NP and Ramsar Site
Guayaquil
Manglares Churute ER and Ramsar Site
Machala
Tumbes
Mancora
Talara
PERU
0 60 120 180 240 300 km
Serranilla Bank
Bajo Nuevo
CARIBBEAN SEA
Quitasueño Bank
Serrana Bank
COLOMBIA
Cayo de Roncador
Isla de Providencia
Old Providence McBean Lagoon NatNP
Isla de San Andrés
Cayos del E.S.E. (Courtown Cays)
Cayos Haines a Cayos Cotton INR
Cayos de Albuquerque
0 50 100 150 km
PACIFIC OCEAN
Galapagos Islands
ECUADOR
I. Culpepper
I. Wenman
Galápagos MRR
Galápagos NP (land area only)
Galapagos Islands World Heritage Site
I. Pinta
I. Marchena
I. Genovesa
Archipiélago de Colón (Galápagos) Biosphere Reserve
I. Fernandina
I. San Salvador
I. Santa Cruz
I. Isabela
I. San Cristobal
I. Santa Maria
I. Española
0 50 100 150 km

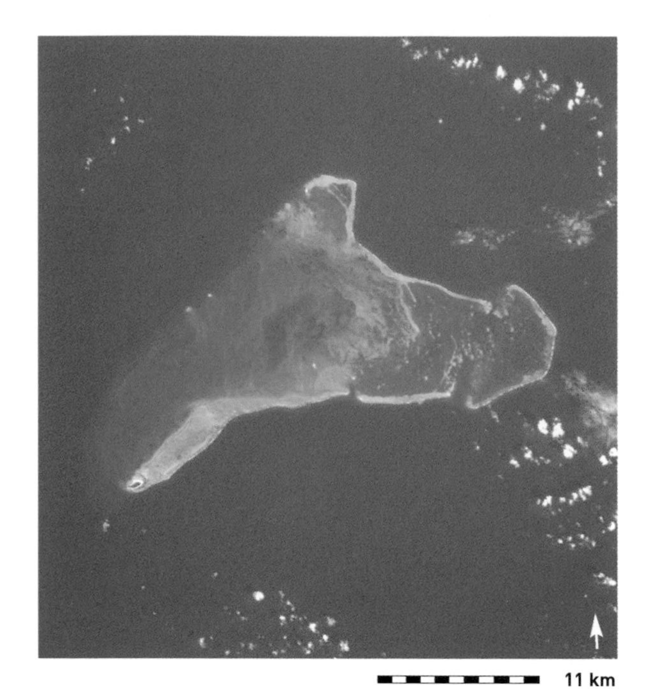

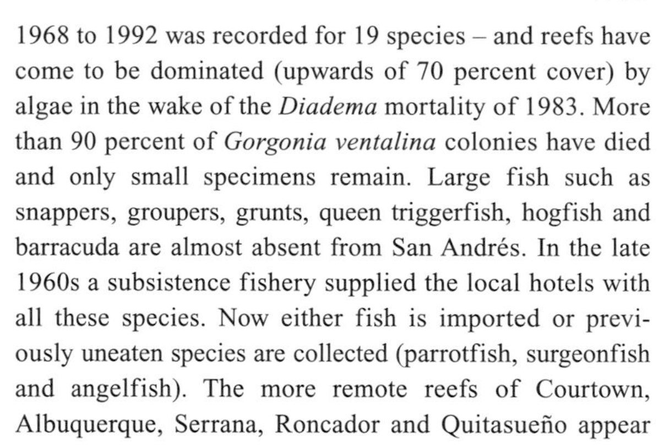

1968 to 1992 was recorded for 19 species – and reefs have come to be dominated (upwards of 70 percent cover) by algae in the wake of the *Diadema* mortality of 1983. More than 90 percent of *Gorgonia ventalina* colonies have died and only small specimens remain. Large fish such as snappers, groupers, grunts, queen triggerfish, hogfish and barracuda are almost absent from San Andrés. In the late 1960s a subsistence fishery supplied the local hotels with all these species. Now either fish is imported or previously uneaten species are collected (parrotfish, surgeonfish and angelfish). The more remote reefs of Courtown, Albuquerque, Serrana, Roncador and Quitasueño appear to be showing signs of a similar decline, with a low proportion of live coral cover, and depletion of a number of commercially important fish stocks. While overfishing may be causing the latter, the demise of the corals themselves may be indicative of a Caribbean-wide decline.

There are a few small reef developments along the Pacific coast of Colombia, most notably at Tebada and Ensenada de Utría. These are small fringing and patch reefs and are relatively young, developed over the basalt rocks of the Cordillera del Baudó. They are formed by no more than a half dozen scleractinian corals, mostly *Pocillopora* spp. The reefs of the Ensenada de Utría are protected in a national park. Given the remote location, human influence in this area is low. However, the reefs were heavily impacted, with coral bleaching and mortality caused by the recent El Niño. In addition to this area, there are fringing and patch reefs around the coast of Isla Gorgona, particularly on its eastern shores. These reefs are protected within a national park. Far offshore, the oceanic island of Malpelo also has some important coral communities down to depths of 35 meters.

Coral cover is very high on some of these Pacific reefs. Despite the 1982-83 El Niño event, which caused widespread mortality, including the decline of live coral cover in Isla Gorgona from 70 percent to 15 percent, there has been a rapid and nearly complete recovery, and in 1998 coral cover was estimated to be almost 60 percent. At one site on Malpelo island, live coral cover was 65 percent in 1972 and is now at 45 percent.

Colombia has designated a number of protected areas containing coral reefs. The Caribbean ones are generally larger, but these suffer more notable problems of management and some illegal activities continue.

Colombia

General Data	
Population (thousands)	39 686
GDP (million US$)	51 800
Land area (km²)	1 141 957
Marine area (thousand km²)	750
Per capita fish consumption (kg/year)	5
Status and Threats	
Reefs at risk (%)	44
Recorded coral diseases	6
Biodiversity	
Reef area (km²)	940
Coral diversity	49 / 77
Mangrove area (km²)	3 659
No. of mangrove species	11
No. of seagrass species	na

Left: The Serrana Bank, Colombia. An isolated reef structure in the Caribbean Sea (STS080-718-46, 1996). Right: Algae are now forming a dominant part of the community in many of Colombia's offshore reefs.

Ecuador

A few coral communities occur on the mainland coast of Ecuador and one true reef at Machalilla. However, it is in the Galapagos Islands that reefs are best developed. This archipelago is influenced by a major surface current, the South Equatorial Current, which flows from the east, largely fed by the cool Peru Oceanic Current (20-24°C) and the colder Peru Coastal Current (15°C). This current is strongly driven by the nearly constant southeast trade winds, while additional impetus is given by the Panama Current which flows south from the Panama Bight in December to January. Below the South Equatorial Current, an easterly Equatorial Undercurrent is generated at a depth of 100 meters, which is deflected to the surface by Fernandina and Isabela. Cool nutrient-rich water is therefore present all year round (except during El Niño events) and this restricts coral growth and reef development to the eastern sides of Isabela, Santa Cruz and the northern coasts of San Cristóbal.

For the most part these reefs are poorly developed patches and do not form true fringing structures. Species diversity is also low. Although the reefs are well protected there have been some impacts from bleaching and bioerosion. Fishing pressures have recently increased dramatically in a few areas, notably for the export trade in sea cucumbers and shark. Significant bleaching also impacted these reefs, both in 1982-83 and in 1997-98, with both events causing considerable coral mortality.

Although the overall human population is low in the Galapagos, the fishing lobby is significant and powerful. In nearshore waters, the most important industrial fisheries include lobster and sea cucumber, while numbers of fishers have grown considerably. The number of lobster fishers alone grew from 500 in 1999 to nearly 1 000 in 2000. Efforts to place restrictions on these industries have led to considerable hostility and violence by the fishers, but also to some weakening of catch limits as a form of appeasement.

Ecuador

General Data	
Population (thousands)	12 920
GDP (million US$)	13 008
Land area (km²)	256 925
Marine area (thousand km²)	1 064
Per capita fish consumption (kg/year)	8
Status and Threats	
Reefs at risk (%)	16
Recorded coral diseases	0
Biodiversity	
Reef area (km²)	<50
Coral diversity	25 / 23
Mangrove area (km²)	2 469
No. of mangrove species	7
No. of seagrass species	na

Protected areas with coral reefs

Site name	Designation	Abbreviation	IUCN cat.	Size (km²)	Year
Colombia					
Corales del Rosario	Natural National Park	NatNP	II	1 200.00	1977
Ensenada de Utría	Natural National Park	NatNP	II	543.00	1987
Isla de Malpelo	Fauna and Flora Sanctuary	FFS	Ia	na	1995
Isla Gorgona	Natural National Park	NatNP	II	492.00	1984
Old Providence McBean Lagoon	Natural National Park	NatNP	II	9.95	1996
Sierra Nevada de Santa Marta	Natural National Park	NatNP	II	3 830.00	1959
Tayrona	Natural National Park	NatNP	II	150.00	1964
Ecuador					
Galápagos	Marine Resource Reserve	MRR	IV	79 900.00	1986
ARCHIPIÉLAGO DE COLÓN (GALÁPAGOS)	UNESCO BIOSPHERE RESERVE			7 665.14	1984
GALAPAGOS ISLANDS	WORLD HERITAGE SITE			7 665.14	1978
MACHALILLA	RAMSAR SITE			550.95	1990

MAP 5g

Florida
USA
Gulf of Mexico
Cay Sal
Bank
BAHAMAS
BAHAMAS
BAHAMAS
ATLANTIC
OCEAN
Great Bahama Bank
Rincón de
Guanabo
Puerto
Escondido
Archipiélago
de Sabana
Havana Bay
HAVANA
Mariel
Matanzas
Archipiélago
de los Colorados
Península
Guanahacabibes HR
El Veral NR
Jibacoa Bacunayagua ETC
CUBA
Puerto
de Sagua
Subarchipiélago de Sabana
- Camaguey IMA
Cayo Santa
Maria TNA
Cayo Guillermo-
Santa Maria PN
Cayo Centro y Oeste
de Cayo Coco RE
Cayo Coco/Cayo
Guillermo TNA
Cayo Romano NP
Lanzanillo Pajonal RF
Pinar del Río
Golfo de Batabanó
Ciénaga de Zapata
Biosphere Reserve
Cabo
Corrientes
NR
Ciénaga de
Zapata NP
and ETC
Las Salinas
WRef
Cienfuegos
Caguanes PN
Cayo Caguanes/
Cayos de Piedra MFIR
Buenavista
Biosphere
Reserve
Archipiélago de
Camaguey
Desembocadura del Río
Máximo-Laguna Sabinal RF
Los Indios -
San Felipe PN
Isla de
la Juventud
Cayo Cantiles WRef
Cazones Gulf
Trinidad
Casilda
Península de
Guanahacabibes
Biosphere Reserve
Cayo Largo - Cayo
Rosario TNA
Golfo de Ana Maria
Cayos de Ana
Maria WRef
Cayo
Sabinal
TNA
La Isleta - Nuevas
Grandes IMA
Camaguey
Nuevitas
Sur de Guanahacabibes PN
Archipiélago
de los Canarreos
Subarchipiélago de
los Canarreos IMA
Cayo Blanco de
Casilida TNA
Cayo Algodon Grande TNA
Puerto Padre
Bahia del Naranjo MNP
Peninsula de Saetia ETC
Punta Francés -
Punta Pederales PNM
Punta del Este
RE and TNA
Sur Isla de
la Juventud NP
Laberinto de
Doce Leguas
Santa
Cruz
del Sur
Guayabal
Holguin
Cayo Saetia TNA
Alejandro de
Humboldt PN
Golfo de
Guacanayabo
Delta del
Cauto WRef
Subarchipiélago de
Jardines de la Reina IMA
Cuchillas del Toa
Biosphere Reserve
Baracoa
Hatibonico
WRef
Guantanamo
Gran Parque
Sierra Maestra IMA
Cayo Caguama MNP
Baitiquirí ETC
Guantanamo Bay
CARIBBEAN SEA
Cabo Cruz
Little Cayman
Cayman Brac
Desembarco del
Granma National Park
World Heritage Site
Desembarco del Granma PN
Baconao
Biosphere
Reserve
Baconao HR
Grand Cayman
CAYMAN ISLANDS
(UK)
N
0 60 120 180 240 300 km
84° 82° 80° 78° 76° 74°
24° 22° 20°
HAITI

Cuba

MAP 5g

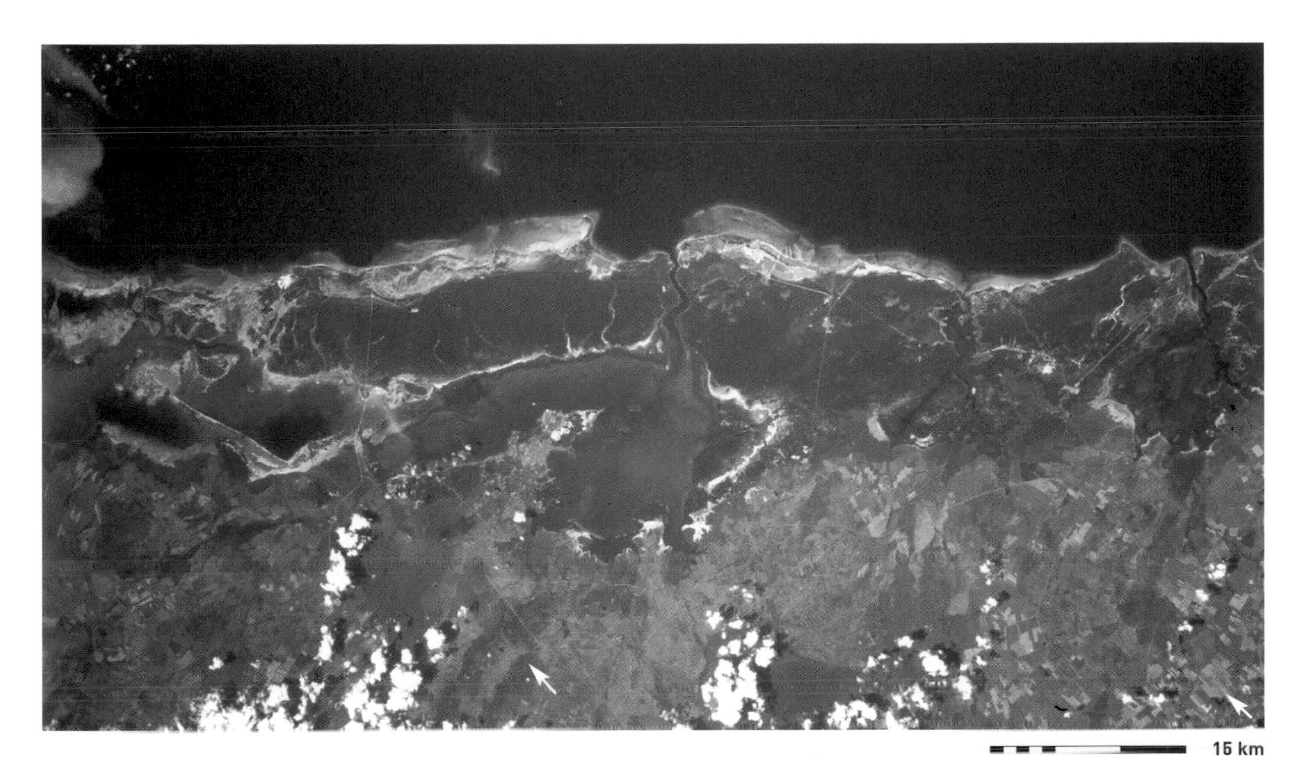

Cuba is the largest of the Caribbean islands, with a long, complex coastline and considerable chains of offshore islands and coral cays. Coral reefs stretch along virtually the entire border of the Cuban shelf. The majority of these lie offshore in long tracts which resemble barrier reefs, separated from the main island by broad lagoons. The longest runs for some 400 kilometers along the north coast from the Archipiélago de Sabana to the Archipiélago de Camaguey. On the south coast a similar reef tract stretches for over 350 kilometers from Trinidad to Cabo Cruz. Unlike true barrier reefs, the lagoons behind these reef tracts are very shallow. In most cases these wide lagoons, together with the long archipelagos of small coral cays which lie on their outer edges, have protected the reefs from adverse anthropogenic impacts. Hurricanes are more frequent in the south and west where the reef communities are dominated by species resistant to sedimentation and water movement, especially in the Gulf of Batabanó.

Only short stretches of coast have been heavily urbanized or industrialized. For these reasons, pollution tends to be localized: less than 3 percent of reefs in Cuba are believed to be affected by a significant degree of organic pollution. Many of the reefs appear to have shown a general increase in algal cover, probably associated with the *Diadema* die-off which has affected the rest of the region. Populations of the urchin in Cuba show no signs of recovery, so algal species such as *Cladophora catenata*, *Microdictyon marinum*, *Lobophora variegata*, *Dictyota* spp., *Sargassum* spp. and *Halimeda* spp. achieve biomass figures as high as 3 kilos per square meter. This occurs on reefs which are far from sources of organic pollution and may indicate that the changes are part of the Caribbean-wide impacts of *Diadema* die-off and loss of *Acropora* spp. to disease, rather than a result of direct anthropogenic impacts. In terms of reef fish, Cuban populations have higher biomass, species richness and average size than many other countries in the region, but these parameters were declining in the 1980s and 1990s due to overfishing. In 1998, coral bleaching was reported to have been severe on all coasts, although bleaching-related mortality was low.

Levels of sewage, organic and inorganic pollution are high in Havana Bay and this has caused the diversity of scleractinians, sponges and gorgonians to decline severely. These reefs are now dominated by just a few species of scleractinian corals, mainly *Siderastrea radians*, by the sponges *Clathria venosa* and *Iotrochota birotulata*, and by the gorgonians *Plexaura homomalla*, *P. flexuosa* or *Pseudoplexaura* spp.

Nuevitas Bay in northeast Cuba. The fringing reefs offshore are clearly visible, while there are important mangrove communities around the lagoon (NM23-729-782, 1997).

Cuba

General Data	
Population (thousands)	11 142
GDP (million US$)	14 694
Land area (km^2)	110 437
Marine area (thousand km^2)	345
Per capita fish consumption (kg/year)	13
Status and Threats	
Reefs at risk (%)	46
Recorded coral diseases	1
Biodiversity	
Reef area (km^2)	3 020
Coral diversity	29 / 57
Mangrove area (km^2)	7 848
No. of mangrove species	5
No. of seagrass species	4

A trumpetfish Aulostomus maculatus *amongst soft corals.*

Fishing plays a very important role in the Cuban economy, both as a generator of foreign exchange and as a source of protein. In general, catches rose systematically during 1960-75, leading to the overfishing of species such as the lane snapper *Lutjanus synagris* in the Gulf of Batabanó, the Nassau grouper *Epinephelus striatus* and queen conch *Strombus gigas* across the whole Cuban shelf, and shrimps (*Penaeus* spp.) on the southern shelf. The decline in lane snappers led to their replacement by grunts in the Gulf of Batabanó, which are of lower food quality and commercial value. The proliferation of grunts subsequently prevented recuperation of the lane snapper stocks, in spite of the imposition of severe administrative and protective measures. The overfishing of queen conch was based mainly on the illegal extraction of the meat as bait (with a rough estimate of more than 1 500 tons per year), or for selling the shells as curios. Closed seasons during the breeding period (April-September), the prohibition of catching juveniles, and quotas have all been implemented. Two assessments suggested a slight recovery in the south of Cuba stock in Cabo Cruz in 1990, and another in the south of the Gulf of Batabanó in 1991. The spiny lobster *Panulirus argus* is another resource closely linked to coral reefs and since 1978 catches have varied between 11 000 and 13 000 tons per year. However, this harvest is mainly based on lobsters in the seagrass beds in the Gulf of Batabanó rather than in the reefs themselves, where an important reproductive population remains. Annual lobster exports were valued at US$100 million in the late 1990s.

The collection of black coral for ornamental jewellery has continued for four decades and, as a consequence, stocks have been depleted in some places, especially the shallower waters along the north of Pinar del Río Province, in Matanzas Bay, Puerto de Sagua and Cazones Gulf.

The Cuban government announced in 1995 that gross domestic product had declined by 35 percent during 1989-93, a decline closely related to the loss of aid from the former USSR and economic sanctions imposed by the USA. Although there has been some economic growth since then, living standards remain at a depressed level compared with 1990. Fluctuations in the price of nickel and sugar have compelled the state to open up areas for tourism development, and this industry now plays a key role in generating foreign currency earnings. However, regulations for the protection of coral reefs directed at both tourists and tour guides are not yet fully implemented. Physical damage and the extraction of stony corals and other organisms are degrading the reefs in some tourist areas, such as the reefs of Rincón de Guanabo and Puerto Escondido to the northeast of Havana Province. The effects of coastal construction are generally unmonitored. At the end of the 1990s the tourist industry largely catered for the European market, but enormous expansion is to be expected if the political situation between Cuba and the USA changes.

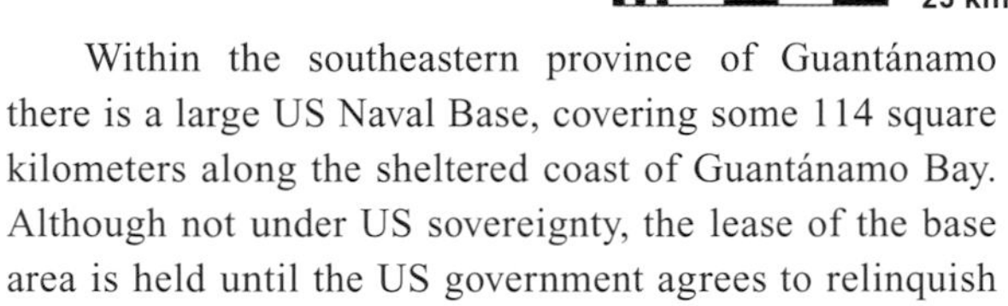

Within the southeastern province of Guantánamo there is a large US Naval Base, covering some 114 square kilometers along the sheltered coast of Guantánamo Bay. Although not under US sovereignty, the lease of the base area is held until the US government agrees to relinquish it. The area includes considerable military developments and associated recreational facilities, and much of the bay area has been dredged and degraded. Despite this, there are some coral communities, and recreational diving is practiced by personnel. The beach and waters of Cuzco Beach have been declared a preserve, dredging is prohibited and visitor numbers restricted.

Protected areas with coral reefs

Site name	Designation	Abbreviation	IUCN cat.	Size (km²)	Year
Cuba					
Cayo Coco/Cayo Guillermo	Touristic Natural Area	TNA	V	320.00	1986
Cayo Romano	National Park	NP	V	920.00	1986
Cayo Sabinal	Touristic Natural Area	TNA	V	335.00	na
Cayos de Ana Maria	Wildlife Refuge	WRef	IV	69.00	na
Ciénaga de Zapata	National Park	NP	V	na	na
Punta Francés – Punta Pederales	Parque Nacional Marino	PNM	II	174.24	1985
Subarchipiélago de Jardines de la Reina	Integrated Management Area	IMA	V	305.80	na
Subarchipiélago de los Canarreos	Integrated Management Area	IMA	V	331.10	na
Subarchipiélago de Sabana – Camaguey	Integrated Management Area	IMA	V	1 789.08	na
Sur Isla de la Juventud	National Park	NP	V	800.00	1992
Buenavista	**UNESCO Biosphere Reserve**			**3 135.00**	**2000**
Ciénaga de Zapata	**UNESCO Biosphere Reserve**			**6 253.54**	**2000**
Cuchillas del Toa	**UNESCO Biosphere Reserve**			**1 275.00**	**1987**
Desembarco del Granma National Park	**World Heritage Site**			**418.63**	**1999**
Península de Guanahacabibes	**UNESCO Biosphere Reserve**			**1 015.00**	**1987**

Left: Golfo de Guacanayabo in southeast Cuba. Reef development in this shallow water has formed complex reticulated structures (NM23-729-780, 1997). Right: Schoolmasters Lutjanus apodus. *These snappers are more regularly observed on remote reefs which are less heavily fished.*

MAP 5h

78°30'
78°00'
77°30'
77°00'
76°30'
76°00'
18°30'
Pedro Pt.
Montego Bay
Montego Bay
Falmouth
Discovery Bay
Ocho Rios PA
Galina Pt.
N
Lucea
Bogue FS
Montego Bay MP
Montpelier
Port Maria
Sandy Reef
Maroon Town
Annotto Bay
Negril MP
Little London
South Negril Pt.
JAMAICA
Ewarton
Savanna la Mar
Port Antonio
Black River
Santa Cruz
Manchioneal
18°00'
KINGSTON
May Pen
Morant Pt.
Parottee Pt.
South-east Pt.
Port Morant
Southfield
Long Bay
Portland Bight
Alligator Reef
South Jamaica Shelf
Albatross Bank
Portland Bight PA
17°30'
Morant Bank
Morant Cays
CARIBBEAN SEA
Pedro Cays
Portland Rock
Pedro Bank
Northeast Cay
Blower Rock
Middle Morant Cay NR/SciR NR
Middle Cay
17°00'
Southwest Cay
South Cay
SE Pedro Cay NR/SciR NR
Banner Reef
0 10 20 30 40 50 km

Jamaica

MAP 5h

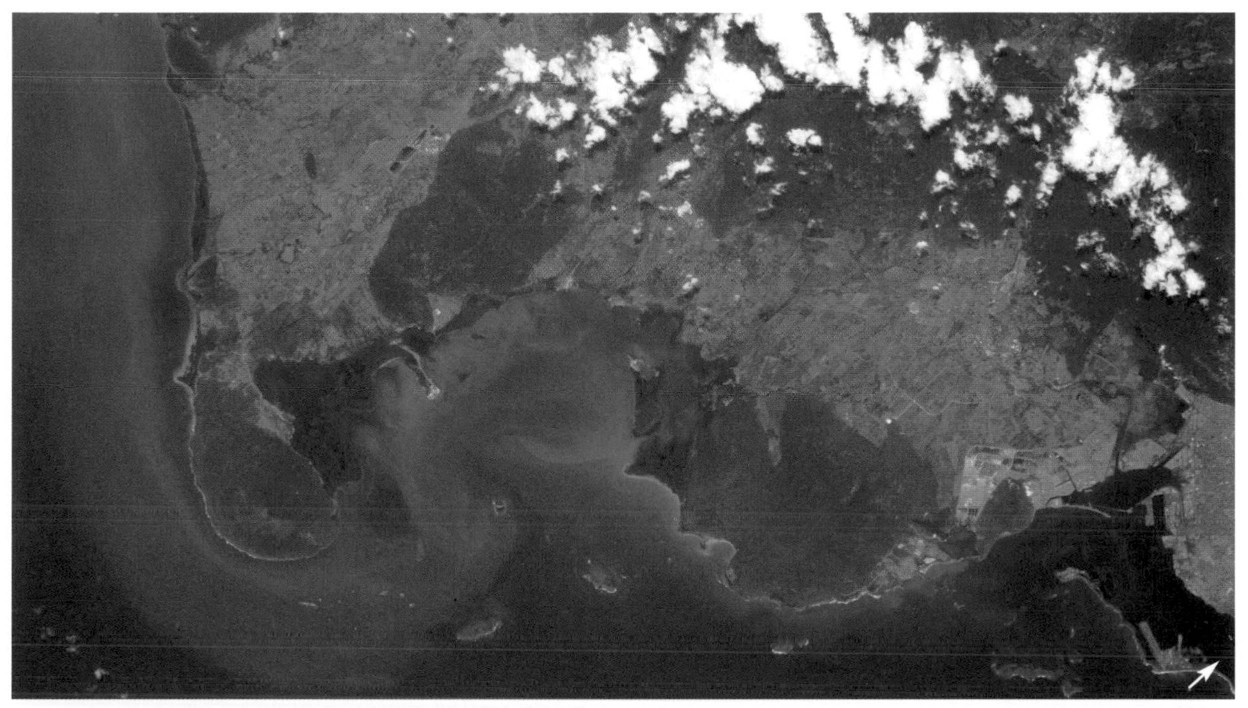

Jamaica is the third largest island in the Greater Antilles and is located in the center of the Caribbean Sea. Cuba, 150 kilometers north, moderates the effects of the northeast trade winds on the fringing reefs of the north coast, which grow on a narrow shelf. Patchy reef formations on the south coast, punctuated by rivers and sediment slopes, grow on a wider shelf extending up to 20 kilometers offshore. Reefs and corals also grow on nine offshore banks, notably at the Pedro Cays, 70 kilometers south, and the Morant Cays, 50 kilometers to the southeast. Coral cover on the mainland fringing reefs is low, although this was not always the case (see below). Cretaceous basement rocks are covered by Tertiary limestone, and on the north coast by Pleistocene reef deposits. Past changes in sea level have created terraces above and below present sea level to form raised or drowned cliffs. There are two wet seasons, in October and May, and two dry seasons. The water temperature on the north coast ranges from 26 to 30°C. The weather, particularly on the north coast, is dominated by the northeast trade winds, occasionally interrupted by cold fronts from North America in winter. Two of the most severe hurricanes on record, Allen and Gilbert, hit Jamaica in the 1980s, with significant impacts on the coral reefs.

Jamaica has a long history of exploiting its marine resources. Since early colonial days there was a substantial import of fish to feed the growing population, including turtle meat from the Cayman Islands, and dried fish from North America. Fishing the immediate offshore waters was also undertaken, but the maximum yield of some 11 000 tons of fish per year in the 1960s was clearly unsustainable and fish stocks have now collapsed. Overfishing is particularly bad on the north coast, where the narrow coastal shelf concentrates fishing into a smaller area, while making the shallow reef communities more accessible. Many of the fish now caught have not yet attained reproductive maturity, and it has been suggested that reef fish stocks in Jamaica may be being supplemented by fish larvae from other parts of the Caribbean. The offshore banks are also heavily fished, and there is a large conch fishery on Pedro Bank.

Jamaican reefs are further stressed by human impacts resulting from terrestrial activities, including sedimentation caused by soil erosion, but more particularly from nutrient pollution. Coastal development has been rapid in many parts of Jamaica, encouraged by massive tourism developments. In many areas sewage receives little or no treatment.

Portland Bight in southern Jamaica. This important area for coral reefs and mangroves was declared a protected area in 1999, and has full community involvement in its management (STS065-95-82, 1994).

Protected areas with coral reefs

Site name	Designation	Abbreviation	IUCN cat.	Size (km²)	Year
Jamaica					
Bogue	Fisheries Sanctuary	FS	IV	na	1979
Middle Morant Cay	Nature Reserve	NR	na	na	na
Montego Bay	Marine Park	MP	II	15.30	1991
Negril	Marine Park	MP	na	na	1998
Ocho Rios	Protected Area	PA	V	na	1966
Portland Bight	Protected Area	PA	V	1 876.15	1999

The combined effects of hurricanes and the regional die-off of *Diadema* are particularly well understood in Jamaica. Here, as elsewhere, the urchin was a major herbivore on reefs before 1983, and its disappearance, combined with severe hurricanes and white band disease, has resulted in a dramatic shift in Jamaica's reef from coral-dominated communities to those dominated by algae. Hurricane Allen, in 1980, caused the destruction of most of the dominant shallow reef-building colonies of *Acropora palmata* and *A. cervicornis*, which led to the temporary abundance of species such as *Agaricia agaricites*, with smaller encrusting and plate-like growth forms. There was a partial recovery from the impacts of this hurricane and new coral growth was recorded. However, the *Diadema* die-off led to considerable algal growth, with small ephemeral species being replaced by large macroalgae, leading to mortality amongst the *Agaricia* species and preventing further recruitment by juvenile corals. Hurricane Gilbert did destroy large amounts of algae but the bare spaces created were more rapidly recolonized by algae than by the slow-growing corals, suggesting that a new stable ecosystem had been created which is algal rather than coral-dominated. White band disease has further decimated *Acropora* populations and, over the last ten years, black band and yellow band disease have impacted some areas. Live coral cover at multiple sites along the coastline declined from over 50 percent in the late 1970s to less than 5 percent in the early 1990s. By the late 1990s, coral cover increased again to 10-15 percent at depths of 5-15 meters, partly due to increases in the abundance of *Diadema* in these shallow waters. Deeper reefs have been less severely impacted by these various events, particularly on the southern shelf edge where some active *Acropora* growth continues. It is postulated that increased herbivory could bring back the former high cover of coral, but herbivorous fish remain chronically overfished.

Jamaica's reefs have been well studied by scientists for several decades, notably through the work undertaken from the Discovery Bay Marine Laboratory. Efforts to reverse some of the many problems facing the country are beginning in some areas, and a number of marine protected areas have been declared. Active management, with full community involvement, is being pursued in a number of these, notably Montego Bay, Negril and the recently declared Portland Bight Protected Area.

Jamaica

General Data	
Population (thousands)	2 653
GDP (million US$)	4 383
Land area (km²)	11 044
Marine area (thousand km²)	251
Per capita fish consumption (kg/year)	17
Status and Threats	
Reefs at risk (%)	99
Recorded coral diseases	5
Biodiversity	
Reef area (km²)	1 240
Coral diversity	36 / 57
Mangrove area (km²)	106
No. of mangrove species	5
No. of seagrass species	3

Cayman Islands

MAP 5i

The Cayman Islands are Overseas Territories of the UK and consist of three islands: Grand Cayman, Cayman Brac and Little Cayman. All are very low-lying, with a maximum elevation of only 42 meters. Jamaica and the Cayman Islands are situated on either side of the Oriente Transform Fault, which also separates the south coast of Cuba from Jamaica. The Cayman Trough, to the east of Jamaica and the southeast of Grand Cayman, is actively spreading at the mid-Cayman Rise. Weather patterns in the Caymans are broadly similar to those in Jamaica, but the eastern islands are generally drier than Grand Cayman.

The reefs of the Cayman Islands are similar to each other. Grand Cayman has a narrow carbonate shelf which rarely exceeds 1.5 kilometers in width, and which is frequently much narrower. The fringing reefs often have well defined spur and groove formations, below which there are two distinct terraces: one at about 9 meters deep, the other at 12-16 meters. This second deeper terrace plunges vertically into the abyssal depths of the Cayman Trench. Coral cover is generally high. Historically, Grand Cayman had one of the largest green turtle rookeries in the Caribbean. The turtles were exported to Jamaica as a major food source during the early days of colonial rule, and it has been estimated that some 13 000 individuals were exported annually in the early 18th century, before the fishery collapsed through overfishing towards the end of the century.

There has been a remarkable expansion in the Cayman Islands over the last 30 years. The resident population has grown from 8 500 to 30 000, while the economy has boomed. The Islands are a thriving offshore financial center, with more than 40 000 registered companies in 1997, including almost 600 banks and trust companies whose assets currently exceed US$500 billion. Despite this, tourism is the mainstay of the economy, accounting for about 70 percent of gross domestic product and 75 percent of foreign currency earnings. The tourist industry is aimed at the luxury market and caters mainly for visitors from North America. Numbers are very high, with some 1.4 million visitors arriving annually. About 40 percent of these visitors go diving, attracted by the easy access to clear waters and sheer drop-offs.

Most of the pressure on the reefs arises from the massive, tourist-focussed development. Pollution and the contamination of groundwater by sewage are potential problems, as is overfishing. The deeper reefs off George Town have been destroyed by the continual anchoring of

Dive tourism in the Cayman Islands is a critical part of the island's economy.

Cayman Islands

General Data	
Population (thousands)	35
GDP (million US$)	612
Land area (km²)	277
Marine area (thousand km²)	119
Per capita fish consumption (kg/year)	na
Status and Threats	
Reefs at risk (%)	100
Recorded coral diseases	3
Biodiversity	
Reef area (km²)	230
Coral diversity	35 / 57
Mangrove area (km²)	71
No. of mangrove species	3
No. of seagrass species	na

MAP 5i

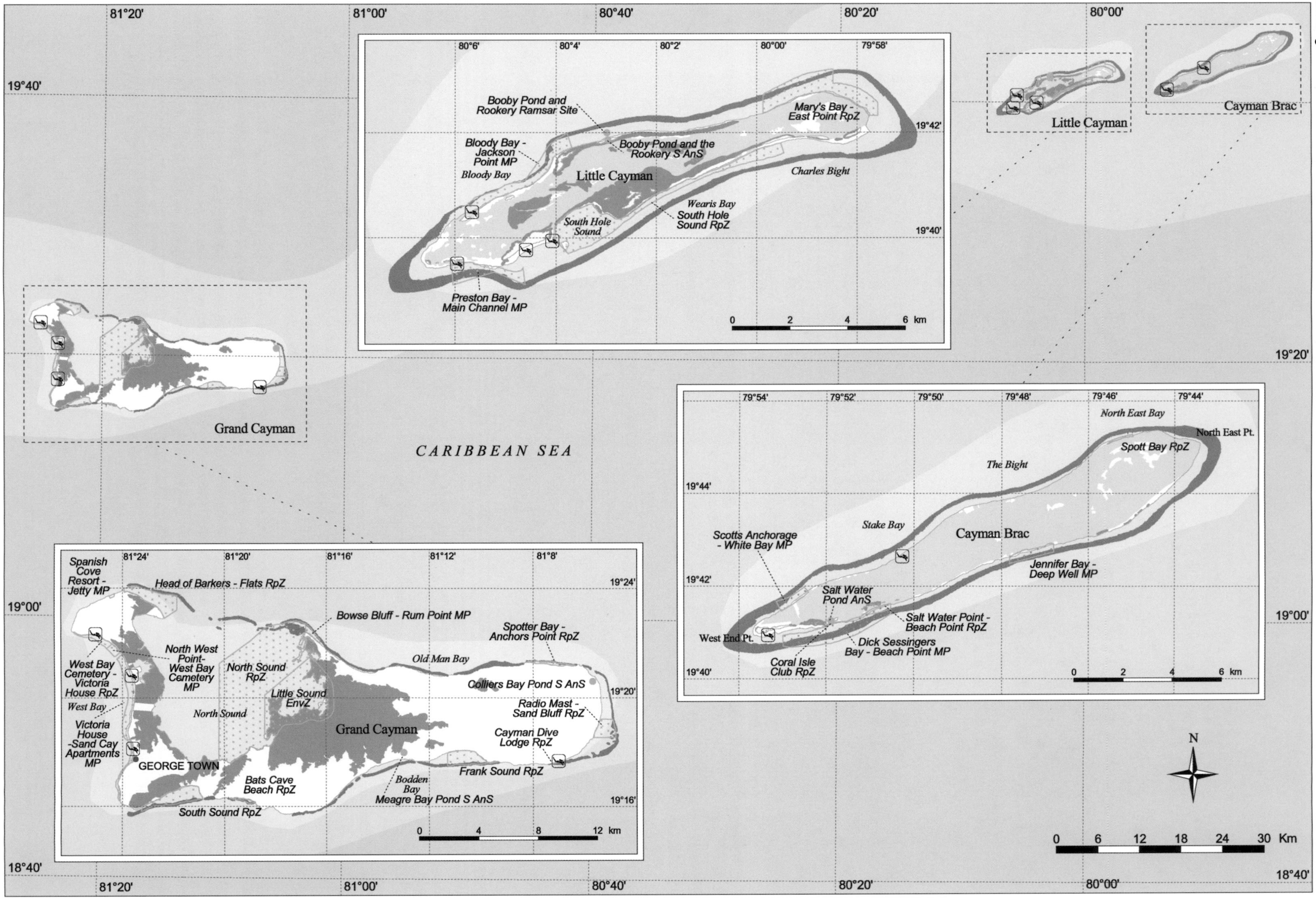

81°20'
81°00'
80°40'
80°20'
80°00'
19°40'
19°20'
19°00'
18°40'
CARIBBEAN SEA
Grand Cayman
Little Cayman
Cayman Brac
0 6 12 18 24 30 Km
N
80°6'
80°4'
80°2'
80°00'
79°58'
19°42'
19°40'
Booby Pond and Rookery Ramsar Site
Mary's Bay - East Point RpZ
Bloody Bay - Jackson Point MP
Booby Pond and the Rookery S AnS
Bloody Bay
Little Cayman
Charles Bight
Wearis Bay
South Hole Sound RpZ
South Hole Sound
Preston Bay - Main Channel MP
0 2 4 6 km
79°54'
79°52'
79°50'
79°48'
79°46'
79°44'
19°44'
19°42'
19°40'
North East Bay
North East Pt.
Spott Bay RpZ
The Bight
Stake Bay
Cayman Brac
Scotts Anchorage - White Bay MP
Jennifer Bay - Deep Well MP
Salt Water Pond AnS
Salt Water Point - Beach Point RpZ
West End Pt.
Dick Sessingers Bay - Beach Point MP
Coral Isle Club RpZ
0 2 4 6 km
81°24'
81°20'
81°16'
81°12'
81°8'
19°24'
19°20'
19°16'
Spanish Cove Resort - Jetty MP
Head of Barkers - Flats RpZ
Bowse Bluff - Rum Point MP
Spotter Bay - Anchors Point RpZ
North West Point- West Bay Cemetery MP
North Sound RpZ
Old Man Bay
West Bay Cemetery - Victoria House RpZ
Little Sound EnvZ
Colliers Bay Pond S AnS
West Bay
North Sound
Radio Mast - Sand Bluff RpZ
Victoria House -Sand Cay Apartments MP
Grand Cayman
Cayman Dive Lodge RpZ
GEORGE TOWN
Bats Cave Beach RpZ
Bodden Bay
Frank Sound RpZ
Meagre Bay Pond S AnS
South Sound RpZ
0 4 8 12 km

Protected areas with coral reefs

Site name	Designation	Abbreviation	IUCN cat.	Size (km²)	Year
Cayman Islands					
Bats Cave Beach (Grand Cayman)	Replenishment Zone	RpZ	IV	0.31	1986
Bloody Bay – Jackson Point (Little Cayman)	Marine Park	MP	II	1.61	1986
Bowse Bluff – Rum Point (Grand Cayman)	Marine Park	MP	II	0.60	1986
Cayman Dive Lodge (Grand Cayman)	Replenishment Zone	RpZ	IV	0.04	1986
Coral Isle Club (Cayman Brac)	Replenishment Zone	RpZ	IV	0.01	1986
Dick Sessingers Bay – Beach Point (Cayman Brac)	Marine Park	MP	II	1.43	1986
Frank Sound (Grand Cayman)	Replenishment Zone	RpZ	IV	2.24	1986
Head of Barkers – Flats (Grand Cayman)	Replenishment Zone	RpZ	IV	3.65	1986
Jennifer Bay – Deep Well (Cayman Brac)	Marine Park	MP	II	0.61	1986
Little Sound (Grand Cayman)	Environmental Zone	EnvZ	Ib	17.31	1986
Mary's Bay – East Point (Little Cayman)	Replenishment Zone	RpZ	IV	1.80	1986
North Sound (Grand Cayman)	Replenishment Zone	RpZ	IV	33.10	1986
North West Point – West Bay Cemetery (Grand Cayman)	Marine Park	MP	II	1.55	1986
Preston Bay – Main Channel (Little Cayman)	Marine Park	MP	II	0.81	1986
Radio Mast – Sand Bluff (Grand Cayman)	Replenishment Zone	RpZ	IV	1.77	1986
Salt Water Point – Beach Point (Cayman Brac)	Replenishment Zone	RpZ	IV	0.26	1986
South Hole Sound (Little Cayman)	Replenishment Zone	RpZ	IV	3.16	1986
South Sound (Grand Cayman)	Replenishment Zone	RpZ	IV	3.17	1986
Spott Bay (Cayman Brac)	Replenishment Zone	RpZ	IV	0.33	1986
Victoria House – Sand Cay Apartments (Grand Cayman)	Marine Park	MP	II	8.01	1986
West Bay Cemetery – Victoria House (Grand Cayman)	Replenishment Zone	RpZ	IV	0.69	1986

cruiseships, and nearby shallow reefs have been damaged by the resulting sedimentation. There are also direct concerns about the carrying capacity of dive sites and there have been some declines in fish stocks associated with overfishing. There was large-scale bleaching of corals in 1987, and even more severe bleaching in 1995-96 and 1998. In addition, white band disease has been reported.

There is reason for optimism, however. *Diadema* died out in 1983 but this did not result in an algal bloom because grazing fish were still abundant, and in 1998 the *Diadema* seemed to be recovering in areas on west Grand Cayman. *Acropora* species are still common, although impacted by storms. A comprehensive system of marine protected areas has been established covering 34 percent of the coastal waters of the islands, enforced by a number of guards, and also regularly subject to detailed monitoring.

Selected bibliography

MEXICO

Carriquiry JD, Reyes Bonilla H (1997). Community structure and geographic distribution of the coral reefs of Nayarit, Mexican Pacific. *Ciencias Marinas* 23(2): 227-248.

Fenner DP (1988). Some leeward reefs and corals of Cozumel, Mexico. *Bull Mar Sci* 42(1): 133-144.

Fenner DP (1991). Effects of Hurricane Gilbert on coral reefs, fishes and sponges at Cozumel, Mexico. *Bull Mar Sci* 48(3): 719-730.

Glynn PW, Morales GEL (1997). Coral reefs of Huatulco, West Mexico: reef development in upwelling Gulf of Tehuantepec. *Revista de Biología Tropical* 45(3): 1033-1047.

Grigg PW, Ault JS (2000). A biogeographic analysis and review of the far eastern Pacific coral reef region. *Coral Reefs* 19: 1-23.

Gutiérrez D, García-Saez C, Lara M, Padilla C (1993). Comparación de arrecifes coralinos: Veracruz y Quintana Roo. In: Salazar-Vallejo SI, González NE (eds). *Biodiversidad Marina y Costera de México.* CONABIO/CIQRO, México. 787-806.

Ketchum JT, Reyes Bonilla H (1997). Biogeography of hermatypic corals of the Archipiélago Revillagigedo, Mexico. *Proc 8th Int Coral Reef Symp* 1: 471-476.

Lara M, Padilla C, García C, Espejel JJ (1992). Coral reef of Veracruz, Mexico I. Zonation and community. *Proc 7th Int Coral Reef Symp* 1: 535-544.

Reyes Bonilla H, Lopez Perez A (1998). Biogeography of the stony corals (Scleractinia) of the Mexican Pacific. *Ciencias Marinas* 24(2): 211-224.

Salazar-Vallejo SI, González NE (eds) (1993). *Biodiversidad Marina y Costera de México.* Centro de Investigaciones de Quintana Roo, Chetumal, Mexico.

Tunnell JW Jr (1988). Regional comparison of southwestern Gulf of Mexico to Caribbean Sea coral reefs. *Proc 6th Int Coral Reef Symp* 3: 303-308.

BELIZE

Carter J, Gibson J, Carr A III, Azueta J (1994). Creation of the Hol Chan Marine Reserve in Belize: a grass-roots approach to barrier reef conservation. *Env Professional* 16(3): 220-231.

Gischler E, Hudson JH (1998). Holocene development of three isolated carbonate platforms, Belize, central America. *Mar Geol* 144(4): 333-347.

Littler DS, Littler MM (1997). An illustrated marine flora of the Pelican Cays, Belize. *Bull Biol Soc Washington* 9: 149.

McClanahan TR, Muthiga NA (1998). An ecological shift in a remote coral atoll of Belize over 25 years. *Env Cons* 25(2): 122-130.

Polunin NVC, Roberts CM (1993). Greater biomass and value of target coral-reef fishes in two small Caribbean marine reserves. *Mar Ecol Prog Ser* 100: 167-176.

Price ARG, Heinanen AP, Gibson JP, Young ER (1992). *A Marine Conservation and Development Report: Guidelines for Developing a Coastal Zone Management Plan for Belize.* IUCN, Gland, Switzerland.

Rützler K, Macintyre IG (eds) (1982). *Smithsonian Contributions to the Marine Sciences, 12: The Atlantic Barrier Reef Ecosystem at Carrie Bow Cay, Belize, I: Structure and Communities.* Smithsonian Institution Press, Washington DC, USA.

HONDURAS, NICARAGUA, GUATEMALA AND EL SALVADOR

Fenner DP (1993). Some reefs and corals of Roatan (Honduras), Cayman Brac, and Little Cayman. *Atoll Res Bull* 388: 1-30.

Guzman HM (ed) (1998). Marine-terrestrial flora and fauna of Cayos Cochinos Archipelago, Honduras: The Smithsonian Tropical Research Institute 1995-1997 project. *Revista de Biología Tropical* 46 (suppl.1).

Jameson SC, Trott LB, Marshall MJ, Childress MJ (2000). Nicaragua in the Caribbean Sea Large Marine Ecosystem. In: Sheppard C (ed). *Seas at the Millennium: An Environmental Evaluation.* Elsevier Science Ltd, Oxford, UK.

Ryan JD, Miller LJ, Zapata Y, Downs O, Chan R (1998). Great Corn Island, Nicaragua. In: UNESCO. *CARICOMP – Caribbean Coral Reef, Seagrass and Mangrove Sites.* Coastal Region and Small Island Papers 3, UNESCO, Paris, France. 95-105.

Tortora LR, Keith DE (1980). Scleractinian corals of the Swan Islands, Honduras. *Carib J Sci* 16(1-4): 65-72.

COSTA RICA AND PANAMA

Clifton KE, Kim K, Wulff JL (1997). A field guide to the reefs of Caribbean Panamá with an emphasis on western San Blas. *Proc 8th Int Coral Reef Symp* 1: 167-184.

Cortés J (1990). The coral reefs of Golfo Dulce, Costa Rica. Distribution and community structure. *Atoll Res Bull* 344: 1-37.

Cortés J (1997). Biology and geology of eastern Pacific coral reefs. *Proc 8th Int Coral Reef Symp* 1: 57-64.

Cortés J, Guzman H (1998). Organisms of Costa Rican coral reefs: description, geographic distribution and natural history of Pacific zooxanthellate corals (Anthozoa: Scleractinia). *Revista de Biología Tropical* 46(1): 55-92.

Glynn PW, Maté JL (1997). Field Guide to the Pacific coral reefs of Panamá. *Proc 8th Int Coral Reef Symp* 1: 145-166.

Guzmán HM, Cortés J (1992). Cocos Island (Pacific Costa Rica) coral reefs after the 1982-83 El Niño disturbance. *Revista de Biología Tropical* 40: 309-324.

Guzmán HM, Jimenez CE (1992). Contamination of coral reefs by heavy-metals along the Caribbean coast of Central-America (Costa Rica and Panama). *Mar Poll Bull* 24(11): 554-561.

Guzmán HM, Burns KA, Jackson JBC (1994). Injury, regeneration and growth of Caribbean reef corals after a major oil spill in Panama. *Mar Ecol Prog Ser* 105(3): 231-241.

Guzmán HM, Holst I (1994). Biological inventory and present status of coral reef at both ends of the Panama Canal. *Revista de Biología Tropical* 42(3): 493-514.

Guzmán HM, Guevara C (1998). Arrecifes coralinos de Bocas del Toro, Panamá: II. Distribución, estructura y estado de conservación de los arrecifes de las islas Bastimentos, Solarte, Carenero y Colón. *Revista de Biología Tropical* 46: 893-916.

Nunez JC (1992). The coral reefs of the Gandoca-Manzanillo National Wildlife Refuge, Limon, Costa-Rica. *Revista de Biología Tropical* 40(3): 325-333.

Ogden JC, Ogden NB (1994). The coral reefs of the San Blas Islands: revisited after 20 years. In: Ginsburg RN (ed). *Proceedings of the Colloquium on Global Aspects of Coral Reefs: Health, hazards and history, 1993.* Rosenstiel School of Marine and Atmospheric Science, Miami, USA. 267-272.

Shulman MJ, Robertson DR (1996). Changes in the coral reefs of San-Blas, Caribbean Panama – 1983 to 1990. *Coral Reefs* 15(4): 231-236.

COLOMBIA AND ECUADOR

Acosta A (1994). Contamination gradient and its effect on the coral community structure in the Santa Marta Area, Colombian Caribbean. In: Ginsburg RN (ed). *Proceedings of the Colloquium on Global Aspects of Coral Reefs: Health, Hazards and History, 1993.* University of Miami, Miami, Florida, USA. 233-239.

Díaz JM (ed) (2000). *Areas Coralinas de Colombia.* Instituto de Investigaciones Marinas y Costeras (INVEMAR), Santa Marta, Colombia.

Díaz JM, Garzón-Ferreira J, Zea S (1995a). *Los Arrecifes Coralinos de la Isla de San Andrés, Colombia: Estado actual y perspecitvas para su conservación.* Academia Colombiana de Ciencias Exactas, Físicas y Naturales. Santafé de Bogatá.

Díaz JM, Sánchez JA, Zea S, Garzón-Ferreira J (1995b). Morphology and marine habitats of two southwestern Caribbean atolls: Albuquerque and Courtown. *Atoll Res Bull* 435: 1-33.

Garzon-Ferreira J, Kielman M (1994). Extensive mortality of corals in the Colombian Caribbean during the past two decades. In: Ginsburg RN (ed). *Proceedings of the Colloquium on Global Aspects of Coral Reefs: Health, Hazards and History, 1993.* University of Miami, Miami, Florida, USA. 247-253.

Glynn PW, Colley SB et al (1994). Reef coral reproduction in the Eastern Pacific – Costa Rica, Panama, and Galapagos-Islands (Ecuador), 2. Poritidae. *Mar Biol* 118(2): 191-208.

Zea S, Geister J, Garzon-Ferreira J, Diaz JM (1999). Biotic changes in the reef complex of San Andres Island (southwestern Caribbean Sea, Colombia) occurring over nearly three decades. *Atoll Res Bull*: 1-16.

CUBA

Alcolado PM, Herrera-Moreno A, Martínez-Estalella N (1994). Sessile communities as environmental bio-monitors in Cuban coral reefs. In: Ginsburg RN (ed). *Proceedings of the Colloquium on Global Aspects of Coral Reefs: Health, Hazards and History, 1993.* University of Miami, Miami, Florida, USA. 27-33.

Alcolado PM, Claro R, Menendez G, Martinez-Daranas B (1997). General status of Cuban coral reefs. *Proc 8th Int Coral Reef Symp*: 341-344.

Williams D (1999). *Diving and Snorkelling Cuba.* Lonely Planet Publications, Hawthorn, Australia.

JAMAICA

Bruckner AW, Bruckner RJ et al (1997). Spread of a black-band disease epizootic through the coral reef system in St Ann's Bay, Jamaica. *Bull Mar Sci* 61(3): 919-928.

Goreau TF (1959). The ecology of Jamaican coral reefs. I. Species composition and zonation. *Ecology* 40: 67-90.

Hughes TP (1994). Catastrophes, phase-shifts, and large-scale degradation of a Caribbean coral reef. *Science* 265(5178): 1547-1551.

Sary Z, Oxenford HA et al (1997). Effects of an increase in trap mesh size on an overexploited coral reef fishery at Discovery Bay, Jamaica. *Mar Ecol Prog Ser* 154: 107-120.

THE CAYMAN ISLANDS

Blanchon P, Jones B et al (1997). Anatomy of a fringing reef around Grand Cayman: storm rubble, not coral framework. *J Sedimentary Res* 67(1 PtA): 1-16.

Ghiold J, Smith SH (1990). Bleaching and recovery of deep-water, reef-dwelling invertebrates in the Cayman Islands, BWI. *Carib J Sci* 26(1-2): 52-61.

Jones B, Hunter KC, Hunter NG, Hunter IG (1997). Geology and hydrogeology of the Cayman Islands. In: Vacher HL, Quinn T (eds). *Developments in Sedimentology, 54: Geology and Hydrology of Carbonate Islands.* Elsevier Science BV, Amsterdam, Netherlands.

Map sources

Maps 5a and 5b

For the Yucatan Peninsula, reefs and coastline have been combined from Hydrographic Office (1995) and Jordán-Dahlgren (1993). Sources for the former include various finer resolution charts and surveys from 1820-47 and 1980-1989, while the latter is based on multiple high resolution sources, combined with expert knowledge, transferred onto maps at a scale of 1:250 000 base maps for the mainland coast, but not offshore reefs. Further reefs have been added, largely for the Gulf of Mexico, using maps at various scales presented in Bezaury-Creel et al (1997). Additional point data for small named reefs in the eastern Bahía de Campeche and the Campeche Bank have been added using geographic co-ordinates from the same report. The reefs off Cozumel island are approximate, and are based on a tourist map of the island. Data for the Pacific Coast are largely taken from UNEP/IUCN (1988a)*, which was prepared at a scale of 1:10 000 000.

Bezaury-Creel J, Macias Ordoñez R, García Beltrán G, Castillo Arenas G, Pardo Caicedo N, Ibarra Navarro R, Loreto Viruel A (1997). Implementation of the International Coral Reef Initiative (ICRI) in Mexico. Commission for Environmental Cooperation (CEC). In: *The International Coral Reef Initiative – The Status of Coral Reefs in Mexico and the United States Gulf of Mexico.* Amigos de Sian Ka'an AC, CINVESTAV, NOAA, CEC, and The Nature Conservancy. http://benthos.cox.miami.edu/mexico/icri/home.htm

Hydrographic Office (1995). Gulf of Honduras and Yucatan Channel. *British Admiralty Chart No. 1220.* 1:1 000 000. May 1995. Taunton, UK.

Jordán-Dahlgren E (1993). *Atlas de los Arrecifes Coralinos del Caribe Mexicano. Parte 1. El Sistema Continental.* CIQROO-UNAM.

Map 5c

Coral reef and mangrove data are taken from Gibson et al (1993), data originally prepared at the World Conservation Monitoring Centre. Some further detail of reefs and coastline has been appended from Hydrographic Office (1989a, 1989b, and 1989c).

Gibson JP, Price ARG, Young E (1993). *Guidelines for Developing a Coastal Zone Management Plan for Belize: The GIS Database.* A Marine Conservation and Development Report. IUCN, Gland, Switzerland.

Hydrographic Office (1989a). Belize, Colson Point to Belize City, including Lighthouse Reef and Turneffe Islands. *British Admiralty Chart No. 959.* 1:125 000. October 1989. Taunton, UK.

Hydrographic Office (1989b). Gulf of Honduras. *British Admiralty Chart No. 1573.* 1:125 000. October 1989. Taunton, UK.

Hydrographic Office (1989c). Belize, Monkey River to Colson Point. *British Admiralty Chart No. 1797.* 1:125 000. October 1989. Taunton, UK.

Hydrographic Office (1995). Gulf of Honduras and Yucatan Channel. *British Admiralty Chart No. 1220.* 1:1 000 000. May 1995. Taunton, UK.

Map 5d

For the Bay Islands coral reef areas were estimated from 1:150 000 prints of Landsat-5 TM images (Path/Row 17/49, 15/4/94. Bands 2, 3, 4), from 1994. There was no ground-truthing on this work.

Offshore reefs and islands around the Cayos Miskitos have been prepared from Hydrographic Office (1964). Most of data from this chart actually date from a hydrographic survey of 1830-43, with additions from a US Government chart of 1927. Although this chart does not feature large areas of reef directly, reefs have been broadly interpreted from those few features which are marked as reefs and from shallow submerged rocks where these occur in areas of active reef development. Further reef data are take from from Petroconsultants SA (1990)*, and from UNEP/IUCN (1988a)*.

Hydrographic Office (1964). River Hueson to False Cape, including Morrison and Mosquito Cays. *British Admiralty Chart No. 2425.* August 1929 (minor corrections to 1964). Taunton, UK.

Map 5e

Coral reefs have been prepared for Costa Rica from IGN (various dates). For Panama, Caribbean reefs have been taken from UNEP/IUCN (1988a)* at an approximate scale of 1:1 600 000, while Pacific reefs have been gathered at various scales from Glynn and Maté (1997).

Glynn PW, Maté JL (1997). Field guide to the Pacific coral reefs of Panamá. *Proc 8th Int Coral Reef Symp* 1: 145-166.

IGN (various dates). *Costa Rica. 1:200 000, 9-map series.* Instituto Geográfico Nacional, San José, Costa Rica.

Map 5f

Locations of reefs for the north coast of Colombia are based on Hydrographic Office (1990, 1991a, 1991b). Sources for these are US Government charts of 1938, 1977, 1986 and 1987, most of which are largely based on earlier surveys (1935-38) with corrections from the 1970s and 1980s. For the offshore Isla de San Andrés, reefs have been estimated from an original map in Díaz et al (1995a). Similar maps, presented in Díaz et al (1995b) were utilized to estimate reef areas for Courtown and Albuquerque atolls. All three maps give detailed habitat summaries and reef areas have been interpreted as all areas dominated by hermatypic corals. For Isla Providencia reefs and coastline were estimated from Hydrographic Office (1912). This chart does not mark reefs, but reef areas have been estimated from reef features marked as very shallow waters and submerged rocks which clearly demarcate a reef structure. For the Galápagos, locations of coral reefs have been taken from Glynn and Wellington (1983). These are mostly very small structures.

Díaz JM, Garzón-Ferreira J, Zea S (1995a). *Los Arrecifes Coralinos de la Isla de San Andrés, Colombia: Estado actual y perspecitvas para su conservación.* Academia Colombiana de Ciencias Exactas, Físicas y Naturales. Santafé de Bogatá.

Díaz JM, Sánchez JA, Zea S, Garzón-Ferreira J (1995b). Morphology and marine habitats of two southwestern caribbean atolls: Albuquerque and Courtown. *Atoll Res Bull* 435: 1-33.

Glynn PW, Wellington GM (1983). *Corals and Coral Reefs of the Galápagos Islands.* University of California Press, Berkeley, USA.

Hydrographic Office (1912). Old Providence Island and Coral Bank. *British Admiralty Chart No. 1334.* 1:55 000. June 1912 (minor corrections to 1960). Taunton, UK.

Hydrographic Office (1990). Colombia – North Coast: Isla Fuerte to Cabo Tiburón including Golfo de Urabá. *British Admiralty Chart No. 1278.* 1:200 000. September 1990. Taunton, UK.

Hydrographic Office (1991a). Colombia – North Coast: Bahía Santa Marta to Punta Canoas. *British Admiralty Chart No. 1276.* 1:200 000. March 1991. Taunton, UK.

Hydrographic Office (1991b). Colombia – North Coast: Punta Canoas to Isla Fuerte. *British Admiralty Chart No. 1277.* 1:200 000. March 1991. Taunton, UK.

Map 5g

Coral reef areas are based on Petroconsultants SA (1990)*.

Map 5h

Coral reef and mangrove data were kindly provided by Tommy Lindell. They are largely based on four Landsat TM images, from 1985 and 1995. These have been extensively checked against topographic charts, nautical charts, aerial photographs and ground-truthing work. For the present map, coral reefs are taken from the layers described as "corals" and "coral reefs". In some areas it was difficult to differentiate corals from vegetation. These areas have been omitted. Further details about this work are provided in Lindell (1997, 1999).

Lindell T (1997). Mapping of the coastal zone of Jamaica. *Proc Fourth International Conference on Remote Sensing for Marine and Coastal Environments, Orlando, Florida, 17-19 March, 1997.*

Lindell T (1999). Coastal zone mapping of Jamaica for planning and management. *Proc Pecora 14/Land Satellite Information III, Dec 1999, Denver CO, USA.*

Map 5i

For Cayman Brac and Little Cayman reef data are taken from Logan (1983). These maps are based on aerial photographs flown in 1958, 1971, 1977 and field studies in 1981 and 1983. For Grand Cayman coral reefs are taken from DOS (1978a, 1978b).

DOS (1978a). *Cayman Islands 1:25 000: Grand Cayman.* Series E821 (DOS 328), Sheet 1, 2nd edn-DOS 1978. Directorate of Overseas Surveys, UK and Survey Department, Cayman Islands.

DOS (1978b). *Cayman Islands 1:25 000: Grand Cayman.* Series E821 (DOS 328), Sheet 2, 3rd edn-DOS 1978. Directorate of Overseas Surveys, UK and Survey Department, Cayman Islands.

Logan A (1983). *Shallow Marine Substrates of the Lesser Caymans,* BWI Monochrome maps at 1:12 500, prepared by A Logan, Department of Geology, University of New Brunswick, Canada.

* See Technical notes, page 401

Chapter 6
Eastern Caribbean and Atlantic

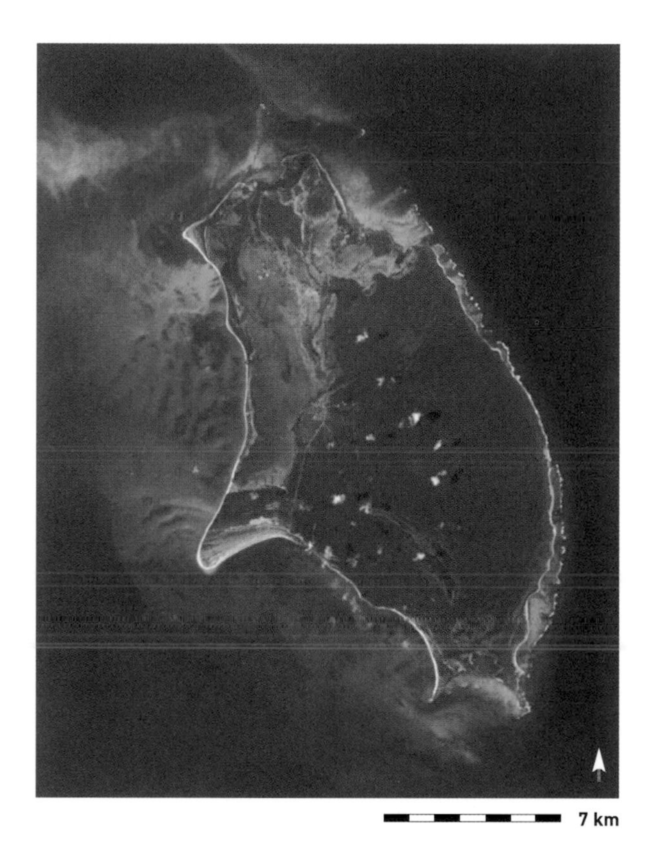

The Eastern Caribbean region is dominated by small islands, lying in a broad arc around the eastern end of the Caribbean Sea. Fringing coral reef communities are found in places on the shores of most islands, although their condition varies considerably. The region also includes the long coastline of Venezuela and, although there is virtually no reef development along this shore, there are important reefs associated with the offshore chain of islands immediately to the north.

As with other areas of the Caribbean, the reefs of the Eastern Caribbean have suffered considerably from the combined impacts of the *Diadema* die-off and coral disease. Many of the more northerly islands have also been swept by major hurricanes in recent years, greatly reducing coral cover over wide areas.

Tourism is the largest industry in this region and vast numbers of visitors have driven rapid and often poorly planned coastal development, with associated problems of sedimentation and pollution. Patterns of fishing, and of overfishing, vary considerably between islands. The region also contains some important protected areas, notably those off Saba, Bonaire and St. Lucia, which have been particularly well managed, leading to the maintenance or recovery of healthy reef ecosystems in localized areas. These provide a model for reef management throughout the region.

Beyond the waters of the Caribbean there are some considerable reefs and coral communities along the coastline of Brazil. Although still poorly known, these reefs are receiving increasing attention and house important and unusual communities, with a high proportion of endemic species. In the few scattered islands of the Central Atlantic and along the less turbid areas of the West African coastline some coral communities are also recorded, although there is no significant development of large reef structures.

Left: The island of Barbuda, Lesser Antilles has extensive fringing reefs (STS026-35-11, 1988). Right: The butter hamlet Hypoplectrus unicolor *is a distinctive Caribbean species, with a range of highly distinctive color morphs. This one is the barred variety var.* puella.

MAP 6

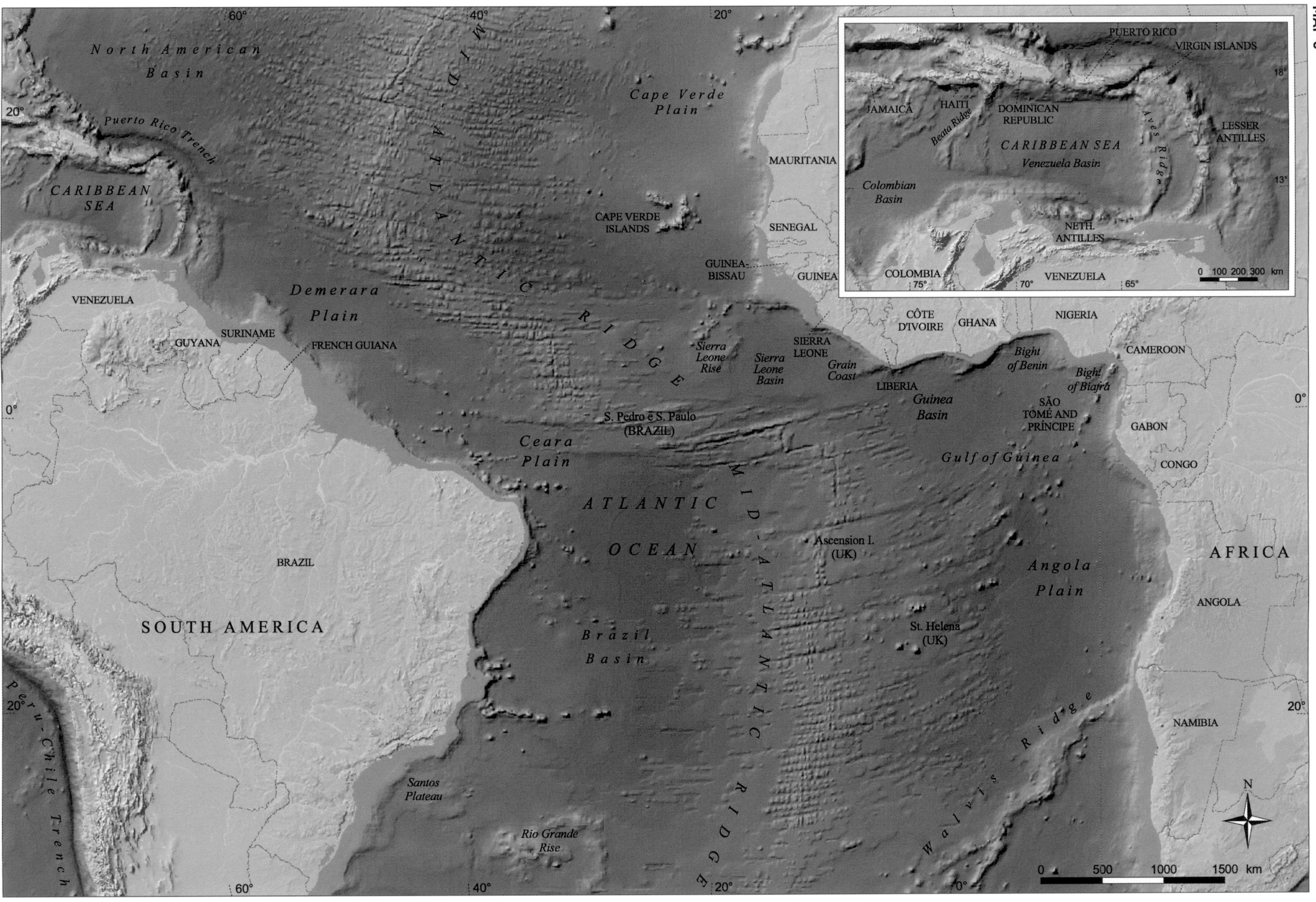

North American Basin
Puerto Rico Trench
CARIBBEAN SEA
MID-ATLANTIC RIDGE
Cape Verde Plain
CAPE VERDE ISLANDS
MAURITANIA
SENEGAL
GUINEA-BISSAU
GUINEA
VENEZUELA
Demerara Plain
GUYANA
SURINAME
FRENCH GUIANA
Sierra Leone Rise
Sierra Leone Basin
SIERRA LEONE
Grain Coast
CÔTE D'IVOIRE
GHANA
NIGERIA
LIBERIA
Bight of Benin
Bight of Biafra
CAMEROON
Guinea Basin
SÃO TOMÉ AND PRÍNCIPE
GABON
S. Pedro e S. Paulo (BRAZIL)
Ceara Plain
Gulf of Guinea
CONGO
ATLANTIC OCEAN
Ascension I. (UK)
AFRICA
BRAZIL
Angola Plain
ANGOLA
SOUTH AMERICA
Brazil Basin
St. Helena (UK)
NAMIBIA
Walvis Ridge
Peru-Chile Trench
Santos Plateau
Rio Grande Rise
PUERTO RICO
VIRGIN ISLANDS
JAMAICA
HAITI
DOMINICAN REPUBLIC
Beata Ridge
CARIBBEAN SEA
Venezuela Basin
Aves Ridge
LESSER ANTILLES
Colombian Basin
NETH. ANTILLES
COLOMBIA
VENEZUELA
0 100 200 300 km
0 500 1000 1500 km
N
0°
20°
40°
60°
18°
13°
65°
70°
75°

Haiti, the Dominican Republic and Navassa Island

MAP 6a

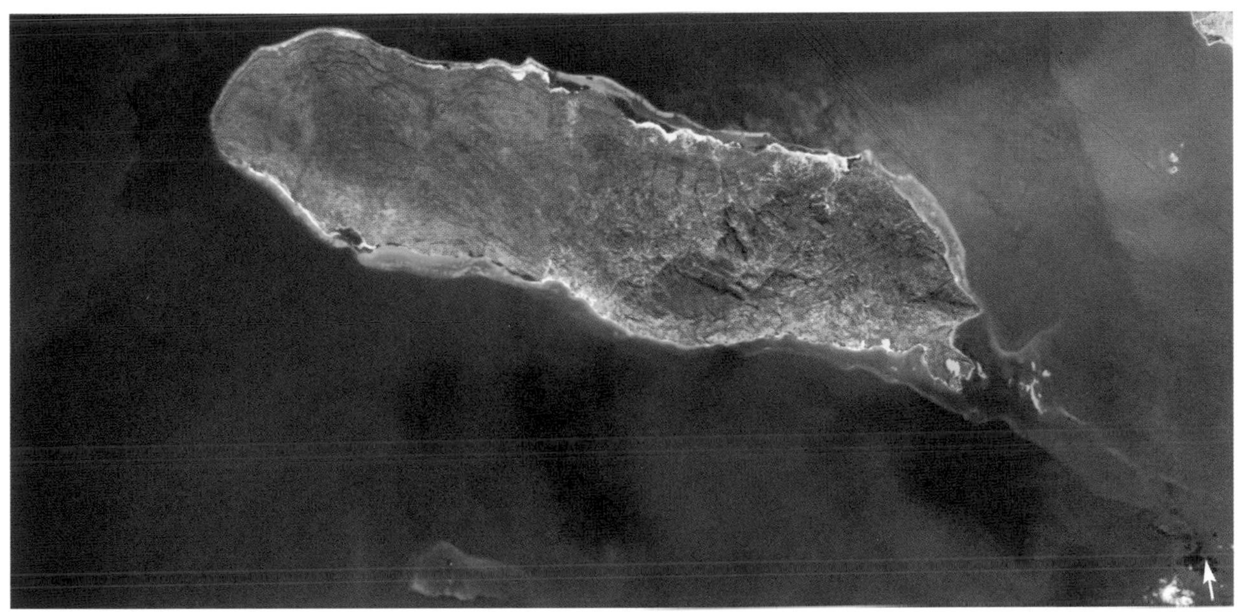

Haiti makes up the western part of the island of Hispaniola, the second largest island in the Caribbean. It is a mountainous country with a central plain enclosed by mountain ranges to the north and south which extend out into two long peninsulas enclosing the Golfe de la Gonâve. Offshore there are a number of islands, including the large Île de la Tortue in the north and the central Île de la Gonâve. Very little indeed is known about the coral reefs in Haiti. What information exists is largely for the area round the capital Port-au-Prince and Les Arcadins islands. Coral reefs are also known to occur all around Île de la Gonâve; on the Rochelois Bank and Les Îles Cayemites; around Île à Vache on the south coast; and also on the north coast between Cap Haitien and the border with the Dominican Republic. Marine benthic surveys were carried out in Les Arcadins in the 1980s. The reef profile was found to be similar to other Caribbean fringing reefs: a reef crest dominated by *Millepora complanata*, an *Acropora palmata* zone (with 100 percent live cover in 1989) and a shallow fore reef dominated by *Montastrea annularis*. Extensive seagrass beds occur here – shallow beds (2-4 meters) generally have more algae species with a higher biomass, whereas the deeper beds (12-14 meters) are more sparsely populated with algae. In total, 35 species of scleractinian coral have been recorded in Les Arcadins, as well as 12 gorgonians and 54 species of sponges. There are two unusual aspects to these reefs. Firstly the soft coral *Icillogorgia* spp., which normally occurs in cryptic habitats or deeper water elsewhere, is common in the open shallow waters of Les Arcadins. Secondly, the sponge *Niphates digitalis*, which rarely exhibits gigantism in shallow water elsewhere in the Caribbean, does so here.

Haiti is the most impoverished country in the western hemisphere, so the reef resources, although poorly documented, are likely to be under intense pressure. Perhaps less than 1 percent of the native terrestrial vegetation remains intact. The steep relief and high rainfall have caused widespread and severe soil erosion, which is likely to have placed a high sediment load on coastal reefs. About 75 percent of the population live in poverty and so almost all fishing activities are carried out at the subsistence level, and anecdotal reports suggest that this is so intense that few fish reach reproductive size. There are no sewage treatment plants or sanitary landfills and the levels of nutrients flowing from settlements into coastal waters are probably high. This has been linked to an abundance of fleshy algae in the reefs off Les Irois and the

The Île de la Gonâve, in Haiti, has a number of important reefs (STS060-84-56, 1994).

Haiti

General Data	
Population (thousands)	6 868
GDP (million US$)	2 183
Land area (km2)	27 156
Marine area (thousand km2)	127
Per capita fish consumption (kg/year)	3
Status and Threats	
Reefs at risk (%)	100
Recorded coral diseases	0
Biodiversity	
Reef area (km2)	450
Coral diversity	na / 57
Mangrove area (km2)	134
No. of mangrove species	na
No. of seagrass species	na

Baie de Port-au-Prince. This bay is also severely polluted with oil, industrial chemicals and solid waste. Although a monitoring station has been established there by the Fondation pour la Protection de la Biodiversité Marine, it is yet to implement a full monitoring protocol. There are currently no marine protected areas in Haiti, although a park including Les Arcadins was proposed in 1989.

Dominican Republic

The Dominican Republic makes up the larger, eastern part of the island of Hispaniola. Like Haiti, it too is mountainous, with considerable riverine runoff. Fringing reefs and small barrier reefs are scattered along some 170 kilometers of its coastline, and there are important reef communities on the offshore banks of Navidad and La Plata (Silver Bank), which lie to the north of the country. The best developed reefs include a small barrier reef in Montecristi in the northwest, some narrow fringing communities along the central north coast and a further barrier-type development in the far east. Reef development is less extensive on the south coast, but there is some in the east on the mainland and the neighboring Isla Saona. Around Santo Domingo there are small reefs on narrow platforms, while there are also some in the far south of the country around Jaragua National Park.

Coral cover has declined considerably in most near-shore areas, and algae have proliferated at the expense of reef corals at many localities. High coral cover is now largely restricted to deeper reefs, and to those lying further offshore. In 2000, average coral cover was 35 percent on the Montecristi barrier reef and 40 percent on parts of the offshore banks. In the Del Este National Park the diversity of the main reef groups is high, with 22 octocorals, 26 scleractinians and 36 sponges on the shallow spur and groove formations. Here most of the algae are calcareous, although *Dictyota* is also abundant. Reef flat communities occur on low-relief consolidated carbonate platforms which are exposed to strong wave action from the Mona Passage. Algae (36 species) provide the dominant cover (over 70 percent in some cases), but 14 species of corals also occur, principally *Acropora palmata*, *Diploria clivosa*, *Porites asteroides* and *P. porites*. Closer to shore, patch reefs occur in amongst seagrass beds. Algae again are dominant, some 21 species accounting for more than 50 percent of the benthic cover. Information on the status of coral reefs in the Jaragua National Park is more scarce, but this park protects many different coastal ecosystems. There are large and regionally important populations of manatees, crocodiles, turtles (leatherback, green, hawksbill and loggerhead turtles all nest there) and flamingos.

Many of the reefs in the north and around Santo Domingo have been severely affected by a variety of human impacts. Degradation is probably due to increases

Above: The whitespotted filefish, Cantherhines macrocerus, *somewhat confusingly has an orange color phase without white spots. Below: A classic Caribbean reef scene, with a massive brain coral alongside soft corals.*

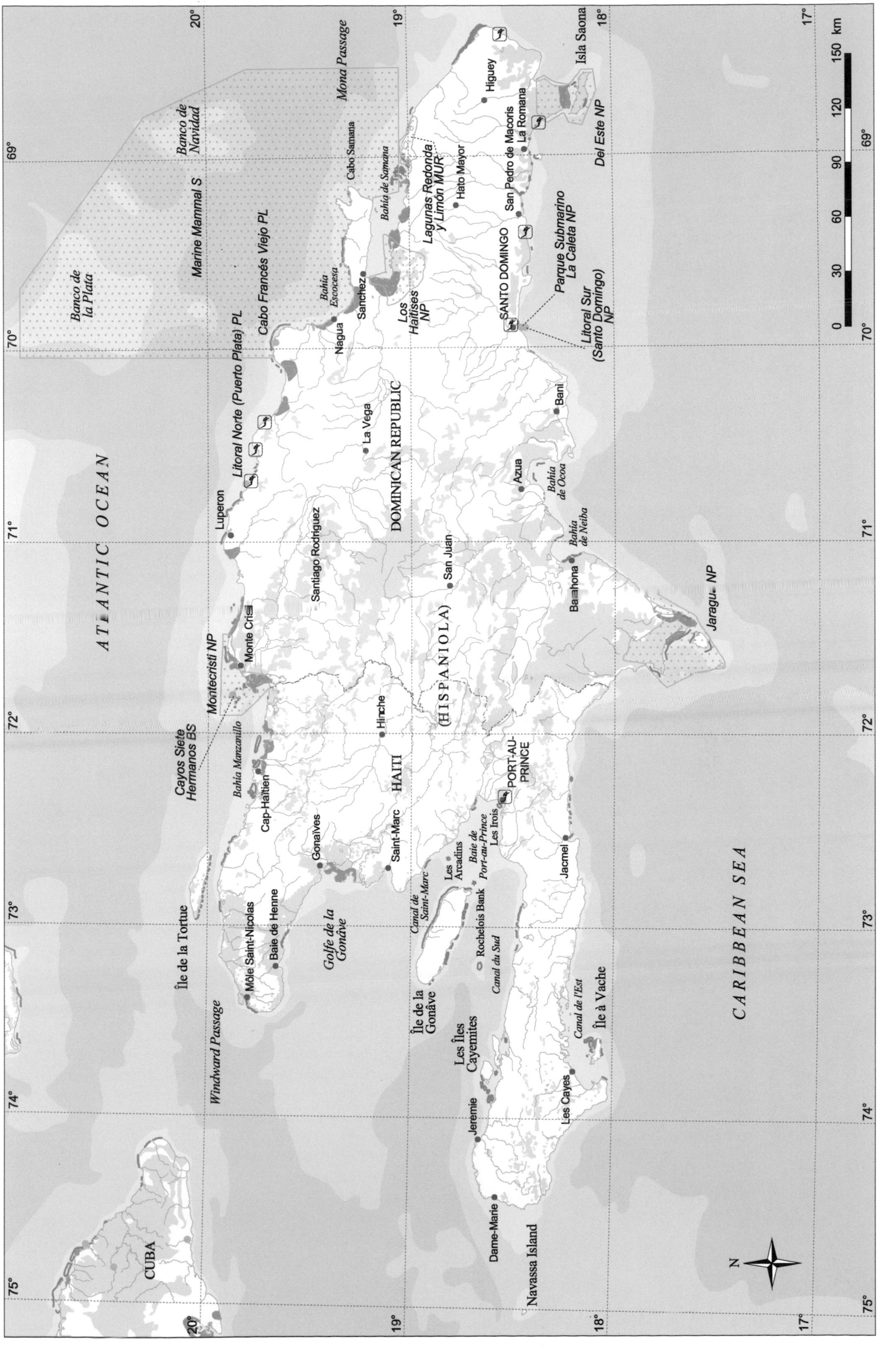

AT ANTIC OCEAN
CARIBBEAN SEA
CUBA
Île de la Tortue
Windward Passage
Cayos Siete Hermanos BS
Montecristi NP
Bahía Manzanillo
Môle Saint-Nicolas
Baie de Henne
Cap-Haïtien
Monte Cri
Luperon
Litoral Norte (Puerto Plata) PL
Cabo Francés Viejo PL
Banco de la Plata
Marine Mammal S
Banco de Navidad
Mona Passage
Gonaïves
Golfe de la Gonâve
Santiago Rodriguez
Bahía Escocesa
Nagua
Cabo Samana
Sanchez
La Vega
Hinche
Saint-Marc
HAITI
DOMINICAN REPUBLIC
Bahía de Samana
Los Haitises NP
Île de la Gonâve
Canal de Saint-Marc
Les Arcadins
(HISPANIOLA)
San Juan
Lagunas Redonda y Limón MUR
Hato Mayor
Les Îles Cayemites
Jeremie
Dame-Marie
Rochelois Bank
Baie de Port-au-Prince
Canal du Sud
Les Irois
PORT-AU-PRINCE
SANTO DOMINGO
San Pedro de Macoris
La Romana
Higuey
Navassa Island
Azua
Bani
Bahía de Ocoa
Parque Submarino La Caleta NP
Les Cayes
Canal de l'Est
Île à Vache
Jacmel
Ba hona
Bahía de Neiba
Litoral Sur (Santo Domingo) NP
Isla Saona
Del Este NP
Jaragu NP
N
75°
74°
73°
72°
71°
70°
69°
20°
19°
18°
17°
0 30 60 90 120 150 km

Protected areas with coral reefs

Site name	Designation	Abbreviation	IUCN cat.	Size (km²)	Year
Dominican Republic					
Del Este	National Park	NP	II	808.00	1975
Marine Mammal	Sanctuary	S	na	38 000.00	1996
Jaragua	National Park	NP	II	1 374.00	1983
Litoral Sur (Santo Domingo)	National Park	NP	Unassigned	10.75	1968
Montecristi	National Park	NP	II	1 309.50	1983
Parque Submarino La Caleta	National Park	NP	II	10.10	1986

in sediments (from upland deforestation, wetland removal, soil erosion and coastal construction for the tourist industry), nutrients (from fertilizers as well as domestic wastewater) and pesticides (in agricultural runoff). Coral disease and the *Diadema* die-off have undoubtedly exacerbated the effects of these direct human impacts. Reefs in the southeast and southwest have generally suffered less. There is an important artisanal fishery in the country, with catches estimated at 13 000 tons in 1998. Overfishing, particularly of conch and lobster, is a problem, although there may now be a reduction in pressure as some fishers are turning to work in tourism, while the use of fish aggregating devices is directing the focus of a number of fishers towards pelagic fisheries.

Approximately 20 percent of the coral reefs in the Dominican Republic occur within marine parks and sanctuaries, and the majority of these occur in the Jaragua and Del Este National Parks. Both cover large areas and are far from the centers of human activity. Management levels in the parks as a whole are limited and many continue to be heavily fished. The parks near Haiti are regularly and heavily poached by vessels from that country.

The offshore banks of Navidad and La Plata have significant areas of reef where coral cover remains high. These are also important breeding grounds for the largest population of humpback whales in the region, some 3 000 individuals, and support an important whale-watching industry in Samana Bay. They are at the center of the large Marine Mammal Sanctuary, which incorporates both banks and extends to include the northeastern coast between Cabo Samana and Cabo Francés.

Dominican Republic

General Data	
Population (thousands)	8 443
GDP (million US$)	9 945
Land area (km²)	48 444
Marine area (thousand km²)	261
Per capita fish consumption (kg/year)	12
Status and Threats	
Reefs at risk (%)	89
Recorded coral diseases	2
Biodiversity	
Reef area (km²)	610
Coral diversity	na / 57
Mangrove area (km²)	325
No. of mangrove species	6
No. of seagrass species	4

Navassa Island

Navassa Island was claimed as a US territory in 1857 for the exploitation of guano resources, but has no permanent inhabitants. An uplifted limestone structure of around 5 square kilometers, it is located in the Jamaica Passage between Jamaica and Haiti, some 50 kilometers from Haiti's Cap Dame-Marie. There are important coral reef communities on all sides, with live coral cover of 20-25 percent on the leeward (west) coast. Some 36 hard coral species have been recorded to date. The reefs were also observed to have considerable structural complexity, with high levels of coral recruitment. Incidence of coral disease was low, and reasonable numbers of *Diadema* were reported. Although the surrounding waters have yet to be documented, it has been suggested that significant coral communities may also occur on shallow seamounts nearby. There are few human impacts on these reefs, although there is some artisanal fishing by Haitian fishers. The island is now controlled by the US Department of the Interior and there are strict controls on access and entry.

Puerto Rico and the Virgin Islands

MAP 6b

The US Commonwealth of Puerto Rico is a large mountainous island on the northern edge of the Caribbean Sea, lying to the east of the Dominican Republic. Mona Island is an uplifted carbonate island to the west of the main island, while to the east there are two other significant islands, those of Culebra and Vieques. The Virgin Islands form an archipelago of about 100 islands to the east of Puerto Rico. The most western islands are an "unincorporated territory" of the USA, the US Virgin Islands, while those in the east are an overseas territory of the UK, the British Virgin Islands. The majority of these islands are on a single shallow platform which is an extension of the Puerto Rico shelf. St. Croix to the south lies on the separate Cruzan platform, separated by the 4 500 meter deep Virgin Islands Trough. To the northeast this forms the deep, narrow Anegada Passage which separates the Virgin Islands from the Lesser Antilles. North of the Virgin Islands and Puerto Rico, the seismically active Puerto Rico Trench constitutes the northern boundary of the Caribbean tectonic plate, which is moving at a rate of 2-4 centimeters per year in an easterly direction with respect to the North American plate. The strike-slip motion of this plate boundary has created a deep trench, forming the deepest point in the Caribbean (9 200 meters) at the Milwaukee Depth, some 150 kilometers north of Mona Island. Puerto Rico and the rest of the Virgin Islands are predominantly volcanic in origin, with the exception of St. Croix and Anegada which, like Mona, were formed by uplifted sedimentary rocks.

The dominant current flow is from east to west, with water movement being driven by the Atlantic Northern Equatorial Current. These islands also lie within the trade wind belt, with wind blowing mainly from the east and southeast during the summer months, and from the east and northeast during the winter. As a result of all these factors, the prevailing swell is from the east, and sediment transport is along the north and south margins.

Puerto Rico

Coral reefs are discontinuous around the main island of Puerto Rico, and most abundant along the east, south and west coasts. The offshore islands are more continuously fringed by reefs. Coral cover is highly varied, and the island includes some of the best developed and most diverse coral reefs in the US Caribbean territories. As elsewhere, coral disease has had a significant impact on the total coral cover. The *Diadema* die-off was also

St. Croix, US Virgin Islands. Wide areas of shallower water can clearly be seen around the island (STS054-74-49, 1993).

MAP 6b

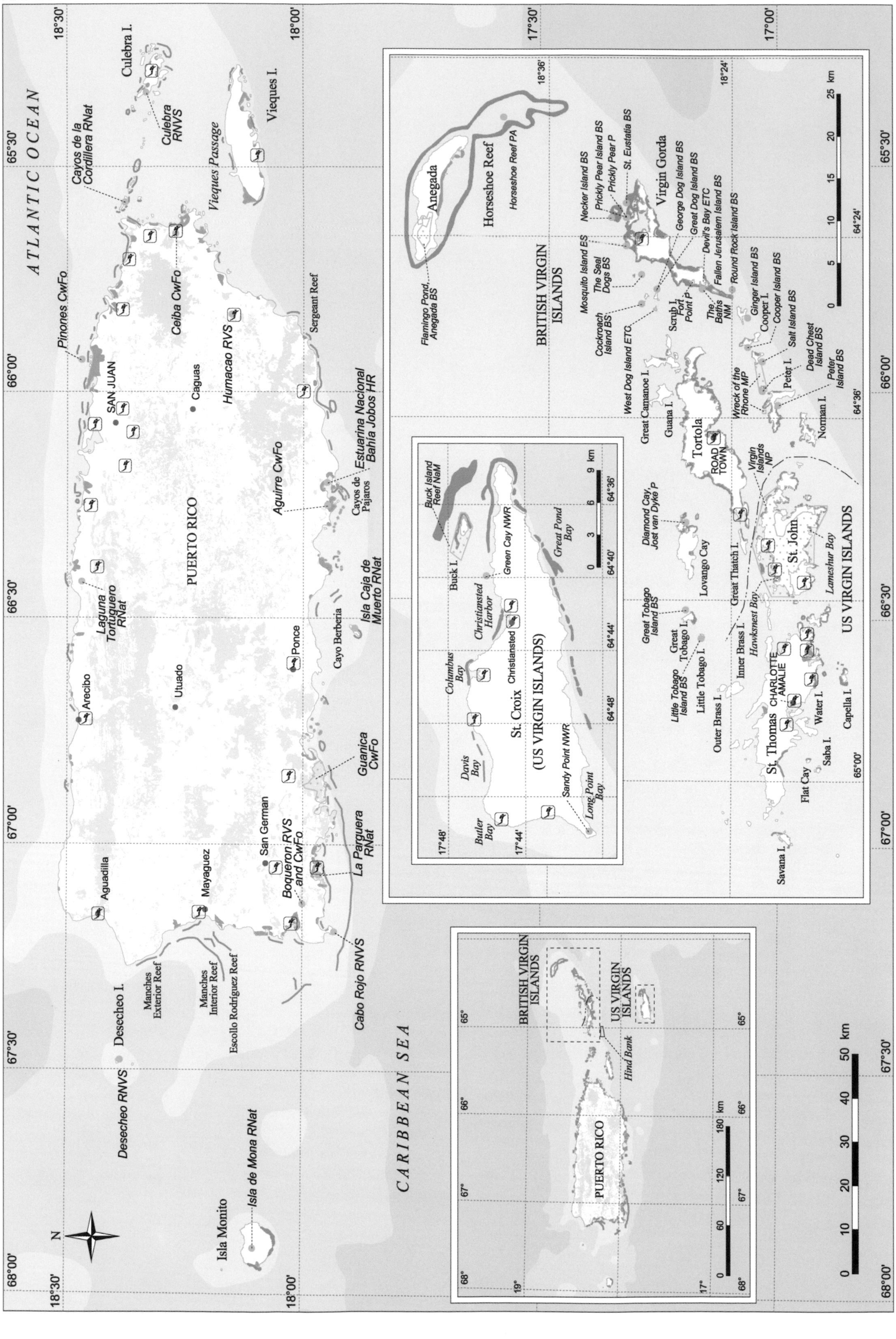

ATLANTIC OCEAN
CARIBBEAN SEA
PUERTO RICO
SAN JUAN
Aguadilla
Arecibo
Laguna Tortuguero RNat
Pinones CwFo
Cayos de la Cordillera RNat
Culebra I.
Culebra RNVS
Ceiba CwFo
Vieques Passage
Vieques I.
Humacao RVS
Caguas
Utuado
Mayaguez
San German
Ponce
Desecheo RNVS
Desecheo I.
Manches Exterior Reef
Manches Interior Reef
Escollo Rodríguez Reef
Isla Monito
Isla de Mona RNat
Boqueron RVS and CwFo
Cabo Rojo RNVS
La Parguera RNat
Guanica CwFo
Cayo Berberia
Isla Caja de Muerto RNat
Aguirre CwFo
Cayos de Pajaros
Estuarina Nacional Bahía Jobos HR
Sergeant Reef
68°00' 67°30' 67°00' 66°30' 66°00' 65°30'
18°30' 18°00' 17°30' 17°00'

PUERTO RICO
BRITISH VIRGIN ISLANDS
US VIRGIN ISLANDS
Hind Bank
68° 67° 66° 65°
19° 17°
0 60 120 180 km

St. Croix
(US VIRGIN ISLANDS)
Buck Island Reef NaM
Buck I.
Columbus Bay
Davis Bay
Butler Bay
Christiansted Harbor
Christiansted
Green Cay NWR
Great Pond Bay
Sandy Point NWR
Long Point Bay
17°48' 17°44'
64°48' 64°44' 64°40' 64°36'
0 3 6 9 km

BRITISH VIRGIN ISLANDS
Anegada
Flamingo Pond, Anegada BS
Horseshoe Reef
Horseshoe Reef PA
Mosquito Island BS
Necker Island BS
Prickly Pear Island BS
Prickly Pear P
St. Eustatia BS
Cockroach Island BS
The Seal Dogs BS
West Dog Island ETC
Virgin Gorda
George Dog Island BS
Great Dog Island BS
Devil's Bay ETC
Fallen Jerusalem Island BS
Round Rock Island BS
Scrub I.
Fort Point P
The Baths NM
Great Camanoe I.
Guana I.
Tortola
ROAD TOWN
Great Tobago Island BS
Diamond Cay, Jost van Dyke P
Little Tobago Island BS
Great Tobago I.
Little Tobago I.
Lovango Cay
Outer Brass I.
Inner Brass I.
Great Thatch I.
Hawksnest Bay
Virgin Islands NP
Wreck of the Rhone MP
Ginger Island BS
Cooper I.
Cooper Island BS
Salt Island BS
Peter I.
Dead Chest Island BS
Peter Island BS
Norman I.
St. Thomas
CHARLOTTE AMALIE
St. John
Savana I.
Flat Cay
Saba I.
Water I.
Capella I.
Lameshur Bay
US VIRGIN ISLANDS
18°36' 18°24'
65°00' 64°36' 64°24'
0 5 10 15 20 25 km

0 10 20 30 40 50 km
68°00' 67°30' 67°00' 66°30' 66°00' 65°30'

considerable here, but numbers are now reported to be increasing. Coral bleaching in the late 1980s caused significant mortality, and a major bleaching event was also observed in 1998, though little associated mortality appears to have occurred.

Puerto Rico has one of the most dynamic economies in the Caribbean and tourism has traditionally been an important source of income, with estimated arrivals of nearly 4 million people in 1993. Construction and tourism were the leading sectors in the economy in 1998, and this has had considerable impacts on the reefs. Clearance of over 75 percent of Puerto Rico's mangroves, combined with dredging, agricultural runoff, pollution from untreated sewage, and sedimentation from forest clearance have had a considerable impact on most coastal reefs. Although there are no big commercial fisheries, small-scale fisheries are significant, with a total catch of over 1 600 tons in 1996. Overfishing of large predators, parrotfishes and spiny lobsters is widely reported. Oil spills have further impacted reefs in some areas. The offshore island of Vieques is used by the US military as a bombing range, resulting in many craters on the reefs measuring some 5-13 meters in diameter. Whether the positive impacts associated with the exclusion of fishers and tourists from these reefs by the military compensates for this destruction is still a matter of some controversy.

Efforts to control some of the more damaging activities and protect some of the reefs from further decline are now underway. A number of marine protected areas have been designated, together with seasonal fishery on some spawning aggregations. New legislation is being developed to begin to address some of the pollution problems of the area.

US Virgin Islands

Coral reefs are widespread around all of the main islands. These are mostly fringing reefs, but there is a small barrier reef off St. Croix, and there are a number of offshore patch reefs and bank structures.

Nowhere else in the Caribbean have the combined effects of hurricanes and disease on coral reef population structure been more pronounced than in the US Virgin Islands. In 1976 live coral cover on the fore reef at Buck Island, dominated by *Acropora palmata*, was 85 percent. Since then, eight hurricanes have caused serious physical damage to these reefs. Hurricane Hugo, in 1989, was undoubtedly the most severe, but in 1995 Hurricanes Luis and Marilyn hit the islands within a ten-day period and caused extensive damage in some areas. Others were less affected, either because of the uneven impacts of the storms themselves or because there was so little coral remaining to be damaged. White band disease has also greatly impacted the region and killed many acroporid corals, with as many as 64 percent of all colonies being affected. Other diseases have also hit less abundant species in the Virgin Islands, such as *Agaricia agaricites* and *Stephanocoenia michelinii*. The situation in St. John is similar, with 80 percent of *Acropora palmata* colonies in Hawksnest Bay being lost in just seven months. Coral cover around St. John was about 30 percent before Hurricane Hugo reduced it to some 8-18 percent. In Lameshur Bay the dominant coral, *Montastrea annularis*, declined by about 35 percent and there has been no substantial recovery even though coral recruitment is occurring. Despite extensive bleaching in 1998, there was little associated mortality.

Tourism is the islands' primary economic activity, accounting for more than 70 percent of gross domestic

Left: The spiny lobster industry is of great economic importance in the Caribbean. In Puerto Rico, as in many areas, there are reports of overharvesting. Right: A grey angelfish Pomacanthus arcuatus.

product and 70 percent of employment. Damage to reefs associated with tourism and recreation includes significant harm caused by boat anchors and ship groundings. The Virgin Islands National Park on St. John attracts a million visitors a year, mostly arriving on cruiseships or smaller boats, and an estimated 30 000 anchors are dropped in a single year. In 1989 the cruiseship *Windspirit* destroyed some 300 square meters of reef with its anchor and chain and there has been little recovery since. This resulted in the successful prosecution of the boat owners by the park authorities, and remains one of the few examples of such action for damages incurred to coral reefs. Direct damage by divers and snorkellers has also been recorded at the most heavily used sites. Mooring buoys were installed following a survey which found that 33 percent of boats anchored in seagrass beds and 14 percent on coral reefs. Unfortunately, there is now little coral left to protect and no limits have been set on the size of vessels allowed in park waters.

Overfishing is widespread throughout the islands, even within protected areas. This is further exacerbated by the widespread loss of fish habitats, including seagrass and mangrove areas, such that fish stocks are highly depleted in most areas.

Other threats to the reefs include sedimentation, land clearance, coastal development and sewage discharge (the eutrophication of some reefs in the Virgin Islands has been attributed to leaching from septic tanks during heavy rain). One of the world's largest petroleum refineries is at St. Croix, which also represents a significant potential threat to reefs as well as other ecosystems. In 1999 a Marine Conservation District was declared to the southwest of St. Thomas, in cooperation with fishers, divers and local and federal government. Known as Hind Bank, the area is closed to all fishing and anchoring, and represents an important step towards more comprehensive fisheries management.

British Virgin Islands

Coral reefs are widespread throughout the British Virgin Islands, including fringing reefs close to most islands, patch reefs in offshore areas, and a long barrier-type structure, Horseshoe Reef, extending to the southeast of Anegada. As with the US Virgin Islands, the reefs have been severely impacted by the passage of several hurricanes in recent years and, although not all areas were equally affected, some sites lost up to 100 percent of their live coral cover. There are now reports of a partial recovery in most places. The reefs have also suffered from coral disease and from the 1998 bleaching event. Although less well studied, it can generally be assumed that many of these impacts have had consequences similar to those in the nearby US Virgin Islands.

Human impacts vary across the islands, but significant deterioration or loss of reef habitats has been noted close to the more heavily populated areas. Coastal development has been particularly severe on Tortola and Virgin Gorda, with the clearance of almost all mangroves. Considerable increases in coastal sedimentation have resulted from road building and other construction projects. Large amounts of sewage pass into the sea untreated, although newer developments tend to include sewage treatment facilities. These islands have the

	Puerto Rico	US Virgin Islands	British Virgin Islands
General Data			
Population (thousands)	3 916	121	20
GDP (million US$)	40 865	na	210
Land area (km²)	9 063	350	161
Marine area (thousand km²)	205	6	81
Per capita fish consumption (kg/year)	1	10	na
Status and Threats			
Reefs at risk (%)	100	100	100
Recorded coral diseases	11	8	5
Biodiversity			
Reef area (km²)	480	200	330
Coral diversity	31 / 57	34 / 57	28 / 57
Mangrove area (km²)	92	10	4
No. of mangrove species	4	na	na
No. of seagrass species	4	5	na

greatest concentration of charter yachts in the world and anchor damage is widespread, particularly in the more popular anchorages where wide areas of benthic communities have been destroyed. There is some eutrophication in the more enclosed bays which is at least in part related to these vessels. Although several hundred moorings are available these are clearly insufficient for the numbers of boats. There are only relatively few commercial fishers (less than 200), with a total catch of some 800 tons in 1998. Despite this, the impacts of commercial and recreational fishing remain substantial, notably on lobster, conch, groupers and snappers.

Although a number of marine protected areas have been declared, active management is limited. The pressures of tourist numbers in existing sites, together with the impacts of legitimate activities within their boundaries, including fishing, further reduce the effectiveness of many of these sites.

Protected areas with coral reefs

Site name	Designation	Abbreviation	IUCN cat.	Size (km²)	Year
Puerto Rico					
Boqueron	Wildlife Refuge	RVS	IV	2.37	1964
Cayos de la Cordillera	Nature Reserve	RNat	IV	0.88	1980
Estuarina Nacional Bahía Jobos	Hunting Reserve	HR	IV	11.33	1981
Isla Caja de Muerto	Nature Reserve	RNat	IV	1.88	1988
Isla de Mona	Nature Reserve	RNat	IV	55.54	1986
La Parguera	Nature Reserve	RNat	IV	49.73	1979
US Virgin Islands					
Buck Island Reef	National Monument	NaM	III	3.56	1961
Green Cay	National Wildlife Refuge	NWR	IV	0.06	1977
Hind Bank	Marine Conservation District	MarCD	IV	41.00	1999
Virgin Islands	National Park	NP	II	53.08	1956
British Virgin Islands					
Cooper Island	Bird Sanctuary	BS	IV	1.38	1959
Dead Chest Island	Bird Sanctuary	BS	IV	0.14	1959
Fallen Jerusalem Island	Bird Sanctuary	BS	IV	0.12	1959
Fort Point	Park	P	IV	0.15	1978
Horseshoe Reef	Protected Area	PA	na	30.00	1990
Mosquito Island	Bird Sanctuary	BS	IV	0.50	1959
Necker Island	Bird Sanctuary	BS	IV	0.30	1959
Peter Island	Bird Sanctuary	BS	IV	4.30	1959
Prickly Pear	Park	P	na	0.95	1988
Prickly Pear Island	Bird Sanctuary	BS	IV	0.70	1959
Round Rock Island	Bird Sanctuary	BS	IV	0.08	1959
Salt Island	Bird Sanctuary	BS	IV	0.78	1959
St. Eustatia	Bird Sanctuary	BS	IV	0.11	1959
The Baths	Natural Monument	NM	III	0.03	1990
The Seal Dogs	Bird Sanctuary	BS	IV	0.03	1959
Wreck of the Rhone	Marine Park	MP	III	3.24	1980

The Lesser Antilles, Trinidad and Tobago

MAPS 6c and d

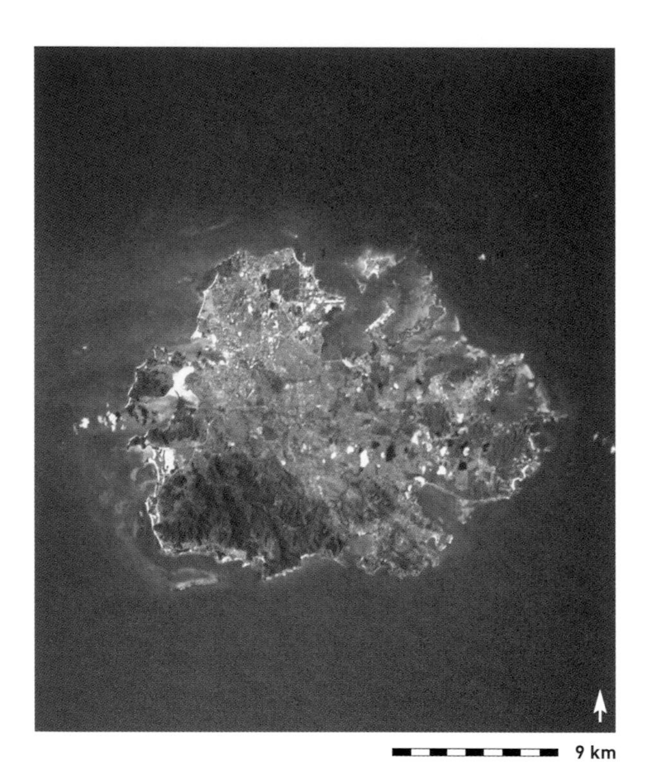

The Lesser Antilles are a group of islands lying in an 800 kilometer long arc, stretching from the Anegada Passage due east of the Virgin Islands, southwards to Grenada which lies close to the South American continental shelf. These islands form the eastern margin of the Caribbean Sea, with the oceanic waters of the Atlantic to the east. The deep waters of the Puerto Rico Trench lie to the north and northeast. This trench is the result of the subduction of the Atlantic plate under the Caribbean plate. Towards the south, these deep waters rise up towards the island of Barbados, which lies about 150 kilometers east of the main island chain. Geologically the islands are quite varied, but are dominated by two types, older sedimentary islands and more recent volcanic ones.

There has been progressive degradation of reefs throughout the Lesser Antilles over the last two decades. Many have lost large amounts of live coral cover and there have been considerable increases in algal cover. Fish numbers have also decreased, and the average size of many species is now much reduced as large individuals rarely survive the intensive fishing which takes place across the region. Recent hurricanes have been very damaging to important reef-building species such as *Montastrea annularis*, and coral diseases have severely affected populations of the two shallow species of *Acropora*.

In the following account, brief descriptions are given of the islands of the Lesser Antilles, following a sequence from north to south. Trinidad and Tobago, although distinct from the Lesser Antilles, is also considered here.

Anguilla

Anguilla is an overseas territory of the UK, and consists of a small carbonate island together with a number of smaller offshore cays. Fringing reefs are widespread, particularly on the south coast, and there are other reefs on the offshore cays. Anguilla has suffered fewer impacts than most other islands in the region. With no rivers there is little point-source runoff, and although tourism development has been significant, it has had little direct impact. Dog Island, some distance off to the northwest, has reefs which are among the least impacted and visitors are actively discouraged. Plans to establish a satellite launching pad on the small island of Sombrero have recently been overturned, largely due to environmental concerns. It harbors an important bird colony, and its surrounding reefs are considered important, although they remain poorly documented.

Left: Antigua is surrounded by intermittent bank barrier reef structures, a number of which fall within protected areas (STS064-76-BB, 1994). Right: St. Eustatius, Netherlands Antilles, clearly showing the volcanic origin common to many islands in the Lesser Antilles.

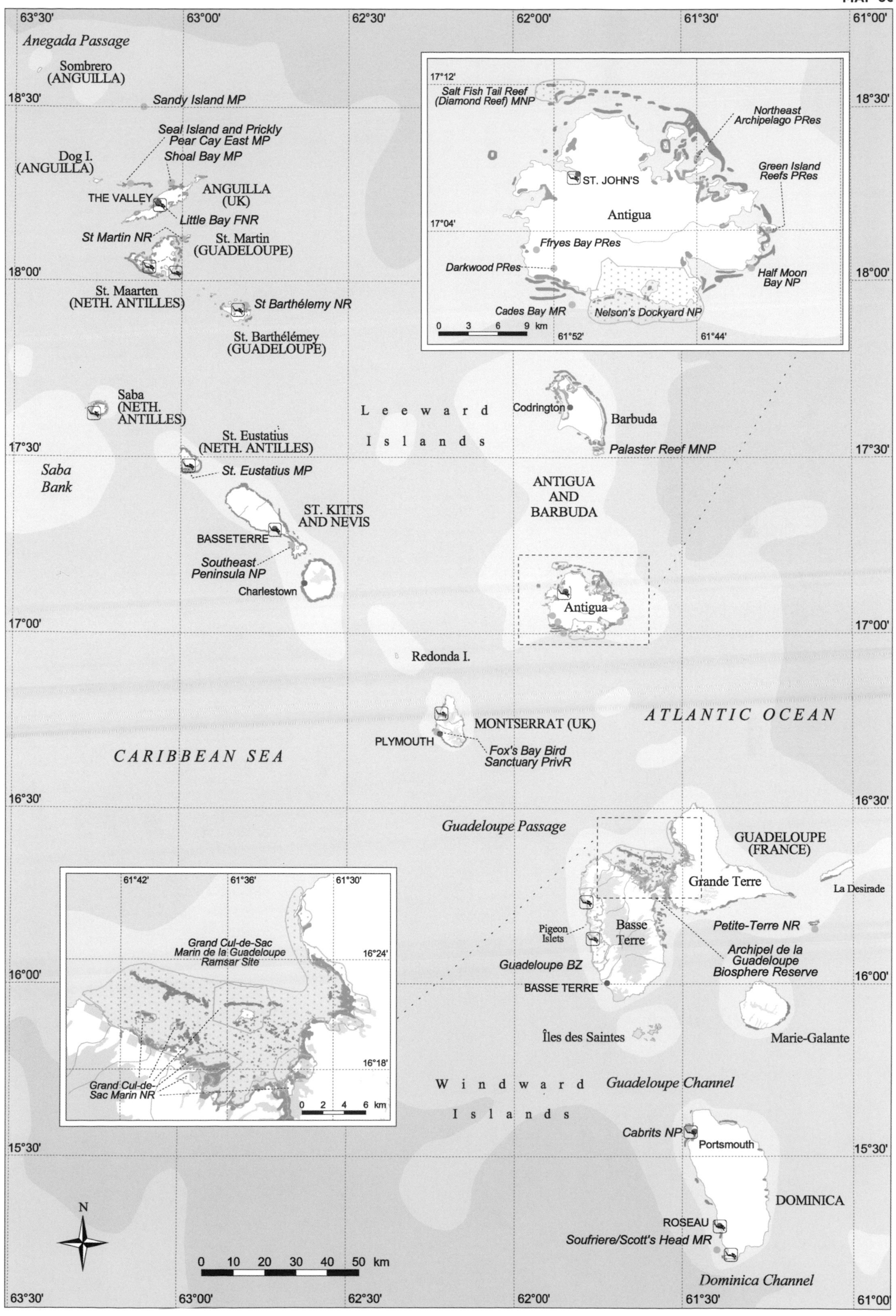

Anegada Passage
Sombrero (ANGUILLA)
Sandy Island MP
Seal Island and Prickly Pear Cay East MP
Shoal Bay MP
Dog I. (ANGUILLA)
THE VALLEY
ANGUILLA (UK)
Little Bay FNR
St Martin NR
St. Martin (GUADELOUPE)
St. Maarten (NETH. ANTILLES)
St Barthélemy NR
St. Barthélémey (GUADELOUPE)
Saba (NETH. ANTILLES)
St. Eustatius (NETH. ANTILLES)
St. Eustatius MP
Saba Bank
ST. KITTS AND NEVIS
BASSETERRE
Southeast Peninsula NP
Charlestown
Leeward Islands
Codrington
Barbuda
Palaster Reef MNP
ANTIGUA AND BARBUDA
Antigua
Redonda I.
MONTSERRAT (UK)
PLYMOUTH
Fox's Bay Bird Sanctuary PrivR
ATLANTIC OCEAN
CARIBBEAN SEA
Salt Fish Tail Reef (Diamond Reef) MNP
Northeast Archipelago PRes
Green Island Reefs PRes
ST. JOHN'S
Antigua
Ffryes Bay PRes
Darkwood PRes
Half Moon Bay NP
Cades Bay MR
Nelson's Dockyard NP
0 3 6 9 km
17°12'
17°04'
61°52'
61°44'
Guadeloupe Passage
GUADELOUPE (FRANCE)
Grande Terre
La Desirade
Pigeon Islets
Basse Terre
Petite-Terre NR
Guadeloupe BZ
Archipel de la Guadeloupe Biosphere Reserve
BASSE TERRE
Îles des Saintes
Marie-Galante
Grand Cul-de-Sac Marin de la Guadeloupe Ramsar Site
Grand Cul-de-Sac Marin NR
61°42'
61°36'
61°30'
16°24'
16°18'
0 2 4 6 km
Windward Islands
Guadeloupe Channel
Cabrits NP
Portsmouth
DOMINICA
ROSEAU
Soufriere/Scott's Head MR
Dominica Channel
N
0 10 20 30 40 50 km
63°30'
63°00'
62°30'
62°00'
61°30'
61°00'
18°30'
18°00'
17°30'
17°00'
16°30'
16°00'
15°30'

MAP 6d

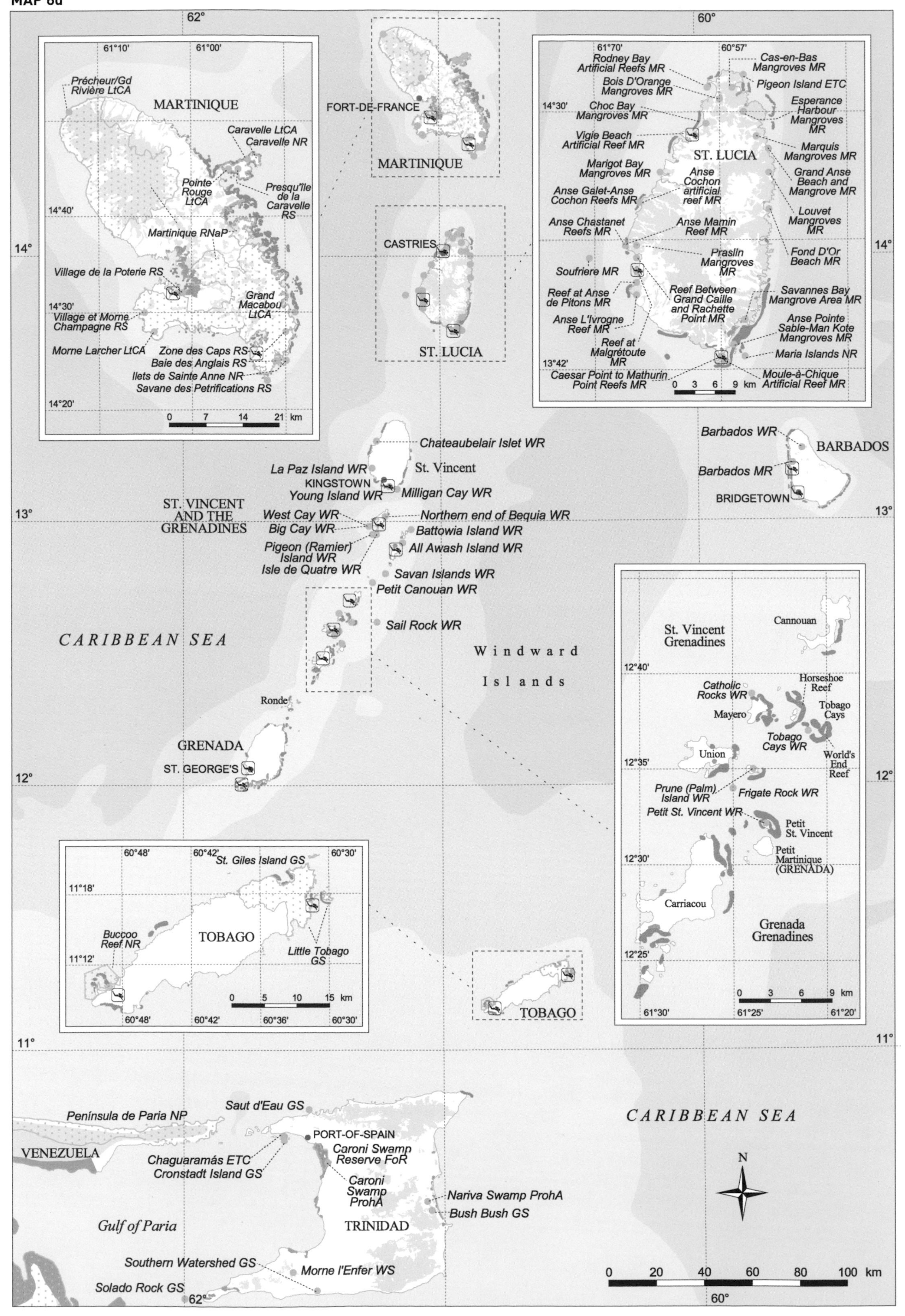

MARTINIQUE
Précheur/Gd Rivière LtCA
Caravelle LtCA
Caravelle NR
Pointe Rouge LtCA
Presqu'île de la Caravelle RS
Martinique RNaP
Village de la Poterie RS
Grand Macabou LtCA
Village et Morne Champagne RS
Morne Larcher LtCA
Zone des Caps RS
Baie des Anglais RS
Ilets de Sainte Anne NR
Savane des Petrifications RS
FORT-DE-FRANCE
CASTRIES
ST. LUCIA
Rodney Bay Artificial Reefs MR
Cas-en-Bas Mangroves MR
Bois D'Orange Mangroves MR
Pigeon Island ETC
Choc Bay Mangroves MR
Esperance Harbour Mangroves MR
Vigie Beach Artificial Reef MR
Marquis Mangroves MR
Marigot Bay Mangroves MR
Anse Cochon artificial reef MR
Grand Anse Beach and Mangrove MR
Anse Galet-Anse Cochon Reefs MR
Louvet Mangroves MR
Anse Chastanet Reefs MR
Anse Mamin Reef MR
Praslin Mangroves MR
Fond D'Or Beach MR
Soufriere MR
Reef at Anse de Pitons MR
Reef Between Grand Caille and Rachette Point MR
Savannes Bay Mangrove Area MR
Anse L'Ivrogne Reef MR
Anse Pointe Sable-Man Kote Mangroves MR
Reef at Malgrétoute MR
Maria Islands NR
Caesar Point to Mathurin Point Reefs MR
Moule-à-Chique Artificial Reef MR
Barbados WR
BARBADOS
Barbados MR
BRIDGETOWN
Chateaubelair Islet WR
St. Vincent
La Paz Island WR
KINGSTOWN
Milligan Cay WR
Young Island WR
ST. VINCENT AND THE GRENADINES
West Cay WR
Northern end of Bequia WR
Big Cay WR
Battowia Island WR
Pigeon (Ramier) Island WR
All Awash Island WR
Isle de Quatre WR
Savan Islands WR
Petit Canouan WR
Sail Rock WR
CARIBBEAN SEA
Windward Islands
Ronde
GRENADA
ST. GEORGE'S
St. Vincent Grenadines
Cannouan
Catholic Rocks WR
Horseshoe Reef
Mayero
Tobago Cays
Tobago Cays WR
Union
World's End Reef
Prune (Palm) Island WR
Frigate Rock WR
Petit St. Vincent WR
Petit St. Vincent
Petit Martinique (GRENADA)
Carriacou
Grenada Grenadines
TOBAGO
St. Giles Island GS
Buccoo Reef NR
Little Tobago GS
Península de Paria NP
Saut d'Eau GS
VENEZUELA
PORT-OF-SPAIN
Caroni Swamp Reserve FoR
Chaguaramás ETC
Cronstadt Island GS
Caroni Swamp ProhA
Nariva Swamp ProhA
Bush Bush GS
Gulf of Paria
TRINIDAD
Southern Watershed GS
Morne l'Enfer WS
Solado Rock GS
N
0 20 40 60 80 100 km

Antigua and Barbuda

Antigua and Barbuda, together with the tiny uninhabited island of Redonda, are an independent Caribbean nation. Coral reefs are relatively widespread in the coastal waters. Antigua has some fringing reefs, but also more extensive, though intermittent, bank barrier reef structures offshore. Barbuda has extensive fringing reefs, particularly along its eastern coastline, topped by a well developed algal ridge. The reefs, particularly in nearshore areas, are reported to have been degraded in recent years, possibly due to increasing sedimentation and nutrient enrichment associated with coastal development. Offshore reefs and those to the north of Barbuda generally have higher coral cover and species richness. Hurricanes Luis and Marilyn caused further damage when they struck the islands in 1995.

Netherlands Antilles
(Windward Group)

A number of islands in the region make up the dependency of the Netherlands Antilles. These include two islands close to Venezuela (Bonaire and Curaçao) and the islands of Saba, St. Eustatius and the southern half of St. Maarten (the northern part of St. Maarten (St. Martin) is a part of the French Antilles). Saba and St. Eustatius are both volcanoes, with steep cliffs and little structural reef development, but important coral communities. The St. Eustatius Marine Park was established in 1998 to protect four areas, including coral reefs and wrecks along the coast. Visitors to the park are required to pay a small user fee which helps to offset management costs. Visitor numbers are growing rapidly, from 3 000 in 1997 to 8 300 in 1999.

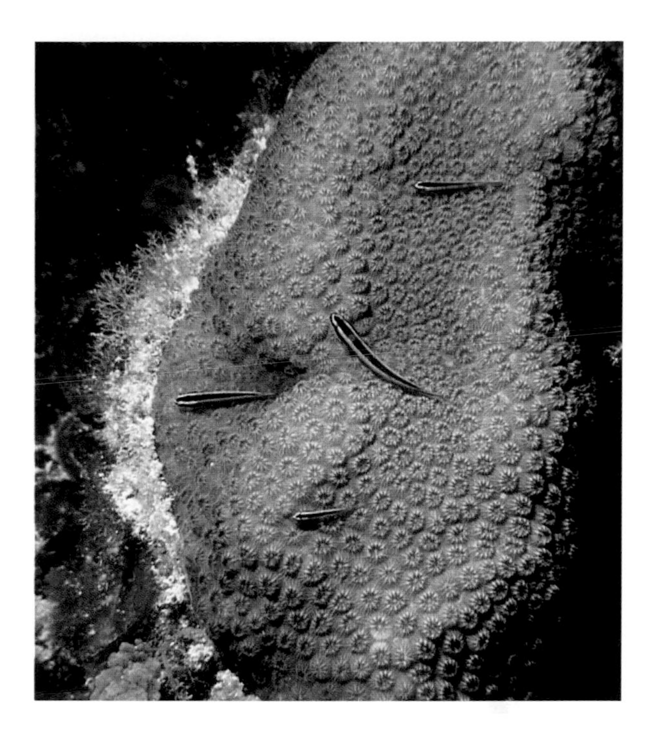

Offshore from Saba, there are again extensive coral communities in many areas. The precipitous coastline limits coastal development and, although this is a very popular diving destination, human impacts are minimal. All benthic communities down to a depth of 60 meters in Saba are protected in the Saba Marine Park. There were some 5 000 visitors to the island in 1997, and the park user fee (US$3 per dive in 1998) together with souvenir sales and yacht fees generated the majority of the income for management of the park.

	Anguilla	Antigua and Barbuda	Netherlands Antilles*	St. Kitts and Nevis
General Data				
Population (thousands)	12	66	210	39
GDP (million US$)	64	450	1 813	171
Land area (km^2)	86	462	810	275
Marine area (thousand km^2)	90	110	79	10
Per capita fish consumption (kg/year)	na	37	22	37
Status and Threats				
Reefs at risk (%)	100	100	100	100
Recorded coral diseases	0	1	10	0
Biodiversity				
Reef area (km^2)	<50	240	420	180
Coral diversity	na / 57	na / 57	40 / 57	na / 57
Mangrove area (km^2)	5	13	11	>0.71
No. of mangrove species	na	na	2	na
No. of seagrass species	na	na	na	na

* Including Bonaire and Curaçao

Small gobies Gobiosoma *sp. on a boulder star coral* Montastrea annularis.

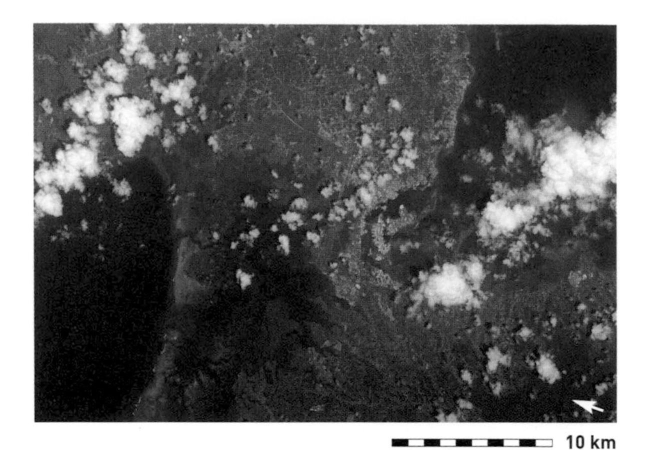

About 7 kilometers west of Saba is a large shallow platform, the Saba Bank, which may be a submerged atoll. Although only recently studied, it appears to have high coral cover in places, and is important for lobster and snapper fisheries.

The entire island of St. Maarten has shown rapid coastal development in recent years, paralleling fast population growth and a dramatic expansion of tourism. The reefs of the south and west coasts are seriously threatened by sewage pollution and siltation, while there is also much recreational boating and anchor damage. To date, no protected areas have been created.

St. Kitts and Nevis

This small independent state consists of two volcanic islands with steep mountainous slopes. There are fringing reefs along much of the coastline, and a number of deeper submerged reef structures. There is little published information about these reefs. Tourism is an important industry and there are now a number of dive operators.

Montserrat

This small island – an overseas territory of the UK – is mountainous and includes considerable forest cover. Since 1995 however, the Soufriere Hills volcano has been active almost continuously, with major pyroclastic flows into the sea. Small scattered reefs and coral communities were originally described along much of the coastline, but it seems likely that they have been severely impacted by the massive inputs of sediment, and possibly chemical influences, associated with volcanic activity. Most of the island's people have now been evacuated and the capital Plymouth was itself destroyed in 1997.

Guadeloupe and dependencies

Guadeloupe is an overseas territory of France, consisting of the twin islands of Grande Terre and Basse Terre which make up Guadeloupe proper, together with the nearby Îles des Saintes and Marie-Galante. The territory also administers the island of St. Barthélémy (St. Barths) and the northern half of St. Martin (see above). Basse Terre is high and volcanic, while Grande Terre is flat and calcareous. The western coast of Basse Terre has coral communities but no major reef structures. There are some fringing and bank barrier reef structures, particularly on the southern coastline of Grande Terre, while the northern and eastern coast of this island have well developed algal ridges. The best developed reefs are in the Grand Cul-de-Sac Marin, a shallow embayment fringed with extensive mangrove areas and dominated by seagrasses. There are several patch reefs within this bay, while its outer edge is bounded by a barrier reef with spur and groove formations and a reef slope with coral growth down to a depth of 55 meters. Discontinuous fringing reefs are found in a few parts of the other islands, notably on the southern shores of Marie-Galante. St. Barthélémy and St. Martin have limited coral reef development. Mass bleaching was reported in 1998 at Guadeloupe, but some bleaching occurs every year in September when water temperatures reach 29°C.

Fishing is an important activity in Guadeloupe, and in 1998 there were more than 2 000 professional fishermen, with a further 1 000 also thought to be fishing regularly. These have a considerable impact on the nearshore communities, most of which are considered to be overexploited. The annual catch was about 8 500 tons from these islands in 2000. Tourism, a major activity for the islands, is further driving the problems caused by coastal development and pollution. Diving is a popular activity. The Pigeon Islets (to the west of Grande Terre) are a popular dive site, but there is evidence of damage being caused by an estimated 80 000 divers per year.

Dominica

Dominica, a high volcanic island with steep topography, is an independent state. There is only limited reef development on the narrow coastal shelves, although there

Left: The Grand Cul-de-Sac Marin in Guadeloupe has important mangrove, seagrass and patch and barrier reef communities (STS092-316-12, 2000). Right: A school of yellow goatfish Mulloidichthys martinicus, *among corals and sponges in the Saba Marine Park.*

are several important coral communities, particularly on the south, west and northwest coasts. Several species of whale and dolphin are found in the waters around Dominica, which is fast positioning itself as the leading whale-watching destination in the region. The small population and minimal coastal development mean that the corals have not been severely impacted by human activities, and Dominica has been spared from a direct hurricane since Hurricane David in 1979.

Martinique

Martinique, like Guadeloupe, is an overseas territory of France. Reefs are absent on the leeward northern, northwest and west coasts, because the shelf is narrow and there is a high sediment load from the erosion of Mount Pelée. There are, however, some coral communities along this coastline. Similarly, there are no true reefs along the northeast coast, although south of Presqu'île de la Caravelle a barrier reef continues along the shore for about 25 kilometers. The lagoon behind this reef is up to 30 meters deep in places and there are extensive seagrass communities. Fringing reefs have developed along the coast behind the barrier reef. Algae, including *Sargassum*, *Turbinaria* and *Dictyota*, have proliferated on the reefs of Martinique since the *Diadema* die-off. Eutrophication from the city of Fort-de-France may be combining with the lack of grazing organisms in maintaining this situation. Overfishing is a problem, with about 900 registered fishers in 1997, but many others operating. There were an estimated 50 000 wire-mesh fish traps around the island in 2000.

St. Lucia

St. Lucia is another high volcanic island. Coral reefs are generally poorly developed, often only forming a thin veneer over the underlying volcanic substrates. The best developed reefs are in the south and east, although the best studied and most heavily utilized coral communities occur along the west coast. Certain reefs around Soufriere showed up to 50 percent live coral cover, but these sites were strongly impacted by Hurricane Lenny in 1999, which brought strong wave action on the leeward coast. Fishing is a very important activity around the island and overexploitation is a problem. Concerted efforts have recently been undertaken to manage the nearshore fisheries, and in the Soufriere Marine Management Area a number of no-take reserves have been established, interspersed with other use zones. Studies have shown huge increases in fish biomass in the reserves, while

	Montserrat	Guadeloupe*	Dominica	Martinique
General Data				
Population (thousands)	6	426	72	415
GDP (million US$)	40	2 085	191	2 654
Land area (km²)	105	1 735	732	1 101
Marine area (thousand km²)	7	90	29	45
Per capita fish consumption (kg/year)	na	28	35	26
Status and Threats				
Reefs at risk (%)	na	100	100	100
Recorded coral diseases	0	1	0	1
Biodiversity				
Reef area (km²)		250	<100	240
Coral diversity	na / na	na / 57	na / 57	34 / 57
Mangrove area (km²)	>0.02	40	2	16
No. of mangrove species	na	na	na	na
No. of seagrass species	na	na	na	na

* Including St. Martin and St. Barthélémy

A view of Simpson Bay Lagoon, from the French St. Martin to the Dutch St. Maarten, showing the significant coastal development on this island.

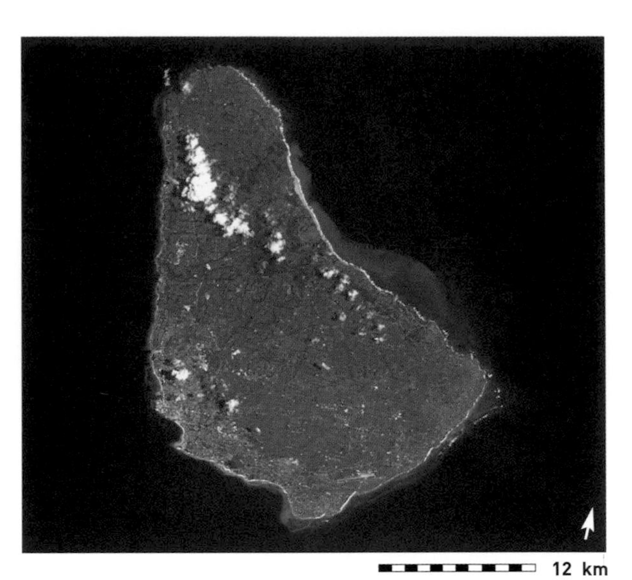

fishermen have reported significant increases in their catches from adjacent fishing priority areas. Studies have shown a tripling of fish biomass in the marine reserves, while fishermen have reported significant increases in their catches from adjacent areas. Tourism is also popular in the islands, and diving is increasingly focussed towards sites in the marine management area. Fees from divers, and anchor fees from yachts, mean that the management authority is now self-financing. This provides perhaps the best example in the region of reef management for multiple uses with full community participation. A new marine management area is now under development further north on the same coast.

St. Vincent and the Grenadines

St. Vincent is a relatively young volcanic island. To the north, Mount Soufriere most recently erupted in 1979. The relatively young coastline, together with new volcanic sediments, have prevented the development of extensive reefs. There are no reef developments around the north and east coasts, and only a few coral communities are found on rocky headlands along the west coast. Small areas of fringing reefs occur on the south and southeast coasts. Running south from the main island is the chain of the Grenadines, where there are considerable areas of reef. Large bank barrier reef complexes have developed on the windward side of some islands. Among the best developed

	St. Lucia	St. Vincent and the Grenadines	Barbados	Grenada
General Data				
Population (thousands)	156	115	274	89
GDP (million US$)	478	237	1 768	223
Land area (km²)	605	390	440	367
Marine area (thousand km²)	15	38	186	25
Per capita fish consumption (kg/year)	22	20	40	28
Status and Threats				
Reefs at risk (%)	100	96	100	100
Recorded coral diseases	2	2	0	1
Biodiversity				
Reef area (km²)	160	140	<100	150
Coral diversity	na / 57	na / 57	33 / 57	na / 57
Mangrove area (km²)	1	>0.45	>0.07	2
No. of mangrove species	na	na	na	na
No. of seagrass species	1	na	1	1

Left: A banded butterflyfish Chaetodon striatus *amidst gorgonians and soft corals. Right: Fringing reefs around Barbados have declined over many decades although there are still submerged reefs off the west and southeast coasts (STS051-72-95, 1993).*

reefs are those around the small islands of the Tobago Cays. Each island has a fringing reef, the larger Horseshoe Reef encircles them to the east, while beyond this there is the larger World's End Reef. The reefs of St. Vincent, and particularly the Grenadines, support important fishing and tourism, while large numbers of yachts visit these waters. The Tobago Cays are particularly important, but their condition has deteriorated recently because of storm damage, white band disease, physical damage from fishing gear and boat anchors, and pollution from visiting yachts.

Barbados

Barbados is, in many ways, an anomaly. It lies east of the main Lesser Antilles chain in the Atlantic Ocean. Fringing reefs are largely absent, although there is a small fringing structure near Folkestone on the leeward west coast. There are also sub-surface reefs along this coast, where a gently sloping shelf extends about 300 meters seaward to a depth of 10 meters. At the edge of the shelf, the sea floor drops evenly to a depth of about 20 meters. Seaward from this there are further submerged patch reef structures, together with two larger bank barrier reefs, 12-20 meters deep and up to 100 meters wide. Offshore, submerged bank barriers are also found off the southeast coast. The eastern, Atlantic, coast is subject to very high wave energies throughout the year, and much of this coastline is a bare carbonate platform extending out to deep water. Nearshore reefs in Barbados have suffered considerably. Reef flat corals disappeared over 100 years ago with the intensification of agriculture, while considerable declines in coral cover and diversity have been reported on offshore reefs since the 1980s, linked to eutrophication from urbanization and tourism developments.

Grenada

Grenada is the most southerly of the Lesser Antilles, and the country also governs the southernmost islands of the Grenadines. There are some fringing and patch reefs around all the coasts of Grenada itself, although the total

Trinidad and Tobago

General Data	
Population (thousands)	1 176
GDP (million US$)	5 499
Land area (km²)	5 152
Marine area (thousand km²)	74
Per capita fish consumption (kg/year)	14
Status and Threats	
Reefs at risk (%)	100
Recorded coral diseases	5
Biodiversity	
Reef area (km²)	<100
Coral diversity	na / 57
Mangrove area (km²)	>70
No. of mangrove species	7
No. of seagrass species	2

French grunts Haemulon flavolineatum *against a thriving colony of* Acropora cervicornis. *Over wide areas of the Caribbean such scenes are now rare as a result of overfishing and coral disease.*

area of reef is not great. Off the eastern coasts of Carriacou and Petit Martinique relatively large bank barrier reefs have been formed. Many of the shallow reefs were reported to have become overgrown with algae during the 1980s, probably linked to the *Diadema* die-off, but possibly exacerbated by sewage and agrochemical pollution and increased sedimentation.

Trinidad and Tobago

The large island of Trinidad and the nearby Tobago lie well south of the chain of the Lesser Antilles, on the continental shelf of South America. Reef development around Trinidad is severely restricted. The Orinoco River lies to the south and discharges huge volumes of sediment into the sea, creating turbid conditions which predominate along the south and east coastlines of the island. The western coastline faces the Gulf of Paria which, along with high levels of sediments, has near estuarine conditions arising from the high freshwater inputs and semi-enclosed nature of this gulf. There are small, low diversity coral communities in places on the north shore. Tobago lies close to the edge of the continental shelf, and here reef development is much better, with a number of fringing reefs, particularly on the north shore and in the southwest. Tobago has a considerable tourism industry, and the impacts of tourism have undoubtedly led to the degradation of some coastal reefs.

Protected areas with coral reefs

Site name	Designation	Abbreviation	IUCN cat.	Size (km²)	Year
Anguilla					
Little Bay	Fish Nursery Reserve	FNR	na	na	na
Sandy Island	Marine Park	MP	na	na	na
Seal Island and Prickly Pear Cay East	Marine Park	MP	na	na	na
Shoal Bay	Marine Park	MP	na	na	na
Antigua and Barbuda					
Green Island Reefs	Park Reserve	PRes	IV	na	na
Northeast Archipelago	Park Reserve	PRes	IV	na	na
Palaster Reef	Marine National Park	MNP	II	5.00	1973
Salt Fish Tail Reef (Diamond Reef)	Marine National Park	MNP	II	20.00	1973
Cades Bay	Marine Reserve	MR	ETC	na	1999
Barbados					
Barbados	Marine Reserve	MR	II	2.30	1980
Dominica					
Cabrits	National Park	NP	II	5.31	1986
Soufriere/Scott's Head	Marine Reserve	MR	V	na	na
Guadeloupe					
Grand Cul-de-Sac Marin	Nature Reserve	NR	IV	37.36	1987
Petite-Terre	Nature Reserve	NR	IV	9.90	1998
St. Barthélémy	Nature Reserve	NR	IV	12.00	1996
St. Martin	Nature Reserve	NR	IV	30.60	1998
GRAND CUL-DE-SAC MARIN DE LA GUADELOUPE	RAMSAR SITE			**200.00**	**1993**
ARCHIPEL DE LA GUADELOUPE	UNESCO BIOSPHERE RESERVE			**697.00**	**1992**

Protected areas with coral reefs

Site name	Designation	Abbreviation	IUCN cat.	Size (km²)	Year
Martinique					
Caravelle	Littoral Conservation Area	LtCA	IV	2.57	1988
Caravelle	Nature Reserve	NR	IV	4.22	1976
Grand Macabou	Littoral Conservation Area	LtCA	Unassigned	1.13	1982
Pointe Rouge	Littoral Conservation Area	LtCA	Unassigned	0.54	1985
Netherlands Antilles (Windward)					
St. Eustatius	Marine Park	MP	na	na	1998
Saba	Marine Park	MP	na	8.20	1987
St. Kitts and Nevis					
Southeast Peninsula	National Park	NP	II	26.10	na
St. Lucia					
Anse Chastanet Reefs	Marine Reserve	MR	IV	na	1990
Anse Cochon Artificial Reef	Marine Reserve	MR	IV	na	1990
Anse Galet – Anse Cochon Reefs	Marine Reserve	MR	IV	na	1990
Anse L'Ivrogne Reef	Marine Reserve	MR	IV	na	1986
Anse Mamin Reef	Marine Reserve	MR	IV	na	1986
Anse Pointe Sable – Man Kote Mangroves	Marine Reserve	MR	IV	na	1986
Caesar Point – Mathurin Point Reefs	Marine Reserve	MR	IV	na	1990
Maria Islands	Nature Reserve	NR	IV	0.12	1982
Moule-à-Chique Artificial Reef	Marine Reserve	MR	IV	na	1990
Pigeon Island	Other Area	ETC	III	0.20	1978
Reef at Anse de Pitons	Marine Reserve	MR	IV	na	1986
Reef at Malgrétoute	Marine Reserve	MR	IV	na	1986
Reef between Grand Caille and Rachette Point	Marine Reserve	MR	IV	na	1986
Rodney Bay Artificial Reefs	Marine Reserve	MR	IV	na	1986
Soufriere	Marine Reserve	MR	na	na	na
Vigie Beach Artificial Reef	Marine Reserve	MR	IV	na	1990
St. Vincent					
Frigate Rock	Wildlife Reserve	WR	IV	na	1987
Isle de Quatre	Wildlife Reserve	WR	IV	na	1987
Prune (Palm) Island	Wildlife Reserve	WR	IV	na	1987
Tobago Cays	Wildlife Reserve	WR	IV	38.85	1987
West Cay	Wildlife Reserve	WR	IV	na	1987
Trinidad and Tobago					
Buccoo Reef	Nature Reserve	NR	Ia	6.50	1973
Little Tobago	Game Sanctuary	GS	IV	1.01	1928

Venezuela and Aruba, Bonaire and Curaçao

MAP 6e

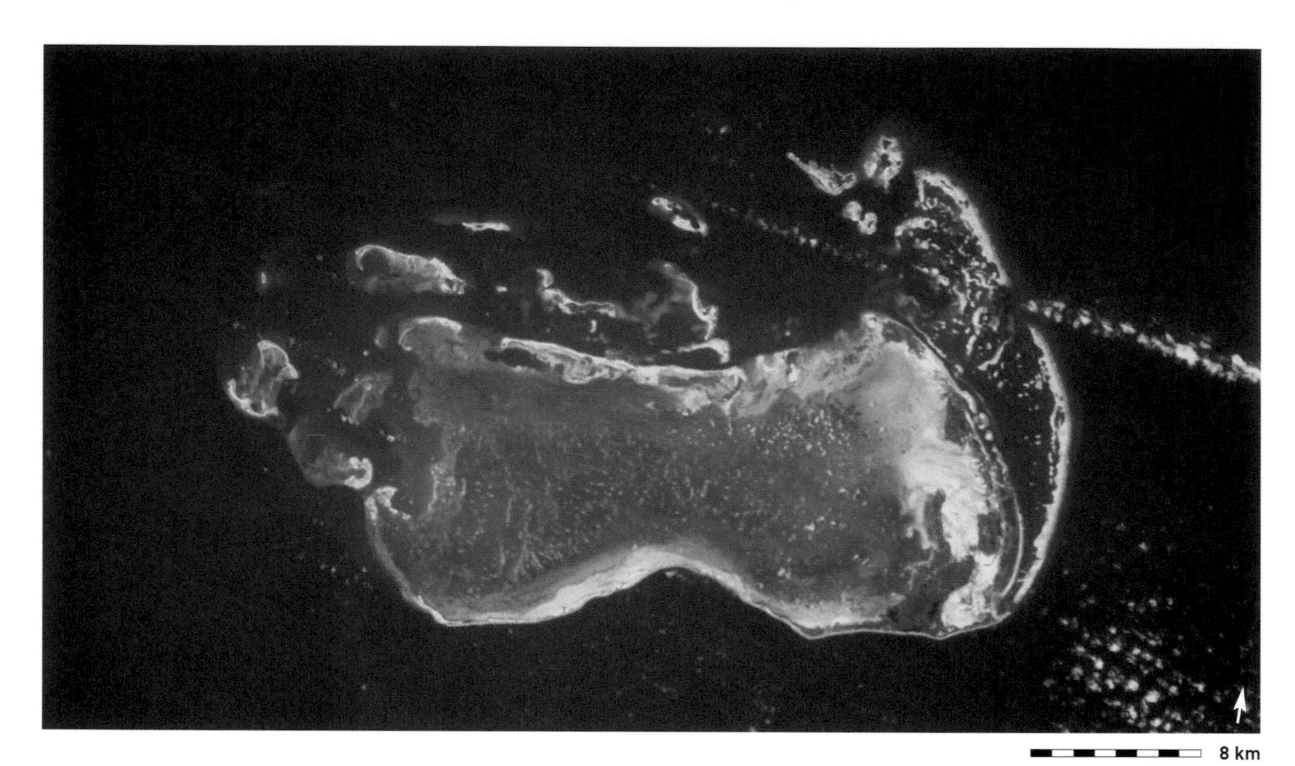

Venezuela is a large country with a long, north-facing coastline delimiting the southeastern edge of the Caribbean Sea. In the east this coastline is dominated by the vast delta of the Orinoco River, which carries considerable quantities of freshwater into the Western Atlantic, just south of the island of Trinidad. Further west, the coastline generally has higher relief, and there are numerous smaller rivers. Coral reef development is thus highly limited by freshwater and sediment runoff, and nearshore coral reefs are scarce. Small reef systems exist at Morrocoy and coral communities in Mochima. Between these two locations there are a few other small reef developments, for example in San Esteban, Turiamo Bay and Ciénaga de Ocumare Bay. The reefs in the Parque Nacional Morrocoy occur along the seaward margins of small cays at the mouth of the Golfete de Guare (Borracho and Cayo Sombrero) and to the south of Punta Tucacas. This is a generally low energy area with moderate to low wave activity, and hurricanes are very rare. Mangroves, mainly *Rhizophora mangle*, grow on the leeward side of these islands, which are separated from the mainland by extensive seagrass beds. The reef platforms are approximately 50 meters wide and typically slope down to a depth of 12 meters. Until recently, they were dominated by *Montastrea cavernosa*, *M. annularis* and several species of soft coral (*Pseudopterogorgia* spp., *Plexaura* spp. and *Eunicea* spp.). Further reefs are located on the continental coastline around the Mochima National Park, although diversity is lower here, with only about 25 scleractinian coral species recorded.

In January 1996 there was mass coral mortality at Morrocoy, which left less than 5 percent live coral cover. All corals except *Porites porites*, *Siderastrea siderea* and *Millepora alcicornis* at the main monitoring station were killed. In addition to corals, mass mortalities were recorded amongst fish, crustaceans, molluscs, echinoderms and sponges. The ultimate cause of this event remains unclear. The more protected reefs, in the lee of coral cays and away from open water, appeared to show greater levels of survival. Given the proximity of these reefs to an oil refinery, petrochemical plant and various other industries, it has been suggested that an unreported anthropogenic impact such as a chemical spill may have been responsible.

Venezuela also holds jurisdiction over a number of offshore islands, most lying in oceanic water at some distance from the continental shelf. These include Las Aves, Los Roques, Isla la Orchilla and La Blanquilla,

The reefs of Los Roques in Venezuela, a large marine protected area where coral cover remains high (STS077-719-105, 1996).

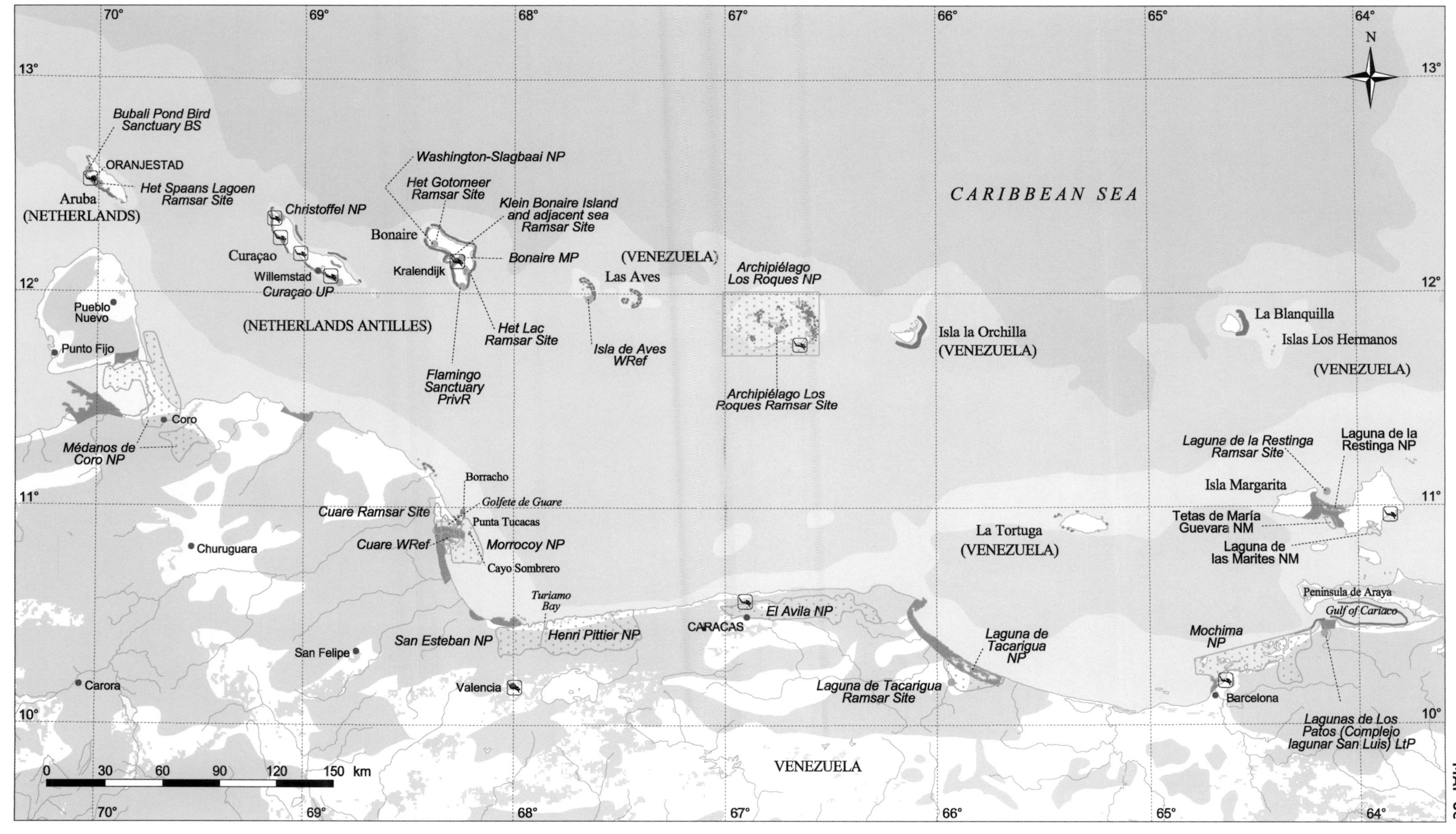

CARIBBEAN SEA
Bubali Pond Bird Sanctuary BS
ORANJESTAD
Het Spaans Lagoen Ramsar Site
Aruba (NETHERLANDS)
Christoffel NP
Curaçao
Willemstad
Curaçao UP
Washington-Slagbaai NP
Het Gotomeer Ramsar Site
Klein Bonaire Island and adjacent sea Ramsar Site
Bonaire
Kralendijk
Bonaire MP
Het Lac Ramsar Site
Flamingo Sanctuary PrivR
(NETHERLANDS ANTILLES)
(VENEZUELA)
Las Aves
Isla de Aves WRef
Archipiélago Los Roques NP
Archipiélago Los Roques Ramsar Site
Isla la Orchilla (VENEZUELA)
La Blanquilla
Islas Los Hermanos
(VENEZUELA)
Pueblo Nuevo
Punto Fijo
Coro
Médanos de Coro NP
Laguna de la Restinga Ramsar Site
Laguna de la Restinga NP
Isla Margarita
Tetas de María Guevara NM
Laguna de las Marites NM
Borracho
Golfete de Guare
Cuare Ramsar Site
Punta Tucacas
Cuare WRef
Morrocoy NP
Cayo Sombrero
Churuguara
La Tortuga (VENEZUELA)
Peninsula de Araya
Gulf of Cariaco
Turiamo Bay
El Avila NP
CARACAS
San Esteban NP
Henri Pittier NP
San Felipe
Valencia
Carora
Laguna de Tacarigua NP
Laguna de Tacarigua Ramsar Site
Mochima NP
Barcelona
Lagunas de Los Patos (Complejo lagunar San Luis) LtP
VENEZUELA
0 30 60 90 120 150 km
70° 69° 68° 67° 66° 65° 64°
13° 12° 11° 10°
N

which lie in a chain parallel to the coast. These reefs have a high species diversity, including some 270 species of coral reef fish. Los Roques is an archipelago of 40 small islands, including one rocky island and 39 coral cays in an atoll-like formation. The continental shelf is narrow to the south but nearly 1 kilometer wide to the north. Coral cover remains high at this site, averaging 27 percent in 1999/2000, but reaching 60 percent in some localities. A total of 51 hermatypic coral species have been recorded. The whole archipelago was declared a Venezuelan national park in 1972 and is one of the largest marine national parks in the Caribbean.

Water moves through these offshore islands in a westerly direction, the current being a branch of the Caribbean Current. This probably protects the offshore reefs from most of the terrestrial runoff from the mainland. The principal threat is intensive fishing, particularly on the fringing reefs of Los Roques. Reef-based tourism is not intensively developed. The military control many of the smaller islands and the exclusion of fishermen and tourists may be the most effective protection for reefs in the country.

More remote from these is the Isla de Aves, a small and extremely remote islet in the Caribbean Sea, over 200 kilometers west of Dominica in the Lesser Antilles, and about 550 kilometers north of mainland Venezuela. There is very little information describing the marine communities around this island.

Aruba, Bonaire and Curaçao

Politically Bonaire and Curaçao are part of the Netherlands Antilles, and are sometimes referred to as the leeward islands of the Netherlands Antilles. Aruba maintains a separate constitution, but still forms a part of the Kingdom of the Netherlands. Bonaire and Curaçao are oceanic islands surrounded by deep water, but Aruba is located on the South American continental shelf only 27 kilometers north of Venezuela. The easterly trade winds create markedly different physical regimes between the

Venezuela	
General Data	
Population (thousands)	23 543
GDP (million US$)	56 042
Land area (km²)	916 560
Marine area (thousand km²)	522
Per capita fish consumption (kg/year)	20
Status and Threats	
Reefs at risk (%)	44
Recorded coral diseases	1
Biodiversity	
Reef area (km²)	480
Coral diversity	23 / 57
Mangrove area (km²)	2 500
No. of mangrove species	7
No. of seagrass species	4

Aruba	
General Data	
Population (thousands)	70
GDP (million US$)	na
Land area (km²)	183
Marine area (thousand km²)	6
Per capita fish consumption (kg/year)	9
Status and Threats	
Reefs at risk (%)	94
Recorded coral diseases	0
Biodiversity	
Reef area (km²)	<50
Coral diversity	na / 57
Mangrove area (km²)	4
No. of mangrove species	2
No. of seagrass species	na

A coney Cephalopholis fulva *with sponges behind.*

Protected areas with coral reefs

Site name	Designation	Abbreviation	IUCN cat.	Size (km²)	Year
Netherlands Antilles (Leeward)					
Bonaire	Marine Park	MP	na	26.00	1979
Curaçao	Underwater Park	UP	na	10.36	1983
KLEIN BONAIRE ISLAND AND ADJACENT SEA	RAMSAR SITE			6.00	1980
Venezuela					
Archipiélago Los Roques	National Park	NP	II	2 211.20	1972
Mochima	National Park	NP	II	949.35	1973
Morrocoy	National Park	NP	II	320.90	1974
San Esteban	National Park	NP	II	435.00	1987
ARCHIPIÉLAGO LOS ROQUES	RAMSAR SITE			2 132.20	1996
CUARE	RAMSAR SITE			99.68	1988

leeward and windward sides of these islands. The reef profile of Bonaire and Curaçao is generally similar: a submarine terrace extending between 50 and 100 meters offshore, and ending in a drop-off at a depth of 8-12 meters which slopes steeply to 50-60 meters. There is a second drop-off at 80-100 meters ending in a sandy plain. Prolific coral growth occurs across this terrace and on the shallower slope. Conspicuous spur and grooves are a feature of the Bonaire reef slope, especially along the northwestern shore. Along the eastern windward shore there is little coral growth in any water less than 12 meters in depth. Shallower waters harbor an abundance of crustose coralline algae and dense *Sargassum platycarpum*, though in places these also grow down to a depth of 40 meters. Being located on the continental shelf, Aruba does not have sharply sloping underwater relief.

Coral cover at depths of 10-20 meters at four sites on Curaçao and Bonaire decreased from 50-55 percent to 25-30 percent between 1973 and 1992, but was mostly unchanged at a 30-40 meter depth. Bonaire is widely cited as one of the region's best examples of a self-financing marine park. Divers are charged a fee of US$10 per year to dive on the reefs, contributing about 60 percent of the running costs of the park, with a significant proportion of the remainder being generated from the sale of souvenirs. Studies have shown that the user fee is seen as a positive thing by the majority of visitors, raising awareness of conservation issues while giving them some sense of participation or ownership. The deterioration in coral cover around this island is linked to the Caribbean-wide declines, perhaps slightly exacerbated by tourism development. This pressure is increasing: 57 000 visitors arrived in 1994 (25 000 of whom were divers), rising to 70 000 in 1999. Despite this pressure, direct physical damage to the coral by divers remains low, with less than 3 percent of colonies affected.

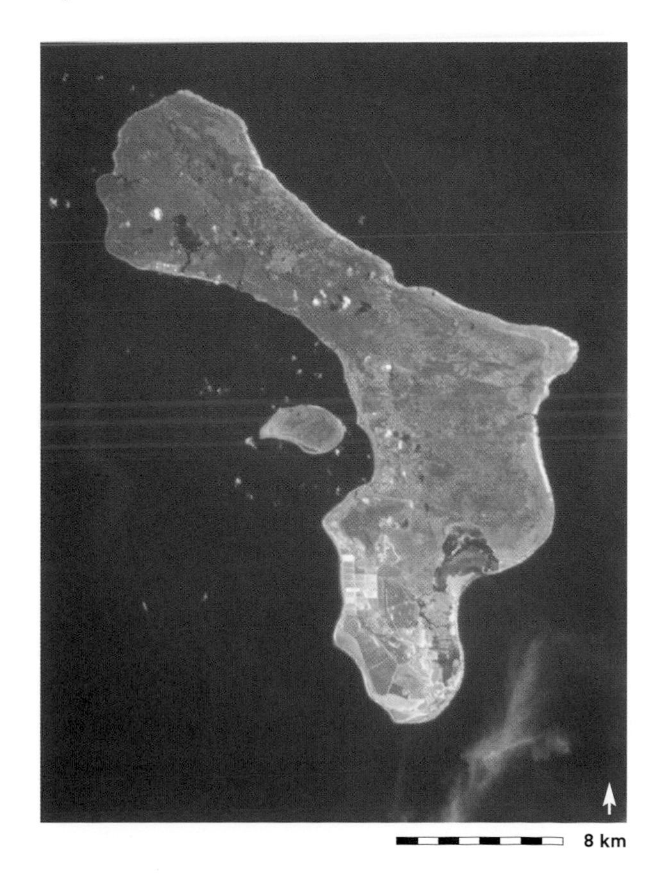

The waters around Bonaire are one of the best known marine parks in the Caribbean (STS075-706-41, 1996).

MAP 6f

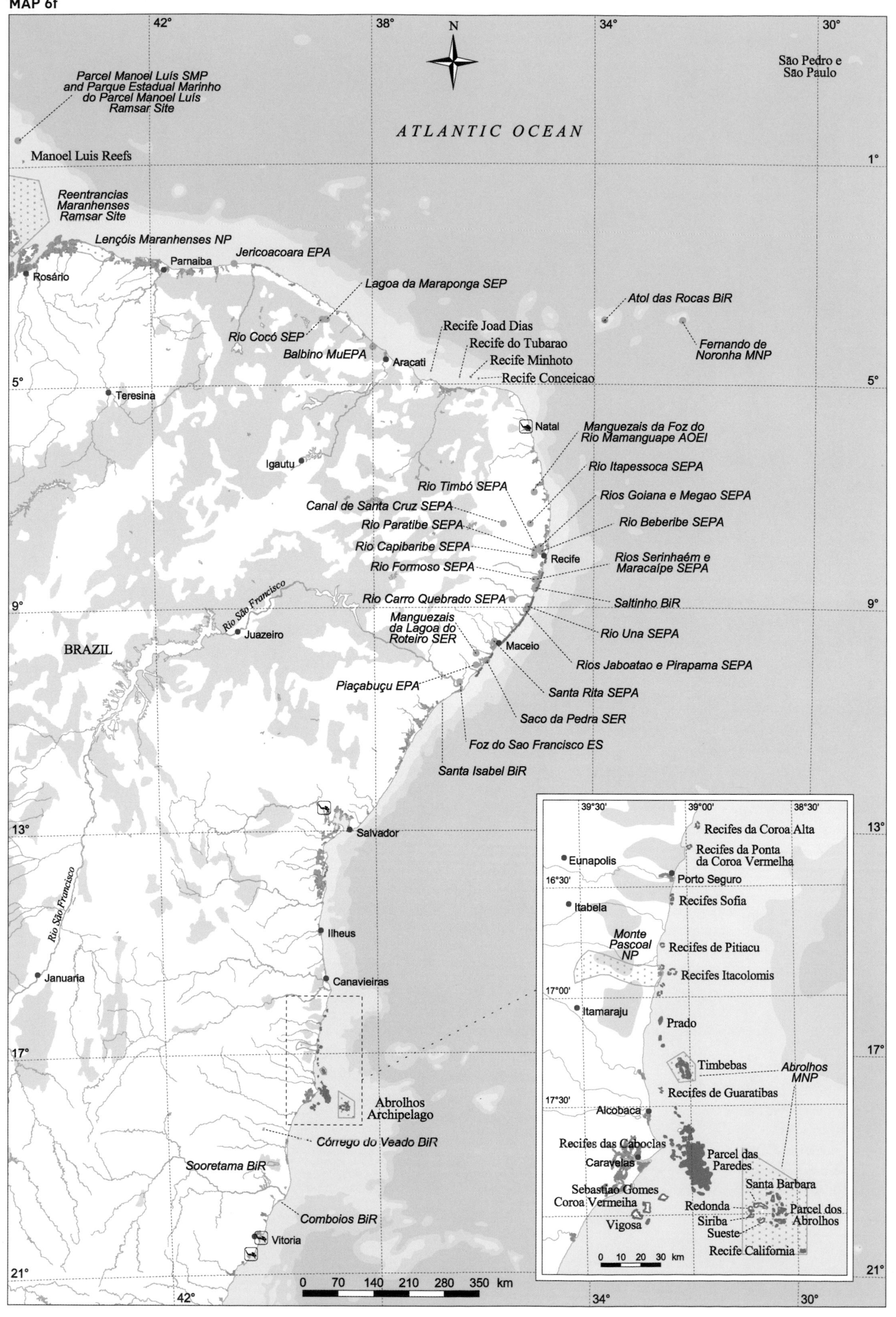

N
42°
38°
34°
30°
1°
5°
9°
13°
17°
21°
ATLANTIC OCEAN
São Pedro e São Paulo
Parcel Manoel Luís SMP and Parque Estadual Marinho do Parcel Manoel Luís Ramsar Site
Manoel Luis Reefs
Reentrancias Maranhenses Ramsar Site
Lençóis Maranhenses NP
Jericoacoara EPA
Parnaiba
Rosário
Lagoa da Maraponga SEP
Rio Cocó SEP
Balbino MuEPA
Aracati
Recife Joad Dias
Recife do Tubarao
Recife Minhoto
Recife Conceicao
Atol das Rocas BiR
Fernando de Noronha MNP
Teresina
Natal
Manguezais da Foz do Rio Mamanguape AOEI
Igautu
Rio Itapessoca SEPA
Rio Timbó SEPA
Rios Goiana e Megao SEPA
Canal de Santa Cruz SEPA
Rio Beberibe SEPA
Rio Paratibe SEPA
Rio Capibaribe SEPA
Recife
Rios Serinhaém e Maracaípe SEPA
Rio Formoso SEPA
Rio São Francisco
Rio Carro Quebrado SEPA
Saltinho BiR
Juazeiro
Manguezais da Lagoa do Roteiro SER
Rio Una SEPA
Maceio
BRAZIL
Rios Jaboatao e Pirapama SEPA
Piaçabuçu EPA
Santa Rita SEPA
Saco da Pedra SER
Foz do Sao Francisco ES
Santa Isabel BiR
Salvador
Rio São Francisco
Ilheus
Januaria
Canavieiras
Abrolhos Archipelago
Córrego do Veado BiR
Sooretama BiR
Comboios BiR
Vitoria
0 70 140 210 280 350 km
39°30'
39°00'
38°30'
16°30'
17°00'
17°30'
Recifes da Coroa Alta
Recifes da Ponta da Coroa Vermelha
Eunapolis
Porto Seguro
Recifes Sofia
Itabela
Monte Pascoal NP
Recifes de Pitiacu
Recifes Itacolomis
Itamaraju
Prado
Timbebas
Abrolhos MNP
Recifes de Guaratibas
Alcobaca
Recifes das Caboclas
Caravelas
Parcel das Paredes
Santa Barbara
Sebastiao Gomes
Coroa Vermeiha
Redonda
Parcel dos Abrolhos
Vigosa
Siriba
Sueste
Recife California
0 10 20 30 km

Brazil and West Africa

MAP 6f

The waters of both Brazil and West Africa are separated from the Caribbean reefs by vast barriers inimical to reef growth. For the Brazilian reefs these barriers include the huge river mouths of the Amazon and Orinoco, as well as the intervening sediment-rich coastline of the Guyanas. The coastlines of West Africa (and the intervening Atlantic islands) are separated from the Caribbean center of diversity by large expanses of open ocean. For these reasons there is virtually no supply of larval recruits from the Caribbean either to Brazil or West Africa. Hence the coral reef organisms found in these two areas are ecologically isolated.

Brazil

Coral reef growth in Brazil is limited to the northeast and eastern shores. Most of the northern coastline of Brazil is dominated by areas of massive riverine input, with freshwater and sediments dominating the continental shelf over wide areas to the east of the Amazon. This coastline is also swept by the west and northward flow of the northern arm of the South Equatorial Current, and these factors combine to isolate Brazil from the Caribbean. A result of this is that Brazil's coral fauna is notable for having a low species diversity yet a high degree of endemism. Just 19 species of reef-building coral are recognized, of which at least six (including all three species of the genus *Mussismilia*), and possibly as many as ten, are found nowhere else. Another interesting feature of Brazilian coral communities is that there are no acroporid corals, which are the major shallow water corals elsewhere in the world.

The westernmost reef systems, in closest proximity to the Caribbean, are the recently described Manoel Luis Reefs, lying relatively close to the Amazon river mouth. These reefs are some 10 kilometers in length and consist of numerous pinnacles rising from a depth of 25-30 meters up to the surface waters. Some 16 hermatypic corals have been recorded, including 12 scleractinian species. These reefs are still poorly known, but their location, as the closest reefs to the Caribbean, may be important for any movements of species between these regions.

There are a few oceanic islands to the northeast of Brazil. Coral communities of 12 species form dense structures, but not true reefs, on the islands of Fernando de Noronha. The nearby Atol das Rocas is a true atoll some 3.7 kilometers across, encircling a shallow lagoon. The carbonate deposits, which are some 10 meters

Left: Mussismilia harttii, *one of several species endemic to Brazil (photo: JEN Veron). Right:* Madracis decactis *is a truly Atlantic species, found in the Caribbean, Brazil and West Africa. In Brazil it typically forms tall grey columns (photo: JEN Veron).*

thick, are predominantly the result of coralline algal deposits. Only eight coral species have been recorded, of which *Siderastrea stellata* is dominant in all areas. Nearly 1 000 kilometers northeast of Brazil, São Pedro e São Paulo is a group of some 15 rocks and islets. They lie in the westward flowing South Equatorial Current and hence there is little or no migration of coral larvae to these rocks. Only two species of hermatypic coral (*Scolymia wellsi* and *Madracis decactis*) have been recorded.

The eastern continental shelf of Brazil is of irregular and limited width (about 50 kilometers) in most places. Narrow reefs, formed by pinnacles of *Siderastrea stellata* and *Millepora alcicornis*, are found along the coast to the north of Natal. Further south there are many reefs parallel to the coast. These are characterized by an emergent reef crest with only two species of coral (*Favia gravida* and *Siderastrea stellata*) and an algal ridge of *Melobesiacea* and *Dendropoma* spp. There are typically three zones on the seaward slope dominated respectively by *Millepora alcicornis*, *Mussismilia harttii* and, at depth, *Montastrea cavernosa*. Gorgonian corals are particularly abundant on these reefs.

In the State of Bahia, the continental shelf widens considerably, extending from 5 to 65 kilometers offshore, and reaching 200 kilometers in the far south around the Abrolhos Archipelago. This is the largest and richest area of coral reefs in the South Atlantic. Sixteen species of stony coral are recorded, and coral cover approaches 20 percent in some areas of shallow reef. Reefs include fringing reefs and offshore banks. A common growth form is the development of mushroom shaped pinnacles called chapeiroes, highly characteristic of Brazilian reefs. They are typically 2-50 meters in diameter, and extend vertically to a height of between 1 and 25 meters. The tops of chapeiroes close to shore frequently fuse together with open spaces beneath the coalesced surface. Channels between individual chapeiroes sometimes fill up with sediment. The tops of some of these inshore reefs are often completely exposed at low tides. Further out to sea the chapeiroes do not fuse together and the reefs consist of very large individual chapeiroes in water about 15-20 meters deep. The Abrolhos Archipelago incorporates the most extensive reefs, and also includes some small islands and sand cays, with some areas of mangrove. Coral bleaching was reported from northern Bahia and the Abrolhos reefs in 1998, but levels of mortality were low.

Many of the coastal reefs of Brazil exhibit signs of degradation, particularly close to human settlement. The major concern for the coral reefs of Abrolhos is the expansion in tourism, increased sedimentation from inland deforestation for agriculture and rapid coastal development. The number of visitors to the Abrolhos Marine Park increased fourfold between 1988 and 1993. Associated problems such as anchor damage, litter, collection of souvenirs and reef walking are of considerable concern.

Brazil

General Data	
Population (thousands)	172 860
GDP (million US$)	503 484
Land area (km²)	8 507 080
Marine area (thousand km²)	3 661
Per capita fish consumption (kg/year)	7
Status and Threats	
Reefs at risk (%)	84
Recorded coral diseases	0
Biodiversity	
Reef area (km²)	1 200
Coral diversity	na / 17
Mangrove area (km²)	13 400
No. of mangrove species	7
No. of seagrass species	1

West Africa

True reefs do not occur along the West African coast or the Cape Verde and Gulf of Guinea archipelagos, although mature coral communities are found at various locations. In all some 15 species of hermatypic and ahermatypic corals have been recorded. The region's heavy rainfall drains through several major rivers, principally the Niger, and creates a large freshwater input to the Gulf of Guinea.

Some of the islands and reefs of the Abrolhos Archipelago. Additional structures, including the marine park, lie further offshore (STS054-86-1, 1993).

Protected areas with coral reefs

Site name	Designation	Abbreviation	IUCN cat.	Size (km²)	Year
Brazil					
Abrolhos	Marine National Park	MNP	II	913.00	1983
Atol das Rocas	Biological Reserve	BiR	Ia	362.49	1979
Fernando de Noronha	Marine National Park	MNP	II	112.70	1988
Parcel Manoel Luís	State Marine Park	SMP	II	452.37	1991
Recife de Fora	State Marine Park	SMP	IV	17.00	na
PARQUE ESTADUAL MARINHO DO PARCEL MANOEL LUÍS	RAMSAR SITE			**452.37**	**2000**

This warm, low salinity water is a permanent feature of the Grain Coast and the Bight of Biafra, and a seasonal feature of the whole coast from Mauritania to Angola. Outside this region the marine waters are generally much colder, the result of currents or upwellings. These oceanographic factors combine to restrict significant coral growth to shallow protected bays, outside which the number of species and size of coral colonies rapidly decrease. In open water, hermatypic corals are generally temperature limited to depths shallower than 20 meters with some exceptions in the offshore archipelagos.

Two different types of coral community have been described. The more common one comprises *Millepora alcicornis* and three species of *Porites*, two species of *Siderastrea*, *Favia* and *Madracis*, as well as *Montastrea cavernosa*, with three species of ahermatypic scleractinian coral (*Phyllangia americana*, *Tubastrea* sp. and *Dendrophyllia dilatata*). This type of community is found mainly in the islands, though it also occurs in more brackish coastal waters. The second community type consists of colonies of the monospecifc genus *Schizoculina* which is endemic to the Gulf of Guinea. Various theories exist as to the evolutionary origin of the West African coral communities. It has been proposed that they have developed either as a result of long distance dispersion from the Caribbean via Bermuda and the Azores, or from Brazil, or even that they could include some relict species from the ancient Mediterranean-Tethys Sea. Very little is known about sub-tidal benthic communities over wide areas of West Africa, and it is quite possible that there are important and diverse coral communities in a number of areas which are yet to be documented.

The great barracuda, Sphyraena barracuda, *can grow to nearly 2 meters in length. Brazil is the southernmost portion of its Atlantic range.*

Selected bibliography

HAITI, THE DOMINICAN REPUBLIC AND NAVASSA ISLAND

Geraldes FX (1998). Parque Nacional del Este, Dominican Republic. In: Kjerfve B (ed). *CARICOMP – Caribbean Coral Reef, Seagrass and Mangrove Sites.* UNESCO, Paris, France.

Luczkovich JJ, Wagner TW et al (1993). Discrimination of coral reefs, seagrass meadows, and sand bottom types from space – a Dominican Republic case-study. *Photogrammetric Engineering and Remote Sensing* 59(3): 385-389.

UNDP (1995). *Creation of Les Arcadins Marine Park and Fisheries Project.* UNDP Project Document.

UNESCO (1997). *Coasts of Haiti – Resource Assessment and Management Needs. Results of a Seminar and Related Field Activities.* Coastal Region and Small Island Papers 2. UNESCO, Paris, France.

Williams EH Jr, Clavijo I, Kimmel JJ, Colin PL, Diaz Carela C, Bardales AT, Armstrong RA, Bunkley-Williams L, Boulon RH, Garcia JR (1983). A checklist of marine plants and animals of the south coast of the Dominican Republic. *Carib J Sci* 19: 39-53.

PUERTO RICO AND THE VIRGIN ISLANDS

Bruckner AW, Bruckner RJ (1997). Outbreak of coral disease in Puerto Rico. *Coral Reefs* 16(4): 260.

Bythell JC, Bythell M et al (1993). Initial results of a long-term coral reef monitoring program – impact of Hurricane Hugo at Buck Island Reef National Monument, St-Croix, United States Virgin Islands. *J Exp Mar Biol Ecol* 172(1-2): 171-183.

Edmunds PJ (1991). Extent and effect of black band disease on a Caribbean reef. *Coral Reefs* 10(3): 161-165.

Lirman D, Fong P (1997). Patterns of damage to the branching coral *Acropora palmata* following Hurricane Andrew: damage and survivorship of hurricane-generated asexual recruits. *J Coast Res* 13(1): 67-72.

Macintyre IG, Raymond B, Stuckenrath R (1983). Recent history of a fringing reef, Bahia Salina del Sur, Vieques Island, Puerto Rico. *Atoll Res Bull* 268: 1-6.

Rogers CS, McLain LN et al (1991). Effects of Hurricane Hugo (1989) on a coral reef in St. John, USVI. *Mar Ecol Prog Ser* 78(2): 189-199.

THE LESSER ANTILLES, TRINIDAD AND TOBAGO

Bouchon-Navaro Y, Louis M, Bouchon C (1997). Trends in fish species distribution in the West Indies. *Proc 8th Int Coral Reef Symp* 1: 987-992.

Gabrié C (2000). *State of Coral Reefs in French Overseas Departements and Territories.* Ministry of Spatial Planning and Environment and State Secretariat for Overseas Affairs, Paris, France.

Humphrey JD (1997). Geology and hydrogeology of Barbados. In: Vacher HL, Quinn T (eds). *Developments in Sedimentology, 54: Geology and Hydrology of Carbonate Islands.* Elsevier Science BV, Amsterdam, Netherlands.

Nowlis JS, Roberts CM, Smith AH, Siirila E (1997). Human-enhanced impacts of a tropical storm on nearshore coral reefs. *Ambio* 26/8: 515-521.

Polunin NVC, Roberts CM (1993). Greater biomass and value of target coral-reef fishes in two small Caribbean marine reserves. *Mar Ecol Prog Ser* 100: 167-176.

Rakitin A, Kramer DL (1996). Effect of a marine reserve on the distribution of coral reef fishes in Barbados. *Mar Ecol Prog Ser* 131: 97-113.

Sheppard CRC, Matheson K, Bythell JC, Blair Myers C, Blake B (1995). Habitat mapping in the Caribbean for management and conservation: use and assessment of aerial photography. *Aquatic Conservation: Marine and Freshwater Ecosystems* 5: 277-298.

VENEZUELA AND ARUBA, BONAIRE AND CURAÇAO

Bone D, Perez D, Villamizar A, Penchaszadeh P, Klein E (1998). Parque Nacional Morrocoy, Venezuela. In: Kjerfve B (ed). *CARICOMP – Caribbean Coral Reef, Seagrass and Mangrove Sites.* UNESCO, Paris, France.

De Meyer K (1998). Bonaire, Netherlands Antilles. In: Kjerfve B (ed). *CARICOMP – Caribbean Coral Reef, Seagrass and Mangrove Sites.* UNESCO, Paris, France.

Leendert PJ, Pors J, Nagelkerken IA (1998). Curaçao, Netherlands Antilles. In: Kjerfve B (ed). *CARICOMP – Caribbean Coral Reef, Seagrass and Mangrove Sites.* UNESCO, Paris, France.

Meesters EH, Knijn R, Willemsen P, Pennartz R, Roebers G, van Soest RMW (1991). Sub-rubble communities of Curaçao and Bonaire coral reefs. *Coral Reefs* 10: 189-197.

BRAZIL AND WEST AFRICA

Amaral FD (1994). Morphological variation in the reef coral *Montastrea cavernosa* in Brazil. *Coral Reefs* 13: 113-117.

Amaral FD, Hudson MM, Coura MF (1998). Levantamento preliminar dos corais e hidrocorais do Parque Estadual Marinho do Parcel do Manuel Luiz (MA). *Resumos do XIII Simpósio de Biologia Marinha.* São Sebastio, Cebimar-USP. 13.

Laborel J (1974). West African corals: an hypothesis on their origin. *Proc 2nd Int Coral Reef Symp* 1: 452-443.

Leão ZMAN, Tellas MD, Sforza R, Bulhoes HA, Kikuchi RKP (1994). Impact of tourism development on the coral reefs of the Abrolhos area, Brazil. In: Ginsburg RN (ed). *Proceedings of the Colloquium on Global Aspects of Coral Reefs: Health, Hazards and History, 1993.* University of Miami, Miami, Florida, USA. 255-260.

Leão ZMAN, Ginsburg RN (1997). Living reefs surrounded by siliciclastics sediments: the Abrolhos coastal reefs, Bahia, Brazil. *Proc 8th Int Coral Reef Symp* 2: 1767-1772.

Leão de Moura R, Martins Rodrigues MC, Francini-Filho RB, Sazima I (1999). Unexpected richness of reef corals near the southern Amazon river mouth. *Coral Reefs* 18: 170.

Maida M, Ferreira BP (1997). Coral reefs of Brazil: an overview. *Proc 8th Int Coral Reef Symp* 1: 263-274.

Testa V (1996). Calcareous algae and corals in the inner shelf of Rio Grande do Norte, NE Brazil. *Proc 8th Int Coral Reef Symp*: 737-742.

Werner TB, Pinto LP, Dutra GF, Pereira PG do P (2000). Abrolhos 2000: conserving the Southern Atlantic's richest coastal biodiversity into the next century. *Coastal Management* 28: 99-108.

Map sources

Map 6a

For the Dominican Republic coral reefs are taken from Hydrographic Office (1970, 1985, 1986, 1990, 1991). Most of this information is derived from data gathered during the 1980s, although some surveys were conducted in the 1940s. For Haiti coral reef data is taken from Petroconsultants SA (1990)*, with some additional reef areas added from UNEP/IUCN (1988a)*.

Hydrographic Office (1970). Eastern Part of Haiti to Puerto Rico including Mona Passage. *British Admiralty Chart No. 3689.* 1:614 000. Taunton, UK.

Hydrographic Office (1985). West Indies Plans on the North Coast of the Dominican Republic. Punta Mangle to Pointe Yaquezi and Bahia de Samana and Approaches. *British Admiralty Chart No. 463.* 1:200 000. Taunton, UK.

Hydrographic Office (1986). West Indies Dominican Republic. Bayajibe to Haina. *British Admiralty Chart No. 467.* 1:200 000. Taunton, UK.

Hydrographic Office (1990). West Indies Dominican Republic and Puerto Rico. Mona Passage. *British Admiralty Chart No. 472.* 1:200 000. Taunton, UK.

Hydrographic Office (1991). West Indies Dominican Republic – South Coast. Cabo Caucedo to Isla Alto Velo. *British Admiralty Chart No. 471.* 1:200 000. Taunton, UK.

Map 6b

For Puerto Rico and for the US Virgin Islands coral reefs have been taken from UNEP/IUCN (1988a)*. at scales of 1:700 000 and 1:100 000 respectively. For the British Virgin Islands reefs are based on DOS (1982).

DOS (1982). *British Virgin Islands.* 1:63 360. Directorate of Overseas Surveys, UK.

Maps 6c and 6d

Coral reef data were taken from UNEP/IUCN (1988a)* for the following countries: Antigua and Barbuda at 1:150 000; Barbados at 1:90 000; Dominica at 1:90 000; Netherlands Antilles at 1:300 000; St. Lucia at 1:150 000 (and below).

For Guadeloupe coral reefs are derived from IGN (1988). For Martinique coral reefs are derived from Hydrographic Office (1991a, 1991b), which are based on French Government charts of 1984 to 1988 with later corrections. For Montserrat, coral reefs are derived from Hydrographic Office (1986). For Saba, reefs were digitized from a sketch map at c.1:30 000 prepared by K Buchan (Park manager, Saba Marine Park). For St. Kitts and Nevis reefs are derived from DOS (1979), which is based on 1:25 000 DOS maps prepared from 1968 air photography and field surveys to 1972. Additional coral reef data for St. Lucia are taken from Hydrographic Office (1995a). For St. Vincent reefs are taken from Hydrographic Office (1995a, 1995b), which is mostly based on admiralty surveys from 1858-89 and 1933-35.

DOS (1979). *Saint Christopher and Nevis.* 1:50 000. Department of Overseas Surveys, London, UK.

Hydrographic Office (1986). Montserrat and Barbuda. *British Admiralty Chart No. 254.* 1:50 000. July 1986. Taunton, UK.

Hydrographic Office (1991a). Northern Martinique: Pointe Caracoli to Fort-de-France. *British Admiralty Chart No. 371.* 1:75 000. April 1991. Taunton, UK.

Hydrographic Office (1991b). Northern Martinique: Fort-de-France to Pointe Caracoli. *British Admiralty Chart No. 494.* 1:75 000. April 1991. Taunton, UK.

Hydrographic Office (1995a). West Indies: Southern Martinique to Saint Vincent. *British Admiralty Chart No. 596.* 1:175 000. January 1995. Taunton, UK.

Hydrographic Office (1995b). West Indies: Saint Vincent to Grenada. *British Admiralty Chart No. 597.* 1:175 000. September 1995. Taunton, UK.

IGN (1988). Guadeloupe. *Carte 510, Edition 5.* 1:100 000. Institut Géographique National, Paris, France.

Map 6e

For Curaçao and Bonaire, coral reefs have been taken from UNEP/IUCN (1988a)* at 1:550 000. For Aruba, coral reefs are taken from Hydrographic Office (1987). For Venezuela, coral reef data have been taken as arcs from Petroconsultants SA (1990)*, with some additional reef areas for Morrocoy, Isla la Orchilla and La Blanquilla added from UNEP/IUCN (1988a).

Hydrographic Office (1987). Aruba and Curaçao. *British Admiralty Chart No. 702.* 1:100 000. August 1987. Taunton, UK.

Map 6f

Coral reefs are largely taken from UNEP/IUCN (1988a)* at an approximate scale of 1:10 000 000 (and 1:2 000 000 for parts of northeast Brazil). Further detail has been added for the Manoel Luis reefs based on sketch maps in Leão de Moura et al (1999) and for the Abrolhos region based on a 1:1 000 000 sketch map in Leão et al (1988).

Leão ZMAN, Araujo TMF, Nolasco MC (1988). The coral reefs off the coast of eastern Brazil. *Proc 6th Int Coral Reef Symp*: 339-347.

Leão de Moura R, Martins Rodrigues MC, Francini-Filho RB and Sazima I (1999). Unexpected richness of reef corals near the southern Amazon River mouth. *Coral Reefs* 18: 170.

* See Technical notes, page 401

Part III

The Indian Ocean and Southeast Asia

The Indian Ocean is the third largest ocean. It is closed to the north, and a large proportion of its waters are tropical or near tropical. Unlike the Atlantic it is largely bounded by relatively arid countries and does not receive particularly high inputs of freshwater or terrestrial sediments. The great exception to this is the Bay of Bengal in the northeast, which is fed by massive riverine discharge from a number of rivers, leading to conditions of high sediments and low, fluctuating salinities – inimical to coral reef development. To the northwest are the enclosed sea areas of the Red Sea and the Arabian Gulf, with very different tectonic histories, but both occurring in highly arid regions with little terrestrial runoff. The coast of East Africa is also relatively dry. Continental shelf areas are generally narrow, although there are a few nearshore island groups which are important for coral reef development. There are also several oceanic island groups, notably in the west and central parts of this ocean. The largest chain of islands follows the Chagos-Laccadive Ridge, a volcanic trace which has formed the Lakshadweep Islands in India, the Maldives and the Chagos Archipelago. Réunion and Mauritius are high islands lying at the most recent end of this volcanic trace, with active vulcanism on Réunion. The Seychelles form a more complex group of islands with varied origins. To the east there are fewer remote oceanic islands, and the region blends into the reefs of Southeast Asia with the island chains of the Andaman and Nicobar Islands, and the Mentawai Islands to the west of Sumatra in Indonesia.

There are large areas of coral reef right across this region, making up nearly 20 percent of the world total. Fringing reefs predominate along much of the Red Sea, particularly northern and central parts. Further south in the Red Sea and Arabian Gulf coastal sediments and high salinities restrict fringing reef development, though there are extensive offshore patch reefs. Cool upwellings limit the development of true reefs along parts of southern Arabia and Pakistan. Further south there are fringing communities on the coasts of East Africa, and particularly along the shores of continental islands. Some of the best developed reef structures occur in isolated oceanic locations. There are numerous atolls and platform structures in the west and central regions of the oceans, and the Maldives and Chagos Archipelago include the largest atoll structures in the world. The continental coastlines of India and Sri Lanka have very limited reef development as there are various adverse conditions, including high turbidity, fluctuating salinity and high wave energy. There are important though little known reefs around the Andaman and Nicobar Islands and to the north of Sumatra. Australia also has significant reef communities, including extensive fringing reefs, offshore platform and barrier structures and high latitude communities (these are considered further in Chapter 11).

Species diversity is high across the region, following a narrow band of high diversity in the Central Indian Ocean and forming two distinctive sub-centers of diversity in the Western Indian Ocean and the Red Sea. Elsewhere there is greatly reduced diversity, notably in the Arabian Gulf and along the shores of mainland India. Despite their high latitude there is relatively little diminution of diversity in the reefs of the northern Red Sea. By contrast there are latitudinal declines in species numbers, and as a consequence in the development of reefs themselves, in both southern Africa and Western Australia.

Wide areas of this region were affected by the 1997-98 El Niño Southern Oscillation event, and in 1998 warm waters swept across wide areas of the Indian Ocean, leading to bleaching and massive levels of coral mortality on reefs from Western Australia to the shores of East Africa. In the Maldives, Chagos and Seychelles (which together make up over 5 percent of the world's coral reefs), more than 60 percent of corals died in all areas, and up to 100 percent of corals were lost in some places. Recovery has now begun in most areas, but the overall scale of this event was so large that full recovery may take years or decades, while there are concerns that such events may be repeated with global climate change.

Direct human pressures on coral reefs in the Indian Ocean are highly varied. The Arabian Gulf contains the largest concentrations of oil reserves in the world, and there is chronic oil pollution in this sea, exacerbated by occasional massive oil spills. Tanker traffic also carries the threat of oil pollution to other areas, notably the narrow straits at the mouth of the Arabian Gulf, and at the northern and southern ends of the Red Sea. Coastal development is sporadic – there are vast areas of the Arabian coastline with little development, but in others, such as around the major ports and some of the tourist areas of East Africa, coastal development is having a direct impact on reefs. Tourism too is sporadic, but is critical to the economies of Egypt, Kenya, Tanzania and the islands of the Indian Ocean.

Southeast Asia

Southeast Asia is one of the most important areas in the world for coral reefs. Over 30 percent of the world's reefs are found in this complex region which straddles the waters between the Indian and Pacific Oceans. The region includes the continental coastlines of Myanmar and Thailand, Malaysia, Cambodia, Vietnam and China. The most extensive coastlines, however, follow the great complex of islands which are dominated by the Philippines and Indonesia. Indonesia alone has over 50 000 square kilometers of coral reefs, nearly 18 percent of the world total. Japan lies on the edges of this region, and also has a considerable area of reefs surrounding island chains which follow natural clines of diminishing species diversity. The coastlines define a large number of partially enclosed seas. While the waters around the continental shores are generally shallow, deep oceanic waters come in close contact with offshore islands in many areas.

Fringing reefs predominate, although there are also extensive barrier reef systems, and a number of atolls and near-atolls. This region is the great center of coral reef biodiversity, and there are more species here, in almost all animal groups, than anywhere else. To some degree this diversity is encouraged or maintained by the complexity of coastline and the great range of habitats found in the region, but its ultimate origins can be traced back over geological timescales. While extinctions were occurring in other regions it would appear that species were able to survive in this region, and even to diversify as sea levels fluctuated and areas became isolated and then reconnected with one another.

Unfortunately this region is also the most threatened and disturbed by human activities. Some 82 percent of the region's reefs were considered to be threatened by human activities in the 1998 *Reefs at Risk* study. Most of these threats are linked to the rapidly growing economies and populations in this region. These are driving massive changes in the landscape, with forest clearance and agricultural intensification leading to increased sedimentation and pollution from agricultural chemicals. Massive urban expansion has also led to enormous pollution problems close to urban centers. Fishing pressures are ubiquitous, from chronic overfishing for local consumption and the highest rates of blast fishing in the world, to target species overfishing even in many of the remotest parts of the region.

Chapter 7 Western Indian Ocean

The Western Indian Ocean is distinctive in terms of coral reefs. Its eastern boundary is marked by the Seychelles and the shallow Mascarene Ridge which extends down towards Mauritius. East of these is a considerable expanse of deep ocean separating the reefs of this region from those of the Maldives and Chagos. The region's southern margins are the oceanic waters to the south of Mauritius, Réunion and Madagascar, while on the continental coastlines of southern Mozambique and northern South Africa the gradual cooling of water temperatures is mirrored by a diminution in coral diversity. The growth of reefs ceases close to this international border. The northern edge of the region ends along the eastern coast of Somalia where coral growth again becomes highly restricted, here by cold water upwellings associated with regional patterns in oceanic currents.

The reefs of mainland East Africa are predominantly fringing, closely following the coastlines of the mainland and islands on the continental shelf. Madagascar has some discontinuous fringing reef development, as well as some barrier reef systems off its west coast. The remainder of the region is dominated by oceanic islands. The northern Seychelles are actually a remnant of continental crust, with high islands and fringing reef systems. There are also two volcanic island chains, the Comoros and the Réunion-Rodrigues chain. Both show classic reef development, with limited fringing reefs on the most recent islands, but wide fringing and barrier reef development on the older ones. The Réunion hotspot also produced the vast reef areas of the Chagos-Laccadive ridge across the Central Indian Ocean (Chapter 8).

The reefs of this region have high levels of species diversity. Although they are similar to those of the Central Indian Ocean, there are distinctive and endemic species which have led to the recognition by some authors of a Western Indian Ocean center of diversity. Detailed knowledge of the reef biotas is lacking for much of the region, a factor which is related to a lack of infrastructure and indigenous expertise combined with problems of national security in some areas. This region was badly damaged by the coral bleaching event of 1998, with many areas suffering over 50 percent mortality.

Human populations along much of the coastline are rapidly increasing. Most of the coastal populations are very poor, and heavily dependent on the adjacent reefs for food. Unfortunately there is little control over the utilization of these resources, either through traditional or formal management regimes, and large areas of reefs have been degraded through overfishing or destructive fishing techniques. Growing interest in the reefs for tourism is leading to new pressures in some areas, however it is also providing an economically powerful incentive for protecting reefs, and there is considerable potential for environmentally sustainable tourism developments.

Left: The Seychelles anemonefish Amphiprion fuscocaudatus *with dominos* Dacyllus trimaculatus *in a giant anemone. The anemonefish are often restricted to particular countries or regions. Right: The brightly colored* Fromia monilis *starfish is common on reefs throughout the Indian Ocean and is a regionally important grazer of algae.*

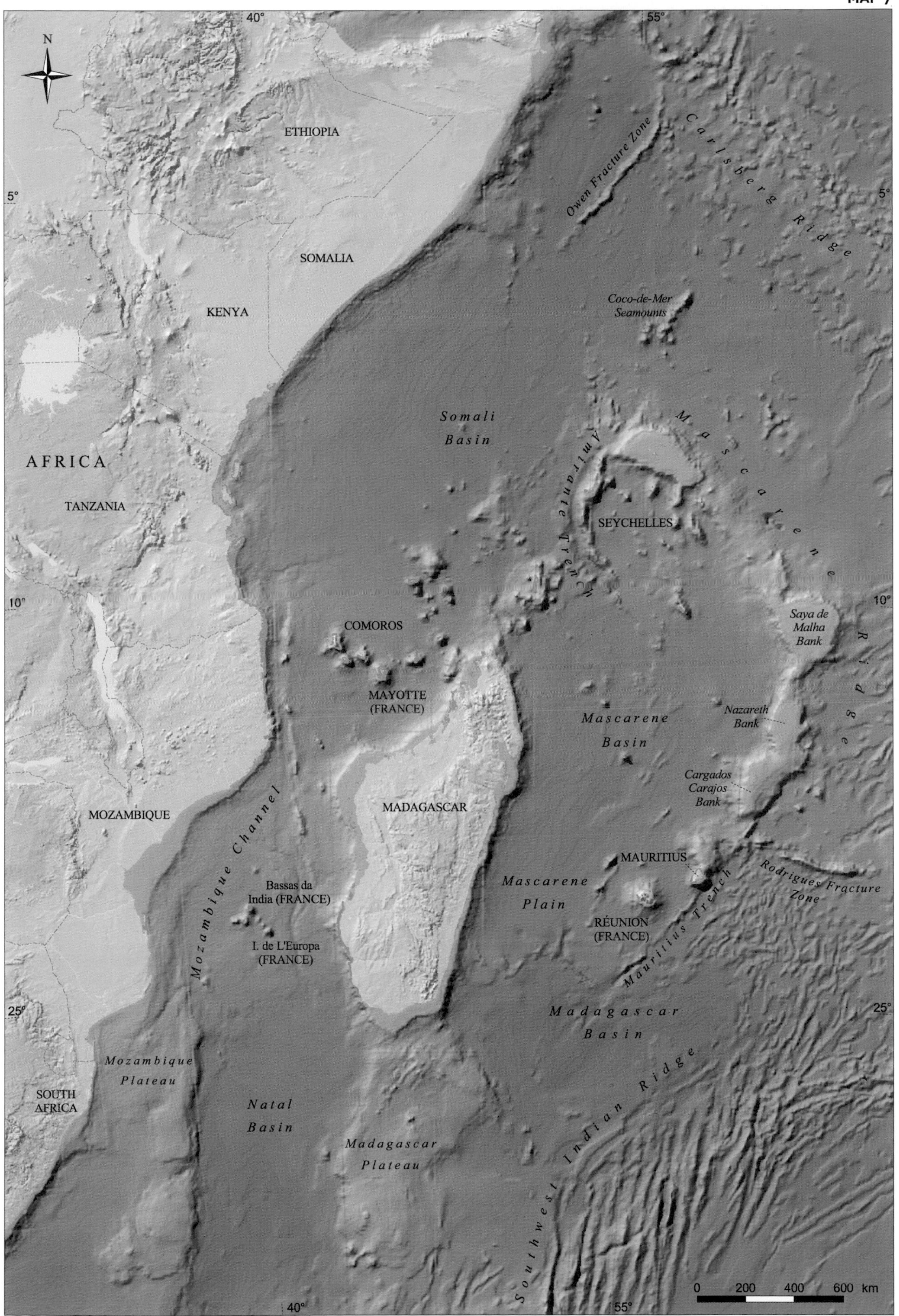

N
40°
55°
5°
10°
25°
ETHIOPIA
SOMALIA
KENYA
AFRICA
TANZANIA
MOZAMBIQUE
SOUTH AFRICA
Owen Fracture Zone
Carlsberg Ridge
Coco-de-Mer Seamounts
Somali Basin
Amirante Trench
Mascarene Ridge
SEYCHELLES
Saya de Malha Bank
COMOROS
MAYOTTE (FRANCE)
Nazareth Bank
Mascarene Basin
Cargados Carajos Bank
MADAGASCAR
Mozambique Channel
MAURITIUS
Rodrigues Fracture Zone
Mascarene Plain
Bassas da India (FRANCE)
I. de L'Europa (FRANCE)
RÉUNION (FRANCE)
Mauritius Trench
Madagascar Basin
Mozambique Plateau
Natal Basin
Madagascar Plateau
Southwest Indian Ridge
0 200 400 600 km

MAP 7a

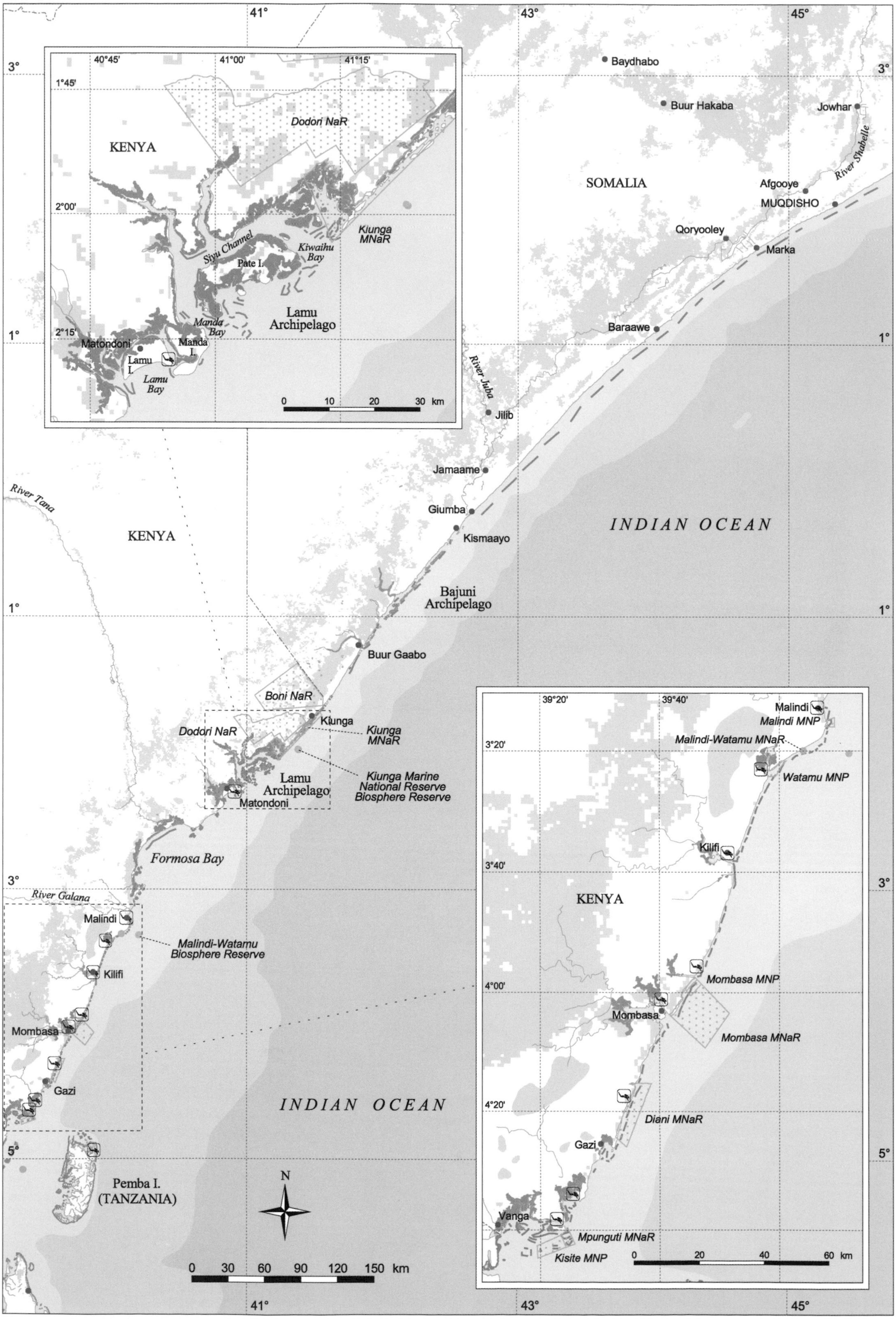

41° 43° 45°
3° 3°
1° 1°
5° 5°

KENYA

40°45' 41°00' 41°15'
1°45'
2°00'
2°15'
Dodori NaR
Siyu Channel
Pate I.
Kiwaihu Bay
Kiunga MNaR
Lamu Archipelago
Manda Bay
Manda I.
Matondoni
Lamu I.
Lamu Bay
0 10 20 30 km

SOMALIA
Baydhabo
Buur Hakaba
Jowhar
River Shabelle
Afgooye
MUQDISHO
Qoryooley
Marka
Baraawe
River Juba
Jilib
Jamaame
Giumba
Kismaayo
INDIAN OCEAN
Bajuni Archipelago
Buur Gaabo
River Tana
KENYA
Boni NaR
Kiunga
Dodori NaR
Kiunga MNaR
Lamu Archipelago
Matondoni
Kiunga Marine National Reserve Biosphere Reserve
Formosa Bay
River Galana
Malindi
Malindi-Watamu Biosphere Reserve
Kilifi
Mombasa
Gazi
INDIAN OCEAN
Pemba I. (TANZANIA)
N
0 30 60 90 120 150 km

39°20' 39°40'
3°20'
3°40'
4°00'
4°20'
Malindi
Malindi MNP
Malindi-Watamu MNaR
Watamu MNP
Kilifi
KENYA
Mombasa MNP
Mombasa
Mombasa MNaR
Diani MNaR
Gazi
Vanga
Mpunguti MNaR
Kisite MNP
0 20 40 60 km

Kenya and southern Somalia

MAP 7a

Kenya has a relatively narrow coastal plain to the south, with a series of raised Pleistocene reef platforms above the present day intertidal platform. North of Malindi the coastal plain becomes much broader and is dominated by older sedimentary plains. There is a relatively narrow continental shelf, only extending about 5 kilometers offshore south of Malindi, broadening to 60 kilometers offshore in the north. There are two major permanent rivers, the Athi-Galana-Sabaki which reaches the coast in an estuary just north of Malindi, and the Tana which lies further north again and reaches the coast in a large delta with associated swamps, mangrove communities and shifting dunes. There are several nearshore islands, notably those of the Lamu Archipelago in the mouth of Lamu and Manda Bays, but also a chain of about 50 calcareous barrier islands further north around Kiunga. Patterns of coastal currents are largely driven by the major oceanographic currents. South of Malindi, the East African Coastal Current flows northeast throughout the year coming up from Tanzania and originally driven by the South Equatorial Current. North of Malindi this same East African Coastal Current continues to flow for part of the year during the Southeast Monsoon (April-October). During the Northeast Monsoon (December-March), however, it is reversed, countered by the southward flowing Somali Current. Around Malindi the two currents meet and flow out to sea, forming the North Equatorial Counter Current.

Fringing reefs are well developed in southern Kenya. However, to the north, where there are large areas of loose sediment and significant freshwater influences, levels of development are lower. There are fringing reefs in places off the Lamu islands, and also along many of the barrier islands further north.

Patterns of biodiversity appear to follow patterns of reef development, with generally higher diversity in the south. Active coral growth is not continuous along the fringing reefs, but is interspersed with extensive seagrass and algal beds. Where hard substrates occur, live coral cover was typically about 30 percent prior to 1998. Some 55 coral genera and up to 200 species have been recorded in Kenya. Mangroves are widespread in creeks and inlets as well as the larger estuaries. Important mangrove communities are also found on the leeward shores of offshore islands and on their corresponding mainland coasts. There are very important nesting communities of terns and gulls on a number of offshore islands, notably the barrier islands in Kiunga. Much of the Kenyan

A saddleback butterflyfish Chaetodon falcula *over a shallow reef scene.*

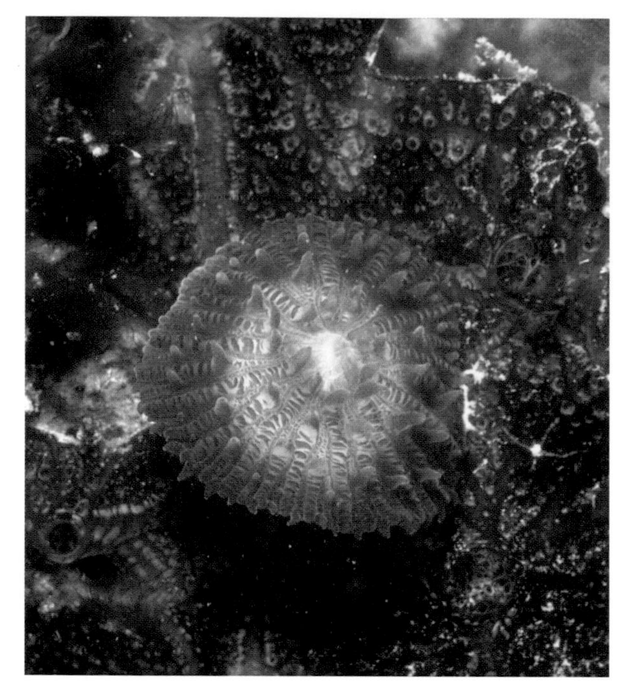

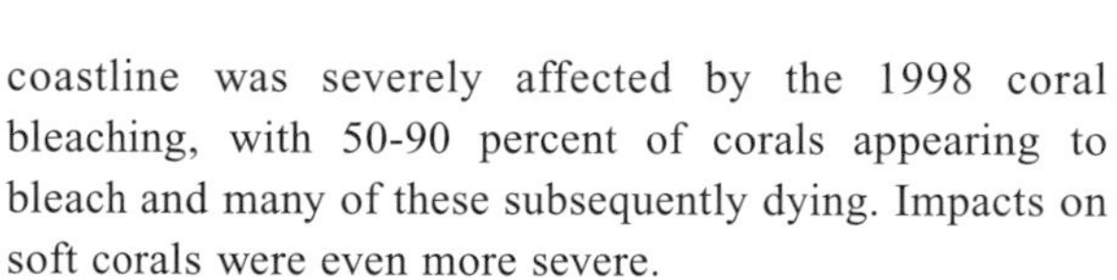

coastline was severely affected by the 1998 coral bleaching, with 50-90 percent of corals appearing to bleach and many of these subsequently dying. Impacts on soft corals were even more severe.

Coastal areas of Kenya are densely populated and there are large-scale artisanal and commercial fisheries. Fishing using handlines, traps, spearguns and gill and seine nets is common, with artisanal fishing concentrated in lagoons and commercial fishing also operating from sail-powered dhows. Other fisheries, including netting for aquarium fish and sport fishing in offshore waters, are increasing. Exploitation is heavy and stocks in several localities are considered to be overexploited. A number of marine parks and reserves have been established, however. Fishing is prohibited in the parks and only "traditional" methods of handlines and traps are permitted in the reserves. Protection of these areas has had clear impacts, with increases in fish abundance and diversity as well as live coral cover. Sea urchin densities are notably higher in non-protected reefs, and this may be impacting coral cover and topographic complexity.

Tourism is a major industry for Kenya, and of the 750 000 holiday makers visiting the country annually, 70 percent spend at least part of their time on the coast. Coastal tourism is particularly focussed in the southern areas, including Malindi, Mombasa and Diani. Many of the hotels have direct frontage onto the marine parks and so visitor numbers are high. Diving is a popular activity in many of the southern reefs, but there is only a limited infrastructure for recreational diving in the north. Diving activities peak from October until April when water conditions tend to be clearer and calmer.

Overexploitation is a continuing problem on many Kenyan reefs, including illegal activities in protected areas, although policing is increasingly effective in places. There has been local opposition to the establishment of the Diani Marine Reserve due to a perceived loss of benefits – efforts are being made to address this. Increasing levels of sediment load arising from *ex situ* changes in landuse are a problem, particularly in the Athi-Galana-Sabaki River, and are probably affecting reefs near Malindi. Direct physical damage by divers (primarily coral breakage) has been clearly demonstrated, but only affects small areas

Kenya

GENERAL DATA	
Population (thousands)	30 340
GDP (million US$)	9 621
Land area (km²)	587 709
Marine area (thousand km²)	117
Per capita fish consumption (kg/year)	5
STATUS AND THREATS	
Reefs at risk (%)	91
Recorded coral diseases	0
BIODIVERSITY	
Reef area (km²)	630
Coral diversity	na / 237
Mangrove area (km²)	530
No. of mangrove species	9
No. of seagrass species	13

Left: Zoanthids, like this one in Kenya, are closely related to scleractinian corals and sea anemones. Right: A nalolo blenny Ecsenius nalolo *shelters under the protective spines of a sea urchin.*

Protected areas with coral reefs

Site name	Designation	Abbreviation	IUCN cat.	Size (km²)	Year
Kenya					
Diani	Marine National Reserve	MNaR	VI	75.00	1993
Kisite	Marine National Park	MNP	II	28.00	1978
Kiunga	Marine National Reserve	MNaR	VI	250.00	1979
Malindi	Marine National Park	MNP	II	6.30	1968
Malindi-Watamu	Marine National Reserve	MNaR	VI	177.00	1968
Mombasa	Marine National Park	MNP	II	10.00	1986
Mombasa	Marine National Reserve	MNaR	VI	200.00	1986
Mpunguti	Marine National Reserve	MNaR	VI	11.00	1978
Watamu	Marine National Park	MNP	II	32.00	1968
Kiunga Marine National Reserve	**UNESCO Biosphere Reserve**			**600.00**	**1980**
Malindi-Watamu Biosphere Reserve	**UNESCO Biosphere Reserve**			**196.00**	**1979**

and is probably countered by the increased protection given to important tourist sites. Anchor damage in the protected areas is limited through a system of user moorings, however these are rarely used outside protected areas. There do not appear to be major impacts associated with eutrophication.

Southern Somalia

The coastline of southern Somalia in many ways continues the patterns of northern Kenya. The continental shelf begins to narrow again. The Shabelle and Juba Rivers converge and enter the sea near Kismaayo where there is an estuarine and mangrove system. Close to the Kenyan border there is a continuation of the chain of small barrier islands, known as the Bajuni Archipelago. The same patterns of ocean currents are found in this region as in northern Kenya. During the Northeast Monsoon the Somali Current flows from the northeast while during the Southeast Monsoon the East African Coastal Current reaches considerable strengths, generating cold upwellings which are responsible for inhibiting reef development further north along this coastline. Fringing reefs are relatively well developed in the south and around the islands of the Bajuni Archipelago, but further north the diversity and abundance of living coral decreases, although fossil structures remain. Data on biodiversity are unavailable for southern Somalia, but it is likely that the trend of decreasing diversity towards the north continues.

Political instability in southern Somalia has prevented the gathering of information about the reefs, or their utilization by local human populations in Somalia for a number of years. The instability is clearly a problem for biodiversity protection, particularly in the south, where security is weakest. There are currently no effective legal controls on the exploitation of natural resources, and these are clearly not a priority. In some areas natural resources may actually be protected by such instability, but over-exploitation is likely to be an issue.

Somalia

General Data	
Population (thousands)	7 253
GDP (million US$)	686
Land area (km²)	639 129
Marine area (thousand km²)	828
Per capita fish consumption (kg/year)	2
Status and Threats	
Reefs at risk (%)	95
Recorded coral diseases	0
Biodiversity	
Reef area (km²)	710
Coral diversity*	59 / 308
Mangrove area (km²)	910
No. of mangrove species	6
No. of seagrass species	4

* The higher coral diversity figure is likely to be a considerable overestimate as it is based on the biogeographic region which includes the Gulf of Aden and Socotra

Tanzania

MAP 7b

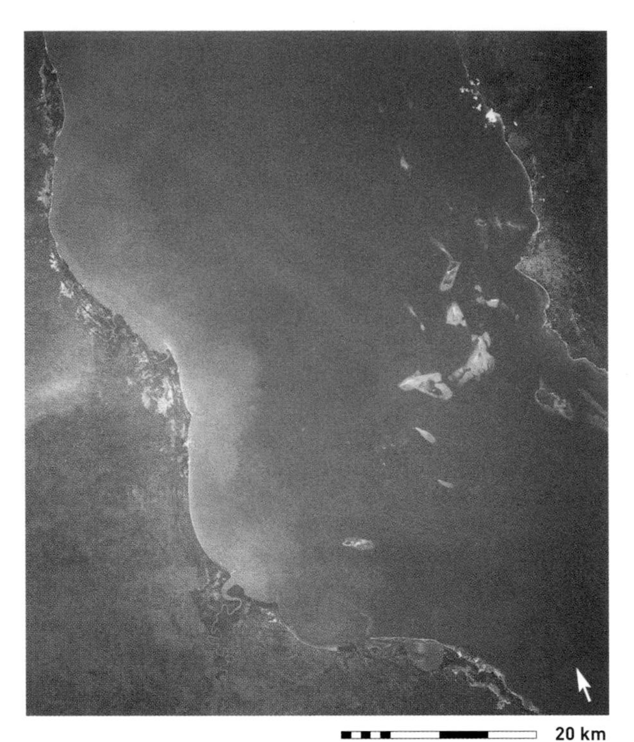

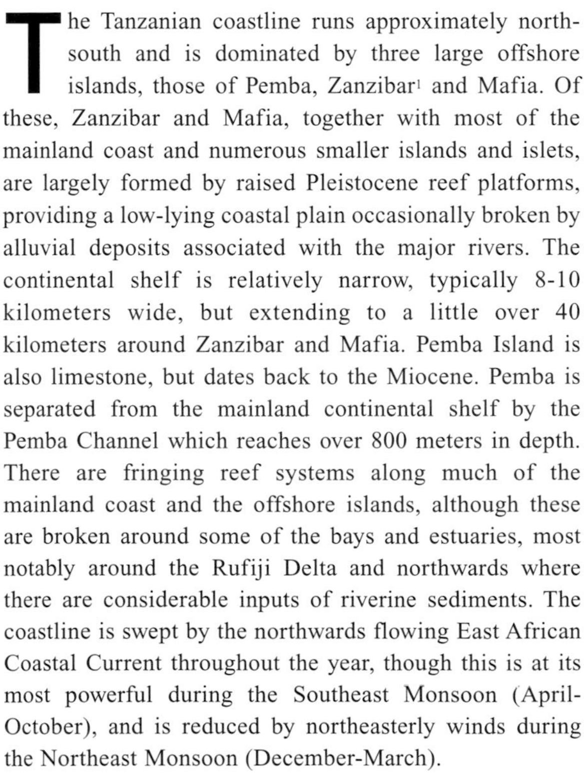

The Tanzanian coastline runs approximately north-south and is dominated by three large offshore islands, those of Pemba, Zanzibar[1] and Mafia. Of these, Zanzibar and Mafia, together with most of the mainland coast and numerous smaller islands and islets, are largely formed by raised Pleistocene reef platforms, providing a low-lying coastal plain occasionally broken by alluvial deposits associated with the major rivers. The continental shelf is relatively narrow, typically 8-10 kilometers wide, but extending to a little over 40 kilometers around Zanzibar and Mafia. Pemba Island is also limestone, but dates back to the Miocene. Pemba is separated from the mainland continental shelf by the Pemba Channel which reaches over 800 meters in depth. There are fringing reef systems along much of the mainland coast and the offshore islands, although these are broken around some of the bays and estuaries, most notably around the Rufiji Delta and northwards where there are considerable inputs of riverine sediments. The coastline is swept by the northwards flowing East African Coastal Current throughout the year, though this is at its most powerful during the Southeast Monsoon (April-October), and is reduced by northeasterly winds during the Northeast Monsoon (December-March).

Coral reefs are well developed in many places. Close to the mainland there are fringing and patch reefs along much of the coast to the north of the Pangani River, with a wide lagoon with only occasional patch reefs further south around Dar es Salaam. Coral cover is highly varied, with estimates on different patch reefs varying between 1 and 80 percent. Clearly in some areas reefs are not actively developing, and represent little more than occasional coral growth on Pleistocene reef deposits. Coral diversity increases with distance from the coast, and up to 39 genera of coral have been reported from individual patch reefs off the Tanga coast. Fringing reefs begin again off the mainland coast south of the Rufiji Delta and are very well developed, with deep spur and groove formations on outer slopes. These are particularly well represented in the areas around and to the south of Mtwara where undamaged reefs, especially those further offshore, often show over 50 percent live coral cover.

Offshore reefs are highly developed around the main three islands, their associated islets, and the Songo Songo Archipelago in the south. Reefs around parts of Pemba are prolific, with corals recorded to 64 meters, and cover on western reef slopes at 21-60 percent. Cover tends to be low (rarely above 15 percent) on eastern shores of all the

Left: A fishing dhow just beyond the reef, with an uplifted coralline shore of Chumbe Island, Zanzibar. Right: The Zanzibar Channel in Tanzania, with numerous important patch reefs (STS026-42-85, 1988).

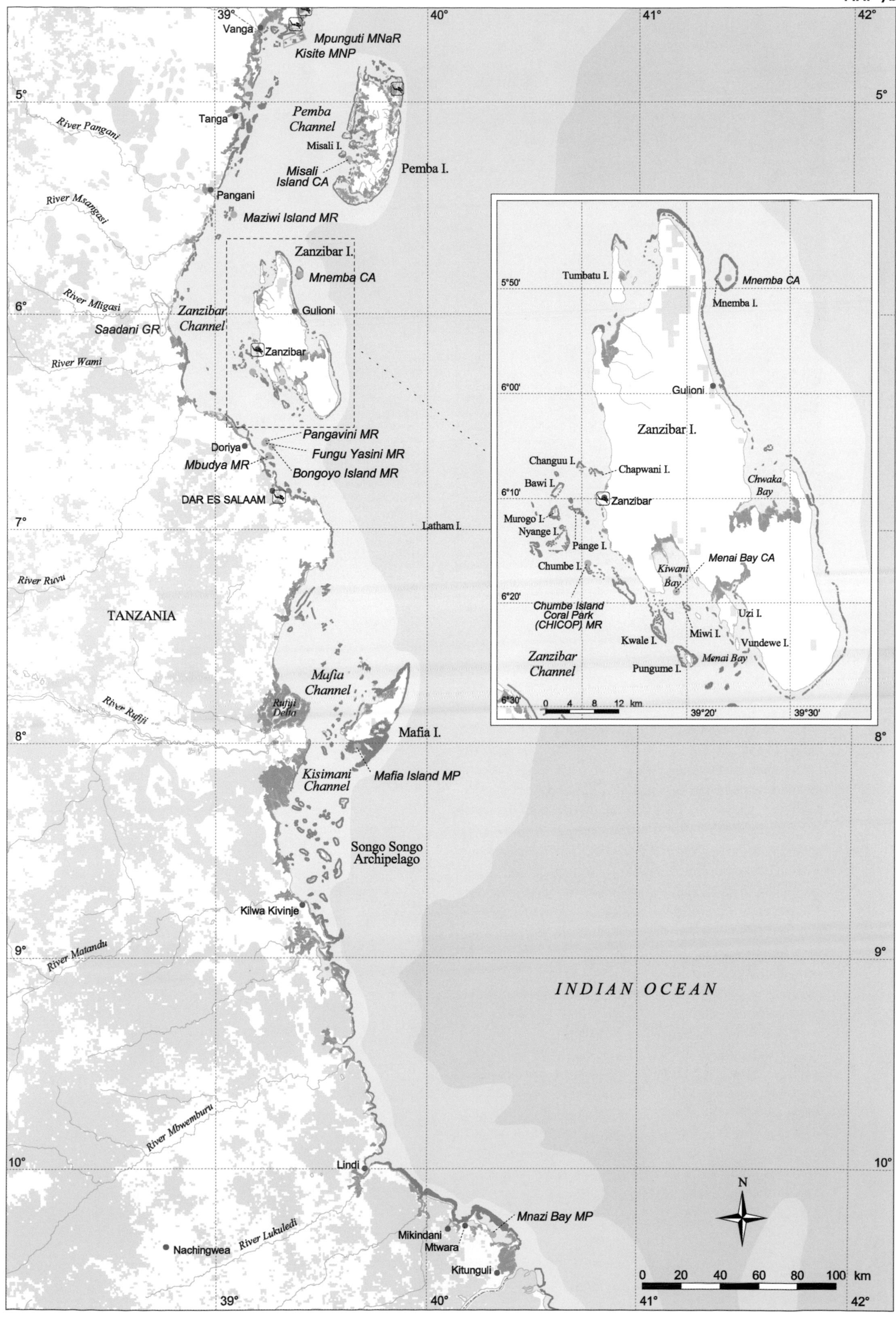

Vanga
Mpunguti MNaR
Kisite MNP
Tanga
River Pangani
Pemba Channel
Misali I.
Misali Island CA
Pemba I.
Pangani
River Msangasi
Maziwi Island MR
Zanzibar I.
Mnemba CA
River Mligasi
Gulioni
Zanzibar Channel
Saadani GR
Zanzibar
River Wami
Pangavini MR
Doriya
Fungu Yasini MR
Mbudya MR
Bongoyo Island MR
DAR ES SALAAM
Latham I.
River Ruvu
TANZANIA
Mufia Channel
Rufiji Delta
River Rufiji
Mafia I.
Kisimani Channel
Mafia Island MP
Songo Songo Archipelago
Kilwa Kivinje
River Matandu
INDIAN OCEAN
River Mbwemburu
Lindi
Mnazi Bay MP
Mikindani
Mtwara
Nachingwea
River Lukuledi
Kitunguli
N
0 20 40 60 80 100 km
Tumbatu I.
Mnemba I.
Zanzibar I.
Changuu I.
Chapwani I.
Bawi I.
Chwaka Bay
Murogo I.
Nyange I.
Pange I.
Chumbe I.
Menai Bay CA
Kiwani Bay
Chumbe Island Coral Park (CHICOP) MR
Uzi I.
Kwale I.
Miwi I.
Vundewe I.
Menai Bay
Pungume I.
0 4 8 12 km
5°50'
6°00'
6°10'
6°20'
6°30'
39°20'
39°30'
39°
40°
41°
42°
5°
6°
7°
8°
9°
10°

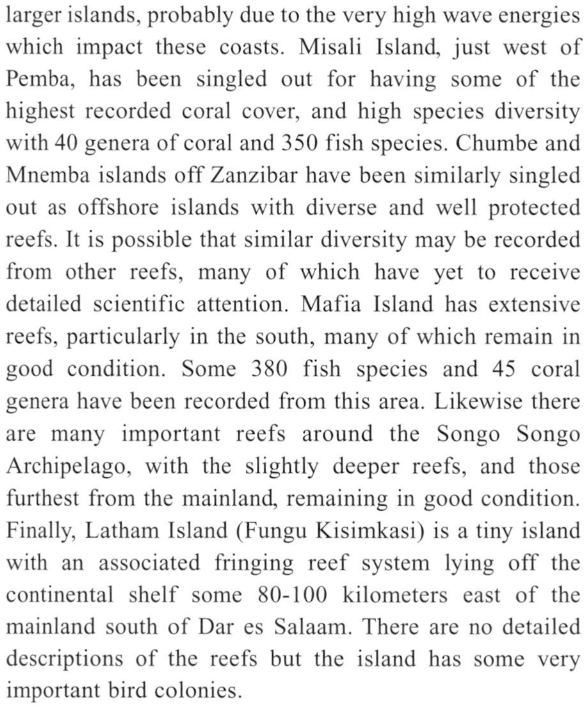
larger islands, probably due to the very high wave energies which impact these coasts. Misali Island, just west of Pemba, has been singled out for having some of the highest recorded coral cover, and high species diversity with 40 genera of coral and 350 fish species. Chumbe and Mnemba islands off Zanzibar have been similarly singled out as offshore islands with diverse and well protected reefs. It is possible that similar diversity may be recorded from other reefs, many of which have yet to receive detailed scientific attention. Mafia Island has extensive reefs, particularly in the south, many of which remain in good condition. Some 380 fish species and 45 coral genera have been recorded from this area. Likewise there are many important reefs around the Songo Songo Archipelago, with the slightly deeper reefs, and those furthest from the mainland, remaining in good condition. Finally, Latham Island (Fungu Kisimkasi) is a tiny island with an associated fringing reef system lying off the continental shelf some 80-100 kilometers east of the mainland south of Dar es Salaam. There are no detailed descriptions of the reefs but the island has some very important bird colonies.

Mangroves are well developed in most river mouths, and seagrass ecosystems are widespread, particularly in the shallow waters around the Mafia and Songo Songo Archipelagos. The 1998 coral bleaching event had a significant impact on most reefs, although this was far from uniform. Around Mafia Island reefs dominated by *Acropora* suffered 70-90 percent mortality, but those with less *Acropora* were far less affected. Similar local variation in the degree of impact between reefs was noted in Zanzibar.

The coastal population in Tanzania is very large, mostly concentrated in Tanga, Zanzibar, Dar es Salaam and Mtwara. Rapid population growth along the coasts, combined with poverty and poor management and understanding of coastal resources, has led to the rapid and

Tanzania

General Data	
Population (thousands)	35 306
GDP (million US$)	na
Land area (km²)	944 983
Marine area (thousand km²)	241
Per capita fish consumption (kg/year)	10
Status and Threats	
Reefs at risk (%)	99
Recorded coral diseases	0
Biodiversity	
Reef area (km²)	3 580
Coral diversity*	na / 314
Mangrove area (km²)	1 155
No. of mangrove species	10
No. of seagrass species	10

* The higher coral diversity figure is an estimate for Mozambique and Tanzania combined

Left: A hawksbill turtle Eretmochelys imbricata. *Right: The Rufiji Delta, showing the large inputs of sediments, but also the important areas of mangrove forest (STS026-42-87, 1988).*

Protected areas with coral reefs

Site name	Designation	Abbreviation	IUCN cat.	Size (km²)	Year
Tanzania					
Bongoyo Island	Marine Reserve	MR	II	na	1975
Chumbe Island Coral Park	Marine Sanctuary	MS	II	0.30	1994
Fungu Yasini	Marine Reserve	MR	II	na	1975
Mafia Island	Marine Park	MP	VI	822.00	1995
Maziwi Island	Marine Reserve	MR	II	na	1981
Mbudya	Marine Reserve	MR	II	na	1975
Menai Bay	Conservation Area	CA	VI	470.00	1997
Misali Island	Conservation Area	CA	VI	21.58	1998
Mnazi Bay	Marine Park	MP	VI	650.00	2000
Mnemba	Conservation Area	CA	VI	0.15	1997
Pangavini	Marine Reserve	MR	II	na	1975

extreme degradation of coral reefs and other coastal communities along large sectors of the coast. Fishing is a critical activity, providing a major protein source for much of the coastal population. Overfishing is a problem on most reefs, and has been exacerbated by destructive fishing practices. Most notable among these are various seine-net fishing techniques in which a small mesh (2-8 centimeters) net with a weighted foot rope is dragged through the benthos, either onto the beach or directly into a boat. Some techniques additionally involve beating the substrate with poles to frighten fish into the net and/or use of a very small mesh scoop to haul fish from the water. Dynamite fishing was also once widespread, but its use has been reduced drastically throughout the country following a nationwide campaign in 1996-97. This involved major community-driven action which included naming culprits, but also an amnesty for all those who surrendered their dynamite and made a public statement not to re-offend. Coral mining is another highly destructive activity which is also widespread along the entire coast. In 2000 it was estimated that 1 500 tons of coral were being mined every year from the Mikindani Bay area in southern Tanzania alone. Some 12 percent of Tanga's reefs are believed to be totally destroyed, largely through destructive fishing, and a further 64 percent are in poor to moderate condition.

There is only primary sewage treatment in Zanzibar Town, and little or no treatment on any of the mainland coast. Tourism is a growing and important sector of the economy, but there are few environmental controls and there may be increasing impacts on the reefs. Nonetheless, tourism is also providing impetus for further reef protection measures in a number of areas. The Chumbe Island Coral Park provides the best example of "low impact" tourism in the region, and tourism here provides support not only for reef management, but also for an important education program with schools and local communities in Zanzibar. One further coastal activity that has grown rapidly since 1989 is commercial seaweed farming, now practiced along the majority of the coastline of Zanzibar and increasing on Mafia, Pemba and the mainland coast. This activity is low technology and hence is being taken up at the community and individual family level and may be reducing pressure on fish resources.

Although a number of marine reserves were designated in 1975 none of these was fully implemented. Subsequent legislation under the Marine Parks and Reserves Act in 1994 has rectified this situation and there are now five marine reserves and two marine parks designated under this act. The latter are large areas, incorporating reefs and other ecosystems, with zoning systems and focussing towards sustainable use. Protected areas are declared under separate legislation in Zanzibar and Pemba. The Menai Bay Conservation Area off the south coast of Zanzibar was established in 1997 and is one of a number of new marine protected areas being operated at the local level, with local government and community involvement in park utilization and management.

1. Officially this island is known as Unguja, while the term Zanzibar refers to the administrative state which includes both this island and Pemba. Despite this, the term Zanzibar is most commonly used in relation to the single island, and this is the usage applied here.

MAP 7c

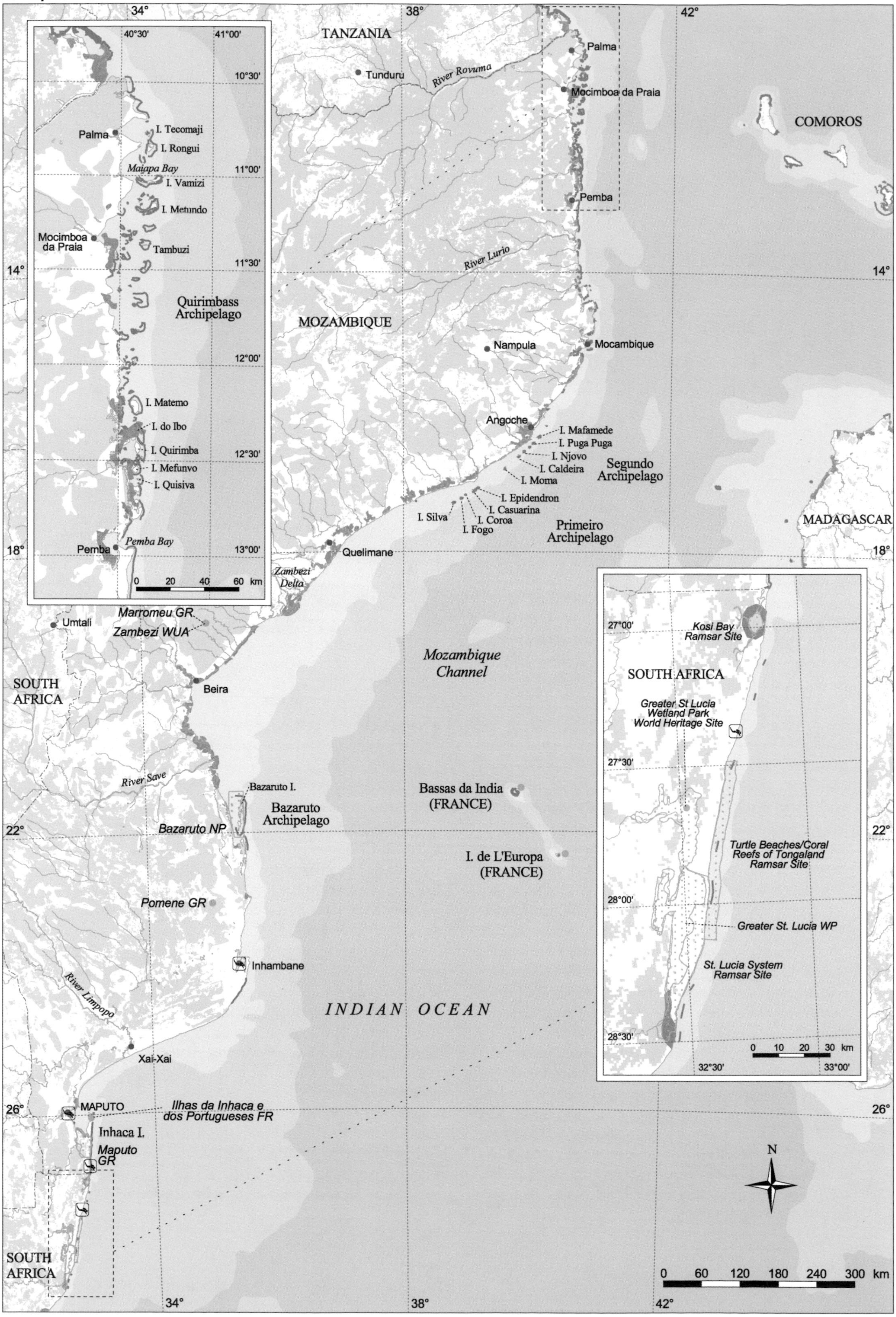

TANZANIA
Tunduru
River Rovuma
Palma
Mocimboa da Praia
COMOROS
Pemba
River Lurio
MOZAMBIQUE
Nampula
Mocambique
Angoche
I. Mafamede
I. Puga Puga
I. Njovo
I. Caldeira
I. Moma
Segundo Archipelago
I. Epidendron
I. Casuarina
I. Coroa
I. Fogo
I. Silva
Primeiro Archipelago
MADAGASCAR
Quelimane
Zambezi Delta
Marromeu GR
Zambezi WUA
Umtali
Mozambique Channel
Beira
SOUTH AFRICA
River Save
Bazaruto I.
Bazaruto Archipelago
Bazaruto NP
Bassas da India (FRANCE)
I. de L'Europa (FRANCE)
Pomene GR
Inhambane
River Limpopo
INDIAN OCEAN
Xai-Xai
MAPUTO
Ilhas da Inhaca e dos Portugueses FR
Inhaca I.
Maputo GR
SOUTH AFRICA
N
0 60 120 180 240 300 km
34° 38° 42°
14° 18° 22° 26°

Quirimbass Archipelago
40°30' 41°00'
10°30' 11°00' 11°30' 12°00' 12°30' 13°00'
Palma
I. Tecomaji
I. Rongui
Maiapa Bay
I. Vamizi
I. Metundo
Mocimboa da Praia
Tambuzi
I. Matemo
I. do Ibo
I. Quirimba
I. Mefunvo
I. Quisiva
Pemba
Pemba Bay
0 20 40 60 km

27°00' 27°30' 28°00' 28°30'
Kosi Bay Ramsar Site
SOUTH AFRICA
Greater St Lucia Wetland Park World Heritage Site
Turtle Beaches/Coral Reefs of Tongaland Ramsar Site
Greater St. Lucia WP
St. Lucia System Ramsar Site
0 10 20 30 km
32°30' 33°00'

Mozambique and South Africa

MAP 7c

Mozambique has a long coastline facing the Mozambique Channel and Madagascar. In the north, heavily faulted Cretaceous to Tertiary sediments line the coast. South of Angoche the coastline is dominated by Quaternary to Recent sediments, largely sands interspersed with heavy alluvial deposits, particularly in the central region between Angoche and Bazaruto Island where some 24 rivers meet the coast, including the large delta areas of the Zambezi and Save Rivers. In many areas the sands form flat plains, although high dune systems are also common, particularly in the southern third of the country where they often lie in front of coastal barrier lakes and swamps. There are several offshore island groups, including a number of small coralline islands directly south of the border with Tanzania, the Quirimbass Archipelago, and two short island chains due south of Angoche – the Primeiro and Segundo Archipelagos. Larger islands include those of the Bazaruto Archipelago and Inhaca Island in the far south. The continental shelf is less than 20 kilometers wide in the north, broadening to a maximum of about 130 kilometers in the center of the country and then narrowing again in the south.

The South Equatorial Current which flows westwards across the Indian Ocean meets the East African coastline in the region of the Tanzania/Mozambique border where it then divides, with one branch forming the constant south-flowing Mozambique Current. Part is deflected eastwards south of Inhambane and forms a gyre, circulating in the Mozambique Channel, and the remainder flows on to join the Agulhas Current off South Africa. Aside from these main current patterns there are a number of counter currents associated with the larger embayments along this coastline, and these create quite strong north-flowing coastal currents in places.

Fringing reefs are numerous along the northern coastline away from river mouths and around the offshore islands, with mangroves and patch reefs on the western shores and simpler reef profiles along exposed eastern shores. Over 50 hard coral genera and over 300 reef fish species have been recorded from the Quirimbass Archipelago. Diversity and coral cover appear to be lower in the Primeiro and Segundo Archipelagos, which may be related to cold water upwellings, but this region has extensive seagrass and important dugong and turtle populations. There are also reported to be large seabird nesting colonies on some of these islands.

The central section of the coastline has been called

A large school of convict surgeon Acanthurus triostegus, *which are important algal grazers on some shallow reefs.*

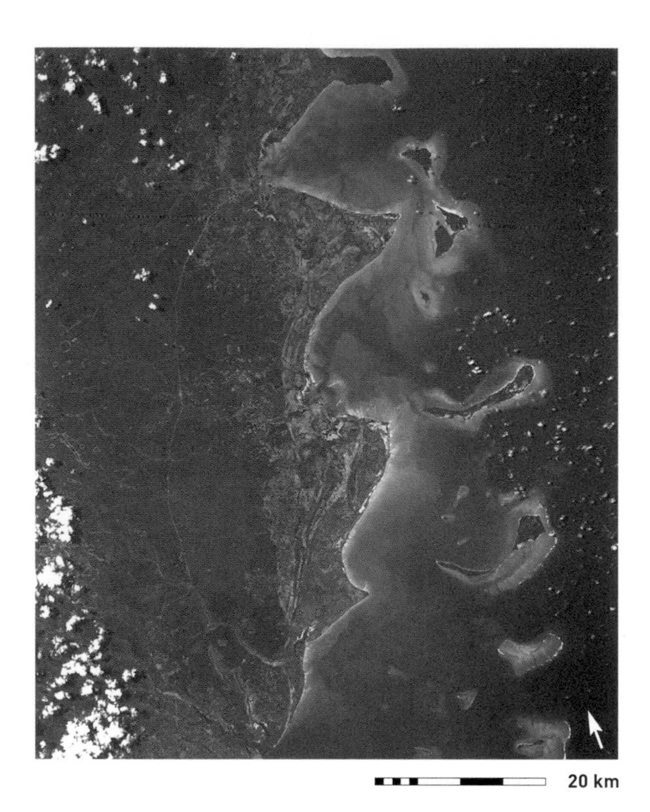

the swamp coast, and is dominated by riverine sediments. These prevent any major reef developments, while there are extensive mangroves inshore. Further south again, reef development is limited, but there are true reefs as well as rocky structures with coral communities around both the Bazaruto and Inhaca Islands. At the former, reefs are mostly patch reef structures, and active growth is restricted to the shallowest waters, but here coral cover can reach 90 percent, and some 27 scleractinian coral genera have been recorded. Three small fringing reefs have developed off Inhaca and Portugueses Islands, as well as a number of smaller patch reefs. These have been relatively well studied and have a high diversity of corals and fish. Elsewhere there are anecdotal reports of extensive coral and gorgonian communities in offshore areas, but there have been few, if any, surveys of the region. Probably the largest remaining population of dugongs is found in the Bazaruto Archipelago, estimated at 150 individuals in the early 1990s, but thought to have declined to 60-80 animals by 1999. Crown-of-thorns starfish are reported to have destroyed a number of reef areas off Bazaruto and Inhambane.

The 1998 coral bleaching event appears to have caused significant mortality of corals in Mozambique, particularly in the north, but with considerable variation in the degree of impact between localities.

Mozambique has a large coastal population. The majority of these people moved to urban areas during the civil unrest which ended in 1992. These are a source of considerable pollutants to nearby coastal waters as most sewage is untreated. Away from these urban areas much of the coastline is dominated by slash-and-burn agriculture, which releases sediments and nutrients into nearby waters. Tourism is growing, particularly in the south, and is generally considered detrimental to the environment, especially vehicular or camping-based tourism from South Africa which brings few benefits to the country, and may lead to unsustainable levels of recreational fishing and

Mozambique

General Data	
Population (thousands)	19 105
GDP (million US$)	2 089
Land area (km²)	788 629
Marine area (thousand km²)	565
Per capita fish consumption (kg/year)	2
Status and Threats	
Reefs at risk (%)	76
Recorded coral diseases	0
Biodiversity	
Reef area (km²)	1 860
Coral diversity*	194 / 314
Mangrove area (km²)	925
No. of mangrove species	10
No. of seagrass species	8

* The higher coral diversity figure is an estimate for Mozambique and Tanzania combined

Left: The Quirimbass Archipelago has some of the most important reef areas in Mozambique (STS51I-31-45, 1985). Right: The shore crab Grapsus *is widespread on rocky shores throughout the region.*

Protected areas with coral reefs

Site name	Designation	Abbreviation	IUCN cat.	Size (km²)	Year
Mozambique					
Bazaruto	National Park	NP	II	150.00	1971
Ilhas da Inhaca e dos Portugueses	Faunal Reserve	FR	VI	20.00	1965
South Africa					
Greater St. Lucia (South Africa)	Wetland Park	WP	II	2 586.86	1895
GREATER ST. LUCIA WETLAND PARK	WORLD HERITAGE SITE			2 395.66	1999
TURTLE BEACHES/ CORAL REEFS OF TONGALAND	RAMSAR SITE			395.00	1986

damage to turtle nesting beaches. Efforts to develop coastal resorts have also been poorly controlled to date, although this may be changing. Most reef-based tourism operates around the Bazaruto Archipelago and there is evidence of significant damage to reefs caused by divers and boats.

Fishing is an important activity in Mozambique. Trawling for prawns dominates the commercial fishery and generates 40 percent of the country's foreign exchange earnings. This catch is highly dependent on the mangroves and estuaries which act as nursery areas. There is little agreement about the overall size of the artisanal catch, with estimates that these may make up anywhere between 20 and 70 percent of total landings (estimates of total landings similarly vary between 18 500 and 90 000 tons per year). Fishing is considerable on the Quirimbass reefs and seagrass systems, and there are now migrant fishermen coming to the region, bringing the potential for overfishing. Exploitation of the Primeiro and Segundo Archipelagos is relatively low due to the lack of permanent human settlement and often rough seas. Removal of molluscs for the curio trade is reported to be of significance on a number of reefs.

Mozambique still has many reefs which have escaped heavy human impact, however this is changing, and quite rapidly in some areas. There are only two protected areas which incorporate reefs and, while there are active management measures in place at these sites, there are no immediate proposals for any protected areas on the important reefs in the north of the country.

South Africa

GENERAL DATA	
Population (thousands)	43 421
GDP (million US$)	114 585
Land area (km²)	1 223 124
Marine area (thousand km²)	1 525
Per capita fish consumption (kg/year)	8
STATUS AND THREATS	
Reefs at risk (%)	na
Recorded coral diseases	0
BIODIVERSITY	
Reef area (km²)	<50
Coral diversity	na / na
Mangrove area (km²)	11
No. of mangrove species	6
No. of seagrass species	3

South Africa

Although reef communities extend into the waters of South Africa it is arguable whether these are true reef structures. There are three main areas: the northern, central and southern reef complexes. All are submerged communities growing over late Pleistocene dune and beach sequences, and reaching a minimum depth of about 8 meters. Diversity is lower than the reefs of more northern countries, with only 43 species of scleractinian coral recorded. Coral cover (hard and soft) is high, however, making up almost 50 percent of benthic cover (and 95 percent of the live cover). These reefs were largely unaffected by the 1998 coral bleaching event. Large numbers of divers visit the reefs, with over 90 000 recreational dives per year, mostly visiting Two Mile Reef in the central complex. Lying offshore, these areas are not threatened by terrestrial sources of pollution or sedimentation, and they are protected within the St. Lucia Marine Reserve (a part of the wetland park). Artisanal fishing is permitted in much of the reserve.

Madagascar

MAP 7d

11 km

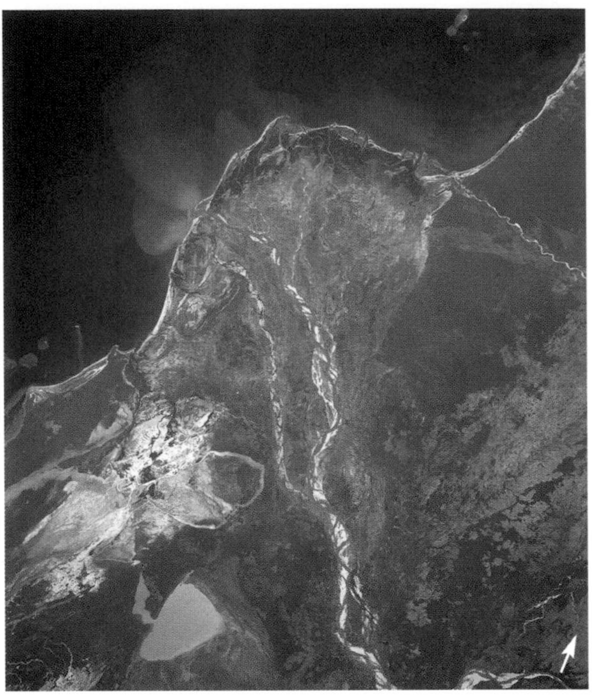

15 km

Madagascar is one of the world's largest islands. Along with the Indian sub-continent it was separated off from the rest of Africa during the Jurassic, and was then separated from the Indian sub-continent (and the granitic Seychelles) during the late Jurassic/early Cretaceous. There are clear differences between the physical conditions and resulting ecological communities on the east and west coasts. The east coast is steep and, in places, mountainous. This is matched by a steeply shelving bathymetry and narrow continental shelf. The central and southern sections of this coast are dominated by vast sandy beaches and barrier islands and there is no offshore reef development. Further north the coastline becomes more complex, with a number of embayments and rocky headlands as well as offshore islands. There are a number of emergent fossil reefs along the more northerly sections of this coastline. Active coral growth is also widespread in the north, often growing on fossil structures offshore, although not always contributing to active reef accretion. There is a submerged and fragmented barrier reef described off Toamasina, although the recent status of this is unclear. Discontinuous fringing reefs also occur off the coast around Foulpointe and Mananara, Nosy Boraha (Sainte Marie Island), and the Masoala Peninsula.

The west coast consists of a wide coastal plain, with numerous rivers, and also a wider continental shelf. This coast is swept by the northward flowing currents of the Mozambique Gyre and is further affected by large tidal ranges. Reef development is extensive in both the northern and southern parts of this coast. The southernmost reefs are offshore around Banc de l'Etoile and Nosy Manitsa. There are extensive fringing reefs along much of the coast north of Androka as far as Cap St. Vincent, varying between 500 meters and a few kilometers offshore, and separated from the shore by a generally shallow channel. Around Tulear a more complex system of offshore reefs is present, with shoreline fringing communities, a series of inner lagoon reefs, and a well developed barrier reef – the Grand Récif – which runs continuously for 18 kilometers. Between the Baie des Assassins and Morombe a sequence of reefs, many with associated sand cays, has developed offshore, forming a fragmented barrier reef system. This same barrier system reappears north of the Mangoky Delta with a series of submerged banks and emergent reefs with sand cays. Along most of the central section of the west coast there is no reef development, probably due to the terrigenous sediments discharged from rivers. Offshore, however, there are reefs towards the edge of the continental shelf associated with the

Left: One of the best known reefs in Madagascar is the Grand Récif, a barrier reef close to Tulear (STS065-84-92, 1994). Right: Terrigenous sediments impact or inhibit reef development along considerable lengths of Madagascar's coast, as here at the Mangoky Delta. Sedimentation has been greatly increased by poor landuse practices often far inland (STS033-71-94, 1989).

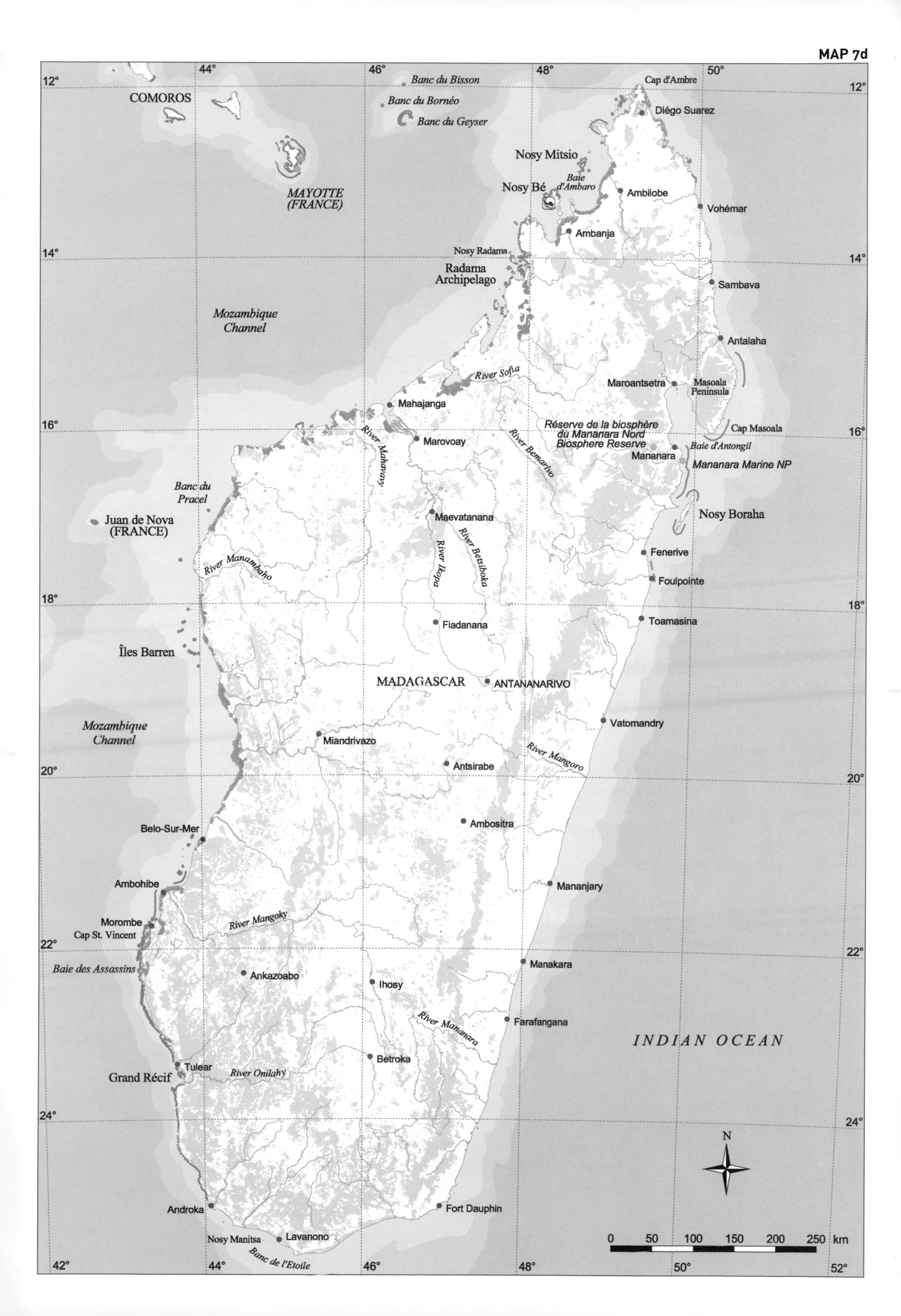

COMOROS
Banc du Bisson
Banc du Bornéo
Banc du Geyser
MAYOTTE (FRANCE)
Cap d'Ambre
Diégo Suarez
Nosy Mitsio
Baie d'Ambaro
Nosy Bé
Ambilobe
Vohémar
Ambanja
Nosy Radama
Radama Archipelago
Sambava
Mozambique Channel
Antalaha
River Sofia
Maroantsetra
Masoala Peninsula
Mahajanga
Réserve de la biosphère du Mananara Nord Biosphere Reserve
Cap Masoala
River Mahavavy
Marovoay
River Bemarivo
Mananara
Baie d'Antongil
Mananara Marine NP
Banc du Pracel
Juan de Nova (FRANCE)
Maevatanana
Nosy Boraha
River Betsiboka
River Ikopa
River Manambaho
Fenerive
Foulpointe
Toamasina
Fiadanana
Îles Barren
MADAGASCAR
ANTANANARIVO
Vatomandry
Mozambique Channel
Miandrivazo
River Mangoro
Antsirabe
Ambositra
Belo-Sur-Mer
Ambohibe
Mananjary
Morombe
Cap St. Vincent
River Mangoky
Baie des Assassins
Manakara
Ankazoabo
Ihosy
River Mananara
Farafangana
INDIAN OCEAN
Betroka
Tulear
River Onilahy
Grand Récif
N
Androka
Fort Dauphin
Nosy Manitsa
Lavanono
Banc de l'Etoile
0 50 100 150 200 250 km
12° 14° 16° 18° 20° 22° 24°
42° 44° 46° 48° 50° 52°

Protected areas with coral reefs

Site name	Designation	Abbreviation	IUCN cat.	Size (km²)	Year
Madagascar					
Mananara Marine	National Park	NP	II	10.00	1989
RÉSERVE DE LA BIOSPHÈRE DU MANANARA NORD	UNESCO BIOSPHERE RESERVE			1 400.00	1990

Îles Barren and the Banc du Pracel, although these remain poorly documented. In the northeast, fringing reefs reappear along the coast and the offshore islands, notably Nosy Bé and the Mitsio and Radama Archipelagos, although their distribution is discontinuous around the many rivers and bays. On the outer edge of the continental shelf in the far north, there is another series of raised banks, actually forming a near continuous ridge which may be the remains of a large barrier reef system. Coral cover is reported to be very high along the outer slopes, heavily dominated by formations of the sheet coral *Pachyseris speciosa*.

Most research has been centered around Nosy Bé in the north and Tulear in the south, and very little is known about the intervening reef areas. Some 130 species of scleractinian coral and 700 fish species have been recorded on the reefs off Tulear, but it has been estimated that there may be 200 coral species and 1 500 fish species in the whole country. Along the western coastline, mangroves form a major community and seagrasses are widespread, often forming the dominant communities in the channels behind fringing reefs. It would appear that most reefs were hit by the 1998 bleaching event, although data on the impacts are only available for a few sites. Some 30 percent bleaching was observed on the mid-west coast, for example at Belo-sur-Mer, but bleaching-related mortality was relatively low.

For its size, Madagascar is relatively sparsely populated. The majority of the coastal population is concentrated on the eastern coast, while the western coast is less developed, aside from the larger cities of Tulear and Mahajanga. It is this west coast, however, that also supports the majority of fishing and tourism-based activities. Artisanal fishing is a critical activity, accounting for an estimated 55 percent of all fishery production from an estimated 1 250 fishing villages operating over 20 000 small vessels (pirogues, mostly without engines). Reef-associated species are heavily relied upon, accounting for 43 percent of total production. It remains a largely traditional fishery, although there are increasing numbers of migrant fishers who do not observe existing customs and taboos. Larger-scale commercial and export fisheries make up the remainder of the fishery and, together with aquaculture, provide critical foreign exchange earnings. Tourism is another important and relatively rapidly developing activity, with at least 50 percent of arrivals visiting the coast.

One of the greatest threats to Madagascar's reefs is siltation from inappropriate landuse practices. Most of Madagascar's land area has been converted from natural systems and soil erosion affects nearly 80 percent of the island, with massive sedimentation offshore. Urban and industrial waste is poorly controlled and a problem near major cities. Overfishing may be significant – fishing levels have greatly increased in recent years and there is evidence of reduced yields. Despite the considerable potential for ecotourism most developments seem to have been poorly planned and contribute to pollution, while also causing conflicts with local fishing communities. There is only one marine protected area with coral reefs, the Mananara Marine National Park on the northeast coast which incorporates three coral islets, including Nosy Antafana. This site has two rangers and there is some community involvement in its management. There are a number of proposals for new parks.

Madagascar

GENERAL DATA	
Population (thousands)	15 506
GDP (million US$)	3 264
Land area (km²)	594 854
Marine area (thousand km²)	1 205
Per capita fish consumption (kg/year)	7
STATUS AND THREATS	
Reefs at risk (%)	87
Recorded coral diseases	0
BIODIVERSITY	
Reef area (km²)	2 230
Coral diversity	135 / 315
Mangrove area (km²)	3 403
No. of mangrove species	9
No. of seagrass species	10

Mayotte, Comoros and outlying islands

MAP 7e

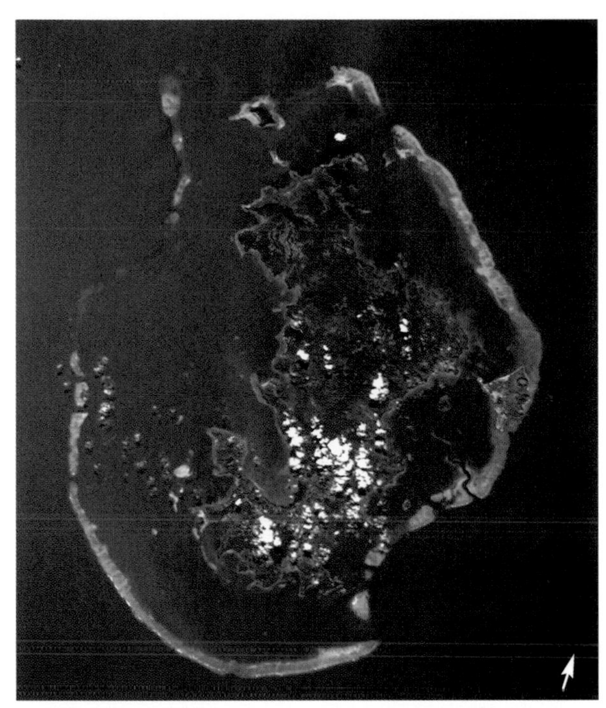

There are a number of small oceanic islands lying between Mozambique and Madagascar. The most important of these are the four large volcanic islands of the Comoros Archipelago situated at the northern entrance to this channel. Mayotte, the easternmost of these islands is a *collectivité territoriale* under French control, while the remaining islands form the Federal Islamic Republic of the Comoros. Mayotte is geologically the oldest and is surrounded by a wide lagoon which reaches 70 meters in depth before a barrier reef some 3-15 kilometers offshore. The remaining islands are surrounded by fringing reefs, although Ngazidja (Grande Comore), the youngest island which is still volcanically active, has very steep and barren shores and fringing reefs are restricted to only a few parts of the coastline. Mwali (Mohéli) has the most extensive reef systems, with fringing reefs on all coasts. East of these islands lies the Banc du Geyser/Zélée (Map 7d), a horseshoe-shaped reef which is probably part of the same volcanic system. This reef breaks the surface during low tides, and lies between Mayotte and the French territory of Îles Glorieuses. All of these reefs lie in the path of the westward flowing Equatorial Counter Current, which coincides with the northern leg of the Mozambique Gyre.

Mayotte

Mayotte's reefs are relatively well studied, and harbor more than 200 species of coral. They were adversely affected by a bleaching event in 1982-83, which apparently caused mortality and degradation on about 36 percent of the fringing reefs. Crown-of-thorns starfish outbreaks since 1983 have been a major problem, although a bounty system resulted in fishermen collecting large numbers, with a peak of some 8 000 collected in 1998. The 1998 bleaching event caused even more widespread mortality, with greater than 90 percent mortality recorded on the outer slopes. Recovery from this event is now being noted, particularly on the inshore reefs. Fisheries and tourism are important activities, with some 3 600 fishermen and 9 000 visitor arrivals per year in the late 1990s. Two protected areas have been established, although these only cover some 2 percent of the total area of the lagoon. A comprehensive management plan for the lagoon was under development in late 2000.

Comoros

The densely populated Comoros is one of the world's poorest countries. Deforestation and conversion of land to agriculture are creating massive problems of soil erosion,

Left: Mayotte has a number of fringing reefs and is almost completely encircled by its barrier reef (STS51D-41-3, 1985). Right: Algae, including fleshy green varieties such as this, were quick to colonize many of the bare surfaces following the massive coral mortalities of 1998.

MAP 7e

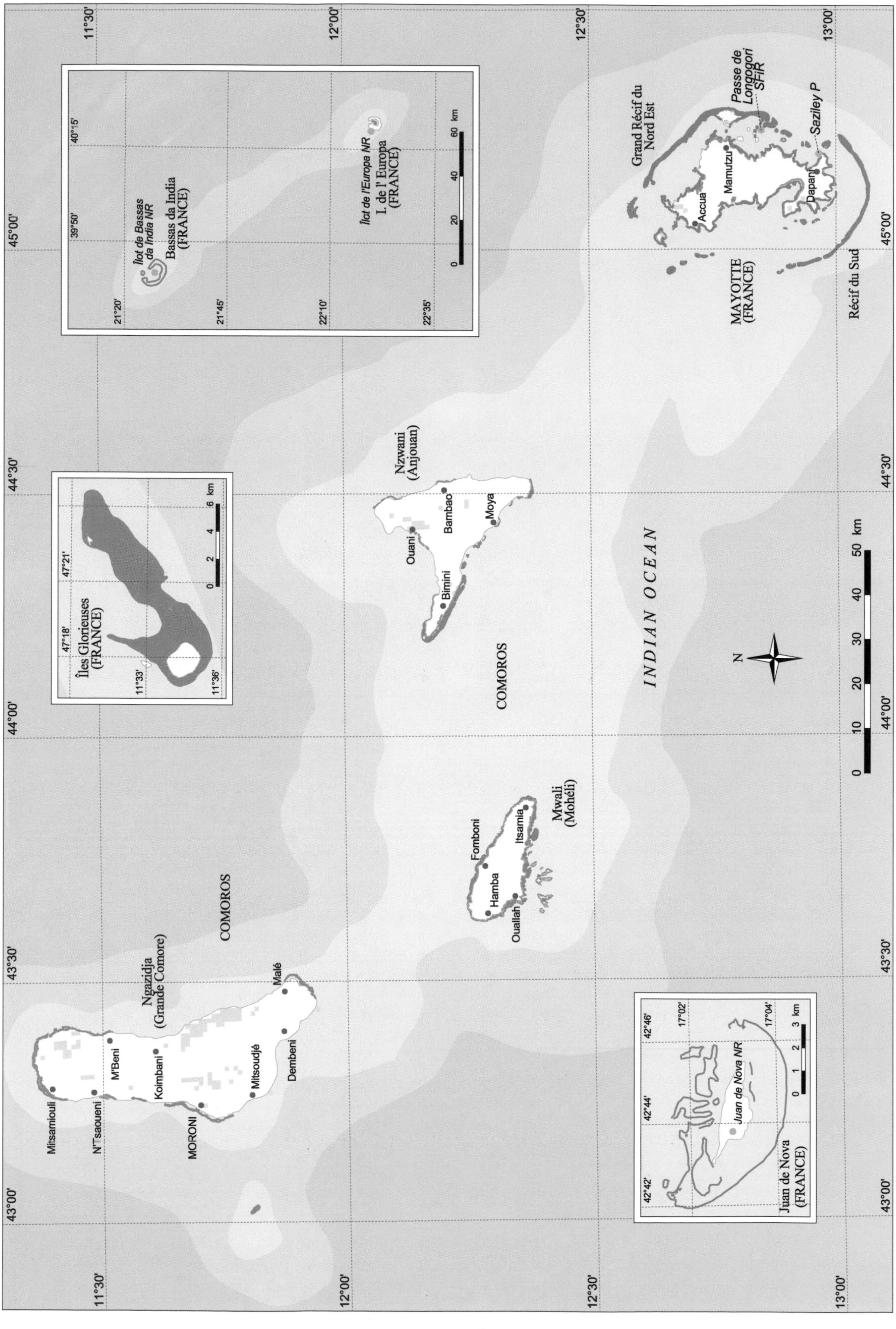

43°00'
43°30'
44°00'
44°30'
45°00'
11°30'
12°00'
12°30'
13°00'
Mitsamiouli
N'Tsaoueni
M'Beni
Koimbani
MORONI
Mitsoudjé
Dembeni
Malé
Ngazidja
(Grande Comore)
COMOROS
Îles Glorieuses
(FRANCE)
47°18'
47°21'
11°33'
11°36'
0 2 4 6 km
39°50'
40°15'
21°20'
21°45'
22°10'
22°35'
Îlot de Bassas da India NR
Bassas da India
(FRANCE)
Îlot de l'Europa NR
I. de l' Europa
(FRANCE)
0 20 40 60 km
Fomboni
Hamba
Ouallah
Itsamia
Mwali
(Mohéli)
COMOROS
Ouani
Bimini
Bambao
Moya
Nzwani
(Anjouan)
INDIAN OCEAN
N
0 10 20 30 40 50 km
42°42'
42°44'
42°46'
17°02'
17°04'
Juan de Nova NR
Juan de Nova
(FRANCE)
0 1 2 3 km
Grand Récif du Nord Est
Accua
Mamutzu
Passe de Longogori SFiR
Dapani
Sazíley P
MAYOTTE
(FRANCE)
Récif du Sud

Protected areas with coral reefs

Site name	Designation	Abbreviation	IUCN cat.	Size (km²)	Year
Mayotte					
Passe de Longogori	Strict Fishing Reserve	SFiR	VI	4.50	1990
Saziley	Park	P	II	41.80	1991

particularly in Nzwani (Anjouan) and Mwali. The subsequent heavy siltation may be affecting large areas of the reefs offshore. Fisheries are important, with over 4 500 registered fishermen operating from traditional boats in nearshore waters. Reef walking by fishers gathering octopus and small fish is causing some degradation of reef flats. Blast fishing is also reported to be a problem on Mwali. There is little or no information regarding overfishing problems in the Comoros, although as population densities continue to rise this may create significant problems.

Mayotte

General Data	
Population (thousands)	156
GDP (million US$)	na
Land area (km²)	375
Marine area (thousand km²)	74
Per capita fish consumption (kg/year)	na
Status and Threats	
Reefs at risk (%)	100
Recorded coral diseases	0
Biodiversity	
Reef area (km²)	570
Coral diversity	na / 313
Mangrove area (km²)	10
No. of mangrove species	na
No. of seagrass species	na

Comoros

General Data	
Population (thousands)	578
GDP (million US$)	235
Land area (km²)	1 660
Marine area (thousand km²)	175
Per capita fish consumption (kg/year)	20
Status and Threats	
Reefs at risk (%)	99
Recorded coral diseases	0
Biodiversity	
Reef area (km²)	430
Coral diversity	na / 314
Mangrove area (km²)	26
No. of mangrove species	na
No. of seagrass species	4

A bicolor cleaner wrasse Labroides bicolor *follows a coral grouper* Cephalopholis miniata. *Cleaner fishes play a critical role in removing parasites and other material from many reef fish.*

Seychelles

MAPS 7f and g

The Seychelles is a very large archipelagic nation in the Western Indian Ocean. The 115 named islands and atolls together with their associated reef systems can be clearly divided into two distinct regions: the high islands to the north and the low coralline islands spread over wide areas to the south and southwest.

The Seychelles Bank lies at the northernmost point of the Mascarene Ridge and is a large, shallow area (some 31 000 square kilometers) of water, mostly above a depth of 100 meters. In its center are a number of high granitic islands of continental origin. These have been described as a "micro-continent", having been left behind during the northwards migration of the Indian sub-continent about 135 million years ago. These are surrounded by widespread but discontinuous, fringing reefs. Along the east coast of Mahé and the west coast of Praslin such fringing reefs are well developed. Reef flats reaching over 2 kilometers in width and terminating in a high algal ridge are followed by a reef slope descending to a floor typically at 8-12 meters. Such clearly zoned reefs are less apparent in more sheltered locations where more complex reef formations have developed. Coral cover varies, being virtually absent from some former reef structures, but abundant in other areas, including non-reefal slopes and granitic surfaces.

The low coralline islands to the south and west of the Seychelles Bank fall into a number of geographic groups. The largest is that of the Amirante Islands, which extend along a shallow north-south ridge, with the Alphonse group forming a slightly separate southern section of this chain. Further south are two small and more disparate island groups, those of Providence-Farquhar and the Aldabra group. Finally, directly to the south of the Seychelles Bank are the isolated islands of Platte and Coetivy. The reefs in these outer island areas are highly varied, and include true atolls (St. Joseph, Alphonse, Farquhar), raised atolls (Aldabra), submerged or partially submerged atolls (Desroches, Coetivy), and platform or bank structures (African Banks, Providence-Cerf). Coral cover varies considerably between localities, ranging from close to zero on some banks and reef slopes (notably the large Providence-Cerf Bank), to 60-70 percent on some atoll slopes.

The Seychelles lie in an area of relatively high faunal diversity. Some 101 hermatypic coral species and 920 fish species have been listed. The reef fauna is fairly typical of the Western Indian Ocean, as exemplified by the reef fish: many are widespread across the ocean basin or wider areas of the Indo-Pacific, however about 15 percent are confined to the western part. The coral reefs of the

Aldabra Atoll is a raised atoll in the southwest Seychelles and a World Heritage Site. There are many unique species on land, including the last giant tortoises in the region, while the reefs are important and relatively pristine (STS068-248-44, 1994).

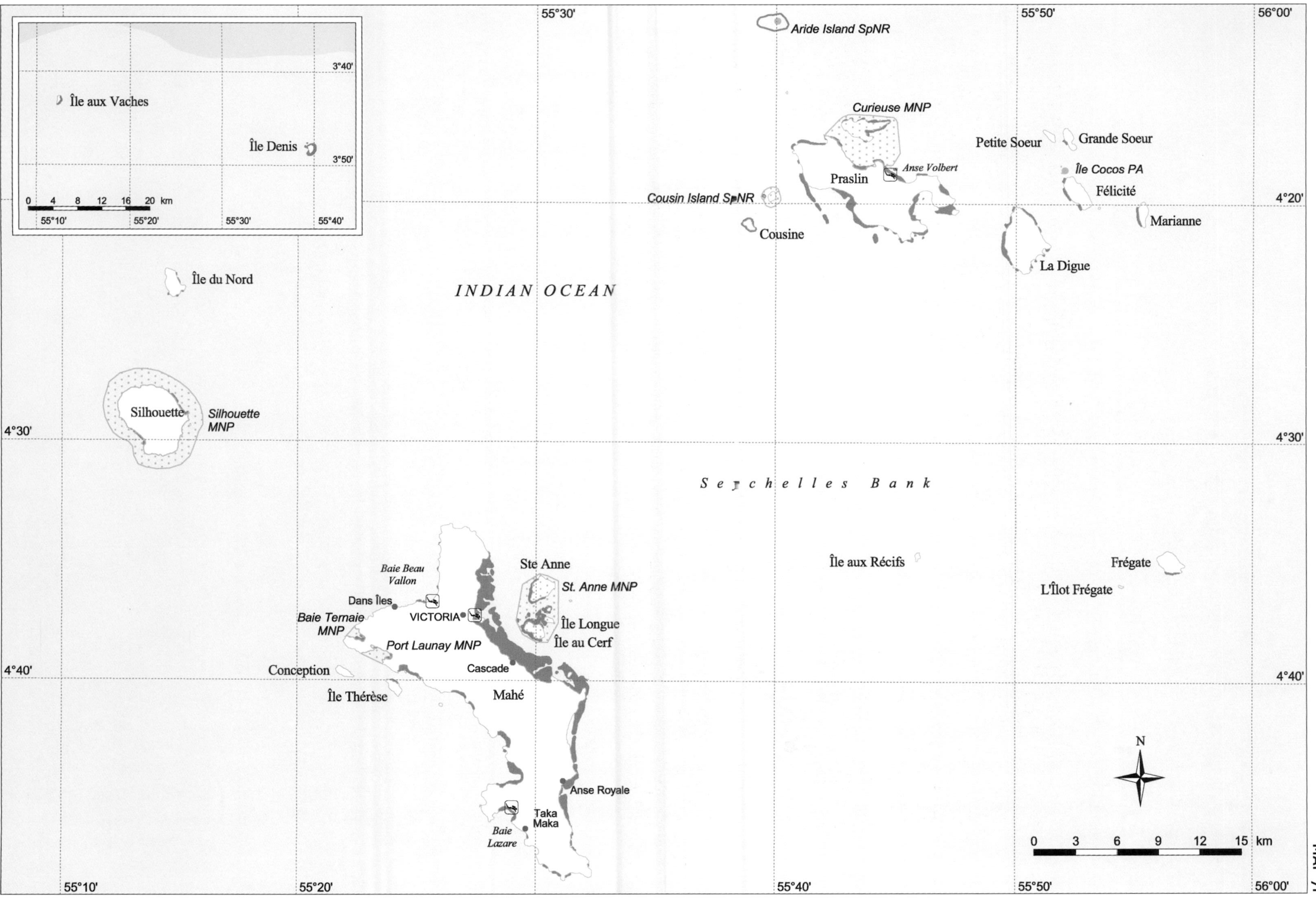

55°10'
55°20'
55°30'
55°40'
55°50'
56°00'
4°20'
4°30'
4°40'
3°40'
3°50'
0 4 8 12 16 20 km
Île aux Vaches
Île Denis
Aride Island SpNR
Curieuse MNP
Petite Soeur
Grande Soeur
Île Cocos PA
Félicité
Marianne
Praslin
Anse Volbert
Cousin Island SpNR
Cousine
La Digue
Île du Nord
INDIAN OCEAN
Silhouette
Silhouette MNP
Seychelles Bank
Île aux Récifs
Frégate
L'Îlot Frégate
Baie Beau Vallon
Dans Îles
Baie Ternaie MNP
VICTORIA
Port Launay MNP
Ste Anne
St. Anne MNP
Île Longue
Île au Cerf
Cascade
Conception
Île Thérèse
Mahé
Anse Royale
Taka Maka
Baie Lazare
N
0 3 6 9 12 15 km

MAP 7g

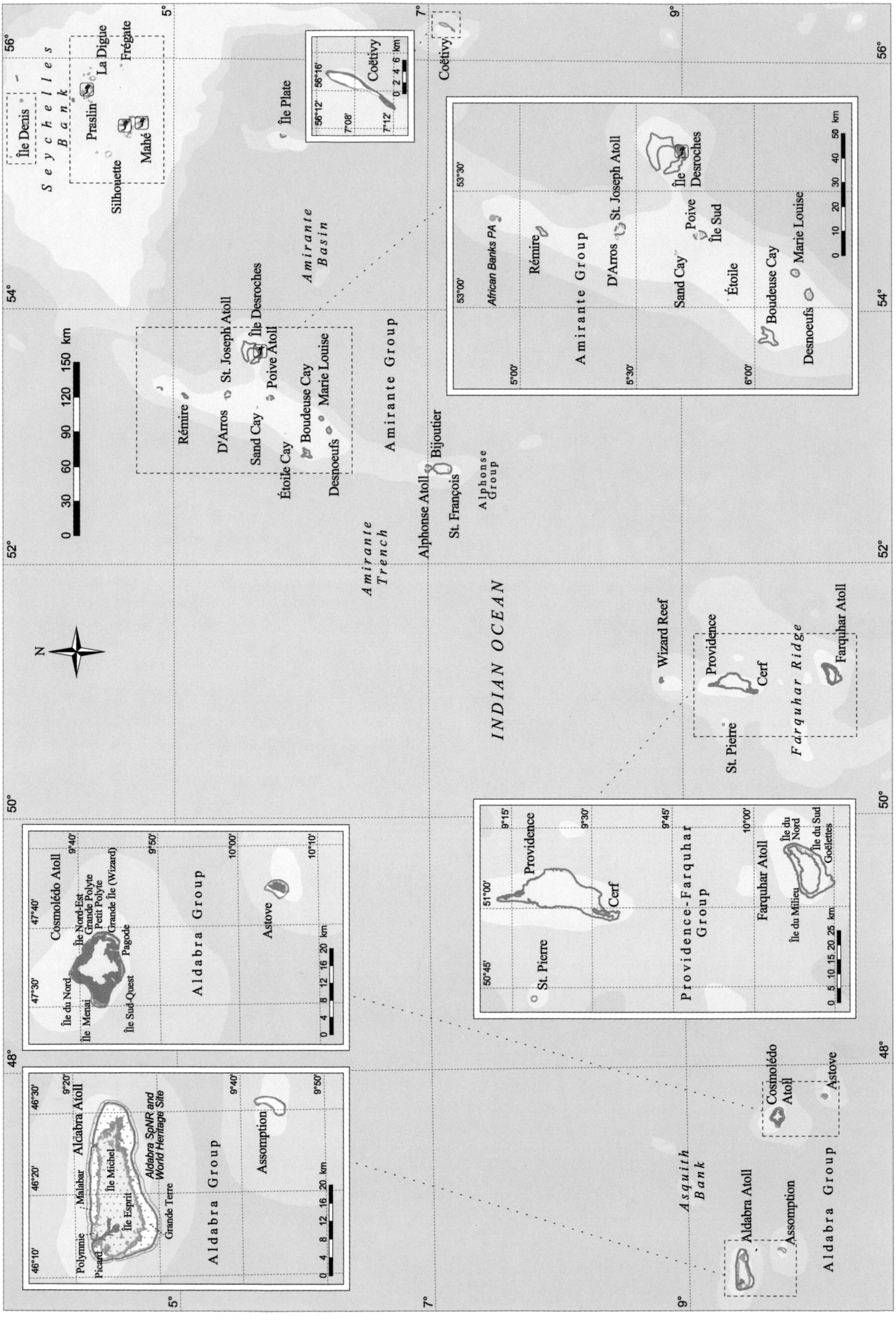

Seychelles Bank
Île Denis
Praslin
La Digue
Frégate
Silhouette
Mahé
Île Plate
Coëtivy
56°12'
56°16'
7°08'
7°12'
0 2 4 6 km
Amirante Basin
Rémire
D'Arros
St. Joseph Atoll
Île Desroches
Poive Atoll
Sand Cay
Étoile Cay
Boudeuse Cay
Marie Louise
Desnoeufs
Amirante Group
Bijoutier
Alphonse Atoll
St. François
Alphonse Group
Amirante Trench
0 30 60 90 120 150 km
N
INDIAN OCEAN
Wizard Reef
Providence
Cerf
St. Pierre
Farquhar Ridge
Farquhar Atoll
African Banks PA
Rémire
Amirante Group
D'Arros
St. Joseph Atoll
Île Desroches
Poive
Île Sud
Sand Cay
Étoile
Boudeuse Cay
Marie Louise
Desnoeufs
53°00'
53°30'
5°00'
5°30'
6°00'
0 10 20 30 40 50 km
Providence
Cerf
St. Pierre
Providence-Farquhar Group
Farquhar Atoll
Île du Nord
Île du Sud
Goëlettes
Île du Milieu
50°45'
51°00'
9°15'
9°30'
9°45'
10°00'
0 5 10 15 20 25 km
Cosmolédo Atoll
Île du Nord
Île Menai
Île Nord-Est
Grande Polyte
Petit Polyte
Grande Île (Wizard)
Pagode
Île Sud-Ouest
Aldabra Group
Astove
47°30'
47°40'
9°40'
9°50'
10°00'
10°10'
0 4 8 12 16 20 km
Aldabra Atoll
Polymnie
Picard
Malabar
Île Michel
Île Esprit
Grande Terre
Aldabra SpNR and World Heritage Site
Aldabra Group
Assomption
46°10'
46°20'
46°30'
9°20'
9°40'
9°50'
0 4 8 12 16 20 km
Asquith Bank
Aldabra Atoll
Assomption
Cosmolédo Atoll
Astove
Aldabra Group
5°
7°
9°
48°
50°
52°
54°
56°

entire archipelago were very heavily impacted by the 1997-98 El Niño event, with bleaching occurring on 60-95 percent of corals, and subsequent coral mortalities of 50-90 percent. The longer-term impacts on the reef communities are somewhat unpredictable given the scale of damage in all areas.

Human impacts on the reefs are varied, but clearly significant in the granitic islands. Most of the national fish consumption is of nearshore fishes, a large proportion of which are reef associated. The reefs of the Seychelles Bank are thus quite heavily utilized, and there are clearly documented examples of overfishing from a few localities.

Fishing pressure in the southern islands, by contrast, is relatively low. There is some fishing by vessels visiting from the granitic islands, but also some small commercial fisheries operations run from a few of the inhabited islands. The offshore tuna populations are the center of a major export fishery, with a tuna cannery in Mahé serving a large number of vessels from the Indian Ocean.

Tourism is a critical industry in the Seychelles, being one of the main providers of employment and the main foreign exchange earner. In 1996 there were some 131 000 visitors, with receipts totalling US$147 million. Virtually all tourism is coastal and beach orientated, with a large

Above: Shrimpfish Aeoliscus strigatus *in their unusual head-down swimming motion over a shallow reef. Left: The massive coral mortalities associated with the bleaching event in 1998 allowed many former corals to be overgrown with algal species. Right: Île Cocos and its surrounding shallow reefs are one of a number of protected areas in the Seychelles.*

Seychelles

General Data	
Population (thousands)	79
GDP (million US$)	449
Land area (km²)	489
Marine area (thousand km²)	1 334
Per capita fish consumption (kg/year)	65
Status and Threats	
Reefs at risk (%)	17
Recorded coral diseases	0
Biodiversity	
Reef area (km²)	1 690
Coral diversity	206 / 310
Mangrove area (km²)	29
No. of mangrove species	9
No. of seagrass species	8

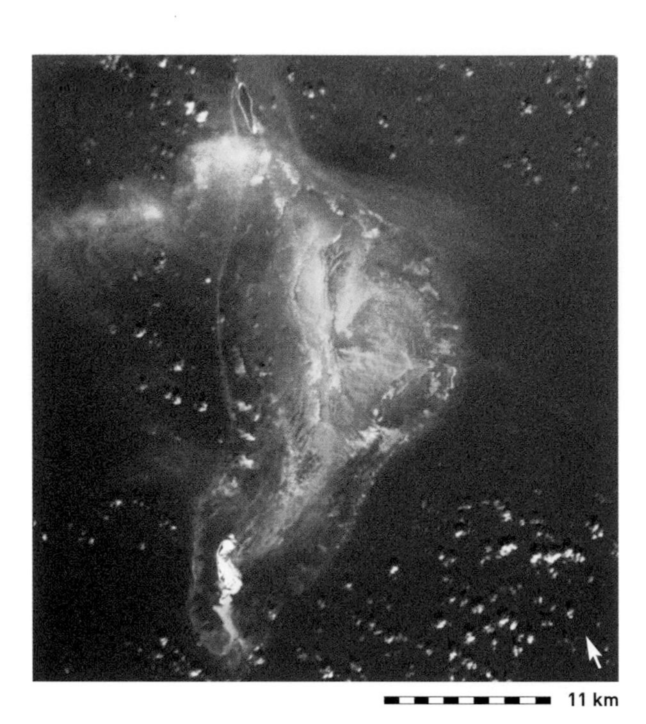

proportion of visitors on diving holidays, and many others making day trips to reefs. Most tourists remain on the granitic islands, but there are now also some exclusive developments on the outer islands, while boat-based holidays take tourists to most areas.

Land reclamation has built upon, and destroyed, a large area of the fringing reefs of east Mahé, which were once the best developed fringing reefs in the country. This work has also had impacts on the adjacent reefs through heavy sedimentation. Elsewhere terrestrial sources of sewage pollution, sediments, and solid waste are problematic, while the increases in tourism are bringing these problems to new areas.

There is a clear awareness of environmental issues at the governmental level and efforts are being made to improve sewage treatment in some areas. A number of marine protected areas have been established and active management is underway. The remote island of Aldabra has long been recognized for its unique flora and fauna and is well protected, with a research station and permanent staff.

Protected areas with coral reefs

Site name	Designation	Abbreviation	IUCN cat.	Size (km²)	Year
Seychelles					
Aldabra	Special Nature Reserve	SpNR	Ia	350.00	1981
Aride Island	Special Nature Reserve	SpNR	Ia	0.70	1973
Baie Ternaie	Marine National Park	MNP	II	0.80	1979
Cousin Island	Special Nature Reserve	SpNR	Ia	0.28	1975
Curieuse	Marine National Park	MNP	II	14.70	1979
Île Cocos	Protected Area	PA	Unassigned	0.01	1987
Port Launay	Marine National Park	MNP	II	1.58	1979
Silhouette	Marine National Park	MNP	II	30.45	1987
St. Anne	Marine National Park	MNP	II	14.23	1973
ALDABRA ATOLL	WORLD HERITAGE SITE			**350.00**	**1982**

The bank of shallow water around Providence and Cerf Islands has the appearance of a true platform reef, although recent studies have shown that there is very little living coral on its seaward margins, while the surface is dominated by seagrass (STS033-76-43, 1989).

Mauritius and Réunion

MAP 7h

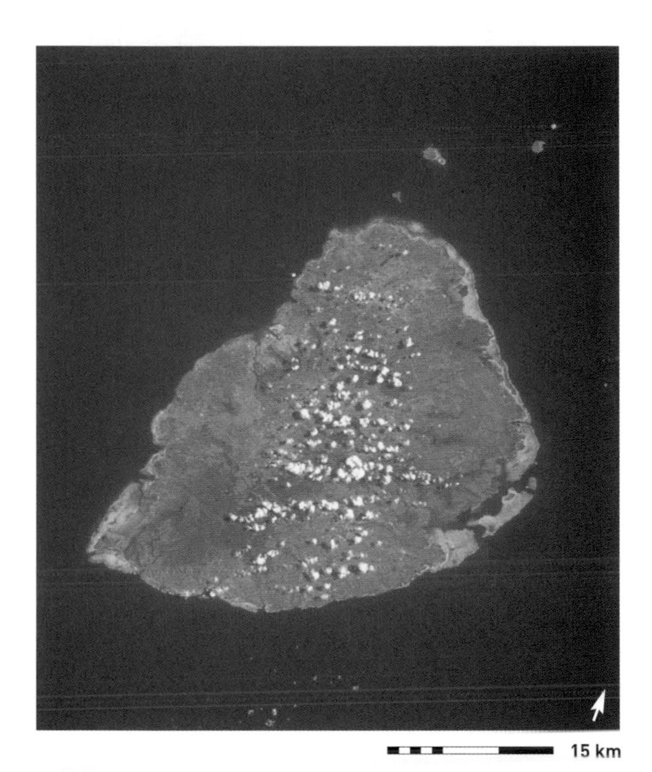

The Mascarene islands of Mauritius, Rodrigues and Réunion lie at the southern end of the Mascarene Ridge and are geologically relatively young. All three islands are of volcanic origin and show a clear sequence of reef development. Réunion is the youngest. Still volcanically active, it lies directly over the hot spot which has provided the geological origin of the entire Chagos-Laccadive Ridge (Chapter 8), as well as the southerly parts of the Mascarene Ridge. Moving eastwards the islands become comparatively older, and with this the coral reefs are both better developed and further offshore.

In addition to the main islands listed here there are several others which fall under the jurisdiction of Mauritius and of Réunion. These lie at some distance from the main islands and are considered separately, below.

Mauritius and Rodrigues

Mauritius is almost completely encircled by fringing reefs, with substantial lagoon and barrier reef development on the east and southwest coasts. The lagoons are dominated by algae, but with some areas of seagrass. The reef slopes have a clear spur and groove zone. Below about 20 meters there is usually only a thin veneer of coral rock overlying volcanic rocks. Rodrigues is the oldest of the volcanic islands and has a highly developed reef structure, although a true barrier reef has not formed. The island is totally encircled by reefs, with wide shallow reef flats extending out from the shore – in the east this narrows to 50 meters in places but is more typically 1-2 kilometers wide, while at its widest extent in the west it reaches 10 kilometers. Seagrasses are widespread in the lagoon, and reef flats and mangrove communities are reported to be increasing. The outer slopes are steep, and have 50-70 percent coral cover. In Mauritius the 1998 bleaching event affected 30-40 percent of corals, though very few died. The high rates of survival have, in part, been related to overcast and windy conditions for much of February and March, which were associated with cyclone Anacelle and which mitigated the warming impacts observed elsewhere in the region.

Many of the reefs around Mauritius have been degraded by human activities. Problems include high levels of sedimentation and pollution arising from the clearance of the forest and subsequent agricultural runoff. Further pollution comes from domestic and light industrial effluents. There has also been direct damage to the

Left: Mauritius has fringing reefs on most of its coastline, but also a barrier reef in the southeast (STS103-731-80, 1999).
Right: The schooling bannerfish Heniochus diphreutes *is a butterflyfish which feeds on zooplankton above the reef.*

MAP 7h

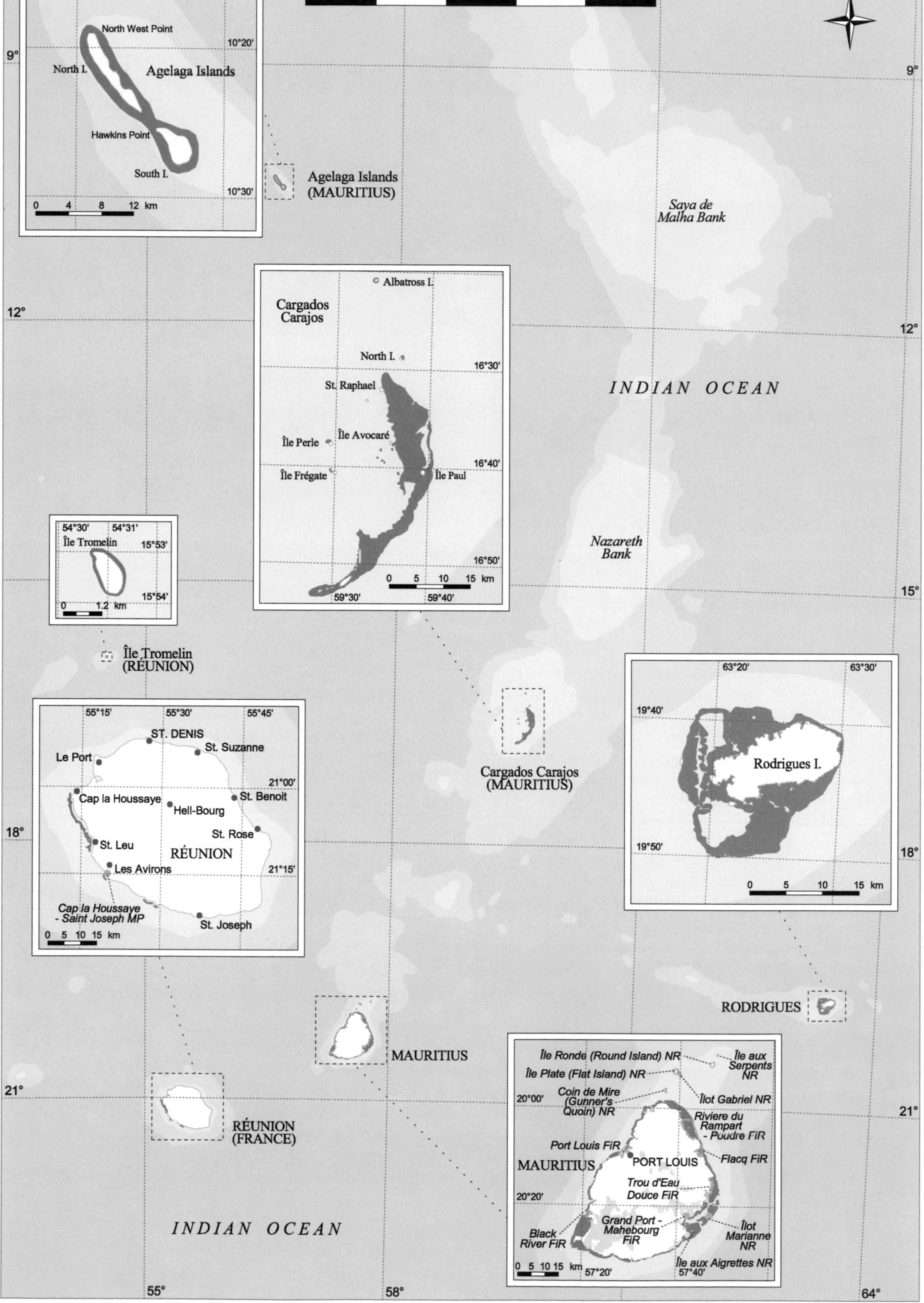

55°
58°
61°
0 90 180 270 360 450 km
N
9°
12°
15°
18°
21°
64°
56°30'
56°40'
North West Point
10°20'
North I.
Agelaga Islands
Hawkins Point
South I.
10°30'
0 4 8 12 km
Agelaga Islands
(MAURITIUS)
Saya de
Malha Bank
Albatross I.
Cargados
Carajos
North I.
16°30'
St. Raphael
Île Avocaré
Île Perle
16°40'
Île Frégate
Île Paul
16°50'
0 5 10 15 km
59°30'
59°40'
INDIAN OCEAN
Nazareth
Bank
54°30'
54°31'
Île Tromelin
15°53'
15°54'
0 1.2 km
Île Tromelin
(RÉUNION)
55°15'
55°30'
55°45'
ST. DENIS
St. Suzanne
Le Port
21°00'
Cap la Houssaye
St. Benoit
Hell-Bourg
St. Rose
St. Leu
RÉUNION
Les Avirons
21°15'
Cap la Houssaye
- Saint Joseph MP
St. Joseph
0 5 10 15 km
Cargados Carajos
(MAURITIUS)
63°20'
63°30'
19°40'
Rodrigues I.
19°50'
0 5 10 15 km
RODRIGUES
MAURITIUS
Île Ronde (Round Island) NR
Île aux
Serpents
NR
Île Plate (Flat Island) NR
Coin de Mire
(Gunner's
Quoin) NR
20°00'
Îlot Gabriel NR
Riviere du
Rampart
- Poudre FiR
Port Louis FiR
MAURITIUS
PORT LOUIS
Flacq FiR
Trou d'Eau
Douce FiR
20°20'
Grand Port -
Mahebourg
FiR
Black
River FiR
Îlot
Marianne
NR
Île aux Aigrettes NR
0 5 10 15 km
57°20'
57°40'
RÉUNION
(FRANCE)
INDIAN OCEAN

reefs – blast fishing was a problem in the past and anchor damage continues. Large areas are also affected by crown-of-thorns starfish, which have undergone population explosions since the early 1980s. Tourism is a critical sector of the economy, and Mauritius had 487 000 arrivals in 1996. Coastal development to cater for this industry has added significant impacts, notably through pollution, but also through coral and shell collection for sale to tourists as well as direct diver impacts.

By contrast the island of Rodrigues remains relatively undeveloped, with a small human population. Fisheries are an important industry and there is a well developed octopus fishery which exports to Mauritius. Tourism is a small, but growing, sector of the economy, with some 26 000 visitors in 1997. Soil erosion and sedimentation are still a problem around this island, but overall the reefs, which are further offshore, remain in relatively healthy condition.

Mauritius holds jurisdiction over a string of islands and reefs running north along the Mascarene Ridge – the northernmost island is Albatross, although there are reef communities on the Nazareth Bank some 240 kilometers further north (and still within Mauritian waters). The main group of islands and reefs in this area lie on a long reef structure on the Cargados Carajos Bank. These include St. Brandon (North Island), St. Raphael, Île Perle, Île Frégate and Île Paul, plus a chain of over a dozen islands in the south. There is little published information about these reefs, however they are thought to include a broad reef flat and possibly the largest continuous algal ridge in the Indian Ocean. There are large and important seabird colonies on a number of islands. The islands are leased to a private fishing company which is based, along with a meteorological station, on St. Raphael.

Also administered by Mauritius is the isolated Agelaga, a complex of two islands (North and South Island) and a substantial reef area. Again there is very little published literature describing this island.

Réunion

Réunion, a territory of France, only has a few fringing reef communities restricted to its leeward western shores, although corals are found growing directly on volcanic substrates in the southeast. Although not extensive, these reefs have been well studied. An estimated 1 000 species of fish occur in the surrounding waters, including 250-300 reef-associated species, and 149 recorded coral species. The 1998 coral bleaching had some impact, particularly where corals were already stressed by other factors, but recovery was good in almost all areas. Coral cover at a number of survey sites on the outer reef slopes and lagoons was at 30-50 percent after the bleaching.

The majority of the people of Réunion live close to the coast and have had a major impact on it. There were some 641 registered commercial fishers in 1996, most of whom were operating in nearshore areas. Overexploitation of coastal fishes has been occurring for some time, and destructive fishing practices have been reported.

Tourism is the main source of income on Réunion. There were 347 000 visitors in 1996 and, although diving and snorkelling are not the major attractions, over 50 percent of hotel bookings are on the west coast, close to the coral reefs. The impacts of overfishing and coastal development together with sewage pollution are reported

Mauritius

GENERAL DATA	
Population (thousands)	1 179
GDP (million US$)	3 544
Land area (km²)	2 035
Marine area (thousand km²)	1 291
Per capita fish consumption (kg/year)	21
STATUS AND THREATS	
Reefs at risk (%)	81
Recorded coral diseases	0
BIODIVERSITY	
Reef area (km²)	870
Coral diversity	161 / 294
Mangrove area (km²)	na
No. of mangrove species	2
No. of seagrass species	7

In addition to the main islands of Mauritius and Rodrigues, Mauritius also administers a large area of remote reefs, notably the Cargados Carajos Bank (STS033-75-92, 1989).

to have severely degraded nearly 30 percent of the reef flats. Efforts are underway to reduce impacts on the coral reefs, including tighter controls on land-based sources of pollution.

Most of the coral reefs are formally protected within a marine park, which had 11 park rangers in 2000. The area incorporates a number of fishing reserves, while consideration is being given to the designation of nature reserves or other additional forms of protection within the park boundaries.

Outlying French territories

France administers a number of islands (Maps 7e and h) around Madagascar (administered alongside Madagascar prior to its independence). Sometimes known by the collective title of Îles Éparses (scattered islands), these islands are all administered from Réunion, although their ownership remains disputed with Madagascar. Most of these are located in the Mozambique Channel. Recent information on the status of any of these reefs is unavailable.

On the same latitude as Grande Comore, but close to

Réunion

General Data	
Population (thousands)	721
GDP (million US$)	6 148
Land area (km²)	2 576
Marine area (thousand km²)	318
Per capita fish consumption (kg/year)	10
Status and Threats	
Reefs at risk (%)	100
Recorded coral diseases	0
Biodiversity	
Reef area* (km²)	<50
Coral diversity	134 / 295
Mangrove area (km²)	na
No. of mangrove species	na
No. of seagrass species	na

* The Îles Éparses have a further 243 km² of reef, with a land area of 23 km² in a marine area of 640 000 km²

Above: Corals do not dominate in all areas, but reef fish, such as these humpback unicornfish Naso brachycentron *are often still found around rocky reefs. Below: Young hawksbill turtles* Eretmochelys imbricata. *There are important turtle nesting beaches on a number of the isolated Indian Ocean islands.*

the northern tip of Madagascar, lie the Îles Glorieuses, a group of four small raised coral islands on a 17 kilometer long coralline platform. At the narrowest point of the Mozambique Channel lies Juan de Nova, another coralline island which was mined for phosphates and had a resident population until 1972. The island lies on a coralline platform almost 12 kilometers in length. Towards the southern end of the Mozambique Channel there are two more islands and reef systems. Bassas da India is a nearly perfect circular atoll about 12 kilometers across with little or no emergent land at high tide. Europa is another atoll structure, about 14 kilometers across, but with a significant land area and a shallow, mangrove-fringed lagoon. This is also one of the most important sites in the world for the breeding of green turtles, with 8 000-15 000 breeding females. Tromelin lies much closer to Réunion, east of northern Madagascar, and its ownership is disputed by Mauritius. This is a raised coralline islet surrounded by fringing reefs on all sides, with a reef flat of about 150 meters width. Some 15 scleractinian coral genera have been described from this island, and there are also 1 500-2 000 breeding green turtles. The island has an airstrip and a meteorological station, although there are no permanent inhabitants.

Although uninhabited, the islands have military barracks and all except Bassas da India also have meteorological stations. These islands have all been declared nature reserves and, although there is little active management, they are perhaps better protected by their remote locations and the presence of military personnel on the islands.

Protected areas with coral reefs

Site name	Designation	Abbreviation	IUCN cat.	Size (km²)	Year
Réunion					
Cap la Houssaye – St. Joseph	Marine Park	MP	VI	na	1998
L'Etang	Fishing Reserve	FiR	VI	na	1992
Pointe de Bretagne – Pointe de l'Etang Sale	Fishing Reserve	FiR	VI	na	1978
Ravine Trois Bassins – Pointe de Bretagne	Fishing Reserve	FiR	VI	na	1978
St. Leu	Fishing Reserve	FiR	VI	na	1992
Saline l'Hermitage (lagoon)	Fishing Reserve	FiR	VI	na	1992
Saline l'Hermitage (reef)	Fishing Reserve	FiR	VI	na	1992
St. Pierre	Fishing Reserve	FiR	VI	na	1992
Îles Éparses					
Juan de Nova	Nature Reserve	NR	IV	na	1975
Îles Glorieuses	Nature Reserve	NR	IV	na	1975
Île Tromelin	Nature Reserve	NR	IV	na	1975
Îlot de Bassas da India	Nature Reserve	NR	IV	na	1975
Îlot d'Europa	Nature Reserve	NR	IV	na	1975
Mauritius					
Balaclava	Marine Park	MP	II	na	1997
Black River	Fishing Reserve	FiR	IV	9.00	1983
Flacq	Fishing Reserve	FiR	IV	6.00	1983
Grand Port – Mahebourg	Fishing Reserve	FiR	IV	22.00	1983
Port Louis	Fishing Reserve	FiR	IV	5.00	1983
Rivière du Rampart – Poudre d'Or	Fishing Reserve	FiR	IV	35.00	1983
Trou d'Eau Douce	Fishing Reserve	FiR	IV	7.00	1983

Selected bibliography

REGIONAL SOURCES

Aleem AA (1984). Distribution and ecology of seagrass communities in the Western Indian Ocean. *Deep Sea Res* Part A 31: 919-922.

Gabrié C (2000). *State of Coral Reefs in French Overseas Départements and Territories.* Ministry of Spatial Planning and Environment and State Secretariat for Overseas Affairs, Paris, France.

Lindén O, Lundin CG (eds) (1997). *The Journey from Arusha to Seychelles: Successes and Failures of Integrated Coastal Zone Management in Eastern Africa and Island States.* The World Bank, Washington DC, USA.

Lindén O, Sporrong N (eds) (1999). *Coral Reef Degradation in the Indian Ocean: Status Reports and Project Presentations.* CORDIO Programme, Stockholm, Sweden.

McClanahan T, Sheppard C, Obura D (eds) (2000). *Coral Reefs of the Indian Ocean: Their Ecology and Conservation.* Oxford University Press, Oxford, UK and New York, USA.

Richmond MD (ed) (1997). *A Guide to the Seashores of Eastern Africa and the Western Indian Ocean Islands.* SIDA/ Department for Research Cooperation, SAREC, Stockholm, Sweden.

Scheer G (1984). The distribution of reef corals in the Indian Ocean with a historical review of its investigation. *Deep Sea Res* Part A 31: 885-900.

Sheppard CRC (1987). Coral species of the Indian Ocean and adjacent seas: a synonymized compilation and some regional distributional patterns. *Atoll Res Bull* 307: 1-32.

Souter D, Obura D, Lindén O (2000). *Coral Reef Degradation in the Indian Ocean. Status Report, 2000.* CORDIO-SIDA/SAREC Marine Science Programme, Stockholm, Sweden.

Stoddart DR, Yonge M (eds) (1971). *Symposia of the Zoological Society of London, 28: Regional Variation in Indian Ocean Coral Reefs.* Academic Press, London, UK.

Wilkinson C, Lindén O, Cesar H, Hodgson G, Rubens J, Strong AE (1999). Ecological and socioeconomic impacts of 1998 coral mortality in the Indian Ocean and ENSO impact and a warning of future change. *Ambio* 28: 188-196.

KENYA AND SOUTHERN SOMALIA

van der Elst R, Salm RV (1999). *Overview of the Biodiversity of the Somali Coastal and Marine Environment.* Report prepared for IUCN Eastern Africa Programme and Somali Natural Resources Management Programme.

McClanahan TR, Obura D (1994). Status of Kenyan coral reefs. In: Ginsburg RN (ed). *Proceedings of the Colloquium on Global Aspects of Coral Reefs: Health, Hazards and History, 1993.* Rosenstiel School of Marine and Atmospheric Science, University of Miami, USA.

Muthiga NA, Bigot L, Nilsson A (2000). Regional report: coral reef programs of Eastern Africa and the Western Indian Ocean. *Proc Int Tropical Marine Ecosystems Management Symp*: 114-143.

Obura DO, Muthiga NA, Watson M (2000). Kenya. In: McClanahan T, Sheppard C, Obura D (eds). *Coral Reefs of the Western Indian Ocean: Their Ecology and Conservation.* Oxford University Press, Oxford, UK and New York, USA.

Sommer C, Schneider W, Poutiers J-M (1996). *FAO Species Identification Guide for Fishery Purposes: The Living Marine Resources of Somalia.* Food and Agriculture Organization of the United Nations, Rome, Italy.

UNEP (1998). *Eastern Africa Atlas of Coastal Resources. 1: Kenya.* UNEP, Nairobi, Kenya.

TANZANIA

Darwall WRT, Guard M (2000). Southern Tanzania. In: McClanahan TR, Obura DO, Sheppard CRC (eds). *Coral Reefs of the Western Indian Ocean: Ecology and Conservation.* Oxford University Press, New York, USA.

Dulvy NK, Stanwell-Smith D, Darwall WRT, Horrill CJ (1995). Coral Mining at Mafia Island, Tanzania: a management dilemma. *Ambio* 24: 358-365.

Guard M, Mmochi AJ, Horrill C (2000). Tanzania. In: Sheppard C (ed). *Seas at the Millennium: An Environmental Evaluation.* Elsevier Science Ltd, Oxford, UK.

Horrill JC, Darwall WRT, Ngoile M (1996). Development of a marine protected area: Mafia Island, Tanzania. *Ambio* 25: 50-57.

Horrill JC, Kamukuru AT, Mgaya YD, Risk M (2000). Northern Tanzania and its major islands. In: McClanahan T, Sheppard C, Obura D (eds). *Coral Reefs of the Western Indian Ocean: Their Ecology and Conservation.* Oxford University Press, Oxford, UK and New York, USA.

Lindén O, Lundin CG (eds) (1995). *Integrated Coastal Zone Management in Tanzania.* The World Bank and SIDA, Washington DC, USA.

MOZAMBIQUE AND SOUTH AFRICA

Gell F, Rodrigues MJ (1998). The reefs of Mozambique. *Reef Encounter* 24: 24-27.

Turpie JK, Beckley LE, Katua SM (1999). Biogeography and the selection of priority areas for conservation of South African coastal fishes. *Biol Cons* 92: 59-72.

MADGASCAR

Gabrié C, Vasseur P, Randriamiarana H, Maharavo J, Mara E (2000). The coral reefs of Madagascar. In: McClanahan T, Sheppard C, Obura D (eds). *Coral Reefs of the Western Indian Ocean: Their Ecology and Conservation.* Oxford University Press, Oxford, UK and New York, USA.

Pichon M (1972). The coral reefs of Madagascar. In: Richard-Vindard G, Battistini R (eds). *Biogeography and Ecology of Madagascar.* Dr W Junk Publishers, The Hague.

Rajonson J (1995). Mangroves and coral reefs of Madagascar. In: Lindén O (ed). *Proceedings of the Workshop and Policy Conference on Integrated Coastal Zone Management in Eastern Africa including the Island States.* Swedish Agency for Research Cooperation with Developing Countries, Sweden.

COMOROS, MAYOTTE AND OUTLYING ISLANDS

Dossar MBA (1997). Integrated coastal zone management in Comoros. In: Lindén O, Lundin CG (eds). *The Journey from Arusha to Seychelles: Successes and Failures of Integrated Coastal Zone Management in Eastern Africa and island states.* The World Bank, Washington DC, USA.

Naim O, Quod J-P (1999). The coral reefs of French Indian Ocean Territories (FIOT). *Reef Encounter* 26: 33-36.

SEYCHELLES

Jennings S, Marshall SS, Polunin NVC (1996). Seychelles' marine protected areas: comparative structure and status of reef fish communities. *Biol Cons* 75: 201-209.

van der Land J (1994). The 'Oceanic Reefs' expedition to the Seychelles (1992-1993). *Zool Verh Leiden* 297: 5-36.

Lundin CG, Lindén O (eds) (1995). *Integrated Coastal Zone Management in the Seychelles.* The World Bank and SAREC-SIDA, Washington DC, USA.

Smith JLB, Smith MM (1969). *Fishes of the Seychelles,* 2nd edn. JLB Smith Institute of Ichthyology, Grahamstown, South Africa.

Vine PJ (1989). *Seychelles.* Immel Publishing, London, UK.

MAURITIUS AND RÉUNION

Fagoonee I (1990). Coastal marine ecosystems of Mauritius. *Hydrobiologia* 208: 55-62.

Lindén O, Lundin CG (eds) (1997). *The Journey from Arusha to Seychelles: Successes and Failures of Integrated Coastal Zone Management in Eastern Africa and Island States.* The World Bank, Washington DC, USA.

Naim O, Quod J-P (1999). The coral reefs of French Indian Ocean Territories (FIOT). *Reef Encounter* 26: 33-36.

Map sources

Map 7a

For Kenya, coral reef data have been combined from Petroconsultants SA (1990)* and UNEP/IUCN (1988b)*. The latter data were only available at a scale of 1:2 500 000. Mangrove data were generously supplied by UNEP and were derived from a more detailed (1:25 000) coverage originally prepared for the Kenya Wildlife Service by W Ferguson in 1995. For Somalia coral reef data are taken from UNEP/IUCN (1988b) at a scale of 1:5 000 000.

Map 7b

Detailed coral reef and mangrove data were generously supplied by UNEP. These data were prepared by Christopher A Muhando at the Institute for Marine Sciences in Zanzibar (with support from the Swedish Agency for Research Cooperation (SAREC)) at a scale of 1:250 000.

Map 7c

Coral reef and mangrove data were generously supplied by UNEP. Coral reef data were originally prepared from MND (1986). Mangrove data are based on 1:1 000 000 maps created under project: FAO/PNUD MOZ/86/003 C MOZ/92/013. For South Africa, coral reefs have been taken from UNEP/IUCN (1988b)* at an approximate scale of 1:2 000 000.

MND (1986). *1:200 000 map series.* Maps: 42621-M, 42622-M, 42623-M, 42624-M, 42625-M, 42626-M, 42627-M, 42628-M, 42629-M, 42630-M and 42630-M. Ministry of National Defense of the Republic of Mozambique, 1st edn 15-X-1986. Division of Navigation and oceanography, Ministry of Defense of Russia.

Map 7d

Coral reef data have been combined from Petroconsultants SA (1990)* and UNEP/IUCN (1988b)*. The latter data were only available at a scale of 1:2 500 000.

Map 7e

For the Comoros, coral reef data were generously supplied by UNEP and are derived from IGN (1995a, 1995b, 1995c). For Mayotte, coral reef data are taken from Hydrographic Office (1978). For the outlying reefs and islands, reef data are from Petroconsultants SA (1990)*.

Hydrographic Office (1978). Comoros Islands. *British Admiralty Chart No. 563.* 1:300 000. Taunton, UK.

IGN (1995a). *Archipel des Comores. Anjouan.* 3615. 1:50 000. Institut Géographique National, Paris, France.

IGN (1995b). *Archipel des Comores. Grande Comore.* 3615. 1:50 000. Institut Géographique National, Paris, France.

IGN (1995c). *Archipel des Comores. Mohéli.* 3615. 1:50 000. Institut Géographique National, Paris, France.

Map 7f

Coral reef and island boundaries are based on Hydrographic Office (1980).

Hydrographic Office (1980). Mahé, Praslin and adjacent islands. *British Admiralty Chart No. 742.* 1:25 000. Taunton, UK.

Map 7g

Coral reef and island boundaries have been derived from several sources, outlined below.

DOS (1978a). *Aldabra Island East* 1:25 000 Series Y852 (DOS 304P) 3rd edn. Department of Overseas Surveys, UK.

DOS (1978b). *Aldabra Island West* 1:25 000 Series Y852 (DOS 304P) 3rd edn. Department of Overseas Surveys, UK.

DOS (1978c). *Farquhar Group* 1:25 000 Series 304P 1st edn. Department of Overseas Surveys, UK.

DOS (1979). *Cosmoledo Group* 1:25 000 Series 304P 1st edn. Department of Overseas Surveys, UK.

DOS (1993a). *Providence Group (North)* 1:25 000 Series 304P 3rd edn. Department of Overseas Surveys, UK.

DOS (1993b). *Providence Group (South)* 1:25 000 Series 304P 3rd edn. Department of Overseas Surveys, UK.

Hydrographic Office (1978). Anchorages in the Seychelles group and outlying islands. *British Admiralty Chart No. 724.* Various scales. Taunton, UK.

Hydrographic Office (1994). Islands North of Madagascar. *British Admiralty Chart No. 718.* Various scales. Taunton, UK.

Map 7h

For the main island of Mauritius coastline, reefs and bathymetry were obtained from Hydrographic Office (1984). For Rodrigues coastline, reefs and bathymetry were obtained from Hydrographic Office (1914). These data were largely derived from a survey undertaken in 1874, however comparisons with a 1983 Department of Overseas Surveys map showed minimal differences in general coastal morphology and reef area. For the Cargados Carajos Shoals and Agalega, islands and reefs were obtained from Hydrographic Office (1969). These are largely based on a survey undertaken in 1846 for Cargados Carajos, and a sketch survey of 1934 for Agalega, but with additions.

For Réunion, coastline and reefs are based on UNEP/IUCN (1988b)*, which was originally prepared at a scale of 1:300 000. For Tromelin, coastline and reefs were obtained from Hydrographic Office (1969). For this island these are largely based on French Government survey of 1959, with additions to 1968.

Hydrographic Office (1914). Rodriguez Island. *British Admiralty Chart No. 715.* February 1914. Taunton, UK.

Hydrographic Office (1969). Cargados Carajos Shoals. *British Admiralty Chart No. 1818.* January 1969 (last major corrections 1941). Taunton, UK.

Hydrographic Office (1984). Mauritius. *British Admiralty Chart No. 711.* 1:125 000. October 1984. Taunton, UK.

* See Technical notes, page 401

Chapter 8
Central Indian Ocean

The southern continental coastline of Central Asia, stretching from Pakistan to Bangladesh, has remarkably little reef development. There are no true reefs recorded off Pakistan, while most of the western and eastern coastlines of India are dominated by high levels of sediments, preventing reef formation. In the far southeast of India there is some reef development and there are a few important reefs around Sri Lanka. In stark contrast to these continental shores, the oceanic waters to the south, and around the Andaman and Nicobar Islands in the east, abound with reefs. The dominant formation is a single arc stretching from the Indian islands of Lakshadweep, along the chain of the Maldives, to the Chagos Archipelago. This follows the Chagos-Laccadive Ridge, a volcanic structure left by the northward movement of the oceanic crust over the Réunion hotspot. These reefs include the world's largest atoll structures.

Biogeographically this is a region of transition. India's Andaman and Nicobar Islands lie on the edges of insular Southeast Asia, the region of highest reef biodiversity in the world. The fauna on these reefs includes many species restricted to Southeast Asia, or which have the Andaman and Nicobar Islands as the westernmost edge of their range. To the west, the reefs from India south to the Chagos Archipelago include certain elements of more typically Indian Ocean species, as well as small numbers of species which characterize the Western Indian Ocean. Maps showing patterns of species diversity on coral reefs (see Chapter 1) clearly show how diversity is restricted to a narrow path of maximum diversity – the "Chagos Stricture" – centered on the southern Maldives and Chagos Archipelago.

Similar levels of diversity again become widespread further west along the coast of East Africa and the Arabian Peninsula. It is considered, from these patterns and others, that the Central Indian Ocean reefs could provide a critical link between the eastern and western margins of the Indian Ocean.

Human pressures on the reefs in the region vary considerably. The reefs of the Chagos and parts of the Andaman and Nicobar Islands are among the least impacted coral reefs worldwide. Studies on water quality in the Chagos Archipelago suggest that these may be some of the least polluted waters in the world, and that even persistent organic pollutants from remote sources may be lower here than elsewhere. By contrast the coral reefs in Sri Lanka and parts of mainland India are under enormous pressure. Although historical data on the distribution of reefs is scarce, it seems highly probable that some reefs have already been lost from these areas. The importance of reefs to the social and economic well being of the region's people is widely recognized, and there are a number of efforts at the national level to restrict damaging activities and set aside areas for conservation.

Left: A red-footed booby Sula sula *and chick on one of the important seabird nesting islands in the Chagos Archipelago.*
Right: A school of bengal snapper Lutjanus kasmira *swims over bleached corals during the 1998 coral bleaching event.*

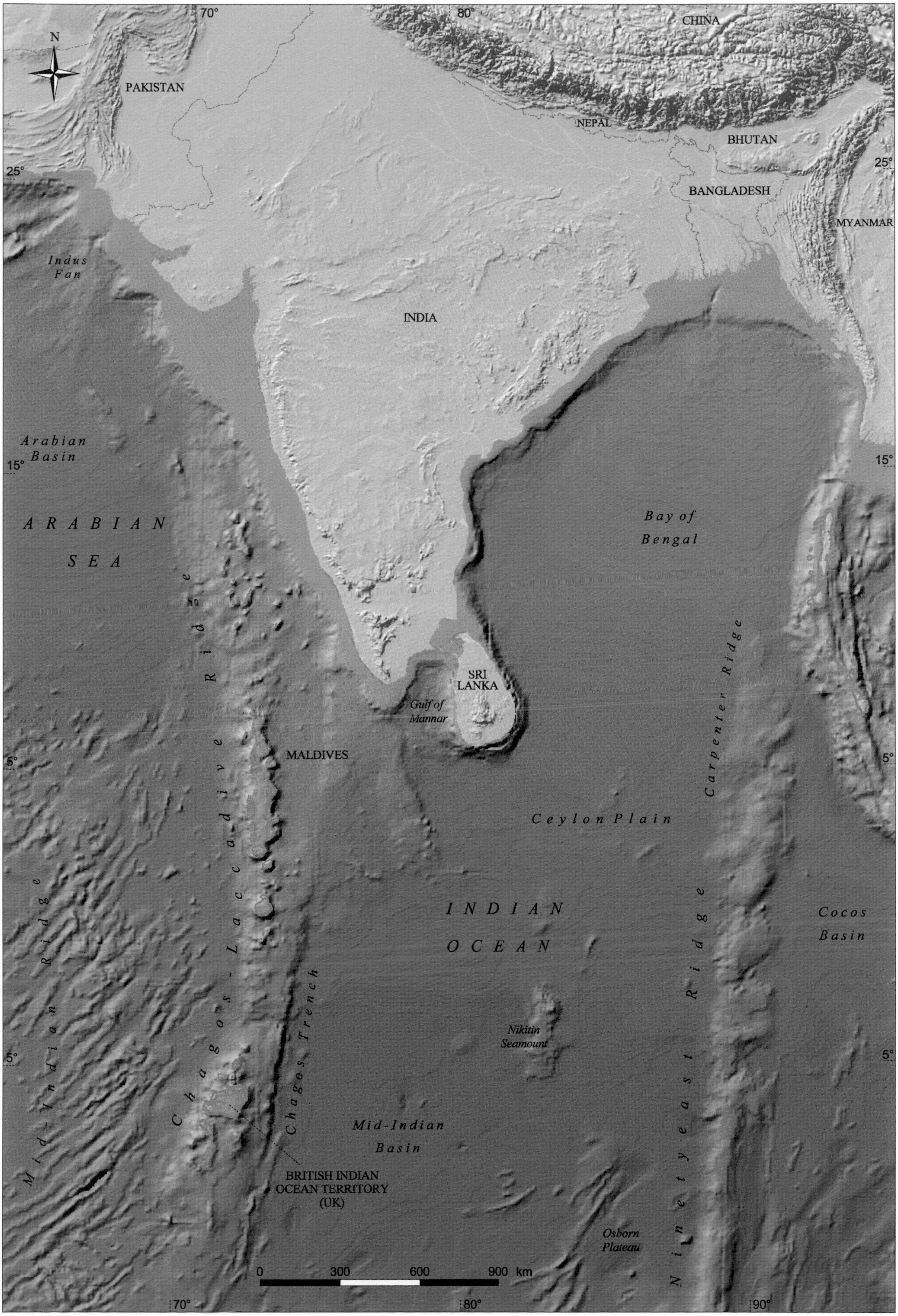

N

70°
80°
CHINA
PAKISTAN
NEPAL
BHUTAN
25°
25°
BANGLADESH
MYANMAR
Indus Fan
INDIA
Arabian Basin
15°
15°
Bay of Bengal
A R A B I A N
S E A
Chagos - Laccadive Ridge
Carpenter Ridge
SRI LANKA
Gulf of Mannar
MALDIVES
5°
5°
Ceylon Plain
Mid-Indian Ridge
I N D I A N
O C E A N
Cocos Basin
Chagos Trench
Nikitin Seamount
5°
5°
Mid-Indian Basin
BRITISH INDIAN OCEAN TERRITORY (UK)
Ninetyeast Ridge
Osborn Plateau
0 300 600 900 km
70°
80°
90°

MAP 8a

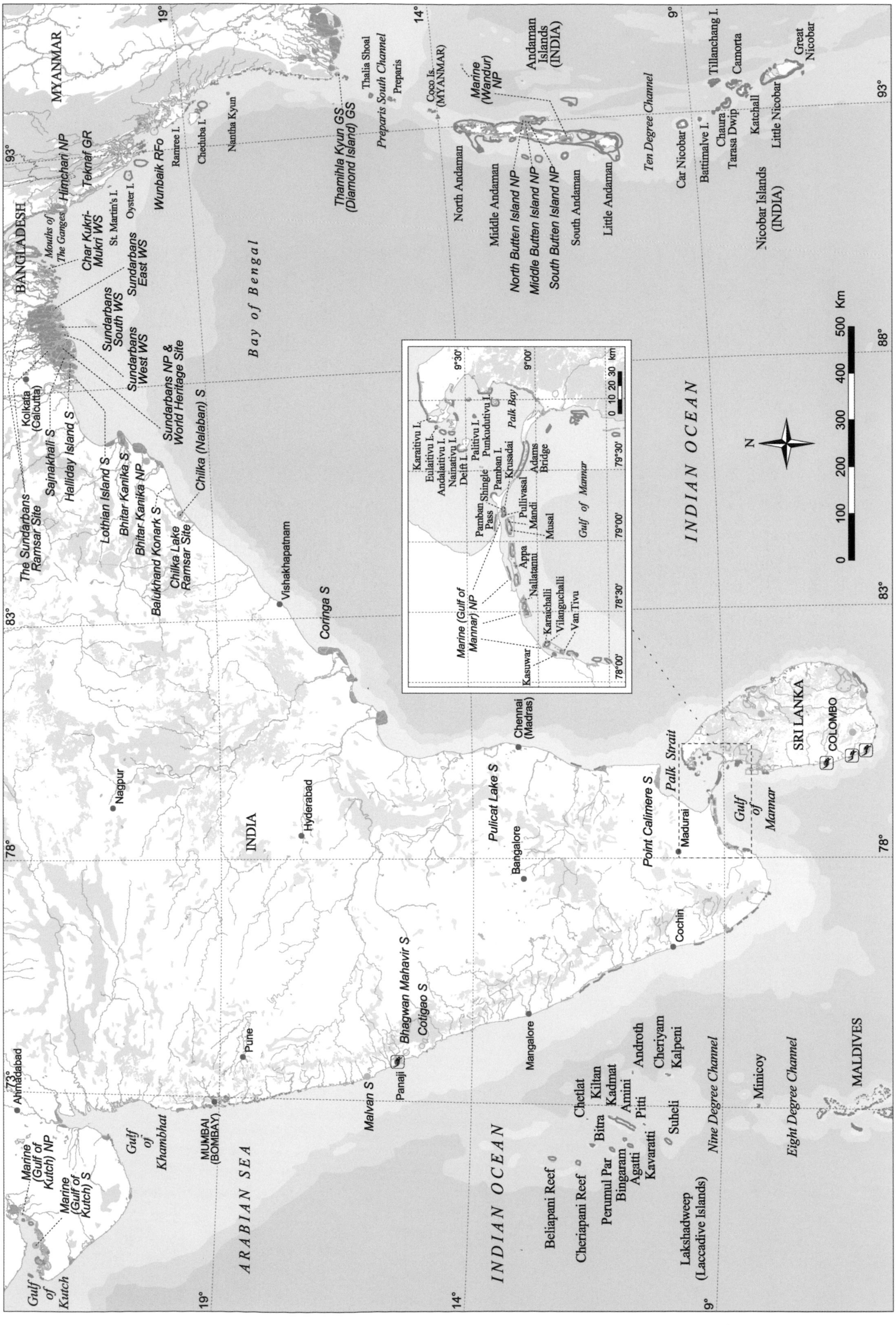

BANGLADESH
MYANMAR
INDIA
SRI LANKA
MALDIVES
Andaman Islands (INDIA)
Nicobar Islands (INDIA)
Coco Is. (MYANMAR)
Lakshadweep (Laccadive Islands)

Bay of Bengal
INDIAN OCEAN
ARABIAN SEA
Gulf of Kutch
Gulf of Khambhat
Gulf of Mannar
Palk Strait
Palk Bay
Preparis South Channel
Ten Degree Channel
Nine Degree Channel
Eight Degree Channel
Mouths of The Ganges

Kolkata (Calcutta)
Ahmadabad
MUMBAI (BOMBAY)
Pune
Panaji
Mangalore
Cochin
Madurai
Bangalore
Chennai (Madras)
Hyderabad
Nagpur
Vishakhapatnam
COLOMBO

Marine (Gulf of Kutch) NP
Marine (Gulf of Kutch) S
The Sundarbans Ramsar Site
Sajnakhali S
Halliday Island S
Lothian Island S
Bhitar Kanika S
Bhitar Kanika NP
Balukhand Konark S
Chilka Lake Ramsar Site
Chilka (Nalaban) S
Sundarbans NP & World Heritage Site
Sundarbans West WS
Sundarbans South WS
Sundarbans East WS
Char Kukri-Mukri WS
Himchari NP
Teknaf GR
Wunbaik RFo
Thamihla Kyun GS (Diamond Island) GS
Marine (Wandur) NP
North Butten Island NP
Middle Butten Island NP
South Butten Island NP
Coringa S
Pulicat Lake S
Point Calimere S
Bhagwan Mahavir S
Cotigao S
Malvan S

St. Martin's I.
Oyster I.
Ramree I.
Cheduba I.
Nantha Kyun
Thalia Shoal
Preparis
North Andaman
Middle Andaman
South Andaman
Little Andaman
Car Nicobar
Battimalve I.
Chaura
Tarasa Dwip
Tillanchang I.
Camorta
Katchall
Little Nicobar
Great Nicobar

Beliapani Reef
Cheriapani Reef
Perumul Par
Bingaram
Agatti
Kavaratti
Bitra
Chetlat
Kiltan
Kadmat
Amini
Pitti
Androth
Cheriyam
Kalpeni
Suheli
Minicoy

Marine (Gulf of Mannar) NP
Karaitivu I.
Eulaitivu I.
Andalaitivu I.
Nainativu I.
Delft I.
Palitivu I.
Punkudutivu I.
Pamban Pass
Shingle
Pamban I.
Krusadai
Adams Bridge
Pullivasal
Mandi
Musal
Appa
Nallatanni
Karaichalli
Vilanguchalli
Van Tivu
Kasuwar
78°00'
78°30'
79°00'
79°30'
9°00'
9°30'
0 10 20 30 km

0 100 200 300 400 500 Km
N
73° 78° 83° 88° 93°
9° 14° 19°

India, Pakistan and Bangladesh

MAP 8a

India, despite its vast size, has only a few coral reefs off its mainland coast, mostly concentrated around the Gulf of Kutch to the northwest, and the Gulf of Mannar near Sri Lanka in the southeast. Reefs are highly developed in the more remote archipelagos of Lakshadweep and the Andaman and Nicobar islands. The distribution and status of any reefs outside these areas remains largely unknown.

The reefs and coral communities of the Gulf of Kutch are predominantly patchy structures built up on sandstone or other banks or around the small islands on the southern side of the gulf. They have adapted to extreme environmental conditions of high temperatures, fluctuating and high salinities, large tidal ranges and heavy sediment loads. As a result diversity is low, with only 37 hard coral species recorded and no branching species. Coral sand mining was a significant industry in the Gulf of Kutch in the early 1980s and may have added to already difficult conditions. Chronic oil pollution in the area may also be affecting the reefs. There is an oil pipeline right through the national park, parts of which were impacted by a major oil spill in 1999. Industrial pollution is a further concern, and the clearance of mangroves may have increased levels of sedimentation. The impacts of the 1998 coral bleaching were quite varied within this area, but on average were much lower than on reefs to the south, with about 30 percent mortality. Further down the coast there are some small, low diversity communities, but conditions here are quite harsh, with low salinities during the monsoon and high turbidity and wave action. Corals are also reported from the Gaveshani Bank some 100 kilometers off the coast from Mangalore.

The best developed mainland reef structures are located in the southeast, with fringing reefs occurring off Palk Bay, and on the coasts and islands of the Gulf of Mannar, including Adams Bridge, a string of reefs stretching across towards Sri Lanka. Diversity is high in this area, with 117 hard coral species recorded, as well as a number of ecosystems including seagrass and mangrove communities. Unfortunately reefs in this region were recorded as rapidly deteriorating as early as 1971, associated with high levels of siltation and the removal of coral rock combined with cyclone impacts. Coral rubble mining still occurs in the region, and mining of sand from the beaches is ongoing. Fisheries are thought to have a considerable impact, with some 47 fishing villages comprising a total of 50 000 people. Apart from overexploitation of general reef fish stocks there are concerns about other fisheries including sea fans, sea cucumbers, spiny lobsters, seahorses and

Pamban Island in the Gulf of Mannar. This region has some of the most important coral reefs off the mainland coast of India (STS033-76-60, 1989).

shells for mother-of-pearl. About 1 000 marine turtles are taken annually and dugongs are also hunted. The 1998 coral bleaching event appears to have severely impacted the reefs in the Gulf of Mannar, with 60-80 percent mortality.

A large proportion of the reefs in both the Gulf of Kutch and the Gulf of Mannar now fall within legally gazetted protected areas, but these suffer from both weak management and virtually no monitoring. There are concerns that the Gulf of Kutch Marine National Park will be rescinded to allow for industrial development.

The Lakshadweep Islands (Laccadives) are located about 300 kilometers west of the southernmost tip of India. They are true atolls and related reef structures, built up over a volcanic base, marking the northernmost and oldest trace of the Réunion hot spot which went on to form the entire Chagos-Laccadive Ridge. There are 12 coral atolls with about 36 islands (with a total land area of 32 square kilometers), about a third of which are inhabited, and also four major submerged reefs and five major submerged banks. Typically the atolls have shallow lagoons, averaging a depth of 3-5 meters, with islands mostly occurring on the eastern rims. The outer slopes of the atolls descend steeply and have prolific coral growth. The local population on these islands numbers some 51 000, and fishing is an important activity, although largely focussed on offshore (non-reef) stocks. There has been sand mining in some lagoons which is likely to have impacted areas of reef. Tourism is a small but

	India	Pakistan	Bangladesh
General Data			
Population (thousands)	1 014 004	141 554	129 194
GDP (million US$)	418 720	62 915	31 838
Land area (km²)	3 089 857	877 664	138 470
Marine area (thousand km²)	2 297	233	80
Per capita fish consumption (kg/year)	5	2	10
Status and Threats			
Reefs at risk (%)	61	na	100
Recorded coral diseases	3	0	0
Biodiversity			
Reef area (km²)	5 790	<50	<50
Coral diversity	208 / 345	na / na	na / na
Mangrove area (km²)	6 700	1 683	5 767
No. of mangrove species	28	4	21
No. of seagrass species	15	na	na

Left: A shallow scene of branching Acropora. *Right: Dense mangrove forests dominate the Sundarbans in the northern Bay of Bengal.*

Protected areas with coral reefs

Site name	Designation	Abbreviation	IUCN cat.	Size (km²)	Year
India					
Great Nicobar	Biosphere Reserve (National)	BR	VI	885.00	1989
Gulf of Kutch	Marine National Park	NP	II	162.89	1980
Gulf of Kutch	Marine Sanctuary	S	IV	293.03	1980
Gulf of Mannar	Marine National Park	NP	II	6.23	1986
Gulf of Mannar	Biosphere Reserve (National)	BR	VI	10 500.00	1989
Wandur (Mahatma Gandhi)	Marine National Park	NP	II	281.50	1983

growing activity: access requires a permit and tourist numbers are currently below 1 000 per year. The 1998 El Niño warming event caused dramatic coral bleaching, with significant subsequent coral mortality of 43-87 percent. This is probably slightly lower than that experienced further south in the Chagos-Laccadive chain.

The Andaman and Nicobar group consists of some 500 islands. Many are the high peaks of a submerged mountain range, a continuation of the Arakan mountains of Myanmar. The islands fall into two clear districts: Andaman to the north and Nicobar to the south, separated by the 160 kilometer wide Ten Degree Channel. There are fringing reefs along the coastlines of many of these islands. Their location is far closer to Indonesia and the Southeast Asian center of biodiversity than to India, and species diversity is higher than at any other reefs in India, with some 219 coral species recorded and around 571 species of reef fish. Although only 38 islands are inhabited, the population has been rising rapidly, largely through immigration, especially in the Andaman District. Close to these areas there may now be some human impacts on the reef communities, while sedimentation is expected to increase as further areas are opened up to logging. At the present time, however, many of the reefs are still largely free from human impacts, and pollution generally remains low. Despite access difficulties, tourist numbers are growing, and dive operators are now taking divers to the islands on "live-aboards", usually departing from Thailand. The reefs were apparently very badly affected by the 1997-98 bleaching, with up to 80 percent mortality reported in some areas. Recent surveys have nonetheless shown an average of 56 percent live coral cover, suggesting a varied impact among the reefs. A detailed network of protected areas has been established in the islands. The majority of these are terrestrial but extend to the coastline, offering at least partial protection to adjacent reef communities.

Pakistan

While there is little published information describing the sub-littoral marine communities of Pakistan, this country is not believed to have any true coral reefs. However, coral communities on hard substrates are suspected, particularly in the west. Any such communities may be very similar to those described for southern Arabia.

Bangladesh

In Bangladesh, as with much of the Indian coast of the Bay of Bengal, the high levels of turbidity and freshwater influx prevent reef development. There is one small area of reef development off the coast of St. Martin's Island or Jinjiradwip, where some 66 hard coral species from 22 genera have been recorded. These small reef areas are considered seriously threatened by sedimentation, cyclone damage, overfishing and anchor damage. Despite the small area of corals, branching *Acropora* are harvested for the curio trade and are now reported to be rare.

Clear oceanic waters with high coral cover are found around Lakshadweep and the Andaman and Nicobar Islands, although both areas were affected by the 1998 coral bleaching and mortality.

MAP 8b

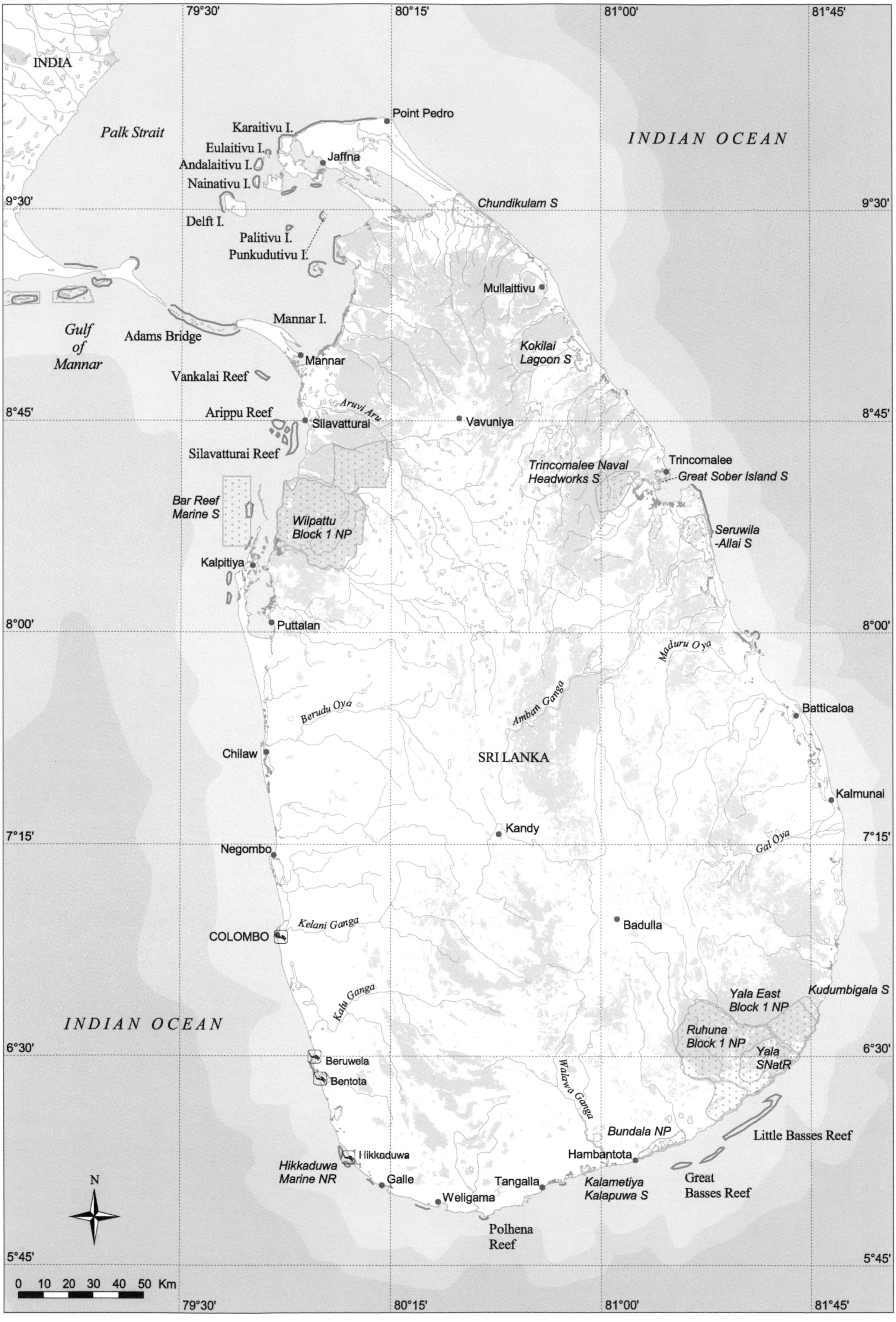

79°30'
80°15'
81°00'
81°45'
INDIA
Palk Strait
Point Pedro
INDIAN OCEAN
Karaitivu I.
Eulaitivu I.
Andalaitivu I.
Jaffna
Nainativu I.
9°30'
Chundikulam S
Delft I.
Palitivu I.
Punkudutivu I.
Mullaittivu
Gulf of Mannar
Adams Bridge
Mannar I.
Mannar
Kokilai Lagoon S
Vankalai Reef
8°45'
Arippu Reef
Aruvi Aru
Silavatturai
Vavuniya
Silavatturai Reef
Trincomalee
Trincomalee Naval Headworks S
Great Sober Island S
Bar Reef Marine S
Wilpattu Block 1 NP
Seruwila -Allai S
Kalpitiya
8°00'
Puttalan
Maduru Oya
Batticaloa
Berudu Oya
Amban Ganga
Chilaw
SRI LANKA
Kalmunai
Kandy
7°15'
Negombo
Gal Oya
Kelani Ganga
COLOMBO
Badulla
Kalu Ganga
Yala East Block 1 NP
Kudumbigala S
INDIAN OCEAN
Ruhuna Block 1 NP
Yala SNatR
6°30'
Beruwela
Bentota
Walawa Ganga
Bundala NP
Little Basses Reef
Hikkaduwa
Hambantota
Hikkaduwa Marine NR
Galle
Tangalla
Kalametiya Kalapuwa S
Great Basses Reef
Weligama
N
Polhena Reef
5°45'
0 10 20 30 40 50 Km

Sri Lanka

MAP 8b

Sri Lanka is a large continental island off the southeast coast of the Indian sub-continent. About 30 percent of the land area is low-lying (less than 30 meters elevation). Offshore the continental shelf is particularly narrow to the south and east, widening to the northwest to join that of India. Much of the coastline is dominated by high wave energy, while the southern and western coasts are further affected by considerable turbidity associated with numerous river mouths. Largely as a consequence of this, coral reefs are not abundant in the coastal waters.

It has been estimated that fringing reefs of varying quality occur along about 2 percent of the coastline, mostly along northwestern and eastern coasts. This statistic includes many coral communities which have developed on non-coral, or fossil reef, platforms. Most reefs could be described as fringing-type formations (although not all are mature structures with clear zonation patterns). Additionally there are some barrier reefs on the northwest coast at Vankalai, Silavatturai and Bar Reef, while in the southeast corals have colonized offshore ridges at Great Basses and Little Basses. The reefs around the Jaffna Peninsula in the north are mainly fringing reefs, but not very well developed. The greatest reef development is in the northwest between Mannar Island and the Kalpitiya Peninsula.

Marine diversity is not as high as among the reefs of the oceanic areas of the Indian Ocean. Coral cover is relatively low, although it reaches more than 50 percent (live hermatypic coral cover) on Bar Reef and reefs in the northwest. During the 1998 El Niño warming event some of the reefs underwent significant bleaching, notably in the south, with bleaching reported to depths of 42 meters near Batticaloa on the east coast. Shallow corals down to 3-5 meters were reported to have died in almost all areas except Trincomalee in the northeast, where bleaching did not take place. Significant reductions in butterflyfish and other coral-dependent species have already been recorded.

Nearshore fisheries are a critical activity in Sri Lanka, providing food, employment and income. Marine fisheries account for 90-95 percent of the total landings, and nearshore fisheries some 70-80 percent of these. Although coral reefs are not widespread, one estimate has suggested that up to 50 percent of the nearshore capture fishery depends directly on coral reef ecosystems. One other important economic activity is the collection of live fish for the aquarium trade. This has grown considerably over the past two decades: some 250 species of reef fish and 50 invertebrates have been exported, in an industry

Bennett's butterflyfish is widespread across the Indo-Pacific. Feeding primarily on coral polyps, it is one of the species which has been impacted by mass mortality of corals.

Protected areas with coral reefs

Site name	Designation	Abbreviation	IUCN cat.	Size (km²)	Year
Sri Lanka					
Bar Reef Marine	Sanctuary	S	IV	306.70	1992
Hikkaduwa Marine	Nature Reserve	NR	IV	1.01	1979

valued at approximately US$3 million in 1998. Other species harvested for export in 1998 included 260 tons of sea cucumbers, and over 800 tons of molluscs.

Tourism plays an important part in the national economy, with coastal tourism estimated to contribute around US$200 million per year. Although reef-related tourism is only a very small fraction of this, it is important in the southwest, particularly around Hikkaduwa where the reef received over 10 000 visitors in 1994.

The threats to Sri Lanka's reefs are numerous and it is likely that the total reef area of this nation may once have been much larger. Many of the remaining reefs are highly degraded. Principle causes of degradation include very high levels of sedimentation arising from erosion of deforested land, poor agricultural practices and construction. Historically, coral mining has led to almost complete destruction of many reefs along the south and southwest coast and may have had similar impacts in the east. Although officially banned in 1983, mining in the sea continues in many areas where it is a traditional activity providing relatively high income employment. Coral rock, taken from living and fossil reefs, is used as a raw material in lime production. In addition to direct destruction, coral mining leads to increased erosion and high turbidity over wide areas of the coastline. Further threats to the remaining reefs arise from destructive fishing practices, including dynamite fishing, uncontrolled exploitation of resources, and pollution arising from sewage and industrial effluent. The combination of threats and current state of degradation of many reefs may slow recovery from the 1998 bleaching event. Although some legislation is in place controlling such activities as coral mining, enforcement is clearly a problem. Only two protected areas (Bar Reef and Hikkaduwa) are specifically designated for the protection of coral reefs, and management is either extremely weak or absent.

Sri Lanka

General Data	
Population (thousands)	19 239
GDP (million US$)	10 738
Land area (km²)	66 580
Marine area (thousand km²)	531
Per capita fish consumption (kg/year)	21
Status and Threats	
Reefs at risk (%)	86
Recorded coral diseases	0
Biodiversity	
Reef area (km²)	680
Coral diversity	100 / 318
Mangrove area (km²)	89
No. of mangrove species	23
No. of seagrass species	7

The highly camouflaged devil scorpionfish Scorpaenopsis diabolis *can be almost invisible when resting on the bottom of the reef. Its dorsal spines are venemous.*

Maldives

MAPS 8c and d

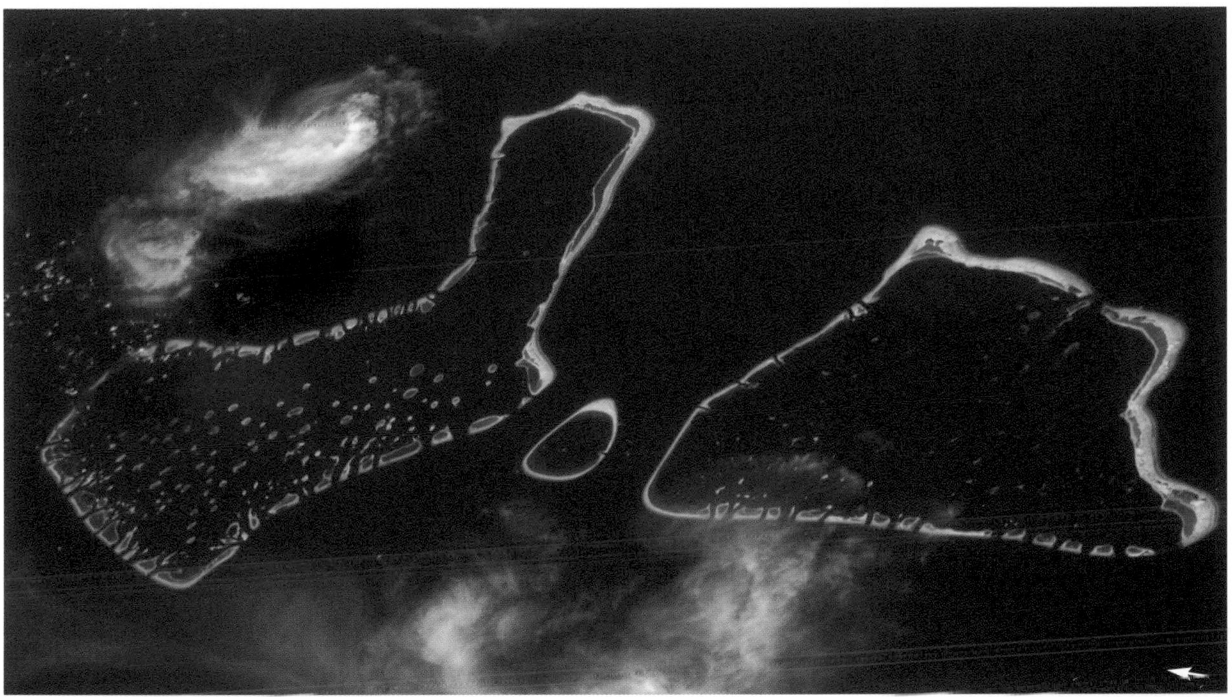

The Maldives are a spectacular chain of 22 coral atolls which run for some 800 kilometers north to south in the Central Indian Ocean. These include the largest surface-level atolls in the world: the area of Thiladhunmathi and Miladhunmadulu Atolls (with two names, but a single atoll structure) is some 3 680 square kilometers, while Huvadhoo Atoll in the south is over 3 200 square kilometers. (The Great Chagos Bank to the south occupies an even greater area, but is now largely submerged.)

There are an estimated 1 200 coralline islands, 199 of which are inhabited (although only three of these are larger than 3 square kilometers). The maximum altitude is only 5 meters above sea level. These islands and reefs make up the central and largest sector of the Chagos-Laccadive Ridge, which marks a volcanic trace left by the Réunion hotspot. The atolls rise steeply from the base of the ridge, and are aligned in two parallel chains. The atoll rims are not unusual, with a wide reef flat, typically bearing a number of islands and sand cays broken by deep channels. The atoll lagoons range from 18 to 55 meters in depth, and within these are a number of patch reefs and knolls, but also some reef structures known as faros which are common in the Maldives, but very unusual elsewhere. These have the appearance of miniature atolls, with a central lagoon, and often bear small islands on their rim.

In terms of their biodiversity, the Maldivian atolls form part of the "Chagos stricture" and so are an important link or stepping stone between the reefs of the Eastern Indian Ocean and those of the East African region. The fauna therefore combines elements of both eastern and western assemblages. Diversity is very high. At least 209 scleractinian corals are recorded, with maximum diversity reported in the south. Over 1 000 epipelagic and shore fishes are recorded from the Maldives, a large proportion of which are reef associated. Coral cover on the atoll edges and on the faros and lagoon knolls was prolific, over 60 percent to depths of at least 20 meters. During the 1998 El Niño warming event some of the worst coral bleaching was recorded in this region and up to 90 percent of hermatypic corals were reported to have died in some areas. Impacts of this mortality are the subject of continuing study – while some growth of new corals is now occurring, the impacts on the wider ecology may continue for decades, even assuming no further extreme events.

More than any other nation outside the Western Pacific, the Maldives is dependant on coral reefs for the

The atolls of Felidu, Wataru and Malaku typify the many atolls of this coral reef nation. The lagoons include numerous patch reefs and circular "faros" (STS081-ESC-5863, 1997).

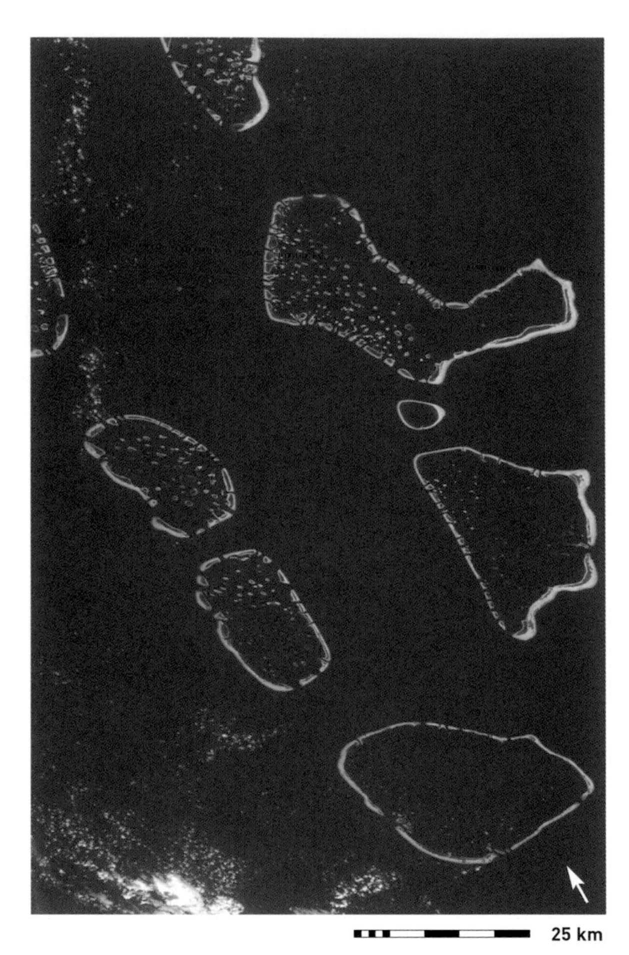

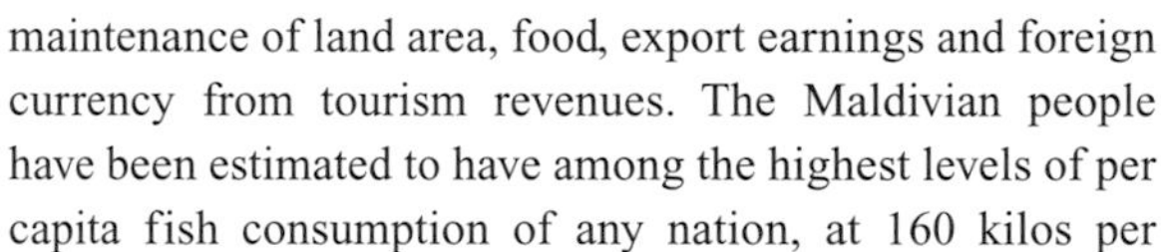

maintenance of land area, food, export earnings and foreign currency from tourism revenues. The Maldivian people have been estimated to have among the highest levels of per capita fish consumption of any nation, at 160 kilos per person per year. The majority of this consumption is of tuna and other pelagic species, while the majority of export fisheries are also centered on tuna. Some reef fish are taken for local consumption, but the most important reef fishery is the capture of live bait for the offshore tuna fishery. Fish exports for the live fish markets of East and Southeast Asia have also been significant through the late 1990s, and this is having an impact on grouper stocks.

Land reclamation has occurred on a number of reefs, while others have been severely impacted by coral mining. Given the geography of this coral reef nation, this latter activity has traditionally been the only means of acquiring natural building material. In the early 1990s it was estimated that between 200 000 and a million cubic meters were mined annually. Mining is now restricted to a few specified reef areas. The first legally gazetted protected areas were designated in 1995 as "protected dive sites"; more sites were established in 1999.

Tourism is restricted to particular resort islands (88 in 1999), which are usually distinct from the local population centers. In 1998 there were almost 400 000 visitors, and diving and snorkelling were a major attraction for almost all. The islands benefit from a relatively stable climate all year round, as well as easily accessible reefs with a high abundance of fish (including many of the larger species

Maldives

General Data	
Population (thousands)	301
GDP (million US$)	215
Land area (km²)	210
Marine area (thousand km²)	996
Per capita fish consumption (kg/year)	160
Status and Threats	
Reefs at risk (%)	11
Recorded coral diseases	1
Biodiversity	
Reef area (km²)	8 920
Coral diversity	212 / 244
Mangrove area (km²)	na
No. of mangrove species	9
No. of seagrass species	1

Left: A wide shallow reef flat on an atoll perimeter. Right: A broad view of the tight arrangement of atolls in the central Maldives, clearly following two parallel chains (STS056-152-160, 1993).

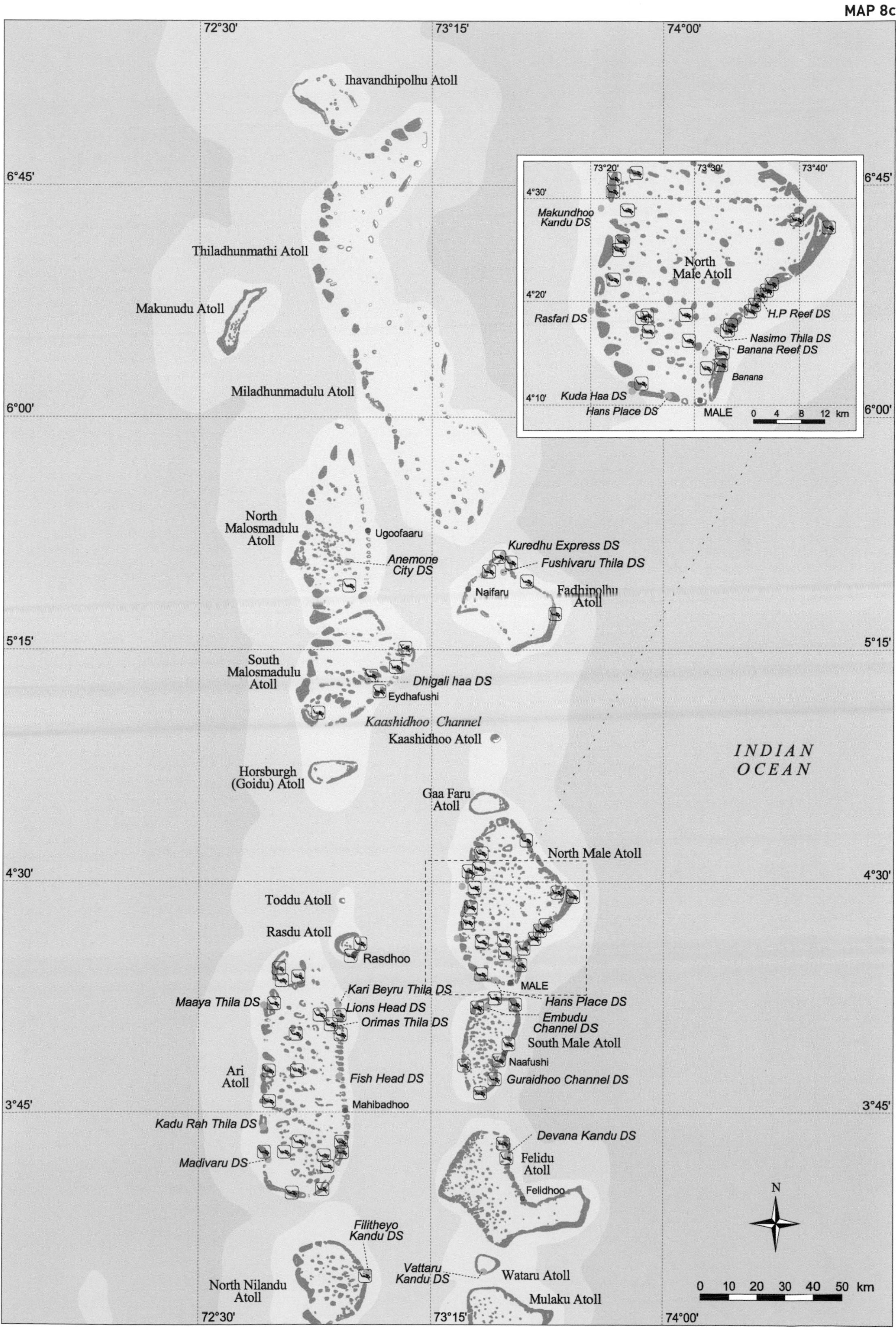
72°30'
73°15'
74°00'
Ihavandhipolhu Atoll
Thiladhunmathi Atoll
Makunudu Atoll
Miladhunmadulu Atoll
6°45'
6°00'
5°15'
4°30'
3°45'
North Malosmadulu Atoll
Ugoofaaru
Anemone City DS
Kuredhu Express DS
Fushivaru Thila DS
Naifaru
Fadhipolhu Atoll
South Malosmadulu Atoll
Dhigali haa DS
Eydhafushi
Kaashidhoo Channel
Kaashidhoo Atoll
Horsburgh (Goidu) Atoll
INDIAN OCEAN
Gaa Faru Atoll
North Male Atoll
Toddu Atoll
Rasdu Atoll
Rasdhoo
MALE
Kari Beyru Thila DS
Maaya Thila DS
Lions Head DS
Orimas Thila DS
Hans Place DS
Embudu Channel DS
South Male Atoll
Naafushi
Ari Atoll
Fish Head DS
Guraidhoo Channel DS
Mahibadhoo
Kadu Rah Thila DS
Devana Kandu DS
Madivaru DS
Felidu Atoll
Felidhoo
Filitheyo Kandu DS
Vattaru Kandu DS
Wataru Atoll
North Nilandu Atoll
Mulaku Atoll
N
0 10 20 30 40 50 km
73°20'
73°30'
73°40'
4°30'
4°20'
4°10'
Makundhoo Kandu DS
North Male Atoll
Rasfari DS
H.P Reef DS
Nasimo Thila DS
Banana Reef DS
Banana
Kuda Haa DS
Hans Place DS
MALE
0 4 8 12 km

MAP 8d

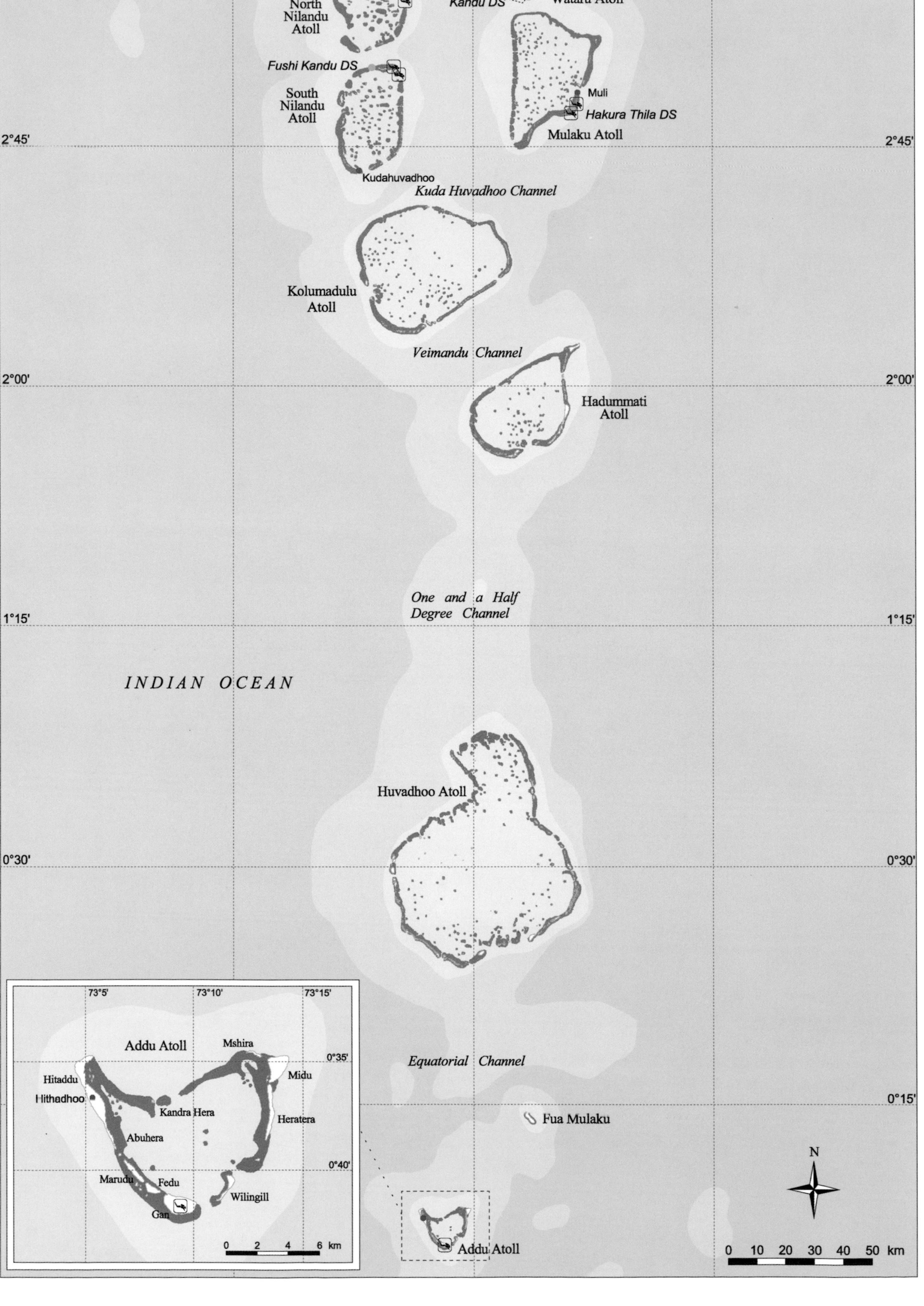

72°30'
73°15'
74°00'
Filitheyo Kandu DS
Felidu Atoll
Vattaru Kandu DS
Wataru Atoll
North Nilandu Atoll
Fushi Kandu DS
South Nilandu Atoll
Muli
Hakura Thila DS
Mulaku Atoll
2°45'
Kudahuvadhoo
Kuda Huvadhoo Channel
Kolumadulu Atoll
Veimandu Channel
2°00'
Hadummati Atoll
One and a Half Degree Channel
1°15'
INDIAN OCEAN
Huvadhoo Atoll
0°30'
Equatorial Channel
0°15'
Fua Mulaku
N
Addu Atoll
0 10 20 30 40 50 km
73°5'
73°10'
73°15'
Addu Atoll
Mshira
0°35'
Hitaddu
Midu
Hithadhoo
Kandra Hera
Heratera
Abuhera
0°40'
Marudu
Fedu
Wilingill
Gan
0 2 4 6 km

such as sharks and manta rays). Whale and dolphin watching is beginning in some areas. The impacts of tourism are localized, but may be significant in certain sites. Impacts include direct diver and anchor damage, interruption of sand movements through the building of groynes or jetties, localized eutrophication from direct sewage discharge into the lagoons, and thermal pollution from desalination plants. Solid waste disposal is a significant problem in most areas. Undoubtedly the greatest concern for this entire nation is the impact of climate change. Coral bleaching and mortality have already caused significant problems: in the future such events will be exacerbated by sea-level rise, and may be further compounded by reduced calcification rates on surviving corals.

Protected areas with coral reefs

Site name	Designation	Abbreviation	IUCN cat.	Size (km²)	Year
Maldives					
Anemone City	Dive Site	DS	Unassigned	na	1999
Banana Reef	Dive Site	DS	Unassigned	na	1995
Devana Kandu	Dive Site	DS	Unassigned	na	1995
Dhigali Haa	Dive Site	DS	Unassigned	na	1999
Embudu Channel	Dive Site	DS	Unassigned	na	1995
Filitheyo Kandu	Dive Site	DS	Unassigned	na	1999
Fish Head	Dive Site	DS	Unassigned	na	1995
Fushi Kandu	Dive Site	DS	Unassigned	na	1999
Fushivaru Thila	Dive Site	DS	Unassigned	na	1995
Guraidhoo Channel	Dive Site	DS	Unassigned	na	1995
HP Reef	Dive Site	DS	Unassigned	na	1995
Hakura Thila	Dive Site	DS	Unassigned	na	1999
Hans Place	Dive Site	DS	Unassigned	na	1995
Kadu Rah Thila	Dive Site	DS	Unassigned	na	1995
Kari Beyru Thila	Dive Site	DS	Unassigned	na	1999
Kuda Haa	Dive Site	DS	Unassigned	na	1995
Kuredhu Express	Dive Site	DS	Unassigned	na	1999
Lions Head	Dive Site	DS	Unassigned	na	1995
Maaya Thila	Dive Site	DS	Unassigned	na	1995
Madivaru	Dive Site	DS	Unassigned	na	1999
Makundhoo Kandu	Dive Site	DS	Unassigned	na	1995
Nasimo Thila	Dive Site	DS	Unassigned	na	1999
Orimas Thila	Dive Site	DS	Unassigned	na	1995
Rasfari	Dive Site	DS	Unassigned	na	1995
Vattaru Kandu	Dive Site	DS	Unassigned	na	1999

During the 1998 coral bleaching event the majority of corals died. The darker branches on this colony have already died and become overgrown with filamentous algae.

British Indian Ocean Territory

MAP 8e

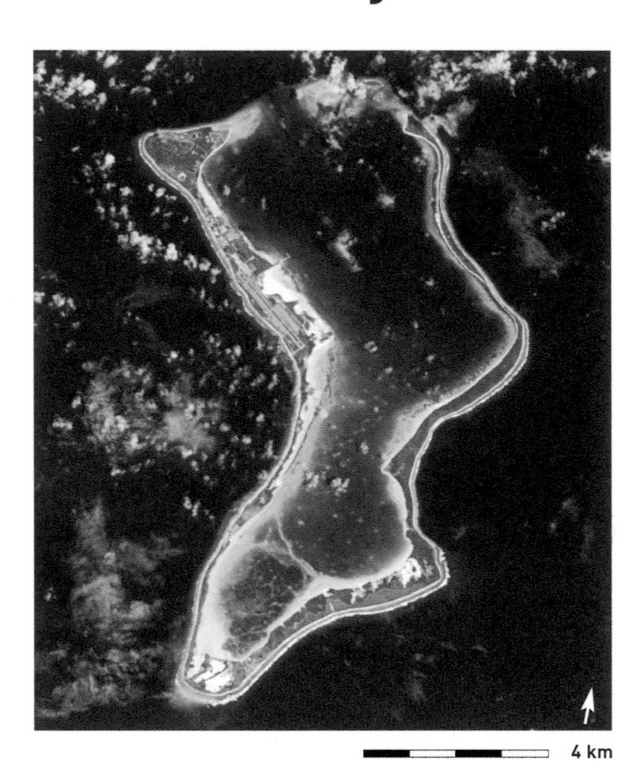

The British Indian Ocean Territory (BIOT) covers a very large area of reefs and islands, also known as the Chagos Archipelago. There are some 50 islands and islets and, although the total land area is only 60 square kilometers, there is a vast area of reefs. These include five true atolls (Blenheim Reef, Diego Garcia, Egmont, Peros Banhos and Salomon), a mostly submerged atoll (Great Chagos Bank, the largest atoll structure in the world at some 13 000 square kilometers), and a number of submerged banks (including Speakers Bank, Pitt Bank and Centurion Bank). The southernmost atoll, Diego Garcia, is unusual in having a narrow but continuous land rim extending around 90 percent of the atoll's circumference. The northerly atolls, by contrast, have only small islands scattered around them. As with the Maldives, the Chagos Archipelago has grown up over the volcanic trace of the Réunion hotspot, and forms the newest and southernmost extension of the Chagos-Laccadive Ridge. The reefs and islands are highly isolated – the nearest reef structures are those of the Maldives, some 500 kilometers to the north, while the nearest continental land mass is that of Sri Lanka, more than 1 500 kilometers away.

With some 220 scleractinian species, the reefs of the Chagos are among the most diverse known for hermatypic corals in the Indian Ocean. While recorded fish faunas are currently lower than those for the Maldives, it is likely that many more have yet to be recorded. Like the Maldives, the reefs of the Chagos lie close to the mid-point between the eastern and western faunas of the Indian Ocean. This fact, combined with their high diversity, lends support to their role as an important biogeographic stepping stone in the so-called Chagos stricture. The faunal characteristics of the Chagos have close affinities to both the Indonesian high diversity faunas and the East African faunas. Further interesting biodiversity features, including a small number of endemic or near endemic species, may be associated with the isolation of the Chagos. Undoubtedly the most interesting of these is the coral *Ctenella chagius* which may be unique to the Chagos, although there is one reported observation from Mauritius. This species is the only extant representative of the family Meandrinidae in the entire Indo-Pacific, although this family was widespread in the Cretaceous (and is widespread in the Caribbean). The Chagos goby *Trimmatom offucius* is endemic to the area and the related *T. nanus* was first reported from these reefs. The latter is the smallest fish species in the world, reaching maturity at only 8 millimeters in length.

Prior to 1998, coral cover was high on both seaward

Left: The southernmost atoll of Diego Garcia includes a major US military base. This atoll is also notable for the narrow but nearly continuous island following the atoll rim (STS038-86-105, 1990). Right: Coralline algae, rather than scleractinian corals, dominate the reef crest on many of the reefs in the Central Indian Ocean, such as this on Peros Banhos.

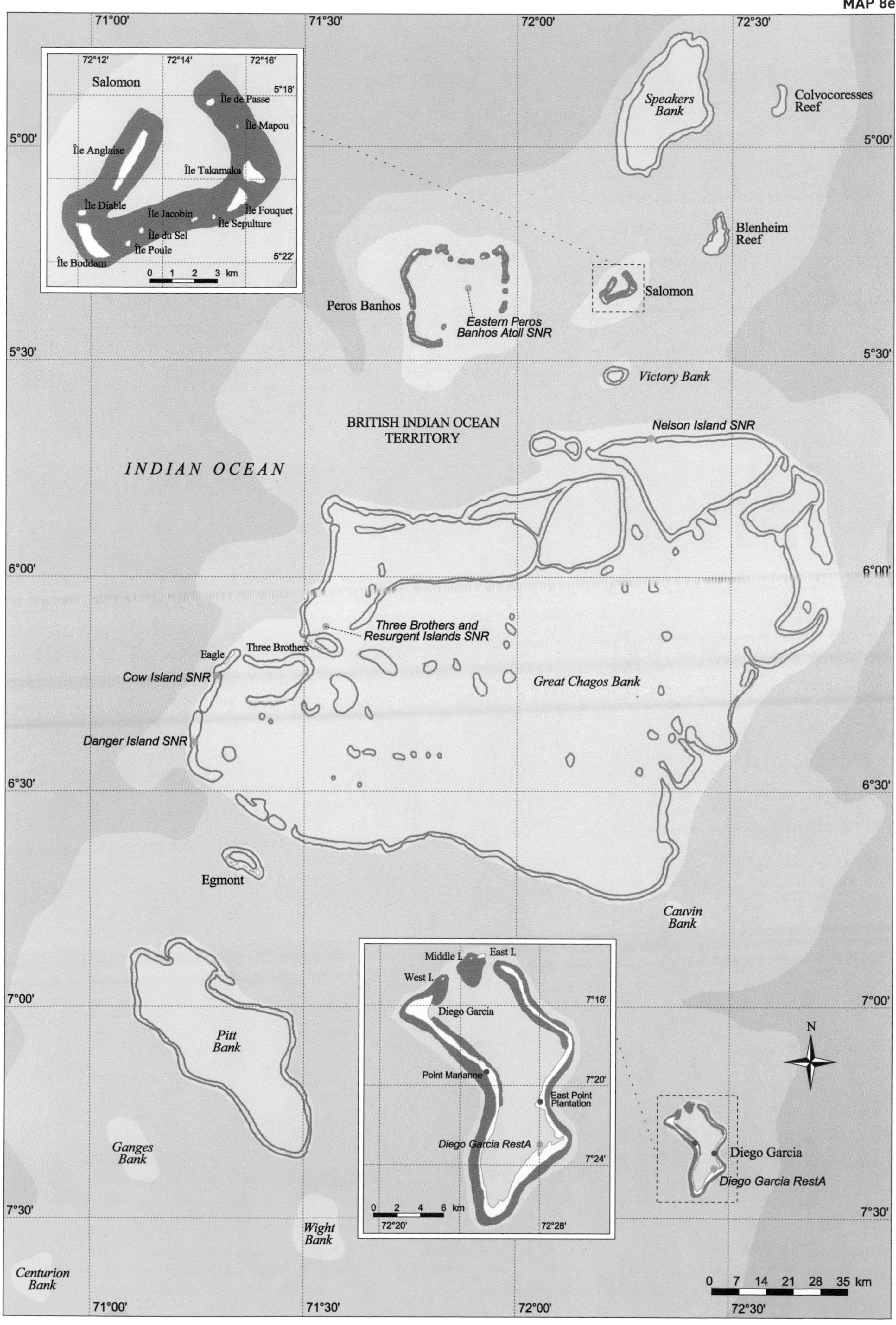

71°00'
71°30'
72°00'
72°30'
5°00'
5°30'
6°00'
6°30'
7°00'
7°30'
Salomon
72°12'
72°14'
72°16'
5°18'
5°22'
Île de Passe
Île Mapou
Île Anglaise
Île Takamaka
Île Diable
Île Jacobin
Île Fouquet
Île Sepulture
Île du Sel
Île Poule
Île Boddam
0 1 2 3 km
Speakers Bank
Colvocoresses Reef
Blenheim Reef
Salomon
Peros Banhos
Eastern Peros Banhos Atoll SNR
Victory Bank
BRITISH INDIAN OCEAN TERRITORY
Nelson Island SNR
INDIAN OCEAN
Three Brothers and Resurgent Islands SNR
Three Brothers
Eagle
Cow Island SNR
Great Chagos Bank
Danger Island SNR
Egmont
Cauvin Bank
Middle I.
East I.
West I.
7°16'
Diego Garcia
Point Marianne
7°20'
East Point Plantation
Diego Garcia RestA
7°24'
0 2 4 6 km
72°20'
72°28'
Pitt Bank
Ganges Bank
Wight Bank
Centurion Bank
N
Diego Garcia
Diego Garcia RestA
0 7 14 21 28 35 km

British Indian Ocean Territory

General Data	
Population*	0
GDP (million US$)	0
Land area (km²)	72
Marine area (thousand km²)	554
Per capita fish consumption (kg/year)	0
Status and Threats	
Reefs at risk (%)	3
Recorded coral diseases	0
Biodiversity	
Reef area (km²)	3 770
Coral diversity	172 / 329
Mangrove area (km²)	na
No. of mangrove species	2
No. of seagrass species	1

* There is a non-resident population of some 3 000 military and civilian personnel on Diego Garcia

and lagoonal reef slopes, typically 50-80 percent of the substrate down to a depth of about 40 meters. Unfortunately this area was heavily damaged during the 1998 coral bleaching event and, although no records of bleaching intensity were made at the time, coral loss has been estimated as averaging 80-85 percent on seaward slopes, and was close to 100 percent in some areas. In addition to its important marine fauna, the Chagos is home to the most diverse and one of the largest populations of breeding seabirds in the Indian Ocean. In 1996, 167 000 breeding pairs of 17 species were observed, including critical populations of the red-footed booby *Sula sula*.

A number of the islands in the Chagos Archipelago, inhabited from the late 18th century, were transformed by the development of coconut plantations and the introduction of rats and other animals. However, it is unlikely that this had a major influence on the marine environment as there was no major export fishery. There was a forced evacuation of the islands in the early 1970s when the military base on the southernmost island of Diego Garcia was established. This has some 3 000 personnel and large vessels permanently at anchor in the lagoon. The impacts of this base have included dredging in the lagoon and some mining of the reef flat, as well as a substantial recreational fishery. There are, however, strict environmental controls on many activities. Personnel are not permitted to dive, and snorkelling is also forbidden on the outer reef slopes. The remaining islands are now uninhabited, although there are a number of visiting yachts and other vessels (commercial tourist-carrying vessels are not permitted). These may be causing localized impacts through anchor damage and sewage pollution, notably in the enclosed lagoon of Salomon Atoll.

Left: A red-footed booby at rest in a palm tree. The northern atolls of the Chagos are a major stronghold for this species in the Indian Ocean. Right, above: A coconut or robber crab Birgus latro. *This land crab can reach 4 kilos in weight, and is found on remote Indo-Pacific islands where it has not been hunted. Right, below: A black-spotted pufferfish* Arothron nigropunctatus.

There is a large offshore tuna fishery as well as a small licensed inshore fishery operated by Mauritian fishermen who visit the reefs for a few months each year. There have also been reports of illegal fishing, notably for sharks and sea cucumbers, although the BIOT Administration has run a fisheries protection vessel for part or all of the year over recent years. A number of the islands and their associated reefs have been declared protected areas. These cover substantial areas of reef. They are occasionally patrolled by military personnel, although the licensed fishing vessels are allowed to operate within their borders. Overall, partly as a result of their history and continuing isolation, but further supported by current management measures, the reefs of the Chagos probably represent some of the most pristine and best protected in the Indian Ocean.

Protected areas with coral reefs

Site name	Designation	Abbreviation	IUCN cat.	Size (km²)	Year
British Indian Ocean Territory					
Cow Island	Strict Nature Reserve	SNR	II	na	1998
Danger Island	Strict Nature Reserve	SNR	II	na	1998
Diego Garcia	Restricted Area	RestA	V	na	1994
Eastern Peros Banhos Atoll	Strict Nature Reserve	SNR	II	na	1998
Nelson Island	Strict Nature Reserve	SNR	II	na	1998
Three Brothers and Resurgent Islands	Strict Nature Reserve	SNR	II	na	1998

A shallow lagoon scene in Salomon Atoll in 1996. These reefs were devastated by the coral bleaching and mortality which occurred in 1998.

Selected bibliography

REGIONAL SOURCES

Brown BE (1997). *Integrated Coastal Management: South Asia.* University of Newcastle, Newcastle upon Tyne, UK.

Debelius H (1993). *Indian Ocean Tropical Fish Guide.* Aquaprint Verlags GmbH, Neu Isenburg, Germany.

GBRMPA, The World Bank, IUCN (1995). *A Global Representative System of Marine Protected Areas. Volume 3: Central Indian Ocean, Arabian Seas, East Africa and East Asian Seas.* The World Bank, Washington DC, USA.

Lindén O, Sporrong N (eds) (1999). *Coral Reef Degradation in the Indian Ocean: Status Reports and Project Presentations.* CORDIO Programme, Stockholm, Sweden.

ODA (ed) (1996). *Proceedings of the International Coral Reef Initiative South Asia Workshop.* Overseas Development Administration, London, UK.

Rajasuriya A, Zahir H, Muley EV, Subramanian BR, Venkataraman K, Wafar MVM, Khan MSM, Whittingham E (2000). Status of coral reefs in South Asia: Bangladesh, India, Maldives and Sri Lanka. In: Wilkinson CR (ed). *Status of Coral Reefs of the World: 2000.* Australian Institute of Marine Science, Cape Ferguson, Australia.

Scheer G (1984). The distribution of reef corals in the Indian Ocean with a historical review of its investigation. *Deep Sea Res* Part A 31: 885-900.

Sheppard CRC (1987). Coral species of the Indian Ocean and adjacent seas: a synonymized compilation and some regional distributional patterns. *Atoll Res Bull* 307: 1-32.

Sheppard CRC (1998). Biodiversity patterns in Indian Ocean corals, and effects of taxonomic error in data. *Biodiversity and Conservation* 7: 847-868.

Stanley Gardiner J (1909). The Percy Sladen Trust Expedition to the Indian Ocean. *The Transactions of the Linnean Society of London Second Series – Zoology* XII: 1-419.

Stanley Gardiner J (1936). The Percy Sladen Trust Expedition to the Indian Ocean. *The Transactions of the Linnean Society of London Second Series – Zoology* XIX: 393-486.

Stoddart DR, Yonge M (eds) (1971). *Symposia of the Zoological Society of London, 28: Regional Variation in Indian Ocean Coral Reefs.* Academic Press, London, UK.

UNEP/IUCN (1988). *UNEP Regional Seas Directories and Bibliographies: Coral Reefs of the World. Volume 2: Indian Ocean.* UNEP and IUCN, Nairobi, Kenya, Gland, Switzerland and Cambridge, UK.

White AT, Fouda MM, Rajasuriya A (1997). Status of coral reefs in South Asia, Indian Ocean and Middle East seas (Red Sea and Persian Gulf). *Proc 8th Int Coral Reef Symp* 1: 301-306.

Wilkinson C, Lindén O, Cesar H, Hodgson G, Rubens J, Strong AE (1999). Ecological and socioeconomic impacts of 1998 coral mortality in the Indian Ocean and ENSO impact and a warning of future change. *Ambio* 28: 188-196.

INDIA, PAKISTAN AND BANGLADESH

Ahmed M (1995). *Coral Reef Ecosystem of Bangladesh – an Overview.* Paper presented at International Coral Reef Initiative South Asia Regional Workshop, Male, Maldives, 1995.

Bahuguna A, Nayak S (1994a). *Coral Reef Mapping of the Lakshadweep Islands.* Space Applications Centre (ISRO), Ahmedabad, India.

Bahuguna A, Nayak S (1994b). *Mapping the Coral Reefs of Tamil Nadu Using Satellite Data.* Space Applications Centre (ISRO), Ahmedabad, India.

Bahunguna A, Nayek S, Patel A, Aggarwal JP, Patel GA (1993). *Coral Reefs of the Gulf of Kachchh, Gujarat.* Space Applications Centre (ISRO), Ahmedabad, India.

Gopinadha Pillai CS (1971). Composition of the coral fauna of the southeastern coast of India and the Laccadives. In: Stoddart DR, Yonge M (eds). *Symposia of the Zoological Society of London, 28: Regional Variation in Indian Ocean Coral Reefs.* Published for the Zoological Society of London by Academic Press, London, UK.

Nayak S, Bahuguna A, Ghosh A (1994). *Coral Reef Mapping of the Andaman and Nicobar Group of Islands.* Space Applications Centre (ISRO), Ahmedabad, India.

Pande P, Kothari A, Singh S (eds) (1991). *Directory of National Parks and Sanctuaries in Andaman and Nicobar Islands: Management Status and Profiles.* Indian Institute of Public Administration, New Delhi, India.

Pernetta JC (1993). *A Marine Conservation and Development Report: Marine Protected Area Needs in the South Asian Seas Region. Volume 2: India.* IUCN, Gland, Switzerland.

Saldanha CJ (1989). *Andaman, Nicobar and Lakshadweep: an Environmental Impact Assessment.* Oxford and IBH Publishing Co., New Delhi, India.

Wafar MVM, Whitaker R (1992). Coral reef surveys in India. *Proc 7th Int Coral Reef Symp* 1: 134-137.

SRI LANKA

De Bruin GHP, Russel BC, Bogusch A (1995). *FAO Species Identification Field Guide for Fishery Purposes: The Marine Fishery Resources of Sri Lanka.* Food and Agriculture Organization of the United Nations, Rome, Italy.

Maldeniya R (1997). The coastal fisheries of Sri Lanka: resources, exploitation and management. In: Silvestre GT, Pauly D (eds). *ICLARM Conference Proceedings, 53: Status and Management of Tropical Coastal Fisheries in Asia.* International Center for Living Aquatic Resources Management, Manila, Philippines.

Öhman MC, Rajasuriya A, Lindén O (1993). Human disturbances on coral reefs in Sri Lanka: a case study. *Ambio* 22: 474-480.

Öhman MC, Rajasuriya A, Olafsson E (1997). Reef fish assemblages in north-western Sri Lanka: distribution patterns and influences of fishing practices. *Environmental Biology of Fishes* 49: 45-61.

Rajasuriya A, De Silva MWRN, Öhman MC (1995). Coral reefs of Sri Lanka: human disturbance and management issues. *Ambio* 24: 428-437.

Rajasuriya A, Öhman MC, Johnstone R (1998). Coral and sandstone reef-habitats in north-western Sri Lanka: patterns in the distribution of coral communities. *Hydrobiologia* 362: 31-43.

Rajasuriya A, Öhman MC, Svensson S (1998). Coral and rock reef habitats in southern Sri Lanka: patterns in the distribution of coral communities. *Ambio* 27: 723-728.

Rajasuriya A, Premaratne A (2000). Sri Lanka. In: Sheppard C (ed). *Seas at the Millennium: An Environmental Evaluation* Elsvier Science Ltd, Oxford, UK.

MALDIVES

Anderson RC, Randall JE, Kuiter RH (1998). Additions to the fish fauna of the Maldive Islands. Part 2: New records of fishes from the Maldive Islands, with notes on other species. *Ichth Bull JLB Smith Inst Ichth* 67: 20-32.

Edwards AJ, Dawson Shepherd A (1992). Environmental implications of aquarium-fish collection in the Maldives, with proposals for regulation. *Env Cons* 19: 61-72.

NIO (1991). *Scientific Report on Status of Atoll Mangroves from the Republic of Maldives.* Report submitted to Ministry of External Affairs, New Delhi, December 1991. National Institute of Oceanography, Goa, India.

Pernetta JC (1993). *A Marine Conservation and Development Report: Marine Protected Area Needs in the South Asian Seas Region. Volume 3: Maldives.* IUCN, Gland, Switzerland.

Randall JE (1992). *Diver's Guide to Fishes of the Maldives.* Immel Publishing, London, UK.

Randall JE, Anderson RC (1993). Annotated checklist of the epipelagic and shore fishes of the Maldive Islands. *Ichth Bull JLB Smith Inst Ichth* 59: 1-47.

Sluka RD, Reichenbach N (1996). Grouper density and diversity at two sites in the Republic of Maldives. *Atoll Res Bull* 438: 1-16.

BRITISH INDIAN OCEAN TERRITORY

Anderson RC, Sheppard CRC, Spalding MD, Crosby R (1998). Shortage of sharks at Chagos. *Shark News* (newsletter of the IUCN Shark Specialist Group) 10: 1-3.

BIOT Administration (1997). *The British Indian Ocean Territory Conservation Policy, October 1997.* British Indian Ocean Territory Administration, Foreign and Commonwealth Office, London, UK.

Sheppard C, Topp J (1999). *Natural History of the Chagos Archipelago, 3: Birds of Chagos.* Friends of the Chagos, London, UK.

Sheppard CRC (1999). *Coral Decline and Weather Patterns over 20 years in the Chagos Archipelago, Central Indian Ocean.* A report commissioned by the Government of the British Indian Ocean Territory. School of Biological Sciences, University of Warwick.

Sheppard CRC, Seaward MRD (eds) (1999). *Linnean Society Occasional Publications, 2: Ecology of the Chagos Archipelago.* Westbury Academic and Scientific Publishing and Linnean Society of London, Otley and London, UK.

Spalding MD, Anderson RC (1997). *Natural History of the Chagos Archipelago, 2: Reef Fishes of Chagos.* Friends of the Chagos, London, UK.

Stoddart DR, Taylor JD (1971). Geography and ecology of Diego Garcia Atoll, Chagos Archipelago. *Atoll Res Bull* 149: 1-237.

Topp J, Seaward M (1999). *Natural History of the Chagos Archipelago, 4: Plants of Chagos.* Friends of the Chagos, London, UK.

Winterbottom R, Anderson RC (1997). A revised checklist of the epipelagic and shore fishes of the Chagos Archipelago, Central Indian Ocean. *Ichth Bull JLB Smith Inst Ichth* 66: 1-28.

Map sources

Map 8a

Coral reefs of India were derived from relatively low resolution (1:10 000 000 to 1:2 000 000) UNEP/IUCN (1988b)*, and from Petroconsultants SA (1990)*, with additional higher resolution data for the Laccadive Islands from Hydrographic Office (1989a and b). For Bangladesh, small areas of coral were added from Ahmed (1995), which includes a sketch map at an approximate scale of 1:33 000.

Ahmed M (1995). Coral Reef Ecosystem of Bangladesh – an Overview. Paper presented at International Coral Reef Initiative South Asia Regional Workshop, Male, Maldives, 1995.

Hydrographic Office (1989a). Islands of Lakshadweep. *British Admiralty Chart No. 705.* Various scales. Taunton, UK.

Hydrographic Office (1989b). Lakshadweep Sea northern part. *British Admiralty Chart No. 2738.* 1:750 000. Taunton, UK.

Map 8b

Coral reef data are largely based on UNEP/IUCN (1988b)*, with original data at a scale of approximately 1:200 000. Small additional polygons were added for individual sites for the Buona Vista Reef from Karunaratne and Weerakkody (1995) and for Kalpitiya Peninsula based on Ohman et al (1993).

Karunaratne L, Weerakkody P (1995). Report on the Status and Bio-Diversity of the Buona-Vista Coral Reef. Draft report.

Ohman MC, Rajasuriya A, Lindén O (1993). Human disturbances on coral reefs in Sri Lanka: a case study. *Ambio* 22(7): 474-480.

Maps 8c and 8d

Atoll names are provided on the map, these are the "traditional" or geographic names, and do not always equate with the names of the administrative units which are used in some sources. The spelling of Maldivian names varies considerably between sources. Coral reef and island boundaries are based on Hydrographic Office (1992a, b, c, d), much of which was based on satellite imagery of 1984 and 1986, supplemented by aerial photography of 1969.

Hydrographic Office (1992a). Addoo Atoll to North Huvadhoo Atoll. *British Admiralty Chart No. 1011.* 1:300 000. October 1992. Taunton, UK.

Hydrographic Office (1992b). North Huvadhoo Atoll to Mulaku Atoll. *British Admiralty Chart No. 1012.* 1:300 000. October 1992. Taunton, UK.

Hydrographic Office (1992c). Mulaku Atoll to South Maalhosmadula Atoll. *British Admiralty Chart No. 1013.* 1:300 000. October 1992. Taunton, UK.

Hydrographic Office (1992d). South Maalhosmadula Atoll to Ihavandhippolhu Atoll. *British Admiralty Chart No. 1014.* 1:300 000. October 1992. Taunton, UK.

Map 8e

Coral reef and islands are based on USDMA (1976). Source data for this map include previous editions (original 1906, large corrections 1971) with amendments to the area boundaries derived from 1976 Landsat data.

USDMA (1976). Indian Ocean, Chagos Archipelago. *Chart No. 61610.* 1:360 000. US Defense Mapping Agency Hydrographic Center.

* See Technical notes, page 401

MAP 9

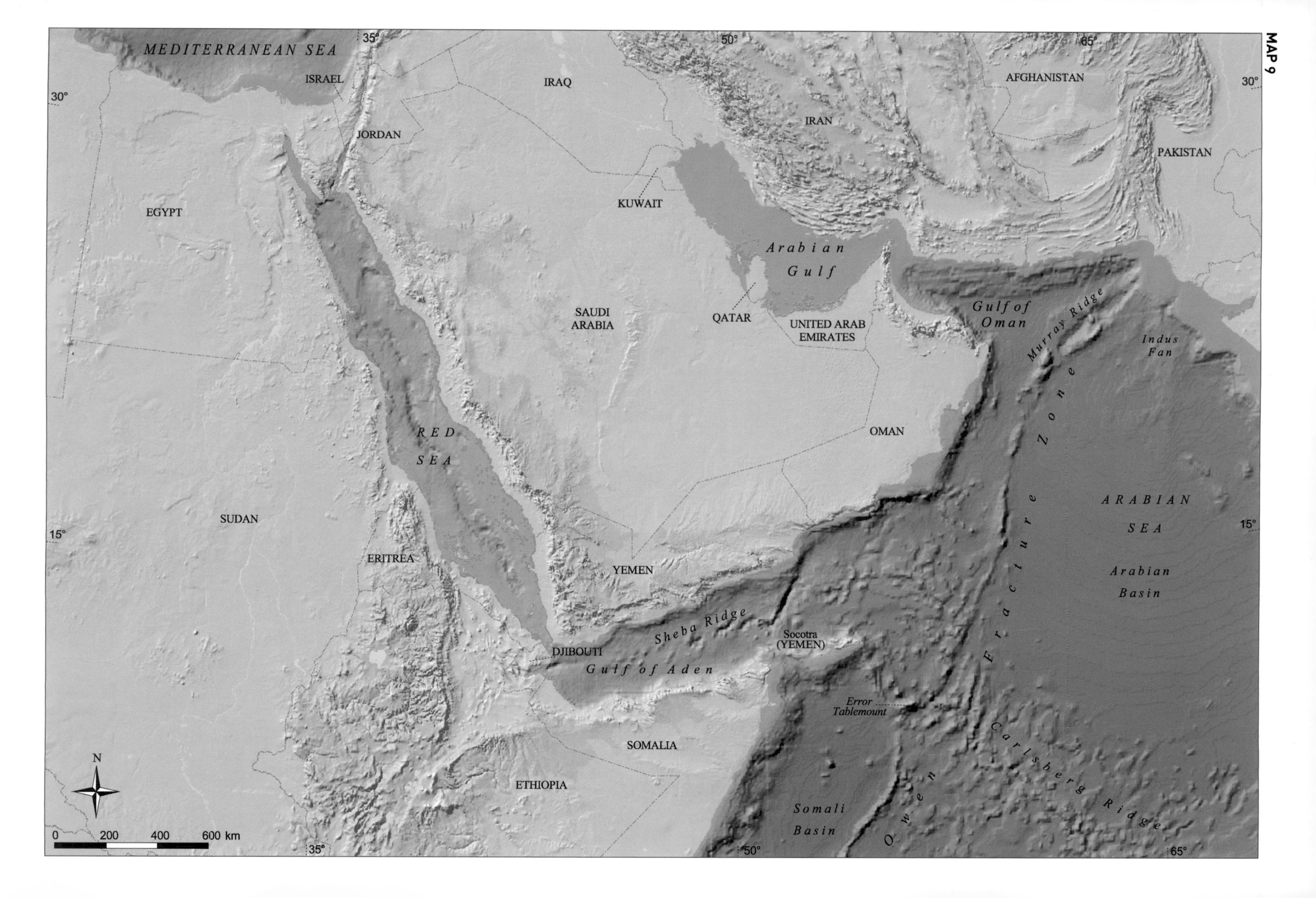

MEDITERRANEAN SEA
ISRAEL
JORDAN
IRAQ
KUWAIT
IRAN
AFGHANISTAN
PAKISTAN
EGYPT
SAUDI ARABIA
QATAR
Arabian Gulf
UNITED ARAB EMIRATES
Gulf of Oman
Murray Ridge
Indus Fan
OMAN
RED SEA
SUDAN
ERITREA
YEMEN
ARABIAN SEA
Arabian Basin
Fracture Zone
Sheba Ridge
Socotra (YEMEN)
DJIBOUTI
Gulf of Aden
Error Tablemount
SOMALIA
ETHIOPIA
Somali Basin
Owen
Carlsberg Ridge
30°
35°
50°
65°
15°
N
0
200
400
600 km

CHAPTER 9
Middle Eastern Seas

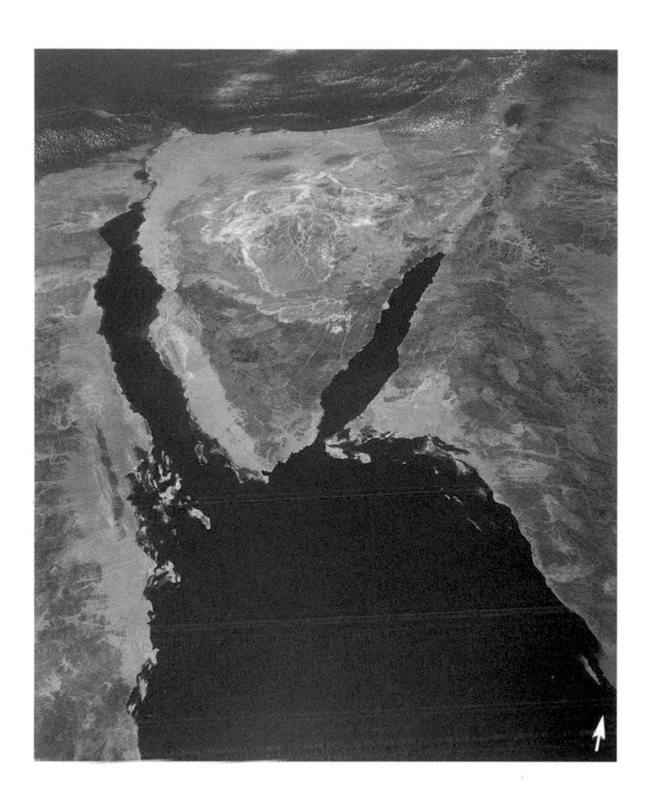

The seas surrounding the Arabian Peninsula are an area of striking contrasts, in their geology and their biology, and in their status in relation to man. They are bordered by some of the world's richest and poorest countries. The vast majority of reefs are very little known, while others have been the focus of study for decades. They include some of the most northerly reef communities in the world, subject to harsh climatic extremes including high and low temperatures, but also to high levels of solar insolation. Despite this, the Red Sea and Gulf of Aden may be the most biologically diverse coral reef area away from the Southeast Asian center of diversity. Biologically the area is relatively isolated as there are no true reefs along much of the coasts of Pakistan or eastern Somalia, which might be seen as the edges of the region.

The region is clearly divided into five major waterbodies: the Arabian Gulf (alternatively referred to as the Gulf or Persian Gulf), the Gulf of Oman, the Arabian Sea, the Gulf of Aden and the Red Sea, each with distinctive ecological and oceanographic characteristics. Until the mid to late 1980s however, only the Red Sea and Arabian Gulf had been studied in any detail. The Arabian Sea coast was almost entirely unknown until the late 1980s, and the Gulf of Aden until the mid to late 1990s. As a consequence, many of the earlier reviews of this region have ignored wide areas where there are highly distinctive and important marine communities.

Both the Red Sea and the Arabian Gulf are partially isolated from the Indian Ocean, and both have inflowing surface currents for much of the year. Geologically speaking, the Red Sea is an ocean, defined by its sea floor which is a basaltic and spreading rift system. It has been separating Africa from the Arabian Peninsula for the last 70 million years. Although remarkably deep and steep-sided along the northern two thirds of this coastline, the continental shelf south of about 19°N becomes very wide, giving rise to quite different conditions and ecological communities. The connection to the Indian Ocean is very shallow, and has been closed many times in its history, each time causing massive changes in salinity and loss of most or all species living in its waters. The latest phase of isolation from

Left: The northern Red Sea and Sinai Peninsula. Fringing reefs line many of these coastlines but are often too narrow to be visible at this scale. Shallow platform reefs are visible in the mouth of the Gulf of Suez (STS040-78-88, 1991). Right: Pocillopora *corals. Large monospecific communities are widespread in the Gulf of Aden and southern Arabia (photo: Jerry Kemp).*

the Indian Ocean took place during the Pleistocene, and reconnection with the Indian Ocean probably only occurred some 17 000 years ago. It remains unclear whether this latest phase led to the total extinction of coral reef species, or whether some survived in certain refugia in the southern Red Sea and/or the Gulf of Aqaba. Many of the reefs that are visible today are actually relatively thin modern veneers of reef deposits which have recolonized older Pleistocene reefs. The Red Sea and Gulf of Aden have large numbers of endemic species, and it may well have been these same climatic shifts and periods of isolation and reconnection that drove the development of new species. Areas in the Gulf of Aden may have acted as refuges for these species when the Red Sea itself was devoid of life.

Reefs are poorly developed along the southern and eastern shores of Arabia, due to the regular cold upwellings associated with the Somali Current. Along the areas of coastline most exposed to these upwellings, in southern Oman and eastern Yemen, macroalgal rather than coral communities predominate, but in more sheltered areas such as the leeward sides of islands, extensive and high-cover coral communities are found. Recent work in the eastern Gulf of Aden has revealed unexpectedly extensive and diverse coral communities along both northern and southern shores, including some of the most diverse fish communities in the wider region. This is in marked contrast to the previously held view that this body of water was devoid of such coral communities.

The Arabian Gulf is a vast shallow sea with little in common with the Red Sea, other than the fact that it too has been subject to periodic drying out over recent geological history. The natural environment is one of harsh climatic extremes, relating to both the high latitude and shallow waters. As a consequence reef development is somewhat restricted and biological diversity is very low.

Human pressures on the reefs in the region vary considerably. Fisheries are an important activity in some countries, although there are few detailed records of catch sizes. Overfishing may not yet be as widespread a problem as in many other regions, but occurs in some areas, including around Yemen and the Gulf of Aden where lobster and shark fisheries are having particular impacts. The region is the principal world petroleum producer and exporter, and a major global shipping route, with related risks of pollution, collisions, groundings, ballast and other discharges. Chronic oil pollution is higher in the Arabian Gulf than in any other coral reef area. Massive development has occurred in parts of the Saudi Arabian Red Sea and in the Arabian Gulf, leading to direct impacts from land reclamation and sedimentation, and also more widespread degradation associated with urban and industrial pollution. Coastal and reef-based tourism has only really developed in the northern Red Sea, but here the rates of growth have been massive, with significant negative impacts in some areas and important examples of successful management in others.

Many of the physical and biological features of this region are best explained and understood from the perspective of natural rather than political sub-divisions, so the main sections of this chapter follow natural sub-divisions. Saudi Arabia, which dominates the region in terms of reef area, is treated separately. However, detailed information about the biology and oceanography of this country can also be found in the other sub-regional sections.

Above: A grey reef shark Carcharhinus amblyrhynchos. *Sharks are now being heavily fished in the southern Red Sea and Gulf of Aden. Below: Red Sea racoon butterflyfish* Chaetodon fasciatus, *one of many species endemic to the Red Sea and Gulf of Aden* (photo: Jerry Kemp).

Northern Red Sea: Egypt, Israel, Jordan

MAP 9a

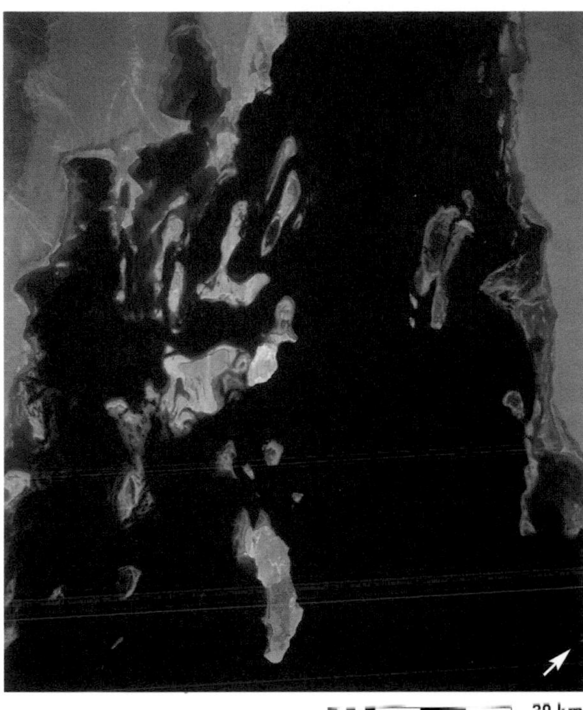

The northern Red Sea includes the coastlines of Egypt, Israel and Jordan, and a substantial part of Saudi Arabia's Red Sea coastline. This section begins with a description of the coral reefs, together with many of the physical and biogeographic characteristics which are continuous between these countries. This is followed by individual descriptions of each country along with the relevant human interactions and uses of coral reefs. For this purpose, Saudi Arabia, which spans several regions, is considered independently.

The northern Red Sea enjoys a number of important and interesting geological and biogeographic features. In the far north, the Red Sea rift system splits into the Gulfs of Suez and Aqaba. Both of these are also rift systems, but have markedly different morphologies. The Gulf of Suez is a spreading rift, but has remained very shallow, averaging a depth of about 30 meters. This area is subjected to considerable climatic extremes associated with its northerly latitude and shallow waters, and hence species diversity is generally low compared to the rest of the Red Sea. There are intermittent fringing reefs along most of the western side, while the eastern side has smaller discontinuous patch reefs.

The Gulf of Aqaba is quite different. It has been formed by a strike-slip rift system as the Arabian Peninsula has moved both in parallel and apart from the Sinai. The same faulting goes on into the Dead Sea rift. The gulf is actually very deep, reaching about 2 000 meters and remaining deep right up to its northern shores. At its southern "mouth" there is a shallow sill and some relatively extensive areas of high quality shallow water reefs, particularly on the eastern side. Inside the gulf itself only narrow fringing reefs have developed on the steeply shelving coastline. Reef flats are often only a few tens of meters wide, while reef slopes are characteristically steep to vertiginous.

South of the Gulfs of Aqaba and Suez, both eastern and western shores of the Red Sea are lined by fringing reefs. These are continuous, often for tens of kilometers, and typically have relatively narrow reef flats. Offshore reefs are well developed at the mouths of the Gulfs of both Aqaba and Suez where there are a number of platform reefs and islands. To the south of the region, in an area known as Gebel Elba, the coastal reefs lie as far as 70 kilometers offshore. There are likely to be some interesting and important areas of shallow water communities in this area, although it has not yet been studied.

The reefs in this region extend into high latitudes and have adapted to relatively low temperatures. Mean surface

Left: Reefs and islands in the southern Gulf of Suez (STS026-41-59, 1988). Right: A steep reef slope, typical of the northern and central Red Sea fringing reefs (photo: Jerry Kemp).

MAP 9a

N

ISRAEL
JORDAN
Sinai Peninsula
EGYPT
SAUDI ARABIA
EGYPT
RED SEA

El Suweis (Suez)
Qalet el Nakhl
El Thamad
Eilat
Eilat Coral R
Aqaba
Aqaba MP
Taba Coast PCo
Haql
Zafarana
Abu Zenîma
Gulf of Suez
Gulf of Aqaba
Nuweiba
Al Bir
Abu Gallum MRPA
Dahab
Ras Mohammed NP
Dahab PCo
Al Bad
Tabuk
Ras Gharib
El Tûr
Ras Shu Kheir
Nabq MRPA
Ash Sharmah
Ras Mohammed NP
Jemsa
Al Muwaylih
Duba
Shaghab
Hurghada
An Nu'man
Ras Abu Soma
Safaga
Al Uwaynidhiyah
The Brothers (El Akhawein)
Al Wadj
Quseir
Umm Urumah
Mashabih
Al Wadj Bank

33° 34° 35° 36°
30° 29° 28° 27° 26° 25°
0 20 40 60 80 100 km

Gulf of Aqaba
Arabian Penninsula
Sinai Penninsula
Nabq MRPA
Tiran-Senafir NP
Nabq
Strait of Tiran
Ras al Qasbah
Tiran I.
Sinafir I.
Sharm el Sheikh
Ras Mohammed NP
RED SEA
34°15' 34°30' 34°45'
28°15' 28°00' 27°45'
0 8 16 24 km

Ghânim I.
Sha'ab Ali
Sha'ab Mahmoud
Ashrâfi Is.
Straits of Gubal
Gûbal Is.
Jemsa
Tawîla Is.
Siyûl Is.
Shâdwan I.
S. Abu Shiban
S. el Erg
S. Abu Nigara
Umm Qamar
Sha'ab Abu Qalawa
EGYPT
Gifatin I.
Hurghada
Sheraton Reef
33°30' 33°45' 34°00'
27°45' 27°30' 27°15'
0 7 14 21 km

temperatures at Suez are 17.5°C, with low extremes dropping to below 10°C. Salinities are also very high in the north, typically about 40.5‰ (parts per thousand), but reaching 42.5‰ in the northern Gulf of Suez.

Although biodiversity remains high even at the northernmost tip of the Gulf of Aqaba, there is clearly some reduction in species numbers with increasing latitude in both of the northerly gulfs. This may be associated with the extreme winter cold, although high salinity probably also plays a role, particularly in the Gulf of Suez. Despite this general picture, a number of other coral and fish species are found in the northern regions which become rare or absent in similarly shallow waters further south. Some 218 hard corals have been recorded in the Gulf of Aqaba. Live coral cover is generally high throughout the region, reaching 60-80 percent on many reef slopes. The most northerly mangroves in the Indian Ocean region are located along the Egyptian coastline of Sinai and these, together with most other mangrove areas in the northern and central Red Sea, are composed of monospecific stands of *Avicennia marina*.

No reefs were observed to be bleached in 1998. Crown-of-thorns starfish have generally been rare in the region, though outbreaks have affected a number of reefs in the southern Gulf of Aqaba since 1998. These were quite localized and temporary, and large numbers were removed, including 70 000 from Tiran and a further 27 000 from Gordon Reef in the Straits of Tiran. Similar outbreaks in more remote locations may have gone unnoticed.

Generally, human population densities are low along most of the coastal zone and are almost entirely confined to urban centers. Fisheries are generally not a major industry. Most are artisanal, undertaken particularly by Bedouin peoples using traditional techniques. Tourism, by contrast, is a major industry, particularly in Egypt. Further details of these and other impacts are considered for the separate countries below.

Egypt

Egypt's extensive coastline incorporates a significant proportion and a considerable range of the coral reefs found in the Red Sea, including a small number of reefs and islands lying in deep water at some distance from the continental shelf. Human activities along this coastline are highly varied, and include areas of quite intensive use and considerable reef degradation, but also areas which remain relatively remote and inaccessible, and which are largely unimpacted by humans.

Marine fishing is not a major industry in Egypt. There is a small amount of commercial fishing in the southern reef areas, and heavy trawling activity was reported in the Gulf of Suez in the late 1990s. However, many reefs are only lightly fished. In contrast, pollution from shipping and oil spillage are a significant threat, notably along the coastline of the Gulfs of Suez and Aqaba. Ship groundings have also been a problem, causing direct physical destruction to some reefs, and raising concerns about the potential economic repercussions arising from any damage to the major tourist beaches and dive sites. The Suez Canal also provides an additional threat. The canal itself was first opened in 1869 and provides a direct sea-level connection between the Red Sea and the Mediterranean. Such a connection allows species to move between these two seas

Bedouins in the Sinai Peninsula, with the fringing reef clearly visible behind.

and to invade areas where they have not previously been recorded (although in fact conditions in the canal are very harsh and highly saline, making the transfer more difficult). Thus far there has been a quite considerable flow of species from the Red Sea into the Mediterranean, but relatively few have made the reverse journey and their impacts on reefs are insignificant.

The greatest impact on the reefs has been the explosion of coastal tourism since the 1980s, with massive growth of resort towns in Sinai and along the mainland Red Sea coast. The latter areas, especially around Hurghada and Safaga, have been particularly poorly planned, leading to the degradation or loss of many of the nearshore fringing reefs. New developments are continuing, notably at Ras Abu Soma but also at localities further south. On the Aqaba coast of Sinai there has again been a massive expansion of the tourist industry – in the Ras Mohammed and Nabq areas tourist rooms increased from nearly 600 in 1988 to over 6 000 by 1995 and 16 000 in 1999 – while massive new hotel developments are being planned at Nabq Bay and close to the Israeli border at Taba. The international airport at Sharm el Sheikh was receiving more than 30 European charter flights per week in the late 1990s. Despite this boom, relatively strict planning measures have been

	Egypt	Israel	Jordan
General Data			
Population (thousands)	68 360	5 842	4 999
GDP (million US$)	55 680	79 610	6 108
Land area (km²)	982 940	20 744	90 177
Marine area (thousand km²)	242*	4.1*	0.2
Per capita fish consumption (kg/year)	7	23	4
Status and Threats			
Reefs at risk (%)	61	100	75
Recorded coral diseases	5	0	0
Biodiversity			
Reef area (km²)	3 800	<10	<50
Coral diversity	126 / 318	145 / na	na / na
Mangrove area (km²)	861	0	0
No. of mangrove species	2	0	0
No. of seagrass species	9	4	na

*Marine area includes Mediterranean Sea

Left: Manta rays Manta birostris *are often seen where reefs lie adjacent to deeper water. Right: The two-banded anemonefish* Amphiprion bicinctus, *endemic to the Red Sea and Gulf of Aden (photo: Jerry Kemp).*

Protected areas with coral reefs

Site name	Designation	Abbreviation	IUCN cat.	Size (km²)	Year
Egypt					
Abu Gallum	Managed Resource Protected Area	MRPA	VI	458.00	1992
Dahab	Protected Coastline	PCo	VI	75.00	1992
Gebel Elba	Conservation Area	CA	IV	4 800.00	1986
Nabq	Managed Resource Protected Area	MRPA	VI	587.00	1992
Ras Mohammed	National Park	NP	II	460.00	1983
Red Sea Islands	Protected Area	PA	VI	na	1983
Sharm el Sheikh	Protected Coastline	PCo	VI	75.00	1992
Taba Coast	Protected Coastline	PCo	VI	735.00	1996
Tiran – Senafir	National Park	NP	II	371.00	1983
Israel					
Eilat Coral	Reserve	R	IV	0.50	na
Jordan					
Aqaba	Marine Park	MP	Unassigned	2.00	na

adopted and enforced in the south Sinai area and the direct impact on the reefs has been low.

A very substantial proportion of Egypt's coral reefs are protected, including all those in the Gulf of Aqaba and all the fringing reefs around islands in the Red Sea itself. There are 22 islands covered by this legislation, including the important and remote offshore islands of the Brothers (El Akhawein), Daedalus (Abu El Kizan), Zabargad and Rocky. The reefs of the Sinai Peninsula have undergone active management since the early 1990s. Mooring buoys have been installed and restrictions are enforced at the sites. A user fee system, (US$5 per day in 2000) helps to support these activities. The significant value of reefs in the national economy has led to the recognition and establishment of a fine system for damage to the reef substrate (from ship groundings and other activities). This has been calculated at US$300 per square meter for each year until estimated recovery (up to 100 years if large, slow-growing *Porites* colonies are damaged).

Israel

Israel has only about 12 kilometers of Red Sea coastline, which is now entirely taken up with urban and industrial development. Nearshore there is a small area of reef, however the stresses in these waters are considerable, including poorly treated sewage discharge, mariculture effluents, bilge and ballast water discharges, and other chemical discharges (including phosphates, detergents, pesticides, hydrocarbons). On the coast itself there has been sand nourishment of the beaches (the addition of sand from elsewhere to satisfy tourist requirements), and solid waste is a problem. Although protected, the reef is further subject to some of the highest diver densities in the world, with an estimated 200 000 dives per year in the late 1990s, largely taking place in the nature reserve. Declines in the reef are notable – even in a less heavily dived location live coral cover dropped from 70 percent in 1996 to 30 percent in 2000, and coral recruitment was also reported to be declining. Direct damage by divers is high, although it has been reduced following the introduction of diver education programs.

Jordan

Jordan has a short coastline, with considerable urban and industrial development in the north but relatively little further south, although it is likely to expand to these areas in the future. Diving tourism is a significant part of the economy and most of the reefs are protected. Enforcement has been a problem, although a number of staff have now been trained in the Ras Mohammed National Park in Egypt. Pollution from the fertilizer industry and sewage are problems in the north of the country.

Saudi Arabia

MAPS 9b, c, d, and e

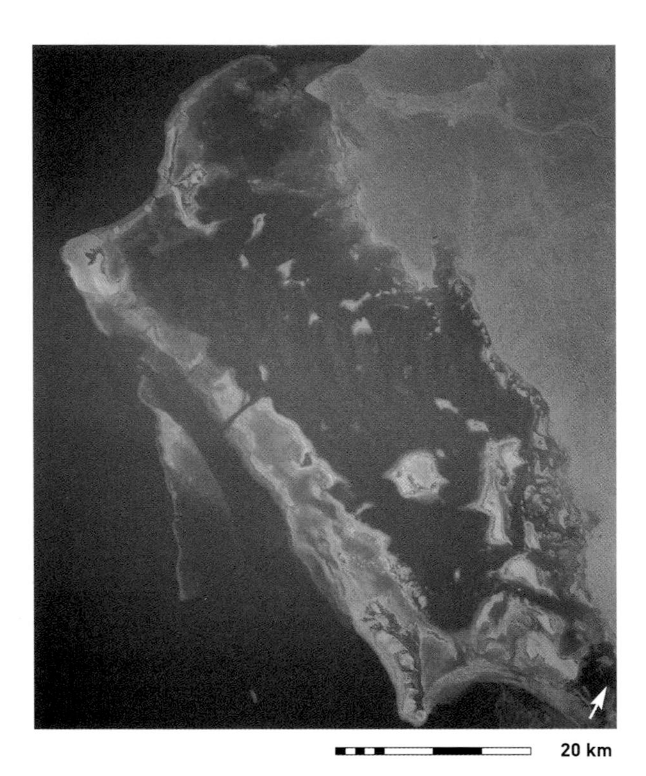

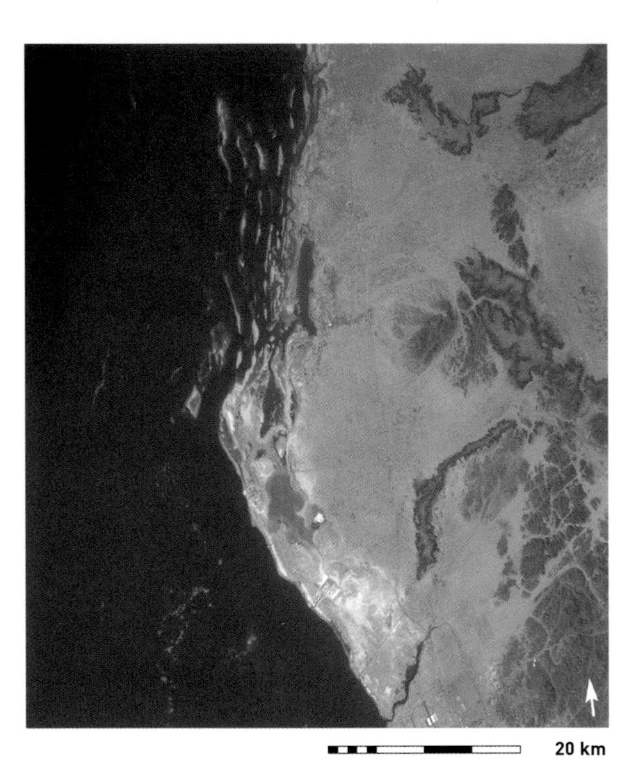

Left: The Al Wadj Bank. In addition to fringing and barrier reefs, this area includes important seagrass and mangrove communities (STS038-77-11, 1990). Right: The Red Sea coastline running north from Jeddah. Although reefs have been badly disrupted as this city has grown, important fringing and patch reefs remain to north and south (STS062-90-81, 1994).

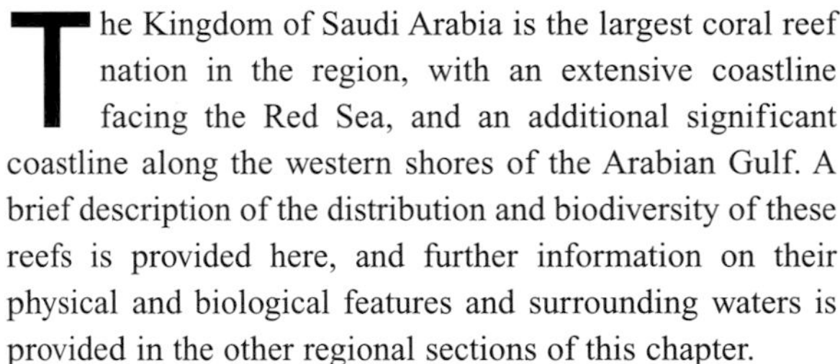

The Kingdom of Saudi Arabia is the largest coral reef nation in the region, with an extensive coastline facing the Red Sea, and an additional significant coastline along the western shores of the Arabian Gulf. A brief description of the distribution and biodiversity of these reefs is provided here, and further information on their physical and biological features and surrounding waters is provided in the other regional sections of this chapter.

The Red Sea coastline extends from the border with Jordan in the northern Gulf of Aqaba all the way to the border with Yemen in the southern Red Sea, following the clear climatic and physical gradients described elsewhere. This region is arid and dominated by high relief along much of its length. Offshore the waters mirror the patterns of the western shores of the Red Sea. In the north there is little or no continental shelf, reef flats are narrow, and the reef profiles are often steep to vertiginous. Further south the continental shelf widens, and in the far south becomes very wide, with extensive, shallow, and turbid inshore waters.

In terms of biodiversity, this coastline incorporates the full wealth of Red Sea species, including those endemic to northern regions, but also the communities and species which are more abundant further south. Surveys from 1997 to 1999 revealed some 260 species of hard coral.

Fringing reefs form a near continuous strip along much of the northern coastline. Further south there is a complex series of fringing, patch and barrier reefs and small islands near the Saudi Arabian coastline on the Al Wadj Bank. This area also houses important and extensive seagrass and mangrove communities. South of this, a discontinuous barrier-type structure has been described running from Al Wadj to Jeddah and termed the Little Barrier Reef. In the far south of the country physical conditions inhibit the development of extensive reef areas close to the continental coastline. However, in a direct parallel with conditions on the coastline of Eritrea, there is extensive mangrove and seagrass development along this coastline, while offshore there is important reef development around the Farasan Islands. Reefs at some locations from Yanbu to Rabigh were observed to be heavily bleached in August/September 1998, associated with elevated sea surface temperatures.

Large parts of Saudi Arabia's Red Sea coastline are undeveloped, particularly away from the central towns of Jeddah and Yanbu. Sewage pollution and land reclamation are concerns around many of the larger towns, including Al Wadj, Yanbu, Jeddah and Jizan. Close to these there are an estimated 18 desalination plants along the Red Sea coast,

Protected areas with coral reefs

Site name	Designation	Abbreviation	IUCN cat.	Size (km²)	Year
Saudi Arabia					
Asir	National Park	NP	V	4 500.00	1981
Dawat Ad-Dafl/Dawat Al-Musallamiyah/Coral Islands	Protected Area	PA	Unassigned	2 100.00	na
Farasan Islands	Protected Area	PA	Ia	600.00	1989
Umm al-Qamari Islands	Protected Area	PA	Ib	1.60	1978

creating localized problems through the return of warm, highly saline waters together with chemicals such as chlorine and anti-scaling compounds. Oil pollution is a threat to reefs around some of the major ports and the refinery in Yanbu. Jeddah is the largest of the Red Sea ports and has undergone massive expansion in recent decades, including large amounts of reclamation and building work directly on the fringing reef flats. Intensive industrial and urban development now extends over more than 100 kilometers of this coastline, and many of the nearshore reefs (together with associated seagrass and mangrove areas) have been severely degraded or destroyed, with pollution and sedimentation combining with the direct impacts of reclamation. Away from these urban areas coastal development remains limited and the reefs are in relatively good condition.

Fishing is not a major industry in the country. There is significant fishing for food and recreation on the nearshore reefs close to the towns, threatening local populations of target species such as large groupers, but there is little or no artisanal fishing. Some commercial fishing activities operate out of Jeddah and Jizan, mostly in the shallow bank areas to south of the country, where there is trawling for prawns and some fishing for pelagic species. There are no detailed statistics describing the size of this fishery.

Tourism is largely unknown, and there is no active promotion of diving or snorkelling, although a number of dive centers cater for local needs, which include significant numbers of expatriate workers. Such recreational activities are most significant on the reefs around Jeddah. A large number of marine protected areas have been proposed along this coastline, though few have been declared.

Arabian Gulf coast

Saudi Arabia probably has some of the most extensive and diverse coral reefs in the Gulf. There are fringing reefs around a number of the offshore islands, with coral growth extending to depths of about 18 meters. Closer to the mainland there are smaller patches and pinnacles. Up to 50 coral species and over 200 fish species have been recorded, with the greatest diversity found in offshore areas. Live coral cover decreased considerably through the 1990s, and extensive coral mortality linked to the 1998 bleaching event was reported on nearshore reefs.

Extensive sections of this coastline are developed and there are large numbers of offshore oil platforms. Impacts on the reefs include those arising from oil pollution, solid waste, and industrial and sewage effluents. There have also been more direct impacts from land reclamation. A large area of reefs have legal protection in one of the only marine protected areas in the Arabian Gulf, although it is unclear to what degree this site is actively managed. Some of the Gulf's more general biological and physical features, together with major human impacts, are discussed more fully in the final section of this chapter.

Saudi Arabia

General Data	
Population (thousands)	22 024
GDP (million US$)	102 677
Land area (km²)	1 948 734
Marine area (thousand km²)	82
Per capita fish consumption (kg/year)	7
Status and Threats	
Reefs at risk (%)	60
Recorded coral diseases	3
Biodiversity	
Reef area (km²)	6 660
Coral diversity*	187 / 314
Mangrove area (km²)	292
No. of mangrove species	3
No. of seagrass species	5

*The higher figure is an estimate for the Red Sea (possibly an underestimate); some 68 species are estimated for the Arabian Gulf

MAP 9b

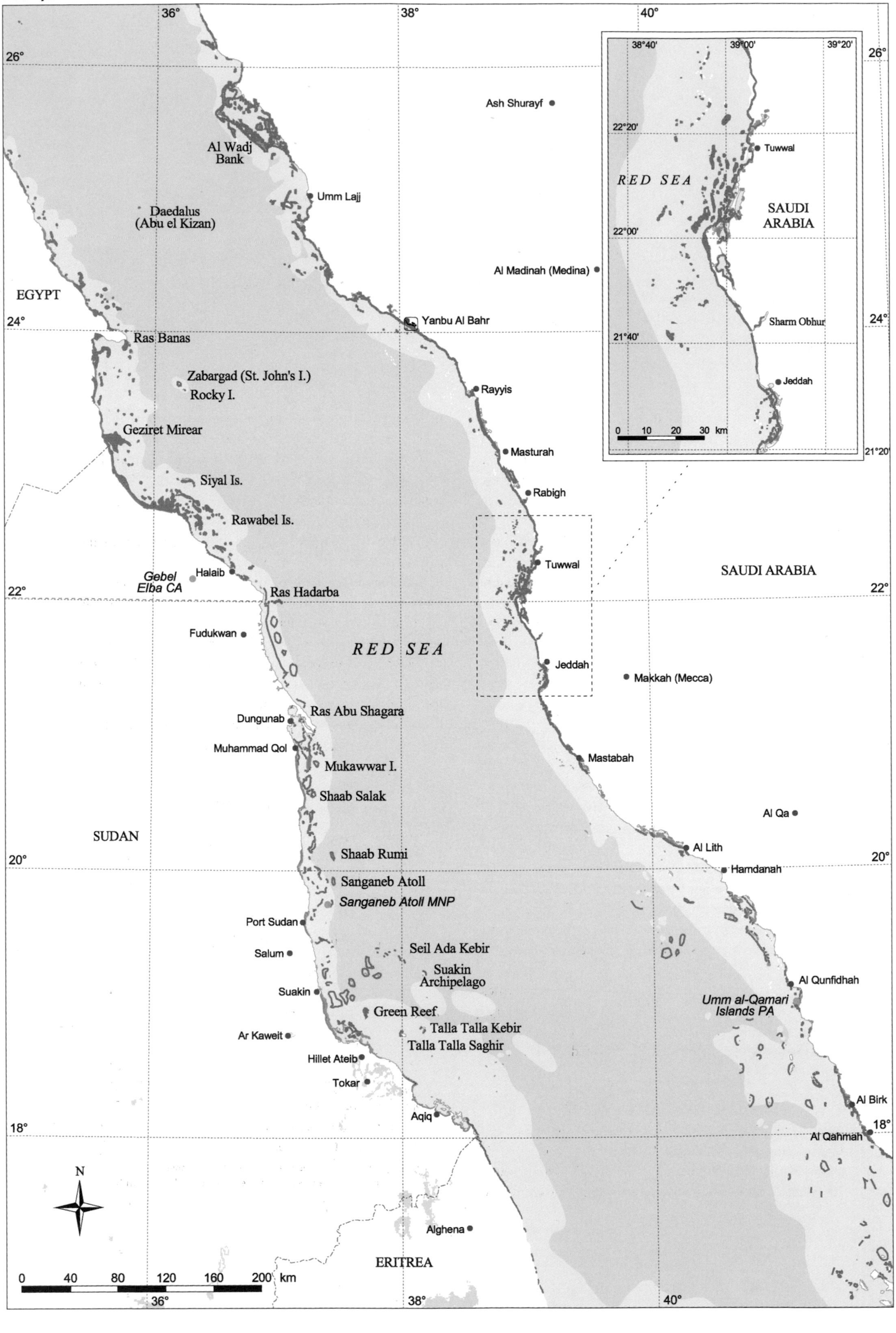

36° 38° 40°
26° 24° 22° 20° 18°

Ash Shurayf
Al Wadj Bank
Umm Lajj
Daedalus (Abu el Kizan)
Al Madinah (Medina)
EGYPT
Yanbu Al Bahr
Ras Banas
Zabargad (St. John's I.)
Rocky I.
Rayyis
Geziret Mirear
Masturah
Siyal Is.
Rabigh
Rawabel Is.
Tuwwal
SAUDI ARABIA
Halaib
Gebel Elba CA
Ras Hadarba
Fudukwan
RED SEA
Jeddah
Makkah (Mecca)
Ras Abu Shagara
Dungunab
Muhammad Qol
Mukawwar I.
Mastabah
Shaab Salak
Al Qa
SUDAN
Al Lith
Shaab Rumi
Hamdanah
Sanganeb Atoll
Sanganeb Atoll MNP
Port Sudan
Seil Ada Kebir
Salum
Suakin Archipelago
Al Qunfidhah
Suakin
Umm al-Qamari Islands PA
Green Reef
Ar Kaweit
Talla Talla Kebir
Talla Talla Saghir
Hillet Ateib
Tokar
Al Birk
Aqiq
Al Qahmah
N
Alghena
ERITREA
0 40 80 120 160 200 km

38°40' 39°00' 39°20'
22°20' 22°00' 21°40' 21°20'
Tuwwal
RED SEA
SAUDI ARABIA
Sharm Obhur
Jeddah
0 10 20 30 km

Central Red Sea: Sudan

MAP 9b

The central Red Sea can be defined politically as the coastlines of Sudan in the west and the central areas of the Saudi Arabian Red Sea Coast in the east. Details of the latter, particularly the human interactions with reefs, can be found in the section on Saudi Arabia.

Geomorphologically the region is again characterized by steeply shelving coastlines to the north. However, heading south from about 20°N is a relatively rapid transition as the coastlines change to a broad and more gently inclined continental shelf. The shores are lined by fringing reefs, mostly with shallow reef flats a few tens of meters wide in the north, although becoming broader further south and stretching out from the coast in areas where there are wide alluvial fans.

In addition to fringing reefs, discontinuous barrier-type structures have been described on both Sudanese and Saudi Arabian coastlines. To the south, the Suakin Archipelago consists of a number of offshore islands rising from relatively deep water. Most of these have significant fringing reefs, although wave action appears to have restricted reef growth on some. Although a number of reef structures on both sides of the Red Sea have some atoll-like features, Sudan has the only true atoll, Sanganeb Atoll, which rises from a depth of 800 meters.

Biologically, this region has many similarities with the more northerly reefs, though it is not affected by such extreme winter cooling and salinities are more stable. These are among the most biologically diverse reefs in the entire Western Indian Ocean region. Coral cover is of course highly varied, but levels of 85 percent (hard and soft coral combined) have been recorded on the reef slope of Sanganeb Atoll. Further south, as the continental shelf widens, there is a relatively rapid transition with the appearance of communities distinctive to the southern Red Sea. In patterns which appear to be mirrored at least among corals and fish, there are considerable changes in dominance and species replacement. A number of endemic Red Sea species are in fact restricted to the northern regions, while others rarely migrate to these parts. Moving south, the area of mangroves also begins to increase and a second species, *Rhizophora mucronata*, occurs.

Port Sudan is a relatively large port, although development here is clearly far less extensive than in Jeddah on the Saudi Arabian coastline, while Suakin is the only other coastal city. Sewage pollution is reported to be a problem close to both cities. Away from these areas, the population densities are low. There is a growing recreational dive industry, almost entirely run from "live-aboard" boats operating our of Egypt and Port Sudan: thus far the total numbers of visitors to these reefs are still low.

Sudan

General Data	
Population (thousands)	35 080
GDP (million US$)	29 761
Land area (km²)	2 490 389
Marine area (thousand km²)	33
Per capita fish consumption (kg/year)	2
Status and Threats	
Reefs at risk (%)	32
Recorded coral diseases	0
Biodiversity	
Reef area (km²)	2 720
Coral diversity	106 / 313
Mangrove area (km²)	937
No. of mangrove species	3
No. of seagrass species	2

Protected areas with coral reefs

Site name	Designation	Abbreviation	IUCN cat.	Size (km²)	Year
Sudan					
Sanganeb Atoll	Marine National Park	MNP	II	260.00	1990

Southern Red Sea: Eritrea and Yemen

MAP 9c

The physical structure of the Red Sea changes significantly in its southern sections. Offshore, the continental shelf broadens and waters remain relatively shallow over wide areas, while the inshore waters become somewhat turbid. Nearshore fringing reefs are less common on the mainland coasts, although some are found around offshore islands. Two major archipelagos are located here: the Farasan Islands off the Saudi Arabian coast which extend into the Kamaran Islands off Yemen, and the Dahlak Archipelago off the coast of Eritrea. There are also several small islands further to the south. The geological origin of these is complex. The Farasan and Dahlak archipelagos and the Kamaran Islands are carbonate platforms which have been uplifted and undergone some further modifications. Some of the islands further south are of volcanic origin, and this is still a tectonically active region. There are large areas of algal reefs – calcareous platforms built almost entirely from coralline red algae – with a probable total area of several hundred square kilometers. These have developed on the sandy sublittoral zones in nearshore waters where conditions of temperature, salinity and turbidity appear inimical to coral growth.

South of about 17°N the coasts of the Red Sea draw together. At the mouth of the Red Sea, known as the Bab el Mandeb (Gate of Lamentations), the water is only about 130 meters deep. Annual precipitation over the entire Red Sea is about 10 millimeters, while evaporation removes something of the order of 2 meters of water per year. There is therefore a net inflow of water through the Bab el Mandeb, although surface currents are more complex with a reversed flow for part of the summer. There is also a deep current of denser more saline water flowing outwards. The water entering the Red Sea from the Gulf of Aden is relatively rich in nutrients and plankton, contributing to the turbidity which appears to restrict reef development in the southern Red Sea.

Temperatures in the southern Red Sea are high, with mean surface values of over 32°C recorded in Yemen, while lagoon waters regularly reach 45°C. By contrast salinities in this region are closer to those of the open ocean.

Biologically, the southern Red Sea is very distinct. Mangroves are well developed along significant stretches of the coastline, as are seagrasses in shallow nearshore waters. In contrast, many of the fringing reefs are poorly developed, and even around the offshore islands partial coral cover or fragmentary reef development is common. There are lower diversities of both fish and corals (and presumably other faunal groups) on almost all reefs, although there is also some species replacement of those found further north. The changes and decreases in diversity

The southern Red Sea, including the narrow straits of Bab el Mandeb, with the Gulf of Aden (STS061-93-12, 1993).

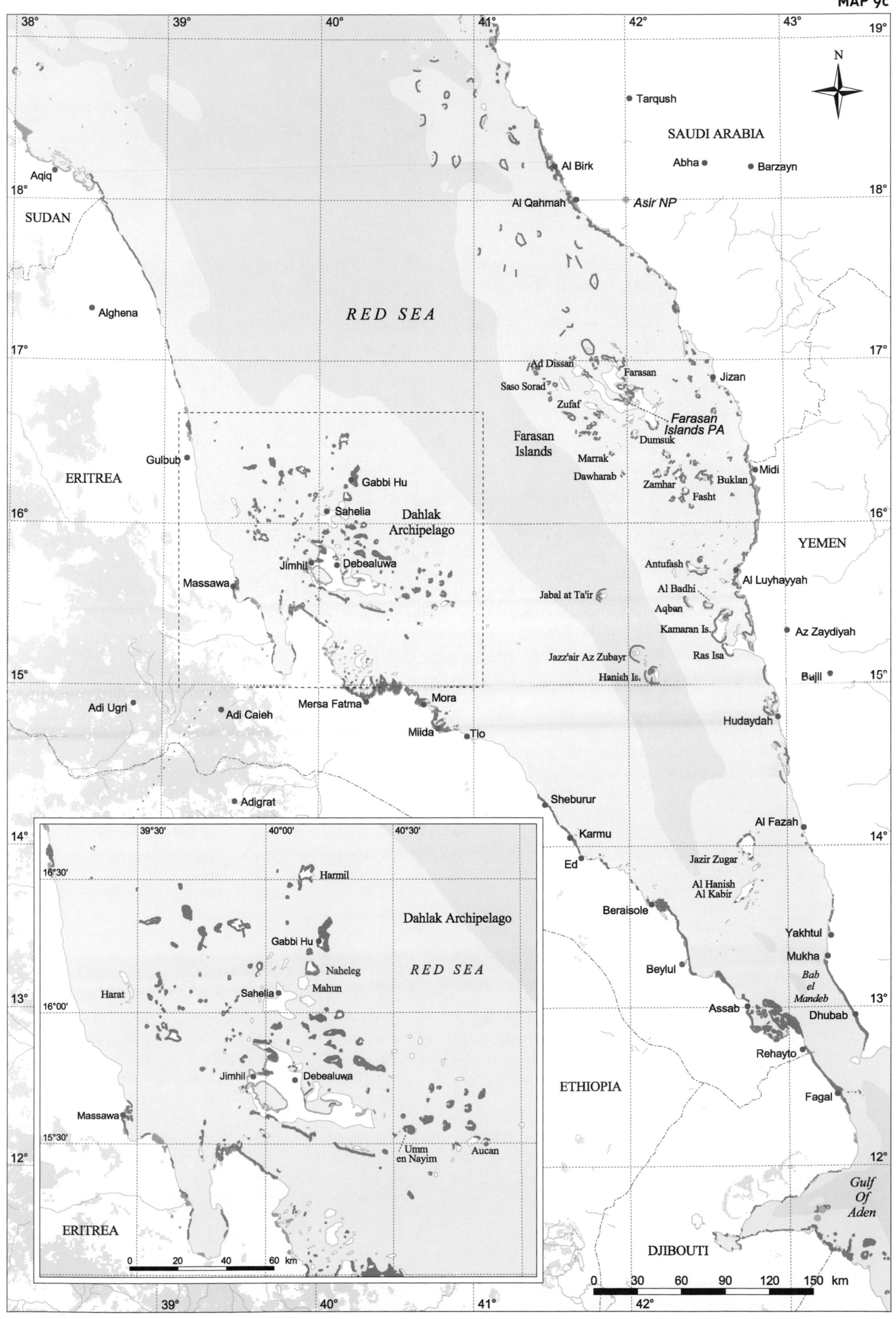

RED SEA
SAUDI ARABIA
SUDAN
ERITREA
YEMEN
ETHIOPIA
DJIBOUTI
Tarqush
Abha
Barzayn
Al Birk
Al Qahmah
Asir NP
Aqiq
Alghena
Ad Dissan
Saso Sorad
Farasan
Jizan
Zufaf
Farasan Islands PA
Farasan Islands
Dumsuk
Marrak
Dawharab
Zamhar
Buklan
Midi
Fasht
Gulbub
Gabbi Hu
Sahelia
Dahlak Archipelago
Jimhil
Debealuwa
Massawa
Antufash
Al Luyhayyah
Al Badhi
Jabal at Ta'ir
Aqban
Kamaran Is.
Az Zaydiyah
Ras Isa
Jazz'air Az Zubayr
Hanish Is.
Bajil
Mersa Fatma
Mora
Adi Ugri
Adi Caieh
Miida
Tio
Hudaydah
Adigrat
Sheburur
Karmu
Ed
Al Fazah
Jazir Zugar
Al Hanish Al Kabir
Beraisole
Yakhtul
Mukha
Beylul
Bab el Mandeb
Assab
Dhubab
Rehayto
Fagal
Gulf Of Aden
Harmil
Naheleg
Mahun
Harat
Umm en Nayim
Aucan
0 20 40 60 km
0 30 60 90 120 150 km

are largely explained by changes in environmental conditions, including increased turbidity and the loss of deeper water species. South of Massawa, reefs support significant growths of *Sargassum* and other macroalgae during the winter, a pattern similar to the coastal communities of southern Arabia. Some of the smaller islands in the southern Red Sea are of regional importance for seabird colonies, and there are important dugong populations in the surrounding waters. Bleaching during the 1998 El Niño event was observed in Eritrea, although mortality was restricted to some shallow water colonies. Such incidents may well have been more widespread in the region, although in Yemen many corals had in fact died in localized bleaching events during a similar warming event in 1995.

Eritrea

The reefs of Eritrea are extensive and suffered little human impact before the 1990s. Since then there have been small increases in both the coastal population and fisheries. Commercial trawlers, including licensed vessels operating from Saudi Arabia, fish mostly in deeper water away from the reefs, although there are thought to be some reef-associated species in their catch, and there is concern that this might indicate they are straying into reef areas. Artisanal fisheries target a broad range of species, including finfish, molluscs, sea cucumbers and pearl oysters. There is also a commercial fishery for the aquarium trade, and around 100 000 fish were exported between 1995 and 1997. The most important and diverse reefs, around the offshore islands including the Dahlak Archipelago, remain in relatively good condition despite the lack of legal protection. By contrast some of the coastal reefs have suffered from development and land reclamation, notably around Massawa.

This was an area of considerable political unrest until separation from Ethiopia was attained in 1993, and there has been sporadic unrest in the south of the country since that time. As a result of these problems there is no significant tourism industry, although it seems likely that this could develop relatively rapidly as and when social and economic stability allow. Considerable efforts have been underway since 1999 to develop a comprehensive management regime for the country's coastal resources, including the designation of protected areas.

Eritrea

General Data	
Population (thousands)	4 136
GDP (million US$)	1 431
Land area (km²)	120 641
Marine area (thousand km²)	39
Per capita fish consumption (kg/year)	<1
Status and Threats	
Reefs at risk (%)	66
Recorded coral diseases	0
Biodiversity	
Reef area (km²)	3 260
Coral diversity	na / 333
Mangrove area (km²)	581
No. of mangrove species	3
No. of seagrass species	na

Yemen

General Data	
Population (thousands)	17 479
GDP (million US$)	15 387
Land area (km²)	733 130
Marine area (thousand km²)	547
Per capita fish consumption (kg/year)	7
Status and Threats	
Reefs at risk (%)	73
Recorded coral diseases	0
Biodiversity	
Reef area (km²)	700
Coral diversity*	na / 344
Mangrove area (km²)	81
No. of mangrove species	2
No. of seagrass species	8

* The higher coral diversity figure is likely to be a considerable overestimate as it is based on the biogeographic region which includes the Gulf of Aden and Socotra

Yemeni Red Sea

Yemen has a long coastline, with a short section facing the Red Sea and a much longer one (described below) facing the Gulf of Aden. In the Red Sea, the Yemen has a more densely populated coastline than many other areas. There are oil terminals in Hudaydah and Mukha, and oil pollution together with sewage and industrial development may be having localized impacts. As with Eritrea, political and military instability have prevented the development of tourism. Fisheries are important, including an offshore trawl fishery, but also line and net fisheries, with reports of overfishing in some areas. A significant sharkfin fishery has been reported in the southern Red Sea and Gulf of Aden, with many fishers coming from Yemen, and often operating illegally in the waters of neighboring countries. Apart from driving a rapid decline in shark stocks there is reported to be a considerable by-catch, including turtles and dolphins.

Southern Arabian Region: Yemen, Djibouti, northern Somalia and Oman

MAP 9d

South of the Bab el Mandeb, the mouth of the Red Sea, the waters rapidly open out into the Gulf of Aden, a wide semi-enclosed sea bordered by Djibouti in the west, Yemen in the North and Somalia and the Yemeni islands of Socotra, Abd al Kiri, Darsa and Semha in the south. This area is of similar tectonic origin to the Red Sea, formed by the spreading of the Sheba Ridge which runs down the center of the Gulf of Aden and out into the Arabian Sea. Onshore, the coastlines are mountainous, while offshore the bathymetry is steep – much of the central part of the gulf is over 2 000 meters deep. To the east the Gulf of Aden joins the Arabian Sea, which is bordered by the coast of Oman in the north and then sweeps northwest forming the Gulf of Oman.

One critical oceanographic feature in this region is that of the seasonally reversing monsoon winds which operate over the entire Northern Indian Ocean. During the summer months there is a sustained strong wind blowing from the southwest along the coast of southern Arabia. This wind drives the surfaces waters away and, in the Arabian Sea, they are replaced by cooler, nutrient-rich waters of about 16-17°C upwelling from the deeper ocean. In the Gulf of Oman the cool water influence is less constant, although occasional upwellings occur and can replace surface waters very rapidly such that falls of up to 10°C over one or two days have been observed. Such upwellings have a significant impact on the ecology, and areas of reef development are few. In the Gulf of Aden, although there are still areas of upwelling, these are not so widespread and have less impact on the shallow inshore communities than is the case in southern Oman.

From an ecological and a biogeographic perspective this is a particularly interesting region. The cool nutrient-rich upwellings in the east have enabled the development of unusual communities dominated by macroalgae. These are found over wide areas of hard substrate, where coral reefs might normally occur and are dominated by *Sargassum* and *Nizamudinnia* in shallow waters and by *Eklonia* in deeper areas. Macroalgal communities such as these are more typically associated with cooler, temperate waters. Small numbers of corals are also sometimes found alongside these algae, while in more sheltered areas, such as the landward sides of some Omani islands, very extensive high cover coral communities are found. Until the mid to late 1990s the Gulf of Aden was almost entirely unknown, but recent studies have revealed extensive and diverse coral communities supporting some of the most diverse fish populations of the entire Arabian region. By contrast, macroalgal communities appear to be much less widespread or dominant in the Gulf of Aden. Another feature, which may

Left: A mixed community of corals and kelp, typical of the areas of upwelling in the Arabian Sea (photo: Jerry Kemp). Right: A coral community in the Strait of Hormuz, Musandam, Oman. Such communities are rich in diversity, but do not feature on most reef maps as they often lack a highly developed physical structure (photo: Jerry Kemp).

MAP 9d

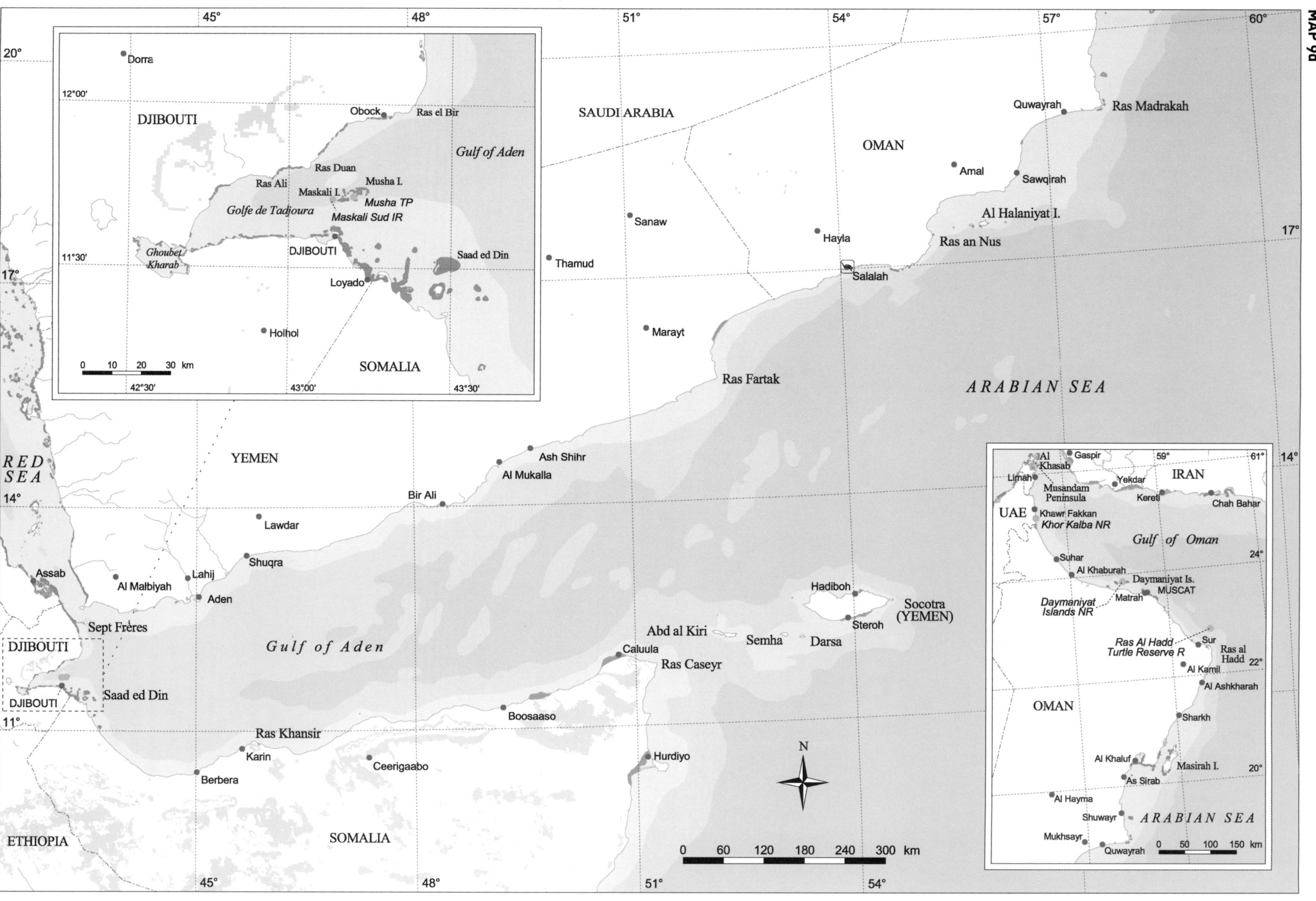

RED SEA
DJIBOUTI
DJIBOUTI
ETHIOPIA
SOMALIA
YEMEN
SAUDI ARABIA
OMAN
Gulf of Aden
ARABIAN SEA
Assab
Al Matbiyah
Lahij
Aden
Sept Frères
Saad ed Din
Shuqra
Lawdar
Bir Ali
Al Mukalla
Ash Shihr
Thamud
Sanaw
Marayt
Ras Fartak
Hayla
Salalah
Ras an Nus
Al Halaniyat I.
Amal
Sawqirah
Quwayrah
Ras Madrakah
Berbera
Karin
Ras Khansir
Ceerigaabo
Boosaaso
Caluula
Ras Caseyr
Hurdiyo
Abd al Kiri
Semha
Darsa
Hadiboh
Steroh
Socotra (YEMEN)
N
0 60 120 180 240 300 km
45° 48° 51° 54° 57° 60°
11° 14° 17° 20°
Dorra
DJIBOUTI
Obock
Ras el Bir
Gulf of Aden
Ras Duan
Ras Ali
Musha I.
Maskali I.
Musha TP
Maskali Sud IR
Golfe de Tadjoura
Ghoubet Kharab
DJIBOUTI
Saad ed Din
Loyado
Holhol
SOMALIA
0 10 20 30 km
12°00' 11°30'
42°30' 43°00' 43°30'
Al Khasab
Gaspir
Limah
Musandam Peninsula
Yekdar
IRAN
Kereti
Chah Bahar
UAE
Khawr Fakkan
Khor Kalba NR
Gulf of Oman
Suhar
Al Khaburah
Daymaniyat Is.
Matrah
MUSCAT
Daymaniyat Islands NR
Ras Al Hadd Turtle Reserve R
Sur
Ras al Hadd
Al Kamil
Al Ashkharah
OMAN
Sharkh
Al Khaluf
Masirah I.
As Sirab
Al Hayma
Shuwayr
ARABIAN SEA
Mukhsayr
Quwayrah
0 50 100 150 km
59° 61° 24° 22° 20°

be linked to the difficult environmental conditions or to unusual processes of coral recruitment or settlement, is the development of low diversity coral communities. In several countries extensive communities, and sometimes even rudimentary reefs, occur which are monospecific or dominated by only two, three or four coral species.

Further interest in this region comes from the biogeographic affinities of the coral communities themselves. The Gulf of Aden and adjacent waters were not subjected to drying out or hypersalinization during the Pleistocene, and it is hypothesized that this area was a critical refuge for some Red Sea species. At present this remains an area of biological transition, or overlap, lying on the biogeographic boundary between the Red Sea, the Arabian Sea and the rest of the Indian Ocean. Many of the species typically regarded as endemic to the Red Sea, or to the Arabian Sea/Arabian Gulf areas, are also found in the Gulf of Aden. While considerable further work is required to understand its evolutionary and genetic processes, this remains a region of considerable interest. In a few localities in the Gulf of Aden species diversity, at least of reef-associated fishes, appears to be extremely high, possibly higher than in any other part of the Arabian region. The impacts of the 1998 bleaching event appear to have been mixed: while widespread mortality has been recorded at some localities, others nearby appear to have been largely unaffected.

Yemen

The Yemeni coast of the Gulf of Aden remains poorly described, but recent studies have found a number of interesting and important coral communities and some true reefs, including around Al Mukalla, Bir Ali and Shuqra. Some of these communities include wide areas of monospecific coral stands, notably of *Pocillopora* and *Montipora*. It was assumed, until the mid-1990s, that there were few significant coral communities off the coast of Socotra, but recent surveys have shown extensive areas of high live coral cover. These are best developed on the northern reaches of both Socotra and the neighboring islands, where some 240 hard coral species have been recorded, making them among the most diverse reefs in the Indian Ocean region. Widespread mortality of corals was reported at some locations following the 1998 bleaching event, but other locations showed little or no impacts and recovery, with rapid growth of new recruits, was reported in 2000.

Human impacts on the reefs in Yemen are still relatively minimal, other than from fishing. Much of the coastline in southern Yemen is undeveloped, although Aden is a major port, with associated pollution and problems of solid waste disposal. The country is quite reliant on fisheries, and has an active offshore pelagic fishery in the Gulf of Aden. Illegal fishing by Yemeni boats is also reported from northern Somalia. Reef fishing has developed and is widespread around Socotra, including an artisanal lobster fishery. Efforts are now underway to develop a tourist industry on Socotra.

Djibouti

This country has some of the best developed reefs outside the Red Sea, including fringing reef communities along parts of the mainland coast, and fringing and platform structures around the reefs and islands of Maskali and Musha and the Sept Frères just south of the Bab el Mandeb. Surveys in 1998 and 1999 described 167 coral species, while

Djibouti

General Data	
Population (thousands)	451
GDP (million US$)	493
Land area (km²)	21 638
Marine area (thousand km²)	7
Per capita fish consumption (kg/year)	3
Status and Threats	
Reefs at risk (%)	100
Recorded coral diseases	0
Biodiversity	
Reef area (km²)	450
Coral diversity	69 / 325
Mangrove area (km²)	10
No. of mangrove species	1
No. of seagrass species	na

Oman

General Data	
Population (thousands)	2 533
GDP (million US$)	16 298
Land area (km²)	2 328
Marine area (thousand km²)	539
Per capita fish consumption (kg/year)	na
Status and Threats	
Reefs at risk (%)	51
Recorded coral diseases	2
Biodiversity	
Reef area (km²)	530
Coral diversity	71 / 128
Mangrove area (km²)	20
No. of mangrove species	1
No. of seagrass species	na

Protected areas with coral reefs

Site name	Designation	Abbreviation	IUCN cat.	Size (km²)	Year
Djibouti					
Maskali Sud	Integral Reserve	IR	Ia	na	1980
Musha	Territorial Park	TP	Unassigned	na	1972
Oman					
Daymaniyat Islands (Oman)	Nature Reserve	NR	IV	200.00	1995

coral cover was highly varied, typically over 20 percent and reaching 90 percent in the Sept Frères. Some of the reefs were reported to have been significantly impacted by the 1998 bleaching, with an estimated 30 percent mortality.

The main economic activity of Djibouti is the operation of the main port, and coral reef areas around it are thought to be heavily degraded. Fishing is not a major activity, although there is an important artisanal fishery, with about 90 small fishing boats, together with about 15 larger (10-14 meters) vessels in the late 1990s. A limited amount of tourism with a coastal focus and some diving on the offshore islands was also taking place in the late 1990s.

Northern Somalia

Along the northern coast, reef development is sporadic, however there are reefs and coral communities in various locations, including wide monospecific stands of *Acropora* in a few areas. Rapid surveys of this coast in 1999 found some 74 scleractinian coral species. The best developed reefs are fringing structures and patch or platform reefs in the area around Saad ed Din and other islands close to Djibouti. Some of the reefs further to the east, near Berbera, were reported to have suffered very extensively from bleaching and mortality as a result of the 1998 bleaching event, while there was also evidence of crown-of-thorns starfish impacts during surveys in 1997 and 1999.

The northern coast of Somalia still has only a sparse human population and coastal resources are regarded as very healthy, although there is a minor nearshore fishery and an opportunistic sharkfin fishery. Further utilization of the fish resources, which may include benthic species, is the result of illegal fishing from Yemen. A more significant offshore trawl fishery has also been reported.

Southern Somalia's reefs are described in Chapter 7.

Oman

Much of Oman's southern coastline and sub-tidal waters are dominated by sand, although there are rocky outcrops, notably around Ras Al Hadd and the offshore islands of Masirah and Al Halaniyat (Kuria Muria). The best developed coral communities and small reef formations are found in four main areas: the Musandam Peninsula; some of the shores and bays of the coast around Muscat and the Daymaniyat Islands; the western coast of Masirah Island and the adjacent mainland; and the sheltered rocky areas of coast around the Al Halaniyat Islands and mainland of Dhofar. Coral growth is restricted both by the cool water upwellings and by the availability of hard substrates. Coral communities with high coral cover but often low diversity have been noted in several areas, including communities dominated by *Porites* spp., *Pocillopora damicornis* and *Acropora* spp. In the Gulf of Masirah near continuous reefs dominated by *Montipora foliosa* have been estimated to cover more than 25 square kilometers. There have been some natural impacts to coral communities in Oman, including storm damage and some predation by crown-of-thorns starfish. Extensive bleaching and associated mortality of shallow corals occurred in Dhofar in 1998, although little or none was observed in other areas.

Human impacts, by contrast, are considerable. Oman has a fairly developed coastline and fishing is widespread. Overfishing is probably only a localized problem on the reef communities, but damage from anchors and fishing gear, together with fishing-related litter, presents much greater problems. One survey found that between 25 and 100 percent of all the coral on *Pocillopora damicornis* reefs surveyed in 1996 was damaged by abandoned nets. There is also a significant abalone fishery operating from the southwest of the country. Abalone are only collected for two months of the year, with total yields of around 35-45 tons per year in the early 1990s. Recreational diving occurs in a few places but remains at low levels. Pollution from terrestrial sources, or indeed from the very high volume of tanker traffic in the region, is minimal and not thought to be impacting reef communities. Oman is one of the few countries in the wider region to have moved towards an integrated system of coastal zone management and has begun to designate a system of marine protected areas.

Arabian Gulf: United Arab Emirates, Qatar, Bahrain, Kuwait, Iran

MAP 9e

The Arabian Gulf is a vast shallow marine basin which has formed on the northeastern and eastern edge of the Arabian tectonic plate. Unlike the Red Sea, it receives considerable riverine input at its northern end from the Shatt al Arab waterway which is formed from the Tigris, Euphrates and Karun Rivers. In addition there are a number of rivers flowing from the Zagros Mountains in Iran. On average the Gulf is only around 35 meters deep, and at its deepest point in the southeast it only reaches about 100 meters. During the last glacial maxima this entire area dried up and all marine life was extinguished. The large rivers at this time would have continued to flow along the eastern edge of the area and out through the Straits of Hormuz. Climatically this is an extremely harsh region. Most of the Gulf is sub-tropical, and the surrounding arid land masses drive extremes of temperatures, with air temperatures frequently reaching 50°C in the summer, but falling to 0°C in the winter. The shallow water does little to ameliorate these fluctuations and most nearshore waters range between 10 and 39°C through the year. Winds further influence temperatures quite considerably, and can also deliver large quantities of sediment. Despite the high riverine input this is a very saline basin – riverine input plus rainfall contribute the equivalent of 15-50 centimeters of water depth per year, but levels of evaporation range from 140 to 500 centimeters so that salinities are typically 40‰, but often reach 70‰ or more in enclosed embayments. The general pattern of current circulation is one of a counter-clockwise flow. Lighter, less saline, waters enter from the Gulf of Oman and flow north along the coast of Iran, then down the eastern shores of the Arabian Peninsula. As evaporation increases the salinity and density of the water, so it sinks, and there is actually an outflow at deeper levels of more saline water through the Straits of Hormuz not unlike that occurring in the Red Sea. Circulation is greatly restricted in the embayments of the Gulf of Salwah and the shallow waters off the United Arab Emirates, driving massive evaporation and even more extreme environmental conditions.

The benthic surfaces of most of the Gulf are featureless soft sediments, dominated by muds in the north and east and carbonate sands in the south and west. There are several areas of rocky shore, both on the mainland and offshore islands. Fringing and patch reefs have developed in a number of places. In many areas the division between true reefs and carbonate structures with little active coral growth is blurred and there are quite a

Left: The Arabian Gulf (STS052-153-131, 1992). Right: A school of yellowspot emperor Gnathodentex aurolineatus. *Coral reef diversity is generally low in the Arabian Gulf.*

MAP 9e

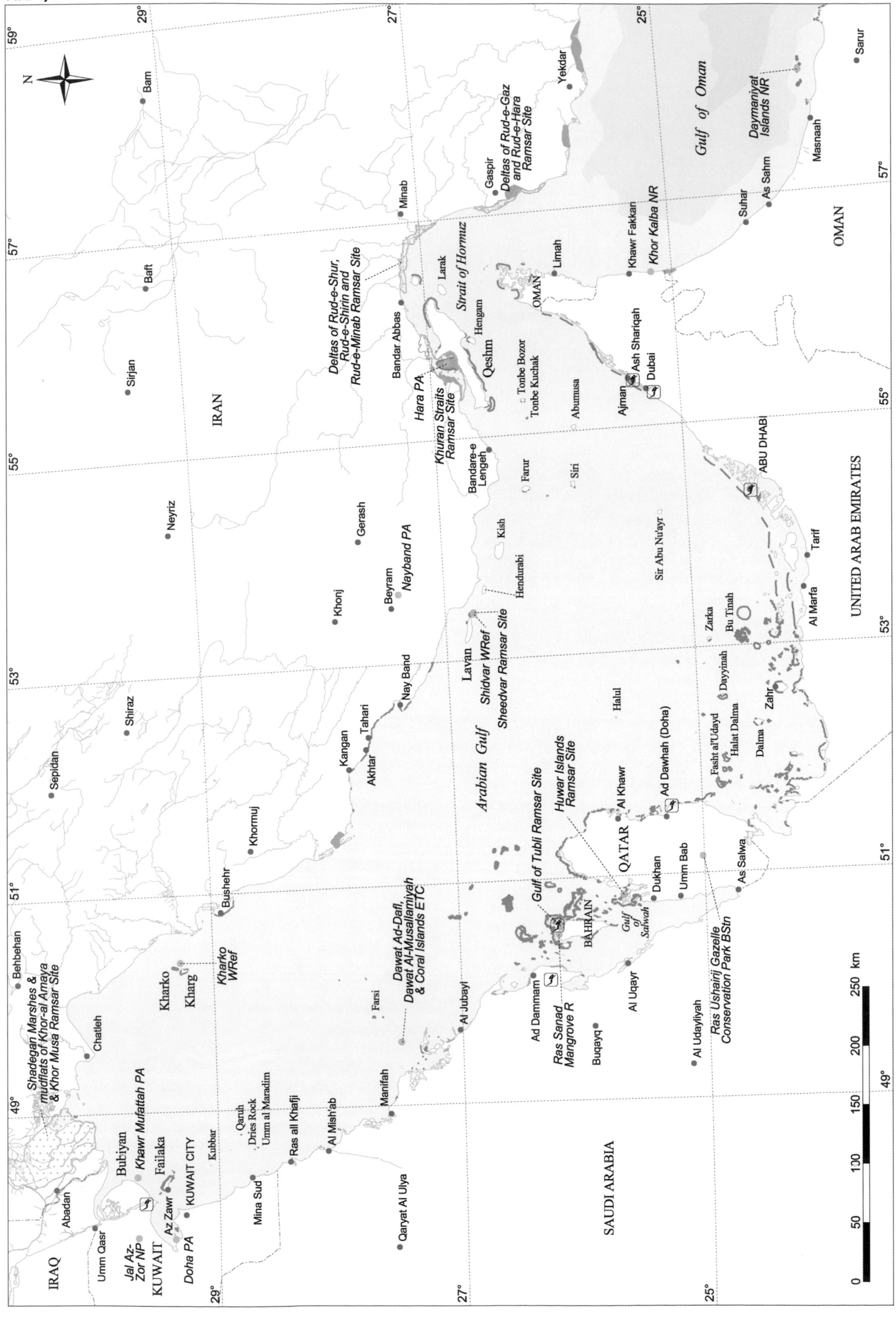

IRAQ
KUWAIT
IRAN
SAUDI ARABIA
QATAR
BAHRAIN
UNITED ARAB EMIRATES
OMAN

Arabian Gulf
Gulf of Oman
Strait of Hormuz
Gulf of Salwah

Umm Qasr
Abadan
Jal Az-Zor NP
Doha PA
Az Zawr
KUWAIT CITY
Bubiyan
Khawr Mufattah PA
Failaka
Kubbar
Qaruh
Dries Rock
Umm al Maradim
Mina Sud
Ras all Khafji
Al Mish'ab
Qaryat Al Ulya
Manifah
Shadegan Marshes & mudflats of Khor-al Amaya & Khor Musa Ramsar Site
Chatleh
Behbehan
Kharko
Kharg
Kharko WRef
Farsi
Dawat Ad-Dafi, Dawat Al-Musallamiyah & Coral Islands ETC
Al Jubayl
Bushehr
Khormuj
Sepidan
Shiraz
Kangan
Akhtar
Tahari
Nay Band
Khonj
Beyram
Nayband PA
Gerash
Neyriz
Sirjan
Baft
Bam
Ad Dammam
Ras Sanad Mangrove R
Buqayq
Al Uqayr
Al Udayliyah
Ras Ushairij Gazelle Conservation Park BStn
Gulf of Tubli Ramsar Site
Huwar Islands Ramsar Site
Dukhan
Umm Bab
As Salwa
Al Khawr
Ad Dawhah (Doha)
Lavan
Shidvar WRef
Sheedvar Ramsar Site
Hendurabi
Kish
Halul
Fasht al'Udayd
Halat Dalma
Dalma
Zahr
Dayyinah
Zarka
Bu Tinah
Al Marfa
Tarif
Sir Abu Nu'ayr
Farur
Siri
Bandare-e Lengeh
Khuran Straits Ramsar Site
Hara PA
Qeshm
Hengam
Tonbe Bozor
Tonbe Kuchak
Abumusa
ABU DHABI
Ajman
Ash Shariqah
Dubai
Bandar Abbas
Deltas of Rud-e-Shur, Rud-e-Shirin and Rud-e-Minab Ramsar Site
Larak
Minab
OMAN
Limah
Khawr Fakkan
Khor Kalba NR
Gaspir
Deltas of Rud-e-Gaz and Rud-e-Hara Ramsar Site
Yekdar
Suhar
As Sahm
Masnaah
Daymaniyat Islands NR
Sarur

29° 27° 25° 59° 57° 55° 53° 51° 49°

0 50 100 150 200 250 km

United Arab Emirates	
General Data	
Population (thousands)	2 369
GDP (million US$)	41 498
Land area (km²)	78 982
Marine area (thousand km²)	52
Per capita fish consumption (kg/year)	29
Status and Threats	
Reefs at risk (%)	65
Recorded coral diseases	3
Biodiversity	
Reef area (km²)	1 190
Coral diversity	30 / 68
Mangrove area (km²)	30
No. of mangrove species	1
No. of seagrass species	1

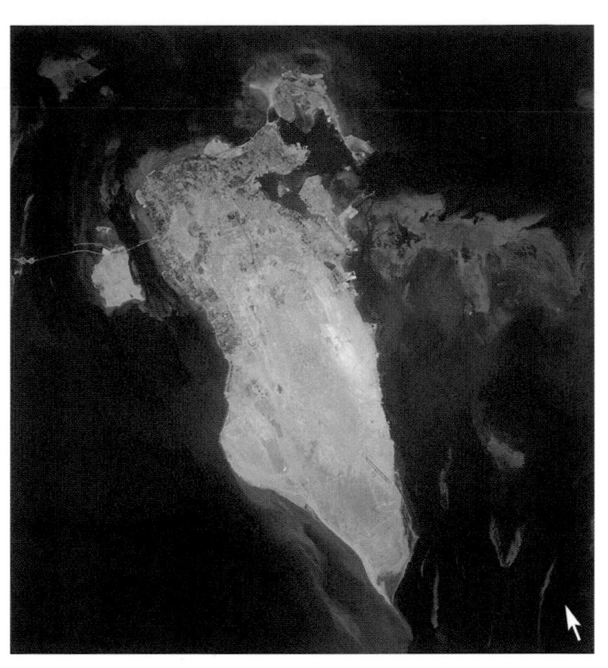

few shallow structures which may be of reefal origin, but do not appear to be accreting structures at the present time. One feature common across much of the region is the periodic decimation of reef communities during occasional cold water events which may arise through a combination of cold upwelling and wind-driven cooling on already cool winter temperatures. During the 1997-98 El Niño event, coral bleaching was recorded at a number of localities, with considerable coral mortality in shallow waters near several countries.

In human terms the most important natural resource in the region is oil, with almost two thirds of the world's proven oil reserves located in the Gulf and adjacent land areas. This has had an enormous influence over the environment of the region, including the construction of numerous oil platforms, but more importantly the release of massive quantities of oil into the marine environment. Prior to the Gulf War in 1991 the waters of the Arabian Gulf already had the world's highest concentrations of hydrocarbons. Much was the result of ballast water discharge from tankers (20 000-35 000 tankers pass through the Straits of Hormuz annually). Further release comes from accidents on the oil platforms and deliberate releases as a result of war. The Nowruz blow-out, a direct result of the Iran-Iraq war, released an estimated 500 000 barrels of oil. Even so, this was greatly surpassed by the Gulf War in 1991, when total releases were of the order of 6-11 million barrels. Somewhat surprisingly the ecological impacts of this pollution may not be as great on coral reefs as might be expected. For the most part there is no direct physical contact with corals, and hence the smothering of corals was somewhat limited. The longer term effects of oil on coral growth and reproduction (see Chapter 2) may be a little more difficult to ascertain however, and may be combined with other environmental stresses.

Fishing is another important industry in the region, with a large industrial trawl fishery which targets shrimp during the open season and finfish for the rest of the year. For the most part these are targeted away from coral areas. The shrimp fisheries were heavily affected by the Gulf War and shrimp stocks collapsed to only 1 percent of their pre-war biomass. Other fisheries are small-scale commercial or artisanal, typically operating from small craft and dhow and using line and trap fishing methods. Although this region was once famous for its pearl oyster fishery this has largely closed with the influx of cultured pearls. There is now a growing recreational fishery which may have an impact on reef communities in some areas.

Reef-based tourism is virtually non-existent in the region, although there is a small amount of recreational diving among local and expatriate residents, some of whom are actively involved in environmental protection and rehabilitation. Although there is a strong interest in the environment in a number of these countries and a significant number of marine protected areas have been proposed, only a few have actually been declared, and only one of these includes coral reefs.

United Arab Emirates

The nearshore waters of the western parts of the United Arab Emirates are shallow, with relatively low water circulation, and some of the highest salinities in the Gulf. Although there are seagrasses, these waters are unsuitable for coral growth. Further offshore there are patch reefs and fringing reefs around many of the islands. Diversity is

The island of Bahrain, together with platform reef structures. Active coral growth is limited to only small areas of these platforms (STS078-748-11, 1996).

low in all areas and many coral communities are dominated by large monospecific stands. Mortality associated with coral bleaching events has been considerable, with over 98 percent loss of *Acropora* in 1996 on reefs near Abu Dhabi, and loss of most of the remaining colonies during the 1998 bleaching event.

Qatar

There are fringing reefs along the north and east coasts, with coral communities growing on the coastal shelf to the east, but no real reef structures. Further offshore there are a number of platform reefs. In the far southwest the Gulf of Salwah is highly saline and unsuitable for coral growth. Shallow reefs to the east of Doha were reported to have undergone mass mortalities as a result of the 1998 coral bleaching event, with losses of up to 100 percent of *Acropora* colonies.

Bahrain

There are no true fringing reefs in this country, but to the north and east there are a number of quite extensive platform reef structures. Diversity and coral cover were generally low, while coral bleaching events in 1996 and 1998 led to mortalities of 85-90 percent of the living coral on many offshore reefs. Over 70 kilometers north of the main island, Abul Thama is a small raised platform with relatively high coral cover of about 25-30 percent. Surrounded by deeper water (50 meters), these corals largely survived these bleaching events.

Bahrain is an industrialized nation. Trawl fisheries undoubtedly had a major impact on offshore ecosystems, and probably impacted a number of coral reefs, until the industry was closed in 1998. Onshore land reclamation has been considerable in the north and west, and there are proposals to reclaim land on Fasht Adham, a large offshore reef area in the east. Industrial effluents are significant, and nearshore waters are routinely dredged, with a major impact of increased sediments on the surrounding reefs.

Kuwait

Kuwait's reefs are largely located in the southern part of the country, and are dominated by platform and patch reefs along the coast from Kuwait City to the border with Saudi Arabia, and with some fringing reefs around offshore islands. Most active reef growth occurs in waters shallower than 10 meters.

There are considerable impacts on these reefs from various human activities. Perhaps the most direct are problems of overfishing, solid waste disposal and widespread anchor damage. These reefs were also among those most directly impacted by the oil spills from the Gulf War, although this did not cause the mass mortalities which were expected by many.

Iran

Very little information is currently available describing reef communities off the coast of Iran. Fringing reefs are known to occur along parts of the mainland and particularly around some of the offshore islands, including Kharg and Kharko Islands in the north and several other small islands to the

	Qatar	Bahrain	Kuwait	Iran
General Data				
Population (thousands)	744	634	1 974	65 620
GDP (million US$)	8 530	5 308	28 111	716 326
Land area (km²)	11 143	612	16 984	1 624 774
Marine area (thousand km²)	31	8	5	206
Per capita fish consumption (kg/year)	10	14	11	5
Status and Threats				
Reefs at risk (%)	66	82	93	88
Recorded coral diseases	0	0	0	0
Biodiversity				
Reef area (km²)	700	570	110	700
Coral diversity	na / 68	na / 68	30 / 68	na / 68
Mangrove area (km²)	<5	1	na	207
No. of mangrove species	1	1	na	2
No. of seagrass species	na	na	2	na

Protected areas with coral reefs

Site name	Designation	Abbreviation	IUCN cat.	Size (km²)	Year
Iran					
Sheedvar Island	Ramsar Site			8.70	1999
Saudi Arabia					
Dawat Ad-Dafl/Dawat Al-Musallamiyah/Coral Islands	Protected Area	PA	Unassigned	2 100.00	na

south. Some 35 coral species have been recorded from around Hormuz Island. As Iran has the deepest and least saline waters of the Gulf, it seems likely that further research may reveal new reef areas and considerable biodiversity.

Fishing is an important industry. From 1989 to 1995 there was a near doubling of the number of fishing vessels to more than 9 000, although fish catches did not show similar increases. Efforts to control the effects of trawling may actually have led to an increase in the impacts on coral reef species. There is also an ornamental fish trade, notably operating from the free trade areas of Kish and Qeshm Islands. Other human impacts, considerable in major industrial areas in the northwest and around Kish and Qeshm, include sedimentation and pollution, together with solid waste and anchor damage. Around Kish Island, algal overgrowth of corals in the late 1990s has been attributed to increased nutrient levels. The narrow waters around the Straits of Hormuz are among the busiest tanker lanes in the world, representing an ongoing threat to the southernmost reefs.

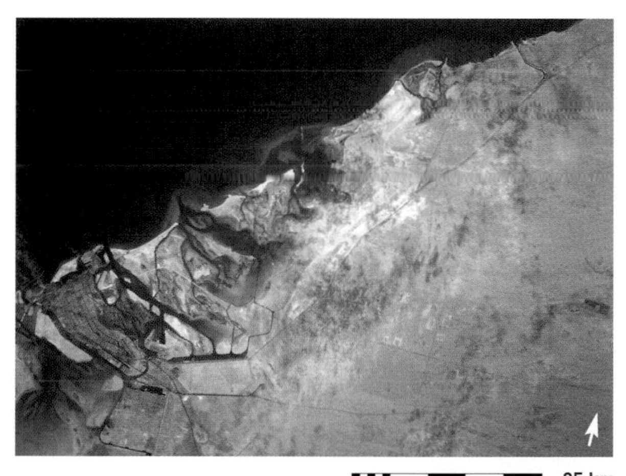

Left: Large groupers are still sold in the fish markets of the Arabian Gulf. Right, above: Abu Dhabi, the United Arab Emirates. These waters have some of the highest salinities in the Gulf, and wide areas are unsuitable for coral growth (STS080-707-77, 1996). Right, below: The port of Dammam in Saudi Arabia, showing intensive coastal development (STS078-748-10, 1996).

Selected bibliography

REGIONAL SOURCES

Coles SL (1988). Limitations on reef coral development in the Arabian Gulf: temperature or algal competition. *Proc 6th Int Coral Reef Symp* 2: 211-216.

Girdler RW (1984). The evolution of the Gulf of Aden and Red Sea in space and time. *Deep Sea Res* Part A 31: 747.

MEPA, IUCN (eds) (1987a). *MEPA Coastal and Marine Management Series, 7: Red Sea and Arabian Gulf. Saudi Arabia: An Assessment of National Coastal Zone Management Requirements.* Meteorology and Environmental Protection Administration, Riyadh, Saudi Arabia.

MEPA, IUCN (eds) (1987b). *MEPA Coastal and Marine Management Series, 1: Red Sea. Saudi Arabia: An Analysis of Coastal and Marine Habitats of the Red Sea.* Meteorology and Environmental Protection Administration, Riyadh, Saudi Arabia.

MEPA, IUCN (eds) (1987c). *MEPA Coastal and Marine Management Series, 3: Red Sea. Saudi Arabia: An Assessment of Coastal Zone Management Requirements for the Red Sea.* Meteorology and Environmental Protection Administration, Riyadh, Saudi Arabia.

Pilcher N, Alsuhaibany A (2000). Regional status of coral reefs in the Red Sea and the Gulf of Aden. In: Wilkinson CR (ed). *Status of Coral Reefs of the World: 2000.* Australian Institute of Marine Science, Cape Ferguson, Australia.

Randall E (1983). *Red Sea Reef Fishes.* Immel Publishing, London, UK.

Roberts CM, Dawson Shepherd AR, Ormond RFG (1992). Large-scale variation in assemblage structure of Red Sea butterflyfishes and angelfishes. *J Biogeog* 19: 239-250.

Roberts CM, Polunin NVC (1992). Effects of marine reserve protection on northern Red Sea fish populations. *Proc 7th Int Coral Reef Symp* 2: 969-977.

Sheppard CRC, Sheppard ALS (1991). Corals and coral communities of Arabia. *Fauna of Saudi Arabia* 12: 1-170.

Sheppard C, Price A, Roberts C (1992). *Marine Ecology of the Arabian Region: Patterns and Processes in Extreme Tropical Environments.* Academic Press, London, UK.

UNEP/IUCN (1988). *UNEP Regional Seas Directories and Bibliographies: Coral Reefs of the World. Volume 2: Indian Ocean.* UNEP and IUCN, Nairobi, Kenya, Gland, Switzerland and Cambridge, UK.

Vine P (1986). *Red Sea Invertebrates.* Immel Publishing, London, UK.

The sources listed above also cover the the following sub-sections of this chapter: Northern Red Sea: Egypt, Israel, Jordan; Saudi Arabia; Central Red Sea: Sudan; Southern Red Sea: Eritrea and Yemen.

SOUTHERN ARABIAN REGION: YEMEN, DJIBOUTI, NORTHERN SOMALIA AND OMAN

Al-Jufaili S, Al-Jabri M, Al-Baluchi A, Baldwin RM, Wilson SC, West F, Matthews AD (1999). Human impacts on coral reefs in the Sultanate of Oman. *Est Coast Shelf Sci* 49 (Supplement A): 65-74.

Coles SL (1995). *Corals of Oman.* Keech, Samdani and Coles (private publication), UK.

DeVantier LM, Turak E, Al-Shaikh KA, Cheung CPS, Abdul-Aziz M, De'ath G, Done TJ (2000). Ecological indicators of status of coral communities for marine protected areas planning: case studies from Arabia. In: Lloyd D, Done TJ, Diop S (eds). *Information Management and Decision Support for Marine Biodiversity Protection and Human Welfare: Coral Reefs.* Australian Institute of Marine Science and United Nations Environment Programme, Townsville, Australia.

DeVantier LM, Turak E, Al-Shaikh KA, De'ath G (in press). Coral communities of the central-northern Saudi Arabian Red Sea. *Fauna of Arabia.*

van der Elst R, Salm RV (1999). *Overview of the Biodiversity of the Somali Coastal and Marine Environment.* Report prepared for IUCN Eastern Africa Programme and Somali Natural Resources Management Programme.

Kemp J (1998). *Marine and Coastal Habitats and Species of the Bir Ali area of Shabwa Province, Republic of Yemen. Recommendations for Protection.* Report to the Environmental Protection Council of the Council of Ministers, Sana'a.

Kemp JM (1998). Zoogeography of the coral reef fishes of the Socotra Archipelago. *J Biogeog* 25: 919-934.

Kemp JM (2000). Zoogeography of the coral reef fishes of the north-eastern Gulf of Aden, with eight new records of coral reef fishes from Arabia. *Fauna of Arabia* 18.

Kemp JM, Benzoni F (1999). Monospecific coral areas on the northern shore of the Gulf of Aden, Yemen. *Coral Reefs* 18: 280.

Kemp JM, Benzoni F (2000). A preliminary study of coral communities of the northern Gulf of Aden. *Fauna of Arabia* 18.

Kemp JM, Obura D (2000). Reefs of the Gulf of Aden and Socotra Archipelago. In: McClanahan T, Sheppard C, Obura D (eds). *Coral Reefs of the Western Indian Ocean: Their Ecology and Conservation.* Oxford University Press, Oxford, UK and New York, USA. 273-275.

McClanahan T, Obura DO (1997). *Preliminary Ecological Assessment of the Saad ed Din, Awdal Region.* IUCN Eastern Africa Programme, Nairobi, Kenya.

Salm RV, Jensen RAC, Papastravou VA (1993). *A Marine Conservation and Development Report: Marine Fauna of Oman: Cetaceans, Turtles, Seabirds, and Shallow Water Corals.* IUCN–The World Conservation Union, Gland, Switzerland.

Schleyer M, Baldwin R (1999). *Biodiversity Assessment of the Northern Somali Coast East of Berbera.* Report prepared for IUCN Eastern Africa Programme and Somali Natural Resources Management Programme.

Sommer C, Schneider W, Poutiers J-M (1996). *FAO Species Identification Guide for Fishery Purposes: The Living Marine Resources of Somalia.* Food and Agriculture Organization of the United Nations, Rome, Italy.

ARABIAN GULF: UNITED ARAB EMIRATES, QATAR, BAHRAIN, KUWAIT, IRAN

Carpenter KE, Harrison PL, Hodgson G, Alsaffar AH, Alhazeem SH (1997a). *The Corals and Coral Reef Fishes of Kuwait.* Kuwait Institute for Scientific Research and Environment Public Authority, Kuwait.

Carpenter KE, Krupp F, Jones DA, Zajonz U (1997b). *FAO Species Identification Guide for Fishery Purposes: Living Marine Resources of Kuwait, Eastern Saudi Arabia, Bahrain, Qatar, and the United Arab Emirates.* Food and Agriculture Organization of the United Nations, Rome, Italy.

Coles SL (1988). Limitations on reef coral development in the Arabian Gulf: temperature or algal competition. *Proc 6th Int Coral Reef Symp* 2: 211-216.

Coles SL, Fadlallah YH (1991). Reef coral survival and mortality at low temperatures in the Arabian Gulf: new species-specific lower temperature limits. *Coral Reefs* 9: 231-237.

Fadlallah YH, Allen KW, Estudillo RA (1994). Damage to shallow reefs in the Gulf is caused by periodic exposure to air during extreme low tides and low water temperatures (Tarut Bay, Eastern Saudi Arabia). In: Ginsburg RN (ed). *Proceedings of the Colloquium on Global Aspects of Coral Reefs: Health, Hazards and History, 1993.* Rosenstiel School of Marine and Atmospheric Sciences, University of Miami, Miami, USA.

McCain JC, Tarr AB, Carpenter KE, Coles SL (1984). Marine ecology of Saudi Arabia: a survey of coral reefs and reef fishes in the Northern Area, Arabian Gulf, Saudi Arabia. *Fauna of Saudi Arabia* 6: 102-126.

MEPA, IUCN (eds) (1987a). *MEPA Coastal and Marine Management Series, 6: Executive Summary – Arabian Gulf. Saudi Arabia: An Assessment of Biotopes and Coastal Zone Management Requirements for the Arabian Gulf.* Meteorology and Environmental Protection Administration, Riyadh, Saudi Arabia.

MEPA, IUCN (eds) (1987b). *MEPA Coastal and Marine Management Series, 7: Red Sea and Arabian Gulf. Saudi Arabia: An Assessment of National Coastal Zone Management Requirements.* Meteorology and Environmental Protection Administration, Riyadh, Saudi Arabia.

MEPA, IUCN (eds) (1987c). *MEPA Coastal and Marine Management Series, 5: Technical Report 5 – Arabian Gulf. Saudi Arabia: An Assessment of Biotopes and Coastal Zone Management Requirements for the Arabian Gulf.* Meteorology and Environmental Protection Administration, Riyadh, Saudi Arabia.

Pilcher NJ, Wilson S, Alhazeem SH, Shokri MR (2000). Status of coral reefs in the Arabian/Persian Gulf and Arabian Sea region. In: Wilkinson CR (ed). *Status of Coral Reefs of the World: 2000.* Australian Institute of Marine Science, Cape Ferguson, Australia.

Price ARG (1990). Rapid assessment of coastal zone management requirements: a case-study from the Arabian Gulf. *Ocean Shore Man* 13: 1-19.

Price ARG, Robinson JH (eds) (1993). *Marine Pollution Bulletin, 27: The 1991 Gulf War: Coastal and Marine Environmental Consequences.* Pergamon Press, Oxford, UK.

Sheppard C, Price A, Roberts C (1992). *Marine Ecology of the Arabian Region: Patterns and Processes in Extreme Tropical Environments.* Academic Press, London, UK.

Map sources

Map 9a

Reefs are taken from Hydrographic Office (1954, 1984, 1994). Sources for these charts include some very old data, but also surveys from the 1980s.

Hydrographic Office (1954). Red Sea: Gezîrat el Dibia to Masamirit Islet. *British Admiralty Chart No. 138.* 1:750 000. December 1954. Taunton, UK.

Hydrographic Office (1984). El Akhawein to Râbigh. *British Admiralty Chart No. 63.* 1:750 000. September 1984. Taunton, UK.

Hydrographic Office (1994). El Suweis (Suez) to El Akhawein (The Brothers) (including the Gulf of Aqaba). *British Admiralty Chart No. 8.* 1:750 000 and 1:300 000 (Aqaba). December 1994. Taunton, UK.

Map 9b

Maps have been prepared for the Saudi Arabian Red Sea from IUCN/MEPA (1984, 1985). Further data have been taken from Hydrographic Office (1955, 1984, 1994) and these same sources, together with Hydrographic Office (1987, 1991) have been used to provide details of reefs in Sudan.

Hydrographic Office (1955). Red Sea: Masamirit Islet to Zubair Islands. *British Admiralty Chart No. 141.* 1:750 000. September 1955. Taunton, UK.

Hydrographic Office (1984). El Akhawein to Râbigh. *British Admiralty Chart No. 63.* 1:750 000. September 1984. Taunton, UK.

Hydrographic Office (1987). Outer Approaches to Port Sudan. *British Admiralty Chart No. 82.* 1:150 000. December 1987. Taunton, UK.

Hydrographic Office (1991). Sawâkin to Ras Qassâr. *British Admiralty Chart No. 81.* 1:300 000. June 1991. Taunton, UK. (Sawâkin inset not utilized.)

Hydrographic Office (1994). El Suweis (Suez) to El Akhawein (The Brothers) (including the Gulf of Aqaba). *British Admiralty Chart No. 8.* 1:750 000 and 1:300 000 (Aqaba). December 1994. Taunton, UK.

IUCN/MEPA (1984). *Report on the Distribution of Habitats and Species in the Saudi Arabian Red Sea: Part 1. Saudi Arabia Marine Conservation Programme, Report No. 4.* IUCN, Gland, Switzerland/Meteorology and Environmental Protection Administration, Jeddah, Kingdom of Saudi Arabia. Includes numerous tables, photos, maps.

IUCN/MEPA (1985). *Distribution of Habitats and Species along the Southern Red Sea Coast of Saudi Arabia. Saudi Arabia Marine Conservation Programme, Report No. 11.* IUCN, Gland, Switzerland/Meteorology and Environmental Protection Administration, Jeddah, Kingdom of Saudi Arabia. Includes numerous tables, photos, maps, annexes.

Map 9c

For Eritrea reefs are derived from Hydrographic Office (1955, 1988, 1991, 1993). For Djibouti reefs are derived from Hydrographic Office (1985, 1991, 1992, 1993). Coral reef data for the former Yemen Arab Republic (North Yemen) were obtained from IUCN (1987). Further data were derived from Hydrographic Office (1985, 1991, 1993).

Hydrographic Office (1955). Red Sea: Masamirit Islet to Zubair Islands. *British Admiralty Chart No. 141.* 1:750 000. September 1955. Taunton, UK.

Hydrographic Office (1985). Straits of Bab el Mandeb to Aden Harbour. *British Admiralty Chart No. 3661.* 1:200 000. November 1985. Taunton, UK.

Hydrographic Office (1988). North and northeast approaches to Mits'iwa. *British Admiralty Chart No. 164.* 1:300 000. March 1988. Taunton, UK.

Hydrographic Office (1991). Jazîrat al Tâ'ir to Bab el Mandeb. *British Admiralty Chart No. 143.* 1:400 000. December 1991. Taunton, UK.

Hydrographic Office (1992). Golfe de Tadjoura and Anchorages. *British Admiralty Chart No. 253.* 1:200 000. September 1992. Taunton, UK.

Hydrographic Office (1993). Gulf of Aden. *British Admiralty Chart No. 6.* 1:750 000. March 1993. Taunton, UK.

IUCN (1987). *The Distribution of Habitats and Species along the YAR Coastline.* IUCN, Gland, Switzerland.

Map 9d

Coral reefs were added to a 1:1 000 000 coastline from IUCN (1986, 1988, 1989). These maps only cover approximately half of the coastline between the Yemen border and the center of Sawqirah Bay and from Ras ad Daffah to Sarimah. For the northern coastline of Somalia, reefs have been taken from Hydrographic Office (1992, 1993).

Hydrographic Office (1992). Golfe de Tadjoura and Anchorages. 1:200 000. *British Admiralty Chart No. 253.* September 1992. Taunton, UK.

Hydrographic Office (1993). Gulf of Aden. 1:750 000. *British Admiralty Chart No. 6.* March 1993. Taunton, UK.

IUCN (1986). *Oman Coastal Zone Management Plan: Greater Capital Area.* Prepared for Ministry of Commerce and Industry, Muscat, Oman. IUCN, Gland, Switzerland.

IUCN (1988). *Oman Coastal Zone Management Plan: Quriyat to Ra's al Hadd.* Prepared for Ministry of Commerce and Industry, Muscat, Oman. IUCN, Gland, Switzerland.

IUCN (1989). *Oman Coastal Zone Management Plan: Dhofar: Volume 2: Resource Atlas.* Prepared for Ministry of Commerce and Industry, Muscat, Oman. IUCN, Gland, Switzerland.

Map 9e

Coral reef areas are based on Hydrographic Office (1989, 1991a, 1991b and 1994). Some additional reef areas are based on Abbott (1994).

Abbott F (1994). Coral Reefs of Bahrain (Arabian Gulf). A draft report, prepared for the World Conservation Monitoring Centre.

Hydrographic Office (1987). Jazireh-ye Lavan to Kalat and Ra's Tannurah. *British Admiralty Chart No. 2883.* 1:350 000. Taunton, UK.

Hydrographic Office (1989). Musay'id to Ra's Laffan. *British Admiralty Chart No. 3950.* 1:150 000. Taunton, UK.

Hydrographic Office (1991a). Ra's Tannurah to Jazirat Faylaka and Jazireh-ye Khark. *British Admiralty Chart No. 2882.* 1:350 000. Taunton, UK.

Hydrographic Office (1991b). Kalat and Ra's al Khafji to Abadan. *British Admiralty Chart No. 2884.* 1:350 000. Taunton, UK.

Hydrographic Office (1994). Jazireh-ye Lavan and Jazirat Das to Ra's Tannurah. *British Admiralty Chart No. 2886.* 1:350 000. Taunton, UK.

Chapter 10 Southeast Asia

Southeast Asia, with its complex coastline and mass of tightly interlocking islands, straddles the world's greatest zone of coral reef biodiversity. Although reef development is rather restricted in a few areas, notably the Gulf of Thailand and the southern coastline of mainland China, for the most part coral reefs are well developed and numerous. Fringing reefs line the coasts of myriad islands, including many of the larger ones, and parts of the mainland. There are also extensive, though often poorly known barrier reefs, while in the deeper waters of the South China Sea and towards the east of the region there is a considerable number of oceanic atolls.

There is a great paucity of information about many areas. Barrier reefs off the coastline of Sumatra, Sulawesi and Palawan as well as the reefs off Myanmar and eastern Indonesia have received little attention from the scientific community.

It has been suggested that biodiversity may have remained, or even accumulated, in this region at the same time that Pleistocene extinctions were occurring in other parts of the world. This is an area which maintained a relatively equitable climate for coral reef development right through the last glaciation, possibly providing a refuge for numerous species. At the same time the massive fluctuations in sea level may have isolated pockets of coral reef diversity, allowing evolution to follow different paths so that, when species reunited, they had diverged, further adding to their diversity. Whatever the causes, this region harbors more species in almost every group of coral reef organism than any other part of the world.

Unfortunately its reefs also face considerable difficulties, with some 82 percent estimated to be threatened by human activities in the recent *Reefs at Risk* report.

Burgeoning human populations are overutilizing the resources in many areas, while wholesale destruction of the forests on land, together with rapid urbanization, is leading to massive loads of sediments and pollution on many reefs. While scientists may have little information about these reefs, most are well known to fishermen, and even the most remote reefs are threatened by overfishing and, particularly, by destructive fishing practices.

Left: Southeast Asia harbors the highest levels of biodiversity of any coral reef region. Here the knobbly branches of Porites *surround a foliose* Montipora *coral. Right: Volcanoes such as Muria on Java are widespread (STS026-41-86, 1998).*

MAP 10

100°
130°
RUSSIA
Kuril Basin
Kuril Trench
MONGOLIA
DEMOCRATIC PEOPLE'S REPUBLIC OF KOREA
SEA OF JAPAN
PACIFIC OCEAN
Izu Trench
YELLOW SEA
JAPAN
REPUBLIC OF KOREA
30° BHUTAN
30°
CHINA
EAST CHINA SEA
Kyushu-Palau Ridge
South Honshu Ridge
INDIA
TAIWAN
Ryukyu Trench
PHILIPPINE SEA
Mariana Trough
VIETNAM
BANGLADESH
Philippine Basin
West Mariana Basin
Bay of Bengal
LAOS
Benham Seamount
MYANMAR
THAILAND
SOUTH CHINA SEA
Philippine Trench
CAMBODIA
Andaman Basin
Gulf of Thailand
PHILIPPINES
Sulu Basin
Palawan Trough
BRUNEI
Celebes Basin
West Caroline Basin
M A L A Y S I A
0° Cocos Basin
0°
I N D O N E S I A
I N D O N E S I A
Java Trench
JAVA SEA
Investigator Ridge
ARAFURA SEA
BANDA SEA
Christmas I. (AUSTRALIA)
Java Ridge
EAST TIMOR
Cocos (Keeling) Is. (AUSTRALIA)
Gulf of Carpentaria
North Australian Basin
Exmouth Plateau
AUSTRALIA
INDIAN OCEAN
N
0 500 1000 1500 km
100°

Thailand, Myanmar and Cambodia

MAP 10a

Thailand is a large country lying in the center of mainland Southeast Asia and extending far south along the Malay Peninsula towards the Malaysian border. The coastline is clearly divided into those sections which border the Gulf of Thailand and a shorter coastline on the Andaman Sea. The Gulf of Thailand is a shallow semi-enclosed sea, generally less than 60 meters in depth. It is heavily sedimented, particularly in the north and west, but highly productive. The tidal systems are diurnal, while monsoonal weather exerts the predominant influence on reef development. In the northeast much of the coast is affected by riverine runoff and there are some large mangrove communities, but fringing reefs have developed away from riverine inputs, and there are quite a number of islands with important fringing reef communities offshore. In the northwest the coastline is indented into the Bight of Bangkok where there is massive riverine input from four major rivers and large reef structures do not occur, although coral communities are reported in a few places, particularly around the offshore islands. More reefs occur on the eastern shores of Thailand (the western shores of the gulf). These are largely around offshore islands, with limited developments around the islands of Prachuab Kirikhan and more extensive ones around the western shores of the islands of Chumphon and on all sides of the islands near Surat Thani. In all areas the relatively harsh physical conditions have restricted reef diversity, and coral numbers in the Gulf of Thailand are far lower than in surrounding regions.

The coastline facing the Andaman Sea is somewhat different. The continental shelf lies about 200 kilometers offshore in the south, although only some 50 kilometers offshore around Phuket. This coastline is heavily affected by the contrasting influences of the strong Southwest Monsoon (May-October) driving onshore winds and rough conditions, while during the Northeast Monsoon (November-April) conditions are generally very calm. This coastline has the largest areas of mangroves in Thailand, and there are also extensive coral reefs, particularly along the shores of the numerous offshore islands. The degree of reef development appears to be related to distance from shore and level of exposure. Fringing communities are generally better developed on the eastern coasts of the islands. Extensive reef structures are reported from the Adang Rawi group in the south and around the Surin Islands, a southerly extension of the Mergui Archipelago. The majority of reef research in this country has been focussed around Phuket and the

Two spot-naped butterflyfish Chaetodon oxycephalus *on a shallow reef. This species feeds primarily on coral polyps.*

MAP 10a

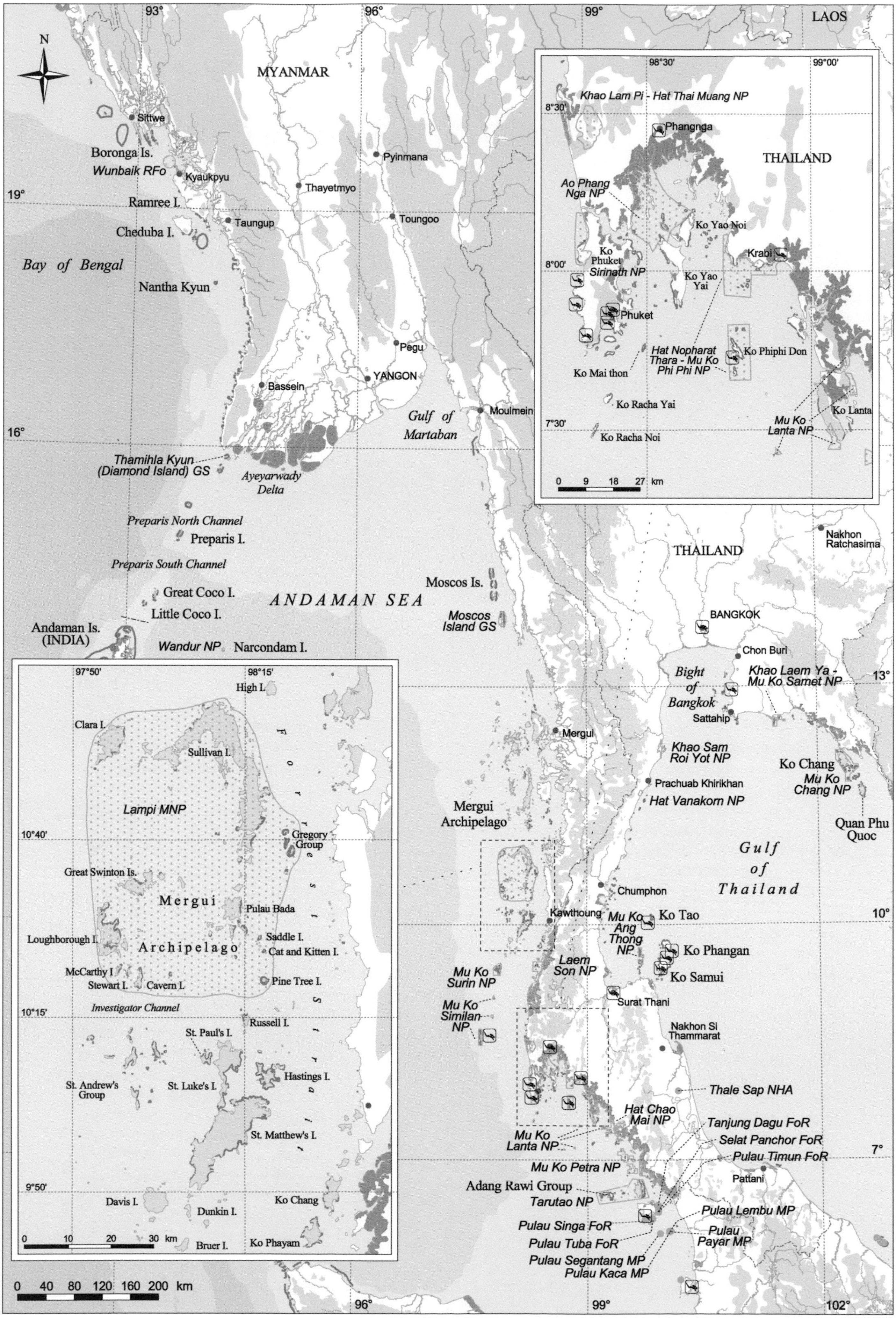

93°
96°
99°
LAOS
N
MYANMAR
Sittwe
Boronga Is.
Wunbaik RFo
Kyaukpyu
Pyinmana
Thayetmyo
19°
Ramree I.
Taungup
Toungoo
Cheduba I.
Bay of Bengal
Nantha Kyun
Pegu
Bassein
YANGON
Gulf of Martaban
Moulmein
16°
Thamihla Kyun (Diamond Island) GS
Ayeyarwady Delta
Preparis North Channel
Preparis I.
Preparis South Channel
Great Coco I.
Little Coco I.
Andaman Is. (INDIA)
Wandur NP
Narcondam I.
ANDAMAN SEA
Moscos Is.
Moscos Island GS
THAILAND
Nakhon Ratchasima
BANGKOK
Chon Buri
Bight of Bangkok
Khao Laem Ya - Mu Ko Samet NP
13°
Sattahip
Mergui
Khao Sam Roi Yot NP
Prachuab Khirikhan
Hat Vanakorn NP
Ko Chang
Mu Ko Chang NP
Quan Phu Quoc
Mergui Archipelago
Gulf of Thailand
Chumphon
Kawthoung
Mu Ko Ang Thong NP
Ko Tao
10°
Ko Phangan
Ko Samui
Laem Son NP
Mu Ko Surin NP
Surat Thani
Mu Ko Similan NP
Nakhon Si Thammarat
Thale Sap NHA
Hat Chao Mai NP
Tanjung Dagu FoR
Mu Ko Lanta NP
Selat Panchor FoR
7°
Pulau Timun FoR
Mu Ko Petra NP
Pattani
Adang Rawi Group
Tarutao NP
Pulau Lembu MP
Pulau Singa FoR
Pulau Payar MP
Pulau Tuba FoR
Pulau Segantang MP
Pulau Kaca MP
0 40 80 120 160 200 km
96°
99°
102°

98°30'
99°00'
Khao Lam Pi - Hat Thai Muang NP
8°30'
Phangnga
THAILAND
Ao Phang Nga NP
Ko Yao Noi
Krabi
8°00'
Ko Phuket
Sirinath NP
Ko Yao Yai
Phuket
Hat Nopharat Thara - Mu Ko Phi Phi NP
Ko Phiphi Don
Ko Mai thon
Ko Racha Yai
7°30'
Ko Racha Noi
Ko Lanta
Mu Ko Lanta NP
0 9 18 27 km

97°50'
98°15'
High I.
Clara I.
Sullivan I.
Lampi MNP
Forest Strait
10°40'
Gregory Group
Great Swinton Is.
Mergui Archipelago
Pulau Bada
Loughborough I.
Saddle I.
Cat and Kitten I.
McCarthy I
Stewart I.
Cavern I.
Pine Tree I.
Investigator Channel
10°15'
Russell I.
St. Paul's I.
Hastings I.
St. Andrew's Group
St. Luke's I.
St. Matthew's I.
9°50'
Davis I.
Dunkin I.
Ko Chang
0 10 20 30 km
Bruer I.
Ko Phayam

coastline of the Andaman Sea. Early reef research into coral bleaching was conducted in these waters, and the reefs of Ko Phuket, which have adapted to high sediment loads, are also of considerable interest. The 1998 bleaching event did not appear to affect reefs in the Andaman Sea, though bleaching was widespread in the Gulf of Thailand where it had not previously been recorded, and up to 60 percent of corals were reported to have been affected.

Pressures on much of Thailand's coastal zone are considerable. Sedimentation is a significant problem for reefs in many areas, but particularly on the mainland coasts. The Gulf of Thailand also has a major trawl fishery. Although this does not directly impact any true reefs, it is likely to have destroyed or degraded small coral communities which may have existed in the open waters of the gulf. Most of Thailand's fisheries are concentrated on offshore stocks, which are believed to have been overfished since the 1970s. In addition, many reefs are utilized for fishing as well as for shell and aquarium fish collection. There have been problems of destructive fishing in some parts, although this is believed to have declined. Some coastal fisherfolk are also now becoming involved in tourism-based activities. In the Andaman Sea, sea gypsies are responsible for some target species fishing and collection for the aquarium trade. It has been estimated that over 50 percent of coastal mangrove forests have been destroyed, largely for conversion to shrimp ponds and for coastal development. Many of the shrimp farms were poorly designed and have since been abandoned, leaving vast areas with neither farms nor the previously highly productive forest areas. Efforts are now underway to restore some of these, although with little success so far. Mangroves are still being cleared, but the rate of loss has now diminished.

Tourism now exerts a considerable influence on reef communities and is probably the most significant reef use in many areas. Unfortunately much of this has been associated with negative impacts. Construction of roads and buildings has led to problems with siltation and pollution. Anchor damage, direct tourist damage and even collection of corals and shells may be having further impacts. It has been estimated that over 40 percent of Thailand's reefs lie within marine national parks, and the Department of Fisheries has been running a coral reef management program since 1995 focussed towards research, training and public eduction to further the protection of reefs outside these areas. Efforts are continuing to establish mooring buoys at all popular dive locations.

Myanmar

Extending from a northern border with Bangladesh to Thailand in the south, Myanmar has a considerable coastline along the Bay of Bengal and the Andaman Sea. The northern coast is bounded by the Arkan mountain range which extends down to the Arkan Peninsula and then

Left: Bangkok is one of many sprawling cities in the region producing vast quantities of sediments and pollution (STS059-235-31, 1994). Right: Thailand's once extensive mangrove forests have been widely replaced by agriculture and prawn farms.

continues below sea level, re-emerging to form a number of small islands north of the Andaman Islands (India). There is a wide level area of coastal plain in the center of the country, dominated by the Ayeyarwady (formerly Irrawaddy) river delta, a large sediment-laden river associated with relatively rapid coastal progradation. In the southeast the coastal plain is again narrow, backed by the Tenasserim Range of mountains. Offshore there are two major island groups: the Moscos Islands to the north and the vast complex of islands forming the Mergui Archipelago in the south.

There is remarkably little information in the scientific literature describing the reef communities of this country, but it seems likely that those on the nearshore islands in

	Thailand	Myanmar	Cambodia
General Data			
Population (thousands)	61 231	41 735	12 212
GDP (million US$)	136 773	33 665	1 187
Land area (km²)	515 139	669 813	182 602
Marine area (thousand km²)	252	513	20
Per capita fish consumption (kg/year)	33	17	9
Status and Threats			
Reefs at risk (%)	96	77	100
Recorded coral diseases	0	0	0
Biodiversity			
Reef area (km²)	2 130	1 870	<50
Coral diversity	238 / 428	77 / 277	na / 337
Mangrove area (km²)	2 641	3 786	851
No. of mangrove species	35	24	5
No. of seagrass species	15	3	1

Snapper taking shelter among mangroves at high tides. Myanmar's Mergui Archipelago includes important areas of relatively undisturbed reef and mangrove.

Protected areas with coral reefs

Site name	Designation	Abbreviation	IUCN cat.	Size (km²)	Year
Myanmar					
Lampi	Marine National Park	MNP	II	3 890.00	1994
Moscos Island	Game Sanctuary	GS	Unassigned	49.21	1927
Thailand					
Ao Phang Nga	National Park	NP	II	400.00	1981
Hat Chao Mai	National Park	NP	II	230.86	1981
Hat Nopharat Thara – Mu Ko Phi Phi	National Park	NP	II	389.96	1983
Khao Laem Ya – Mu Ko Samet	National Park	NP	V	131.00	1981
Khao Sam Roi Yot	National Park	NP	II	98.08	1966
Mu Ko Ang Thong	National Park	NP	Unassigned	102.00	1980
Mu Ko Chang	National Park	NP	II	650.00	1982
Mu Ko Lanta	National Park	NP	II	134.00	1990
Mu Ko Libong	Non Hunting Area	NHA	III	447.49	1979
Mu Ko Petra	National Park	NP	II	494.38	1984
Mu Ko Similan	National Park	NP	II	128.00	1982
Mu Ko Surin	National Park	NP	II	135.00	1981
Sirinath	National Park	NP	II	90.00	1981
Tarutao	National Park	NP	II	1 490.00	1972

the south of the country and around the islands north of the Andamans are extensive and diverse. The Mergui Archipelago consists of over 800 islands, most of which are uninhabited, and many remain forested. Reefs are best developed on the outermost islands, and are thought to be similar to those around the offshore islands of Thailand. Over 100 kilometers offshore from the southern part of the Mergui Archipelago lie the Burma Banks, a series of seamounts which rise up from over 300 meters to flat tops some 15-22 meters below the surface and are reported to have significant hard coral cover. The chain of small islands between the Ayeyarwady Delta and the Andaman Islands is little known but likely to have some interesting and important coral communities. Reefs are also reported at some of the islands off the Bay of Bengal coast and up to the border with Bangladesh.

Myanmar has been a relatively closed country for a number of years and coastal development has been slow, particularly away from the capital. While there is undoubtedly some utilization of reef resources by local people, the pressures are considered to be quite low and the reefs in the south of the country are noted for their significant numbers of large fish, including sharks. At least two marine protected areas have been declared, but there are concerns that resident populations may have been mistreated or displaced for the establishment of these sites. Tourism is growing relatively rapidly since arrangements were made for dive vessels to enter the country in 1997 (via the coastal port of Kawthoung, close to the Thai border), and there are now several vessels operating in the area. Development of the islands themselves has not yet begun.

Cambodia

Cambodia has only a relatively short coastline facing the Gulf of Thailand, though there are several small islands in the adjacent waters. There is very little material available describing the coral reefs off this coastline, but there are known to be coral communities on the mainland coast and some fringing reef structures around the islands. Some 70 hard corals have been recorded at the Koh Tang island group and in a few places coral cover is reported to reach over 50 percent. On the mainland diversity is much lower, and communities are dominated by massive and encrusting corals. Bleaching was reported at a number of localities in 1998, but recovery is thought to have been fairly good.

Malaysia, Singapore and Brunei Darussalam

MAPS 10b and c

Malaysia is a large country split into two land areas: Peninsular Malaysia and east Malaysia. The latter, comprising the states of Sarawak and Sabah, is located along the northern and western edge of the island of Borneo. All of these areas are located on the Sunda Shelf, although the edge of this continental shelf comes relatively close to the land around Sabah.

Although Peninsular Malaysia has a relatively high relief its coastline, particularly in the south and west, is dominated by low-lying land and mangroves or former mangrove areas. Offshore a number of small islands are important for reef development. These include the Pulau Langkawi group in the northwest, Pulau Semblian in the west, and the Pulau Tioman and Pulau Redang groups in the east. East Malaysia also has a very high relief, although in the west there is a generally wide coastal strip with extensive wetlands and mangrove development. Further east, and particularly in Sabah, the coastline is more complex and indented, with a generally narrow coastal strip. Again, a number of offshore island groups are important for reef development, particularly around Sabah.

The region's climate is largely determined by the opposing monsoon systems. During the Northeast Monsoon (November-March) winds are driven from the northeast bearing moist air and typically driving higher rainfalls, as well as high seas, particularly on the north and east coasts of Sabah. During the Southeast Monsoon (May-September), typically drier air flows from the southwest. In the Strait of Malacca there is a permanent current flowing to the northwest, while surface currents are more variable over much of the rest of the region, largely following patterns of wind circulation.

There is relatively little reef development along the mainland coast of Peninsular Malaysia, but reefs occur around all the offshore islands. Conditions for reef development are generally poor in the Strait of Malacca, however there are small low diversity reefs on the mainland close to Port Dickson. There are also reported to be some minor mainland fringing communities on the east coast between Kuala Terengganu and Chukai.

Reef development is highly restricted off the coast of Sarawak, although there are some reefs around the offshore islands of Pulau Talang and Pulau Satar. The most extensive reef development in the country is in the waters around Sabah, which is the region with the highest diversity and optimal conditions for reef development. This is close to the global center of coral reef diversity. Around the southeast coast there are extensive fringing

A feather star on the reef slope in one of east Malaysia's marine parks.

MAP 10b

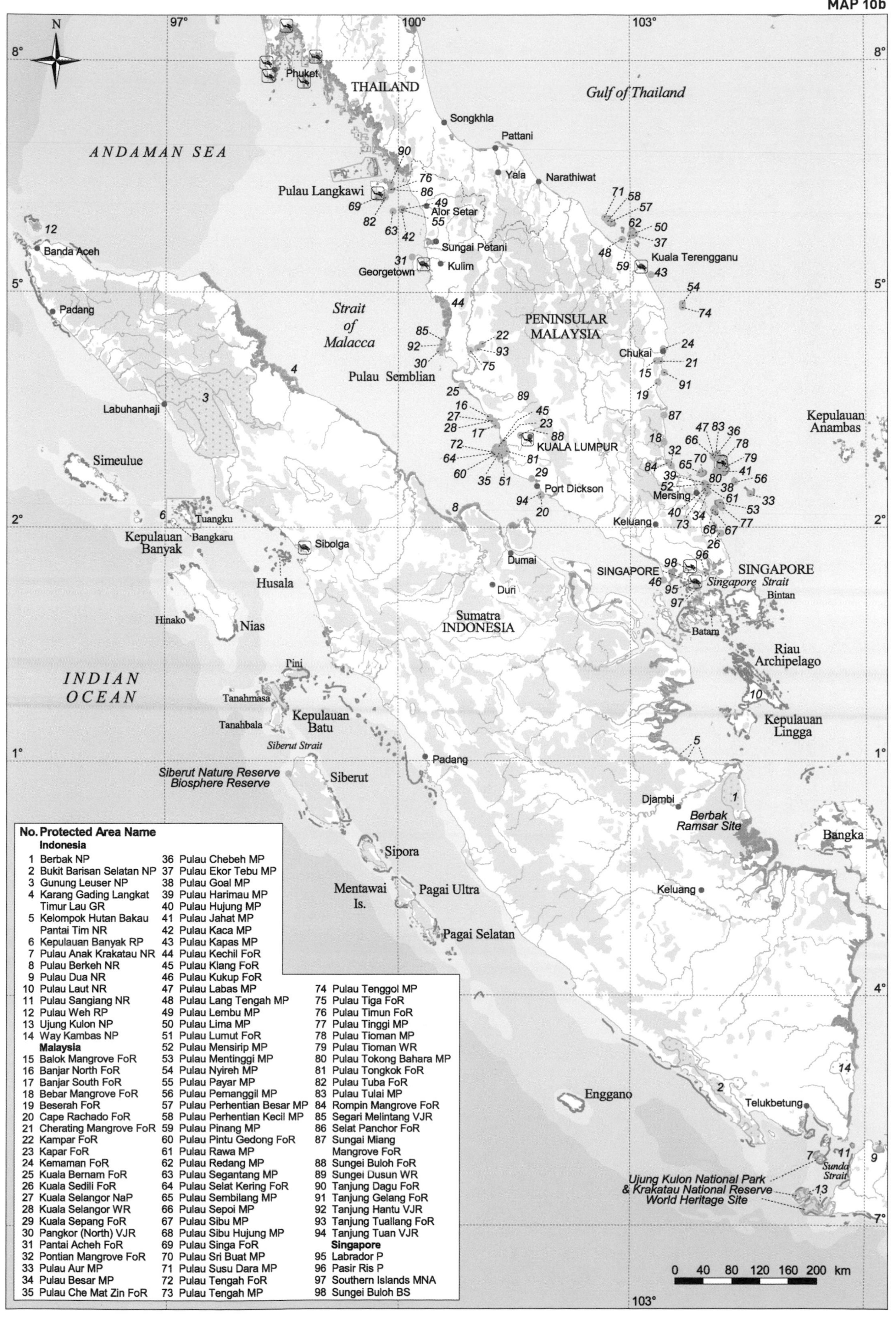

N
97°
100°
103°
8°
5°
2°
1°
4°
7°
Phuket
THAILAND
Gulf of Thailand
Songkhla
Pattani
ANDAMAN SEA
Yala
Narathiwat
Pulau Langkawi
Alor Setar
Sungai Petani
Georgetown
Kulim
Kuala Terengganu
Banda Aceh
Padang
Strait of Malacca
PENINSULAR MALAYSIA
Chukai
Pulau Sembilan
Labuhanhaji
KUALA LUMPUR
Kepulauan Anambas
Simeulue
Port Dickson
Mersing
Tuangku
Bangkaru
Kepulauan Banyak
Keluang
Sibolga
Dumai
SINGAPORE
SINGAPORE
Singapore Strait
Husala
Duri
Bintan
Hinako
Nias
Sumatra INDONESIA
Batam
Riau Archipelago
INDIAN OCEAN
Pini
Tanahmasa
Tanahbala
Kepulauan Batu
Siberut Strait
Kepulauan Lingga
Padang
Siberut Nature Reserve Biosphere Reserve
Siberut
Djambi
Berbak Ramsar Site
Bangka
Sipora
Mentawai Is.
Pagai Ultra
Keluang
Pagai Selatan
Enggano
Telukbetung
Sunda Strait
Ujung Kulon National Park & Krakatau National Reserve World Heritage Site
0 40 80 120 160 200 km
No. Protected Area Name
Indonesia
1 Berbak NP
2 Bukit Barisan Selatan NP
3 Gunung Leuser NP
4 Karang Gading Langkat Timur Lau GR
5 Kelompok Hutan Bakau Pantai Tim NR
6 Kepulauan Banyak RP
7 Pulau Anak Krakatau NR
8 Pulau Berkeh NR
9 Pulau Dua NR
10 Pulau Laut NR
11 Pulau Sangiang NR
12 Pulau Weh RP
13 Ujung Kulon NP
14 Way Kambas NP
Malaysia
15 Balok Mangrove FoR
16 Banjar North FoR
17 Banjar South FoR
18 Bebar Mangrove FoR
19 Beserah FoR
20 Cape Rachado FoR
21 Cherating Mangrove FoR
22 Kampar FoR
23 Kapar FoR
24 Kemaman FoR
25 Kuala Bernam FoR
26 Kuala Sedili FoR
27 Kuala Selangor NaP
28 Kuala Selangor WR
29 Kuala Sepang FoR
30 Pangkor (North) VJR
31 Pantai Acheh FoR
32 Pontian Mangrove FoR
33 Pulau Aur MP
34 Pulau Besar MP
35 Pulau Che Mat Zin FoR
36 Pulau Chebeh MP
37 Pulau Ekor Tebu MP
38 Pulau Goal MP
39 Pulau Harimau MP
40 Pulau Hujung MP
41 Pulau Jahat MP
42 Pulau Kaca MP
43 Pulau Kapas MP
44 Pulau Kechil FoR
45 Pulau Klang FoR
46 Pulau Kukup FoR
47 Pulau Labas MP
48 Pulau Lang Tengah MP
49 Pulau Lembu MP
50 Pulau Lima MP
51 Pulau Lumut FoR
52 Pulau Mensirip MP
53 Pulau Mentinggi MP
54 Pulau Nyireh MP
55 Pulau Payar MP
56 Pulau Pemanggil MP
57 Pulau Perhentian Besar MP
58 Pulau Perhentian Kecil MP
59 Pulau Pinang MP
60 Pulau Pintu Gedong FoR
61 Pulau Rawa MP
62 Pulau Redang MP
63 Pulau Segantang MP
64 Pulau Selat Kering FoR
65 Pulau Sembilang MP
66 Pulau Sepoi MP
67 Pulau Sibu MP
68 Pulau Sibu Hujung MP
69 Pulau Singa FoR
70 Pulau Sri Buat MP
71 Pulau Susu Dara MP
72 Pulau Tengah FoR
73 Pulau Tengah MP
74 Pulau Tenggol MP
75 Pulau Tiga FoR
76 Pulau Timun FoR
77 Pulau Tinggi MP
78 Pulau Tioman MP
79 Pulau Tioman WR
80 Pulau Tokong Bahara MP
81 Pulau Tongkok FoR
82 Pulau Tuba FoR
83 Pulau Tulai MP
84 Rompin Mangrove FoR
85 Segari Melintang VJR
86 Selat Panchor FoR
87 Sungai Miang Mangrove FoR
88 Sungei Buloh FoR
89 Sungei Dusun WR
90 Tanjung Dagu FoR
91 Tanjung Gelang FoR
92 Tanjung Hantu VJR
93 Tanjung Tuallang FoR
94 Tanjung Tuan VJR
Singapore
95 Labrador P
96 Pasir Ris P
97 Southern Islands MNA
98 Sungei Buloh BS

MAP 10c

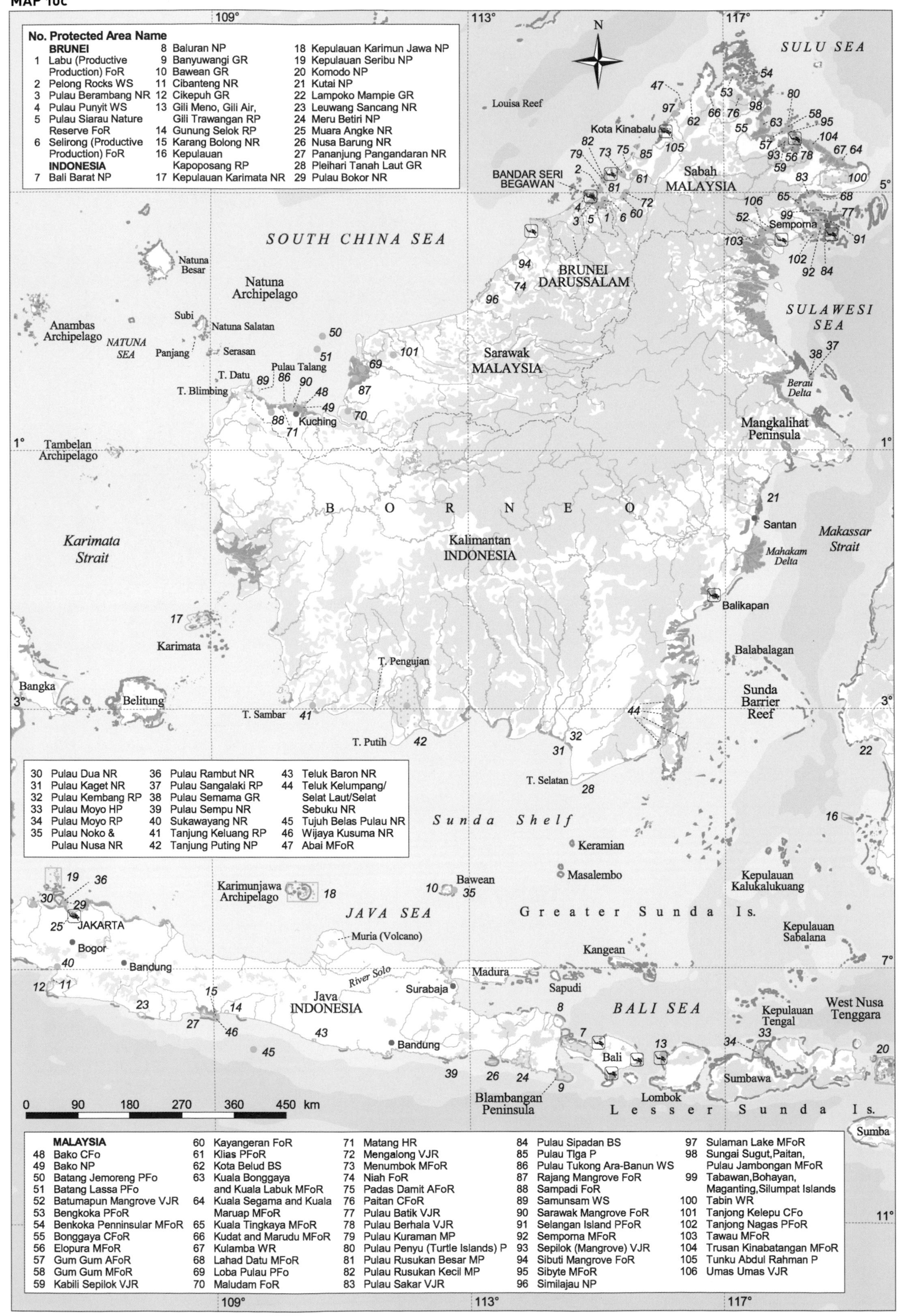

No. Protected Area Name
BRUNEI
1 Labu (Productive Production) FoR
2 Pelong Rocks WS
3 Pulau Berambang NR
4 Pulau Punyit WS
5 Pulau Siarau Nature Reserve FoR
6 Selirong (Productive Production) FoR
INDONESIA
7 Bali Barat NP
8 Baluran NP
9 Banyuwangi GR
10 Bawean GR
11 Cibanteng NR
12 Cikepuh GR
13 Gili Meno, Gili Air, Gili Trawangan RP
14 Gunung Selok RP
15 Karang Bolong NR
16 Kepulauan Kapoposang RP
17 Kepulauan Karimata NR
18 Kepulauan Karimun Jawa NP
19 Kepulauan Seribu NP
20 Komodo NP
21 Kutai NP
22 Lampoko Mampie GR
23 Leuwang Sancang NR
24 Meru Betiri NP
25 Muara Angke NR
26 Nusa Barung NR
27 Pananjung Pangandaran NR
28 Pleihari Tanah Laut GR
29 Pulau Bokor NR
30 Pulau Dua NR
31 Pulau Kaget NR
32 Pulau Kembang RP
33 Pulau Moyo HP
34 Pulau Moyo RP
35 Pulau Noko & Pulau Nusa NR
36 Pulau Rambut NR
37 Pulau Sangalaki RP
38 Pulau Semama GR
39 Pulau Sempu NR
40 Sukawayang NR
41 Tanjung Keluang RP
42 Tanjung Puting NP
43 Teluk Baron NR
44 Teluk Kelumpang/ Selat Laut/Selat Sebuku NR
45 Tujuh Belas Pulau NR
46 Wijaya Kusuma NR
47 Abai MFoR
MALAYSIA
48 Bako CFo
49 Bako NP
50 Batang Jemoreng PFo
51 Batang Lassa PFo
52 Batumapun Mangrove VJR
53 Bengkoka PFoR
54 Benkoka Penninsular MFoR
55 Bonggaya CFoR
56 Elopura MFoR
57 Gum Gum AFoR
58 Gum Gum MFoR
59 Kabili Sepilok VJR
60 Kayangeran FoR
61 Klias PFoR
62 Kota Belud BS
63 Kuala Bonggaya and Kuala Labuk MFoR
64 Kuala Segama and Kuala Maruap MFoR
65 Kuala Tingkaya MFoR
66 Kudat and Marudu MFoR
67 Kulamba WR
68 Lahad Datu MFoR
69 Loba Pulau PFo
70 Maludam FoR
71 Matang HR
72 Mengalong VJR
73 Menumbok MFoR
74 Niah FoR
75 Padas Damit AFoR
76 Paitan CFoR
77 Pulau Batik VJR
78 Pulau Berhala VJR
79 Pulau Kuraman MP
80 Pulau Penyu (Turtle Islands) P
81 Pulau Rusukan Besar MP
82 Pulau Rusukan Kecil MP
83 Pulau Sakar VJR
84 Pulau Sipadan BS
85 Pulau Tiga P
86 Pulau Tukong Ara-Banun WS
87 Rajang Mangrove FoR
88 Sampadi FoR
89 Samunsam WS
90 Sarawak Mangrove FoR
91 Selangan Island PFoR
92 Semporna MFoR
93 Sepilok (Mangrove) VJR
94 Sibuti Mangrove FoR
95 Sibyte MFoR
96 Similajau NP
97 Sulaman Lake MFoR
98 Sungai Sugut, Paitan, Pulau Jambongan MFoR
99 Tabawan, Bohayan, Maganting, Silumpat Islands
100 Tabin WR
101 Tanjong Kelepu CFo
102 Tanjong Nagas PFoR
103 Tawau MFoR
104 Trusan Kinabatangan MFoR
105 Tunku Abdul Rahman P
106 Umas Umas VJR
109° 113° 117°
1° 3° 5° 7° 11°
N
SULU SEA
SOUTH CHINA SEA
SULAWESI SEA
Louisa Reef
Kota Kinabalu
Sabah MALAYSIA
BANDAR SERI BEGAWAN
BRUNEI DARUSSALAM
Semporna
Natuna Besar
Natuna Archipelago
Anambas Archipelago
NATUNA SEA
Subi
Natuna Salatan
Panjang
Serasan
Pulau Talang
T. Datu
T. Blimbing
Kuching
Sarawak MALAYSIA
Berau Delta
Mangkalihat Peninsula
Tambelan Archipelago
B O R N E O
Kalimantan INDONESIA
Santan
Makassar Strait
Mahakam Delta
Karimata Strait
Balikapan
Karimata
Balabalagan
T. Pengujan
Bangka
Belitung
T. Sambar
Sunda Barrier Reef
T. Putih
T. Selatan
Sunda Shelf
Keramian
Masalembo
Bawean
Karimunjawa Archipelago
Kepulauan Kalukalukuang
JAVA SEA
Greater Sunda Is.
JAKARTA
Bogor
Bandung
Muria (Volcano)
Kangean
Kepulauan Sabalana
Madura
River Solo
Surabaja
Sapudi
Java INDONESIA
BALI SEA
Kepulauan Tengal
West Nusa Tenggara
Bandung
Bali
Sumbawa
Blambangan Peninsula
Lombok
Lesser Sunda Is.
Sumba
0 90 180 270 360 450 km

reefs and a small barrier reef. Offshore from the town of Semporna lie a number of islands of volcanic origin with extensive reef developments. Just off the continental shelf lies Pulau Sipadan, a small coral cay with a surrounding reef with high coral cover and diversity. Further north, onshore reef development is restricted, but there are fringing reefs around the Turtle Islands. Off the north and west coasts, and particularly around the offshore islands, there are significant areas of fringing reefs. Over 200 kilometers off the west coast of Sabah there is a coral atoll, Layang Layang, with high biodiversity, although coral cover on the outer slopes was only recorded at 29 percent. Overall some 346 species of scleractinian coral have been identified in Malaysian waters. The impact of the 1998 bleaching appears to have been highly varied, but no widespread mortalities were recorded. At the same time, declines in coral cover were noted throughout eastern Malaysia in the decade up to 1999, linked to various anthropogenic impacts.

Marine fisheries are an important economic activity for Malaysia, with the majority of them commercial and focussed towards non-reef species using trawl and purse seine. Traditional fishing methods account for about a quarter of the total catch, only some of which is reef-dependent, and overfishing is not generally regarded as a major threat. There is significant destructive fishing, notably using explosives, and particularly off the coast of Sabah where more than four blasts per hour have been recorded in several areas. Perhaps the most significant threats to reefs arise from onshore activities, notably the high degree of sedimentation from logging activities and the sedimentation and pollution associated with industry, agriculture and urban development. Tourism development has also had impacts, through the construction of accommodation and associated infrastructure, but also through direct damage caused by anchors and divers. Development on Layang Layang, initially to establish a presence, but subsequently with the construction of a tourist resort, has

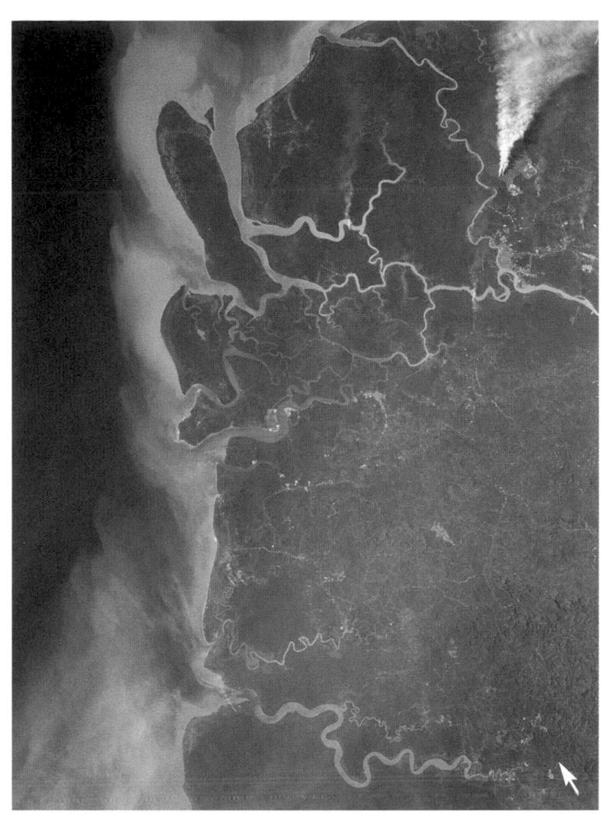

	Malaysia	Singapore	Brunei Darussalam
General Data			
Population (thousands)	21 793	4 152	336
GDP (million US$)	70 402	60 363	4 034
Land area (km2)	330 278	526	5 770
Marine area (thousand km2)	351	1.4	9
Per capita fish consumption (kg/year)	53	na	22
Status and Threats			
Reefs at risk (%)	91	100	100
Recorded coral diseases	0	0	0
Biodiversity			
Reef area (km2)	3 600	<100	210
Coral diversity	281 / 568	176 / 186	na / na
Mangrove area (km2)	6 424	6	171
No. of mangrove species	36	31	29
No. of seagrass species	12	11	4

Deforestation and forest fires have led to large increases in sedimentation in Sarawak and other parts of Malaysia (STS093-708-62, 1999).

Indonesia

MAPS 10b, c, d and e

Indonesia is the world's largest coral reef nation, with over 50 000 square kilometers of reefs (18 percent of the world total), extending nearly 5 000 kilometers from east to west, and harboring over 17 000 islands (including rocks and sandbanks). It touches on both the Indian and Pacific Oceans as well as many seas, including the Andaman, Java, South China, Sulawesi, Banda and Arafura Seas. This same country has a vast array of coral reefs, many poorly described or completely unknown, while it completely straddles the region with the greatest reef biodiversity in the world. For the purposes of this account the physical and biological descriptions are subdivided into a number of geographic sub-units, while human and socio-economic issues are considered together for the entire country.

Sumatra and Java

The western end of the Indonesian islands includes Sumatra and Java which, with Kalimantan, are located on the Sunda Shelf, a vast continental shelf extending across a considerable part of the South China Sea. Both are continental islands, but with the boundary between the Indian and Eurasian tectonic plates lying off their southwestern and southern boundaries, there are numerous volcanoes. The continental shelf lies relatively close to the shore on the western side of Sumatra and south of Java. Some distance off the west coast of Sumatra and off the continental shelf, lies the long chain of the Mentawai Islands. Off the east coast of Sumatra there is a complex of smaller islands at the southern end of the Strait of Malacca, the Riau Archipelago. Further south, towards the Java Sea, Bangka Island lies just off the Sumatra coastline and Belitung Island lies midway between Sumatra and Kalimantan. There are a few small islands north of Java, while Bali lies immediately to the east. Bali, unlike the other islands which continue in a chain to the east, is still located on the Sunda Shelf. The western side of Sumatra is heavily mountainous, with only a narrow coastal plain. In contrast the eastern side is low-lying and there is considerable riverine input all along this coastline. Java is very mountainous in its entirety, although the coastal plain is a little wider to the north and it is here that the most considerable riverine runoff occurs: rates of coastal progradation in the Solo Delta have been measured at 70 meters per year. The coastal waters of both eastern Sumatra and northern Java are generally quite turbid.

Weather and water conditions are largely determined

Left: Jakarta produces considerable quantities of sediment and pollution. The impacts of these on coral cover and diversity decline with distance across the reefs of Kepulauan Seribu (STS056-155-242, 1993). Right: A great diversity of fish and corals in Bali Barat National Park.

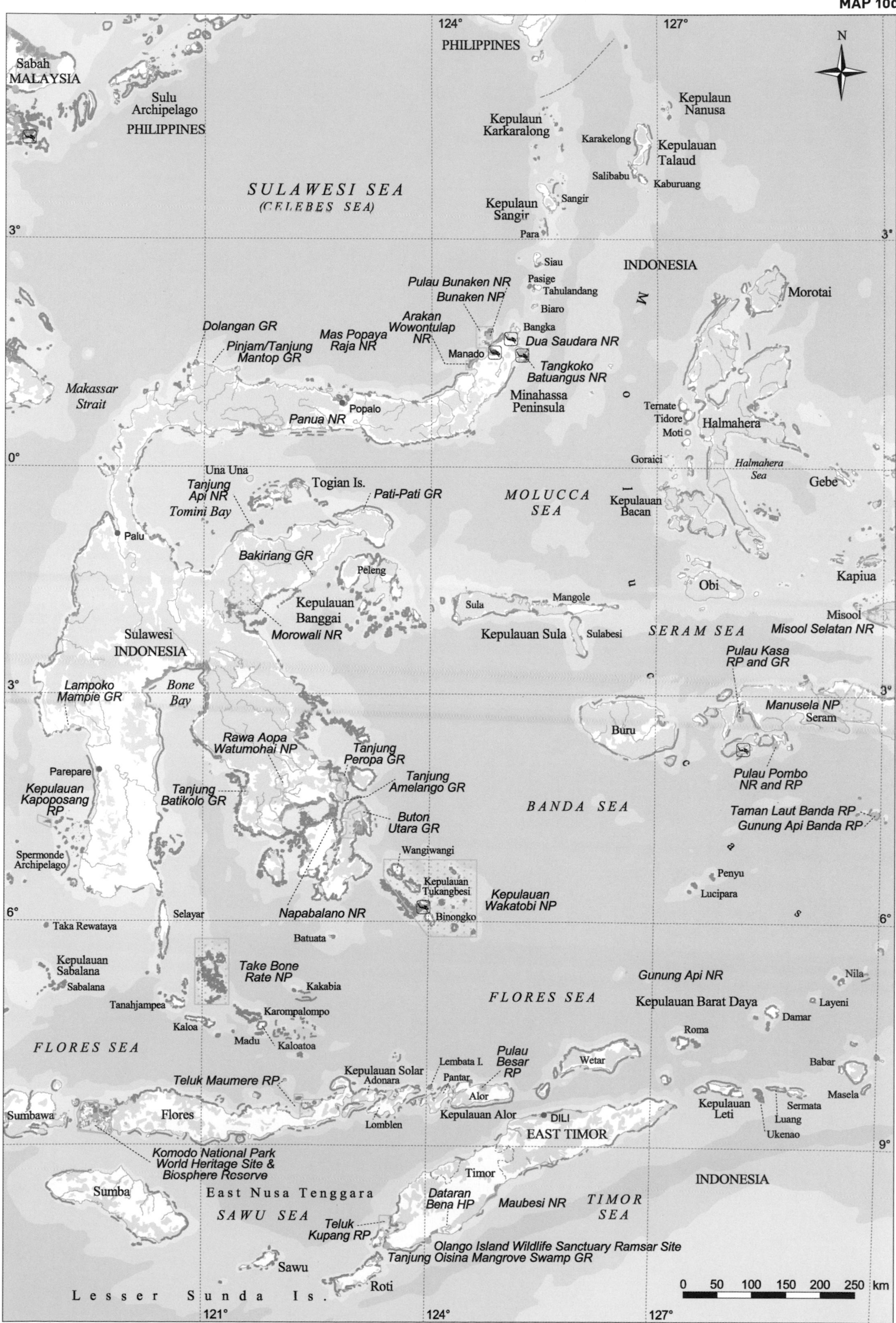

124°
127°
PHILIPPINES
N
Sabah
MALAYSIA
Sulu
Archipelago
PHILIPPINES
Kepulaun
Nanusa
Kepulaun
Karkaralong
Karakelong
Kepulauan
Talaud
Salibabu
Kaburuang
SULAWESI SEA
(CELEBES SEA)
Kepulaun
Sangir
Sangir
Para
3°
Siau
INDONESIA
Pasige
Tahulandang
Biaro
Bangka
Pulau Bunaken NR
Bunaken NP
Arakan
Wowontulap
NR
Mas Popaya
Raja NR
Dolangan GR
Pinjam/Tanjung
Mantop GR
Manado
Dua Saudara NR
Tangkoko
Batuangus NR
Morotai
Makassar
Strait
Popalo
Panua NR
Minahassa
Peninsula
Ternate
Tidore
Moti
Halmahera
0°
Goraici
Halmahera
Sea
Gebe
Una Una
Tanjung
Api NR
Tomini Bay
Togian Is.
Pati-Pati GR
MOLUCCA
SEA
Kepulauan
Bacan
Palu
Bakiriang GR
Peleng
Obi
Kapiua
Kepulauan
Banggai
Sula
Mangole
Misool
Morowali NR
Kepulauan Sula
Sulabesi
SERAM SEA
Misool Selatan NR
Sulawesi
INDONESIA
Pulau Kasa
RP and GR
Lampoko
Mampie GR
Bone
Bay
Manusela NP
Seram
Buru
Rawa Aopa
Watumohai NP
Tanjung
Peropa GR
Parepare
Tanjung
Amelango GR
Pulau Pombo
NR and RP
Kepulauan
Kapoposang
RP
Tanjung
Batikolo GR
BANDA SEA
Taman Laut Banda RP
Gunung Api Banda RP
Buton
Utara GR
Wangiwangi
Spermonde
Archipelago
Kepulauan
Tukangbesi
Kepulauan
Wakatobi NP
Penyu
Lucipara
6°
Selayar
Napabalano NR
Binongko
Taka Rewataya
Batuata
Kepulauan
Sabalana
Sabalana
Take Bone
Rate NP
Kakabia
Gunung Api NR
Nila
FLORES SEA
Kepulauan Barat Daya
Layeni
Tanahjampea
Karompalompo
Damar
Kaloa
Madu
Kaloatoa
Roma
FLORES SEA
Pulau
Besar
RP
Babar
Lembata I.
Wetar
Teluk Maumere RP
Kepulauan Solar
Adonara
Pantar
Alor
Masela
Sermata
Kepulauan
Leti
Sumbawa
Flores
Kepulauan Alor
DILI
Luang
Lomblen
EAST TIMOR
Ukenao
9°
Komodo National Park
World Heritage Site &
Biosphere Reserve
Timor
INDONESIA
Sumba
East Nusa Tenggara
Dataran
Bena HP
Maubesi NR
TIMOR
SEA
SAWU SEA
Teluk
Kupang RP
Olango Island Wildlife Sanctuary Ramsar Site
Tanjung Oisina Mangrove Swamp GR
Sawu
0 50 100 150 200 250 km
Roti
Lesser Sunda Is.
121°
124°
127°

MAP 10e

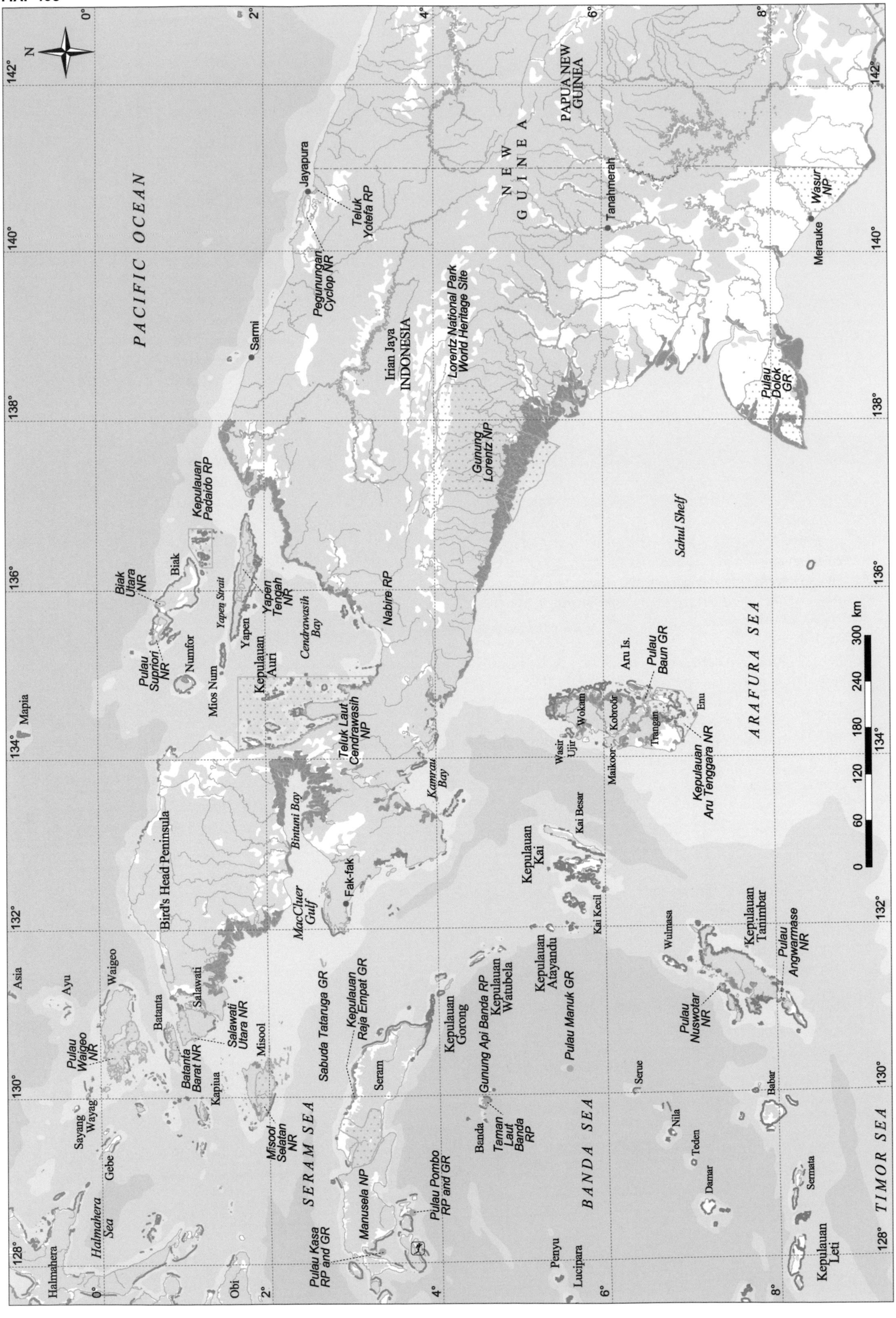

PACIFIC OCEAN
Halmahera
Halmahera Sea
Sayang
Wayag
Gebe
Pulau Waigeo NR
Asia
Ayu
Waigeo
Batanta
Batanta Barat NR
Salawati
Salawati Utara NR
Kapiua
Obi
Misool
Misool Selatan NR
SERAM SEA
Sabuda Tataruga GR
Pulau Kasa RP and GR
Manusela NP
Seram
Kepulauan Raja Empat GR
Pulau Pombo RP and GR
Kepulauan Gorong
Banda
Gunung Api Banda RP
Taman Laut Banda RP
Kepulauan Watubela
Kepulauan Atayandu
Pulau Manuk GR
Penyu
Lucipara
BANDA SEA
Serue
Nila
Teden
Damar
Babar
Sermata
Kepulauan Leti
TIMOR SEA
Wulmasa
Pulau Nuswotar NR
Kepulauan Tanimbar
Pulau Angwarmase NR
Bird's Head Peninsula
MacCluer Gulf
Fak-fak
Bintuni Bay
Kamrau Bay
Kepulauan Kai
Kai Kecil
Kai Besar
Mapia
Pulau Supriori NR
Biak Utara NR
Biak
Numfor
Mios Num
Yapen Strait
Yapen
Kepulauan Padaido RP
Kepulauan Auri
Yapen Tengah NR
Cendrawasih Bay
Teluk Laut Cendrawasih NP
Nabire RP
Wasir
Ujir
Wokam
Maikoor
Kobroor
Aru Is.
Trangan
Pulau Baun GR
Kepulauan Aru Tenggara NR
Enu
ARAFURA SEA
Sahul Shelf
Sarmi
Irian Jaya
INDONESIA
Gunung Lorentz NP
Lorentz National Park World Heritage Site
Jayapura
Pegunungan Cyclop NR
Teluk Yotefa RP
NEW GUINEA
PAPUA NEW GUINEA
Tanahmerah
Pulau Dolok GR
Merauke
Wasur NP
0 60 120 180 240 300 km
N
128° 130° 132° 134° 136° 138° 140° 142°
0° 2° 4° 6° 8°

by the opposing monsoon systems. During the Northeast Monsoon (December-March) winds over Sumatra predominate from the northeast, bearing moist air and typically driving higher rainfalls. This air is deflected in southern Sumatra and out over the Indian Ocean and bears round such that Java is dominated by northwesterly and westerly winds. During the Southeast Monsoon (particularly June-July), typically drier air flows from the southwest across Sumatra, and from the southeast across Java. Patterns of surface water currents are largely driven by these winds and during the Northeast Monsoon currents from the northeast flow in and are largely deflected into southeast and eastward flowing currents along eastern Sumatra and northern Java. These are mirrored by longshore currents flowing south and east along the Indian Ocean shores of these islands. During the Southeast Monsoon some of these patterns are reversed, with strong westward flowing currents along the coasts of Java and deflecting northwards along the east coast of Sumatra. The west coast of Sumatra, by contrast, maintains a southeasterly flowing current all year round. In the Strait of Malacca there is a permanent current flowing to the northwest.

Surprisingly little is known about the development of reefs around Sumatra. Fringing reefs are considered well developed in the north around Aceh and around the islands immediately north of Sumatra. They are are also likely to be widespread along much of the west coast of Sumatra facing the Indian Ocean – and have actually been recorded at the Mentawai Islands – but there is little published material describing the remainder of this coastline. Likewise this region is believed to support some extensive barrier reef systems: an 85 kilometer section is reported in the north, 20 kilometers off the coast of Aceh. This is a submerged or drowned system some 13-20 meters below the surface, but it is not clear to what degree it enjoys active coral growth. Further barrier reefs along the west coast of Sumatra are recorded with a combined length of 660 kilometers, although these have been little studied and rarely mentioned in regional reviews. Reefs are thought to be poorly developed along the east Sumatra coast where there is significant riverine input and the coastline is dominated by large mangrove communities. Fringing reefs are widespread in the Riau Archipelago and 95 species of scleractinian coral have been recorded from Batam Island. Water conditions are highly turbid in this area, however, and coral cover quickly diminishes with depth. Much further south around Belitung Island, fringing reefs have significantly higher diversities, presumably associated with more suitable conditions for reef development – 174 scleractinian species have been recorded.

The fringing reefs around Java have received little attention despite their high accessibility (compared to much of the rest of the country). There are well developed fringing reefs surrounding the volcanic islands in the Sunda Strait. Although not marked on most charts, it has been suggested that there may be extensive reef development off the south coast of Java, but that classic reef flat and reef crest structures have not developed due to the extreme exposure and high energy environment. Fringing reefs are well developed around the Blambangan Peninsula and off the short east coast of Java, with reef flats reaching 200-400 meters in width, but these are again limited off much of the north coast. One of the best known reef complexes in the region is the Kepulauan Seribu patch reef chain, also known as the Thousand Islands. This is a group of almost 700 reefs lying in a chain just northwest of Jakarta Bay. Many have associated islands and most have shallow intertidal reef flats. The reef slopes are quite diverse and there appears to be an increase in diversity with distance from Java – 88 scleractinian species have been recorded at one of the southerly reefs, rising to 190 species in the north. Outbreaks of crown-of-thorns starfish in 1995 may have reduced diversity in these southern islands still further.

Reefs are widely developed around the Karimunjawa Archipelago north of Java, and there are reported to be extensive fringing communities around Bawean Island on its eastern side. Fringing reefs are also well developed along the south coast of Bali and have a deep spur and groove formation associated with the high exposure along this coast. The 1998 bleaching event did affect the reefs around Bali, with over 75 percent bleaching in some areas. North of Java it appeared to be more varied and generally less significant.

Kalimantan

Much of the coastline of Kalimantan, or Indonesian Borneo, is low-lying and subjected to considerable riverine inputs. The Mahakam River, in particular, is noted for its high volume discharge and has been estimated to

A Balinese fishing boat.

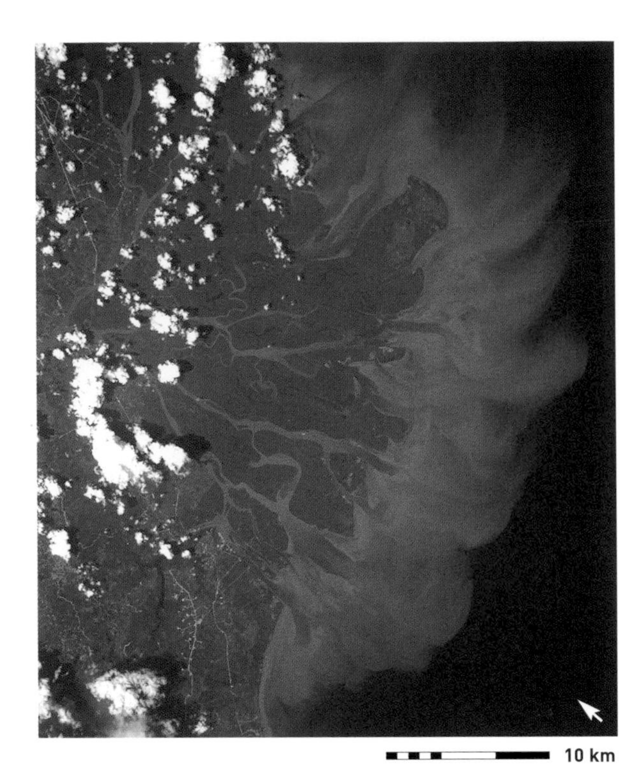

produce 4-10 million tons of sediment annually, with a plume which may extend up to 400 kilometers southeast of the Mahakam Delta. Even between the river mouths, the shores are largely fringed by mudflats and there are extensive mangrove communities. The main island lies on the Sunda Shelf and hence is surrounded by extensive shallow, and often relatively turbid, waters. To the east, however, the continental shelf edge lies relatively close to the mainland. There are several nearshore islands and some much further offshore, notably the Anambas, Natuna and Tambelan Archipelagos on the border between the South China Sea and the Natuna Sea. The patterns of monsoon weather are similar to those described for Sumatra and Java, with Northeast Monsoons bringing a northeasterly airflow which is deflected around the south of Kalimantan such that the south coast actually receives a predominantly westerly flow. Surface water currents at this time mirror these winds. During the Southeast Monsoon airflows are predominantly from the southwest, however the surface water currents are a little different, flowing from the north along the east coast, then deflecting towards the west as they meet the south coast.

Fringing reefs are absent from much of the main Kalimantan coastline, but do occur away from major areas of riverine input. They are thought to be well developed on the offshore continental islands, and also off the large headlands such as Tanjung (headland) Datu and T. Blimbing in the west, and T. Sambar, T. Putih, T. Pengujan and T. Selatan in the south. In the east, extensive reefs are recorded for 140 kilometers between T. Setan and T. Pamerikan, and again around the Mangkalihat Peninsula, while there is also an extensive fringing reef to the north of the Berau Delta. Offshore from the east coast lies Indonesia's longest continuous barrier reef system, the Sunda Barrier Reef, some 630 kilometers long, on the edge of the Sunda Shelf. Despite its size and potential economic, social and biological importance, this reef is largely undescribed. The coral communities of the Anambas, Natuna and Tambelan Archipelagos have not been well studied, although well developed fringing reef communities have been recorded on charts of the area.

Sulawesi and the Nusa Tengarra

This region is sometimes referred to as Wallacea, and encompasses the islands of Sulawesi and the Nusa Tenggara Islands. It is an area of complex oceanography: all of the islands have narrow continental shelves and many are separated from one another by relatively deep waters. The geological history of this region is extremely complex, and there are active volcanoes all along the southern islands and in the northeast peninsula of Sulawesi. All of these islands are mountainous, but their relatively narrow widths mean that there are few major watersheds and riverine input is widely dispersed. Air circulation patterns generally follow those of Kalimantan: during the Northeast Monsoon northerly winds reach the north of Sulawesi, but are rapidly deflected, becoming westerly along the southern coast of Sulawesi and the Nusa Tenggara Islands, while this pattern is almost exactly reversed during the Southeast Monsoon. Surface currents flow permanently eastwards along the north coast of Sulawesi and permanently southwards along the west coast. Between Sulawesi and the Nusa Tenggara there is a strong east flowing current during the Northeast Monsoon, which is reversed during the Southeast Monsoon. South of Nusa Tenggara in the Timor Sea the currents flow permanently westwards.

Conditions in this region are ideal for reef development and there are extensive fringing reefs along the shores of most islands, including some near continuous stretches running for hundreds of kilometers along the coastline of Sulawesi. These are particularly well developed along the eastern arm of Sulawesi where reef flats are typically 100-200 meters wide. In other areas reef flats may be less than 20 meters wide, resulting in their omission from many marine charts. Further offshore a large number of barrier reef systems have been described with a total length of 2 084 kilometers. Among the best known is the Spermonde Barrier Reef, which has a series of reefs leading towards the outer edge in a manner similar to the Great Barrier Reef – some 224 scleractinian corals have been described in this system. South of Peleng Island on the Banggai Platform there is another shelf-edge barrier reef system, the Banggai Barrier Reef. This is of

The Mahakam River produces vast quantities of sediment which inhibit coral reef development over a wide area (STS050-97-65, 1992).

particular interest because of the development of faros, circular atoll-like structures otherwise largely associated with the Maldives (Chapter 8). The Togian Islands, located in the mouth of Tomini Bay in northern Sulawesi, lie in very deep water and boast a number of interesting reef formations including fringing, barrier and atoll reefs. The reefs of the Tomini Bay are some of the most biodiverse in the world, with an estimated 77 species of *Acropora* alone. The 1998 bleaching event appears to have had relatively little impact over much of this region, and little or no bleaching was recorded north and west of Sulawesi.

There is little detailed information describing the reef communities of the Nusa Tenggara Islands, but fringing reefs are again widespread. Studies of Lembata Island in the center of the group show significant variation around the coastline. The northwest fringing reef is well developed with a 200-400 meter wide reef flat rich in seagrasses; this reef flat is even wider on the west coast. By contrast, the south coast has a narrower reef flat, which is fully exposed to Indian Ocean swell and may be further affected by cool water upwellings – a pronounced spur and groove structure is again noted, and a number of deep water species are found which may prefer cooler waters. North of these islands well developed barrier reefs are reported to occur northwest of Sumbawa and north of Flores. At the southern end of the Makassar Strait and in the Flores Sea there are a number of atolls, including the largest in the country: Kalukalukuang, Sabalana and Taka Bone Rate, each over 60 kilometers in length with complex atoll rims formed from individual patch reef structures separated by narrow and deep channels. In the western end of the Banda Sea there are, additionally, many smaller atolls.

The Moluccas and Irian Jaya

This final region, dominated by the coastline of Irian Jaya, also includes the complex island groups of the Moluccas to the west of Irian Jaya and a chain of small archipelagos along the south of the Banda Sea, stretching from Timor in the west to the Aru Islands in the east close to Irian Jaya. Overall this is another region of relatively complex bathymetry. Its waters are very deep, and even islands only a few tens of kilometers apart might be separated by depths of over 1 000 meters. The only areas of relatively extensive shallow water and true continental shelf are a platform west of the Bird's Head (Doberai) Peninsula and the wide expanse of the Arafura Sea, south of Irian Jaya and east of the Aru Islands. The latter largely lies above 100 meters in depth and is quite turbid, in marked contrast to the clear oceanic waters of much of the rest of the region. The coastline of Irian Jaya remains very poorly described. Large areas are low-lying and there is considerable riverine input, particularly along the south coast. The Bird's Head Peninsula is more mountainous.

During the Northeast Monsoon, northwesterly winds cut across most of the region, while during the Southeast Monsoon southeasterly winds come up towards southern Irian Jaya and the southern Moluccas, but these are deflected to become westerly in the more northern areas. Surface currents are somewhat mixed in this region. However, a northward current flows between Irian Jaya and Halmahera and an eastward current flows along the north shore of Irian Jaya during the Northeast Monsoon. This pattern reverses during the Southeast Monsoon.

Along the southeast coast of Irian Jaya wide areas are unsuitable for reef development: this coastline includes some of the largest mangrove forests in the world – those off the central coast and in Bintuni Bay may rival the area occupied by the Sundarbans forest between India and Bangladesh. There are reported to be fringing reefs along much of the higher coastal areas to the west. There is little or no information describing the reef communities around Bird's Head Peninsula. Along the rest of the north coast there are fringing reefs on all islands in Cendrawasih Bay, however the central and eastern coasts of this bay are dominated by mangrove forests and wide mudflats, and fringing reef systems have not developed. Further east, fringing reefs are believed to follow a large proportion of the coastline between Sarmi and the border with Papua New Guinea. For the most part these are poorly described, but reef flats are estimated to reach 300-400 meters wide in places. Further offshore, north of Irian Jaya, and also east of Halmahera, there are several small atolls. Off the east coast of the Aru islands there are vast fringing reefs, with shallow reef flats extending up to 15 kilometers from the coast.

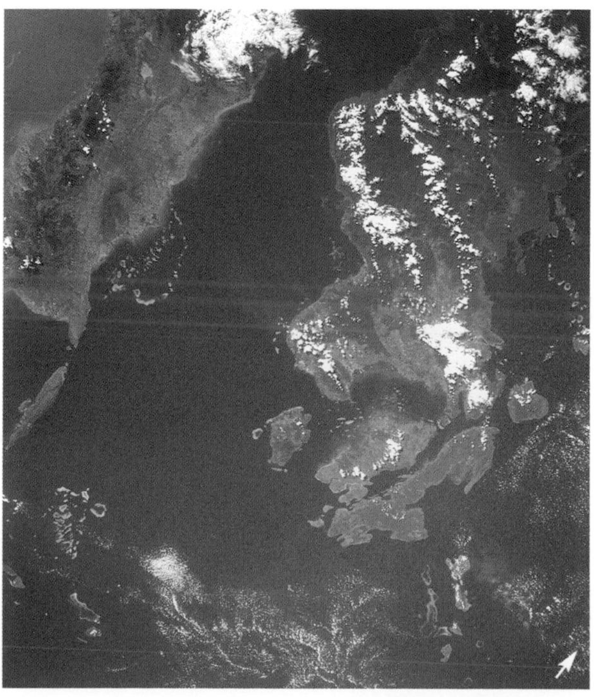

Southern Sulawesi, with a number of reefs clearly visible (STS069-709-42, 1995).

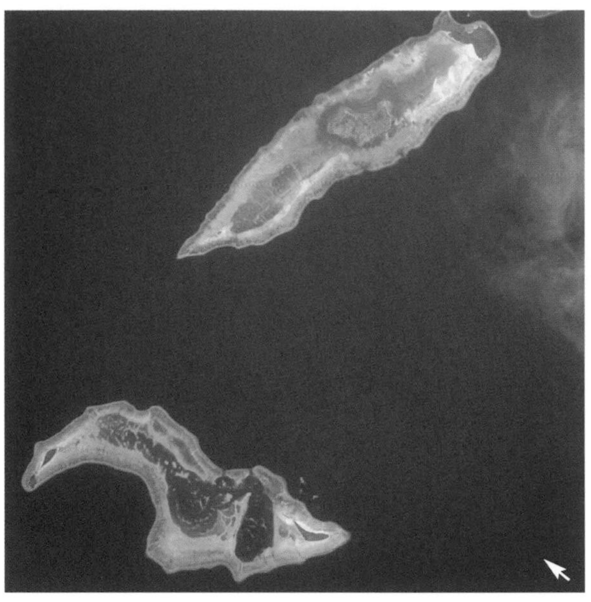

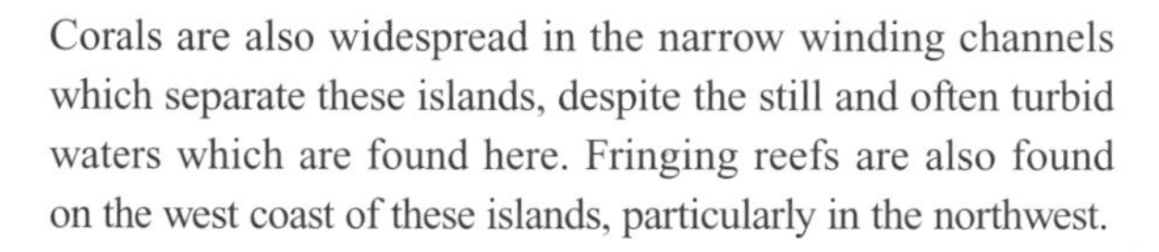
Corals are also widespread in the narrow winding channels which separate these islands, despite the still and often turbid waters which are found here. Fringing reefs are also found on the west coast of these islands, particularly in the northwest.

Socio-economic considerations

Despite the vast area of the Indonesian Archipelago and the lack of detailed information about its reef communities, the majority of its coastal area is already heavily utilized, particularly in the west, and considerable areas are under increasing stress from human activities. About 6 000 of Indonesia's islands are inhabited, and marine and coastal resources and activities generate 25 percent of the country's gross domestic product. One study along the west coast of Lombok made a detailed assessment of coral reef value, particularly looking at fisheries production, but also at tourism, mariculture, ornamental trade and other resources. The estimated value of the reefs in the area was US$5 800 per hectare. This same coastline was utilized by 7 100 fishermen and over 35 percent of their fish catch came from coral reefs.

Fisheries are a major activity, and it has been estimated that 60 percent of protein consumption is derived from fisheries. About 90 percent of all fisheries are artisanal, with products for local consumption or for sale in local markets. Unfortunately overfishing is widespread and is almost continuous in all regions from Sulawesi westwards. In addition a number of destructive fishing practices, blast and cyanide fishing amongst them, are employed in all areas, including many remote reefs and atolls. Blast fishing, in particular, is having an extremely detrimental effect across the country. Although illegal since 1985, few places have escaped it, even in protected areas. The total probable cost of this fishery to the country, in terms of long-term fishery losses and loss of tourist income, has been estimated at US$3 billion over the 20 years from 1999. Indonesia is the largest supplier of live food fish to the Asian markets with large vessels operating among the more remote reefs, and mostly using cyanide (although illegal since 1995). *Muro-ami* fishing has significant impacts in a number of areas, including Kepulauan Seribu. This involves the use of large nets and large groups of fishers, often children, who swim with poles or rocks on ropes and smash the reef surface to frighten fish up into the nets. The impacts of trawling on submerged reef systems are less well known, in part because the location and extent of these reefs is unknown.

Collection of fish and corals for export in the ornamental and aquarium trade is considerable. Indonesia is the world's largest exporter of corals under the regulations of the Convention on International Trade in Endangered Species of Wild Fauna and Flora (CITES). Reaching well over 1 000 tons of coral per year in the early 1990s and now exporting around 500 tons per year, Indonesia has provided approximately 41 percent of all coral exports worldwide since 1985. These exports are relatively low on a unit-area basis because of the very large coral reef area in the country, but they may have localized impacts.

Coastal development causes considerable problems, particularly in the western half of the country. Extensive deforestation has greatly exacerbated the natural influences of freshwater and sediment discharge on reef growth and condition, and these impacts are continually expanding to new areas. Urban and industrial pollution is widespread, entering coastal areas through rivers and discharge pipes. In 1998 it was reported that there was no sewage treatment plant in any major coastal city. Agricultural development

Left: A group of dominos Dascyllus trimaculatus *taking shelter in a large anemone under a gorgonian coral. Right: Luang and Ukenao Atolls to the east of East Timor (STS038-75-43, 1990).*

is leading to increased inputs of nutrients and chemicals, and their effects are now widely apparent. In a gradient across the Spermonde Archipelago, for example, there is a rapid decline in biodiversity and coral cover closely linked with proximity to the highly polluted coastline approaching Makassar. Coral cover at 68 kilometers distance from the town is over 65 percent, dropping to 14 percent at 1.3 kilometers. Mangroves have been widely removed, often for the development of shrimp ponds, but also for commercial woodchip or pulp production, or due to general overexploitation by growing coastal populations. Coral mining is also common, with corals being used for various purposes including building (houses, road foundations, sea walls and jetties), to lime production (for mortar), and decorative use both within the country and for export.

Tourism is now important in many areas, and is itself responsible for a range of problems, particularly associated with the developments on small coral cays. Impacts include land reclamation, dredging of lagoons and mangrove clearance. A large number of the islands in Kepulauan Seribu have been modified in this way. At the same time tourism provides an alternate income source and may lead to the reduction of fishing pressures in some locations. Although there are many protected areas in Indonesia, they do not provide a good network for the vast area of reefs, nor do they yet reach the 300 000 square kilometer goal set by the government for 2000. Most of the existing sites lack comprehensive management and, in many, their conservation value is reported to be rapidly deteriorating.

Indonesia

General Data	
Population (thousands)	224 784
GDP (million US$)	161 324
Land area (km²)	1 909 624
Marine area (thousand km²)	6 121
Per capita fish consumption (kg/year)	18
Status and Threats	
Reefs at risk (%)	82
Recorded coral diseases	0
Biodiversity	
Reef area (km²)	51 020
Coral diversity*	443 / 581-602
Mangrove area (km²)	42 550
No. of mangrove species	45
No. of seagrass species	13

* The range provided for the upper figure is due to uncertain biogeographic boundaries

Protected areas with coral reefs

Site name	Designation	Abbreviation	IUCN cat.	Size (km²)	Year
Indonesia					
Arakan Wowontulap	Nature Reserve	NR	Ia	138.00	1986
Bali Barat	National Park	NP	II	777.27	1982
Baluran	National Park	NP	II	250.00	1980
Bunaken	National Park	NP	II	890.65	1989
Dolangan	Game Reserve	GR	IV	4.63	1981
Gili Meno/Gili Air/Gili Trawangan	Recreation Park	RP	V	29.54	1993
Gunung Api Banda	Recreation Park	RP	V	7.35	1992
Karang Bolong	Nature Reserve	NR	Ia	0.01	1937
Karang Gading Langkat Timur Laut	Game Reserve	GR	IV	157.65	1980
Kepulauan Aru Tenggara	Nature Reserve	NR	Ia	1 140.00	1991
Kepulauan Banyak	Recreation Park	RP	V	2 275.00	na
Kepulauan Kapoposang	Recreation Park	RP	V	500.00	na
Kepulauan Karimata	Nature Reserve	NR	Ia	770.00	1985
Kepulauan Karimun Jawa	National Park	NP	II	1 116.25	1986
Kepulauan Padaido	Recreation Park	RP	V	1 830.00	na

Protected areas with coral reefs

Site name	Designation	Abbreviation	IUCN cat.	Size (km²)	Year
Indonesia cont.					
Kepulauan Seribu	National Park	NP	II	1 080.00	1982
Kepulauan Wakatobi	National Park	NP	II	13 900.00	na
Komodo	National Park	NP	II	1 733.00	1980
Leuwang Sancang	Nature Reserve	NR	Ia	33.07	1978
Morowali	Nature Reserve	NR	Ia	2 250.00	1986
Pananjung Pangandaran	Nature Reserve	NR	Ia	4.19	1934
Pati-Pati	Game Reserve	GR	IV	35.00	1936
Pinjam/Tanjung Mantop	Game Reserve	GR	IV	16.13	1981
Pulau Anak Krakatau	Nature Reserve	NR	Ia	250.35	1990
Pulau Besar	Recreation Park	RP	V	30.00	1986
Pulau Bunaken	Nature Reserve	NR	Ia	752.65	1986
Pulau Dua	Nature Reserve	NR	Ia	0.60	1984
Pulau Kasa	Game Reserve	GR	IV	9.00	1978
Pulau Kasa	Recreation Park	RP	V	11.00	1978
Pulau Moyo	Hunting Park	HP	VI	222.50	1986
Pulau Moyo	Recreation Park	RP	V	60.00	1986
Pulau Pombo	Nature Reserve	NR	Ia	0.02	na
Pulau Pombo	Recreation Park	RP	V	9.98	1973
Pulau Rambut	Nature Reserve	NR	Ia	0.18	1939
Pulau Sangalaki	Recreation Park	RP	V	2.80	na
Pulau Sangiang	Nature Reserve	NR	Ia	7.00	1985
Pulau Semama	Game Reserve	GR	IV	2.20	1982
Pulau Weh	Recreation Park	RP	V	39.00	1982
Sabuda Tataruga	Game Reserve	GR	IV	50.00	1993
Take Bone Rate	National Park	NP	II	5 307.65	1992
Taman Laut Banda	Recreation Park	RP	V	25.00	1977
Tanjung Amelango	Game Reserve	GR	IV	8.50	1975
Teluk Kelumpang/ Selat Laut/Selat Sebuku	Nature Reserve	NR	Ia	666.50	1981
Teluk Kupang	Recreation Park	RP	V	500.00	1993
Teluk Laut Cendrawasih	National Park	NP	II	14 535.00	1990
Teluk Maumere	Recreation Park	RP	V	594.50	1986
Tujuh Belas Pulau	Nature Reserve	NR	Ia	99.00	1987
Ujung Kulon	National Park	NP	II	1 229.56	1992
Komodo National Park	**UNESCO Biosphere Reserve**			**1 735.00**	**1977**
Komodo National Park	**World Heritage Site**			**2 193.22**	**1991**
Ujung Kulon National Park and Krakatau National Reserve	**World Heritage Site**			**1 230.51**	**1991**

Philippines

MAPS 10f and g

The Philippines are a large and complex mass of over 7 000 islands making up the north of insular Southeast Asia. Together with Indonesia to the south, the Philippines lie in the center of global coral reef biodiversity and have a vast area of reefs.

In the far north the archipelago commences with the Batanes and Babuyan Islands in the Luzon Strait, just south of Taiwan. The northern third of Luzon itself is highly mountainous and parts remain heavily forested, while the central parts are predominantly agricultural with large areas of low-lying land. Relatively close to Luzon are the islands of Mindoro and Marinduque, the former mountainous and still largely under forest. South of Luzon lies a complex mass of islands known as the Visayas, including Panay, Negros, Cebu, Bohol, Leyte and Samar, centered around the Visayan Sea which, despite the tight configuration of islands, reaches a depth of more than 200 meters in some places. The southernmost major island is Mindanao, which lies separated from the Visayas by the Bohol Sea. This is another mountainous island, with a narrow shelf on all sides. The Philippines Trench to the east of Mindanao and Samar reaches depths of more than 10 000 meters at a distance of less than 80 kilometers from shore. Stretching to the southwest from Mindanao is a chain of islands known as the Sulu Archipelago, coming close to the coastline of Sabah in Malaysia and separating the Sulawesi (Celebes) Sea in the south from the Sulu Sea to the north. There are several remote islands and atolls in the central Sulu Sea, while its northern edge is marked by the long mountainous island of Palawan as well as various smaller ones.

The eastern side of the country borders the Philippine Sea and the Pacific Ocean and is affected by the ocean currents of the Pacific. The North Equatorial Current reaches this coastline and divides, with a northward branch flowing up the coast of the Visayas and Luzon, becoming the Kuroshio Current as it flows towards Taiwan and Japan. The southward branch flows along the east coast of Mindanao as the Mindanao Current. The western side of the country, facing the South China and Sulu Seas, is more directly affected by the reversing pattern of the monsoon winds.

Fringing reefs are well developed around the Batanes and Babuyan Islands, although live coral cover on the former is reported as low (less than 25 percent). Around Luzon itself reefs are by no means continuous. There are no recorded reefs in the far northwest, and the first to appear on this coast are fringing structures around the Hundred Islands, an area in the Lingayen Gulf. The waters here may

Left: Bongo Island lies sufficiently far from the major riverine sedimentation associated with the adjacent areas of Mindanao to allow fringing reefs to develop (STS61A-40-70, 1985). Right: Banks of fire coral Millepora platyphyllia.

MAP 10f

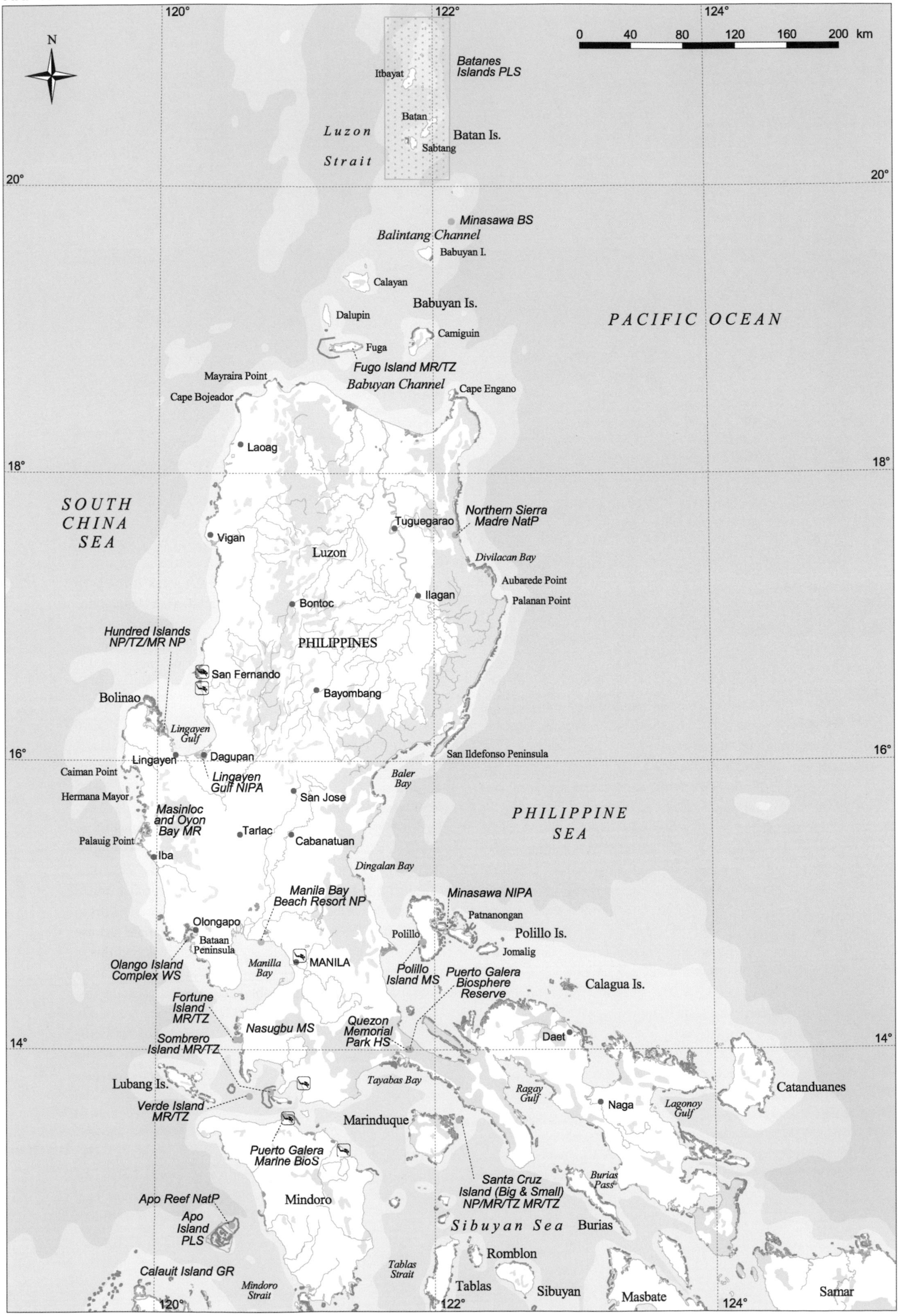

120° 122° 124°

0 40 80 120 160 200 km

N

Itbayat
Batanes Islands PLS
Batan
Luzon Strait
Batan Is.
Sabtang

20°

Minasawa BS
Balintang Channel
Babuyan I.
Calayan
Babuyan Is.
Dalupin
Camiguin
Fuga
Fugo Island MR/TZ
Babuyan Channel
PACIFIC OCEAN
Mayraira Point
Cape Bojeador
Cape Engano
Laoag

18°

SOUTH CHINA SEA
Northern Sierra Madre NatP
Tuguegarao
Vigan
Luzon
Divilacan Bay
Aubarede Point
Ilagan
Palanan Point
Bontoc
Hundred Islands NP/TZ/MR NP
PHILIPPINES
San Fernando
Bayombang
Bolinao
Lingayen Gulf

16°

Lingayen
Dagupan
San Ildefonso Peninsula
Caiman Point
Lingayen Gulf NIPA
Baler Bay
Hermana Mayor
San Jose
Masinloc and Oyon Bay MR
PHILIPPINE SEA
Tarlac
Palauig Point
Cabanatuan
Iba
Dingalan Bay
Manila Bay Beach Resort NP
Minasawa NIPA
Patnanongan
Olongapo
Polillo
Polillo Is.
Bataan Peninsula
Jomalig
Olango Island Complex WS
Manilla Bay
MANILA
Polillo Island MS
Puerto Galera Biosphere Reserve
Calagua Is.
Fortune Island MR/TZ
Nasugbu MS
Quezon Memorial Park HS
Daet

14°

Sombrero Island MR/TZ
Tayabas Bay
Catanduanes
Lubang Is.
Ragay Gulf
Naga
Lagonoy Gulf
Verde Island MR/TZ
Marinduque
Puerto Galera Marine BioS
Santa Cruz Island (Big & Small) NP/MR/TZ MR/TZ
Burias Pass
Apo Reef NatP
Mindoro
Apo Island PLS
Sibuyan Sea
Burias
Romblon
Calauit Island GR
Tablas Strait
Mindoro Strait
Tablas
Sibuyan
Masbate
Samar

be turbid and much of the reef area is reported to have been destroyed by blast fishing. At the mouth of the Lingayen Gulf there are wide fringing reefs around Bolinao and the nearby islands, with discontinuous fringing reefs running south to Manila Bay. The explosion of Mount Pinatubo, with its massive ashfall and mud flows, caused a steep decline in live coral cover from 60-70 percent down to 10-20 percent on the nearest fringing reefs. There is little information about the development of reefs along the east coast of Luzon, although fringing reefs are described at the Polillo Islands and in the Northern Sierra Madre Natural Park. There is little published information describing the reefs around the southern coastline of Luzon, and little for Mindoro and Marinduque, but there are discontinuous fringing reefs in many areas, notably around Puerto Galera in Mindoro. Over 200 kilometers west of Luzon is the atoll-like formation of Scarborough Reef.

Fringing reefs are widespread along much of the coastline of the Visayas, although broken up by areas of soft sediments, particularly close to river mouths. Live cover on some parts of these reefs can exceed 50 percent, and fish diversity is also high, particularly on protected or less heavily fished reefs such as Sumilon and Apo Islands south of Cebu and Negros. Reefs around Mindanao are poorly known, although fringing structures are widespread, and diversity is reportedly high on reefs around Arangasa Island on the east coast.

The Sulu Archipelago has not been described in detail, but includes fringing and barrier reef systems. To the northwest there are two major atoll systems in the Sulu Sea, the Cagayan Islands and Tubbataha, the latter being a structure composed of two atolls. Further west, Palawan has some of the best developed reefs in the country, with fringing and patch reefs along most of the coast and live coral cover reaching between 50 and 90 percent in some places. A number of banks and shoals off the west coast of Palawan are thought to be part of a long, sub-surface barrier reef system. Finally, due west of Palawan lies the large complex of the Spratly Islands, which are disputed between several countries, and so considered in a separate section.

Many of the reefs in this country are severely impacted by human activities. Dense populations utilize fish communities in almost all areas. The vast majority of this fishing is small-scale – coastal waters up to 15 kilometers from the shore fall under local government control and are often closed to larger commercial vessels. It has been estimated that reefs may yield up to 10-15 percent of the total annual fisheries production of the country, and studies have shown that individual reefs may support yields of between 3 and 36 tons of fish per square kilometer per year. Despite this, demersal fish stocks including reef fish, as well as small pelagics, are considered to be biologically and economically overfished in almost all areas other than eastern Luzon, Palawan and the southern Sulu Sea. Such overfishing has significant ecological impacts, including changes in community structure and decreases in diversity. In many areas it has been claimed that there are insufficient adult fish populations to allow local recruitment. Catches of demersal fish have been stable or declining since 1976, while fishing effort has remained stable or increased. It is unclear whether this is solely related to overfishing, or exacerbated by other forms of environmental degradation.

Destructive fishing is also widespread. Although blast fishing is illegal, it continues in nearly every part of the Philippines and causes significant reef loss in many areas. Prior to 1989 blasts were heard at a rate of 10 per hour in a 2-3 kilometer listening radius around Bolinao. Following the introduction of stringent punishments for this illegal activity, these rates have dropped, and there is now little or no blast fishing, but only in this one area. Cyanide fishing is also common for the live fish trade, and there is a significant illegal fishery by vessels from Taiwan, Hong Kong, Singapore, Korea and Japan. The use of cyanide by Philippine fishers is prohibited, and this is monitored for the export fishery, so the majority of legal live fish exports are probably no longer caught in this manner. Live fish are also caught in a few areas to supply a fairly large aquarium trade export, largely to the USA. *Muro-ami* fishing is another method which has been used in the Philippines. Although now illegal it almost certainly continues, while a new method, known as *paaling*, utilizes divers (typically 100 or more at one time) with hoses aiming compressed air at the reef to force the fish into the nets. This may be widespread off the coast of Palawan, and is indiscriminate and destructive to the reef. The Philippines once featured as a major coral exporter. This

Philippines

General Data	
Population (thousands)	81 160
GDP (million US$)	52 072
Land area (km2)	298 120
Marine area (thousand km2)	974
Per capita fish consumption (kg/year)	30
Status and Threats	
Reefs at risk (%)	97
Recorded coral diseases	4
Biodiversity	
Reef area (km2)	25 060
Coral diversity	421 / 577
Mangrove area (km2)	1 607
No. of mangrove species	30
No. of seagrass species	19

formerly legitimate trade has been stopped, although illegal exports may still be considerable.

Sedimentation is another major threat, and loads are high in many rivers as a result of deforestation and poor agricultural practices. Some 60-75 percent of the original mangrove cover has been removed, reducing the role these can play as nursery areas or sediment traps. Urban and industrial effluent is a particular problem in some locations, such as Manila Bay. At Toledo City, Cebu, an estimated 100 000 tons of mine tailings are discharged into the sea daily, with massive losses of fish and coral cover along 7 kilometers of coastline. Similar problems of discharge combined with poor flushing have affected Calancan Bay in Marinduque. Tourism is a growing industry in the Philippines, although diving is not as significant as in other parts of the region, possibly in part related to the degraded nature of so many reefs.

A considerable number of marine protected areas have been declared in the Philippines but few have ever been effectively enforced. Some of the larger sites have failed to win the support of local communities, while in others, the local people have been unable to control the impacts of outsiders. There are a few exceptions to this however, and the two small reserves of Apo Island and Sumilon are globally recognized as examples of good community-based management. In both cases very small no-take zones have been established and actively enforced for a number of years. This has led to increases in fish populations and average sizes, which in turn have led to export of fish from these areas to surrounding waters and an overall increase in fish yields despite the partial closure of reefs. In addition to these benefits, the islands have sold merchandise to tourist divers, and from 1999 received a fee from visiting dive vessels.

Protected areas with coral reefs

Site name	Designation	Abbreviation	IUCN cat.	Size (km²)	Year
Philippines					
Agan-an	Municipal Marine Reserve	MuMR	IV	0.06	1999
Andulay	Municipal Marine Reserve	MuMR	IV	0.06	1999
Apo Island	Protected Landscape/Seascape	PLS	V	6.91	1996
Apo Reef	Natural Park	NatP	II	116.77	1996
Basdiot	Fish Sanctuary	FiS	na	0.01	1988
Batanes	Protected Landscape/Seascape	PLS	V	2 135.78	1994
Bien Unido	Fish Reserve	FishR	na	na	1995
Bio-os	Municipal Marine Reserve	MuMR	IV	0.08	na
Bolisong	Municipal Marine Reserve	MuMR	IV	0.10	1995
Bongalonan	Municipal Marine Reserve	MuMR	IV	0.20	1993
Cabugan	Municipal Marine Reserve	MuMR	IV	0.07	1993
Cabulotan	Municipal Marine Reserve	MuMR	IV	0.06	1993
Cagayan Island	Other Area	ETC	Unassigned	na	1970
Calag-calag	Municipal Marine Reserve	MuMR	IV	0.07	1991
Cangmating	Municipal Marine Reserve	MuMR	IV	0.06	1997
Caohagan	Marine Reserve/Tourist Zone	MR/TZ	na	na	na
Carbin Reef	Municipal Park	MuP	na	2.00	1983
Danjugan Island	Private Reserve	PrivR	Unassigned	0.43	1994
El Nido	Marine Reserve	MR	Unassigned	950.00	1992
Fortune Island	Marine Reserve/Tourist Zone	MR/TZ	Unassigned	na	1978
Fugo Island	Marine Reserve/Tourist Zone	MR/TZ	Unassigned	na	1978

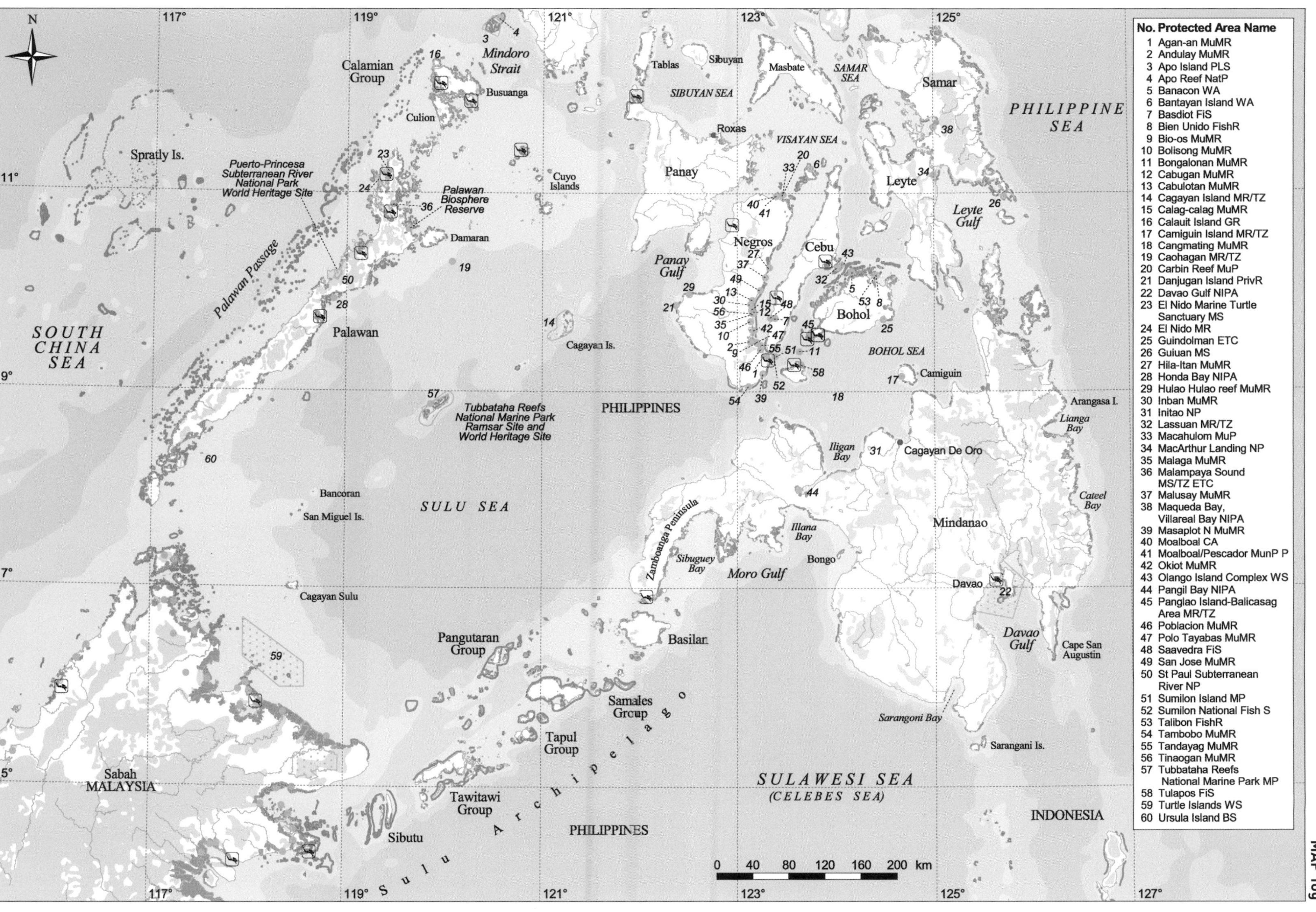
No. Protected Area Name
1 Agan-an MuMR
2 Andulay MuMR
3 Apo Island PLS
4 Apo Reef NatP
5 Banacon WA
6 Bantayan Island WA
7 Basdiot FiS
8 Bien Unido FishR
9 Bio-os MuMR
10 Bolisong MuMR
11 Bongalonan MuMR
12 Cabugan MuMR
13 Cabulotan MuMR
14 Cagayan Island MR/TZ
15 Calag-calag MuMR
16 Calauit Island GR
17 Camiguin Island MR/TZ
18 Cangmating MuMR
19 Caohagan MR/TZ
20 Carbin Reef MuP
21 Danjugan Island PrivR
22 Davao Gulf NIPA
23 El Nido Marine Turtle Sanctuary MS
24 El Nido MR
25 Guindolman ETC
26 Guiuan MS
27 Hila-Itan MuMR
28 Honda Bay NIPA
29 Hulao Hulao reef MuMR
30 Inban MuMR
31 Initao NP
32 Lassuan MR/TZ
33 Macahulom MuP
34 MacArthur Landing NP
35 Malaga MuMR
36 Malampaya Sound MS/TZ ETC
37 Malusay MuMR
38 Maqueda Bay, Villareal Bay NIPA
39 Masaplot N MuMR
40 Moalboal CA
41 Moalboal/Pescador MunP P
42 Okiot MuMR
43 Olango Island Complex WS
44 Pangil Bay NIPA
45 Panglao Island-Balicasag Area MR/TZ
46 Poblacion MuMR
47 Polo Tayabas MuMR
48 Saavedra FiS
49 San Jose MuMR
50 St Paul Subterranean River NP
51 Sumilon Island MP
52 Sumilon National Fish S
53 Talibon FishR
54 Tambobo MuMR
55 Tandayag MuMR
56 Tinaogan MuMR
57 Tubbataha Reefs National Marine Park MP
58 Tulapos FiS
59 Turtle Islands WS
60 Ursula Island BS
N
SOUTH CHINA SEA
PHILIPPINE SEA
SULU SEA
SULAWESI SEA
(CELEBES SEA)
BOHOL SEA
VISAYAN SEA
SIBUYAN SEA
SAMAR SEA
PHILIPPINES
INDONESIA
Sabah
MALAYSIA
Spratly Is.
Calamian Group
Mindoro Strait
Busuanga
Culion
Puerto-Princesa Subterranean River National Park World Heritage Site
Palawan Biosphere Reserve
Damaran
Palawan Passage
Palawan
Cuyo Islands
Cagayan Is.
Tubbataha Reefs National Marine Park Ramsar Site and World Heritage Site
Bancoran
San Miguel Is.
Cagayan Sulu
Pangutaran Group
Tapul Group
Samales Group
Tawitawi Group
Sibutu
Sulu Archipelago
Basilan
Zamboanga Peninsula
Sibuguey Bay
Moro Gulf
Illana Bay
Bongo
Iligan Bay
Cagayan De Oro
Mindanao
Davao
Davao Gulf
Cape San Augustin
Sarangani Bay
Sarangani Is.
Cateel Bay
Lianga Bay
Arangasa I.
Camiguin
Bohol
Cebu
Negros
Panay
Panay Gulf
Roxas
Tablas
Sibuyan
Masbate
Samar
Leyte
Leyte Gulf
0 40 80 120 160 200 km
117° 119° 121° 123° 125° 127°
11° 9° 7° 5°

Protected areas with coral reefs

Site name	Designation	Abbreviation	IUCN cat.	Size (km²)	Year
Philippines cont.					
Guindolman	Other Area	ETC	Unassigned	na	na
Hila-Itan	Municipal Marine Reserve	MuMR	IV	0.06	1996
Hulao Hulao Reef	Municipal Marine Reserve	MuMR	IV	na	1996
Inban	Municipal Marine Reserve	MuMR	IV	0.08	1996
Initao	National Park	NP	Unassigned	0.57	1963
Lassuan	Marine Reserve/Tourist Zone	MR/TZ	na	na	na
Macahulom	Municipal Park	MuP	na	10.00	1983
Malaga	Municipal Marine Reserve	MuMR	IV	0.08	1996
Malusay	Municipal Marine Reserve	MuMR	IV	0.06	1996
Masaplot	Municipal Marine Reserve	MuMR	IV	0.06	1997
Masinloc and Oyon Bay	Marine Reserve	MR	Ia	75.68	1994
Moalboal/Pescador	Park	P	Unassigned	na	na
Northern Sierra Madre	Natural Park	NatP	II	3 195.13	1997
Okiot	Municipal Marine Reserve	MuMR	IV	0.01	1994
Olango Island Complex	Wildlife Sanctuary	WS	Unassigned	9.20	na
Panglao Island – Balicasag Area	Marine Reserve/Tourist Zone	MR/TZ	Unassigned	na	1978
Poblacion	Municipal Marine Reserve	MuMR	IV	0.04	1994
Polo Tayabas	Municipal Marine Reserve	MuMR	IV	0.02	1995
Saavedra	Fish Sanctuary	FiS	na	0.01	1988
San Jose	Municipal Marine Reserve	MuMR	IV	0.10	1996
Sombrero Island	Marine Reserve/Tourist Zone	MR/TZ	Unassigned	na	1977
St. Paul Subterranean River	National Park	NP	II	57.53	1971
Sumilon Island	Marine Park	MP	Unassigned	0.23	1974
Sumilon National Fish	Sanctuary	S	na	0.01	1980
Talibon	Fish Reserve	FishR	na	na	1989
Tambobo	Municipal Marine Reserve	MuMR	IV	0.06	1995
Tandayag	Municipal Marine Reserve	MuMR	IV	0.06	1996
Tinaogan	Municipal Marine Reserve	MuMR	IV	0.25	1996
Tubbataha Reefs	Marine Park	MP	Unassigned	332.00	1988
Tulapos	Fish Sanctuary	FiS	na	0.14	1994
Turtle Islands	Wildlife Sanctuary	WS	VI	2 429.67	1999
OLANGO ISLAND WILDLIFE SANCTUARY	RAMSAR SITE			58.00	1994
PALAWAN BIOSPHERE RESERVE	UNESCO BIOSPHERE RESERVE			11 508.00	1990
PUERTO GALERA BIOSPHERE RESERVE	UNESCO BIOSPHERE RESERVE			235.45	1977
PUERTO PRINCESA SUBTERRANEAN RIVER NATIONAL PARK	WORLD HERITAGE SITE			202.02	1999
TUBBATAHA REEF MARINE PARK	WORLD HERITAGE SITE			332.00	1993
TUBBATAHA REEFS NATIONAL MARINE PARK	RAMSAR SITE			332.20	1999

Spratly Islands, Tung-Sha (Dongsha Qundao) Reefs and the Paracel Islands

MAPS 10g and h

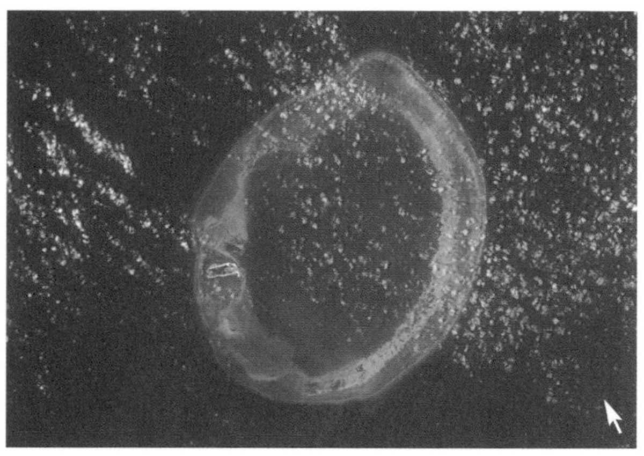

Left: A pink anemonefish *Amphiprion perideraion*. Right: Tung-Sha Atoll (STS055-92-3, 1993).

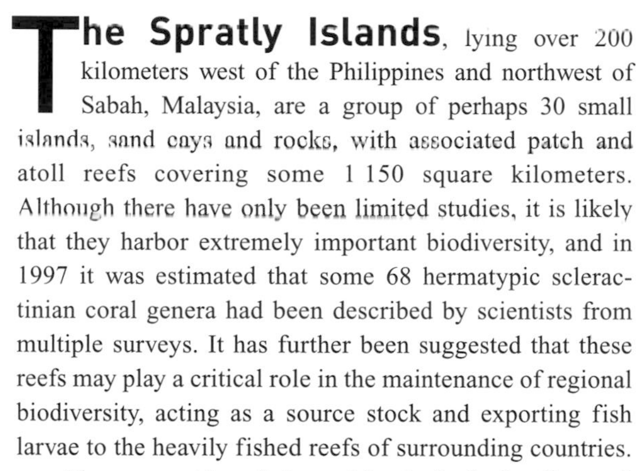

The Spratly Islands, lying over 200 kilometers west of the Philippines and northwest of Sabah, Malaysia, are a group of perhaps 30 small islands, sand cays and rocks, with associated patch and atoll reefs covering some 1 150 square kilometers. Although there have only been limited studies, it is likely that they harbor extremely important biodiversity, and in 1997 it was estimated that some 68 hermatypic scleractinian coral genera had been described by scientists from multiple surveys. It has further been suggested that these reefs may play a critical role in the maintenance of regional biodiversity, acting as a source stock and exporting fish larvae to the heavily fished reefs of surrounding countries.

The ownership of these islands is hotly disputed. China, Taiwan and Vietnam claim all of the islands and reefs, the Philippines claims most of them, while Malaysia claims a southern group and Brunei Darussalam claims one. Vigorous efforts are made by all countries to strengthen these claims and there are numerous military outposts scattered among the reefs. Over 70 people have been killed in active combat in recent years. Because of these military hazards the reefs are not heavily fished, and a significant number of Chinese and Philippine fishers have been arrested for attempting to fish the area. Some fishing does occur, however, and tends to target the largest species, notably sharks, or use explosives to make a rapid catch. Additionally the unspecified number of military personnel (possibly thousands) undoubtedly have some impact through fishing, and have caused wide-scale degradation of terrestrial systems including important seabird nesting colonies. Overall the area is probably in relatively good condition, although the risks of conflict and potential impacts on the environment remain great. One suggestion has been that the area might be declared an international marine park through the development of an agreement similar to the international Antarctic Treaty (where, again, various countries have multiple overlapping claims of sovereignty). There has been little development of such a proposal, although there have been a small number of joint studies of the reefs and islands between Malaysia, the Philippines and Vietnam.

Tung-Sha (Dongsha Qundao) Reefs

The Tung-Sha or Dongsha Qundao reefs are located in the northern reaches of the South China Sea, centered around a large, submerged atoll with a single island. Diversity is not high, although over 70 coral species have been recorded. These reefs are disputed between China and Taiwan, although Taiwan maintains a lighthouse and meteorological station on the island.

Paracel Islands

The Paracel Islands are a group of atolls, atoll complexes and platform reefs, together with some 31 small islands in the South China Sea. Sovereignty over these islands is disputed with Vietnam. A considerable diversity of corals and other groups has been observed, and hermatypic coral cover was reported to be high, particularly on the northeastern reef flats, but low on the reef slopes.

MAP 10h

N

105°
110°
115°
20°
15°
10°
5°

Kinmen NP
Hong Shu Lin NR
CHINA
Ting Kok SSSI
Kin Tsui CoP
Ma On Shan CoP
Tsim Bei Tsui SSSI
Fu Tian-Nei Ling
Ding Dao NR
Lantau North CoP
Lantau South CoP
Hoi Ha Wan SSSI
Shek O CoP
Pat Shin Leng CoP
Yim Tso Ha Egretry SSSI
Kat O Chau SpA
Plover Cove CoP
Gang Kou Hai Gui Wan NR
Da Ya Wan NR
HONG KONG
(CHINA)
Sai Kung West CoP
Sai Kung East CoP
Hok Tsui (Cape d'Aguilar) SSSI
Clear Water Bay CoP
Tung-Sha Reefs
(Dongsha Qundao)
Shanku Mangrove
Biosphere Reserve
Wei Zhou NR
Shan Kou NR
VIETNAM
Ba Mun NR
Ha Long Bay H/CS
& World Heritage Site
HANOI
Cai
Qiao
NR
Wen Lan
He NR
Cat Ba NP
Lin Gao
Jiao NR
Dongzhaigang
Ramsar Site
Dong Zhai Gang NR
Xuan Thuy NR
Xuan Thuy Natural
Wetland Reserve
Ramsar Site
LAOS
Xin Ying
Hong Shu
Lin NR
Qi Lin Cai NR
Tong Gu Ling NR
Qing Lan Gang NR
Qi Lin Cai NR
Gulf of Tonkin
Hainan
Ling Qiang
Shi Dao NR
Xincun
Sanya
Yalong Bay
Dong Dao NR
Da Zhou Dao NR
VIENTIANE
Bao Yu NR
Nan Wan NR
Shan Hu
Jiao NR
Ya Long Wan
Qing Mei Wan NR
San Ya
He NR
Da Dong
Hai NR
North Reef
Amphitrite Group
Lincoln I.
Hue
Crescent Group
Vuladore Reef
THAILAND
Da Nang
Discovery Reef
Bombay Reef
Cu Lao Cham NR
Passau Keah
Macclesfield
Bank
Triton I.
Paracel Is.
Scarborough Reef
(Huangyan Dao)
Cu Lao Re
Quang Ngai
Qui Nhon
SOUTH CHINA SEA
CAMBODIA
Nha Trang
Trident
Shoal
Nares
Bank
Templer Bank
Lys Shoal
Leslie Bank
PHNOM PENH
Subi Reef
Irving
Reef
Wood Bank
Ream NP
Ho Chi Minh
Cu Lao Hon
Western Reef
Foulteron Reef
Jackson Atoll
Phu
Quoc
NR
Binh Chau Phuoc Buu NR
Great Discovery Reef
Sand Cay
Sabina Shoal
Koh Tang
Can Gio Mangrove
Biosphere Reserve
Fiery Cross Reef
Discovery
Small Reef
Johnson
Reef
Bombay
Shoal
Vo Doi NR
Cuarteron Reef
Pearson
Reef
Maralie Reef
Spratly Is.
Palawan
PHILIPPINES
Gulf
of
Thailand
London Reefs
Alison Reef
Cornwallis South Reef
Con Dao NP
Investigator Shoal
Owen Shoal
Barque-
Canada
Reef
Mariveles
Reef
Ardasier Banks
and Reefs
Layang Layang
(Swallow Reef)
Louisa Reef
Sabah
(MALAYSIA)
BRUNEI
Natuna Besar
PENINSULAR
MALAYSIA
INDONESIA
Kalimantan
(INDONESIA)
Kepulauan
Anambas
Sarawak
(MALAYSIA)
0 100 200 300 400 500 km

Vietnam and China

MAPS 10h and i

Vietnam has an extensive coastline encompassing a great latitudinal range. In the far south this coastline is very low-lying and dominated by the Mekong River Delta and the Cau Mau Peninsula. A short section of coastline faces the Gulf of Thailand, while offshore there are a number of islands, including the relatively hilly Phu Quoc Island and associated islands to the south, as well as the islands of Nam Du and Tho Chau, the latter being located about 150 kilometers west of the mainland. Some 80 kilometers offshore from the Mekong Delta there are a number of small hilly islands called the Con Dao (Con Son) Islands. North of the Mekong Delta Vietnam's coastline becomes one of high relief, with little or no coastal plain, and the edge of the continental shelf coming quite close inshore. Further north still the coastline sweeps in to the west and the continental shelf becomes very wide indeed around the Gulf of Tonkin. There are occasional islands along the central coastline, and close to the Chinese border numerous small islands, including the Cat Ba islands and other dramatic limestone islands which rise up vertically from the waters of Ha Long Bay.

Coral reefs have not been described in detail for any locations. There are known to be reefs or coral communities around most of the offshore islands in the southwest, and on the Con Dao Islands. On the east coast, fringing reefs and coral communities have developed along the mainland, and more particularly the offshore islands around Nha Trang. The coastline along much of the Gulf of Tonkin is dominated by soft sediments and there are few reports of reef development. However, there are fringing reef communities further offshore in Ha Long Bay and the Tonkin Gulf.

Biodiversity is greatest in the south-central areas where some 277 coral species have been recorded, while to the north only 165 species are recorded. Typhoon Linda caused some damage to the corals in Con Dao Islands in 1997, and there was some additional mortality in 1998 arising from bleaching. Recovery was reported as slow in Con Dao in 2000, although other affected reefs were reported to be making a good recovery.

Fishing pressure off the southwest coast is thought to be very high, with some 7 000 fishing vessels operating from nearby and probably as many again coming in from other regions. Deforestation has been a major problem in Vietnam, largely linked to the use of defoliants during the Vietnam war. This has caused massive erosion and heavy sedimentation offshore and may be threatening reefs

A lionfish Pterois volitans, *next to the arms of a feather star.*

around Cat Ba Island. Tourism is rapidly increasing in this country, and the Ha Long Bay area receives over a million visitors a year. A small number of marine protected areas have been established, two of which include coral reefs.

Vietnam claims a number of islands in the South China Sea, including the Spratly Islands.

China

Although China has a substantial coastline facing the South China Sea there is little or no true reef development along any of it. Hainan, a large island in the mouth of the Gulf of Tongking, was once reported as having substantial fringing reef communities along parts of its southern coast, but a number of sites originally described in the 1950s were revisited in 1984 and found to have all but disappeared. Significant fringing reef structures around Shalao on the east coast and Xincun Bay in the southeast were visited in 1990 and found to be largely made up of dead coral rubble with only occasional live corals. The most extensive and diverse fringing reef communities are found in the area around Sanya where, in 1978, coral cover was reported as 50-90 percent on the East Reefs and 60 percent on the West Reef. These figures were reported to have fallen to 40-60

Vietnam	
General Data	
Population (thousands)	78 774
GDP (million US$)	10 487
Land area (km²)	327 100
Marine area (thousand km²)	396
Per capita fish consumption (kg/year)	17
Status and Threats	
Reefs at risk (%)	86
Recorded coral diseases	0
Biodiversity	
Reef area (km²)	1 270
Coral diversity	278 / 364
Mangrove area (km²)	2 525
No. of mangrove species	29
No. of seagrass species	9

China	
General Data	
Population (thousands)	1 261 832
GDP (million US$)	101 885
Land area* (km²)	9 291 000
Marine area (thousand km²)	348
Per capita fish consumption (kg/year)	na
Status and Threats	
Reefs at risk (%)	91
Recorded coral diseases	0
Biodiversity	
Reef area (km²)	1 510
Coral diversity	101 / 365
Mangrove area* (km²)	339
No. of mangrove species*	23
No. of seagrass species	na

* Including Taiwan

Large groupers Epinephelus *sp. in holding pens awaiting shipment to the restaurants of Hong Kong.*

Protected areas with coral reefs

Site name	Designation	Abbreviation	IUCN cat.	Size (km²)	Year
Vietnam					
Cat Ba	National Park	NP	II	152.00	1986
Con Dao	National Park	NP	II	150.43	1982
Ha Long Bay	**World Heritage Site**			**1 500.00**	**1994**
China					
Kat O Chau	Special Area	SpA	IV	0.24	1979
Shan Hu Jiao	Nature Reserve	NR	V	85.00	1990

percent and 30-40 percent respectively by 1990, while many species had disappeared. Similarly important and diverse communities have also been reported off the islets in Yalong Bay just southeast of Sanya. The principal threats include coral mining for construction, blast fishing and the collection of corals for handicrafts. There are now reported to be efforts to protect and manage these reefs.

A number of coral communities have been described off the coastline of Hong Kong. These do not form true reefs, and are likely to be mirrored by similar communities in other areas along this coast. They are all undoubtedly threatened by pollution, sedimentation and overfishing on this heavily populated coastline.

China claims a large number of reefs and coral islands scattered across the South China Sea, including all of the Spratly Islands, the Paracel Islands off Vietnam, and the Tung-Sha reefs towards Taiwan. As sovereignty over these areas is disputed they are dealt with separately.

Left: The Moorish idol Zanclus cornutus *is widespread on coral reefs across the Indo-Pacific and feeds primarily on sponges. Right: Mangroves like these were decimated during the Vietnam war, but large areas have now been replanted.*

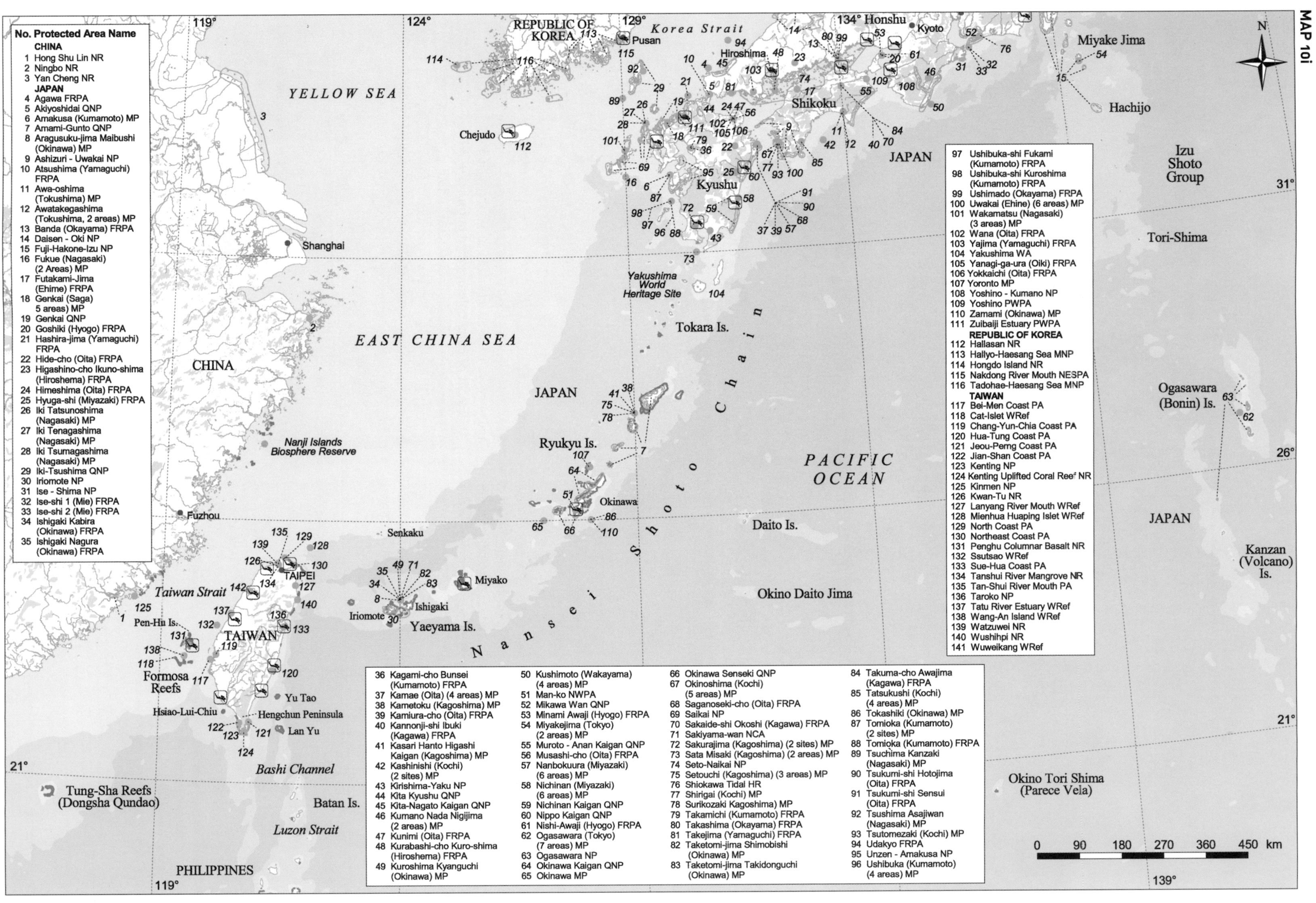

No. Protected Area Name
CHINA
1 Hong Shu Lin NR
2 Ningbo NR
3 Yan Cheng NR
JAPAN
4 Agawa FRPA
5 Akiyoshidai QNP
6 Amakusa (Kumamoto) MP
7 Amami-Gunto QNP
8 Aragusuku-jima Maibushi (Okinawa) MP
9 Ashizuri - Uwakai NP
10 Atsushima (Yamaguchi) FRPA
11 Awa-oshima (Tokushima) MP
12 Awatakegashima (Tokushima, 2 areas) MP
13 Banda (Okayama) FRPA
14 Daisen - Oki NP
15 Fuji-Hakone-Izu NP
16 Fukue (Nagasaki) (2 Areas) MP
17 Futakami-Jima (Ehime) FRPA
18 Genkai (Saga) 5 areas) MP
19 Genkai QNP
20 Goshiki (Hyogo) FRPA
21 Hashira-jima (Yamaguchi) FRPA
22 Hide-cho (Oita) FRPA
23 Higashino-cho Ikuno-shima (Hiroshema) FRPA
24 Himeshima (Oita) FRPA
25 Hyuga-shi (Miyazaki) FRPA
26 Iki Tatsunoshima (Nagasaki) MP
27 Iki Tenagashima (Nagasaki) MP
28 Iki Tsumagashima (Nagasaki) MP
29 Iki-Tsushima QNP
30 Iriomote NP
31 Ise - Shima NP
32 Ise-shi 1 (Mie) FRPA
33 Ise-shi 2 (Mie) FRPA
34 Ishigaki Kabira (Okinawa) FRPA
35 Ishigaki Nagura (Okinawa) FRPA
36 Kagami-cho Bunsei (Kumamoto) FRPA
37 Kamae (Oita) (4 areas) MP
38 Kametoku (Kagoshima) MP
39 Kamiura-cho (Oita) FRPA
40 Kannonji-shi Ibuki (Kagawa) FRPA
41 Kasari Hanto Higashi Kaigan (Kagoshima) MP
42 Kashinishi (Kochi) (2 sites) MP
43 Kirishima-Yaku NP
44 Kita Kyushu QNP
45 Kita-Nagato Kaigan QNP
46 Kumano Nada Nigijima (2 areas) MP
47 Kunimi (Oita) FRPA
48 Kurabashi-cho Kuro-shima (Hiroshema) FRPA
49 Kuroshima Kyanguchi (Okinawa) MP
50 Kushimoto (Wakayama) (4 areas) MP
51 Man-ko NWPA
52 Mikawa Wan QNP
53 Minami Awaji (Hyogo) FRPA
54 Miyakejima (Tokyo) (2 areas) MP
55 Muroto - Anan Kaigan QNP
56 Musashi-cho (Oita) FRPA
57 Nanbokuura (Miyazaki) (6 areas) MP
58 Nichinan (Miyazaki) (6 areas) MP
59 Nichinan Kaigan QNP
60 Nippo Kaigan QNP
61 Nishi-Awaji (Hyogo) FRPA
62 Ogasawara (Tokyo) (7 areas) MP
63 Ogasawara NP
64 Okinawa Kaigan QNP
65 Okinawa MP
66 Okinawa Senseki QNP
67 Okinoshima (Kochi) (5 areas) MP
68 Saganoseki-cho (Oita) FRPA
69 Saikai NP
70 Sakaide-shi Okoshi (Kagawa) FRPA
71 Sakiyama-wan NCA
72 Sakurajima (Kagoshima) (2 sites) MP
73 Sata Misaki (Kagoshima) (2 areas) MP
74 Seto-Naikai NP
75 Setouchi (Kagoshima) (3 areas) MP
76 Shiokawa Tidal HR
77 Shirigai (Kochi) MP
78 Surikozaki Kagoshima) MP
79 Takamichi (Kumamoto) FRPA
80 Takashima (Okayama) FRPA
81 Takejima (Yamaguchi) FRPA
82 Taketomi-jima Shimobishi (Okinawa) MP
83 Taketomi-jima Takidonguchi (Okinawa) MP
84 Takuma-cho Awajima (Kagawa) FRPA
85 Tatsukushi (Kochi) (4 areas) MP
86 Tokashiki (Okinawa) MP
87 Tomioka (Kumamoto) (2 sites) MP
88 Tomioka (Kumamoto) FRPA
89 Tsuchima Kanzaki (Nagasaki) MP
90 Tsukumi-shi Hotojima (Oita) FRPA
91 Tsukumi-shi Sensui (Oita) FRPA
92 Tsushima Asajiwan (Nagasaki) MP
93 Tsutomezaki (Kochi) MP
94 Udakyo FRPA
95 Unzen - Amakusa NP
96 Ushibuka (Kumamoto) (4 areas) MP
97 Ushibuka-shi Fukami (Kumamoto) FRPA
98 Ushibuka-shi Kuroshima (Kumamoto) FRPA
99 Ushimado (Okayama) FRPA
100 Uwakai (Ehine) (6 areas) MP
101 Wakamatsu (Nagasaki) (3 areas) MP
102 Wana (Oita) FRPA
103 Yajima (Yamaguchi) FRPA
104 Yakushima WA
105 Yanagi-ga-ura (Oiki) FRPA
106 Yokkaichi (Oita) FRPA
107 Yoronto MP
108 Yoshino - Kumano NP
109 Yoshino PWPA
110 Zamami (Okinawa) MP
111 Zuibaiji Estuary PWPA
REPUBLIC OF KOREA
112 Hallasan NR
113 Hallyo-Haesang Sea MNP
114 Hongdo Island NR
115 Nakdong River Mouth NESPA
116 Tadohae-Haesang Sea MNP
TAIWAN
117 Bei-Men Coast PA
118 Cat-Islet WRef
119 Chang-Yun-Chia Coast PA
120 Hua-Tung Coast PA
121 Jeou-Perng Coast PA
122 Jian-Shan Coast PA
123 Kenting NP
124 Kenting Uplifted Coral Reef NR
125 Kinmen NP
126 Kwan-Tu NR
127 Lanyang River Mouth WRef
128 Mienhua Huaping Islet WRef
129 North Coast PA
130 Northeast Coast PA
131 Penghu Columnar Basalt NR
132 Ssutsao WRef
133 Sue-Hua Coast PA
134 Tanshui River Mangrove NR
135 Tan-Shui River Mouth PA
136 Taroko NP
137 Tatu River Estuary WRef
138 Wang-An Island WRef
139 Watzuwei NR
140 Wushihpi NR
141 Wuweikang WRef
N
119°
124°
129°
134°
139°
21°
26°
31°
YELLOW SEA
EAST CHINA SEA
PACIFIC OCEAN
Korea Strait
Taiwan Strait
Bashi Channel
Luzon Strait
Nansei Shoto Chain
CHINA
JAPAN
TAIWAN
PHILIPPINES
REPUBLIC OF KOREA
Shanghai
Fuzhou
Pusan
Chejudo
Hiroshima
Kyoto
Honshu
Shikoku
Kyushu
Miyake Jima
Hachijo
Izu Shoto Group
Tori-Shima
Ogasawara (Bonin) Is.
Kanzan (Volcano) Is.
Okino Tori Shima (Parece Vela)
Daito Is.
Okino Daito Jima
Tokara Is.
Yakushima World Heritage Site
Ryukyu Is.
Okinawa
Miyako
Ishigaki
Iriomote
Yaeyama Is.
Senkaku
Nanji Islands Biosphere Reserve
TAIPEI
Pen-Hu Is.
Formosa Reefs
Hsiao-Lui-Chiu
Hengchun Peninsula
Yu Tao
Lan Yu
Batan Is.
Tung-Sha Reefs (Dongsha Qundao)
0 90 180 270 360 450 km

Taiwan and Japan

MAP 10i

Taiwan (China) lies relatively far to the north. Nonetheless, it has a number of well developed coral reef communities at the northern edge of the South China Sea, particularly along its southern edge, and around offshore islands. Taiwan is particularly affected along its southern and eastern edges by the Kuroshio Current, which carries warm water from the south, although its influence is weakened during the winter months by the Northeast Monsoon. An estimated 300 hard coral species have been recorded at the island, along with 1 200 fish species.

Some of the best known and best developed reefs of the mainland are those of the Hengchun Peninsula and the Kenting National Park. These are fringing communities, although they form a discontinuous structure broken up by sand channels. These reefs are further characterized by significant variation in the fauna between localities, with certain areas dominated by alcyonarian coral. Some 250 scleractinian corals from 58 genera have been recorded, together with 39 species (11 genera) of alcyonarian coral. Fringing reefs are also well developed around the offshore islands. In the northern parts of the South China Sea there are diverse fringing reefs around Hsiao-Liu-Chiu, while further north there are patchy coral reefs and occasional fringing reefs around the Pen-Hu (Pescadores) Islands. The islands off the east coast include Lan-Yu and Lu-Tao, both of which are volcanic. These lie in the path of the Kuroshio Current and diverse reef communities have developed. In 1998 there was extensive coral bleaching, and surveys in 1999 and 2000 have suggested that about 20 percent of coral colonies died during this event.

There are thought to be considerable pressures on the reefs in Taiwan, particularly from fishing, coastal development and tourism. Dynamite fishing, trawling and sedimentation are reported to have degraded the reefs around the Pen-Hu Islands, while destructive fishing and tourism are believed to have impacted reefs on the southeast mainland. Aquarium fish collecting and spear fishing are also reported to have affected numbers of reef fish. Nuclear power plants were reported to have been built in the vicinity of a number of reefs, while a nuclear waste disposal site was reportedly established at Lan-Yu.

Japan

The islands of Japan stretch from the edge of the tropics to the mid-temperate regions, and in so doing provide one of the clearest examples of the latitudinal limits to coral

Extensive shallow coral gardens of Acropora abrolhosensis *in the Ryukyu Islands (photo: JEN Veron).*

growth and reef development. The southernmost islands are a long chain, the Nansei Shoto, which clearly subdivides into a series of smaller archipelagos, with the Yaeyama Islands, including the important islands of Iriomote and Ishigaki in the south, followed by the Ryukyu Islands, including the island of Okinawa. Closest to the large island of Kyushu is a final group of small islands, the Tokara Islands. Following on from these the main islands of Japan, including Kyushu, Shikoku and Honshu in the south continue, with numerous offshore islands. One critical factor for reef development in these islands is the Kuroshio Current, which flows northwards along the edge of the continental shelf of the East China Sea, bringing relatively warm waters across the southern islands before passing out into the Pacific Ocean just south of Kyushu.

Away from these islands Japan also has a number of more isolated islands in the Pacific Ocean. The Daito Islands are a small group of three islands some 300 kilometers east of Okinawa. Two are raised atolls, the third a raised platform reef. Coral growth is apparently not well developed on the steeply shelving sides of these islands. South of these there is also reported to be reef development on the isolated reef of Okino Tori Shima lying on the Kyushu-Palau Ridge. Leading southwards from Tokyo there is a sequence of small island groups which follow the volcanic South Honshu Ridge. The Izu Shoto are a widely spaced group of high volcanic islands, lying relatively far north. Further south again the Ogasawara (Bonin) and Kanzan (Volcano) Islands form two groups along a volcanic arc linking Japan and the Mariana Islands to the south. Volcanic activity and a lack of suitable substrates precludes the development of reefs on many of these islands, although rich fringing communities occur in some areas. One of the most isolated reefs, even by Pacific standards, is that of Minami-Torishima (Marcus Island) an atoll lying halfway between the Ogaswara Islands and Wake Island (USA).

The presence of relatively warm waters has enabled hermatypic corals to reach quite high latitudes in Japan and there are records of around 40 coral genera at the larger islands in the north, some reaching into temperate latitudes. In these areas, however, corals are incapable of forming reefs, and it is generally accepted that the northern limit for true reef development in the Nansei chain is the Tokara Islands at around 30°N. The most extensive fringing reef structures are those around the Ryukyu Islands and further south. The remote islands to the east of the country are warmed by the Kuroshio Current and also show a high diversity of coral species. Miyake Jima (34°N) in the north of the Izu Shoto group was reported to have 80 hermatypic corals, while some 156 hermatypic species are recorded from this group as a whole. As in other high latitude reefs around the world, there appears to be considerable interaction between corals and macroalgae, with corals being overgrown during colder winters.

The Ryukyu and Yaeyama Islands have the highest levels of diversity. Altogether approximately 400 coral species have been recorded from Japan, the majority of which are found in the waters around Iriomote and Ishigaki. Coral cover is generally very low – a survey from 1990 to 1992 found that over 60 percent of coral communities in the Nansei chain had less that 5 percent coral cover while only 8 percent had cover of 50 percent or more. It seems highly likely that these low figures relate, at least in part, to the heavy environmental deg-

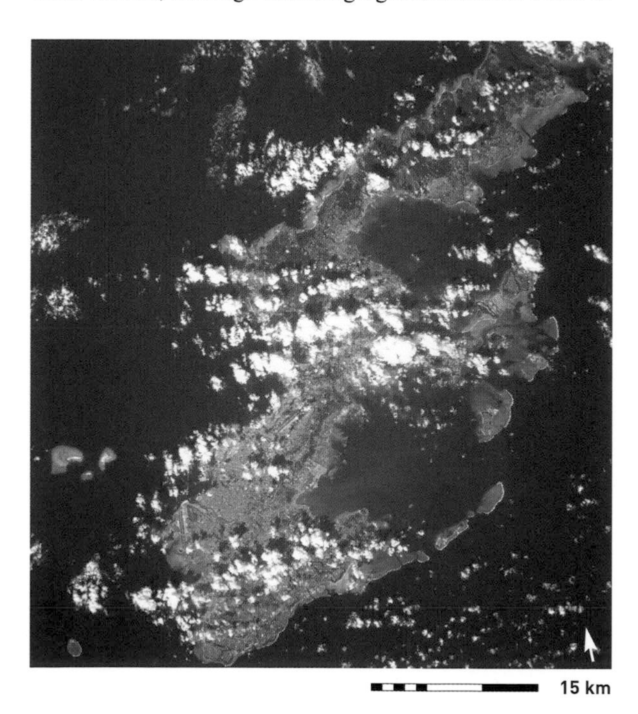

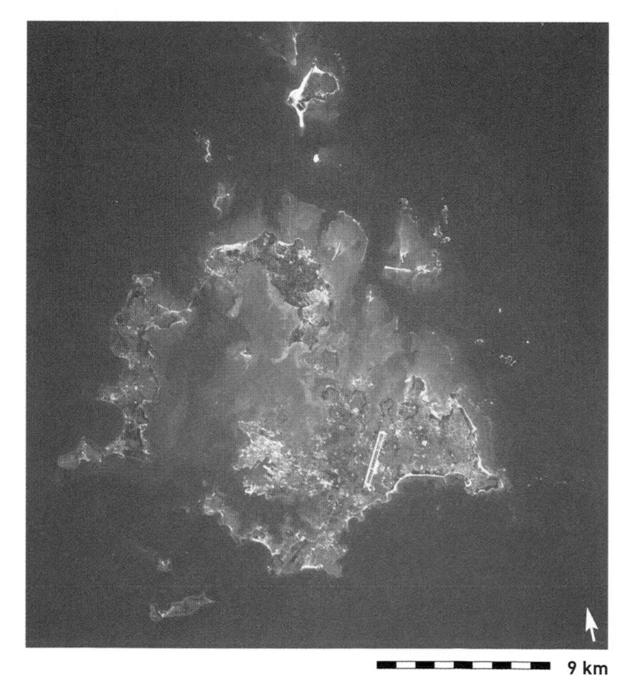

Left: Fringing reefs around Okinawa have been severely damaged or destroyed by sedimentation (STS080-755-79, 1996). Right: Taiwan's Pen-Hu Islands (STS068-239-89, 1994).

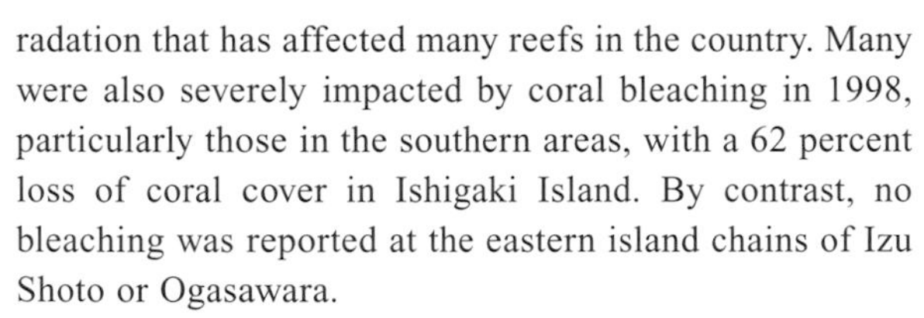

radation that has affected many reefs in the country. Many were also severely impacted by coral bleaching in 1998, particularly those in the southern areas, with a 62 percent loss of coral cover in Ishigaki Island. By contrast, no bleaching was reported at the eastern island chains of Izu Shoto or Ogasawara.

Unfortunately coastal development, forest clearance and poor agricultural practices have led to the rapid demise of many of Japan's most important fringing reefs, and a large number, particularly around the bigger islands such as Okinawa, can now be regarded as totally destroyed. Further death of corals has been caused by crown-of-thorns starfish, which have reached plague proportions in a number of areas since 1970. There is a great deal of tourism in the southern islands of Japan, with the Ryukyu Islands receiving over 4 million tourists per year in the late 1990s. Diving and snorkelling are very popular, and there is some damage caused by reef walking. Coastal development, in part fuelled by tourism, has led to direct destruction, including land reclamation for the major airport in Okinawa, and a new airport in Ishigaki, both built directly on coral reefs. Commercial fishing on the reefs is limited. In Okinawa prefecture total reef fish catch in 1993 was over 6 000 tons, but in 1998 it had dropped to 4 700 tons, a decline which may in part be linked to reef decline, but is further explained by slight decreases in the total numbers of fishers.

While there are many protected areas, estimated to cover nearly 13 percent of the total reef area of the country, it is not clear to what degree these areas provide active protection. Given that some of the greatest threats to the reefs come from external sources a more holistic approach may be required to ensure the survival of the remaining reefs.

Taiwan, China

General Data	
Population (thousands)	22 191
GDP (million US$)	na
Land area (km²)	36 349
Marine area (thousand km²)	285
Per capita fish consumption (kg/year)	na
Status and Threats	
Reefs at risk (%)	88
Recorded coral diseases	0
Biodiversity	
Reef area (km²)	940
Coral diversity	255 / 444
Mangrove area* (km²)	339
No. of mangrove species*	23
No. of seagrass species	5

* For the whole of China

Japan

General Data	
Population (thousands)	126 550
GDP (million US$)	3 300 625
Land area (km²)	373 049
Marine area (thousand km²)	4 022
Per capita fish consumption (kg/year)	67
Status and Threats	
Reefs at risk (%)	91
Recorded coral diseases	0
Biodiversity	
Reef area (km²)	2 900
Coral diversity	420 / 413
Mangrove area (km²)	4
No. of mangrove species	11
No. of seagrass species	8

Left: Blast fishing has reduced vast areas of Southeast Asia's reefs to rubble. Right: Southern Japan is the northernmost part of the range of the white mouth moray eel, Gymnothorax meleagris.

Even the more remote islands appear to have been affected by coastal development. Tourism is a major activity in the Izu Shotu Islands, including dive tourism, and two of the most diverse sites for corals on Miyake Jima have been completely destroyed by port construction. Tourism is also increasing in the Ogasawara Islands, and concerns have been expressed about the rapid increases in development which may occur following the construction of an airport.

Protected areas with coral reefs

Site name	Designation	Abbreviation	IUCN cat.	Size (km²)	Year
Taiwan					
Bei-Men Coast	Protected Area	PA	na	29.80	1987
Jeou-Perng Coast	Protected Area	PA	na	5.30	1987
Kenting	National Park	NP	II	326.31	1982
Kenting Uplifted Coral Reef	Nature Reserve	NR	Ia	1.38	1994
North Coast	Protected Area	PA	VI	56.95	1987
Japan					
Genkai	Quasi National Park	QNP	Unassigned	101.58	1956
Iriomote	National Park	NP	II	125.06	1972
Kamae (Oita) (4 areas)	Marine Park	MP	na	0.34	1974
Kametoku (Kagoshima)	Marine Park	MP	na	0.70	1974
Kasari Hanto Higashi Kaigan (Kagoshima)	Marine Park	MP	na	0.93	1974
Kirishima – Yaku	National Park	NP	II	548.33	1934
Kiyanguchi	Marine Park	MP	na	0.46	1977
Maibishi	Marine Park	MP	na	0.48	1977
Nichinan (Miyazaki) (6 areas)	Marine Park	MP	na	0.56	1970
Nichinan Kaigan	Quasi National Park	QNP	Unassigned	45.42	1955
Ogasawara	National Park	NP	II	60.99	1972
Okinawa	Marine Park	MP	na	1.40	1972
Okinawa Kaigan	Quasi National Park	QNP	Unassigned	103.20	1972
Okinawa Senseki	Quasi National Park	QNP	Unassigned	31.27	1972
Saikai	National Park	NP	V	246.36	1955
Sakiyama-wan	Nature Conservation Area	NCA	Ia	1.28	1983
Sakurajima (Kagoshima) (2 sites)	Marine Park	MP	na	0.15	1970
Sata Misaki (Kagoshima) (2 areas)	Marine Park	MP	na	0.12	1970
Setouchi (Kagoshima) (3 areas)	Marine Park	MP	na	0.58	1974
Shimobishi	Marine Park	MP	na	0.83	1977
Surikozaki (Kagoshima)	Marine Park	MP	na	0.70	1974
Takidunguchi	Marine Park	MP	na	0.37	1977
Tokashiki (Okinawa)	Marine Park	MP	na	1.20	1978
Yoronto (Kagoshima) (3 areas)	Marine Park	MP	na	1.55	1974
Yoshino – Kumano	National Park	NP	V	597.98	1936
Zamami (Okinawa)	Marine Park	MP	na	2.33	1978

Selected bibliography

REGIONAL SOURCES

Barber CV, Pratt VR (1997). *Sullied Seas: Strategies for Combating Cyanide Fishing in Southeast Asia and Beyond.* World Resources Institute and International Marinelife Alliance, Washington DC, USA.

Barber CV, Pratt VR (1998). Policy reform and community-based programs to combat cyanide fishing in the Asia-Pacific region. In: Hatziolos M, Hooten AJ, Fodor M (eds). *Coral Reefs: Challenges and Opportunities for Sustainable Management.* The World Bank, Washington DC, USA.

Benzie JAH (1998). Genetic structure of marine organisms and SE Asian biogeography. In: Hall R, Holloway JD (eds). *Biogeography and Geological Evolution of SE Asia.* Backhuys Publishers, Leiden, Netherlands.

Chou LM (2000). Southeast Asia reefs – status update: Cambodia, Indonesia, Malaysia, Philippines, Singapore, Thailand and Vietnam. In: Wilkinson CR (ed). *Status of Coral Reefs of the World: 2000.* Australian Institute of Marine Science, Cape Ferguson, Australia.

Clough BF (ed) (1993). *Mangrove Ecosystems Technical Reports, 1: The Economic and Environmental Values of Mangrove Forests and their Present State of Conservation in the South-East Asia/Pacific Region.* International Society for Mangrove Ecosystems, Okinawa, Japan, ITTO/ISME/JIAM Project PD71 / 89 Rev.1.

Fujiwara S, Shibuno T, Mito K, Nakai T, Sasaki Y, Dai C-F, Gang C (2000). Status of coral reefs of east and north Asia: China, Japan and Taiwan. In: Wilkinson CR (ed). *Status of Coral Reefs of the World: 2000.* Australian Institute of Marine Science, Cape Ferguson, Australia.

MacKinnon N (1998). Destructive fishing practices in the Asia-Pacific region. In: Hatziolos M, Hooten AJ, Fodor M (eds). *Coral Reefs: Challenges and Opportunities for Sustainable Management.* The World Bank, Washington DC, USA.

McManus JW, Cabanban AS (1992). Coral reef recruitment studies in Southeast Asia: background and implications. In: *Proceedings, Workshop on Coral and Fish Recruitment. Report No. 7, ASEAN-Australia Living Coastal Resources Project, 1-8 June 1992.* Bolinao Marine Laboratory, Bolinao, Philippines.

Silvestre GT, Pauly D (1997). *ICLARM Conference Proceedings, 53: Status and Management of Tropical Coastal Fisheries in Asia.* International Center for Living Aquatic Resources Management, Manila, Philippines.

Sudara S, Wilkinson CR, Chou LM (eds) (1994). *Proceedings, Third ASEAN-Australia Symposium on Living Coastal Resources. Volume 2: Research Papers.* Australian Institute of Marine Science, Townsville, Australia.

Thia-Eng C, Pauly D (eds) (1989). *Coastal Area Management in Southeast Asia: Policies, Management Strategies and Case Studies.* Ministry of Science, Technology and the Environment, Malaysia and International Center for Living Aquatic Resources Management, Manila, Philippines.

Wilkinson CR (ed) (1994). *Living Coastal Resources of Southeast Asia: Status and Management.* Australian Institute of Marine Science, Townsville, Australia.

Wilkinson CR, Sudara S, Chou LM (eds) (1994). *Proceedings, Third ASEAN-Australia Symposium on Living Coastal Resources. Volume 1: Status Reviews.* Australian Institute of Marine Science, Townsville, Australia.

THAILAND, MYANMAR AND CAMBODIA

Eiamsa-Ard M, Amornchairojkul S (1997). The marine fisheries of Thailand, with emphasis on the Gulf of Thailand trawl fishery. In: Silvestre G, Pauly D (eds). *ICLARM Conference Proceedings, 53: Status and Management of Tropical Coastal Fisheries in Asia.* International Center for Living Aquatic Resources Management, Manila, Philippines.

Juntarashote K, Suvanachai P (1999). *A Summary of Coastal Zone Management in the Gulf of Thailand.* Workshop paper, Coastal Zone Management Workshop 1999, University of British Colombia, Canada. http:// www.ire.ubc.ca/czm/gulfthailand.html

Piprell C, Boyd AJ (1994). *Thailand's Coral Reefs. Nature under Threat.* White Lotus, Bangkok.

Sudara S, Nateekarnchanalap S (1988). Impact of tourism development on the reef in Thailand. *Proc 8th Int Coral Reef Symp* 2: 273-278.

Sudara S, Yeemin T (1997). Coral reefs in Thai waters: newest tourist attraction. In: Wilkinson CR, Sudara S, Chou LM (eds). *Proceedings, Third ASEAN-Australia Symposium on Living Coastal Resources. Volume 1: Status Reviews.* Australian Institute of Marine Science, Townsville, Australia.

INDONESIA

Bak RPM, Povel GDE (1989). Ecological variables, including physiognomic-structural attributes, and classification of Indonesian coral reefs. *Neth J Sea Res* 23: 95-106.

Borel Best M, Djohani RH, Noor A, Reksodihardjo G (1992). Coastal marine management programs in Indonesia: components for effective marine conservation. *Proc 7th Int Coral Reef Symp* 2: 1001-1006.

Cesar H (1996). *Economic Analysis of Indonesian Coral Reefs.* The World Bank, Washington DC, USA.

Djohani R (1998). Abatement of destructive fishing practices in Indonesia: who will pay? In: Hatziolos M, Hooten AJ, Fodor M (eds). *Coral Reefs: Challenges and Opportunities for Sustainable Management.* The World Bank, Washington DC, USA.

Hopley D, Suharsono (2000). *The Status of Coral Reefs in Eastern Indonesia.* Australian Institute of Marine Science, Cape Ferguson, Australia.

Priyono BE, Sumiono B (1997). The marine fisheries of Indonesia, with emphasis on the coastal demersal stocks of the Sunda Shelf. In: Silvestre G, Pauly D (eds). *ICLARM Conference Proceedings, 53: Status and Management of Tropical Coastal Fisheries in Asia.* International Center for Living Aquatic Resources Management, Manila, Philippines.

Soekarno (1997). The status of coral reefs in Indonesia. In: Wilkinson CR, Sudara S, Chou LM (eds). *Proceedings, Third ASEAN-Australia Symposium on Living Coastal Resources. Volume 1: Status Reviews.* Australian Institute of Marine Science, Townsville, Australia.

Tomascik T, Mah AJ, Nontji A, Moosa MK (1997a). *The Ecology of Indonesia, VII: The Ecology of the Indonesian Seas. Part One.* Oxford University Press, Oxford, UK.

Tomascik T, Mah AJ, Nontji A, Moosa MK (1997b). *The Ecology of Indonesia, VIII: The Ecology of the Indonesian Seas. Part Two.* Oxford University Press, Oxford, UK.

MALAYSIA, SINGAPORE AND BRUNEI DARUSSALAM

Abu Talib A, Alias M (1997). Status of fisheries in Malaysia – an overview. In: Silvestre G, Pauly D (eds). *ICLARM Conference Proceedings, 53: Status and Management of Tropical Coastal Fisheries in Asia.* International Center for Living Aquatic Resources Management, Manila, Philippines.

Chia LS, Khan H, Chou LM (1988). *ICLARM Technical Reports, 21: The Coastal Environmental Profile of Singapore.* International Center for Living Aquatic Resources Management, Manila, Philippines.

Chou LM (1990). Assessing the coastal living resources of Singapore – a study under the ASEAN-Australia Coastal Living Resources Project. *Wallaceana* 59-60: 7-9.

Chou LM, Low JKY, Loo MGK (1997). The state of coral reefs and coral reef research in Singapore. In: Wilkinson CR, Sudara S, Chou LM (eds). *Proceedings, Third ASEAN-Australia Symposium on Living Coastal Resources. Volume 1: Status Reviews.* Australian Institute of Marine Science, Townsville, Australia.

De Silva MWRN, Wright RAD, Matdanan HJH, Sharifuddin PHY, Agbayani CV (1992). *Coastal Environmental Sensitivity Mapping of Brunei Darussalam.* Department of Fisheries, Ministry of Industry and Primary Resources and Brunei Shell Petroleum Company, Brunei Darussalam.

DOF-MIPR (1992). *ICLARM Technical Reports, 29: The Integrated Management Plan for the Coastal Zone of Brunei Darussalam.* Department of Fisheries, Ministry of Industry and Primary Resources and International Center for Living Aquatic Resources Management, Manila, Philippines.

Rajasuriaya A, De Silva MWRN, Zainin AH (1992). *Survey of Coral Reefs of Brunei Darussalam in Relation to Their Vulnerability to Oil Spills.* Department of Fisheries, Ministry of Industry and Primary Resources, Brunei Darussalam.

Ridzwan AR (1997). The status of coral reefs in Malaysia. In: Wilkinson CR, Sudara S, Chou LM (eds). *Proceedings, Third ASEAN-Australia Symposium on Living Coastal Resources. Volume 1: Status Reviews.* Australian Institute of Marine Science, Townsville, Australia.

PHILIPPINES

Barut NC, Santos MD, Garces LR (1997). Overview of Philippine marine fisheries. In: Silvestre G, Pauly D (eds). *ICLARM Conference Proceedings, 53: Status and Management of Tropical Coastal Fisheries in Asia.* International Center for Living Aquatic Resources Management, Manila, Philippines.

Gomez ED (1997). Reef management in developing countries: the Philippines as a case study. *Proc 8th Int Coral Reef Symp* 1: 123-128.

Gomez ED, Aliño PM, Licuanan WRY, Yap HT (1997). Status report of coral reefs of the Philippines 1994. In: Wilkinson CR, Sudara S, Chou LM (eds). *Proceedings, Third ASEAN-Australia Symposium on Living Coastal Resources. Volume 1: Status Reviews.* Australian Institute of Marine Science, Townsville, Australia.

Hodgson G (1994). Sedimentation damage to coral reefs. In: Ginsburg RN (ed). *Proceedings of the Colloquium on Global Aspects of Coral Reefs: Health, Hazards and History.* University of Miami, Miami, Florida, USA.

Russ GR, Alcala AC (1996). Do marine reserves export adult fish biomass? Evidence from Apo Island, central Philippines. *Mar Ecol Prog Ser* 132: 1-9.

Russ GR, Alcala AC, Cabanban AS (1992). Marine reserves and fisheries management on coral reefs with preliminary modelling of the effects on yield per recruit. *Proc 7th Int Coral Reef Symp* 2: 978-985.

SPRATLY ISLANDS, TUNG-SHA (DONGSHA QUNDAO) REEFS AND THE PARACEL ISLANDS

McManus JW (1994). The Spratly Islands: a marine park? *Ambio* 23: 181-186.

Vo Si Tuan, Nguyen Huy Yet, Aliño PM (1997). Coral and coral reefs in the north of Spratly Archipelago – the results of RP-VN JOMSRE-SCS 1996. *Proc Sci Conf on RP-VN JOMSRE SCS '96.* Ha Noi, Vietnam.

VIETNAM AND CHINA

Fiege D, Neumann V, Li J (1994). Observation on coral reefs of Hainan Island, South China Sea. *Mar Poll Bul* 29: 84-89.

Guozhong W, Bingquan L, Songqing Q (1994). On the severe changes in the ecology and sedimentation of Luweitou fringing coral reefs, Hainan Island, China. In: Ginsburg RN (ed). *Proceedings of the Colloquium on Global Aspects of Coral Reefs: Health, Hazards and History, 1993.* Rosenstiel School of Marine and Atmospheric Science, University of Miami, USA.

Latypov YY (1995). Community structure of scleractinian reefs in the Baitylong Archipelago (South China Sea). *Asian Mar Biol* 13: 27-37.

Latypov YY, Malyutin AN (1996). Structure of coral communities on the eastern part of Baitylong Archipelago, South China Sea. *Asian Mar Biol* 13: 15-24.

Vo Si Tuan (1998). Hermatypic Scleractinia of South Vietnam. *Proc 3rd Int Conf on Marine Biology of Hong Kong and South China Sea.* Hong Kong University Press, Hong Kong. 11-20.

Vo Si Tuan, Hodgson G (1997). Coral reef of Vietnam: recruitment limitation and physical forcing. *Proc 8th Int Coral Reef Symp* 1: 477-482.

TAIWAN AND JAPAN

Dai C-F (1988). Coral communities of southern Taiwan. *Proc 6th Int Coral Reef Symp* 2: 647-652.

Fujiwara S (1997). *Coral Reefs in Japan.* Marine Parks Center of Japan, Tokyo, Japan.

Fujiwara S, Shibuno T, Mito K, Nakai T, Sasaki Y, Dai C-F, Gang C (2000). Status of coral reefs of east and north Asia: China, Japan and Taiwan. In: Wilkinson CR (ed). *Status of Coral Reefs of the World: 2000.* Australian Institute of Marine Science, Cape Ferguson, Australia.

Map sources

Map 10a

For Myanmar, coral reef data for the Mergui Archipelago are taken from Hydrographic Office (1975). Sources for this data include hydrographic surveys undertaken in 1877-1914 and 1930-1939. Only areas marked as coral reefs were included, while it is likely that a number of other submerged rocks and pinnacles are also reef structures. Outside of this area, additional coral reef data have been taken as arcs from Petroconsultants SA (1990)*.

Coral reefs of Thailand are taken from Chansang et al (1999a and b), which include maps for the entire coastline at 1:10 000. These were free-drawn onto World Vector Shoreline (1:250 000). All coral structures were included (fringing, large communities on rocks, small coral communities and patch reefs).

Chansang H, Satapoomin U, Poovachiranon S (eds) (1999a). *Coral Reef Maps of Thailand. Volume 1: Gulf of Thailand.* Coral Reef Management Programme, Department of Fisheries. (In Thai language.)

Chansang H, Satapoomin U, Poovachiranon S (eds) (1999b). *Coral Reef Maps of Thailand. Volume 2: Andaman Sea.* Coral Reef Management Programme, Department of Fisheries. (In Thai language.)

Hydrographic Office (1975). Mergui Archipelago. *British Admiralty Chart No. 216.* 1:300 000. September 1975. Taunton, UK.

Maps 10b, 10c, 10d, 10e

Coral reef data are largely based on arcs from Petroconsultants SA (1990)*. Some additional areas were added by experts during the Regional Reefs at Risk workshop in Manila, 2000. For the Riau Archipelago high resolution data were generously provided through the Regional Reefs at Risk project. These are based on Landsat TM at a working scale of 1:30 000.

Maps 10b and 10c

For large parts of Malaysia, coral reef data have been taken as arcs from Petroconsultants SA (1990)*. For Sabah higher resolution reef polygons were kindly made available through the Regional Reefs at Risk Southeast Asia project, which plot reefs at a scale of 1:200 000. For Brunei coral reef data have been taken from De Silva et al (1992).

De Silva MWRN, Wright RAD, Matdanan HJH, Sharifuddin PHY, Agbayani CV (1992). *Coastal Environmental Sensitivity Mapping of Brunei Darussalam.* Department of Fisheries, Ministry of Industry and Primary Resources and Brunei Shell Petroleum Company, Brunei Darussalam.

Maps 10f and 10g

The coral reef map for the Philippines is largely based on two sources: processed satellite imagery kindly provided by the National Mapping and Resource Information Authority (NAMRIA) and further details from Petroconsultants SA (1990)*. The former data were prepared from SPOT images taken in 1987, at a scale of 1:250 000. Unfortunately expert analysis showed significant reef areas were missing from this analysis, so gaps were filled with the lower resolution Petroconsultants SA data.

In addition, the sub-surface barrier reef system on the western coast of Palawan was mapped based on Hydrographic Office (1985). This data is based on Philippine Government charts to 1976, Admiralty surveys of 1850-54 and US surveys to 1937. As these are sub-surface reefs, they have not been included in the reef area calculations.

Hydrographic Office (1985). South China Sea – Palawan. *British Admiralty Chart No. 967.* 1:725 000. November 1985. Taunton, UK.

NAMRIA (1988). *Land Cover Maps, 1:250 000.* National Mapping and Resources Information Authority, Manila, Republic of the Philippines.

Map 10h

Data showing reef areas in the Spratly Islands were prepared by staff at the University of the Philippines Marine Sciences Institute, using source materials at a scale of 1:250 000. It is possible to distinguish between surface and sub-surface reefs in this dataset, and only the former have been used in the calculation of reef areas.

Map 10i

For Japan, coral reefs and mangrove areas for all the southern islands are taken from Environment Agency (1981-1987). For Taiwan, Petroconsultants SA (1990)* was used as a base map, but further reefs were added based on simple annotation provided by Cheng-feng Dai (Professor, Institute of Oceanography, Taiwan).

For China, coral reef data have been taken as arcs from Petroconsultants SA (1990)*. For Vietnam, coral reef data were provided on a series of hand annotated base maps (various scales between approximately 1:100 000 and 1:750 000), based on the expert knowledge of Vo Si Tuan (Head, Department of Marine Living Resources, Institute of Oceanography, Vietnam). Gaps in this map coverage were filled with data from Petroconsultants SA (1990).

Environment Agency (1981-1987). *Actual Vegetation Map, Okinawa, 1-29.* 1:50 000. The 3rd National Survey on the Natural Environment (Vegetation). Environment Agency, Japan. (29-map series on 26 sheets).

* See Technical notes, page 401

Part IV

The Pacific Ocean

This is a vast region, incorporating the eastern shores of Australia and stretching towards the coastline of the Americas. There are more coral reefs here than in any other part of the world: over 40 percent of the global total, including the most extensive areas of barrier reefs and coral atolls. It was by looking at this region that Darwin developed many of his ideas on the development of reefs.

Although a contiguous ocean, the underlying plate tectonics are a little more complex. Much of the region lies on the main Pacific plate, however the Indo-Australian plate stretches far into the southwest. The boundary between these two plates, and the associated tectonic activity, has driven the development of a number of the island groups including Tonga and Fiji. Further north a similar boundary occurs between the Philippines plate and the Pacific plate. This is the location of the Mariana Islands, but also the deepest point in all the world's oceans – the Challenger Deep at 11 034 meters. Away from plate margins, a number of island groups have formed over mid-plate hotspots. The movement of the ocean crust over these hotspots has led to classic island chains such as Hawai'i, with active volcanoes at one end, and a subsequent line of islands illustrating atoll development as the central volcanic islands subside but the reef structures continue to grow.

Given the size of this ocean, it is dominated by a relatively simple system of ocean currents which are broadly consistent throughout the year. From the southern hemisphere up to the equator there is a westward flowing South Equatorial Current. Immediately north of this is the Equatorial Counter Current, flowing to the east, typically between 3-5° and 10°N. To the north of this, the North Equatorial Current again flows westwards. South of the equator, air movements are dominated by the southeast trade winds which flow from the sub-tropical high pressure areas towards the equator, and tend to be particularly strong from around June to October. The northeast trade winds dominate to the north of the equator, and are particularly strong from November to May. Along the equator itself there are easterly winds. These are strongest in the Eastern Pacific, and tend to be generally light to variable further west.

Tropical storms or cyclones are a regular disturbance in areas away from the equator, and become more numerous in the western parts of the Pacific. The general westward flow of the surface waters, which are heated as they flow, sets up a number of gradients, including an important pressure gradient. Occasionally this system undergoes reversal in a process known as the El Niño Southern Oscillation (an "El Niño event"). Such processes are typified by warm water upwellings in the Eastern Pacific and considerable changes in the "normal" patterns of currents and upwellings across the region and even around the globe. The impacts of such events on coral reefs can be considerable, as witnessed by the mass bleaching events of recent years (see Chapter 2).

In terms of biodiversity this is a very important, though still little studied, region. In the far west it encompasses the edges of the Indonesian-Philippine center of coral diversity, and there is evidence to suggest that biodiversity on reefs in Papua New Guinea may be at least as high as in these countries. Moving east across the region there is a clear gradient of diminishing diversity which appears to be reflected in all of the major groups of coral reef organisms, as well as in mangroves and seagrasses. There are some 45 mangrove species recorded from Australia and Papua New Guinea, but only three from Samoa, with none occurring east of Samoa. Knowledge of the reefs in this region is still extremely limited. Australia's Great Barrier Reef has been extensively studied, but its vast size means that, even here, many reefs are only occasionally visited by scientists. French Polynesia is another relatively well studied territory, but it has been estimated that only about half of the reef systems have even been visited by scientists, and there is published material on less than a quarter of them.

This was one of the last regions on Earth to have been settled by humans. While there is some evidence of early arrivals in Papua New Guinea and the nearby islands up to 30 000 years ago, the movement of peoples out to oceanic islands is largely the result of more recent journeys. The great Polynesian voyages probably began some 3 000-4 000 years ago, and continued until about 1 000 years ago. European arrival in the region had considerable impacts on the native people. In many islands, "new" diseases decimated populations, such that, although there are now high growth rates, many nations still have lower populations than existed before European arrival.

In the *Reefs at Risk* analysis this region was assessed as being one of the least threatened in the world. Population densities are generally low, and there are large areas of coral reefs which are far from any human populations. Despite this, human reliance on the coral reefs of the region is considerable. For many of the small island nations they are a critical source of food, as well as offering protection from storms. Many areas, and some entire nations, are comprised solely of small atoll cays, entirely the product of reef development, and only a few meters above sea level at their highest point.

Western-style development is limited in many countries, and wide areas of reef still fall under some form of customary marine tenure. Artisanal fishing predominates in coastal waters, and traditional systems for controlling this fishing often include relatively complex and effective management regimes (see Chapter 2)

Environmental problems do occur in some areas. There is evidence of target species overfishing in many countries, and populations of clams and trochus have collapsed in several nations, even before export fisheries for these species had begun. Modern fishing methods have allowed access to more remote reefs, and more thorough harvesting. As traditional systems break down, some areas have seen considerable overexploitation, and also destructive fishing. These problems do not affect wide areas, but are important, particularly because they are focussed close to high population densities and are clearly diminishing the potential of the reefs as a renewable source of food.

Tourism is important for the economy of many countries, and is almost entirely focussed on the coastal zone, with diving and snorkelling being key activities. In general, however, tourist numbers remain low compared to other parts of the world, and they are usually restricted to those islands with more developed infrastructures.

Pollution and sedimentation are generally not widespread, but are clearly a concern in localized areas, especially where there is urban development. On the high islands sediment runoff and pollution from agriculture and mining can be a problem.

The transition from traditional to Western society presents a number of interesting difficulties. The desire to establish legally designated marine protected areas has run into conflict with local "owners" of reef resources. In many countries this has prevented the establishment of Western-style protected area systems. While traditional management regimes remain effective this is unproblematic, but as or when such systems are undermined there is considerable potential for overexploitation and damage.

Chapter 11
Australia

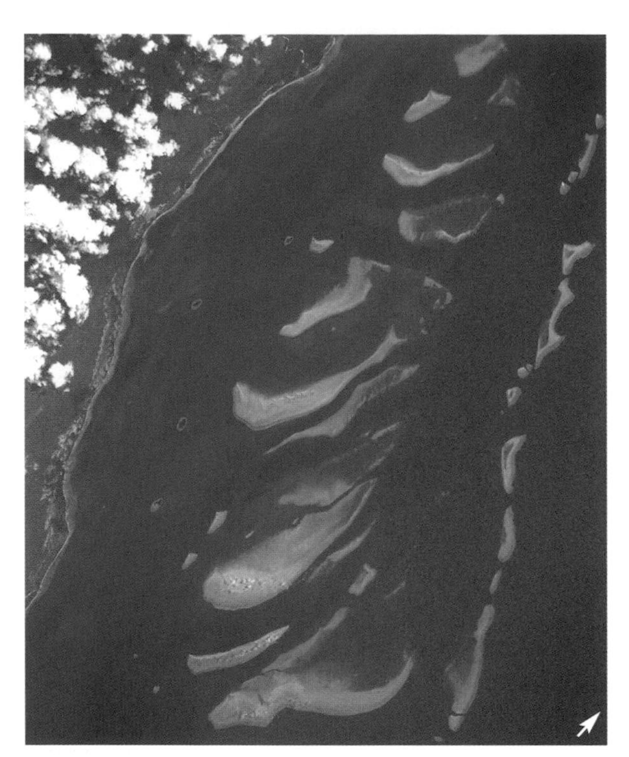

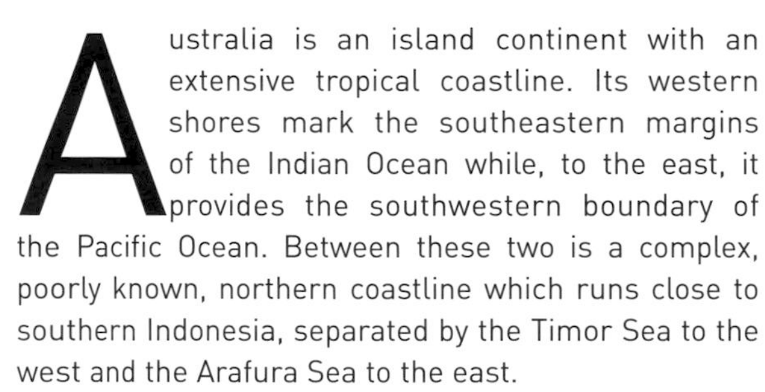

Australia is an island continent with an extensive tropical coastline. Its western shores mark the southeastern margins of the Indian Ocean while, to the east, it provides the southwestern boundary of the Pacific Ocean. Between these two is a complex, poorly known, northern coastline which runs close to southern Indonesia, separated by the Timor Sea to the west and the Arafura Sea to the east.

After Indonesia, Australia has the largest area of coral reefs of any nation, nearly 50 000 square kilometers, or some 17 percent of the world's total area of reefs. Conditions for reef development vary considerably along the coastline. In the far west the climate is dry and there is little terrestrial runoff. Reef development is not continuous, though away from loose coastal sediments there are important areas, including Australia's best developed fringing reefs. The southward flowing Leeuwin Current is also important on this coastline, bringing warm waters to relatively high latitudes and enabling the development of some unique reef communities. Further north there are several reefs on the outer edges of the continental shelf. These include remnants of what may have been a substantial barrier reef structure drowned as a result of rising sea levels over geological time scales. The northern coastline is less known, however this is an area of high terrestrial runoff, and the waters are shallow and turbid, greatly restricting reef development. The eastern boundary of the Arafura Sea is marked by a narrow constriction, the Torres Strait. East of here, the world's largest coral reef complex commences, extending out to the margins of the continental shelf and continuing southwards as the Great Barrier Reef. The warm, southward flowing East Australia Current also supports the development of high latitude reefs along Australia's eastern shores to the south of the Great Barrier Reef. Other reefs are found in Australia's offshore waters. Most notable among these are the extensive reef structures of the Coral Sea, east of the Great Barrier Reef.

Australia also administers the Cocos (Keeling) Islands and Christmas Island in the Indian Ocean, both of which have significant coral reefs.

Australia's original human population, the Australian Aborigines, are thought to have inhabited

Left: The reefs of the northern Great Barrier Reef where the continental shelf is relatively narrow (STS046-77-31, 1992).
Right: The blue starfish Linkia laevigata *is widespread on coral reefs across the Indo-Pacific.*

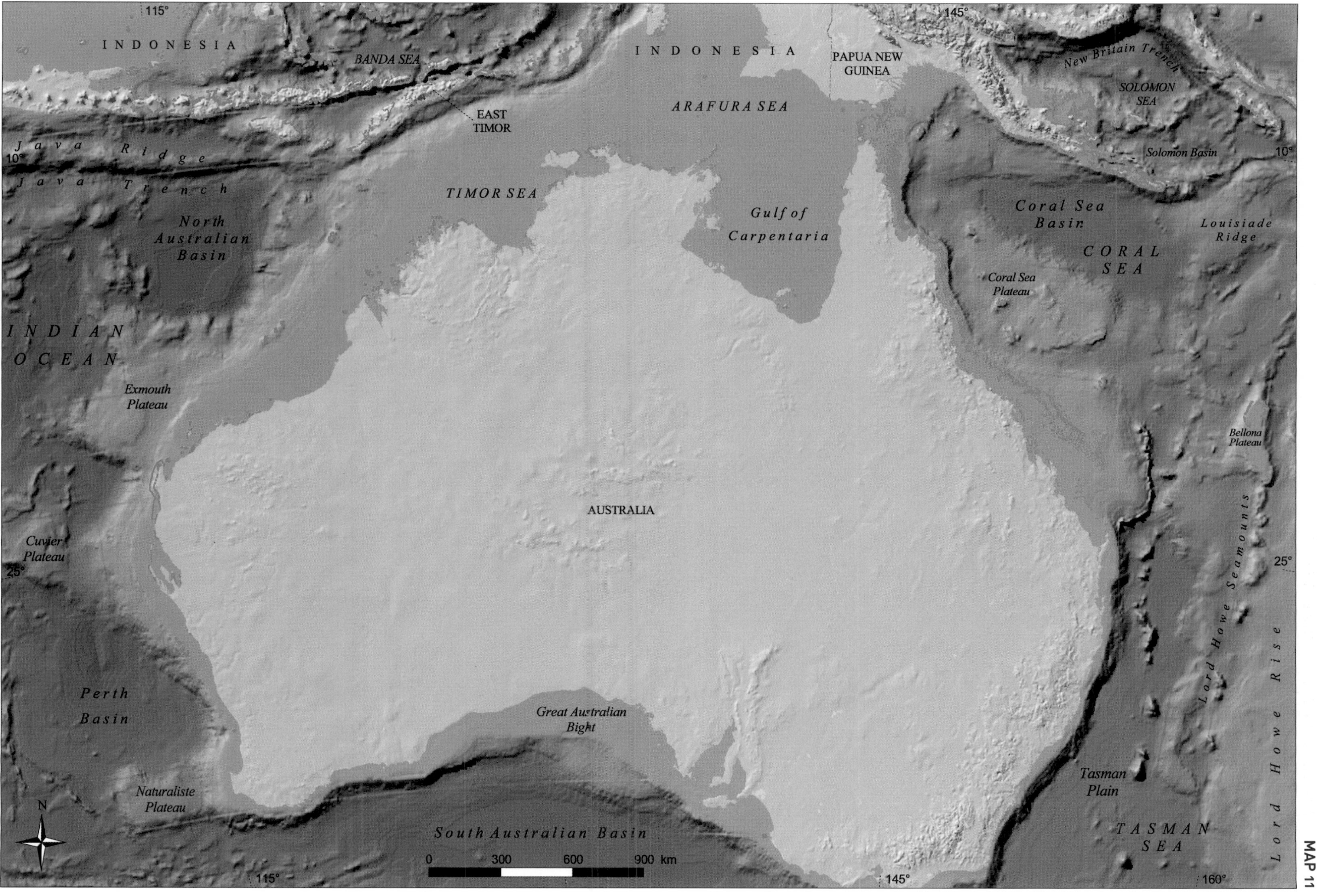

115°
INDONESIA
BANDA SEA
EAST TIMOR
INDONESIA
ARAFURA SEA
PAPUA NEW GUINEA
145°
New Britain Trench
SOLOMON SEA
Solomon Basin
10°
Java Ridge
10°
Java Trench
TIMOR SEA
Gulf of Carpentaria
North Australian Basin
Coral Sea Basin
Louisiade Ridge
CORAL SEA
Coral Sea Plateau
INDIAN OCEAN
Exmouth Plateau
Bellona Plateau
AUSTRALIA
Cuvier Plateau
25°
Lord Howe Seamounts
25°
Lord Howe Rise
Perth Basin
Great Australian Bight
Tasman Plain
Naturaliste Plateau
N
South Australian Basin
TASMAN SEA
0
300
600
900 km
115°
145°
160°

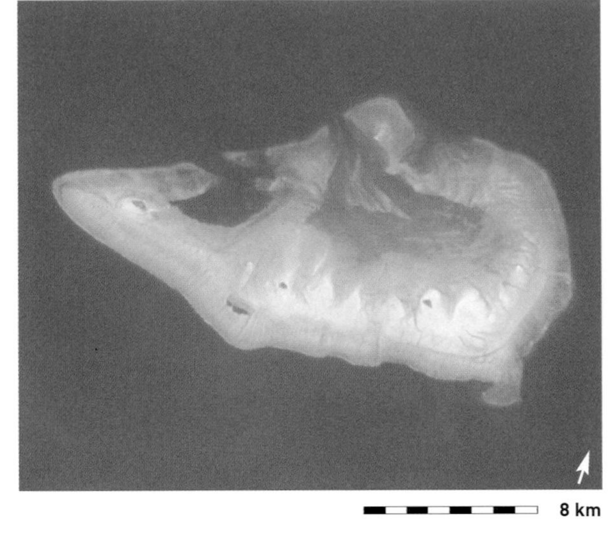

the country for more than 40 000 years. These people, and the Torres Strait Islanders who occupy parts of the far northeast of the country, have traditionally made considerable use of reef resources. It seems likely, however, that their overall impacts remained minimal. Population densities were low, and a large area of offshore reef remained inaccessible to them.

The continent was first described by European travelers in the 17th century. Dampier visited parts of the northwestern coast in 1688 and 1699. Captain James Cook was the first to navigate the waters of the Great Barrier Reef, and indeed ran aground there in 1770. The first British settlement was established in Australia in 1788.

The Aboriginal population has decreased considerably since European occupation, while many of those who remain have been dispossessed of their traditional lands and have ceased to practice their traditional lifestyles. A few remaining coastal populations still have considerable rights regarding their traditional use of the reefs, but their numbers are so low that they are unlikely to have any significant impact except, perhaps, in parts of the Torres Strait region. The dominant human impacts can now be related to fisheries and terrestrial runoff from deforestation, overgrazing and certain agricultural practices. Compared with most countries, however, these impacts remain few. Population densities are low in all coral reef areas, while the location of many reefs at some distance from the shore further protects them from human impacts.

Considerable resources have been put into coral reef research in Australia and, despite the vast area of reefs in the country, there is a good deal of information describing their distribution and biodiversity. Equally importantly, the great majority of Australia's reefs fall within protected areas. The Great Barrier Reef Marine Park is the largest protected reef in the world, and is well managed with a detailed zoning plan, providing areas of strict protection alongside much larger areas of multiple use.

Left: The North West Cape is bordered by Australia's longest fringing reef, the Ningaloo Reef (STS035-76-44, 1990). Right, above: A number of reefs, including Ashmore Reef, lie right on the edge of the continental shelf in the far northwest of Australia (STS060-75-25, 1994). Right, below: Leopard grouper Plectropomus leopardus *amidst branching and soft corals.*

West Australia

MAP 11a

The reefs in the west of Australia encompass a variety of types in a very broad range of oceanographic conditions. For the most part this is a very dry coastline with little terrestrial runoff. It is also, from a human perspective, very sparsely populated and poorly documented. One critical oceanographic feature is the Leeuwin Current which flows south from Indonesia, carrying warm waters to relatively high latitudes, particularly along the continental shelf edge.

Along the mainland coast, reefs are discontinuous but very well developed in places. In the north the continental shelf is very wide and dominated by turbid waters with strong currents. Reef development is little known off the Eighty Mile Beach, though further west there are scattered reefs among the Dampier Archipelago and the Monte Bello Islands. Here, as the continental shelf narrows, there is a great range of oceanographic conditions associated with the gradient between nearshore turbid waters and clear offshore waters, mixed by the complex current regime.

Australia's longest continuous fringing reef system is the Ningaloo Reef which follows some 230 kilometers of coastline running southwards from North West Cape. The reef flats are well developed, lying between 0.5 and 7 kilometers offshore. The continental shelf is narrower here than anywhere else in the country, with the 200 meter contour less than 20 kilometers offshore. These reefs receive the full impact of oceanic waves, so corals tend to be quite low and compact. Biodiversity is relatively high, with some 300 species of coral, nearly 500 species of fish and over 600 molluscs. The area is also noted for the appearance of whale sharks. These giant plankton-feeders occur in considerable numbers between mid-March and mid-May.

The marine areas of the Shark Bay World Heritage Site are of considerable interest, including some of the most extensive seagrass communities in the world and harboring what is probably the largest dugong population in the world (over 10 000). Monkey Mia Bay has become famous for a tame group of bottle-nosed dolphins, but the region is also of considerable importance for other cetaceans, including humpback and southern right whales. Hamelin Pool, within the Shark Bay area, is one of the few places in the world where there are actively growing stromatolites. Hypersaline conditions prevent the survival of most organisms, but photosynthetic bacteria and microalgae survive and form microbial mats as they trap and bind sediments. Over the last 4 000 years these mats have developed into relatively large structures – columns or mounds up to 1.5 meters high. Similar structures have been recorded in fossils dating back 3.5 billion years, some of the earliest known forms of life. Despite this, there are no true coral reefs in the Shark Bay area, although some 80 coral species have been recorded.

The southernmost true reefs in the Indian Ocean are around the Houtman Abrolhos Islands which lie close to 29°S on the edge of the continental shelf. These islands were named by Frederick Houtman in 1619, and the word Abrolhos is derived from the Portuguese expression *abri vossos olhos* ("look out" or "take care") as they were such a navigational hazard. They are located on three carbonate platforms with channels 40 meters deep between them. Lying on the edge of the continental shelf they are directly affected by the Leeuwin Current which moderates the winter temperatures, and may also have a critical role in larval supply. Considering their high latitude, these reefs have a significant diversity, with over 180 coral species and over 230 fish. One of the most interesting ecological features of benthic life around these islands is the occurrence of substantial macroalgae communities dominated by brown algae, including the large kelp *Ecklonia radiata*. Corals dominate the community structure on leeward reef slopes while algal assemblages are predominant on windward slopes and flats, and there is considerable

The Houtman Abrolhos have a very high diversity of species considering their southerly latitude, but also incorporate more temperate species and macroalgal communities (STS093-702-70, 1999).

MAP 11a

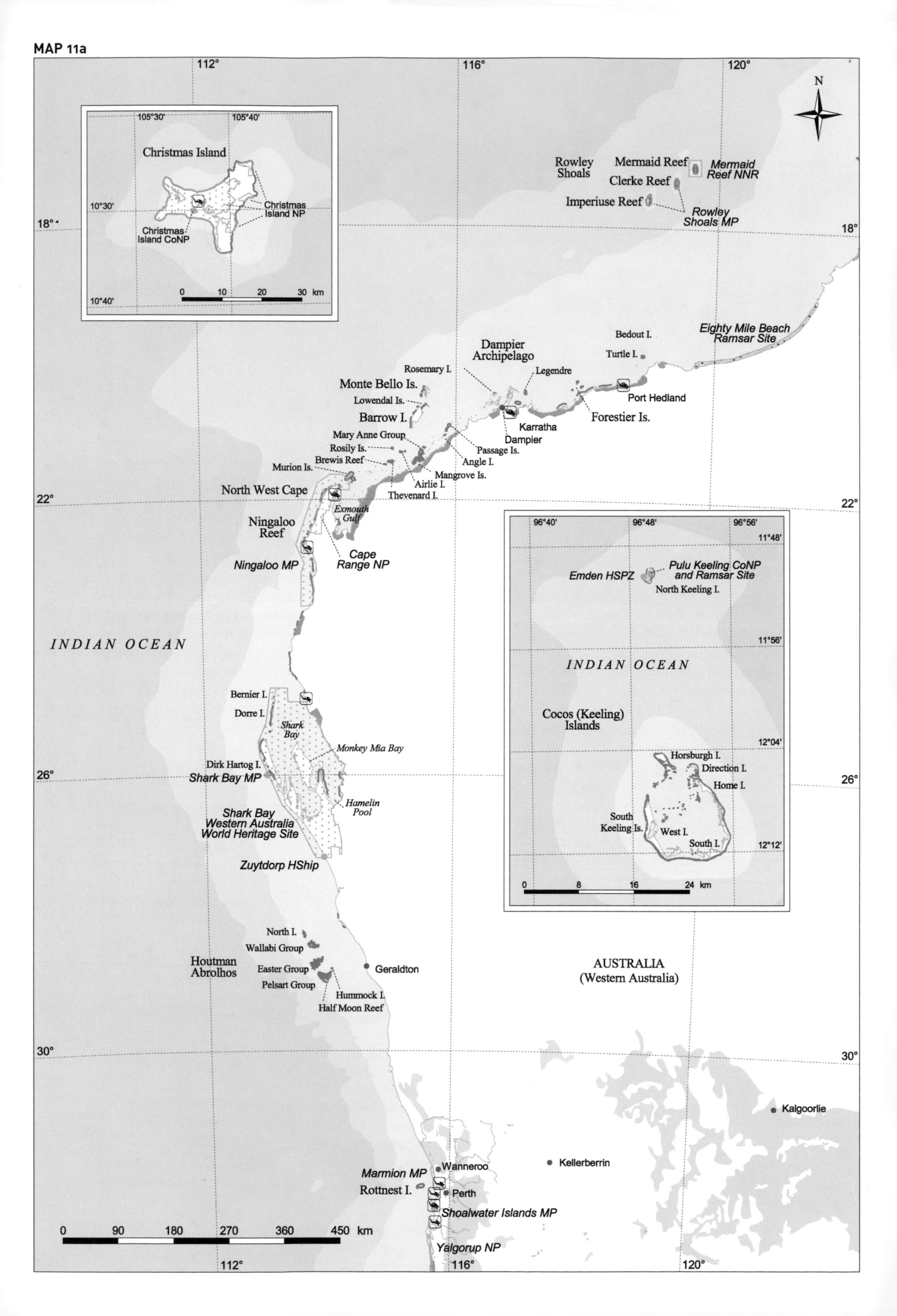

112°
116°
120°
N
Christmas Island
105°30'
105°40'
10°30'
10°40'
Christmas Island NP
Christmas Island CoNP
0 10 20 30 km
Rowley Shoals
Mermaid Reef
Mermaid Reef NNR
Clerke Reef
Imperiuse Reef
Rowley Shoals MP
18°
Eighty Mile Beach Ramsar Site
Bedout I.
Turtle I.
Dampier Archipelago
Legendre
Rosemary I.
Port Hedland
Forestier Is.
Karratha
Dampier
Monte Bello Is.
Lowendal Is.
Barrow I.
Mary Anne Group
Rosily Is.
Brewis Reef
Murion Is.
Passage Is.
Angle I.
Mangrove Is.
Airlie I.
Thevenard I.
North West Cape
22°
Exmouth Gulf
Ningaloo Reef
Ningaloo MP
Cape Range NP
INDIAN OCEAN
96°40'
96°48'
96°56'
11°48'
Emden HSPZ
Pulu Keeling CoNP and Ramsar Site
North Keeling I.
11°56'
INDIAN OCEAN
Cocos (Keeling) Islands
12°04'
Horsburgh I.
Direction I.
Home I.
South Keeling Is.
West I.
South I.
12°12'
0 8 16 24 km
Bernier I.
Dorre I.
Shark Bay
Monkey Mia Bay
Dirk Hartog I.
26°
Shark Bay MP
Hamelin Pool
Shark Bay Western Australia World Heritage Site
Zuytdorp HShip
North I.
Wallabi Group
Houtman Abrolhos
Easter Group
Pelsart Group
Geraldton
Hummock I.
Half Moon Reef
AUSTRALIA (Western Australia)
30°
Kalgoorlie
Kellerberrin
Marmion MP
Wanneroo
Rottnest I.
Perth
Shoalwater Islands MP
Yalgorup NP
0 90 180 270 360 450 km

overlap in some places. These islands thus support a rare combination of sub-tropical and temperate communities in close proximity.

Further south there is no true reef development, though Rottnest Island off the coast near Perth is fringed by shallow platforms where some 25 species of zooxanthellate corals have been recorded. The importance of the warming effect of the Leeuwin Current is equally strong here. As well as corals, about 25-30 percent of the fish and echinoderm populations are generally tropical in their distribution. Despite studies since the 1950s, the first recording of *Acropora* on these reefs was not until 1988. It has been suggested that these and other species may be dependent on larval recruitment from the Houtman Abrolhos reefs.

Low human populations generally restrict impacts on the reefs off the west coast of Australia, although there is some fishing in all areas. Around the Dampier and Monte Bello Islands there is increasing pearl oyster farming, petroleum exploitation and now some tourism, although the impacts of these are still not high. The Monte Bello Islands were used for British nuclear tests in 1952-56. The Ningaloo reefs were heavily fished until recently, however these reefs are now zoned and fishing is restricted to certain areas. These reefs were also reported to have been damaged by outbreaks of the coral eating *Drupella* snails in the 1970s and 1990s, but there now appears to be active recovery in most areas. Levels of tourism are relatively high around Shark Bay. The Houtman Abrolhos Islands were among the first parts of Australia to be settled by Europeans, at least temporarily, following a shipwreck and mutiny in 1629. The islands were heavily mined for guano until the late 1940s and now support a major, well managed commercial rock lobster fishery. Two large protected areas have been declared which provide at least some protection for the reefs in Ningaloo and Shark Bay.

Cocos (Keeling) Islands and Christmas Island

Far out in the Indian Ocean, Australia administers two other territories with important oceanic reef communities. Cocos (Keeling) consists of two atolls on the Cocos Rise, nearly mid-way between Australia and Sri Lanka. They are dominated by the southeast trade winds and swept by the westward flowing equatorial current most of the year, and are occasionally impacted by tropical cyclones. The main atoll of South Keeling is a little over 15 kilometers across, with a near continuous chain of 27 islands along much of its rim. Horsburgh Island in the north lies apart from the others, and holds a particularly important bird nesting colony. North Keeling (Pulu Keeling) is a much smaller atoll, about 3 kilometers across with a single island almost completely encircling a shallow lagoon. The island itself is of considerable interest, being one of the few in the region with its original vegetation, mostly tall hardwood forest. It is also a very important seabird rookery. Some 525 species of fish have been recorded in the waters around the two atolls.

Well developed and distinctive coral reef communities occur in the Houtman Abrolhos. The purple coral is Acropora abrotanoides *which is found in shallow reef areas right across the Indo-Pacific, while the green coral* Acropora seriata *has a disjunct distribution in southwest Australia, insular Southeast Asia and Sri Lanka only* (photo: JEN Veron).

These islands were in fact the only atolls where Charles Darwin ever landed during his voyage on the *Beagle* (in 1836) and thus have a significant place in his development of a theory for atoll development (Chapter 1). There is a small resident population of predominantly Malaysian origin, living on two of the South Keeling Islands, but their impact on the reefs is considered minimal. A national park in North Keeling protects all the island, its surrounding reefs and waters.

Christmas Island is a high, mountainous island some 15 kilometers across, reaching a height of 359 meters. It lies about 300 kilometers south of Java. Fringing reefs surround much of it, with narrow reef flats typically 20-200 meters wide before a steep reef slope down to deep oceanic waters. While the reef faunas clearly contain Indian Ocean elements they show a close affinity to Southeast Asia. Diversity is somewhat limited by a moderate range of reef habitats. The island has important seabird nesting colonies, including the endemic Christmas Island frigatebird. Large numbers of crabs are also noted, including 13 land crabs, the best known of which are the red crabs *Gecarcoidea natalis,* which have a population of some 120 million individuals and undertake a famous annual mass migration to spawn in the sea. The resident population of some 2 000 people originally came to the island to mine its large phosphate deposits, and this continues, although it is strictly regulated. More recently a hotel and casino complex has been developed, drawing tourists from Southeast Asia. Over 60 percent of the island and much of the fringing reef is protected in a national park.

MAP 11b

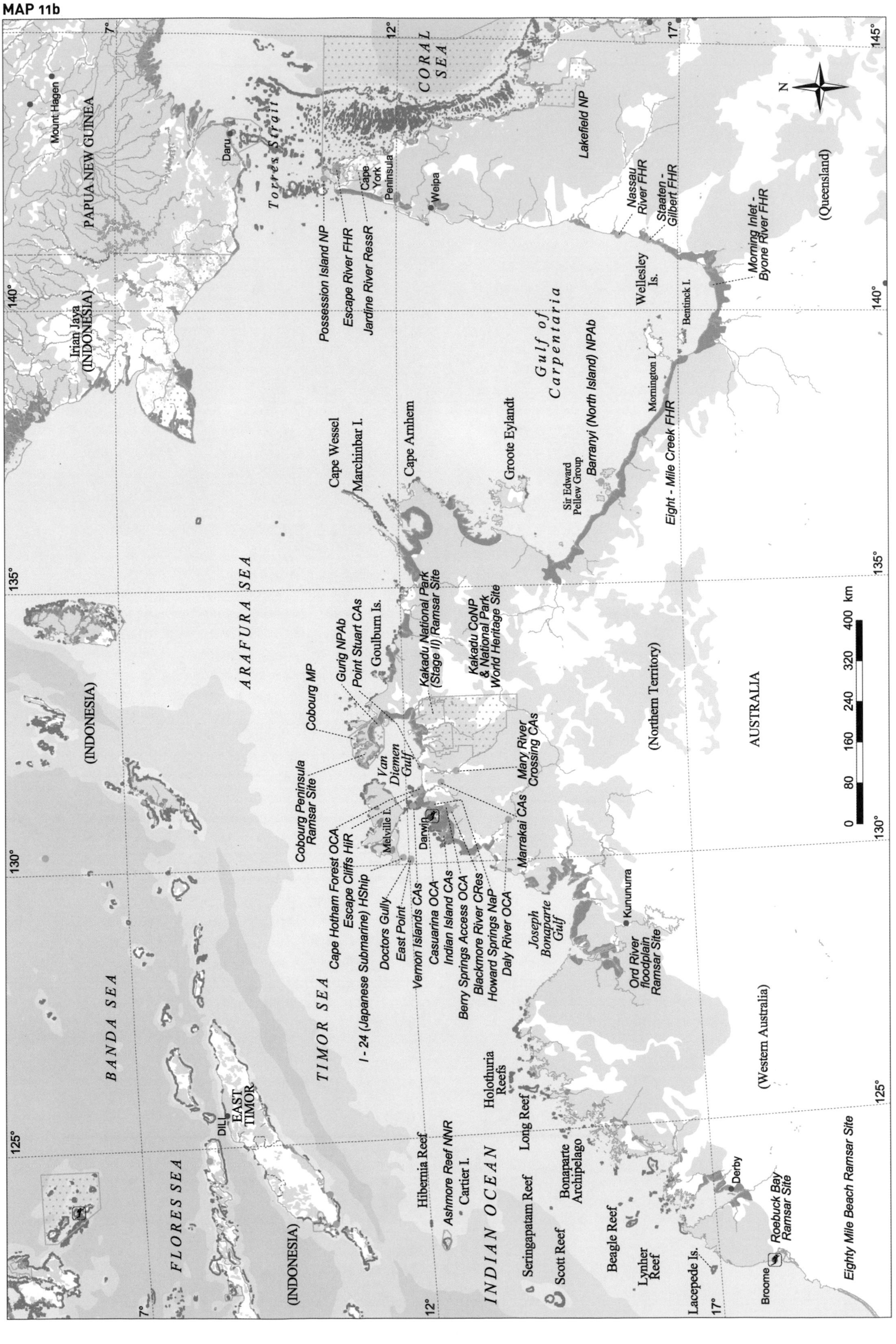

BANDA SEA
FLORES SEA
(INDONESIA)
DILI
EAST TIMOR
(INDONESIA)
TIMOR SEA
ARAFURA SEA
Irian Jaya (INDONESIA)
PAPUA NEW GUINEA
Mount Hagen
Daru
Torres Strait
CORAL SEA
Cobourg Peninsula Ramsar Site
Cobourg MP
Cape Hotham Forest OCA
Escape Cliffs HiR
I - 24 (Japanese Submarine) HShip
Doctors Gully
East Point
Vernon Islands CAs
Casuarina OCA
Indian Island CAs
Berry Springs Access OCA
Blackmore River CRes
Howard Springs NaP
Daly River OCA
Marrakai CAs
Mary River Crossing CAs
Gurig NPAb
Point Stuart CAs
Goulburn Is.
Melville I.
Van Diemen Gulf
Darwin
Kakadu National Park (Stage II) Ramsar Site
Kakadu CoNP & National Park World Heritage Site
Cape Wessel
Marchinbar I.
Cape Arnhem
Groote Eylandt
Gulf of Carpentaria
Sir Edward Pellew Group
Barranyi (North Island) NPAb
Mornington I.
Wellesley Is.
Bentinck I.
Eight - Mile Creek FHR
Morning Inlet - Byone River FHR
Nassau River FHR
Staaten - Gilbert FHR
Lakefield NP
Weipa
Cape York Peninsula
Possession Island NP
Escape River FHR
Jardine River RessR
Hibernia Reef
Ashmore Reef NNR
Cartier I.
INDIAN OCEAN
Holothuria Reefs
Seringapatam Reef
Long Reef
Scott Reef
Bonaparte Archipelago
Joseph Bonaparte Gulf
Beagle Reef
Lynher Reef
Lacepede Is.
Kununurra
Ord River floodplain Ramsar Site
Derby
Broome
Roebuck Bay Ramsar Site
Eighty Mile Beach Ramsar Site
(Western Australia)
(Northern Territory)
AUSTRALIA
(Queensland)
N
0 80 160 240 320 400 km
125° 130° 135° 140° 145°
7° 12° 17°

North Australia

MAP 11b

North of Port Headland and Eighty Mile Beach the continental shelf of Australia widens considerably while the coastline of Indonesia and East Timor forms a northern boundary enclosing the Timor Sea. To the east of Darwin this continental shelf widens further still and connects Australia to New Guinea across the Arafura Sea and the Gulf of Carpentaria. This is Australia's least known and least populated coastline. Wide areas are dominated by an intricate network of rivers and channels with extensive mangrove communities. To the east, in the Arafura Sea, the waters are shallow and turbid and there is little reef development. Fringing reefs are reported further west, but are very poorly described. The only reefs in this region which have received attention are those lying in the northwest on the continental shelf edge or just beyond.

The Rowley Shoals, Scott Reef and Seringapatam Reef are shelf edge atolls lying on the continental slope in clear oceanic waters. An extensive line of other reefs, including Lynher, Cartier, Ashmore and Hibernia, lies just on the continental shelf, and it has been suggested that these may in fact be barrier structures. A number of deeper shoals on the shelf edge indicate that there may have been a more extensive barrier reef along this shelf during recent periods of lower sea level, but that only these structures kept up with rising sea levels. Tidal ranges are very high around these reefs and there is considerable wave energy, so the reef crests are dominated by coralline algae, while only compact coral formations have developed on windward shores. This is also an area regularly affected by cyclones.

Coral cover is typically high. In early 1995 hard coral cover averaged nearly 50 percent on reef slopes in Scott Reef and the Rowley Shoals, but cyclone damage later that same year caused a considerable reduction in this figure. Ashmore Reef has the greatest biodiversity in the region: some 255 species of hermatypic corals have been recorded, 747 fish, 433 molluscs and 192 echinoderms. These compare with 213 hermatypic corals at Scott and Seringapatam Reefs and 184 at the Rowley Shoals. The region also probably has a greater diversity of sea snakes than anywhere else in the world, with 12 species recorded at Ashmore Reef, three of which are thought to be endemic to the Ashmore, Cartier and Hibernia Reefs. Seabird nesting colonies are also extremely important and 17 species (with an estimated 50 000 pairs) have been recorded nesting on the islands of Ashmore. This area was strongly impacted by warm waters associated with the 1998 El Niño event. Widespread coral bleaching was followed by 80 percent mortality at some sites on Scott Reef.

The more northerly reefs lie relatively close to Indonesia and are regularly fished by Indonesians under a joint use agreement. Elsewhere, including near Scott Reef, there is some extraction of natural gas, and further exploratory drilling and the establishment of new oil and gas platforms could bring further human impacts to these otherwise remote reefs. There is also some fishing on all reefs, including collection of trochus, shark and other reef fish, although there is little detailed information available on its impacts. Diving on the Rowley Shoals is increasingly popular, and the reefs are widely regarded as offering some of the best diving in the region. Ashmore Reef and the Rowley Shoals all have some degree of legal protection, and there is ongoing monitoring of Scott Reef and the Rowley Shoals.

Large colonies of blue coral Heliopora coerulea *on Scott Reef, northwest Australia (photo: JEN Veron).*

Torres Strait and the Great Barrier Reef

MAPS 11c, d and e

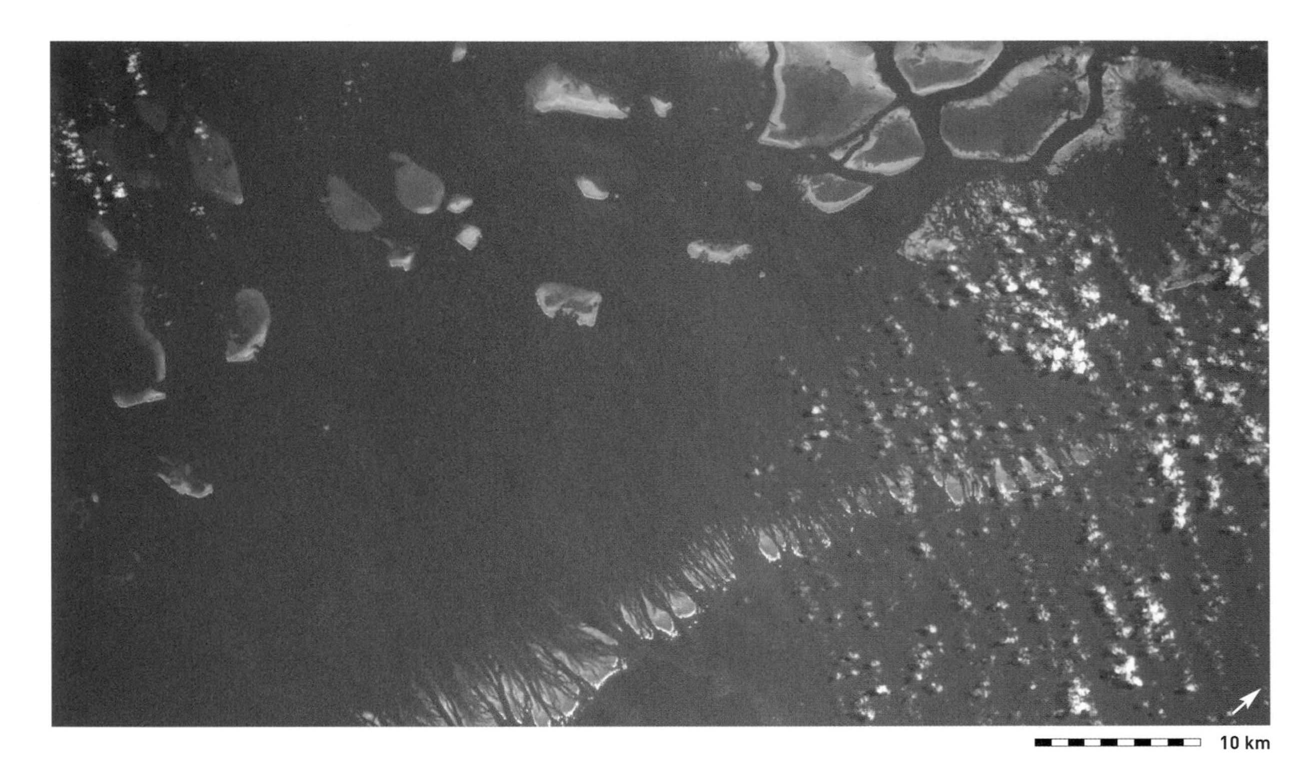

The largest coral reef system in the world runs along the northeastern coastline of Australia, stretching from the Warrior Reefs in the northern Torres Strait for well over 2 000 kilometers to the Capricorn-Bunker group of reefs and islands in the south. Although many of the reefs which make up this system form part of a true barrier reef following the continental shelf on its outermost edge, the Great Barrier Reef is actually a highly complex system including nearly 3 000 separate reefs and coral shoals, as well as high islands with fringing reef systems.

The origins of the Great Barrier Reef can largely be traced back some 2 million years, when continental drift brought the northern coastline of Australia into tropical latitudes and some minor reef development began. Widespread development is thought to be much more recent, however, and can largely be traced back within 500 000 years, making it a much younger structure than many oceanic atolls. As with coral reefs the world over, periods of reef building were continually interrupted by changes in climate and shifting sea levels. Typically reef accretion was confined to relatively short periods of higher sea level when reef structures began to build up on the margins of the continental shelf. As sea levels fell, the reefs died, became land, and were subject to erosion forces which in many places reduced their size again. High sea levels returned and allowed new reef growth, typically most prolific on the remaining structures of the earlier reefs. At the present time in geological history, sea levels are particularly high, such that the base of many of the present reefs lies in depths unsuitable for active reef growth. However, active reef building continues on the ancient structures and the reef continues to thrive. The most recent period of growth is probably only about 8 000 years.

Patterns of currents are complex across the Great Barrier Reef. One of the key driving forces is the South Equatorial Current which flows across the Coral Sea from the east. Where this meets the continental shelf it splits, forming the weak, northward flowing Hiri Current north of about 14ºS and the southward flowing East Australia Current further south. These currents induce localized upwellings onto the shelf, and further have some influence on the current patterns across the continental shelf, although these are predominantly driven by prevailing winds. For much of the year the southeast trade winds predominate, driving northward flowing surface currents in all areas, although most strongly north of about 20ºS. During the Northwest Monsoon (December-February)

The northern edges of the Great Barrier Reef, showing the ribbon reefs with deltaic channels cutting through them (STS049-75-43, 1992).

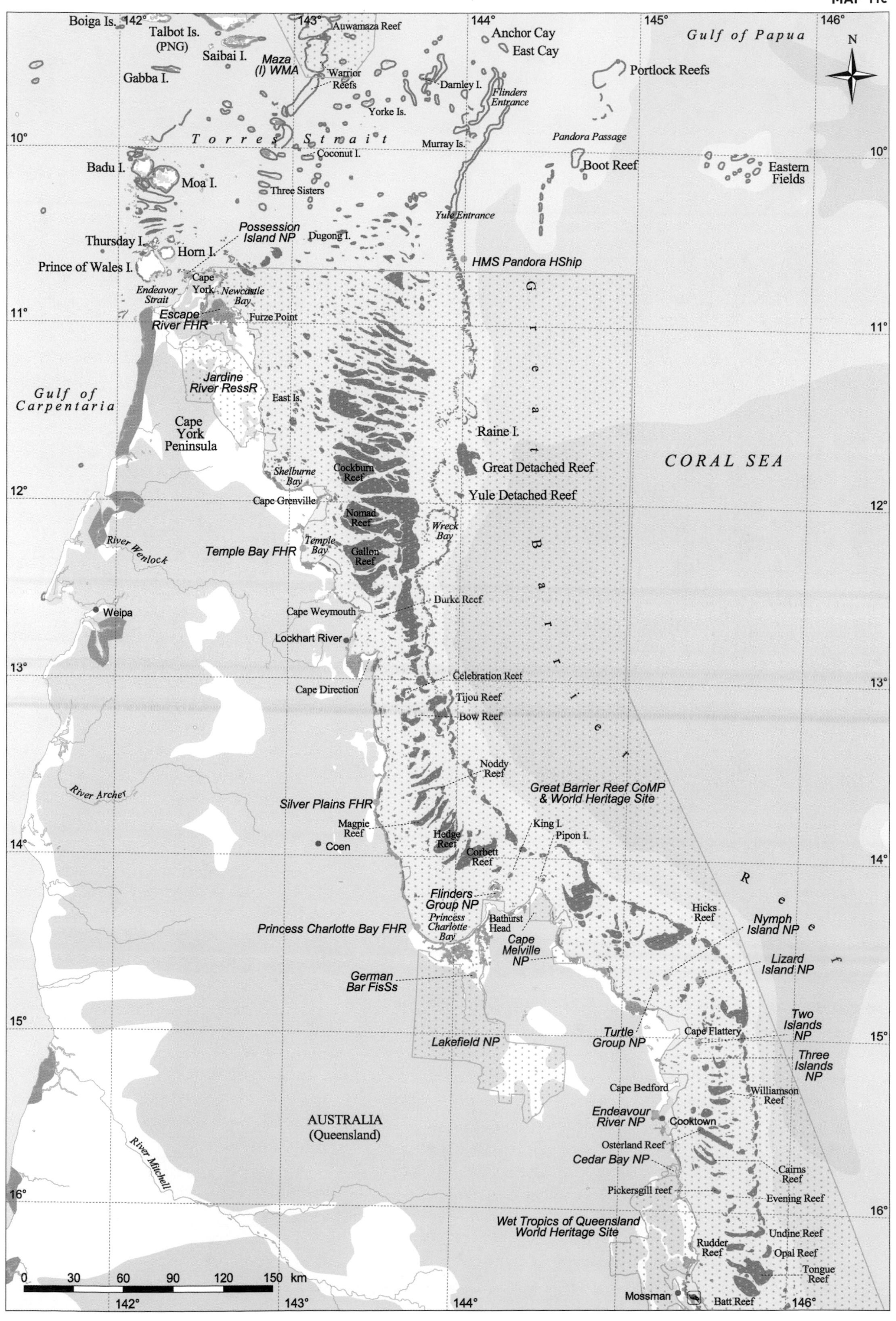

Boiga Is.
Talbot Is. (PNG)
Saibai I.
Gabba I.
Maza (I) WMA
Auwamaza Reef
Warrior Reefs
Anchor Cay
East Cay
Darnley I.
Yorke Is.
Flinders Entrance
Gulf of Papua
Portlock Reefs
Torres Strait
Coconut I.
Murray Is.
Pandora Passage
Boot Reef
Eastern Fields
Badu I.
Moa I.
Three Sisters
Yule Entrance
Thursday I.
Possession Island NP
Dugong I.
Horn I.
Prince of Wales I.
HMS Pandora HShip
Endeavor Strait
Cape York
Newcastle Bay
Escape River FHR
Furze Point
Jardine River RessR
Gulf of Carpentaria
East Is.
Cape York Peninsula
Raine I.
Great Detached Reef
CORAL SEA
Shelburne Bay
Cockburn Reef
Yule Detached Reef
Cape Grenville
Nomad Reef
Wreck Bay
Great Barrier Reef
River Wenlock
Temple Bay FHR
Temple Bay
Gallon Reef
Durke Reef
Weipa
Cape Weymouth
Lockhart River
Celebration Reef
Cape Direction
Tijou Reef
Bow Reef
Noddy Reef
River Archer
Great Barrier Reef CoMP & World Heritage Site
Silver Plains FHR
Magpie Reef
King I.
Pipon I.
Coen
Hedge Reef
Corbett Reef
Flinders Group NP
Princess Charlotte Bay
Bathurst Head
Hicks Reef
Nymph Island NP
Princess Charlotte Bay FHR
Cape Melville NP
Lizard Island NP
German Bar FisSs
Turtle Group NP
Cape Flattery
Two Islands NP
Lakefield NP
Three Islands NP
Cape Bedford
Williamson Reef
AUSTRALIA (Queensland)
Endeavour River NP
Cooktown
River Mitchell
Osterland Reef
Cedar Bay NP
Cairns Reef
Pickersgill reef
Evening Reef
Wet Tropics of Queensland World Heritage Site
Undine Reef
Rudder Reef
Opal Reef
Tongue Reef
Mossman
Batt Reef
0 30 60 90 120 150 km
142° 143° 144° 145° 146°
10° 11° 12° 13° 14° 15° 16°
N

MAP 11d

No.	Protected Area Name
1	Admiralty Island FHR
2	Barr Creek WetR
3	Big Maria Creek EP
4	Bohle River WetR
5	Bowling Green Bay FHR
6	Bowling Green Bay NP
7	Cairns MP
8	Cape Hillsborough NP
9	Cape Upstart NP
10	Cattle Creek WetR
11	Centenary Lakes FisSs
12	Clump Mountain NP
13	Conway NP
14	Dallachy Creek FHR
15	Edmund Kennedy NP
16	Ella Bay NP
17	Foam HShip
18	Great Barrier Reef CoMP
19	Green Island NP
20	Grey Peaks NP
21	Half Moon Creek WetR
22	Halifax WetR
23	Hinchinbrook Island NP
24	Hull River FHR
25	Hull River NP
26	Lindeman Islands NP
27	Magnetic Island NP
28	Maria Creek NP
29	Meunga Creek FHR
30	Midge WetR
31	Murray River FHR
32	Newry Islands NP
33	Orpheus Island NP
34	Palm Creek WetR
35	Repulse Bay FHR
36	Repulse FHR
37	Sand Bay FHR
38	Townsville / Whitsunday M
39	Townsville Town Common CP
40	Trinity Inlet WetR
41	Tully River FHR
42	Whitsunday Islands NP
43	Wreck Creek FHR
44	Yongala HShip
45	Yorkeys Creek WetR

CORAL SEA

Great Barrier Reef

Great Barrier Reef CoMP & World Heritage Site

Wet Tropics of Queensland World Heritage Site

AUSTRALIA (Queensland)

Bougainville Reef, Holmes Reef, Flora Reef, Herald Cays, Dart Reef, Herald Surprise, Flinders Reefs, Agincourt Reefs, Undine Reef, St. Crispin Reef, Rudder Reef, Tongue Reefs, Batt Reef, Trinity Opening, Hastings Reef, Michaelmas Reef, Arlington Reef, Moore Reef, Elford Reef, Sudbury Reef, Gibson Reef, Howie Reef, Wardle Reef, Nathan Reef, Adelaide Reef, Hall-Thompson Reef, Potter Reef, Otter Reef, Britomart Reef, Hinchinbrook I., Truck Reef, Bramble Reef, Bowl Reef, Palm Is., Grub Reef, Centipede Reef, Lynch's Reef, Great Palm I., Davis Reef, Big Broadhurst Reef, Little Broadhurst Reef, Magnetic I., Prawn Reef, Darley Reef, Dingo Reef, Tiger Reef, Kangaroo Reef, Stanley Reef, Old Reef, Wallaby Reef, Cobham Reef, Gould Reef, Line Reef, Hook Reef, Black Reef, Black Reef East, Hardy Reef, Whitsunday Is., Lindeman Is., Cumberland Is.

Cairns, Atherton, Innisfail, Ingham, Townsville, Ayr, Bowen, Proserpine, Mackay

146°, 147°, 148°, 149°, 16°, 17°, 18°, 19°, 20°, 21°

0 20 40 60 80 100 km

currents are reversed, with a weak southward flowing current in the north, but stronger in more southerly latitudes. These patterns of current flow are altered by patterns of tidal flow, particularly in the areas of more complex reef networks, but there remains a predominant pattern of water movements along the continental shelf, with little cross-shelf movement.

Although best seen as a continuous reef complex, it is possible to distinguish a number of ecological regions within the Great Barrier Reef.

Torres Strait

In the far north of Australia the continental shelf forms a wide connecting platform across the Torres Strait to Papua New Guinea. As most of the islands in the Torres Strait fall under Australian jurisdiction so do the reefs and waters of the Strait. There is considerable freshwater and sediment input from the Papua New Guinea coastline, however there are several very extensive platform reefs across the relatively shallow waters of the Strait. The westernmost areas have the shallowest and most turbid waters. A large chain of reefs runs between Prince of Wales Island and Moa Island. Like other reefs in the area, these show a very clear east-west alignment associated with the high velocity tidal currents running through the area. The Warrior Reefs further to the north and east run in a chain towards the coastal town of Daru in Papua New Guinea. Sediment loads are high in this area, and much of the shallow surface of these reefs is dominated by soft muds, although they are fringed by coral on their eastern margins. Finally there is a wide area of platform reefs around Darnley Island, stretching out towards the edge of the continental shelf and the near continuous line of reefs which mark the northern edge of the outer barrier reef.

Northern section

Due east of Cape York the continental shelf remains wide, but it then narrows rapidly towards Raine Island and continues as a platform typically less than 50 kilometers wide. The most distinctive feature of this sector of the Great Barrier Reef is the well developed ribbon-type barrier reefs on the outer edge: long narrow ribbon reefs typically less than 500 meters wide but extending up to 25 kilometers in length and separated by relatively narrow passes. They are located right on the edge of the continental shelf, and depths drop rapidly to over 1 000 meters only a few hundred meters from the eastern edges of some reefs. For about 80 kilometers, in the northernmost sector of these ribbon reefs, there are spectacular deltaic formations in the channels between the reefs. Rather like river deltas, these have been formed in the calmer waters behind the reef by the deposition of sediments from the powerful currents which flow between the reefs. The banks of sediments have then formed a substrate for new reef development.

Inshore of the ribbon reefs there are well developed mid-shelf and inner shelf reefs, while there are also wide areas of submerged *Halimeda*-dominated shoals and banks. This is one of the only areas where there are fringing reefs directly adjacent to the mainland coast, although coral cover and diversity are limited. Raine Island just off the continental shelf has the largest nesting populations of green turtles in the world as well as some of the most important seabird rookeries. There are only a few high islands on the continental shelf, notably the Flinders group and Lizard Island. These have important and extensive fringing reefs.

Central section

This section extends from Mossman in the north to the barrier reef offshore from the Whitsunday and Lindeman Islands. Over this area the continental shelf gradually widens, with reef development largely restricted to its outer third. Closer to the mainland the waters are subject to considerable fluctuations in turbidity and salinity due to the seasonal flooding of rivers. The reefs in this area are younger than those to the north. Many have lower and less extensive reef flats, and coral cays are largely absent, while their outer reef crests are often only clearly developed on the windward southeastern margins. Overall the reefs are less tightly packed and hence do not form such a continuous barrier. The main reefs are also set back a little from the true edge of the continental shelf, although there are several reefal shoals close to the outer shelf margin which rise to within 10 meters or so of the surface and have active coral growth. Over relatively short geological timescales these could evolve into ribbon reef systems similar to those observed to the north. In addition to the barrier reef structures there are important fringing reef communities associated with a number of high island groups, notably the Palm Islands and the Whitsunday and Lindeman Islands to the south.

The Swain and Pompey Complexes

This is the sector of the Great Barrier Reef where the continental shelf is at its widest and the main reefs are furthest from shore. The Pompey Complex has a number of submerged reefs on the edge of the continental shelf. However, about 10 kilometers back from this edge is a vast and complex array of very large reef platforms separated by countless meandering channels making a nearly solid mass of reefs nearly 200 kilometers in length and up to 20 kilometers wide. The high tidal range in this area drives strong currents reaching up to 10 knots, which

scour many of the channels between the reefs. Within the lagoons of individual reefs there are complex patterns of channels and mid-reef coral ridges. To the south the Swain Reefs form a second distinctive complex, dominated by many much smaller and even more closely spaced reefs where large numbers of small coral cays have developed. Inshore of the main Swain and Pompey Complexes reef development is limited, however there are some reefs close to the mainland and associated with island groups such as the Northumberland Islands and Percy Isles.

The Capricorn-Bunker group

South of the Swain Reefs Complex the continental shelf rapidly narrows again and the southernmost reefs of the Great Barrier Reef, the Capricorn-Bunker group, lie a little over 50 kilometers offshore. This is a relatively small complex, well defined with steeply sloping reef edges and deep inter-reefal waters. There are several well developed coral cays, including One Tree Island and Heron Island, which are among the best known reefs of the entire Great Barrier Reef. These reefs actually traverse the tropic of Capricorn and cooler waters are largely responsible for the lower coral diversities found here.

Biodiversity

Levels of diversity are generally very high in the Great Barrier Reef, with some 350 coral species, 1 500-2 000 species of fish, and over 4 000 species of mollusc. While these large numbers may be partly a function of the fact that this area has been intensively studied by scientists for many years, it is further related to the vast area of reefs as well as the great diversity of reef types and physical conditions. Over such a large area it is impossible to generalize about coral cover. However, it should be noted that cyclones and crown-of-thorns starfish clearly impact such statistics. The central areas of the Great Barrier Reef are the most affected by both of these phenomena, and many reefs in these parts have relatively low levels of coral cover when compared to reefs elsewhere in the world. This may well be their natural state, which clearly points to the caution which must be exercised when utilizing measures of coral cover as an expression of reef health.

As might be expected with any reef system traversing such a wide latitudinal range, there is a gradual diminution of species diversity towards higher latitudes. While most of the 350 coral species are recorded in the north, only about 244 species are recorded further south. Even more notable are cross-shelf differences. Close to the mainland there are high levels of nutrient inputs, sediments and freshwater, while offshore such inputs diminish and conditions on the outer reefs can be considered near oceanic, with low levels of nutrients and clear waters. These differences have led to considerable variation in the species assemblages depending on their location on the continental shelf. Such differences are further maintained by the patterns of water movement in the Great Barrier Reef, generally north-south with little cross-shelf transport.

As mentioned in Chapter 1, many corals reproduce in a mass spawning event which takes place once a year. While this is globally widespread, it was first observed and is best documented on the Great Barrier Reef. For a few nights after a particular full moon in the late austral spring (typically November) the majority of scleractinian coral species, together with many other reef organisms including sponges, holothurians, polychaetes and giant clams, undergo a mass spawning event. This is highly

Left: To the south the coastal shelf of the Great Barrier Reef widens considerably around the vast complex of the Swain and Pompey Reefs, before narrowing again around the small Capricorn group (STS043-151-77, 1991). Right: The beaked butterflyfish Chelmon rostratus *is found across Southeast Asia and the Great Barrier Reef.*

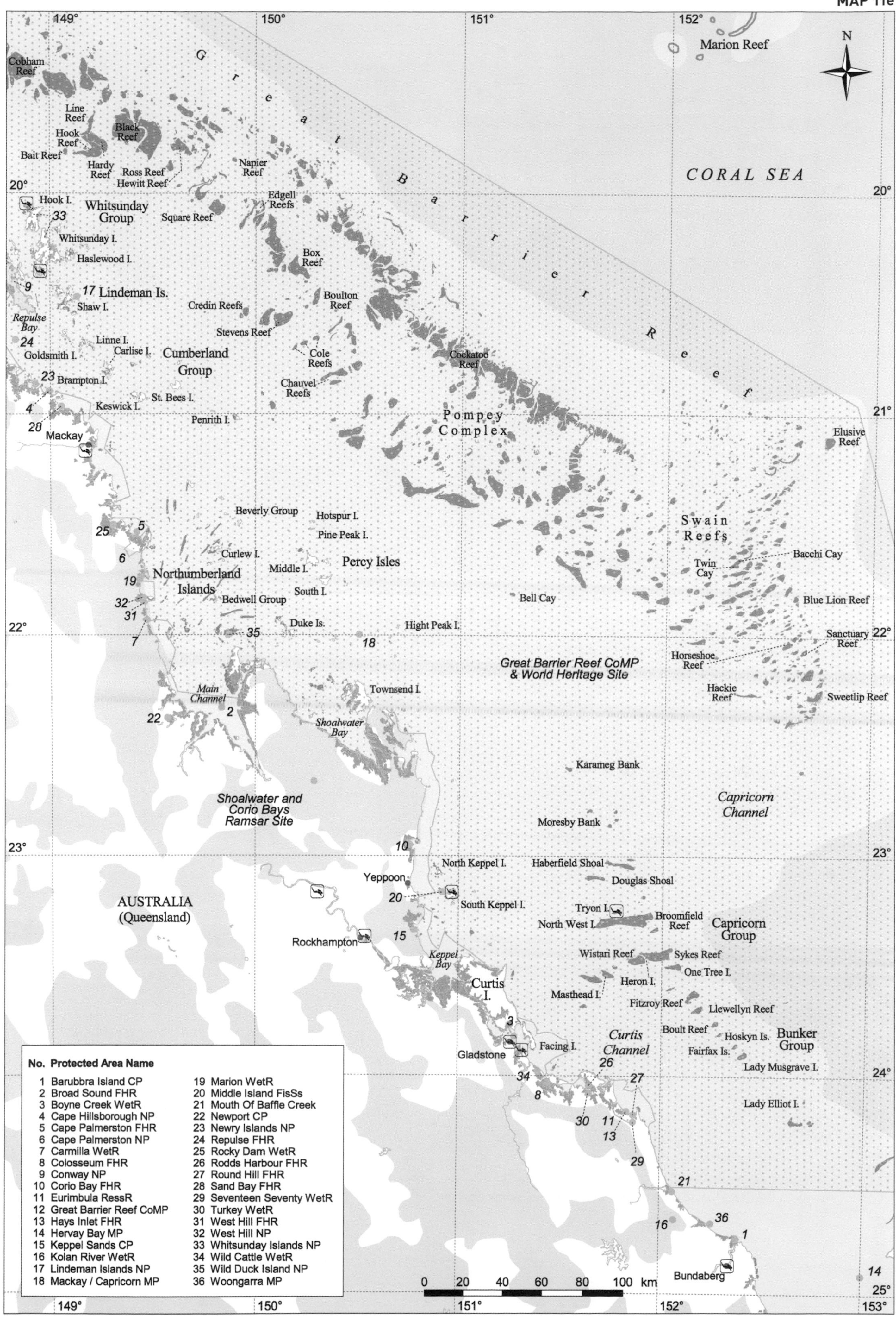

No.	Protected Area Name	No.	Protected Area Name
1	Barubbra Island CP	19	Marion WetR
2	Broad Sound FHR	20	Middle Island FisSs
3	Boyne Creek WetR	21	Mouth Of Baffle Creek
4	Cape Hillsborough NP	22	Newport CP
5	Cape Palmerston FHR	23	Newry Islands NP
6	Cape Palmerston NP	24	Repulse FHR
7	Carmilla WetR	25	Rocky Dam WetR
8	Colosseum FHR	26	Rodds Harbour FHR
9	Conway NP	27	Round Hill FHR
10	Corio Bay FHR	28	Sand Bay FHR
11	Eurimbula RessR	29	Seventeen Seventy WetR
12	Great Barrier Reef CoMP	30	Turkey WetR
13	Hays Inlet FHR	31	West Hill FHR
14	Hervay Bay MP	32	West Hill NP
15	Keppel Sands CP	33	Whitsunday Islands NP
16	Kolan River WetR	34	Wild Cattle WetR
17	Lindeman Islands NP	35	Wild Duck Island NP
18	Mackay / Capricorn MP	36	Woongarra MP

synchronized, with individuals of the same species releasing eggs and sperm often within minutes of one another over wide areas of the reef. The phenomenon was only first observed in November 1982, and yet is one of the most spectacular events on any coral reef. Vast numbers of eggs and sperm are released and form massive slicks on the sea surface. The spawning allows for cross-fertilization between colonies, while the massive scale of such an event ensures that would-be predators are fully satiated, thereby increasing the chances of survival of individual larvae.

In addition to its considerable diversity in terms of coral reef organisms, the Great Barrier Reef is also an extremely important region for other marine and coastal ecosystems, most notably seagrass and mangrove communities. Mangroves generally lie a considerable distance from coral reef communities – with the exception of a few fringing reef systems. But some 37 mangrove species from 19 families have been recorded at the Great Barrier Reef, with the highest levels of diversity in the "wet tropics" north of Cairns. Seagrass communities are also widespread, with some 3 000 square kilometers of mapped shallow seagrasses, and an estimate of at least 2 000 square kilometers of deep (>15 meters) seagrasses. Both seagrass beds and mangroves are extensively used as breeding and nursery grounds by many species, including a number of commercially important species, and some reef species. Seagrass beds are also important for some turtle species as well as large populations of dugongs. Green, hawksbill, loggerhead and flatback turtles all nest in considerable numbers in the region. Unfortunately, with the exception of the flatback turtle, most individuals spend substantial amounts of time in neighboring countries where they are severely threatened by direct hunting and indirect killing, notably as fisheries by-catch. There are globally important populations of dugongs in the region. While there is some traditional hunting of these by Aboriginal and Torres Strait Islander communities the northern population of some 8 000 individuals is considered stable. By contrast, the smaller southern population of about 3 500 individuals is now declining, largely as a result of deaths associated with boat collisions, entanglement in gill nets, and also entanglement in shark nets placed near swimming beaches. In addition some 26 species of cetacean are resident or visitors to the Great Barrier Reef, including significant numbers of humpback whales which breed in the southern and central waters.

There are important seabird communities on the Great Barrier Reef, with over 55 major nesting islands and 1.4-1.7 million breeding birds from some 23 species, with a further 32 non-breeding species. Most of these islands are in the north and south, with around 75 percent of the total seabird biomass in the Capricorn-Bunker group.

The Great Barrier Reef has been one of the regions most extensively impacted by the crown-of-thorns starfish, with the first mass outbreak of this predator observed on Green Island, off Cairns, in 1962. The possible causes of these outbreaks have been debated for some time (see Chapter 2), with much of the work having been conducted on the Great Barrier Reef. While there is still much to learn about these outbreaks, it is clear that they have had a significant impact on the ecology of the region, causing apparently periodic massive losses of live coral cover. Most outbreaks have been recorded in the central sections of the Great Barrier Reef. The 1998 bleaching event also impacted a number of reefs, most notably in the inner shelf areas where some 25 percent of reefs showed bleaching of 60 percent or greater. Overall, bleaching was worst in the central sections of the Great Barrier Reef, while outer reefs generally showed only low levels of bleaching. Mortality was generally low, although some inshore fringing reefs suffered greatly.

Socio-economic considerations

In general the Great Barrier Reef is not heavily affected by human activities, but there are some concerns that deforestation, poor agricultural practices and high concentrations of agricultural chemicals and nutrients in terrestrial runoff may have some impacts, particularly on those reefs closest to the mainland. The majority of reefs, however, are far offshore and this, combined with the prevailing long-shore currents, reduces the effects of land-

Spinner dolphins Stenella longirostris.

based sediments and pollutants. The distance from the mainland of most reefs also makes access more difficult, while the coastal population adjacent to the reef is small overall and does not generally exert a very large direct impact on the reefs, except for some commercial fisheries.

The utilization of marine and coastal resources has a long tradition among the Aboriginal inhabitants of Australia. Further north in Cape York and the Torres Strait, the Torres Strait Islanders, who are of different ethnographic origin, have also been great users of reef resources. Following European colonization the numbers of these peoples diminished and many of their traditional ways of life broke down. There remain some 11 Torres Strait Islander and Aboriginal communities, mostly in the far north, with a population of about 11 000, together with a slightly larger number in urban areas. A small proportion still engage in hunting and fishing on the reef, however their impact, even on such species as dugongs and turtles, is probably still at sustainable levels.

Utilization of marine resources by the wider population is far more significant than that by indigenous communities. Recreational fishing is extremely popular, although it typically targets the reefs closest to the mainland and near the major population centers. The recreational fishing catch has been estimated at 3 500-4 300 tons per year. Commercial reef fish exploitation is predominantly a line fishery concentrated on groupers ("coral trout") and emperors, with a combined annual catch of some 3 000-4 000 tons. Part of this is for the live fish trade, with groupers being air freighted to the Far East, notably Hong Kong. There is also an important lobster fishery to the north, collecting some 50-200 tons annually, and a separate fishery operating in the Torres Strait. In addition there are some fairly small-scale fisheries associated with the aquarium trade, trochus and sea cucumber. The most important commercial fishing within the Great Barrier Reef area is actually trawling, with some 840 licensed vessels, typically landing prawns (5 000-6 500 tons), fish (1 500 tons), scallops (200-1 000 tons) and other crustaceans (500 tons). There are some concerns over the size of the by-catch (typically over 50 percent and sometimes as much as 90 percent of hauls), which includes benthic organisms, fish, and even sea snakes and turtles, and over the wider impacts on benthic communities, particularly in areas of repetitive trawling. Trawling is not permitted over known seagrass communities and in a few other protected areas, however illegal trawling still occurs. There is evidence of overfishing of some target reef fish species, although this is mostly on a small scale and restricted in spatial extent.

The vast majority of the Great Barrier Reef receives protection as the Great Barrier Reef Marine Park, the world's second largest protected area. This covers most of the lagoon and all of the offshore reefs from the Capricorn-Bunker group to the northern tip of Cape York Peninsula. A large proportion of the remaining coastal waters and terrestrial areas of offshore islands which are not covered by the park fall within other protected areas. The park itself is zoned. About 80 percent of its total area is open for general use including commercial fishing and trawling (with permits), and a further 16 percent is also for general use but with trawling prohibited. Only about 5 percent is closed to fishing activities, but this includes over 120 reefs (about 12 percent of the total). The park is managed by a specially designated federal agency, the Great Barrier Reef Marine Park Authority, in collaboration with the Queensland Department of Environment and Heritage. Between them, these organizations employ some 210 staff with an operating budget of Au$27.2 million in 1998-99. About 30 percent of this budget is provided by an environmental management charge levied on all visitors to the park. For administrative purposes the park is divided into four broad sectors. Detailed management plans have been developed for particular localities, while an overall 25-year strategic plan has been developed in collaboration with the major stakeholders.

Active scientific research within the Great Barrier Reef is carried out by a number of organizations, including several universities and the management authorities, however the major research institution which undertakes monitoring and core scientific research is the Australian Institute of Marine Science based in Townsville.

The reefs of the Torres Strait lie outside the Great Barrier Reef Marine Park and do not fall under any strict legal protection, although a fisheries management agreement has been developed with Papua New Guinea. Overfishing is certainly a pressure within the region, while there remains a significant potential threat of pollution, both from the mines in Papua New Guinea and from oil spills associated with the relatively heavy shipping traffic in the strait.

The fringing reef on Orpheus Island in the Palm Islands.

MAP 11f

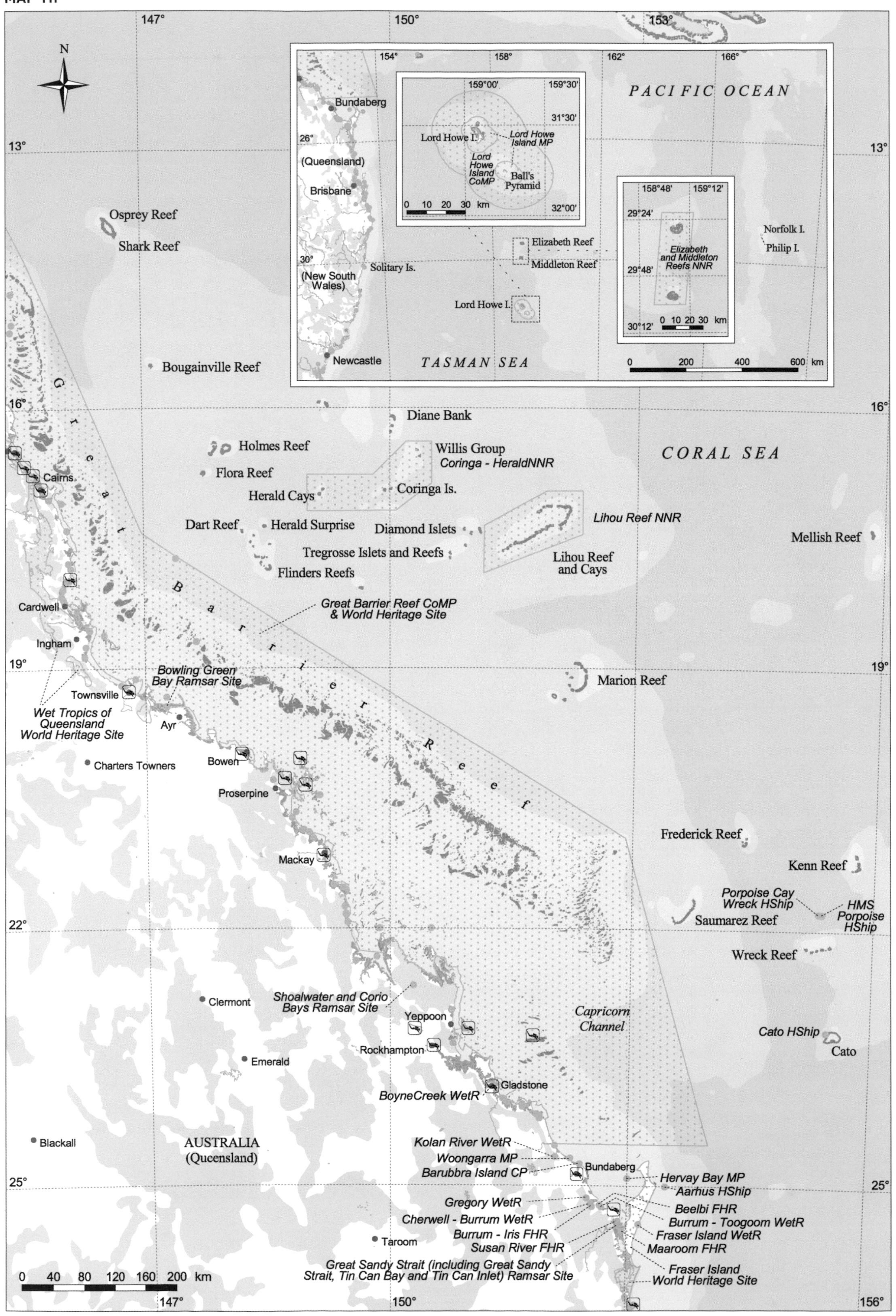

147°
150°
153°
13°
16°
19°
22°
25°
156°
N
PACIFIC OCEAN
TASMAN SEA
CORAL SEA
Bundaberg
(Queensland)
Brisbane
(New South Wales)
Newcastle
Solitary Is.
Lord Howe I.
Lord Howe Island MP
Lord Howe Island CoMP
Ball's Pyramid
Elizabeth Reef
Middleton Reef
Elizabeth and Middleton Reefs NNR
Norfolk I.
Philip I.
Osprey Reef
Shark Reef
Bougainville Reef
Diane Bank
Holmes Reef
Flora Reef
Willis Group
Coringa - HeraldNNR
Herald Cays
Coringa Is.
Dart Reef
Herald Surprise
Diamond Islets
Lihou Reef NNR
Mellish Reef
Tregrosse Islets and Reefs
Flinders Reefs
Lihou Reef and Cays
Great Barrier Reef
Great Barrier Reef CoMP & World Heritage Site
Cairns
Cardwell
Ingham
Townsville
Bowling Green Bay Ramsar Site
Wet Tropics of Queensland World Heritage Site
Ayr
Charters Towners
Bowen
Proserpine
Mackay
Marion Reef
Frederick Reef
Kenn Reef
Porpoise Cay Wreck HShip
HMS Porpoise HShip
Saumarez Reef
Wreck Reef
Clermont
Shoalwater and Corio Bays Ramsar Site
Yeppoon
Rockhampton
Capricorn Channel
Cato HShip
Cato
Emerald
Gladstone
BoyneCreek WetR
Blackall
AUSTRALIA (Queensland)
Kolan River WetR
Woongarra MP
Barubbra Island CP
Hervay Bay MP
Aarhus HShip
Gregory WetR
Beelbi FHR
Cherwell - Burrum WetR
Burrum - Toogoom WetR
Burrum - Iris FHR
Fraser Island WetR
Taroom
Susan River FHR
Maaroom FHR
Great Sandy Strait (including Great Sandy Strait, Tin Can Bay and Tin Can Inlet) Ramsar Site
Fraser Island World Heritage Site
0 40 80 120 160 200 km

The Coral Sea

MAP 11f

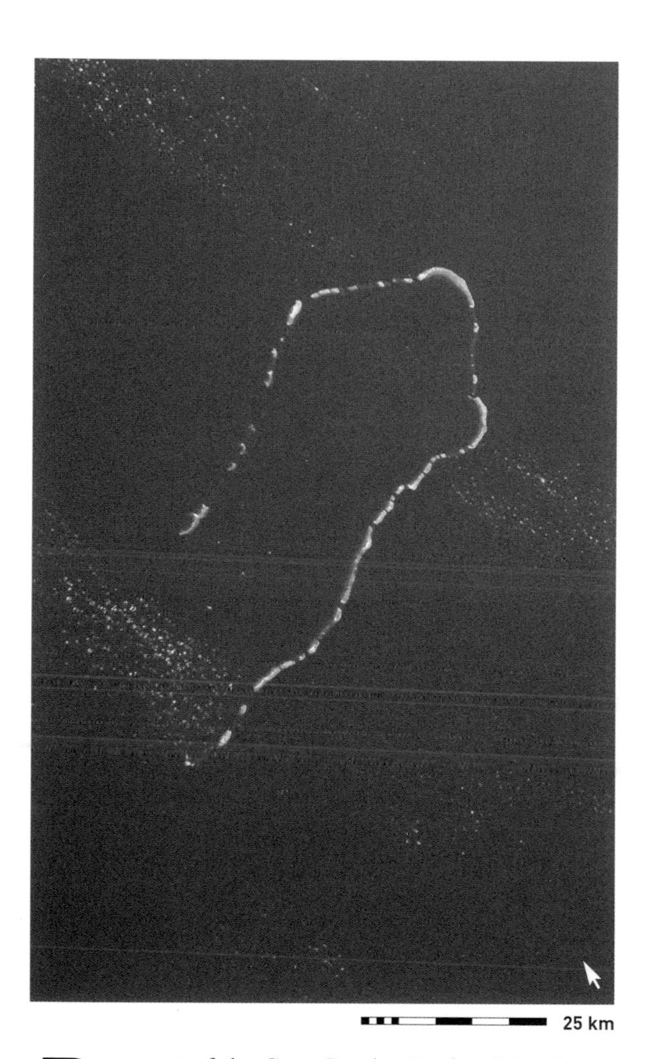

Due east of the Great Barrier Reef and the Torres Strait, Australia holds jurisdiction of a large number of reef formations lying some distance off the continental shelf. In the far north close to Papua New Guinea these include the Portlock Reefs and Eastern Fields. The majority of the remainder are located in an area known as the Coral Sea Plateau. Most are atoll formations, stretching from Osprey Reef in the north to Saumarez Reef in the south. Some are substantial in area – Lihou Reef is a long oval reef structure totalling nearly 2 500 square kilometers. In addition, a number of other reefs lie further south or east, off the Coral Sea Plateau, including Mellish, Frederick, Kenn, Wreck and Cato Reefs.

Information about these reefs remains relatively scant, and they have only been visited by a small number of expeditionary research units. In general they have relatively low coral cover, with maximum estimates of 19-26 percent hard coral cover. In contrast, both coralline algae and sponges make up a considerable proportion of the substrate. Total algal cover is often greater than coral cover. The molluscan fauna is very diverse, with over 730 species listed from the areas around North East Herald Cay alone. Around this same island some 356 fish species have been recorded. There are very important bird nesting colonies on some of the coral cays, while the beaches are widely utilized by nesting turtles.

A small number of dive operators take tourists out to the reefs of the Coral Sea, renowned for the water clarity and wide diversity of near pristine marine life. A number of reefs in the central Coral Sea area are protected. Although not under constant surveillance, they, and many other reefs in the region, benefit from their remote location.

Left: The vast atoll structure of Lihou Reef in the Coral Sea (STS046-90-9, 1992). Right, above: A trumpet emperor Lethrinus miniatus *with extensive branching corals. Right, below: This coral recruit is only 15 millimeters in diameter and probably only a few months old, yet over decades or centuries colonies may reach several meters across.*

High latitude reefs

MAP 11f

South of the Great Barrier Reef there are several reefs and coral communities at high latitudes. The south flowing East Australia Current has an important role to play in maintaining these communities, bringing warm waters as well as the potential for new larval recruits to settle on the reefs.

Lord Howe Island is a high volcanic island with a reef structure extending for some 6 kilometers along its western side. This is the most southerly coral reef in the world, lying beyond 31°S. Elizabeth and Middleton Reefs are platform reefs lying on older volcanic seamounts which form a chain to the north of Lord Howe Island. On the mainland coast there are no true coral reefs. Flinders Reef, east of Brisbane, is a sandstone structure, but has been colonized by a range of tropical corals and other species. Further south, the Solitary Islands also have important coral communities. There are also many smaller benthic communities with coral reef species elsewhere along the coastline of South Queensland and northern New South Wales.

Biodiversity is low in these areas, but they remain of significance as they represent the ecological limits of many species. The offshore reefs are also of interest because of their considerable isolation. Elizabeth and Middleton Reefs have 122 species of reef coral while Lord Howe Island has 65 species. Some of these, particularly at Lord Howe Island, are thought to be temporary populations dependent on recruitment of new individuals from

Australia

General Data	
Population (thousands)	19 165
GDP (million US$)	359 913
Land area (km²)	7 706 304
Marine area (thousand km²)	7 437
Per capita fish consumption (kg/year)	19
Status and Threats	
Reefs at risk (%)	32
Recorded coral diseases	6
Biodiversity	
Reef area (km²)	48 960
Coral diversity	428 / 461
Mangrove area (km²)	11 500
No. of mangrove species	39
No. of seagrass species	21

Interesting and important coral communities have developed at high latitudes around islands to the east of Australia, including the remote Norfolk Island. This species Porites heronensis *is a high latitude species, also recorded on the reefs of Japan, although absent from the reefs of central Southeast Asia (photo: JEN Veron).*

more northerly sources of larvae. Some 477 species of fish have been recorded at Lord Howe Island – for the most part tropical but also including some temperate species. Endemism is relatively high, with about 4 percent of the fish unique to Lord Howe, Elizabeth and Middleton Reefs. An outbreak of the crown-of-thorns starfish in the 1980s caused extensive damage to both Elizabeth and Middleton Reefs, considerably reducing coral cover, particularly on the outer reef slopes.

On the mainland coast, biodiversity in the Flinders Reef is thought to rival that of the Elizabeth and Middleton Reefs, although it remains poorly documented. The Solitary Islands are of more particular interest as they maintain a balance of tropical and temperate species. Only 53 reef coral species are recorded, and some 280 fish of which 80 percent are considered tropical. The islands are also noted for their large populations of sea anemones with their resident clownfishes. Little penguins also nest in the islands making this, with the Galapagos, one of the only places where this group of predominantly Antarctic species may be found near coral reef species.

Most of these reefs and coral communities have some form of legal protection. Elizabeth and Middleton Reefs are a marine reserve, while their isolation protects them from large numbers of visitors. Lord Howe Island has a resident population of about 300, and while tourism provides the mainstay of the economy, total numbers are limited and any impacts on the reef are small.

Protected areas with coral reefs

Site name	Designation	Abbreviation	IUCN cat.	Size (km²)	Year
Australia					
Ashmore Reef	National Nature Reserve	NNR	Ia	583.00	1983
Christmas Island	National Park	NP	II	87.00	1990
Cobourg	Marine Park	MP	VI	2 290.00	1983
Coringa – Herald	National Nature Reserve	NNR	Ia	8 856.00	1983
Elizabeth and Middleton Reefs	National Nature Reserve	NNR	Ia	1 880.00	1987
Emden	Historic Shipwreck Protected Zone	HSPZ	Unassigned	1.00	1982
Great Barrier Reef	Commonwealth Marine Park	CoMP	VI	344 800.00	1979
Lihou Reef	National Nature Reserve	NNR	Ia	8 436.91	1982
Lord Howe Island	Marine Park	MP	VI	480.00	2000
Mermaid Reef	National Nature Reserve	NNR	Ia	539.84	1991
Ningaloo	Marine Park	MP	VI	2 255.64	1987
Pulu Keeling	Commonwealth National Park	CoNP	II	26.02	1995
Rowley Shoals	Marine Park	MP	VI	232.50	1990
Shark Bay	Marine Park	MP	VI	7 487.35	1990
Solitary Islands	Marine Reserve	MR	VI	1 000.00	1991
South West Solitary Island	Nature Reserve	NR	Ia	0.03	1961
Yongala	Historic Shipwreck	HShip	Unassigned	0.78	1982
Cobourg Peninsula	Ramsar Site			2 207.00	1974
Great Barrier Reef	World Heritage Site			348 700.00	1981
Lord Howe Island Group	World Heritage Site			11.76	1982
Moreton Bay	Ramsar Site			1 133.14	1993
Pulu Keeling National Park	Ramsar Site			1.22	1996
Shark Bay Western Australia	World Heritage Site			21 973.00	1991
Shoalwater and Corio Bays	Ramsar Site			2 391.00	1996

Selected bibliography

Collins LB, Zhu ZR, Wyrwoll K-H (1997). Geology of the Houtman Abrolhos. In: Vacher HL, Quinn T (eds). *Developments in Sedimentology, 54: Geology and Hydrology of Carbonate Islands.* Elsevier Science BV, Amsterdam, Netherlands.

Done TJ (1982). Patterns in the distribution of coral communities across the Central Great Barrier Reef. *Coral Reefs* 1: 95-107.

Gladstone W, Dight IJ (1994). Torres Strait baseline study. *Mar Poll Bul* 29: 121-125.

Hatcher BG (1985). Ecological research at the Houtman's Abrolhos: high latitude reefs of Western Australia. *Proc 5th Int Coral Reef Symp* 6: 291-297.

Hearn CJ, Parker IN (1988). Hydrodynamic processes on the Ningaloo coral reef, Western Australia. *Proc 8th Int Coral Reef Symp* 2: 497-502.

Heyward AJ, Halford A, Smith L, Williams DMcB (1998). Coral reefs of north west Australia: baseline monitoring of an oceanic reef ecosystem. *Proc 8th Int Coral Reef Symp* 1: 289-294.

Hopley D (1982). *The Geomorphology of the Great Barrier Reef: Quarternary Development of Coral Reefs.* John Wiley and Sons, New York, USA.

Marsh LM (1992). The occurrence and growth of Acropora in extra-tropical waters off Perth, Western Australia. *Proc 7th Int Coral Reef Symp* 2: 1233-1238.

Playford PE (1997). Geology and hydrogeology of Rottnest Island, Western Australia. In: Vacher HL, Quinn T (eds). *Developments in Sedimentology, 54: Geology and Hydrology of Carbonate Islands.* Elsevier Science BV, Amsterdam, Netherlands.

Randall JE, Allen GR, Steene RC (1997). *Fishes of the Great Barrier Reef and Coral Sea*, 2nd edn. Crawford House Publishing Pty Ltd, Bathurst, Australia.

Stoddart DR, Yonge M (eds) (1978). *The Northern Great Barrier Reef.* The Royal Society, London, UK.

Sudara S, Wilkinson CR, Ming CL (eds) (1994). *Proceedings, Third ASEAN-Australia Symposium on Living Coastal Resources. Volume 2: Research Papers.* Australian Institute of Marine Science, Townsville, Australia.

Sweatman H, Bass D, Cheal A, Coleman G, Miller I, Ninio R, Osborne K, Oxley W, Ryan D, Thompson A, Tomkins P (1998). *Long-Term Monitoring of the Great Barrier Reef.* Australian Institute of Marine Science, Townsville, Australia.

Veron JEN (1986). *Corals of Australia and the Indo-Pacific.* University of Hawai'i Press. 1993 edn. Angus and Robertson, North Ryde, Australia.

Veron JEN (2000). *Corals of the World.* 3 vols. Australian Institute of Marine Science, Townsville, Australia.

Wilkinson CR, Cheshire AC (1988). Cross-shelf variations in coral reef structure and function – influences of land and ocean. *Proc 6th Int Coral Reef Symp* 1: 227-233.

Wilkinson CR, Sudara S, Ming CL (eds) (1994). *Proceedings, Third ASEAN-Australia Symposium on Living Coastal Resources. Volume 1: Status Reviews.* Australian Institute of Marine Science, Townsville, Australia.

Williams DMcB, Hatcher AI (1983). Structure of fish communities on outer slopes of inshore, mid-shelf and outer shelf reefs of the Great Barrier Reef. *Mar Ecol Prog Ser* 10: 239-250.

Woodroffe CD, Falkland AC (1997). Geology and hydrogeology of the Cocos (Keeling) Islands. In: Vacher HL, Quinn T (eds). *Developments in Sedimentology, 54: Geology and Hydrology of Carbonate Islands.* Elsevier Science BV, Amsterdam, Netherlands.

Zann LP (1995). *Our Sea, Our Future. State of the Marine Environment Report, 1995.* Department of the Environment, Sport and Territories, Canberra, Australia.

Zann LP (2000). North Eastern Australia: the Great Barrier Reef region. In: Sheppard C (ed). *Seas at the Millennium: An Environmental Evaluation, Vol 2.* Elsevier Science Ltd, Oxford, UK.

Zell L (1999). *Diving and Snorkelling Australia's Great Barrier Reef.* Lonely Planet Publications, Melbourne, Australia.

Map sources

Map 11a

For Cocos (Keeling) coral reef areas have been copied from a 1:100 000 source map (full reference unavailable, but source was a scanned paper map available on http://www.lib.utexas.edu/Libs/PCL/Map_collection/islands_oceans_poles/Cocos(Keeling)_76.jpg).

The available data for Christmas Island were poor, so reefs have simply been plotted as a line running immediately offshore from the island. In reality this represents an exaggeration of the true reef area. All remaining areas are taken from Petroconsultants SA (1990)*.

Map 11b

Coral features are taken as arcs from Petroconsultants SA (1990)*.

Maps 11c, d and e

For the Great Barrier Reef, coral reef areas were generously supplied (in 1995) by the Great Barrier Reef Marine Park Authority at 1:250 000. For the reefs of the Torres Strait, data are taken from Petroconsultants SA (1990)*.

Map 11f

Coral features are taken as arcs from Petroconsultants SA (1990)*.

* See Technical notes, page 401

Chapter 12 Melanesia

Melanesia occupies a wide swathe of the southwestern Pacific Ocean, stretching from New Guinea in the west to Fiji in the east. This is a region dominated by high islands, with considerable ongoing volcanic activity in the west. A broad range of reef types are found, though atolls are generally not as widespread as the extensive fringing and barrier systems associated with the high islands. Overall this region includes a vast area of reefs, making up about 14 percent of the global total.

Biodiversity is high right across Melanesia, though there is a cline of diminishing diversity towards the east. This trend is hidden in many national statistics by an almost reversed trend of knowledge. The reefs of New Caledonia are the best studied, and a number of reefs in Fiji have also received some scientific attention. Even in these countries, however, there are vast areas which remain unvisited and undescribed by scientists. The remaining countries are very poorly known indeed.

The first peoples to come to this region were Papuans, arriving in New Guinea over 40 000 years ago and reaching the nearby Bismarck Archipelago some 30 000 years ago. Much more recently, about 4 000 years ago, another group, known as Austronesians, arrived by sea from the Southeast Asian areas of what are now the Philippines and Indonesia. They settled in coastal communities from the Bismarck Archipelago to Fiji, and came to dominate most of the region.

The majority of this region remains under traditional stewardship and most reefs are widely utilized by artisanal fishers. Traditional reef management at the level of individual villages, combined with relatively low population densities, has helped to ensure the continued sustainable utilization of most resources, particularly away from towns and centers of Western development.

Attempts to establish marine protected areas along Western lines have only had very limited success, but the rights of villages to manage their own nearshore resources are now quite widely recognized in legal and constitutional systems. This recognition is important in maintaining traditional systems as changes to more Western lifestyles and governance take hold.

Left: Papua New Guinea has a diversity of reef life which rivals the Southeast Asian center of diversity. Right: Ouvéa, New Caledonia, a spectacular atoll formation which has tilted and uplifted along one edge (STS038-74-86, 1990).

MAP 12

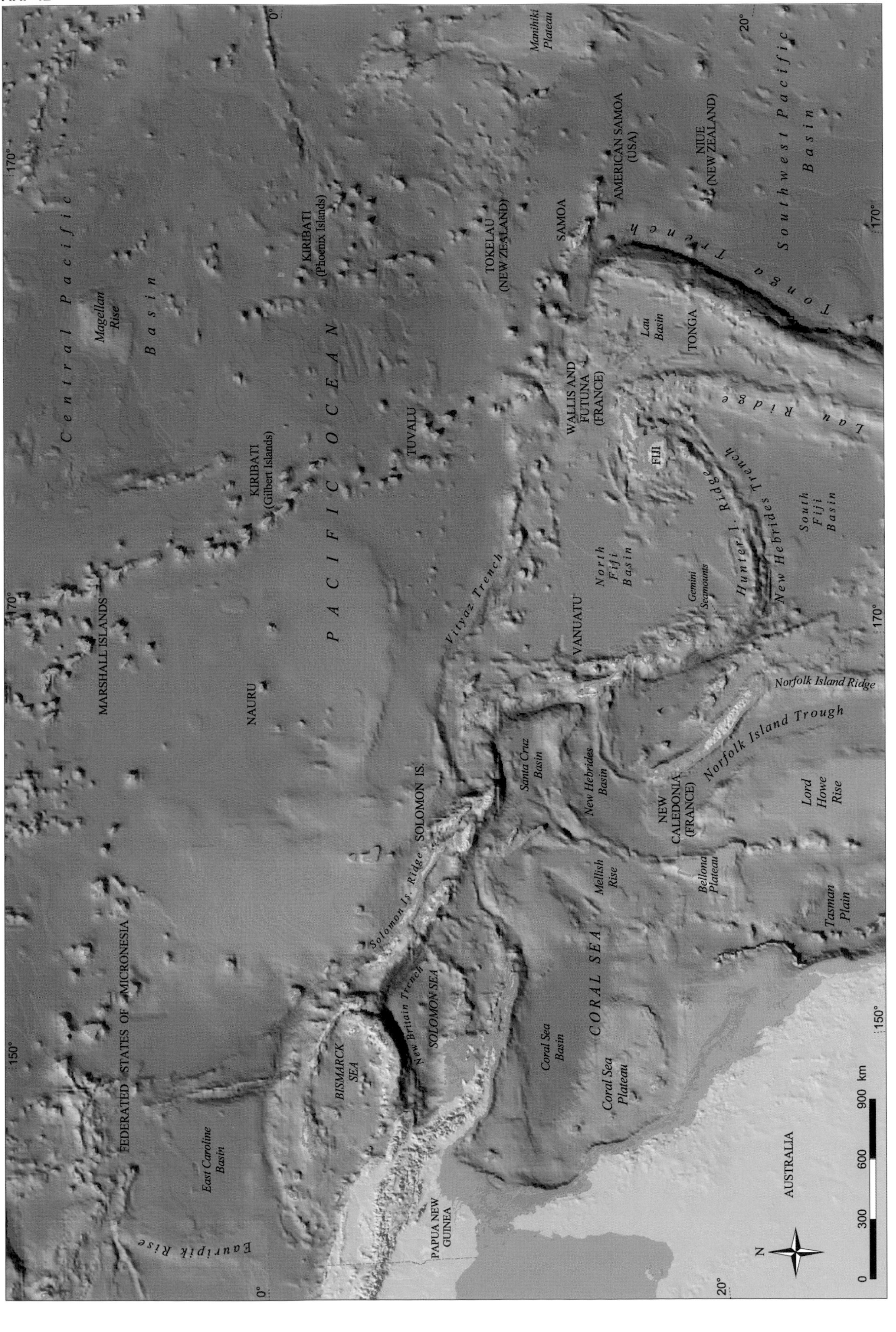

150°
170°
170°
Central Pacific
Magellan Rise
Basin
MARSHALL ISLANDS
FEDERATED STATES OF MICRONESIA
Eauripik Rise
East Caroline Basin
NAURU
KIRIBATI (Gilbert Islands)
KIRIBATI (Phoenix Islands)
0°
0°
PACIFIC OCEAN
BISMARCK SEA
Solomon Is. Ridge
SOLOMON IS.
New Britain Trench
SOLOMON SEA
TUVALU
PAPUA NEW GUINEA
Vityaz Trench
TOKELAU (NEW ZEALAND)
Manihiki Plateau
Santa Cruz Basin
Coral Sea Basin
SAMOA
VANUATU
WALLIS AND FUTUNA (FRANCE)
CORAL SEA
Mellish Rise
New Hebrides Basin
North Fiji Basin
Coral Sea Plateau
AMERICAN SAMOA (USA)
Lau Basin
FIJI
NEW CALEDONIA (FRANCE)
TONGA
Gemini Seamounts
Bellona Plateau
NIUE (NEW ZEALAND)
20°
Norfolk Island Trough
Hunter I. Ridge
20°
New Hebrides Trench
Lau Ridge
Tonga Trench
Southwest Pacific Basin
N
AUSTRALIA
Lord Howe Rise
Norfolk Island Ridge
South Fiji Basin
Tasman Plain
0 300 600 900 km
150°
170°
170°

Papua New Guinea

MAP 12a

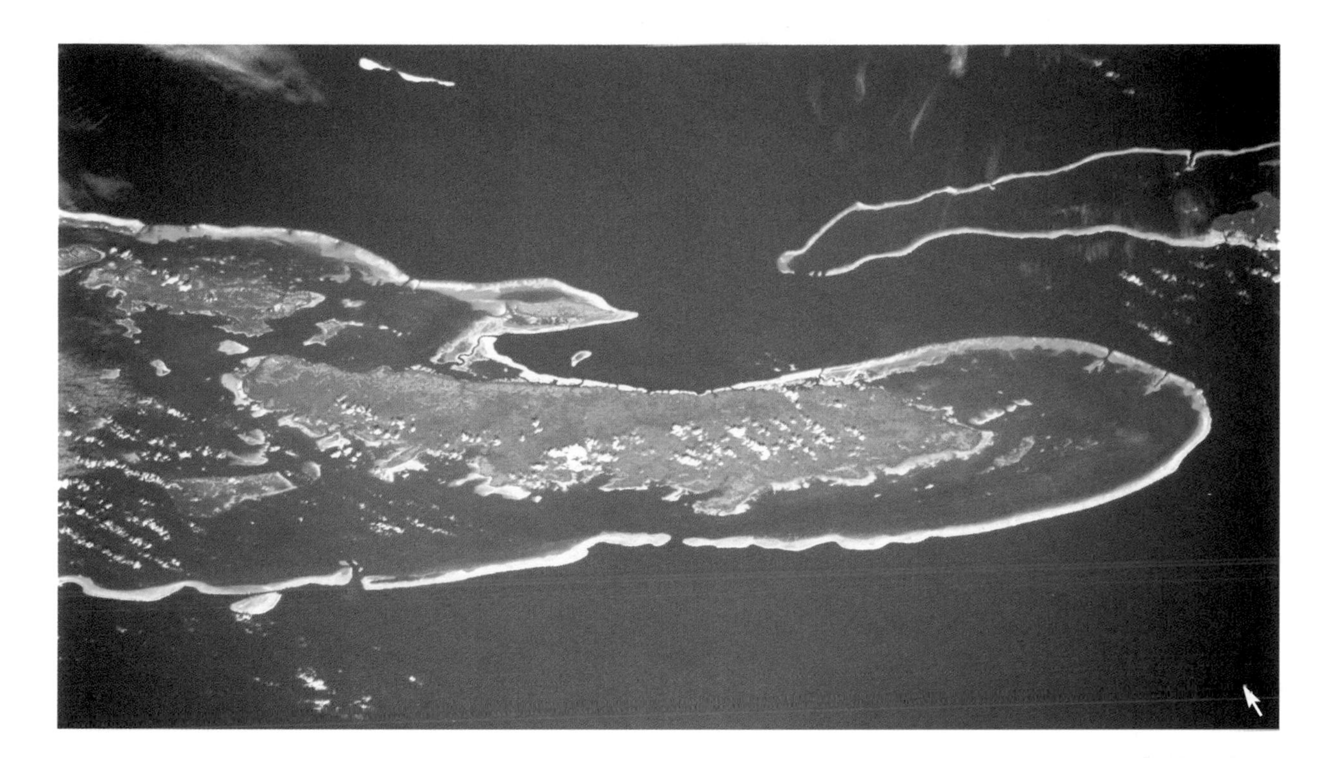

Papua New Guinea is one of the world's major coral reef nations, with a vast area of reefs. The total area is probably considerably larger than the figure of 13 840 square kilometers provided here, as many reefs remain unmapped in the present work. Lying on the eastern edge of the great center of coral reef biodiversity in Southeast Asia, there is every indication that this country enjoys remarkably high levels of biodiversity. It has suffered very little in terms of human impacts and there are great opportunities for continued sustainable management and conservation of its resources.

Papua New Guinea consists of the eastern half of the island of New Guinea together with a large number of smaller islands. To the west the country shares a land border with Irian Jaya (West Papua), Indonesia. To the north of the mainland, the Pacific Ocean becomes enclosed as the Bismarck Sea, bounded by the Bismarck Archipelago to the north and New Britain to the east. East of New Britain and the mainland coast lies the Solomon Sea, further bounded by Bougainville Island in the east and the Louisiade Archipelago in the south. South of the mainland and the Louisiade Archipelago is the Coral Sea, with the Gulf of Papua and the Torres Strait to the west. The islands of the Torres Strait are Australian, but come to within just a few kilometers of the southern coast of Papua New Guinea. In the southwest there are extensive coastal lowlands around the Fly River. Further north the mainland is divided by the long range of mountains known as the Highlands, reaching over 3 000 meters in a number of places and more than 4 500 meters at the highest point (Mount Wilhelm). There are further mountains along much of the north coast, divided at the mouth of the Sepik River. The offshore islands also show considerable relief. The northern coastline and all the islands to the north lie in a region of important tectonic activity where the large Pacific, Australia and Caroline tectonic plates come together, separated by a complex of microplates underlying the Bismarck and Solomon Seas.

Papua New Guinea has a vast area of coral reefs, including fringing, barrier and atoll formations, but there is little information for much of the country and it seems likely that there may still be large areas of unmapped and possibly unknown reefs.

Mainland reefs

The north coast, particularly in the west, is little known, however there are fringing reefs in many areas, including around the nearshore chain of the Schouten Islands. East of

The Calvados Barrier Reef is a spectacular structure, here encircling Sudest Island (STS065-92-50, 1994).

MAP 12a

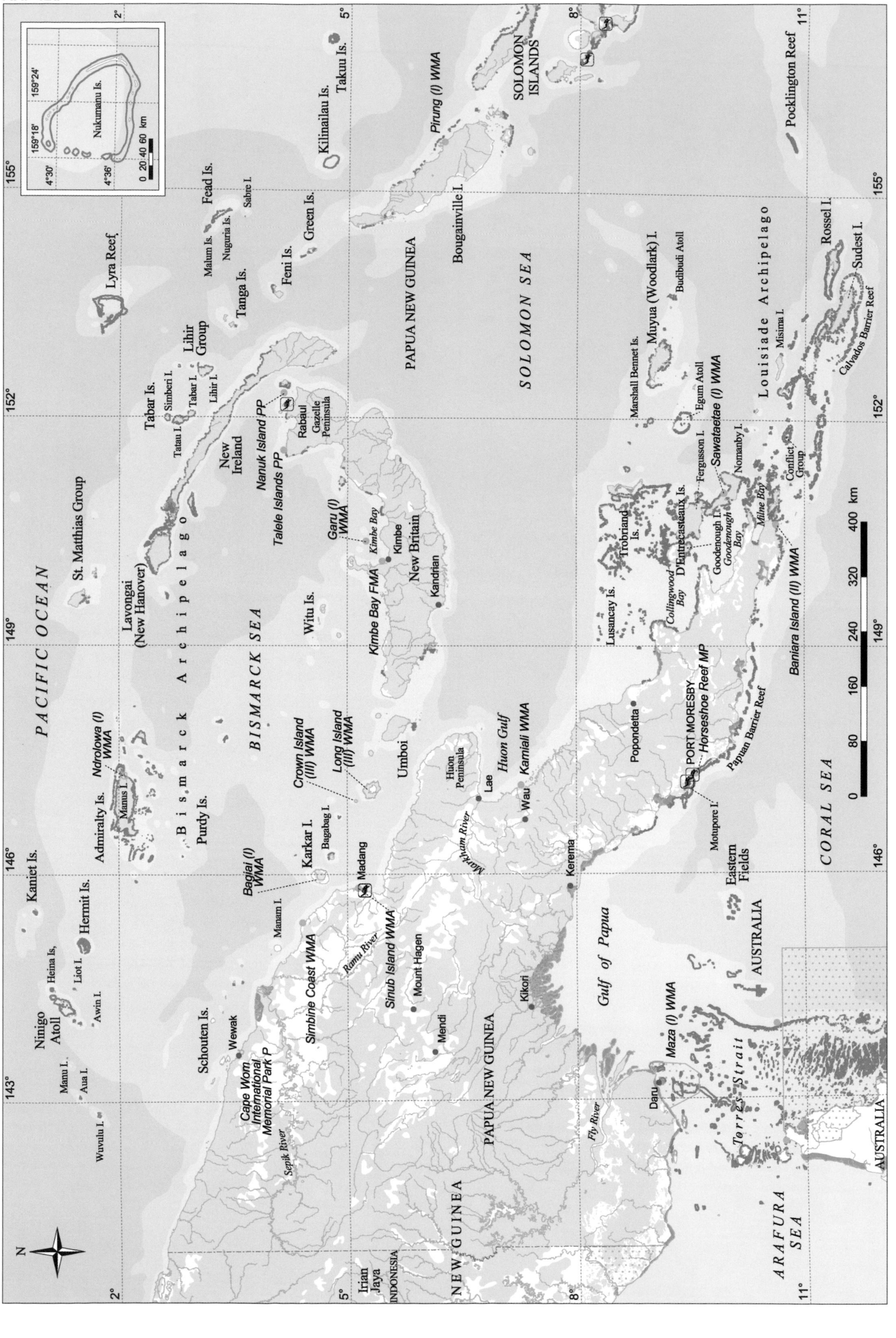

PACIFIC OCEAN
BISMARCK SEA
SOLOMON SEA
CORAL SEA
ARAFURA SEA
Gulf of Papua
Huon Gulf
Torres Strait
Bismarck Archipelago
Louisiade Archipelago
NEW GUINEA
PAPUA NEW GUINEA
SOLOMON ISLANDS
AUSTRALIA
INDONESIA
Irian Jaya
Wuvulu I.
Manu I.
Aua I.
Ninigo Atoll
Heina Is.
Liot I.
Awin I.
Hermit Is.
Kaniet Is.
Admiralty Is.
Manus I.
Ndrolowa (I) WMA
Purdy Is.
St. Matthias Group
Lavongai (New Hanover)
Tabar Is.
Tatau I.
Simberi I.
Tabar I.
Lihir I.
Lihir Group
New Ireland
Lyra Reef
Tanga Is.
Malum Is.
Nuguria Is.
Fead Is.
Sabre I.
Feni Is.
Green Is.
Kilinailau Is.
Takuu Is.
Bougainville I.
Pirung (I) WMA
Nukumanu Is.
Schouten Is.
Wewak
Cape Wom International Memorial Park P
Sepik River
Manam I.
Bagiai (I) WMA
Karkar I.
Bagabag I.
Madang
Crown Island (III) WMA
Long Island (III) WMA
Umboi
Witu Is.
Kimbe Bay FMA
Kimbe Bay
Kimbe
Garu (I) WMA
Talele Islands PP
Nanuk Island PP
Rabaul
Gazelle Peninsula
New Britain
Kandrian
Simbine Coast WMA
Sinub Island WMA
Ramu River
Mount Hagen
Mendi
Markham River
Lae
Huon Peninsula
Wau
Kamiali WMA
Kerema
Kikori
Fly River
Daru
Maza (I) WMA
Eastern Fields
Motupore I.
PORT MORESBY
Horseshoe Reef MP
Papuan Barrier Reef
Popondetta
Lusancay Is.
Collingwood Bay
Trobriand Is.
D'Entrecasteaux Is.
Goodenough I.
Goodenough Bay
Fergusson I.
Sawataetae (I) WMA
Normanby I.
Milne Bay
Baniara Island (II) WMA
Conflict Group
Marshall Bennet Is.
Egum Atoll
Muyua (Woodlark) I.
Budibudi Atoll
Misima I.
Calvados Barrier Reef
Sudest I.
Rossel I.
Pocklington Reef

2° 5° 8° 11° 143° 146° 149° 152° 155°
0 80 160 240 320 400 km

the Sepik and Ramu river mouths fringing reefs continue, often in long unbroken stretches up to the easternmost point of East Cape, while in places barrier reefs run further offshore, notably around Madang where there are about 50 associated offshore islands. It has been estimated that, in all, over half of this coastline may have fringing reefs. There is a major break in the fringing reef around Lae in the Huon Gulf, where the Markham River delivers an estimated 10 million tons of sediment per year. Along the southern coast, reef development is somewhat restricted in the area of the Fly River Delta and the smaller river deltas to the east, where there are extensive mangrove forests, turbidity is high and salinities are variable. Further east, coral reefs are widespread from Port Moresby eastwards. These are sometimes termed the Papuan Barrier Reef as they run some distance offshore, separated by a lagoon about 5 kilometers wide. The total length of this reef is some 560 kilometers, though this is broken by a number of channels.

Northern islands and reefs

To the north of the mainland, the westernmost islands of the Bismarck chain include a number of coralline islands surrounded by fringing reefs, and also a number of atolls, including the large Ninigo Atoll, Liot, Heina and Kaniet (Sae) Islands. The Hermit group is a near-atoll, with two high basaltic islands in the center of its lagoon. East of these lie the Admiralty Islands, dominated by the volcanic Manus Island, but including a number of smaller islands and atoll formations. The large volcanic islands of Lavongai (New Hanover) and New Ireland lie further east, with other smaller high ones to the north, including the St. Matthias group, Tabar and Lihir Islands. Reefs are widespread and include fringing systems as well as platform and atoll structures, though few details are available about these. Equally little is known about the reefs of Bougainville Island: a barrier reef is located about 15 kilometers off the southwest coast and there are other barrier structures with a number of small associated islets off the east coast. Around New Britain the shelf is mostly very narrow and, although there are fringing reefs, they are not continuous. There are also various offshore patch and barrier structures, including around Kimbe Bay and the Gazelle Peninsula in the north.

Southeast

Perhaps the most extensive reef systems in the country are those of Milne Bay Province. The continental shelf is broad and scattered with numerous platform reefs, some with their associated islands (both volcanic and calcareous) between the mainland and the Trobriand Islands to the north. These islands are relatively flat limestone structures. The same shelf continues southwards to the volcanic D'Entrecasteaux Islands. East of the Trobriands are several islands and reefs, including Egum Atoll, the large Muyua (Woodlark) Island with associated fringe and near-barrier reef systems, and Budibudi Atoll in the far east. A long chain of reefs and islands extends southwest from the tip of Papua New Guinea and here there is a vast complex of reefs. The most significant is the Calvados Barrier Reef, extending as a long arm along the southern edge of the continental shelf, right around the tip of Sudest Island to follow the northern edge of the shelf, a total distance of some 640 kilometers. This system encircles many other lagoonal platform reefs and fringing reefs around islands.

Left: Many of Papua New Guinea's fringing reefs remain unexplored. Where the reef flat is narrow they often do not feature on any maps. Right: Reef scene dominated by Porites lichen.

The nearby high island of Rossel is also surrounded by a large barrier reef some 200 kilometers in circumference.

In addition to the reefs described above, quite a number of other systems lie even more remote from the high islands. There are several atolls far off the continental shelf in the Pacific Ocean, including Lyra, Malum and Nuguria east from New Ireland, and Takuu and Nukumanu east from Bougainville. The remote reefs of the northern Coral Sea fall under Australian jurisdiction, but a few are visited by dive vessels operating from Papua New Guinea.

Biodiversity

The reefs of Papua New Guinea are only just being explored in terms of their biodiversity, and studies in the late 1990s revealed extremely diverse communities, including many hitherto undescribed species. A recent survey of multiple sites in the Milne Bay area revealed some 869 reef and nearshore fishes, 637 molluscs and 362 scleractinian corals. When combined with the limited records from previous surveys these totals become even larger, with 1 039 fishes and a predicted 420 coral species from this region alone. While there are affinities with Great Barrier Reef and Coral Sea faunas, the reefs of Milne Bay and by implication all the reefs to the north are thus still closely linked to the Indonesian and Philippine centers of diversity and endemism. With such a variety of reefs it is, of course, not possible to describe anything like a typical reef community. Reefs include a complete range of geomorphological structures, while complex and diverse reef communities have also developed on new volcanic slopes where true reef structures have yet to form. Similarly coral cover, and the dominant species or groups, are highly varied, from low diversity, low coral cover locations, notably close to areas of high sediment loads, to diverse coral slopes with coral cover reaching 100 percent in many areas. Coral bleaching has been observed on a few occasions, with the earliest report describing extensive bleaching at a location in Kimbe Bay in 1983. This led to close to 100 percent mortality, although there was a near complete recovery within 10 years. In 1996-97, bleaching was observed in a number of locations, and was reported to have led to mortalities approaching 80 percent around Motopure Island in Kimbe Bay. In Milne Bay over 50 percent of corals were reported to have bleached in one study in June 1996, however recovery was good. Bleaching was also observed at a number of locations in 2000.

Socio-economic considerations

The most widespread use of coral reefs in Papua New Guinea is for subsistence fisheries. However, few settlements are wholly dependent on fish resources, as fishing is generally second to agriculture for food and income. The dominant commercial offshore fishery is tuna, largely conducted by foreign vessels under license. Inshore commercial fisheries include lobster, sea cucumber, trochus, green snail, pearl shell and some reef fish. The live fish trade has been operating in a few areas since 1991, and numbers of large reef fish are reported to be reduced in the northwest. There have also been reports of blast fishing, particularly around urban centers.

Direct pollution from human settlements is limited to areas close to major towns. Unfortunately there are various other threats which may significantly impact the reefs of Papua New Guinea in the near future. The major commercial industries are logging and mining. Logging is occurring over large areas, although not on the same scale as in much of Southeast Asia, and there is a considerable threat that increased sedimentation will impact nearby nearshore reefs.

Mining, notably for copper, gold and silver, is a major industry, and in the late 1990s environmental controls were still weak. Deliberate or accidental discharge of mine tailings into rivers or directly offshore has caused problems in a number of locations, both by smothering corals and from toxic impacts. The Panguana copper mine on Bougainville was reported to have smothered some 100 square kilometers of sea floor with its tailings prior to its closure in 1989 as a result of civil war. The Ok Tedi mine in the southwest has released tens of millions of tons of tailings into the Ok Tedi and Fly Rivers, causing massive damage to inland forests and possibly affecting offshore reefs. The Misima gold mine in Milne Bay was reported to have caused extensive destruction on nearshore reefs, and there are similar mines at many other localities on the mainland and offshore islands.

Natural factors also affect the status of reefs in Papua New Guinea. The country lies within the cyclone belt, with considerable implications for the reefs. A few have also been severely damaged by volcanic and seismic activity. In

Traditional fishing methods predominate over wide areas of Melanesia.

Protected areas with coral reefs

Site name	Designation	Abbreviation	IUCN cat.	Size (km²)	Year
Papua New Guinea					
Bagiai	Wildlife Management Area	WMA	VI	137.60	1977
Baniara Island	Wildlife Management Area	WMA	Unassigned	0.15	1975
Crown Island	Wildlife Management Area	WMA	VI	59.69	1977
Horseshoe Reef	Marine Park	MP	Unassigned	3.96	1981
Kamiali	Wildlife Management Area	WMA	VI	474.13	1996
Kimbe Bay	Fisheries Management Area	FMA	VI	0.02	1999
Long Island	Wildlife Management Area	WMA	VI	419.22	1977
Maza	Wildlife Management Area	WMA	VI	1 842.30	1978
Nanuk Island	Provincial Park	PP	IV	0.12	1973
Ndrolowa	Wildlife Management Area	WMA	VI	58.50	1985
Pirung	Wildlife Management Area	WMA	VI	442.40	1989
Sawataetae	Wildlife Management Area	WMA	VI	7.00	1977
Simbine Coast	Wildlife Management Area	WMA	VI	0.72	2000
Sinub Island	Wildlife Management Area	WMA	VI	0.12	2000
Talele Islands	Provincial Park	PP	IV	0.40	1973

1998 one of the largest *tsunamis* (tidal waves) on record hit a 25 kilometer section of coastline in the north of the country, with devastating effects on coastal villages. The impact on fringing reefs in the area is unknown.

While tourism as a whole is a relatively small-scale activity, dive tourism is growing fairly rapidly because of the spectacular and unspoiled nature of so many of the reefs. There are now a number of operators, particularly associated with "live-aboard" vessels.

A number of protected areas covering coral reefs have been declared, but the majority of these are simply marine extensions of terrestrial sites, with little or no real provisions for marine protection. Even where they exist, there is little or no local knowledge or application of regulations. In many ways, because of traditional uses and ownership in almost all areas, the application of Western-style national parks and reserves may not be entirely appropriate in this country. Recognizing this, a number of community-run wildlife management areas have been developed. In 2000, the most effective of these included a number of sites which were still awaiting full legal establishment, such as Sinub Island in Madang Lagoon, Simbine Coast (125 kilometers northwest of Madang) and Kimbe Bay. Elsewhere, traditional fisheries combined with relatively low coastal populations spare wide areas of reefs from immediate threat.

Coral reef research has been somewhat limited in this country. Conservation International has been undertaking a number of research expeditions to the reefs in the Milne Bay region, while there are also research facilities on Motupore Island near Port Moresby and in Kimbe Bay.

Papua New Guinea

General Data	
Population (thousands)	4 927
GDP (million US$)	4 730
Land area (km²)	467 498
Marine area (thousand km²)	2 366
Per capita fish consumption (kg/year)	14
Status and Threats	
Reefs at risk (%)	46
Recorded coral diseases	0
Biodiversity	
Reef area (km²)	13 840
Coral diversity	378 / 517
Mangrove area (km²)	5 399
No. of mangrove species	44
No. of seagrass species	7

Solomon Islands

MAP 12b

The Solomon Islands consist of over 900 islands widely distributed in the Western Pacific. The bulk of the land area comprises seven large volcanic islands which form a double chain running from northwest to southeast and converging on the island of Makira (San Cristobal). The Santa Cruz Islands are a second group of three larger volcanic islands lying to the east: Ndenö, Utupua and Vanikolo together with smaller islands, including the Reef Islands and the Duff Islands. In addition to these there are several more remote islands and reefs. Ontong Java is a large atoll of some 1 500 square kilometers lying over 250 kilometers north of Santa Isabel, while nearby there is a smaller atoll, Roncador Reef, which has no associated islands. About 200 kilometers northeast of Malaita is Sikaiana Atoll (Stewart Islands), with a number of small islands around a near-atoll (there is a 45 meter high remnant of the volcano). To the south of the main island chain are two raised atolls, Bellona and Rennell, with fringing reefs around their perimeter. South of these are three large atoll structures with no associated islands – the Indispensable Reefs. The far eastern borders of this island nation are determined by the three small islands of Anuta, Fatutaka and Tikopia.

The Solomon Islands lie on the western margin of the Pacific plate and all are of volcanic origin. There is still volcanic activity in a number of locations, notably on Tinakula in the Santa Cruz Islands and on the submarine volcano of Kavachi, south of New Georgia. The latter is one of the most active volcanoes in the region and has created several new islands in the last century, most recently in May 2000.

Coral reefs are widespread throughout the country. A number of atolls have already been mentioned, and fringing reefs are numerous around most of the islands. Even where they are not marked on maps, such as around Guadalcanal, there are narrow, steeply shelving fringing structures. Barrier reefs are less developed, although there are barrier complexes with associated islands around New Georgia and northeast Choiseul and around Utupua. A complex system occurs around the Reef Islands, including the 25 kilometer Great Reef extending westwards from the main island group. Other shallow platform reefs are found north of the Reef Islands.

Very little is currently known about biodiversity on the reefs of the Solomon Islands, however given their location and the relatively low levels of human impact in many areas, they are likely to include highly diverse and important reef communities. A recent survey of the fish

Left: A sunset wrasse Thalassoma lutescens *takes shelter under a plate* Acropora. *Right: A feather star with a massive coral.*

158°
161°
164°
167°
170°
6°
9°
12°
N
Takuu (Mortlock) Is.
PAPUA NEW GUINEA
Nukumanu Is.
PAPUA NEW GUINEA
Ontong Java Atoll
Roncador Reef
PACIFIC OCEAN
Choiseul
Arnavon MarCA
Shortland Is.
Treasury Is.
Santa Isabel
Vella Lavella
Mbava
Kolombangara
Ranongga
New Georgia
Marovo Lagoon
Simbo
Ghizo
Rendova
Vangunu
Ngatokae
Tetepare
New Georgia Group
Kavachi
Onogou (Ramos) I.
Lau Lagoon
Langalanga Lagoon
Stewart Is.
Sikaiana Atoll
Florida Is.
Nggela
Russell Is.
Malaita
HONIARA
Guadalcanal
Maramasike
SOLOMON ISLANDS
Duff Is.
Great Reef
Reef Is.
Tinakula
SOLOMON SEA
Makira (San Cristobal)
Santa Cruz Is.
Ndenö
Bellona
Utupua
Rennell
East Rennell World Heritage Site
Vanikolo
Anuta I.
Fatutaka I.
Tikopia I.
Indispensable Reefs
CORAL SEA
0 40 80 120 160 200 km
VANUATU

communities in the Santa Cruz Islands identified 725 species (including non-reef species). Some of the most detailed data describing the reefs of the region were gathered during a 1965 Royal Society expedition which visited a large number of the western islands. Overall this expedition concluded that coral reef growth was not well developed, and listed only 87 species of scleractinian coral. But it would appear that these observations were misplaced: little use was made of scuba diving, and it has been further suggested that the reefs may have been impacted by some form of mass mortality just prior to the expedition. Coral bleaching was reported from a wide range of localities in 2000, at the same time as the major bleaching event recorded in Fiji. These include observations from the high islands in the west, but also from Ontong Java Atoll. There is no information about the degree of associated mortality.

The coral reefs of the Solomon Islands include wide areas still largely unimpacted by human activities, although there are also areas where such pressures are large and growing. The islands have one of the fastest population growth rates in the world, and 86 percent of the people are rural. Dependence on coral reefs for protein remains high and subsistence fishing is widespread. In the more populous areas this is leading to overfishing and in certain parts, such as the Lau Lagoon off north Malaita, many of the preferred edible species have been lost. Fishing methods can also be destructive, whether trampling and damaging the reefs with nets, or poison fishing including traditional methods that use coastal plant species to provide the poison. This poison is unselective, killing a number of non-targeted species and reportedly damaging corals.

Traditional management systems are still of considerable importance in the Solomon Islands, as customary marine tenure is widely held and all reefs are "owned" by particular groups who have fishing rights. Christian leaders, traditional *kastom* men, or even the villagers themselves regularly place taboos on particular reefs, usually for a restricted period of time. More complete protection is provided in some areas by other beliefs, such as around Onogou (Ramos) Island, which is believed to house the spirits of the dead and can only be visited after following strict protocols.

Commercial fishing has probably had more far-reaching effects across the islands, notably for selected target species. In 1999 the export of trochus and related snails brought in over US$1 million, with sea cucumbers, shark fins, live fish and spiny lobster also bringing in substantial amounts. Both trochus and sea cucumbers are already overfished and their numbers are declining rapidly in many areas. A significant giant clam fishery peaked in 1983, but overharvesting has depleted these stocks in all areas, exacerbated by illegal poaching by foreign vessels. (A Taiwanese vessel was captured on the Indispensable Reefs in 1986 with 10 tons of frozen adductor muscles on board, representing many tens of thousands of individual clams.) There is some concern that as these different fisheries collapse exploitation of other stocks, such as those used in the live fish trade, will increase.

Efforts to establish giant clam mariculture have been ongoing for about ten years. While this has been interrupted by violence on Guadalcanal, a smaller operation continues near Ghizo. Pearl exports have traditionally been an important industry in the Solomons, and with the export of wild-caught stocks prohibited there are now ongoing efforts to establish a farm near Ghizo. The aquarium trade has been increasing relatively rapidly, much of it around Nggela in the Florida Islands, where

East Rennell, a World Heritage Site, is an uplifted atoll, with the brackish Lake Tegano filling the former lagoon (STS068-244-94, 1994).

Protected areas with coral reefs

Site name	Designation	Abbreviation	IUCN cat.	Size (km²)	Year
Solomon Islands					
Arnavon	Marine Conservation Area	MarCA	VI	82.70	na
EAST RENNELL	WORLD HERITAGE SITE			**370.00**	**1998**

there have been reports of extensive damage. Coral pieces are broken off for collection, damaging methods such as cyanide are used to capture reef fish, and reefs are trampled during capture, resulting in coral breakage.

One unusual but highly significant threat to reefs in the Solomon Islands comes from the use of lime in the habit of chewing betel nuts. The latter are taken from the fruits of a palm and are chewed with a pepper leaf and lime in an addictive habit. The lime is prepared by burning branching corals (typically *Acropora*). Major users may consume 20 kilos of lime per year (derived from over 30 kilos of live coral), and in some areas, such as the lagoon reefs of Malaita, these corals are highly depleted. One estimate suggested that about 6 million kilos of lime are used per year, derived from 10 million kilos of live coral, making this one of the largest single threats to reefs in the country. There are some ongoing efforts to establish coral gardens which might be harvested sustainably, and some communities report that they utilize coral patches on a rotation system.

Although many of the Solomon Islands remain forested, logging is ongoing in many areas and there are few efforts to control sediment runoff. Although there have been no studies it seems highly likely that coral reefs will be impacted in some areas. Particular concern has been expressed about logging activities on the island of Vangunu and the potential impact on the Marovo Lagoon. Previously selectively logged areas on this island are now being clear-felled and converted to oil-palm plantations, and there is concern that the conversion process may produce even higher levels of sedimentation, and that subsequent fertilizer use could create ongoing problems.

There is no sewage treatment in any of the urban centers in the Solomon Islands. As populations grow this will increasingly threaten the health of both humans and reefs. Tourism has never been a major industry, although there are various hotels and "live-aboards" which cater for divers. The establishment of legally gazetted protected areas in the Solomon Islands is complicated by the customary tenure of all reefs. A number of island sanctuaries have recently been repealed. As negotiations on the ownership of at least one of these have been ongoing, there is evidence that a number of villages have been using the confusion to rapidly deplete the surrounding reef resources. The most successful marine protected area is the Arnavon marine conservation area. First established in 1975 there have been a number of disputes and problems, but in 1992 the site was revived and a community-based management committee established. The eastern third of Rennell Island was declared a World Heritage Site in 1998, with boundaries extending seawards for 3 nautical miles.

The current civil unrest in the Solomon Islands is largely confined to the island of Guadalcanal, but general instability is causing considerable disruption, not only to the small tourism industry, but also to development activities, including mariculture. In particular the closure of the Coastal Aquaculture Centre near Honiara in late 1999 has set back aquaculture research considerably, although some of its activities have been transferred to a second center near Ghizo, while other work has relocated to New Caledonia. A new Institute of Marine Resources run by the University of the South Pacific has also been abandoned.

Solomon Islands

GENERAL DATA	
Population (thousands)	466
GDP (million US$)	224
Land area (km²)	27 740
Marine area (thousand km²)	1 630
Per capita fish consumption (kg/year)	33
STATUS AND THREATS	
Reefs at risk (%)	46
Recorded coral diseases	0
BIODIVERSITY	
Reef area (km²)	5 750
Coral diversity	101 / 398
Mangrove area (km²)	642
No. of mangrove species	22
No. of seagrass species	3

New Caledonia

MAP 12c

The Archipelago of New Caledonia is dominated by the large land mass of Grande Terre, the third largest island in the Pacific (after New Guinea and New Zealand). It is of continental origin, having diverged from Australia some 65 million years ago, and has a mountainous interior rising to more than 1 600 meters. The shallow shelf on which the island sits extends a considerable distance to the northwest, and includes the continental Îles Bélep and a number of smaller islands and coral cays further north. To the southeast the shelf continues down to the Île des Pins. The shallow platform on which these islands lie is rimmed by the world's second largest barrier reef, over 1 300 kilometers in length. There are quite regular passes in the reef, largely associated with river mouths on the mainland. In a few locations to the north, notably along the Grand Récif de Koumac and the Récif des Français, a deep lagoon has developed within the single structure of the outer barrier reef flat, forming a rare double barrier structure. Between the barrier reef and the mainland there are many platform structures, while fringing reefs are also widespread in many areas. To the northwest the barrier reef continues beyond the Îles Bélep up to a channel, the Grand Passage. Beyond this there is a group of reefs known as the D'Entrecasteaux Reefs, including Huon Atoll, Surprise Atoll and a number of smaller atoll and barrier-like structures.

Due east of Grande Terre is the low-lying chain of the Loyalty Islands. Maré in the south has some volcanic rocks, while the others are composed primarily of uplifted limestone. Fringing reefs encircle most of Maré and Lifou. Ouvéa to the north is a partially uplifted and tilted atoll with fringing reefs along its eastern (uplifted) coastline, but with a wide reef-fringed lagoon to the west. Moving northwest is the small atoll of Beautemps-Beaupré and then a small group of reefs known as the Astrolabe Reefs. Lying in considerable isolation to the northwest of the Loyalty Islands and to the east of the D'Entrecasteaux Reefs is another significant reef structure, the Petrie Reef. Far to the east of the Loyalty Islands are the two small islands of Matthew and Hunter. Geographically these are a part of the Vanuatu chain, but they are claimed by both countries.

Over 550 kilometers west of Grande Terre are two very large shallow reef areas. The Chesterfield Islands are coral cays along the perimeter of a large atoll. A shallow reef with a very steep outer slope marks its northern and western margins, while to the southeast there is no clear atoll margin, but a gentle slope to considerable depths. To

A reef slope showing high coral cover, with a map puffer Arothron mappa.

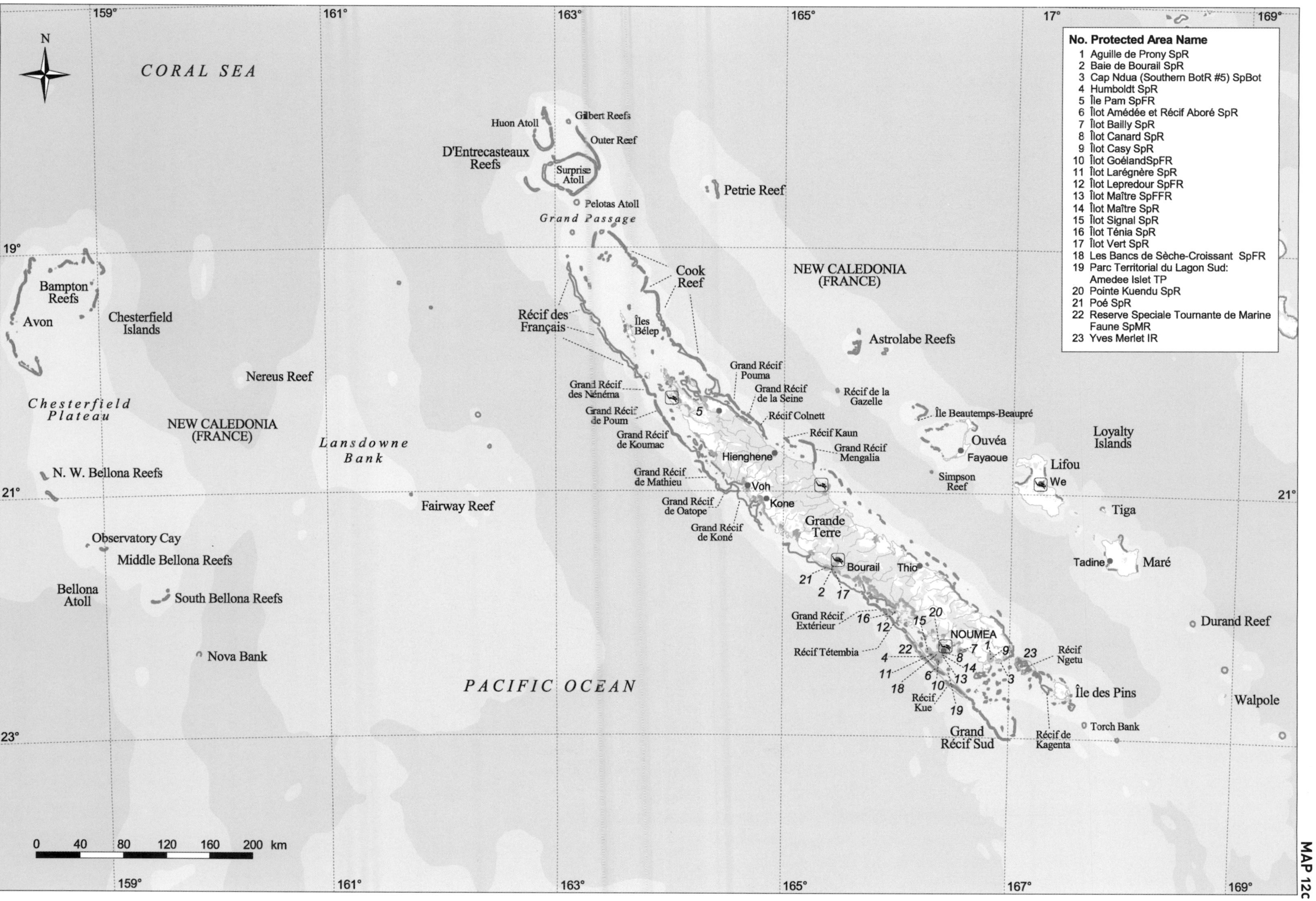
No. Protected Area Name
1 Aguille de Prony SpR
2 Baie de Bourail SpR
3 Cap Ndua (Southern BotR #5) SpBot
4 Humboldt SpR
5 Île Pam SpFR
6 Îlot Amédée et Récif Aboré SpR
7 Îlot Bailly SpR
8 Îlot Canard SpR
9 Îlot Casy SpR
10 Îlot GoélandSpFR
11 Îlot Larégnère SpR
12 Îlot Lepredour SpFR
13 Îlot Maître SpFFR
14 Îlot Maître SpR
15 Îlot Signal SpR
16 Îlot Ténia SpR
17 Îlot Vert SpR
18 Les Bancs de Sèche-Croissant SpFR
19 Parc Territorial du Lagon Sud: Amedee Islet TP
20 Pointe Kuendu SpR
21 Poé SpR
22 Reserve Speciale Tournante de Marine Faune SpMR
23 Yves Merlet IR
CORAL SEA
PACIFIC OCEAN
NEW CALEDONIA (FRANCE)
NEW CALEDONIA (FRANCE)
Bampton Reefs
Avon
Chesterfield Islands
Chesterfield Plateau
N. W. Bellona Reefs
Observatory Cay
Middle Bellona Reefs
Bellona Atoll
South Bellona Reefs
Nova Bank
Nereus Reef
Lansdowne Bank
Fairway Reef
Huon Atoll
Gilbert Reefs
Outer Reef
D'Entrecasteaux Reefs
Surprise Atoll
Pelotas Atoll
Grand Passage
Petrie Reef
Cook Reef
Récif des Français
Îles Bélep
Astrolabe Reefs
Grand Récif Pouma
Grand Récif de la Seine
Récif Colnett
Récif de la Gazelle
Île Beautemps-Beaupré
Grand Récif des Néméma
Grand Récif de Poum
Grand Récif de Koumac
Récif Kaun
Grand Récif Mengalia
Ouvéa
Fayaoue
Loyalty Islands
Lifou
We
Hienghene
Simpson Reef
Grand Récif de Mathieu
Voh
Kone
Grand Récif de Oatope
Grand Récif de Koné
Grande Terre
Tiga
Tadine
Maré
Bourail
Thio
Grand Récif Extérieur
Récif Tétembia
NOUMEA
Récif Ngetu
Île des Pins
Récif Kue
Grand Récif Sud
Récif de Kagenta
Torch Bank
Durand Reef
Walpole
159°
161°
163°
165°
167°
169°
17°
19°
21°
23°
0 40 80 120 160 200 km

the south, Bellona Atoll again has a number of shallow reefs and a few coral islands, notably along its western perimeter.

Between the Chesterfield Plateau and Grande Terre is the wide Landsdowne Bank, which is mostly sandy and 70-80 meters in depth, but includes the small Nereus Reef in the north. To the southeast of this area the Fairway Reef also comes close to the surface and dries at low tide. A number of maps show a large island to the northwest of Nereus Reef which does not actually exist: Île de Sable. However, there may be shallow banks and submerged reefs in this region, which remains poorly charted.

The climate is somewhat seasonal, warm from November to April when it is dominated by frontal systems and when cyclones may occur, followed by a cooler season from June to September when southeast trade winds predominate.

The location of New Caledonia relatively close to the global center of coral reef diversity, combined with the large area and variety of reefs, ensures very high diversity. Unlike many other reefs in the region these have been the subject of considerable study, although many areas in this large archipelago nonetheless remain poorly known and undescribed. Thus far about 1 950 fish species have been recorded, about 5 500 molluscs, 5 000 crustaceans, 600 sponges and 300 corals. Around 5 percent of species are thought to be endemic.

Grande Terre contains about 40 percent of the world's known nickel deposits. Held in the sub-surface rocks of the high mountain areas it is extracted using open-cut techniques involving the removal of about a 30 meter depth of surface soil and rock. Over 300 mines have been dug in the past century, removing 280 million tons of surface rock for the extraction of a further 110 million tons of ore. Sedimentation from these mines has been considerable in many streams and estuaries, and has greatly increased turbidity in some nearshore waters. In the Ouenghi Basin north of Noumea the delta area has extended by 300-400 meters along a 3 kilometer stretch of coastline as a result of this sedimentation over the last 30 years. Much of the sediment flows out onto the east coast in the Thio and Dothio Rivers. Controls reducing sedimentation from new mines have been in place since

The world's second largest barrier reef encircles Grande Terre (STS033-73-61, 1989).

Protected areas with coral reefs

Site name	Designation	Abbreviation	IUCN cat.	Size (km²)	Year
New Caledonia					
Aguille de Prony	Special Reserve	SpR	IV	na	na
Baie de Bourail	Special Reserve	SpR	IV	na	na
Humboldt	Special Reserve	SpR	IV	na	na
Les Bancs de Sèche-Croissant	Special Fauna Reserve	SpFR	IV	na	na
Île Pam	Special Fauna Reserve	SpFR	IV	4.60	1966
Îlot Amédée et Récif Aboré	Special Reserve	SpR	IV	na	na
Îlot Bailly	Special Reserve	SpR	IV	na	na
Îlot Canard	Special Reserve	SpR	IV	na	na
Îlot Casy	Special Reserve	SpR	IV	na	na
Îlot Goéland	Special Fauna Reserve	SpFR	VI	na	na
Îlot Larégnère	Special Reserve	SpR	IV	na	na
Îlot Maître	Special Reserve	SpR	IV	1.54	1981
Îlot Signal	Special Reserve	SpR	IV	na	na
Îlot Ténia	Special Reserve	SpR	IV	na	na
Îlot Vert	Special Reserve	SpR	IV	na	na
Poé	Special Reserve	SpR	IV	na	na
Pointe Kuendu	Special Reserve	SpR	IV	na	na
Tournante de Marine Faune	Special Marine Reserve	SpMR	IV	355.70	1981
Yves Merlet	Integral Reserve	IR	Ia	167.00	1970

the 1970s, however the older abandoned mines will continue to release sediments for many decades. The offshore location of the majority of reefs provides some protection from such impacts, but nearshore areas may suffer considerably owing to the protective nature of the lagoon which holds sediments close to shore.

Aside from sedimentation, most of the human pressures on the coral reefs are centered around the main town of Noumea where there are localized problems of domestic pollution and some overfishing. Here and elsewhere around the southeast there have been significant coastal modifications associated with urbanization and tourism developments. The tourism industry is particularly important to Noumea and there are many hotels, notably in the southeast but also along the west coast and in the Loyalty Islands.

There is a good network of marine protected areas around the southeast of the region, and there are plans to develop a similar network in the north. In addition to these, customary reserves and traditional fishing areas are recognized.

New Caledonia

General Data	
Population (thousands)	202
GDP (million US$)	2 987
Land area (km²)	19 140
Marine area (thousand km²)	1 740
Per capita fish consumption (kg/year)	25
Status and Threats	
Reefs at risk (%)	13
Recorded coral diseases	1
Biodiversity	
Reef area (km²)	5 980
Coral diversity	151 / 359
Mangrove area (km²)	456
No. of mangrove species	16
No. of seagrass species	8

Vanuatu

MAPS 12d and e

Vanuatu represents the main bulk of an island chain which continues into the Santa Cruz Islands of the eastern Solomon Islands. Lying on the western margins of the Pacific plate, the islands are all of volcanic origin, and there is ongoing volcanic activity in a number of locations, including the Banks Islands in the northeast, Lopévi and Ambryn in the central islands, and Tanna in the south. Submarine volcanoes are also active, notably off Épi and Erromango. About 100 kilometers south of Anatom the Gemini Seamounts are another area of volcanic activity – explosions were observed from the eastern seamount in 1996. However the western Gemini Seamount, which rises to about 30 meters below the surface, was reported to have considerable marine life. The Matthew and Hunter Islands, in the far south of the island chain, are disputed between Vanuatu and New Caledonia. All of the islands are volcanic rock or uplifted carbonate structures, or some combination of these. The northern islands form a double chain. Current volcanic activity is generally restricted to the eastern islands and reef development is greatest in the western ones. Fringing reefs predominate, though Cook Reef, north of Éfaté, is a small atoll-like structure with no associated islands. The Reef Islands north of Vanua Lava are also part of a carbonate structure which has undergone a slight uplift. The islands lie in an area particularly prone to tropical cyclones, which cause damage to at least part of the archipelago annually. Cyclone Uma in 1987 was one of the most devastating, causing considerable harm to Éfaté and its reefs. Southeast trade winds predominate between May and October.

Fringing reefs encircle most of the islands from Éfaté southwards. In the central islands fringing reefs are generally not continuous and reef flats can be quite narrow. Typically reefs are best developed on eastern and northern coasts. The eastern coasts of Santo (Espíritu Santo) and Malakula both have wide fringing reefs and a number of coral islands on their coastlines. One interesting phenomenon in relatively recent times has been the observation of significant tectonic uplifts, particularly along the western coastlines of Malakula and Santo. Reefs on the northwest coasts of both islands were uplifted by up to 6 meters in 1965.

Detailed surveys of biodiversity have not been undertaken in most areas, although some 35 locations were visited by divers from the Australian Institute of Marine Sciences in 1988. Typically the reef crests and shallow reef areas are dominated by coralline algae and

Acropora *is the most diverse coral genus, with numerous forms including branching and plate corals.*

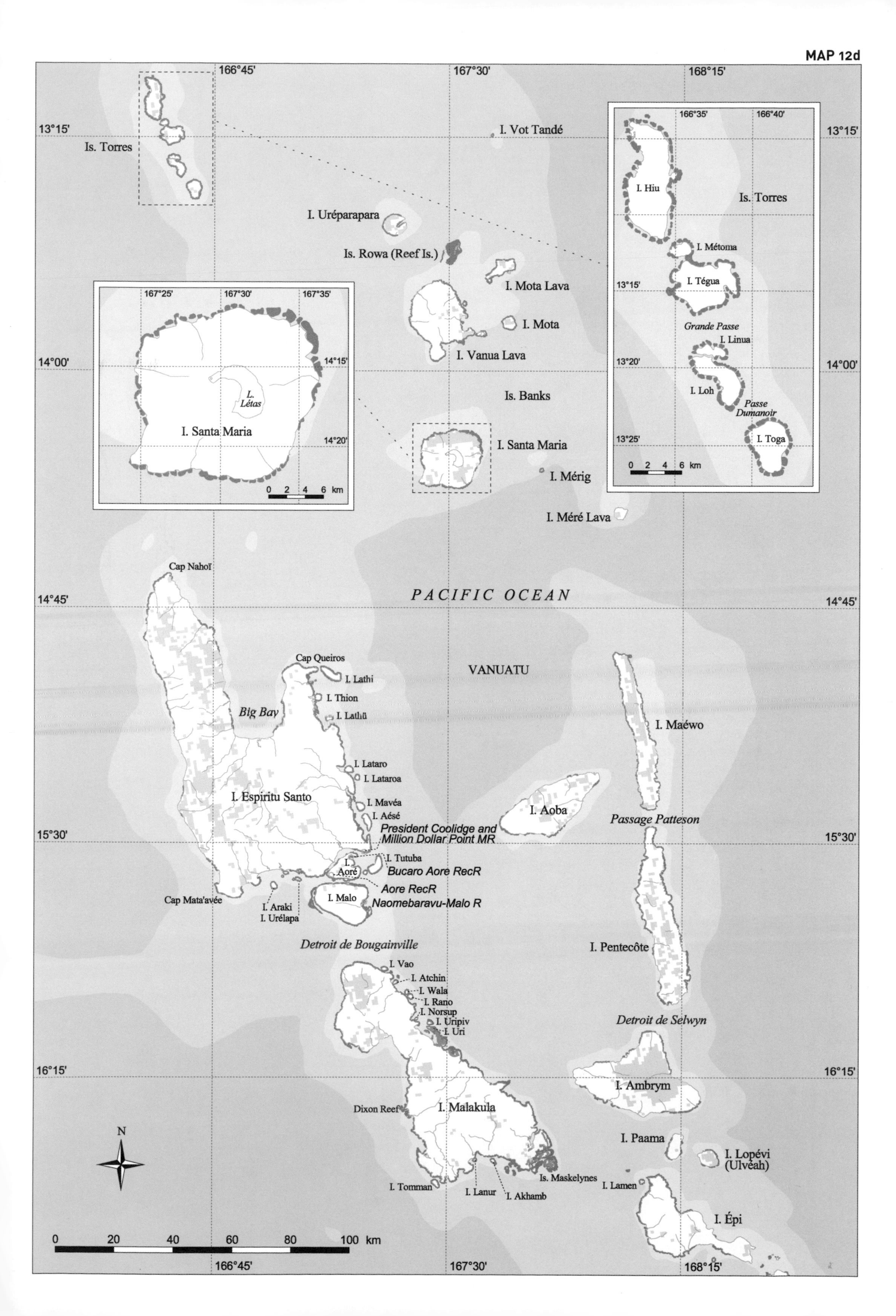

166°45'
167°30'
168°15'
13°15'
14°00'
14°45'
15°30'
16°15'
Is. Torres
I. Vot Tandé
I. Uréparapara
Is. Rowa (Reef Is.)
I. Mota Lava
I. Mota
I. Vanua Lava
Is. Banks
I. Santa Maria
I. Mérig
I. Méré Lava
167°25'
167°30'
167°35'
14°15'
14°20'
L. Létas
0 2 4 6 km
166°35'
166°40'
I. Hiu
I. Métoma
I. Tégua
13°15'
Grande Passe
I. Linua
13°20'
I. Loh
Passe Dumanoir
13°25'
I. Toga
PACIFIC OCEAN
VANUATU
Cap Nahoï
Cap Queiros
I. Lathi
I. Thion
Big Bay
I. Lataro
I. Lataroa
I. Espiritu Santo
I. Mavéa
I. Aésé
President Coolidge and Million Dollar Point MR
I. Tutuba
I. Aoré
Bucaro Aore RecR
Aore RecR
I. Malo
Naomebaravu-Malo R
Cap Mata'avée
I. Araki
I. Urélapa
I. Maéwo
I. Aoba
Passage Patteson
I. Pentecôte
Detroit de Bougainville
I. Vao
I. Atchin
I. Wala
I. Rano
I. Norsup
I. Uripiv
I. Uri
Detroit de Selwyn
I. Ambrym
Dixon Reef
I. Malakula
I. Paama
I. Lopévi (Ulvéah)
Is. Maskelynes
I. Lamen
I. Tomman
I. Lanur
I. Akhamb
I. Épi
N
0 20 40 60 80 100 km

MAP 12e

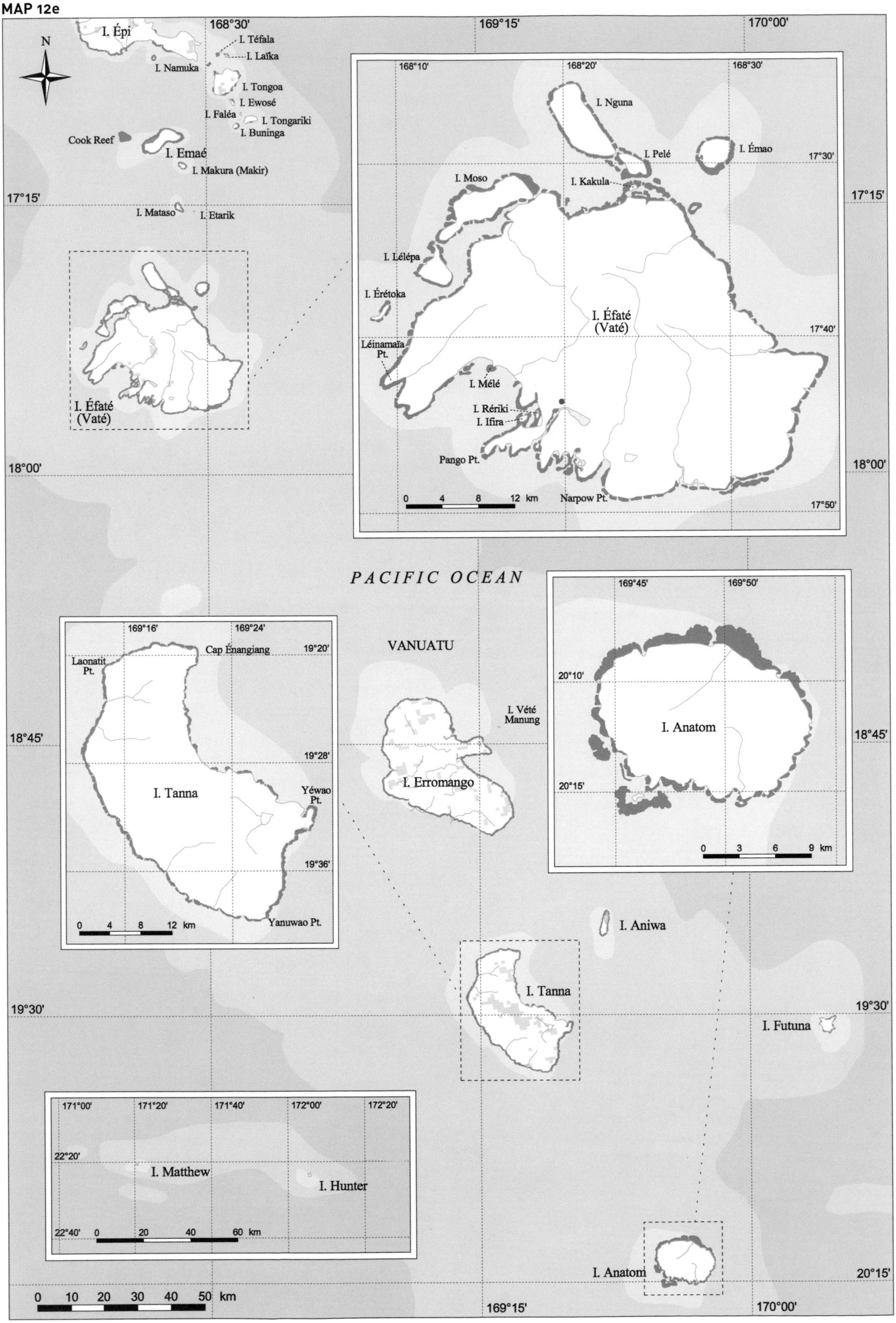

N

I. Épi
168°30'
169°15'
170°00'
I. Téfala
I. Laïka
I. Namuka
I. Tongoa
I. Ewosé
I. Faléa
I. Tongariki
I. Buninga
Cook Reef
I. Emaé
I. Makura (Makir)
17°15'
I. Mataso
I. Etarik
I. Éfaté (Vaté)
18°00'

168°10'
168°20'
168°30'
I. Nguna
I. Pelé
I. Émao
17°30'
I. Moso
I. Kakula
I. Lélépa
I. Érétoka
I. Éfaté (Vaté)
17°40'
Léinamaïa Pt.
I. Mélé
I. Rériki
I. Ifira
Pango Pt.
0 4 8 12 km
Narpow Pt.
17°50'

PACIFIC OCEAN

VANUATU

169°16'
169°24'
Cap Énangiang
19°20'
Laonatit Pt.
19°28'
I. Tanna
Yéwao Pt.
19°36'
0 4 8 12 km
Yanuwao Pt.

I. Vété Manung
18°45'
I. Erromango

169°45'
169°50'
20°10'
I. Anatom
20°15'
0 3 6 9 km

I. Aniwa
I. Tanna
19°30'
I. Futuna

171°00'
171°20'
171°40'
172°00'
172°20'
22°20'
I. Matthew
I. Hunter
22°40'
0 20 40 60 km

I. Anatom
20°15'
0 10 20 30 40 50 km
169°15'
170°00'

Protected areas with coral reefs

Site name	Designation	Abbreviation	IUCN cat.	Size (km²)	Year
Vanuatu					
Aore	Recreation Reserve	RecR	Unassigned	0.37	1984
Bucaro Aore	Recreation Reserve	RecR	Unassigned	0.20	1984
Naomebaravu - Malo	Reserve	R	Unassigned	0.11	1984
President Coolidge and Million Dollar Point	Marine Reserve	MR	Unassigned	1.00	1983

robust plate and branching corals, particularly in exposed locations, with a predominance of massive and branching corals below a depth of 3-5 meters. Massive coral also becomes predominant in embayments, with soft corals in more sheltered locations. In all some 469 fish species and 295 scleractinian corals were observed during this survey, although the complete list, particularly for the fish, is likely to be far longer. Periodic crown-of-thorns starfish outbreaks have been reported, and these combined with the impacts of cyclones and tectonic activity mean that live coral cover and physical state are quite variable across the country.

Vanuatu has a rapidly growing population. While a large number live in the two main towns, over 70 percent live on their traditional lands and remain heavily reliant on subsistence from the land and ocean. Catch methods include gill netting, capture by hand and spear gun and, in more remote areas, traditional techniques including bow and arrow, spears, traps and traditional poisons. Subsistence capture is largely of fish, but also includes substantial amounts of shellfish (34 percent) and lobster (20 percent). Cash income is also provided at the local level through collection of sea cucumbers, trochus, green snails, crustaceans and aquarium fish.

Up to the present time the larger islands of Vanuatu have remained heavily forested. However, there are now increased logging activities in a number of areas which may be impacting coral reefs through sediment runoff. Close to the main urban centers there are considerable concerns about pollution arising from sewage inputs, sediments and storm-water runoff, notably around Port Vila and the airport. Away from these areas concern has been expressed about the overharvesting of some non-motile reef species.

Although relatively restricted in terms of where it operates, tourism is an increasingly important sector of the economy, and diving is a highly popular activity among visitors. Formal protection of reef resources is not widespread, though a number of reserves have been established off Santo. For the most part however these are not respected, or even known about, by local people. The President Coolidge Reserve (a US shipwreck sunk in 1942) is a popular dive location for visitors. Customary tenure of reef resources is legally recognized in the constitution. At the level of villages and local communities a number of more effective management measures have been established, including harvesting restrictions on particular stocks, and sometimes more comprehensive protection of the marine environment. A proposed Environmental and Resource Management Bill currently under consideration might provide an opportunity to put such areas under some form of legal protection as community conservancy areas.

Vanuatu

General Data	
Population (thousands)	190
GDP (million US$)	191
Land area (km²)	12 535
Marine area (thousand km²)	680
Per capita fish consumption (kg/year)	26
Status and Threats	
Reefs at risk (%)	70
Recorded coral diseases	0
Biodiversity	
Reef area (km²)	4 110
Coral diversity	296 / 379
Mangrove area (km²)	16
No. of mangrove species	15
No. of seagrass species	1

Fiji

MAP 12f

Fiji is a vast archipelago centered on two relatively shallow geological features, the Fiji Platform and the Lau Ridge. Geologically, the area lies on the Indo-Pacific plate close to the boundary with the Pacific plate, in an area of relatively complex geology and fracturing. The two largest islands of Viti Levu and Vanua Levu, together with quite a number of smaller ones, lie on the relatively shallow Fiji Platform. Fringing reefs surround most of Viti Levu, with the largest continuous fringing reef running for 100 kilometers along the Coral Coast on its southern shore. Offshore from eastern Viti Levu the Suva Barrier Reef follows the shelf edge up to the island of Ovalau. The northern coast of Viti Levu is dominated by a very complex array of platform reef structures and intervening channels. Running northeast at some distance west of Viti Levu is a string of high islands known as the Yasawa group, again with an associated complex of fringing and patch reefs. These islands lie close to the edge of the Fiji Platform, and part of this shelf-edge is capped by Ethel Reef, a 30 kilometer barrier reef. Immediately south of Viti Levu is the island of Beqa, enclosed to the south and west by the Beqa Barrier Reef. Further south, the large island of Kadavu is separated from the Fiji Platform by the Kadavu Passage. This island has fringing reefs along much of its coastline, but is further dominated by a 95 kilometer long barrier reef running along its southern and eastern coasts and extending into the Great Astrolabe and North Astrolabe Reefs.

The line of the Yasawa group in the west is continued eastwards towards Vanua Levu by Fiji's longest barrier reef structure, the Great Sea Reef which runs along the shelf edge in a near continuous chain for over 200 kilometers, gradually converging towards the coastline of Vanua Levu at its northeastern tip. The Vatu Ira Channel between the two high islands is a tongue of deeper water, also fringed by elongated barrier reef structures including the Vanua Levu Barrier Reef along the eastern edge of this channel and up to the southern shore of Vanua Levu. Much of the southern shores of Vanua Levu are lined by fringing reefs, while the northern edge is marked by a similar complex of platform reefs to that along Viti Levu. Out to the east lies a complex of islands and reefs collectively termed the Ringgold Islands. These include several atolls, and also Budd Reef which is a near-atoll, with a group of high islets located in its lagoon. A group of reefs on the outer edge of thc Ringgold Islands make up the Nukusemanu and Heemskercq Reefs, parts

This broad view shows northern Viti Levu and western Vanua Levu, including the complex platform reef structures along the northern shores of both islands (STS027-32-34, 1988).

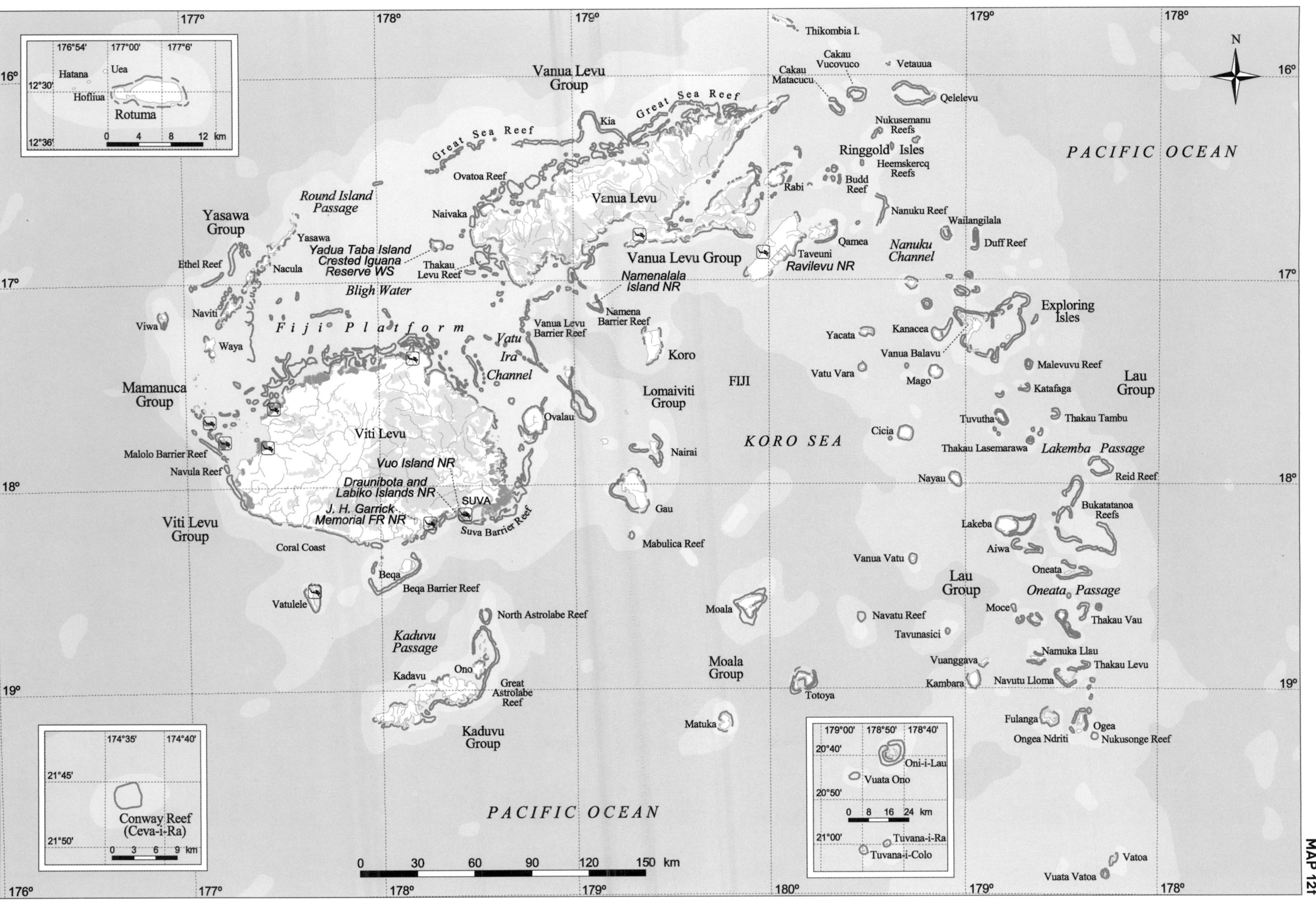

PACIFIC OCEAN
N
Vanua Levu Group
Great Sea Reef
Kia
Vanua Levu
Vanua Levu Group
Thikombia I.
Cakau Matacucu
Cakau Vucovuco
Vetauua
Qelelevu
Nukusemanu Reefs
Ringgold Isles
Heemskercq Reefs
Budd Reef
Rabi
Qamea
Taveuni
Ravilevu NR
Nanuku Reef
Wailangilala
Duff Reef
Nanuku Channel
Exploring Isles
Vanua Balavu
Kanacea
Mago
Malevuvu Reef
Katafaga
Thakau Tambu
Lau Group
Lakemba Passage
Reid Reef
Bukatatanoa Reefs
Yacata
Vatu Vara
Tuvutha
Thakau Lasemarawa
Cicia
Nayau
Lakeba
Aiwa
Oneata
Oneata Passage
Thakau Vau
Moce
Namuka Llau
Thakau Levu
Navutu Lloma
Ogea
Nukusonge Reef
Fulanga
Ongea Ndriti
Vatoa
Vuata Vatoa
Lau Group
Vanua Vatu
Navatu Reef
Tavunasici
Vuanggava
Kambara
Totoya
KORO SEA
FIJI
Koro
Lomaiviti Group
Namenalala Island NR
Namena Barrier Reef
Nairai
Gau
Mabulica Reef
Moala
Moala Group
Matuka
Ovatoa Reef
Naivaka
Thakau Levu Reef
Yadua Taba Island Crested Iguana Reserve WS
Bligh Water
Fiji Platform
Vatu Ira Channel
Vanua Levu Barrier Reef
Ovalau
Round Island Passage
Yasawa
Nacula
Yasawa Group
Ethel Reef
Naviti
Waya
Viwa
Mamanuca Group
Malolo Barrier Reef
Navula Reef
Viti Levu Group
Viti Levu
Vuo Island NR
Draunibota and Labiko Islands NR
J. H. Garrick Memorial FR NR
SUVA
Suva Barrier Reef
Coral Coast
Vatulele
Beqa
Beqa Barrier Reef
Kaduvu Passage
Kadavu
Ono
Great Astrolabe Reef
North Astrolabe Reef
Kaduvu Group
PACIFIC OCEAN
0 30 60 90 120 150 km
Hatana
Hofliua
Uea
Rotuma
176°54'
177°00'
177°6'
12°30'
12°36'
0 4 8 12 km
Conway Reef (Ceva-i-Ra)
174°35'
174°40'
21°45'
21°50'
0 3 6 9 km
179°00'
178°50'
178°40'
20°40'
20°50'
21°00'
Oni-i-Lau
Vuata Ono
Tuvana-i-Ra
Tuvana-i-Colo
0 8 16 24 km
16°
17°
18°
19°
176°
177°
178°
179°
180°

of which are submerged, but may be considered a near-atoll or barrier-type structure.

The Lau Islands make up the eastern edge of the Fiji group and lie at the top of the Lau Ridge, separated from the Fiji Platform by the Nanuku Channel. Most of the northern islands are high and of volcanic origin, but further south carbonate islands predominate. There are a number of atolls and near-atolls throughout the chain. The Exploring Isles make up one of the largest structures in this group, including the high island of Vanua Balavu, as well as a long barrier reef running out to the east and enclosing a number of smaller islands. Towards the center of the group the Bukatatanoa Reefs are another massive barrier reef complex. Lying considerably to the south of the main group of Lau Islands are the smaller islands of Vatoa (a high limestone island with a barrier reef) and the atoll of Vuata Vatoa. Further south again is a complex of four small reef systems including Oni-i-Lau, a small group of islands enclosed by a barrier reef.

The Koro Sea is a relatively enclosed sea between the Lau Islands and Viti Levu. There are a few islands scattered in this area. The Lomaiviti group east of Viti Levu is mostly volcanic and has well developed fringing and barrier structures. Further south, the Moala group is made up of three high volcanic islands with predominantly fringing reefs around them.

Far from the main islands of Fiji are three other reef areas. In the far northwest, the island of Rotuma is volcanic and has wide fringing reefs. A number of smaller islands nearby also have fringing reef structures. In the far southwest Conway Reef or Ceva-i-Ra is a small coral cay of some 200 by 50 meters on a platform reef. Finally, in the southeast, Fiji claims the Minerva Reefs, although these are also claimed by Tonga.

Some of the reefs of the country have been extensively studied in terms of their ecology and biodiversity, but, given the overall extent of Fijian reefs, a vast proportion remain poorly known. Species numbers are high, as might be expected from the location of these reefs in relation to the Indo-Pacific center of diversity as well as from the sheer variety of reef types. Most of the studies have been undertaken close to the University of the South Pacific in Suva or on the Great Astrolabe Reef where there is an associated field study station. Some 298 species of scleractinian coral have been recorded, alongside over 475 species of mollusc (including 253 nudibranchs and 102 bivalves) and some 60 species of ascideans. A total of 1 198 species of fish have been recorded in Fiji's waters, the majority of them reef-associated. The algal flora of these reefs is also well known, and some 422 species have been documented. Early in 2000, a major warming event in the surface waters around Fiji and neighboring countries led to between 50 and 100 percent of corals becoming bleached over wide areas, and extending to depths of 30 meters. Many corals subsequently died, particularly in southern parts of Viti Levu and Vanua Levu.

The rural people of Fiji depend on coral reefs for the vast bulk of their protein, and subsistence catches from reefs are estimated at approximately 17 000 tons per year. Although fishing with hand lines is most common, a vast

Left: Damselfish such as these humbugs Dascyllus aruanus *often take shelter among branching corals. Right: Part of the Beqa Barrier Reef.*

range of techniques are used, including traps, fences, spears, gill nets, hand nets and poisonous plants (notably derris root). Some fishers now also utilize scuba and hookah gear. Gleaning at low tide is also important for shellfish, sea cucumbers, sea urchins and octopus. Customary marine tenure at the level of individual villages has controlled utilization of reefs in many areas, with villages having rights of access to fishing areas or *qoliqolis*. Although such systems are still in place on many islands, there are increasing problems of overexploitation.

Nearshore commercial fisheries probably contribute a further 6 000 tons to the annual fish catch. In many areas target stocks have now declined considerably and this is largely linked to overfishing, although pollution, particularly near urban centers, may play a role. Stocks of emperors, mullets and trevally have declined, while the bump-headed parrotfish *Bolbometopon muracatum* has not been caught in Lau, Kadavu or Vanua Levu for at least ten years and may have been locally extirpated. In a similar manner, collection of bivalves has long been popular as a food source and led to the extermination of the giant clam *Tridacna gigas*, which was last recorded over 50 years ago. Other clam species, including the relatively recently discovered *Tridacna tevoroa* (only recorded from Fiji and Tonga) are also reported to have declined significantly. Black-lip pearl oysters, trochus and the main target species of sea cucumber have also been greatly reduced in recent years, but continue to be collected. Fiji is the major exporter of live coral and fish for aquaria in the Pacific, and there is also a company exporting live fish for the food trade in Hong Kong. Efforts are also underway to establish seaweed farming.

In addition to urban centers, other land-based activities which threaten or degrade reefs in Fiji include mangrove clearance for land reclamation, runoff from mines, agriculture, sugar and timber mills, poorly planned tourist development and solid waste disposal. Although there is some sewage treatment in the larger urban areas this is often inadequate, while solid waste is not only a visual problem but may be a health hazard to both humans and coastal species. Industrial pollution (mainly eutrophication, although there have been recent oil spills in the harbor) is a particular problem close to Suva. On Viti Levu and Vanua Levu intensive commercial farming on steep slopes has led to considerable soil erosion. Similar problems of sedimentation have been observed in some of the more remote and uninhabited islands as a result of overgrazing by goats.

Traditional management of reefs has led to their sustainable use throughout the archipelago for thousands of years. Although the ownership of the seabed now resides with the state, the customary fishing rights of indigenous Fijians remain, under the Fisheries Act of 1942. Traditional management of reefs by the villages continues to some degree, particularly in the outer islands, and includes the setting aside (using taboos) of certain areas such as those which become overfished. Traditional fishing areas have been mapped by the national government. Customary fishing rights have hindered the formal establishment of marine protected areas and, while existing protected areas extend to the shoreline in a number of places, none incorporate sub-littoral elements. Despite this, a number of tourist resorts have established small private sanctuaries through agreements with customary fishing rights holders. Similarly, community-based marine reserves are being established in a few areas, with the support of the government, non-governmental organizations and local communities. These, together with the wider recognition of customary marine tenure, increasing environmental education and the establishment of reef monitoring, may well suffice to protect much larger areas of Fiji's coral reefs in the short and medium term.

Fiji

General Data	
Population (thousands)	832
GDP (million US$)	1 602
Land area (km²)	19 379
Marine area (thousand km²)	1 217
Per capita fish consumption (kg/year)	33
Status and Threats	
Reefs at risk (%)	68
Recorded coral diseases	1
Biodiversity	
Reef area (km²)	10 020
Coral diversity	177 / 398
Mangrove area (km²)	385
No. of mangrove species	9
No. of seagrass species	5

Traditional society and customary marine tenure may have an important role to play in the protection of Fiji's reefs.

Selected bibliography

PAPUA NEW GUINEA

Halstead B, Rock T (1999). *Diving and Snorkelling Papua New Guinea.* Lonely Planet Publications, Melbourne, Australia.

Hoeksema BW (1992). The position of northern New Guinea in the center of marine benthic diversity: a reef coral perspective. *Proc 7th Int Coral Reef Symp* 2: 710-717.

Huber ME (1994). An assessment of the status of the coral reefs of Papua New Guinea. *Mar Poll Bul* 29: 69-73.

Maniwave T, Sweatman H, Marshall P, Munday P, Rei V (2000). Status of coral reefs of Australia and Papua New Guinea. In: Wilkinson CR (ed). *Status of Coral Reefs of the World: 2000.* Australian Institute of Marine Science, Cape Ferguson, Australia.

Munday PL (ed) (2000). *The Status of Coral Reefs of Papua New Guinea.* Australian Institute of Marine Science, Cape Ferguson, Australia.

Pandolfi JM (1992). A review of the tectonic history of New Guinea and its significance for marine biogeography. *Proc 7th Int Coral Reef Symp* 2: 718-728.

Thomas JD (1997). Using marine invertebrates to establish research and conservation priorities. In: Reaka-Kudla ML, Wilson DE, Wilson EO (eds). *Biodiversity II: Understanding and Protecting our Biological Resources.* Joseph Henry Press, Washington DC, USA.

Werner TB, Allen GR (eds) (1998). *RAP Working Papers, 11: A Rapid Biodiversity Assessment of the Coral Reefs of Milne Bay Province, Papua New Guinea.* Conservation International, Washington DC, USA.

SOLOMON ISLANDS

Grano S (ed) (1993). *Solomon Islands: National Environmental Management Strategy.* South Pacific Regional Environmental Programme, Apia, Western Samoa.

Richards AH, Bell LJ, Bell JD (1994). Inshore fisheries resources of Solomon Islands. *Mar Poll Bul* 29: 90-98.

South GR, Skelton PA (2000). Status of coral reefs in the southwest Pacific: Fiji, Nauru, New Caledonia, Samoa, Solomon Islands, Tuvalu and Vanuatu. In: Wilkinson CR (ed). *Status of Coral Reefs of the World: 2000.* Australian Institute of Marine Science, Cape Ferguson, Australia.

Sulu R, Hay C, Ramohia P, Lam M (2002). *The Coral Reefs of the Solomon Islands.* Australian Institute of Marine Science, Cape Ferguson, Australia.

NEW CALEDONIA

Bour W (1988). SPOT images for coral reef mapping in New Caledonia. A fruitful approach for classic and new topics. *Proc 6th Int Coral Reef Symp* 2: 445-448.

Gabrié C (2000). *State of Coral Reefs in French Overseas Départements and Territories.* Ministry of Spatial Planning and Environment and State Secretariat for Overseas Affairs, Paris, France.

South GR, Skelton PA (2000). Status of coral reefs in the southwest Pacific: Fiji, Nauru, New Caledonia, Samoa, Solomon Islands, Tuvalu and Vanuatu. In: Wilkinson CR (ed). *Status of Coral Reefs of the World: 2000.* Australian Institute of Marine Science, Cape Ferguson, Australia.

Zann LP, Vuki V (2000). The south western Pacific Islands region. In: Sheppard C (ed). *Seas at the Millennium: An Environmental Evaluation.* Elsevier Science Ltd, Oxford, UK.

VANUATU

Done TJ, Navin KF (1990). *Vanuatu Marine Resources: Report of a Biological Survey.* Australian Institute of Marine Science, Townsville, Australia

Naviti W, Aston J (2000). Status of coral reef and fish resources of Vanuatu. In: Salvat B, South R, Wilkinson C (eds). Proceedings of the International Coral Reef Initiative Regional Symposium, Noumea, 22-24 May 2000.

South GR, Skelton PA (2000). Status of coral reefs in the southwest Pacific: Fiji, Nauru, New Caledonia, Samoa, Solomon Islands, Tuvalu and Vanuatu. In: Wilkinson CR (ed). *Status of Coral Reefs of the World: 2000.* Australian Institute of Marine Science, Cape Ferguson, Australia.

Zann LP, Vuki V (2000). The south western Pacific Islands region. In: Sheppard C (ed). *Seas at the Millennium: An Environmental Evaluation.* Elsevier Science Ltd, Oxford, UK.

FIJI

Agassiz A (1899). The Islands and coral reefs of Fiji. *Bull Mus Comp Zool* 33: 1-167 and 120 plates.

Ferry J, Kumar PB, Bronders J, Lewis J (1997). Hydrogeology of carbonate islands of Fiji. In: Vacher HL, Quinn T (eds). *Developments in Sedimentology, 54: Geology and Hydrology of Carbonate Islands.* Elsevier Science BV, Amsterdam, Netherlands.

Jennings S, Polunin NVC (1996). Effects of fishing effort and catch rate upon the structure and biomass of Fijian fish communities. *J App Ecol* 33: 400-412.

South GR, Skelton PA (2000). Status of coral reefs in the southwest Pacific: Fiji, Nauru, New Caledonia, Samoa, Solomon Islands, Tuvalu and Vanuatu. In: Wilkinson CR (ed). *Status of Coral Reefs of the World: 2000.* Australian Institute of Marine Science, Cape Ferguson, Australia.

Vuki V, Naqasima M, Vave R (2000). Status of Fiji's coral reefs. In: Salvat B, Wilkinson C, South GR (eds). Proceedings of the International Coral Reef Initiative Regional Symposium, Noumea, 22-24 May 2000.

Zann LP, Vuki V (2000). The south western Pacific Islands region. In: Sheppard C (ed). *Seas at the Millennium: An Environmental Evaluation.* Elsevier Science Ltd, Oxford, UK.

Map sources

Map 12a

Coral reef data have been taken as arcs from Petroconsultants SA (1990)*. Some areas of additional reef have been added for the island groups in the far northwest of the country (western Bismarck Archipelago) from Department of Defence (1971), which were compiled from higher resolution maps, uncontrolled air photography and radar imagery.

Department of Defence (1971). *PNG5 – Vegetation and Timber Resources.* 1:500,000. 1st edn. Department of Defence, Canberra, Australia.

Map 12b

Coral reef data have been taken as arcs from Petroconsultants SA (1990)*.

Map 12c

Coral reef data have been taken as arcs from Petroconsultants SA (1990)*.

Maps 12d and 12e

Mangrove and coral reef data were taken from IGN (1967a and b, 1968a, b and c, and 1971a and b). All these maps are based on aerial photographs taken between 1943 and 1962. Areas of mangrove cover were prepared as polygons and coral reefs as arcs.

IGN (1967a). *Ambrym-Pentecote.* 1:100 000. Series no. 8. Maps 624.041. Institut Géographique National.

IGN (1967b). *Maewo.* 1:100 000. Series no. 7. Maps 624.041. Institut Géographique National.

IGN (1968a). *Aoba.* 1:100 000. Series no. 6. Maps 624.041. Institut Géographique National.

IGN (1968b). *Epi Shepherd.* 1:100 000. Series no. 11. Maps 624.041. Institut Géographique National.

IGN (1968c). *Santo Sud.* 1:100 000. Series no. 5. Maps 624.041. Institut Géographique National.

IGN (1971a). *Lamap.* 1:100 000. Series no. 10. Maps 624.041. Institut Géographique National.

IGN (1971b). *Malekoula.* 1:100 000. Series no. 9. Maps 624.041. Institut Géographiqua National.

Map 12f

Coral reef data have been taken as arcs from Petroconsultants SA (1990)*.

* See Technical notes, page 401

Chapter 13
Micronesia

The northern island areas of the Central and Western Pacific are characterized by widely scattered archipelagos of relatively small islands. The western limits of this region lie along the boundary of the Philippine plate, and there is considerable volcanic activity in the north. While most of the other islands and reefs can be linked to volcanic activity, this largely occurred in the distant geological past. Cores taken through the base of some atolls in the Marshall Islands show reef deposits up to 1.4 kilometers thick, dating back over 50 million years.

Reefs are well developed throughout the region, except on the coastlines of recently active volcanoes. Palau lies closest to the center of reef diversity in the Philippines and Indonesia, and shows very high levels of species diversity. Biodiversity declines to the east.

The peoples of Micronesia have diverse origins, and there is a broad complex of cultures. Palau and the Marianas were probably first populated around 3 500 years ago by peoples from Indonesia and the Philippines. At the same time the western parts of the region were probably settled from eastern Melanesia. At least 15 languages with little in common occur across the region, giving some measure of the independent cultures that exist there. It is believed that, while many of the high island peoples did not travel great distances following their establishment, those of the low islands maintained ocean-going canoes and continued to travel. Early European travelers to the Marshall Islands were shown "stick charts" which consisted of frameworks of sticks, sometimes having other items such as shells bound to them. These charts marked with considerable accuracy the locations of other islands, reefs and even patterns of ocean waves, and clearly played a critical role in oceanic navigation.

At the present time there are considerable differences both in the state of reefs and in the impacts of human cultures. The influence of the USA, associated with rapid Western-style development, is considerable in a number of countries, notably Guam, but also in parts of the Marshall Islands. Urban growth on a few islands has brought with it the breakdown of traditional systems and the sustainable utilization of resources, together with associated problems of pollution. Military activities have also had a considerable impact in this region. Intensive nuclear testing during the 1940s and 1950s impacted a number of atolls in the Marshall Islands, with repercussions to the present. There is also ongoing utilization by the USA of some islands and reefs in the Marshall Islands and the Marianas for military purposes, including target practice. Tourism is a critical and growing economic activity in a few islands, notably Guam, Saipan and Chuuk Atoll. Away from areas of human impact, the region still includes a very large number of islands and reefs in good to excellent condition, where traditional use of the reefs by local peoples remains sustainable and well managed.

Left: The six banded angelfish Pomacanthus sexstriatus. *Right: Coral scene with branching* Acropora. *Diversity in Micronesia diminishes from west to east.*

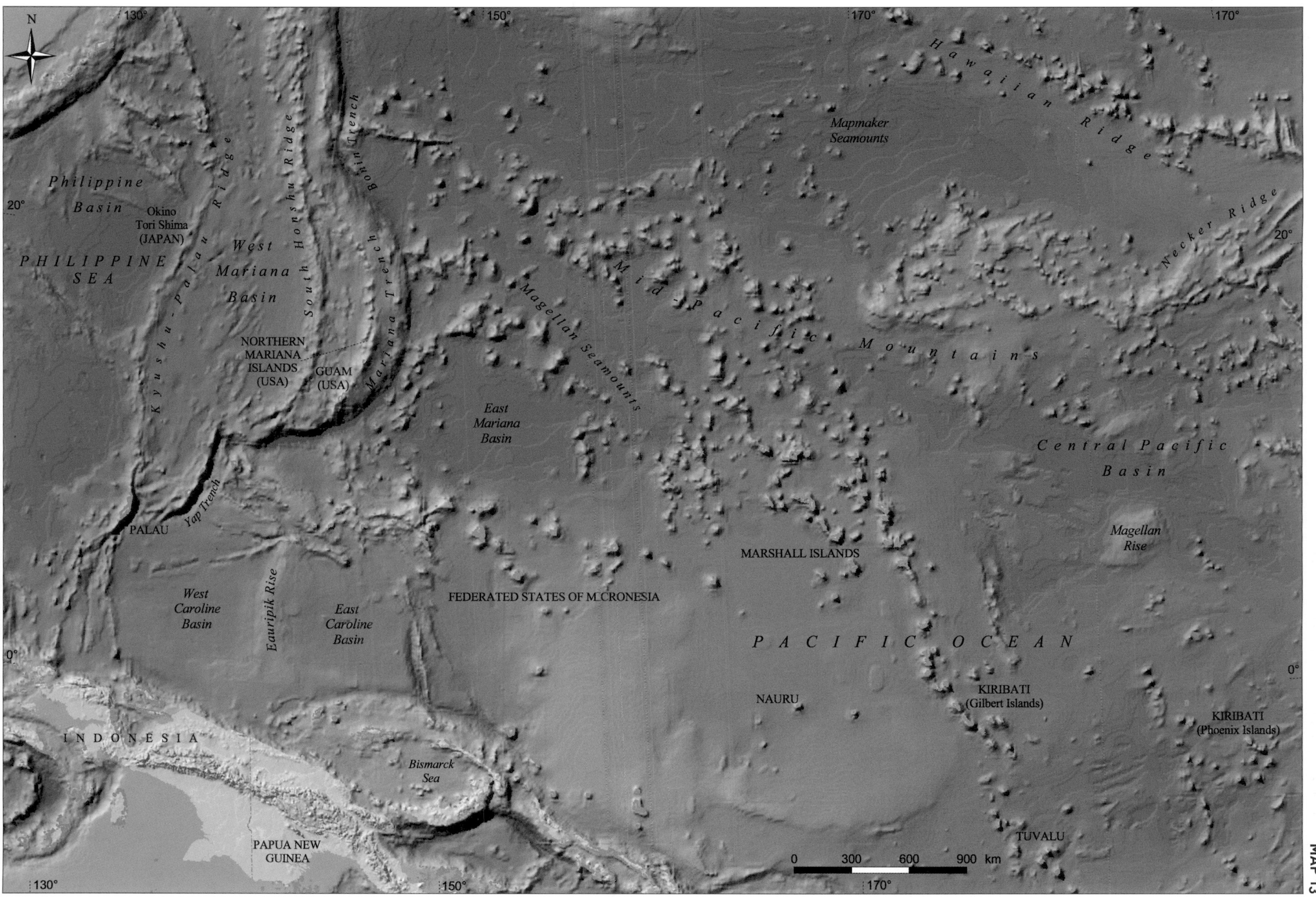

N
130°
150°
170°
170°
20°
0°
Hawaiian Ridge
Mapmaker Seamounts
Philippine Basin
Okino Tori Shima (JAPAN)
PHILIPPINE SEA
Kyushu-Palau Ridge
West Mariana Basin
South Honshu Ridge
Bonin Trench
Mariana Trench
NORTHERN MARIANA ISLANDS (USA)
GUAM (USA)
Magellan Seamounts
Mid-Pacific Mountains
Necker Ridge
East Mariana Basin
Central Pacific Basin
Yap Trench
PALAU
Magellan Rise
MARSHALL ISLANDS
West Caroline Basin
Eauripik Rise
East Caroline Basin
FEDERATED STATES OF M.CRONESIA
PACIFIC OCEAN
NAURU
KIRIBATI (Gilbert Islands)
KIRIBATI (Phoenix Islands)
INDONESIA
Bismarck Sea
PAPUA NEW GUINEA
TUVALU
0 300 600 900 km

MAP 13a

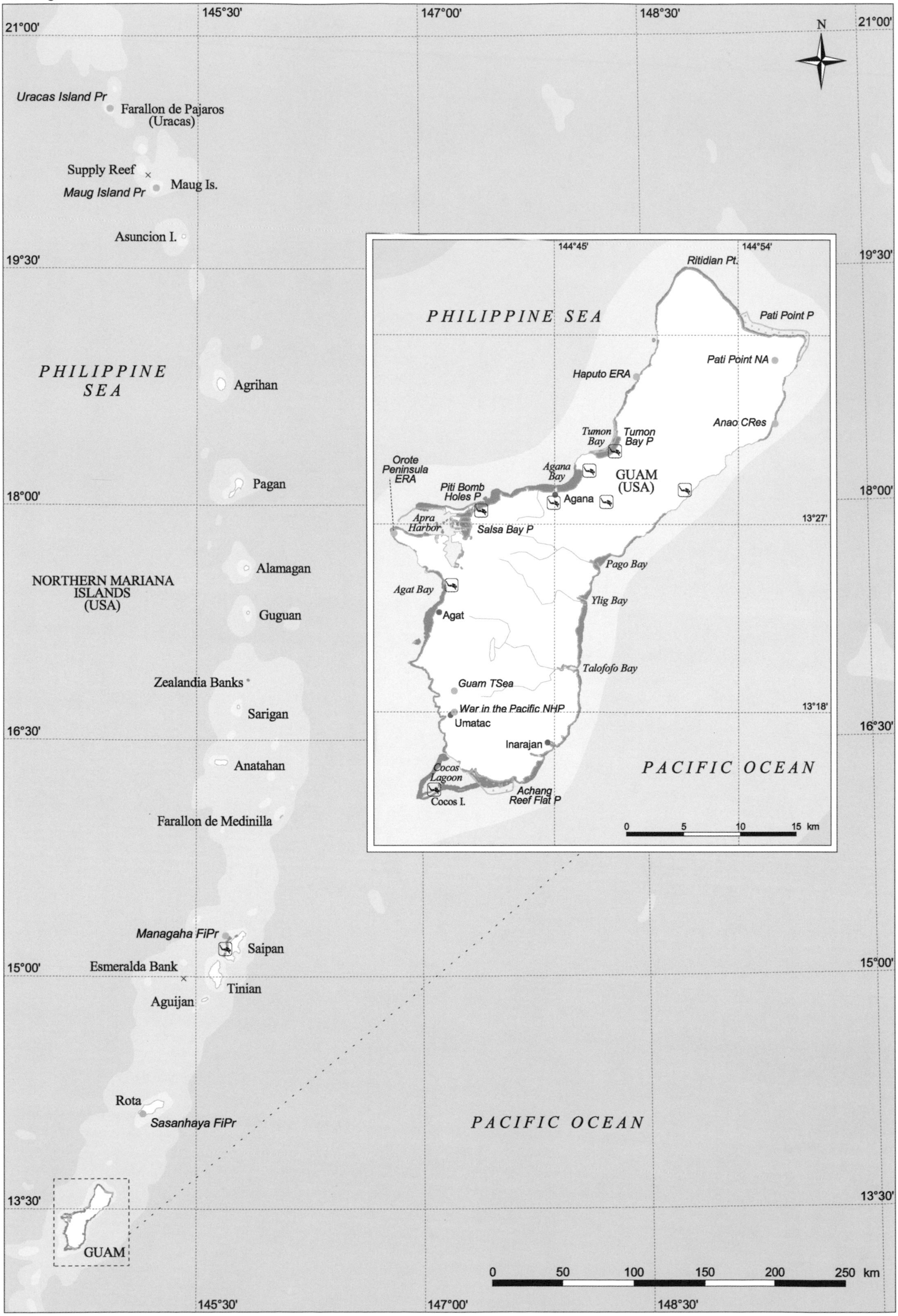

145°30'
147°00'
148°30'
21°00'
N
Uracas Island Pr
Farallon de Pajaros (Uracas)
Supply Reef
Maug Island Pr
Maug Is.
Asuncion I.
19°30'
PHILIPPINE SEA
Agrihan
Pagan
18°00'
Alamagan
NORTHERN MARIANA ISLANDS (USA)
Guguan
Zealandia Banks
Sarigan
16°30'
Anatahan
Farallon de Medinilla
Managaha FiPr
Saipan
15°00'
Esmeralda Bank
Tinian
Aguijan
Rota
Sasanhaya FiPr
PACIFIC OCEAN
13°30'
GUAM
0 50 100 150 200 250 km

144°45'
144°54'
Ritidian Pt.
PHILIPPINE SEA
Pati Point P
Pati Point NA
Haputo ERA
Anao CRes
Tumon Bay
Tumon Bay P
Orote Peninsula ERA
Agana Bay
GUAM (USA)
Piti Bomb Holes P
Agana
13°27'
Apra Harbor
Salsa Bay P
Pago Bay
Agat Bay
Ylig Bay
Agat
Talofofo Bay
Guam TSea
War in the Pacific NHP
13°18'
Umatac
Inarajan
PACIFIC OCEAN
Cocos Lagoon
Achang Reef Flat P
Cocos I.
0 5 10 15 km

Commonwealth of the Northern Mariana Islands and Guam

MAP 13a

The Mariana Islands form a long chain of 15 high islands in the Western Pacific, running approximately 800 kilometers from Farallon de Pajaros (Uracas) in the north to Guam in the south. The southernmost island of Guam is an unincorporated territory of the USA, while the remaining islands, the Commonwealth of the Northern Mariana Islands (CNMI), are a commonwealth in political union with the USA. They are located on the eastern margin of the Philippine tectonic plate. To the east, the Mariana Trench has been formed by the subduction of the Pacific plate and there is considerable volcanic activity, particularly to the north of this chain. The Mariana Trench is the deepest ocean trench in the world, and to the south of Guam, the Challenger Deep is the deepest known point, at 11 034 meters. The climate is fairly stable, with a dry season from January to June when the northeast trade winds predominate, and a wet season from July to November. The region experiences regular typhoons, most recently Supertyphoon Paka, which passed between Guam and Rota in December 1997 with recorded sustained windspeeds on Guam of 185 kilometers per hour and gusts of over 270 kilometers per hour.

Guam is the southernmost and largest of the islands. The northern areas consist of a large uplifted limestone plateau, while the south of the island is dominated by volcanic hills reaching 406 meters. The entire island is encircled by fringing reefs. The five southernmost islands of the CNMI similarly consist of both volcanic and uplifted limestone structures. Rota has a volcanic center with an uplifted limestone terrace all around, and is mostly surrounded by narrow but well developed fringing reefs. Tinian and the nearby Aguijan are uplifted limestone – Tinian has a few areas of narrow fringing reefs, while Aguijan has no clearly developed reef structures, although there are diverse and actively growing coral communities. Saipan also has a volcanic center and reaches nearly 500 meters in elevation, but has a raised limestone perimeter with a gently shelving coastline on the west, and a more dramatic one on the east. Saipan has a well developed barrier reef and lagoon system running off its west coast with fringing reefs in a number of localities. Farallon de Medinilla is another small limestone island, with coral communities in its surrounding waters.

The nine islands north of Farallon de Medinilla are entirely volcanic, including some which are still active and, for the most part, do not have any reef structures.

Two lined rabbitfish Siganus lineatus.

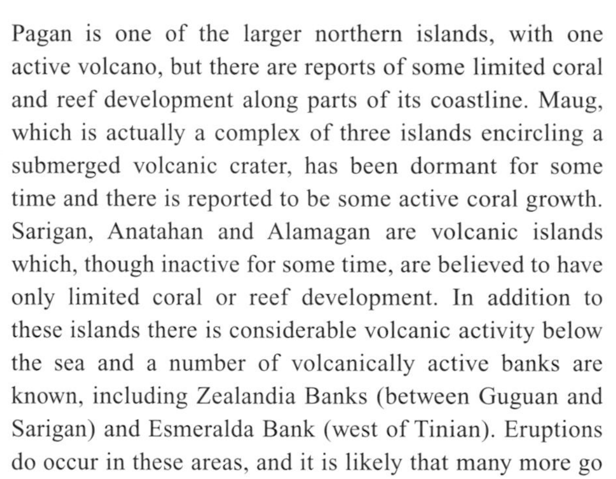

Pagan is one of the larger northern islands, with one active volcano, but there are reports of some limited coral and reef development along parts of its coastline. Maug, which is actually a complex of three islands encircling a submerged volcanic crater, has been dormant for some time and there is reported to be some active coral growth. Sarigan, Anatahan and Alamagan are volcanic islands which, though inactive for some time, are believed to have only limited coral or reef development. In addition to these islands there is considerable volcanic activity below the sea and a number of volcanically active banks are known, including Zealandia Banks (between Guguan and Sarigan) and Esmeralda Bank (west of Tinian). Eruptions do occur in these areas, and it is likely that many more go unrecorded. Supply Reef, near Maug, is an inactive submerged crater reported to have some living coral communities on the crater rim.

The Mariana Islands, lying relatively close to the center of coral reef biodiversity in the Philippines and Indonesia, enjoy high diversity. Guam, which is fairly well studied, has about 300 recorded species of scleractinian corals, 950 species of reef fish, 220 species of benthic algae and more than 1 400 species of molluscs. Live coral cover in Guam reaches 50 percent in some areas, but the majority of sites now show less than 25 percent. Both diversity and cover decrease considerably in the geologically younger northern islands where the volcanic conditions are unfavorable for many species, while in

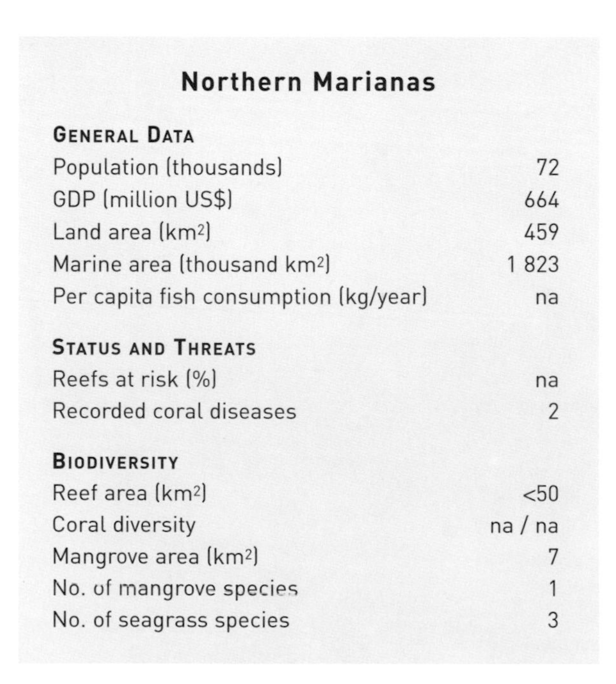

Northern Marianas

General Data	
Population (thousands)	72
GDP (million US$)	664
Land area (km²)	459
Marine area (thousand km²)	1 823
Per capita fish consumption (kg/year)	na
Status and Threats	
Reefs at risk (%)	na
Recorded coral diseases	2
Biodiversity	
Reef area (km²)	<50
Coral diversity	na / na
Mangrove area (km²)	7
No. of mangrove species	1
No. of seagrass species	3

Guam

General Data	
Population (thousands)	155
GDP (million US$)	3 066
Land area (km²)	572
Marine area (thousand km²)	218
Per capita fish consumption (kg/year)	na
Status and Threats	
Reefs at risk (%)	100
Recorded coral diseases	1
Biodiversity	
Reef area (km²)	220
Coral diversity	140 / 220
Mangrove area (km²)	1
No. of mangrove species	10
No. of seagrass species	na

Left: The eastern triangle butterflyfish Chaetodon baronessa *feeds exclusively on* Acropora *polyps. Right: A group of batfish* Platax orbicularis *against a bare volcanic rock.*

Protected areas with coral reefs

Site name	Designation	Abbreviation	IUCN cat.	Size (km²)	Year
Guam					
Anao	Conservation Reserve	CRes	IV	2.63	1953
Guam	Territorial Seashore Park	TSea	VI	61.35	1978
Haputo	Ecological Reserve Area	ERA	IV	1.02	1984
Orote Peninsula	Ecological Reserve Area	ERA	IV	0.66	1984
Pati Point	Natural Area	NA	IV	1.12	1973
War in the Pacific	National Historic Park	NHP	V	7.79	1978
Northern Marianas					
Managaha	Fish Preserve	FiPr	IV	na	2000
Sasanhaya	Fish Preserve	FiPr	IV	na	1994

the far north cooler conditions may further restrict certain species. An expedition to the uninhabited northern islands in 1992 listed 161 species of cnidarians, 520 marine molluscs and 463 fish. Crown-of-thorns starfish had devastating effects, notably in Guam from 1968 to 1970, and again in 1979, killing 90 percent of corals in some areas.

Natural pressures on the reefs around Guam have been greatly exacerbated in recent years by human activities. Agriculture, development and the burning of natural areas have led to increased sedimentation in the surrounding waters, while overfishing is widespread and total catch per unit of effort reportedly fell by 78 percent between 1985 and 1997. The overall impact on the reefs has been considerable. Coral cover is reported to have dropped significantly since the 1970s, when it was over 50 percent in many areas. Algal cover has increased since the crown-of-thorns outbreaks and remains high, possibly due to overexploitation of herbivorous fish. Coral and fish recruitment is also reported to have fallen. Human pressures are largely focussed around the urban areas of the barrier reef system in western Saipan, but also at Rota West Harbor and San Jose Harbor on Tinian where there are problems of pollution and sedimentation. Overfishing is believed to be occurring on Saipan and Tinian, where fisheries data show low average sizes of many reef species. Finally, the island of Farallon de Medinilla has been extensively used for target practice by the US military. Local objections have been raised, with campaigns to have this activity restricted or moved to one of the more active volcanoes where the impacts may be less detectable. Thus far no serious efforts to relocate have been made.

The economy of both territories is highly dependent on tourism. Guam receives more than 1.4 million visitors per year, while the CNMI receives about 500 000, primarily limited to the island of Saipan. Diving and snorkelling are popular tourist activities. Research on the reefs is well advanced, particularly in Guam where there is an active marine laboratory associated with the University of Guam. Several protected areas have been established, including a number of coastal and marine sites in Guam, and some in the CNMI.

A bandcheck wrasse Oxycheilinus digrammus *sheltering below an* Acropora.

Palau and the Federated States of Micronesia

MAPS 13b and c

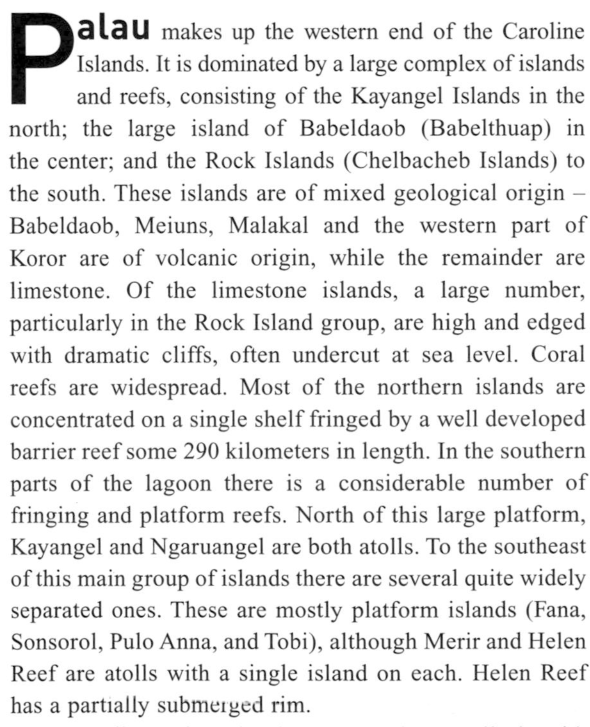

Palau makes up the western end of the Caroline Islands. It is dominated by a large complex of islands and reefs, consisting of the Kayangel Islands in the north; the large island of Babeldaob (Babelthuap) in the center; and the Rock Islands (Chelbacheb Islands) to the south. These islands are of mixed geological origin – Babeldaob, Meiuns, Malakal and the western part of Koror are of volcanic origin, while the remainder are limestone. Of the limestone islands, a large number, particularly in the Rock Island group, are high and edged with dramatic cliffs, often undercut at sea level. Coral reefs are widespread. Most of the northern islands are concentrated on a single shelf fringed by a well developed barrier reef some 290 kilometers in length. In the southern parts of the lagoon there is a considerable number of fringing and platform reefs. North of this large platform, Kayangel and Ngaruangel are both atolls. To the southeast of this main group of islands there are several quite widely separated ones. These are mostly platform islands (Fana, Sonsorol, Pulo Anna, and Tobi), although Merir and Helen Reef are atolls with a single island on each. Helen Reef has a partially submerged rim.

The climate in Palau is warm and generally humid. From November to June the northeasterly trade winds dominate, but for much of the rest of the year winds are lighter and more variable, although occasional typhoons also occur around this time.

Levels of biodiversity are very high. Some 425 species of coral have been reported, including an estimated 300-350 stony corals, together with 1 278 species of reef fish and well over 300 species of sponge. The southern reefs are swept by strong currents and dominated by blue coral *Heliopora coerulea*, however they are also very diverse. Helen Reef was recorded as having 248 stony coral species, perhaps the highest number of any Pacific atoll. Coral cover in all areas was high prior to 1998, typically over 50 percent and reaching 70-80 percent on outer reef slopes in many areas. Marine turtles are relatively common and the estuarine crocodile *Crocodylus porosus* and dugong *Dugong dugon* are found in the lagoon. There are important areas of mangrove and seagrass communities. One interesting and perhaps unique ecosystem in this country is found in its large number of marine lakes. These are inland, but many appear to be connected to the ocean by cave systems and have developed highly distinctive communities which appear to have evolved *in situ* from species that entered the lakes in larval forms. The most distinctive of such organisms are the jellyfish, notably

Left: Ngulu Atoll (STS080-707-26, 1996). Right: Pohnpei, with its fringing and barrier reefs, with adjacent atolls (STS044-93-33, 1991).

PACIFIC OCEAN
Falalop
Ulithi
Fais
Yap Is.
FEDERATED STATES OF MICRONESIA
Ngulu
Sorol
(C A R O L I N E I S L A N D S)

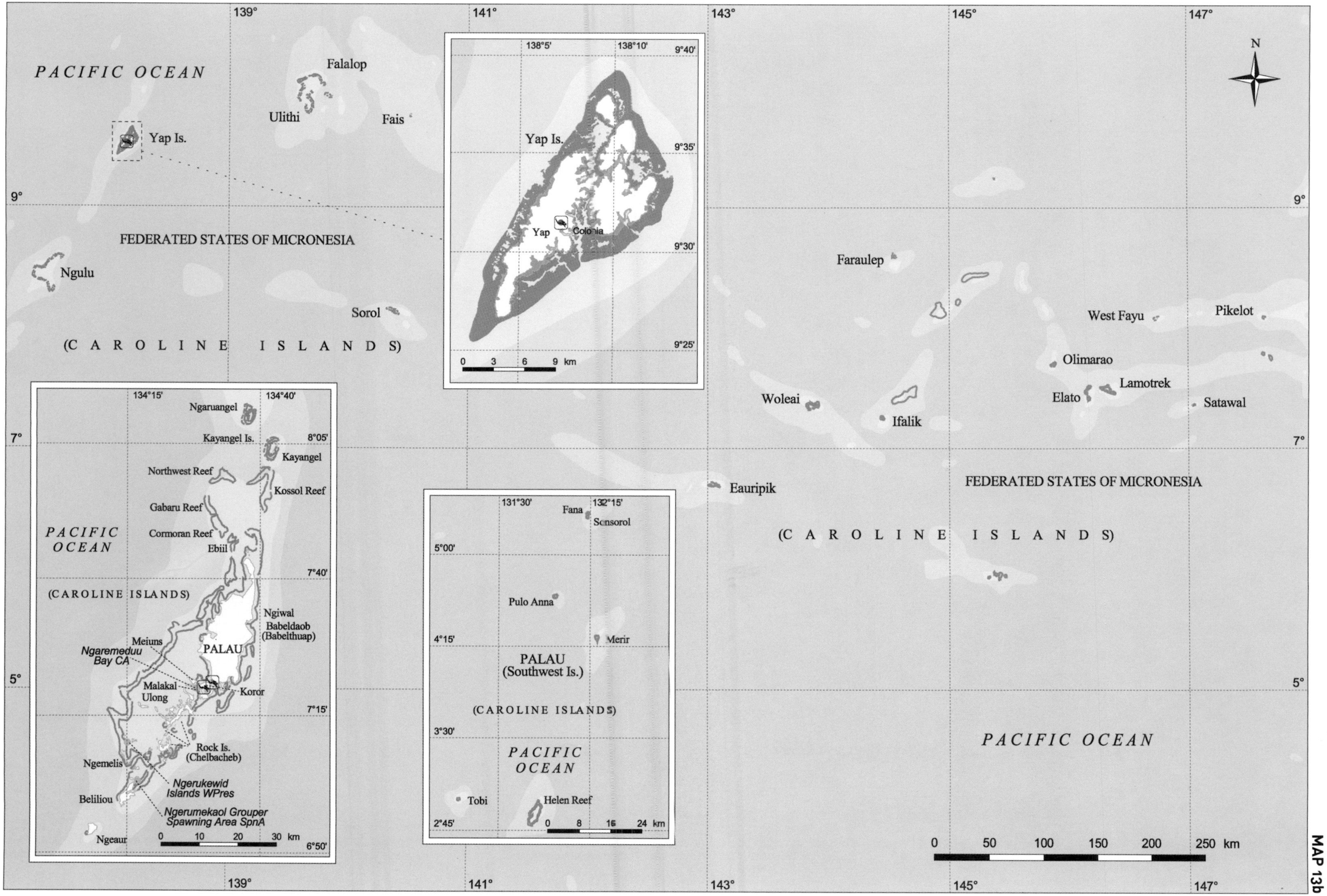

Yap Is.
Yap
Colonia
138°5'
138°10'
9°40'
9°35'
9°30'
9°25'
0 3 6 9 km

PACIFIC OCEAN
(CAROLINE ISLANDS)
134°15'
134°40'
8°05'
7°40'
7°15'
6°50'
Ngaruangel
Kayangel Is.
Kayangel
Northwest Reef
Kossol Reef
Gabaru Reef
Cormoran Reef
Ebiil
Ngiwal
Babeldaob (Babelthuap)
Meiuns
Ngaremeduu Bay CA
PALAU
Malakal
Ulong
Koror
Rock Is. (Chelbacheb)
Ngemelis
Ngerukewid Islands WPres
Beliliou
Ngerumekaol Grouper Spawning Area SpnA
Ngeaur
0 10 20 30 km

131°30'
132°15'
5°00'
4°15'
3°30'
2°45'
Fana
Sonsorol
Pulo Anna
Merir
PALAU (Southwest Is.)
(CAROLINE ISLANDS)
PACIFIC OCEAN
Tobi
Helen Reef
0 8 16 24 km

Faraulep
West Fayu
Pikelot
Olimarao
Lamotrek
Elato
Satawal
Woleai
Ifalik
Eauripik
FEDERATED STATES OF MICRONESIA
(C A R O L I N E I S L A N D S)
PACIFIC OCEAN
0 50 100 150 200 250 km

139° 141° 143° 145° 147°
9° 7° 5°

MAP 13c

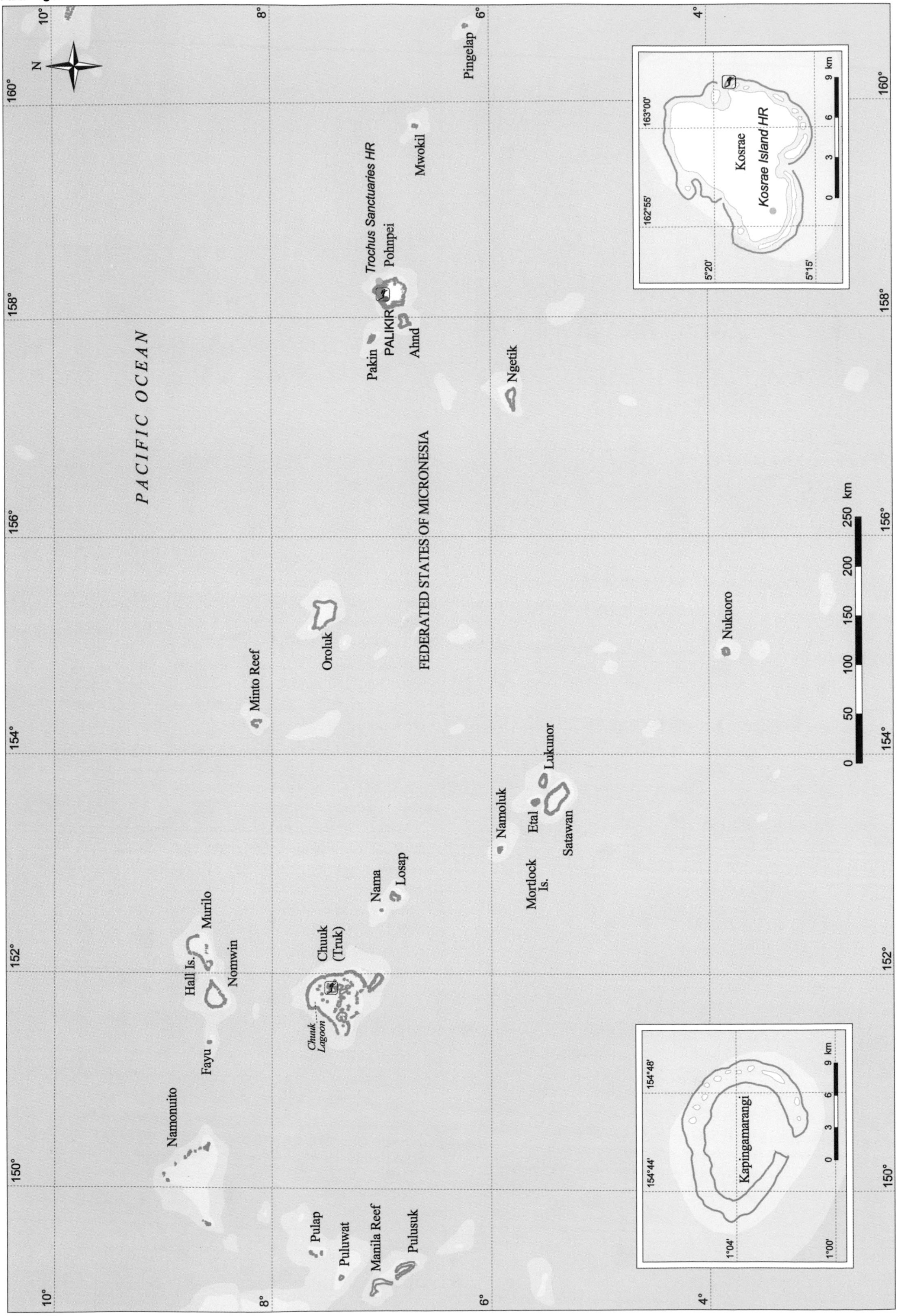

PACIFIC OCEAN

FEDERATED STATES OF MICRONESIA

Namonuito
Fayu
Hall Is.
Murilo
Nomwin
Minto Reef
Pulap
Puluwat
Manila Reef
Pulusuk
Chuuk Lagoon
Chuuk (Truk)
Nama
Losap
Oroluk
Pakin
PALIKIR
Ahnd
Trochus Sanctuaries HR
Pohnpei
Mwokil
Pingelap
Namoluk
Mortlock Is.
Etal
Lukunor
Satawan
Ngetik
Nukuoro

Kapingamarangi
154°44' 154°48' 1°04' 1°00'
0 3 6 9 km

Kosrae
Kosrae Island HR
162°55' 163°00' 5°20' 5°15'
0 3 6 9 km

0 50 100 150 200 250 km

150° 152° 154° 156° 158° 160°
10° 8° 6° 4°
N

Mastigias spp. which have formed vast communities. Marine lakes are most numerous on Koror where there are 58, of which 28 have jellyfish. The reefs of Palau were heavily impacted by crown-of-thorns starfish outbreaks in 1977 and recovery was reported to be poor, even into the 1990s. The reefs were further impacted by the bleaching event of 1998, with mortality reaching over 50 percent in most areas. *Acropora* was devastated almost everywhere, but other corals showed slightly higher levels of survival in nearshore lagoons and fringing reefs. The warm waters of 1998 also had a dramatic impact on jellyfish populations in some of the marine lakes, although they now seem to be recovering. The crown-of-thorns starfish has had a number of outbreaks, and high densities in some areas may be exacerbating the setbacks caused by bleaching-associated mortalities.

The majority of the Palauan population live on Koror, but numbers are growing rapidly and there is some expansion to the other islands. Management of marine resources is devolved to the state level. The individual states (of which there are 16, each typically incorporating several villages) have ownership of all living and non-living resources out to 12 nautical miles, with the exception of highly migratory species. Although traditional law is upheld in the constitution it has been combined with Western statutes, and respect for traditional systems is diminishing. Several of the islands are connected by bridges and causeways which have interfered with natural water circulation in some areas. Sewage and solid waste disposal are a localized problem.

Fishing at the subsistence level is very important, but there are also some export fisheries, including a trochus fishery and a large marine ornamental fishery. Over the past ten years it has been estimated that some 1 800 tons of fish have been taken from the reefs per year: about 1 200 tons for direct consumption, 360 tons for local markets and 250 tons for export. Although such levels may be sustainable, there is considerable evidence of overfishing of certain target species, notably groupers, and a number of these are showing declines in abundance and changes in demographic structure. In the southwestern islands there have been some reports of blast and cyanide fishing.

Mariculture is relatively important. There has been a giant clam mariculture project in existence for many years and coral mariculture is also under development. The dugong population is believed to be in decline, with a 1991 estimate of between 50 and 200 individuals. There is still thought to be some illegal hunting of these animals, and slow rates of reproduction mean that any recovery could take many years. Crocodiles are also now rare and may number fewer than 150 individuals.

There is an active interest in conservation in the country. Protected area legislation is developed at the state level, and a number of sites have been established with regulations ranging from seasonal closures and other fisheries restrictions to strict reserves with no entry permitted. For the most part there is strong community support for these areas. The Ngaremeduu Conservation Area has also recently been established covering parts of three states on the west coast of Babeldaob. Legislation has had to be passed in each state to protect this site.

Left: Palau's southern lagoon (STS106-720-77, 2000). Right: Juvenile giant clams, Tridacna gigas, *in a clam farm.*

Federated States of Micronesia

This country consists of a vast and scattered chain of islands (with Palau also referred to as the Caroline Islands) stretching some 2 900 kilometers from east to west. Politically they are independent, but remain in a "compact of free association" with the USA. Although the total land area is small, there are some 600 islands with diverse geological origins. The total reef area is very large indeed, over 5 000 square kilometers, but remains very poorly known.

There are four states. The State of Yap, in the west, is centered on Yap itself, a large island group with four tightly associated islands formed from uplifted crustal material, including both volcanic and metamorphic rock, and reaching a height of 174 meters above sea level. It is surrounded by a broad reef which is part barrier and part fringing in structure, with lagoon development on the reef flat in some areas. The other islands and reefs in the State of Yap are predominantly atolls with associated islands. They include the two large atolls of Ulithi and Ngulu, and also the small uplifted atoll of Fais. The state of Chuuk (Truk) is dominated by the near-atoll of Chuuk itself, which includes a scattering of high volcanic islands encircled by a barrier reef, the whole structure being some 85 kilometers across its widest axis. This state also incorporates a number of other large atolls, notably Namonuito in the northwest, the Hall Islands (comprising the two atolls of Murilo and Nomwin and the coral platform and island of Fayu), and the Mortlock Islands to the south (a complex of three atolls). Pohnpei is a large volcanic island reaching 798 meters above sea level, and is surrounded by a well developed barrier reef. This state also includes eight other atolls, mostly widely spaced and with only a scattering of small cays on their atoll rims. The easternmost state in the country consists of the single island and reef complex of Kosrae. This is another high volcanic island, surrounded by fringing reefs. In addition to the reefs and islands described there are several areas of relatively shallow banks where reef communities may be well developed, notably between Yap and Chuuk. The climate is similar to that of Palau, with warm moist conditions and northeasterly trade winds dominating from November to June, and more variable conditions the rest of the year.

Biodiversity is slightly lower than in Palau, with decreasing diversity moving from west to east. There is very little information describing the reefs in terms of their biodiversity, but reef status throughout the country is generally thought to be very good. Mangrove communities are particularly well developed around the coastlines of Pohnpei and Yap.

The reefs are of critical importance as a source of food throughout the country. Close to urban areas there is some overfishing and there have been problems with blast fishing. Clams, in particular giant clams, are declining and have been completely eliminated in some areas. There are ongoing efforts to establish giant clam mariculture at a national center in Kosrae. Trochus harvesting is also

Left: A silver gull Larus novaehollandiae *flies over a wide reef flat. Right: Dense branching corals and a black butterflyfish* Chaetodon flavirostris.

Palau

GENERAL DATA	
Population (thousands)	19
GDP (million US$)	92
Land area (km2)	483
Marine area (thousand km2)	601
Per capita fish consumption (kg/year)	108
STATUS AND THREATS	
Reefs at risk (%)	0
Recorded coral diseases	0
BIODIVERSITY	
Reef area (km2)	1 150
Coral diversity	154 / 384
Mangrove area (km2)	na
No. of mangrove species	13
No. of seagrass species	2

Federated States of Micronesia

GENERAL DATA	
Population (thousands)	133
GDP (million US$)	223
Land area (km2)	701
Marine area (thousand km2)	2 980
Per capita fish consumption (kg/year)	73
STATUS AND THREATS	
Reefs at risk (%)	45
Recorded coral diseases	1
BIODIVERSITY	
Reef area (km2)	4 340
Coral diversity	92 / 391
Mangrove area (km2)	86
No. of mangrove species	14
No. of seagrass species	na

an important economic activity in all areas. Coastal development and associated pollution are again localized problems on the largest islands, but for the most part the reefs remain in good condition. Many reefs are owned and managed at the level of the individual villages. There are no permanent protected areas other than a few small trochus sanctuaries.

Tourism is growing with considerable speed in a few islands, although the more remote islands remain largely unvisited. Chuuk Lagoon is widely regarded as one of the world's top dive centers on account of the very large numbers of wrecks which sunk in the lagoon during the Second World War. Around 50 Japanese ships plus numerous Japanese and American aircraft went down during a two-day American attack in February 1944. Although these wrecks are a primary attraction, these same structures can also be seen as artificial reefs, with considerable reef and fish communities. There is also some diving on reefs in the lagoon and on the outer slopes and channels of the barrier reef. Diving is also popular on Yap, Pohnpei and Kosrae. Yap is renowned for its apparently resident populations of manta rays. In Pohnpei there is also some diving on the nearby atolls of Pakin and Ahnd. On Kosrae diving is a relatively new activity; it is well planned and there are already more than 50 buoyed dive sites.

Protected areas with coral reefs

Site name	Designation	Abbreviation	IUCN cat.	Size (km2)	Year
Palau					
Ngerukewid Islands	Wildlife Preserve	WPres	III	12.00	1956
Ngerumekaol Grouper	Spawning Area	SpnA	Unassigned	2.59	na
Ngaremeduu Bay	Conservation Area	CA	VI	na	na
Ngeruangel	Reserve	R	II	na	1996
Ngiwal State	Conservation Area	CA	II	na	1997
Ngemelis Islands	Fishing Reserve	FiR	V	na	1999
Ebiil Channel	Conservation Area	CA	II	na	2000
Ngermach Channel – Bkulachelid	Conservation Area	CA	II	na	1998

Marshall Islands

MAP 13d

The Marshall Islands are a complex of 28 coral atolls and 5 small (non-atoll) islands lying in two broad chains, the eastern Ratak (sunrise) chain and the western Ralik (sunset) chain. The isolated atolls of Enewetak and Ujelang lie to the west of these main chains. Wake Atoll, to the north, is clearly linked to this group biologically and geologically, but is separately administered by the USA (see Chapter 14). In all there are some 1 136 islands dispersed over a vast area of ocean, although the total land area is very small. The atolls are typically circular to elliptical with shallow lagoons. Kwajalein, at some 2 500 square kilometers, is the largest atoll in the Pacific. The two chains were probably formed by plate movement over a volcanic hotspot, although there is no current volcanic activity. Deep drilling into the reefs of Bikini and Enewetak atolls revealed a sequence of reef deposits ranging from 1.3 to 1.4 kilometers thick overlying basalt rock from the original volcanic activity. Dating work on the fossils at the base of these, together with the youngest basalt rock below, indicated that this volcanic activity occurred 50-59 million years ago. All of the islands lie close to sea level with a mean height of about 2 meters.

The climate varies from north to south. In the south, rainfall is relatively high and winds are dominated by the northeast trade winds from December to April and by the southeast trades from May to November. To the north, the northeast trade winds dominate all year round, and typhoons are more common. Six have hit the islands since 1900, some with severe impacts. There is a westerly flowing Equatorial Current north of about 9°N and south of 4°N, but between these latitudes there is the easterly flowing Equatorial Counter Current.

The location of the Marshall Islands places them in an area of fairly high diversity, while the relative lack of pressure on many of the reefs means that there has been little biodiversity loss. Nearly 250 coral species have been recorded at Bikini Atoll, and over 250 species of reef fish are noted. A detailed study of marine algae undertaken in the 1950s revealed some 238 species. Mangroves are also found in the islands, but are not common or diverse. There are important seabird populations, particularly in the northern atolls, with at least 15 breeding species among the 31 recorded in the islands. Some 27 species of whale and dolphin have also been recorded.

Politically the Marshall Islands form an independent state, but exist in a "free association" with the USA. Two thirds of the population live on Majuro and Ebeye where

Bikini Atoll, with the crater in the reef flat from a nuclear test clearly visible in the lagoon edge of the reef flat to the northwest (STS055-96-S, 1993).

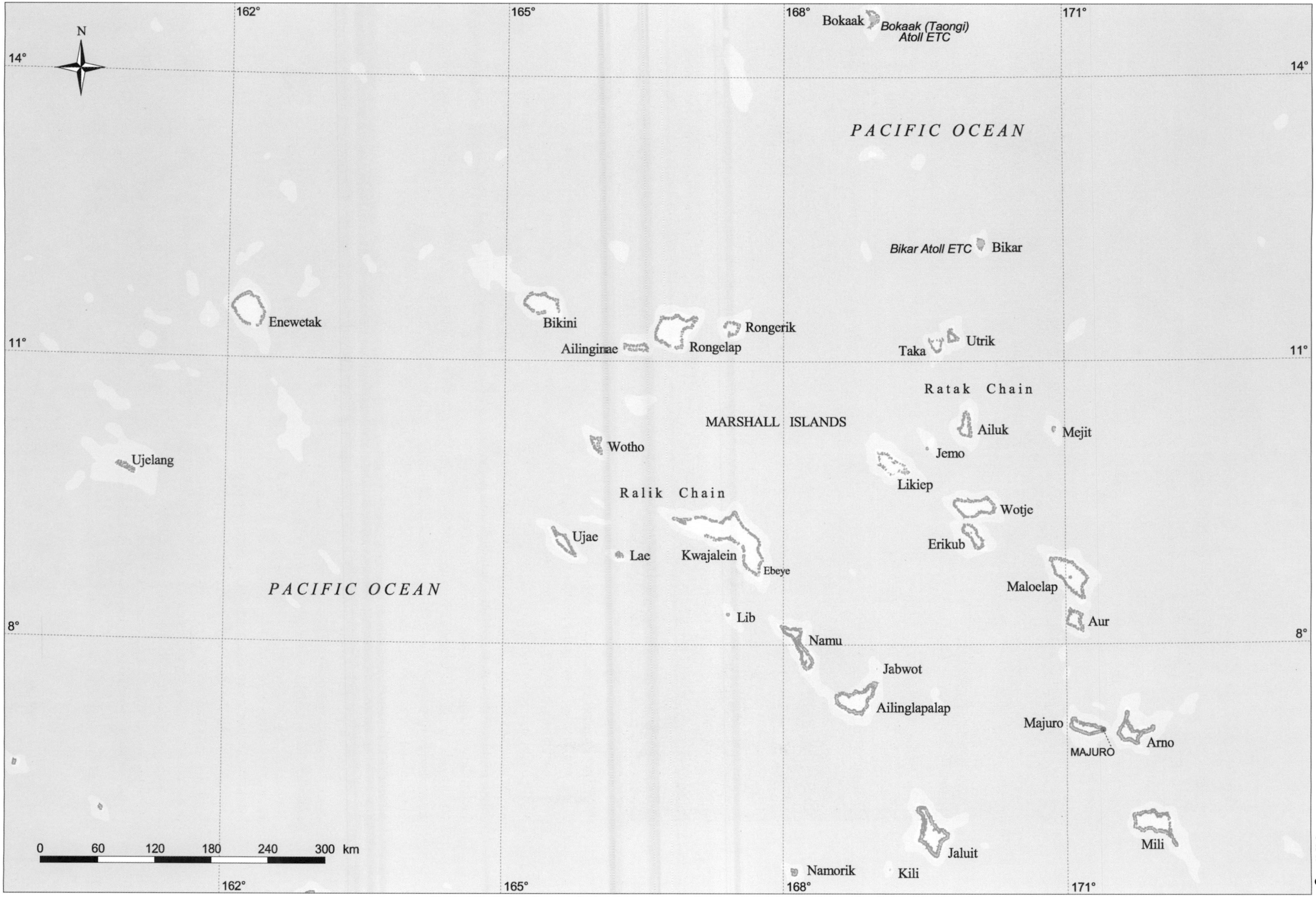

MAP 13d

N

162°
165°
168°
171°
14°
Bokaak
Bokaak (Taongi) Atoll ETC
PACIFIC OCEAN
Bikar Atoll ETC
Bikar
Enewetak
Bikini
Rongerik
Ailinginae
Rongelap
Taka
Utrik
11°
Ratak Chain
MARSHALL ISLANDS
Ailuk
Mejit
Wotho
Jemo
Ujelang
Likiep
Ralik Chain
Wotje
Ujae
Lae
Kwajalein
Ebeye
Erikub
PACIFIC OCEAN
Maloelap
Lib
Aur
8°
Namu
Jabwot
Ailinglapalap
Majuro
MAJURO
Arno
Jaluit
Mili
Namorik
Kili
0 60 120 180 240 300 km

they are concentrated into a relatively small area. Consequently there are various environmental problems, including sewage and solid waste pollution. Much development has taken place with little concern for the environment, and the mining of lagoon sand to obtain building materials is widespread. There has been some breakdown of traditional landuse and cultural systems, exacerbated by the considerable movements forced during the nuclear testing (see below). However, artisanal fishing remains very important, and is now being more widely encouraged. Commercial fishing is largely restricted to foreign-licensed tuna vessels, but there is also an aquarium fishery which has been operating out of Majuro for about 20 years, largely exporting to Hawai'i. Some high value species are also being exploited, including trochus, giant clams and marine turtles. Shark fins are obtained as by-catch from the tuna fisheries. There is also a limited amount of aquaculture development, focussed on clams (for the aquarium trade), pearl oysters and trochus. Sea-level rise associated with climate change is a particular threat to these low-lying islands.

Perhaps the best known "use" of the atolls of the Marshall Islands was nuclear weapons testing by the USA in the 1940s and 1950s, when some 67 nuclear detonations were performed on Bikini and Enewetak atolls. These tests were carried out on land, in the air near ground level or over the water. The largest test, the Bravo hydrogen bomb at Bikini, measured 15 megatons (1 000 times the strength of the Hiroshima bomb): fallout from this explosion was carried to the inhabited atolls of Rongelap, Ailinginae, Rongerik, Utrik and others. A study carried out in 1994 confirmed that some 15 atolls and islands were subject to some radioactive fallout during the 1950s, although most of them are now considered clear. The detailed impacts of these tests on the coral reef environment are still unknown, although there were obviously significant physical affects in the areas of direct impact, while a number of large ships were also sunk in the atoll lagoons. Since the 1960s human pressures on these evacuated atolls (Bikini, Enewetak and Rongelap have been repopulated and re-evacuated on different occasions up to the present day) has been minimal. One positive result of this has been significant increases in some fish groups, including predators such as sharks and jacks (Carangidae) on the surrounding reefs. It may be that these are among the most "pristine" coral reef communities in the region. The USA maintains a military presence in these islands and continues to operate the Kwajalein Missile Range on Kwajalein Atoll.

Although there is a considerable amount of environmental legislation, enforcement is limited. Two protected areas (Bikar and Bokaak Atolls) were established prior to independence but have not been re-established and hence are not officially recognized. Tourism is growing in the islands but remains very low level. There is now some dive tourism to Bikini Atoll where the sharks and wrecks are considered major attractions.

Marshall Islands

General Data	
Population (thousands)	68
GDP (million US$)	75
Land area (km²)	134
Marine area (thousand km²)	2 131
Per capita fish consumption (kg/year)	61
Status and Threats	
Reefs at risk (%)	3
Recorded coral diseases	1
Biodiversity	
Reef area (km²)	6 110
Coral diversity	222 / 340
Mangrove area (km²)	na
No. of mangrove species	4
No. of seagrass species	na

Micronesian reef scene: a coral reef pinnacle rises from deep water nearly to the surface.

Kiribati and Nauru

MAPS 13e and f

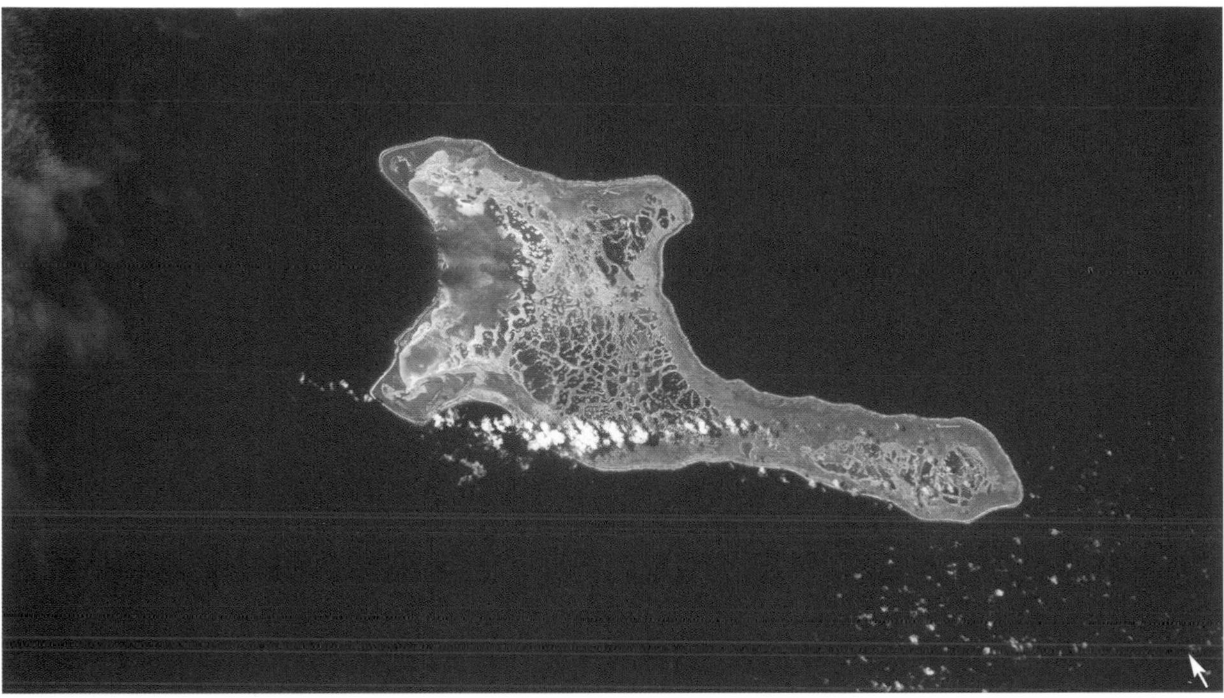

Much of Kiritimati's lagoon is infilled and it has one of the world's largest atoll land areas (STS067-726-94, 1995).

Kiribati's islands and coral reefs straddle a vast swathe of the Pacific Ocean but consist of only some 33 islands or island systems. These are typically divided into three broad groups. Most of the actual islands are now referred to by their Micronesian names, although the island groups are still largely referred to by their European names. In the far west is the long disparate chain of the Gilbert Islands or Tungaru group – this is a chain of 11 atolls and five other islands which have no lagoon but are of similar origin. Often considered alongside this chain is the isolated island of Banaba, a raised atoll reaching a height of some 81 meters similar to Nauru (see below), and the only "high" island in the country. The Phoenix Islands include three atolls and five other islands with fringing reefs. There are also at least two other submerged reef structures (Winslow and Carondelet) which have no associated islands. The Line Islands in the east fall into a northern and southern group. The northern group includes the island of Teraina and the atolls of Tabuaeran and Kiritimati. The latter, pronounced Kirisimas and also known as Christmas Island, has a largely infilled lagoon and the largest land area of any atoll. The southern Line Islands are mostly uninhabited and include the atoll of Millennium Island (formerly Caroline Island) and three other islands with fringing reefs as well as at least one other submerged reef with no associated island. Lying on the equator, the country is largely unaffected by cyclones. The predominant climatic influence comes from the southeast trade winds which create a pronounced windward side to the reefs. The western islands are generally wetter, while the Line Islands lie in the dry equatorial zone. Rainfall is also significantly higher in all areas during El Niño Southern Oscillation events. Although tidal ranges tend to be low (less than 2 meters at spring tides) there is variation in sea level through the year (10-20 centimeters variation in mean monthly levels) which can be further exacerbated by up to 40 centimeters during El Niño years.

The atolls comprise a typical diversity of habitats, including channels, lagoon reefs and shallow reef flats as well as reef slope environments. There is a clear difference between windward and leeward sides, with the windward (eastern) sides typically having a continuous reef margin, narrow reef flat and well developed islands. The leeward reefs are typically much wider, but in some places show a more gradual slope with a less developed reef flat, often submerged at low tide. Spur and groove formations are on all sides, but are usually best developed on lee shores.

Given the wide geographic spread of this country it is possible to follow some of the wider regional trends within the country itself, notably the diminishing species diversity moving from west to east. Some 115 hard coral species have been recorded from Tarawa and Abaiang Atolls in the west, while Tabuaeran in the east has 71. Blue coral *Heliopora coerulea* is reported to be widespread in the west despite being uncommon over nearby areas in the Pacific. Coral cover on the outer reef slopes is typically very high, with measurements on Tarawa and Abaiang of up to 57 percent cover at 3 meters depth and 28-72 percent at 10 meters. Much of the remainder of the benthos is dominated by coralline algae. There are several very important seabird nesting colonies in the Phoenix and Line Islands, with many millions of birds, including the Phoenix petrel and the Polynesian storm-petrel.

In addition to the islands mentioned above, a number of the US Pacific Territories are geographically part of the Kiribati island groups. Baker and Howland Islands lie to the north of the Phoenix Islands, while Jarvis, Kingman and Palmyra lie in the northern Line Islands. Further information about these islands is given in Chapter 14.

The population of Kiribati is low and almost entirely concentrated in the Gilbert Islands. Elsewhere most of the islands are uninhabited, in many areas because freshwater is not always available. Most of the islanders are heavily dependent on fish as a source of protein, and overfishing is a localized problem near population centers. Reports of increasing incidence of ciguatera poisoning have been linked to other environmental disturbance including the dredging of channels and construction of causeways, although these links remain unproven. Locally, notably in the Tarawa lagoon, sewage pollution may be a problem. This has been exacerbated by the construction of causeways in this atoll to link the islands, which has altered circulation patterns in the lagoon and interrupted the migration patterns of spawning fish. Solid waste disposal is a problem in all areas. Despite being low, populations are now rapidly increasing and there are moves to settle some of the uninhabited islands in the Phoenix group.

Milkfish are raised locally in natural ponds in some areas and there is a significant industry in the cultivation of *Eucheuma* algae for export. The possible introduction and cultivation of non-native species is under consideration, although this could have serious implications for the wider environment. Tourism, still at low levels, is increasing. About 4 000 visitors to Kiribati were recorded in 1995.

Phosphate mining has occurred on a number of the islands. This has had a major impact on the terrestrial vegetation, but the effects on coral reefs have been slight, mostly linked to physical damage by access boats during the mining process. Kiritimati (Christmas) and Malden Islands were the location for hydrogen bomb tests by the British and US military in the 1950s and 1960s. While these have had serious medical implications for servicemen there have been no detailed observations concerning the impacts on the natural environment – most of the bombs were exploded in the air some 8-25 kilometers from the islands. Impacts on the reefs were never assessed. As with other low islands in the Pacific, perhaps the greatest threat is that of sea-level rise as a result of climate change.

Despite this list of threats, the majority of the reefs in this coral reef nation are in excellent condition. A number of protected areas have been established. Although they do not incorporate significant marine elements, they do ensure that wider ecosystems are not disturbed. The Fisheries Division intends to establish at least one protected area in each island where fishing is prohibited, as well as seasonally closed areas to protect spawning stocks.

Nauru

Nauru is a single island country lying in considerable isolation to the west of Kiribati. Geologically it is a raised coral atoll reaching a maximum height of 71 meters. The total depth of limestone is some 500 meters over a basalt seamount. The edge of the island has a coastal terrace up to 400 meters wide, perhaps indicating a former sea-level reef. Surrounding the island is a continuous fringing reef with a reef flat up to 300 meters wide. The biodiversity of the reefs has not been extensively surveyed, but the coral fauna is not considered highly diverse. No seagrasses and only one species of mangrove have been recorded.

The branches of acroporid corals greatly increase the reef's structural complexity, providing important shelter for species such as these chromis Chromis viridis.

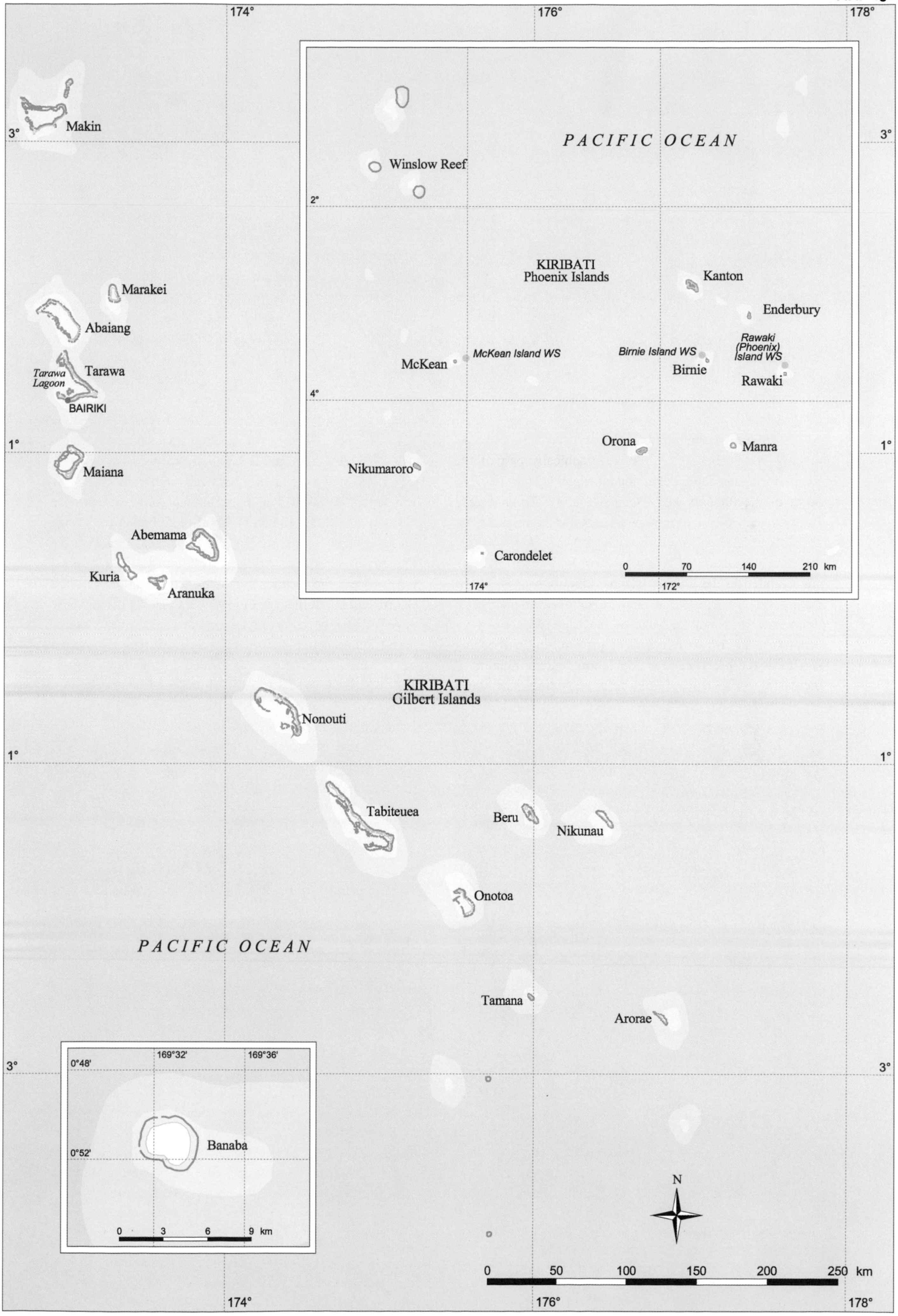

174° 176° 178°
3° 1° 3°

Makin
Marakei
Abaiang
Tarawa
Tarawa Lagoon
BAIRIKI
Maiana
Abemama
Kuria
Aranuka

KIRIBATI
Gilbert Islands

Nonouti
Tabiteuea
Beru
Nikunau
Onotoa
PACIFIC OCEAN
Tamana
Arorae

0°48' 169°32' 169°36' 0°52'
Banaba
0 3 6 9 km

PACIFIC OCEAN
Winslow Reef
2° 4°

KIRIBATI
Phoenix Islands

Kanton
Enderbury
McKean
McKean Island WS
Birnie Island WS
Birnie
Rawaki (Phoenix) Island WS
Rawaki
Orona
Manra
Nikumaroro
Carondelet
0 70 140 210 km
174° 172°

N
0 50 100 150 200 250 km

MAP 13f

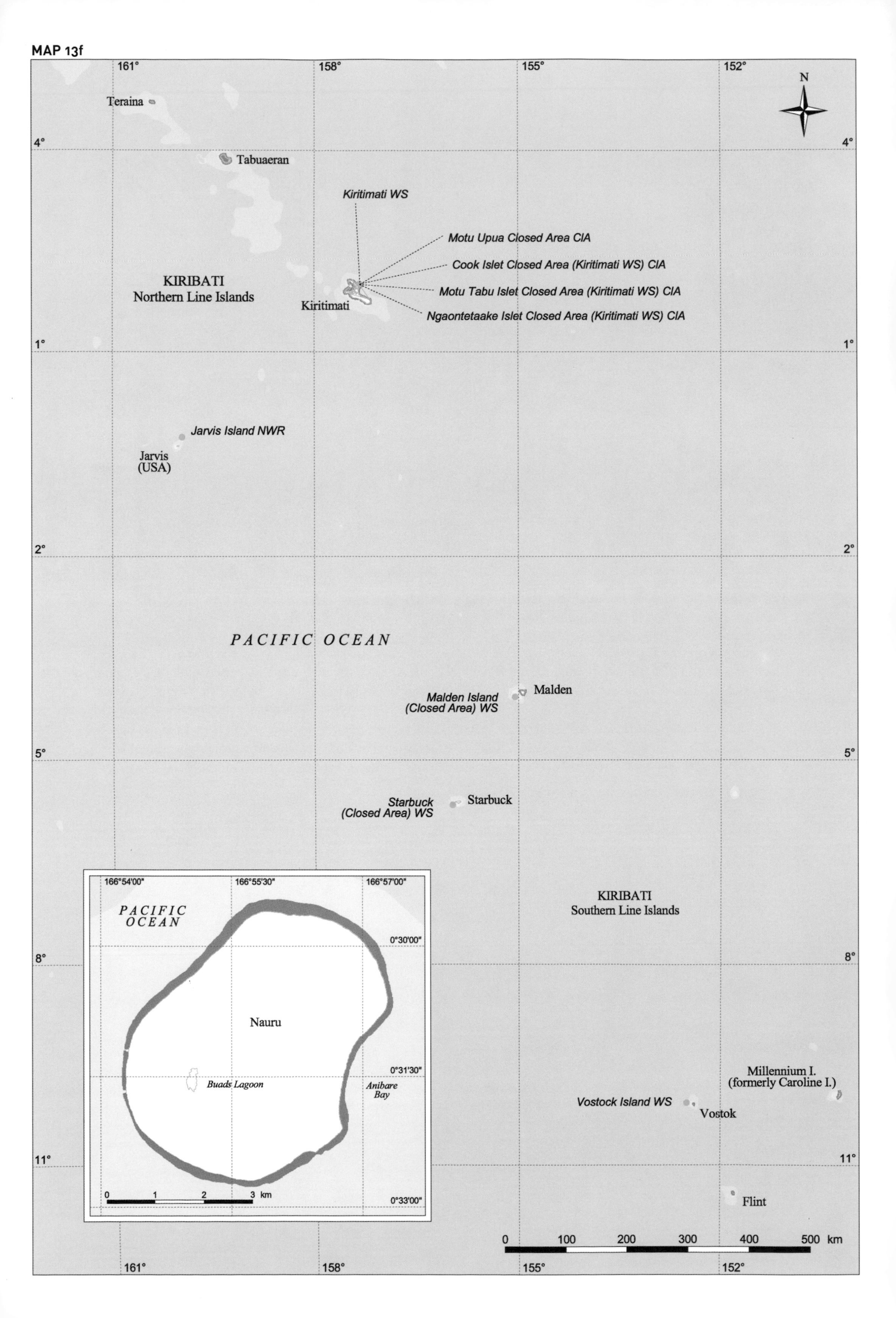
161°
158°
155°
152°
N
Teraina
4°
Tabuaeran
Kiritimati WS
Motu Upua Closed Area ClA
Cook Islet Closed Area (Kiritimati WS) ClA
Motu Tabu Islet Closed Area (Kiritimati WS) ClA
Ngaontetaake Islet Closed Area (Kiritimati WS) ClA
KIRIBATI
Northern Line Islands
Kiritimati
1°
Jarvis Island NWR
Jarvis
(USA)
2°
PACIFIC OCEAN
Malden Island
(Closed Area) WS
Malden
5°
Starbuck
(Closed Area) WS
Starbuck
166°54'00"
166°55'30"
166°57'00"
PACIFIC
OCEAN
0°30'00"
Nauru
0°31'30"
Buads Lagoon
Anibare
Bay
0 1 2 3 km
0°33'00"
KIRIBATI
Southern Line Islands
8°
Millennium I.
(formerly Caroline I.)
Vostock Island WS
Vostok
11°
Flint
0 100 200 300 400 500 km

On land, the entire surface of Nauru has been transformed by phosphate mining, with much of the center of the island now unusable and the population concentrated close to the coast. Although mining is the main source of income the closure of mining operations is planned in the very near future. Fishing is still an important activity, and there are reports of certain species becoming rare as well as average sizes of some fishes diminishing. One impact of the mining industry has been the loss of traditional environmental knowledge. At the same time there are few formal legal controls on reef utilization under the existing fisheries legislation, and there are no protected areas. Sewage pollution is a problem, although there are plans to introduce treatment measures in the near future as part of the environmental rehabilitation of the island. Solid waste is also considered to be a significant problem.

Kiribati

GENERAL DATA	
Population (thousands)	92
GDP (million US$)	43
Land area (km²)	1 050
Marine area (thousand km²)	3 600
Per capita fish consumption (kg/year)	182
STATUS AND THREATS	
Reefs at risk (%)	48
Recorded coral diseases	0
BIODIVERSITY	
Reef area (km²)	2 940
Coral diversity	110 / 365
Mangrove area (km²)	na
No. of mangrove species	4
No. of seagrass species	na

Nauru

GENERAL DATA	
Population (thousands)	12
GDP (million US$)	267
Land area (km²)	28
Marine area (thousand km²)	436
Per capita fish consumption (kg/year)	50
STATUS AND THREATS	
Reefs at risk (%)	100
Recorded coral diseases	0
BIODIVERSITY	
Reef area (km²)	<50
Coral diversity	na / na
Mangrove area (km²)	1
No. of mangrove species	2
No. of seagrass species	na

Cornetfish Fistularia commersonii *are common on Micronesian reefs.*

Selected bibliography

COMMONWEALTH OF THE NORTHERN MARIANA ISLANDS AND GUAM

Jordan J (ed) (1998). *Sensitivity of Coastal Environments and Wildlife to Spilled Oil, Mariana Islands. Volume 2 – Saipan, Tinian, Rota, Aguijan.* Coastal Resources Management Office, Saipan, Northern Marianas.

Mink JF, Vacher HL (1997). Hydrogeology of northern Guam. In: Vacher HL, Quinn T (eds). *Developments in Sedimentology, 54: Geology and Hydrology of Carbonate Islands.* Elsevier Science BV, Amsterdam, Netherlands.

Rock T (1999). *Diving and Snorkelling Guam and Yap,* 2nd edn. Lonely Planet Publications, Melbourne, Australia.

Yamaguchi M (1975). Sea level fluctuations and mass mortalities of reef animals in Guam, Mariana Islands. *Micronesica* 11: 227-243.

PALAU AND FEDERATED STATES OF MICRONESIA

Anthony SS (1997). Hydrogeology of selected islands of the Federated States of Micronesia. In: Vacher HL, Quinn T (eds). *Developments in Sedimentology, 54: Geology and Hydrology of Carbonate Islands.* Elsevier Science BV, Amsterdam, Netherlands.

Golbuu Y (2000). National coral reef status report for Palau. In: Salvat B, Wilkinson C, South GR (eds). Proceedings of the International Coral Reef Initiative Regional Symposium, Noumea, 22-24 May 2000.

Goldman B (1994). Environmental management in Yap, Caroline Islands: can the dream be realized? *Mar Poll Bul* 29: 42-51.

Grano S (ed) (1993). *The Federated States of Micronesia: National Environmental Management Strategy.* South Pacific Regional Environmental Programme, Apia, Western Samoa.

Henson B (ed) (1994). *Republic of Palau: National Environmental Management Strategy.* South Pacific Regional Environmental Programme, Apia, Western Samoa.

Johannes RE (1981). *Words of the Lagoon: Fishing and Marine Lore in the Palau District of Micronesia.* University of California Press, California, USA.

Rock T (1999). *Diving and Snorkelling Guam and Yap.* 2nd edn. Lonely Planet Publications, Melbourne, Australia.

Rock T (2000). *Diving and Snorkelling Chuuk Lagoon, Pohnpei and Kosrae.* Lonely Planet Publications, Melbourne, Australia.

MARSHALL ISLANDS

Emery KO, Tracey JI Jr, Ladd HS (1954). *Geology of Bikini and nearby atolls. Bikini and nearby atolls, part 1, geology. US Geol Survey Prof Pap* 260: 1-265.

Lyons H (1928). The sailing charts of the Marshall Islanders. *The Geographical Journal* 72: 325-328.

Peterson FL (1997). Hydrogeology of the Marshall Islands. In: Vacher HL, Quinn T (eds). *Developments in Sedimentology, 54: Geology and Hydrology of Carbonate Islands.* Elsevier Science BV, Amsterdam, Netherlands.

Price ARG, Maragos JE, Tibon J (2000). The Marshall Islands. In: Sheppard C (ed). *Seas at the Millennium: An Environmental Evaluation.* Elsevier Science Ltd, Oxford, UK.

Wells JW (1954). Recent corals of the Marshall Islands. Bikini and nearby atolls, part 2, oceanography (biologic). *US Geol Survey Prof Pap* 260: 385-486.

Zorpette G (1998). Bikini's nuclear ghosts. *Scientific American.* 9(3).

NAURU AND KIRIBATI

Falkland AC, Woodroffe CD (1997). Geology and hydrogeology of Tarawa and Christmas Island, Kiribati. In: Vacher HL, Quinn T (eds). *Developments in Sedimentology, 54: Geology and Hydrology of Carbonate Islands.* Elsevier Science BV, Amsterdam, Netherlands.

Jacob P (2000). The status of marine resources and coral reefs of Nauru. In: Salvat B, Wilkinson C, South GR (eds) Proceedings of the International Coral Reef Initiative Regional Symposium, Noumea, 22-24 May 2000.

Jacobsen G, Hill PJ, Ghassemi F (1997). Geology and hydrogeology of Nauru Island. In: Vacher HL, Quinn T (eds). *Developments in Sedimentology, 54: Geology and Hydrology of Carbonate Islands.* Elsevier Science BV, Amsterdam, Netherlands.

Lovell ER, Kirata T, Tekinaiti T (2000). National coral reef status report for Kiribati. In: Salvat B, Wilkinson C, South GR (eds). Proceedings of the International Coral Reef Initiative Regional Symposium, Noumea, 22-24 May 2000.

South GR, Skelton PA (2000). Status of coral reefs in the southwest Pacific: Fiji, Nauru, New Caledonia, Samoa, Solomon Islands, Tuvalu and Vanuatu. In: Wilkinson CR (ed). *Status of Coral Reefs of the World: 2000.* Australian Institute of Marine Science, Cape Ferguson, Australia.

Map sources

Map 13a

For Guam, coral reefs and coastline are taken from USGS (1978) (Azimuthal projection). Source data for this map include various hydrographic and topological surveys (from 1945 to 1975). Although these data are old, the location of reefs on this map remains accurate (Charles Birkeland, University of Guam). For the Northern Marianas, coral reefs are taken from Petroconsultants SA (1990)*.

USGS (1978). *Topographic Map of Guam, Mariana Islands.* 1:50 000. US Department of the Interior, Geological Survey.

Maps 13b and 13c

For Palau and most of the Federated States of Micronesia, coral reef data have been taken as arcs from Petroconsultants SA (1990)*. High resolution polygon data were available for Yap, taken from USDI (1983) based on aerial photographs of 1969 with field-checking in 1980. Mangrove data for Yap are also taken from USDI (1983).

USDI (1983). *Topographic map of the Yap Islands (Waqab), Federated States of Micronesia.* 1:25 000. United States Department of the Interior, Geological Survey.

Map 13d

Coral reef data have been taken as arcs from Petroconsultants SA (1990)*.

Maps 13e and 13f

For Nauru, coastline and coral reefs were taken from Hydrographic Office (1955). For Kiribati, coral reef data have been taken as arcs from Petroconsultants SA (1990)*.

Hydrographic Office (1955). Central Pacific Ocean Islands. *British Admiralty Chart No. 979.* 1:55 200. May 1955. Taunton, UK.

* See Technical notes, page 401

CHAPTER 14 Polynesia

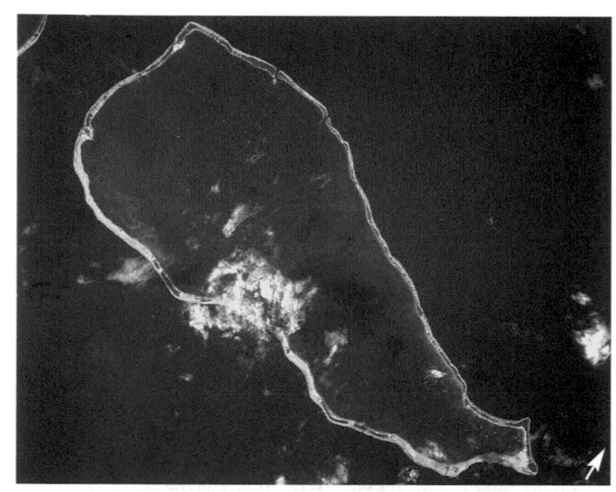

The eastern section of the Pacific, stretching in an easterly direction from Tonga, and to Hawai'i in the north and French Polynesia in the south, constitutes the area known as Polynesia. A very large number of islands and reefs are scattered across this wide region. The westernmost islands of Tonga and Wallis and Futuna lie on the boundary between the Pacific and Indo-Australian plates, with considerable ongoing plate-margin volcanic activity in Tonga. The remainder of the region is located on the Pacific plate, and all of the other islands can be linked to mid-plate volcanic activity. There are many island chains that have been produced by the movement of the plate surface over hotspots, and volcanic activity continues in the Hawaiian Islands.

With over 11 000 square kilometers of coral reefs there is a wide range of reef types harboring highly varied morphologies and ecological communities. The island arcs provide excellent examples of atoll development, from volcanic coastlines with only sporadic coral communities through high islands with fringing reefs, to partially submerged volcanoes with barrier reefs, and finally true atolls. In the north and south the Hawaiian and Austral Archipelagos both provide examples of the latitudinal limits to reef development, with decreases in both diversity and reef constructions in the cooler waters away from the tropics. Polynesia also marks the edge of the Indo-Pacific region. Species diversity is relatively low in almost all species groups, and there are marked declines in diversity as one moves eastwards.

Polynesia's limits are defined by the common features of its peoples. It includes both New Zealand and Easter Island, though these harbor no coral reefs. It was one of the last areas in the world to be reached by humans, probably to central Polynesia some 3 000-4 000 years ago, with secondary migrations spreading out from these first populations. The Samoans are thought to have populated the Marquesas around 300 AD. Hawai'i, by contrast, was probably not settled until between 500 and 1000 AD. Large double canoes some 30-45 meters in length were used for inter-island movement and in the settlement of new areas.

In modern times this region has become an area of contrasts. A number of the countries and states remain remote and isolated with small populations dependent on fisheries. Tourism has become a very important economic resource for a few countries, in particular Tonga, French Polynesia and Hawai'i. Traditional utilization and management of reef resources has largely been lost in the more developed islands, and there are typical problems of overexploitation and pollution associated with the areas of most intense human development. In all, however, the extent of such pressures remains localized, and there are vast areas of reefs in very good condition.

Hawai'i is largely populated by people of non-Polynesian origin and traditional uses of coral reefs have almost completely ceased. Alongside these changes Hawai'i has developed the most extensive network of marine protected areas in the entire Pacific Ocean outside Australia.

Left: Rangiroa Atoll, Tuamotu Islands, French Polynesia, is one of the largest atolls in the Pacific, at 1 800 square kilometers (STS080-750-76, 1996). Right: A peacock grouper Cephalopholis argus. *The species is a target of the live food trade.*

MAP 14

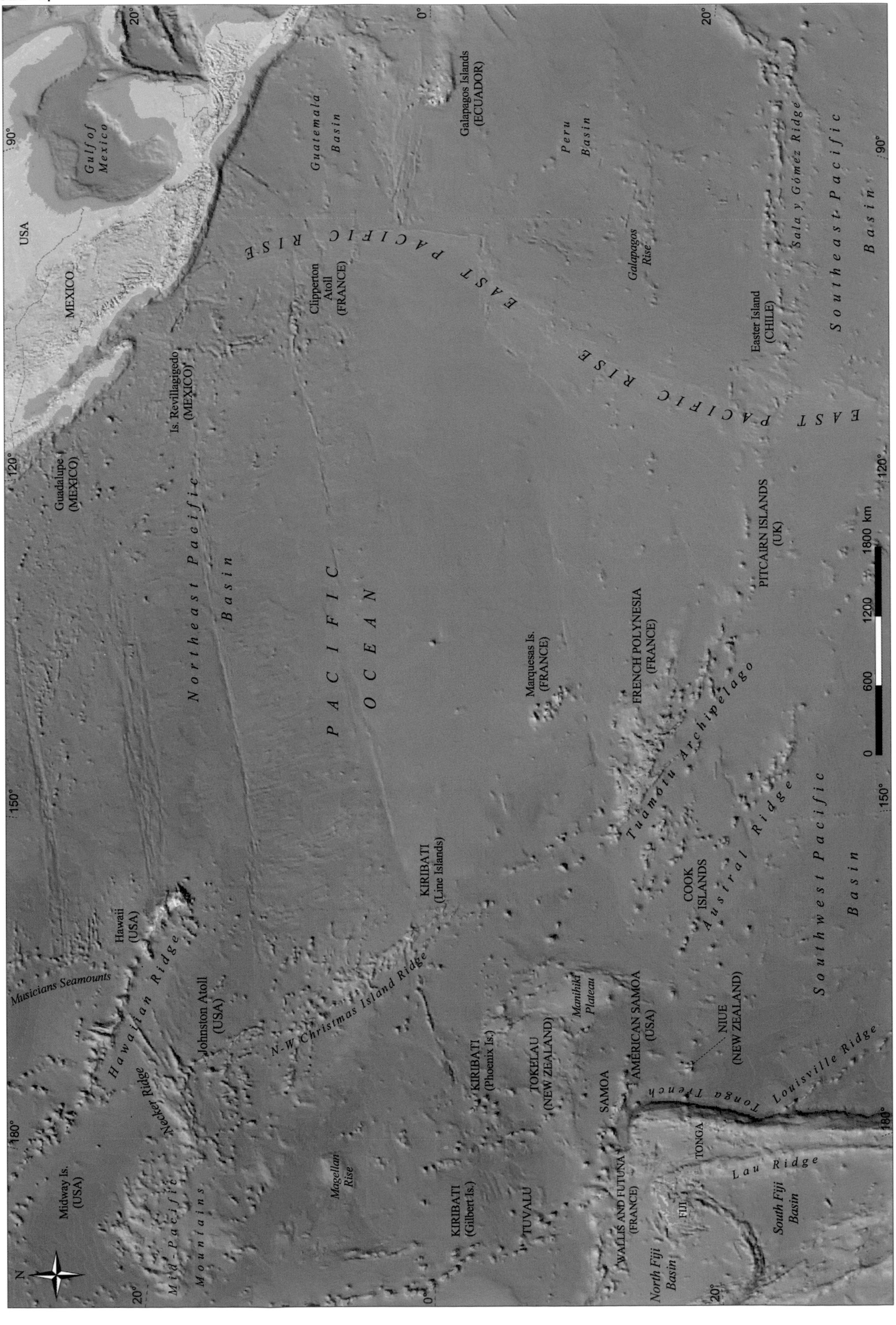

N
Midway Is.
(USA)
Musicians Seamounts
Hawaii
(USA)
Hawaiian Ridge
Necker Ridge
Mid-Pacific
Mountains
Johnston Atoll
(USA)
Guadalupe
(MEXICO)
USA
MEXICO
Gulf of
Mexico
Is. Revillagigedo
(MEXICO)
Northeast Pacific
Basin
N-W Christmas Island Ridge
Magellan
Rise
PACIFIC
OCEAN
Clipperton
Atoll
(FRANCE)
Guatemala
Basin
EAST PACIFIC RISE
KIRIBATI
(Line Islands)
KIRIBATI
(Gilbert Is.)
KIRIBATI
(Phoenix Is.)
Galapagos Islands
(ECUADOR)
TUVALU
TOKELAU
(NEW ZEALAND)
Marquesas Is.
(FRANCE)
Manihiki
Plateau
Peru
Basin
SAMOA
WALLIS AND FUTUNA
(FRANCE)
AMERICAN SAMOA
(USA)
FRENCH POLYNESIA
(FRANCE)
Galapagos
Rise
North Fiji
Basin
FIJI
TONGA
Tonga Trench
Tuamotu Archipelago
COOK
ISLANDS
Austral Ridge
NIUE
(NEW ZEALAND)
Lau Ridge
South Fiji
Basin
Louisville Ridge
PITCAIRN ISLANDS
(UK)
Easter Island
(CHILE)
Sala y Gómez Ridge
Southwest Pacific
Basin
Southeast Pacific
Basin
0 600 1200 1800 km
180° 150° 120° 90°
20° 0° 20°

Tuvalu and Wallis and Futuna

MAP 14a

Tuvalu is a small archipelago (formerly the Ellice Islands) consisting of five true atolls and four other platform islands with encircling fringing reefs. There are also several other seamounts which may reach within 30 meters of the surface.

The lagoons are predominantly sandy with some coral heads. The outer slopes are reported to be rich in both coral cover and diversity, although detailed faunal inventories have not been prepared. Some 400 fish species have been recorded from Funafuti. There are small mangrove stands in a few areas.

Attempts at establishing commercial fisheries have been largely unsuccessful. Some sea cucumbers were exported until 1982, but current densities are not considered high enough to revive the fishery. Efforts to establish a deep water snapper industry for export have largely failed. The country receives its main foreign income through the granting of fishing licenses for offshore pelagics (predominantly tuna). There is no major tourist industry.

Wallis and Futuna

This overseas territory of France consists of three main islands: Wallis (Uvea), Futuna and Alofi. All are high islands of volcanic origin, lying close to the boundary between the Pacific and Indo-Australian tectonic plates. Wallis has fringing reefs around most of its coastline and is further completely encircled by a barrier reef, with a number of sand cays on the reef edge. There is only a small number of deeper channels into the lagoon proper. Futuna has narrow fringing reefs on all coasts, while the uninhabited Alofi has only a few such areas.

There is little in the way of scientific description of the reef communities around these islands. The few studies that have been undertaken show modest levels of diversity, with only 30 coral genera and some 330 species of benthic fish so far described.

Fishing is an important activity, although largely still operating at a subsistence level. However, there have been records of blast fishing. Fringing reefs around Futuna may have been impacted by sediment runoff and are reported to be degraded. There is no significant tourism to the islands, and there are no formal management regimes or protected areas.

Tuvalu

General Data	
Population (thousands)	11
GDP (million US$)	14
Land area (km²)	31
Marine area (thousand km²)	757
Per capita fish consumption (kg/year)	113
Status and Threats	
Reefs at risk (%)	15
Recorded coral diseases	0
Biodiversity	
Reef area (km²)	710
Coral diversity	na / 364
Mangrove area (km²)	na
No. of mangrove species	2
No. of seagrass species	na

Wallis and Futuna

General Data	
Population (thousands)	15
GDP (million US$)	na
Land area (km²)	173
Marine area (thousand km²)	300
Per capita fish consumption (kg/year)	na
Status and Threats	
Reefs at risk (%)	26
Recorded coral diseases	0
Biodiversity	
Reef area (km²)	940
Coral diversity	na / 363
Mangrove area (km²)	na
No. of mangrove species	2
No. of seagrass species	3

MAP 14a

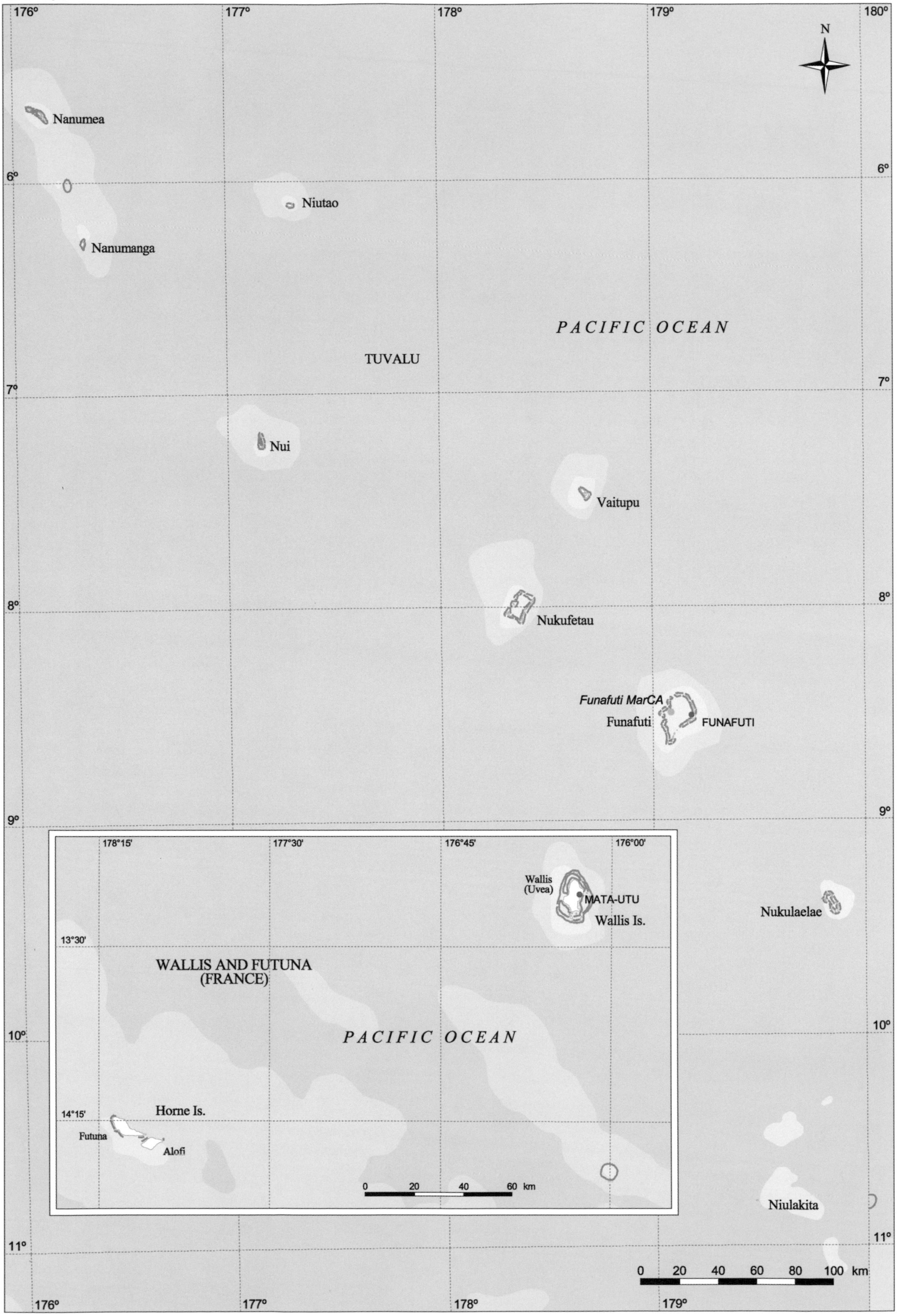

176° 177° 178° 179° 180°

N

Nanumea

6°

Niutao

Nanumanga

PACIFIC OCEAN

TUVALU

7°

Nui

Vaitupu

8°

Nukufetau

Funafuti MarCA

Funafuti

FUNAFUTI

9°

178°15' 177°30' 176°45' 176°00'

Wallis (Uvea)

MATA-UTU

Wallis Is.

Nukulaelae

13°30'

WALLIS AND FUTUNA (FRANCE)

10°

PACIFIC OCEAN

14°15'

Horne Is.

Futuna

Alofi

0 20 40 60 km

Niulakita

11°

0 20 40 60 80 100 km

176° 177° 178° 179°

Tokelau, Samoa and American Samoa

MAP 14b

Tokelau is a group of three small coral atolls, each with numerous islands on its rim. The lagoons are shallow with large numbers of coral outcrops, while the maximum height of the islands is about 4.5 meters. None of these atolls has a deep channel into the lagoon, making boat access difficult. The area has been affected by cyclones on a number of occasions, including 1987, 1990 and 1991. Detailed information about the biodiversity of these reefs is unavailable, but it is likely to be similar to that of the Samoan Islands to the south and Tuvalu to the west.

Tokelau, a territory of New Zealand, is heavily dependent on financial support for development. Concerns about degradation of the natural environment from overfishing and sewage pollution have led to some efforts to improve environmental management. There is a small fish processing plant on Atafu which prepares sundried tuna. At the end of the 1990s the overall threats were very low, although there had been depletion of a number of species such as giant clam and trochus. There is a relatively small population and there are very few visitors.

Samoa

The Samoa Archipelago is a hotspot chain of predominantly high volcanic islands divided into the politically independent western islands of Samoa (formerly Western Samoa) and the eastern islands of American Samoa. Samoa itself is dominated by the two large islands of Upolu and Savai'i with a few very small islands nearby. Savai'i, geologically the youngest, experienced eruptions from two of its volcanoes in the early 1900s. There are fringing reefs around most of the coastline, generally close to the shore, but reaching up to 3 kilometers offshore along the northwestern coast of Upolu.

Information on biodiversity is relatively limited. About 50 species of hard coral have been listed, although this is likely to be a considerable underestimate. Studies of marine algae have been more complete, and some 300 species have so far been described. Some 991 fish species have been recorded in the wider Samoan archipelago, of which at least 890 are shallow reef-dwelling species. There are small areas of seagrass, and mangrove communities are well developed at a few locations around Upolu. The reefs and islands of Samoa were severely impacted by cyclones Ofa and Val in 1990 and 1991, although there is reported to have been good recovery of coral cover and diversity in many areas.

The people of Samoa are generally heavily dependent on the reefs at a subsistence level and for the domestic market. An estimated 4 600 tons of fish were taken for

Decorator crabs cultivate other organisms on their bodies as a form of camouflage or defense. Here a crab has stinging hydroids growing on its antennae, which it uses for defense.

subsistence in 1997, while domestic markets probably add a further 75-80 tons of fish, crustaceans and other invertebrates (1998-99 figures). There has been a noted reduction in biomass and size of fish in shallower and more heavily fished areas. There are two small aquarium trade exporters, while there have also been several attempts at reef restocking with giant clams and green snails. There is also some giant clam aquaculture. As the country develops there are increasing problems of pollution from sewage and solid waste. Poor landuse practices combined with uncontrolled use of agricultural chemicals are creating high loads of nutrients, toxic chemicals and sediments, placing the reefs under increasing levels of stress. Tourism, the fastest growing sector of the economy, is also causing some problems, particularly in the development of hotel facilities.

Samoa has only one major marine protected area, although the Tafua Rainforest Reserve contains some coastal areas, and there are plans to develop others. There are also increasing efforts to involve village participants in conservation awareness. Traditional marine tenure is recognized, maintaining traditional ownership of adjacent lagoon and reef fishery resources by the villages.

American Samoa

The eastern portion of the Samoa Archipelago consists of five high volcanic islands and, in the east, Rose Atoll. It is an unincorporated territory of the USA, and its administration also covers Swains Atoll, a remote atoll which lies between the main islands and those of Tokelau to the north. The high islands are surrounded by fringing reefs, with reef flats typically 50-500 meters wide terminating in a reef slope which drops sharply for 3-6 meters and then descends gradually down to a depth of about 40 meters.

Biodiversity is similar to Samoa's, with 890 species of reef fish recorded around the entire archipelago. Some 200 coral species have been recorded. The reefs were severely impacted by the 1990 and 1991 cyclones, having previously suffered a damaging crown-of-thorns outbreak (1978), but they are reported to have made a considerable recovery from these events.

There are small areas of mangrove on Tutuila and Aunu'u. The reef rim of Rose Atoll is dominated by coralline algae. It has an important green turtle nesting colony, and having been cleared of rats is also a thriving seabird colony. Swains Atoll lies at sea level, but there is a circular island on the reef flat completely enclosing the

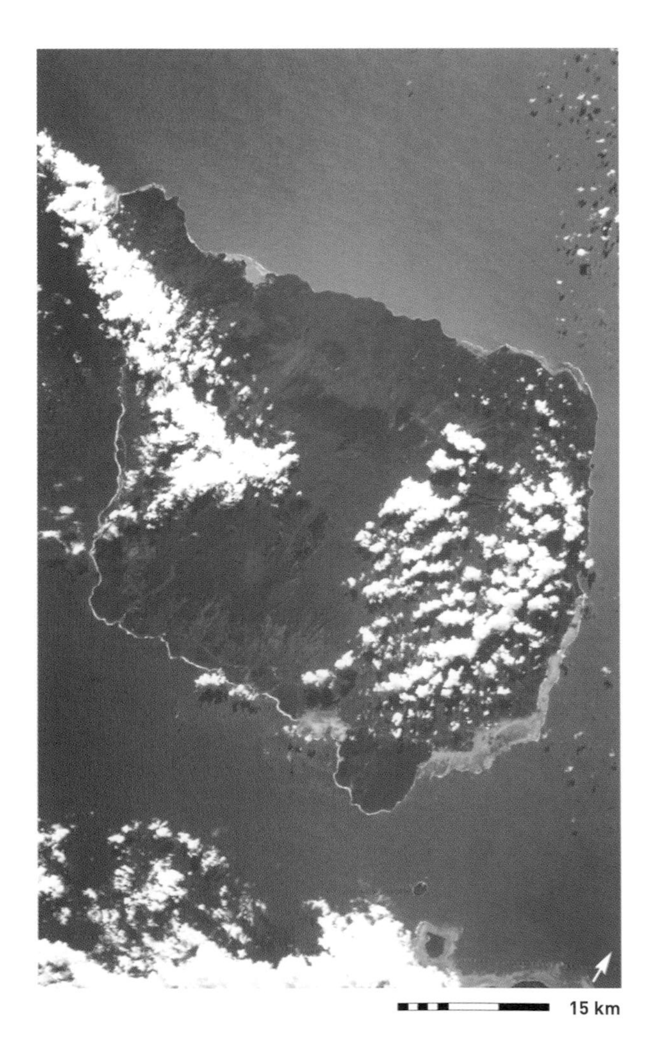

Left: A white-belly damsel Amblyglyphidodon leucogaster *sheltering in a branching* Acropora. *Right: Savai'i Island and reefs with sunlight highlighting ocean surface features* (STS093-716-49, 1999).

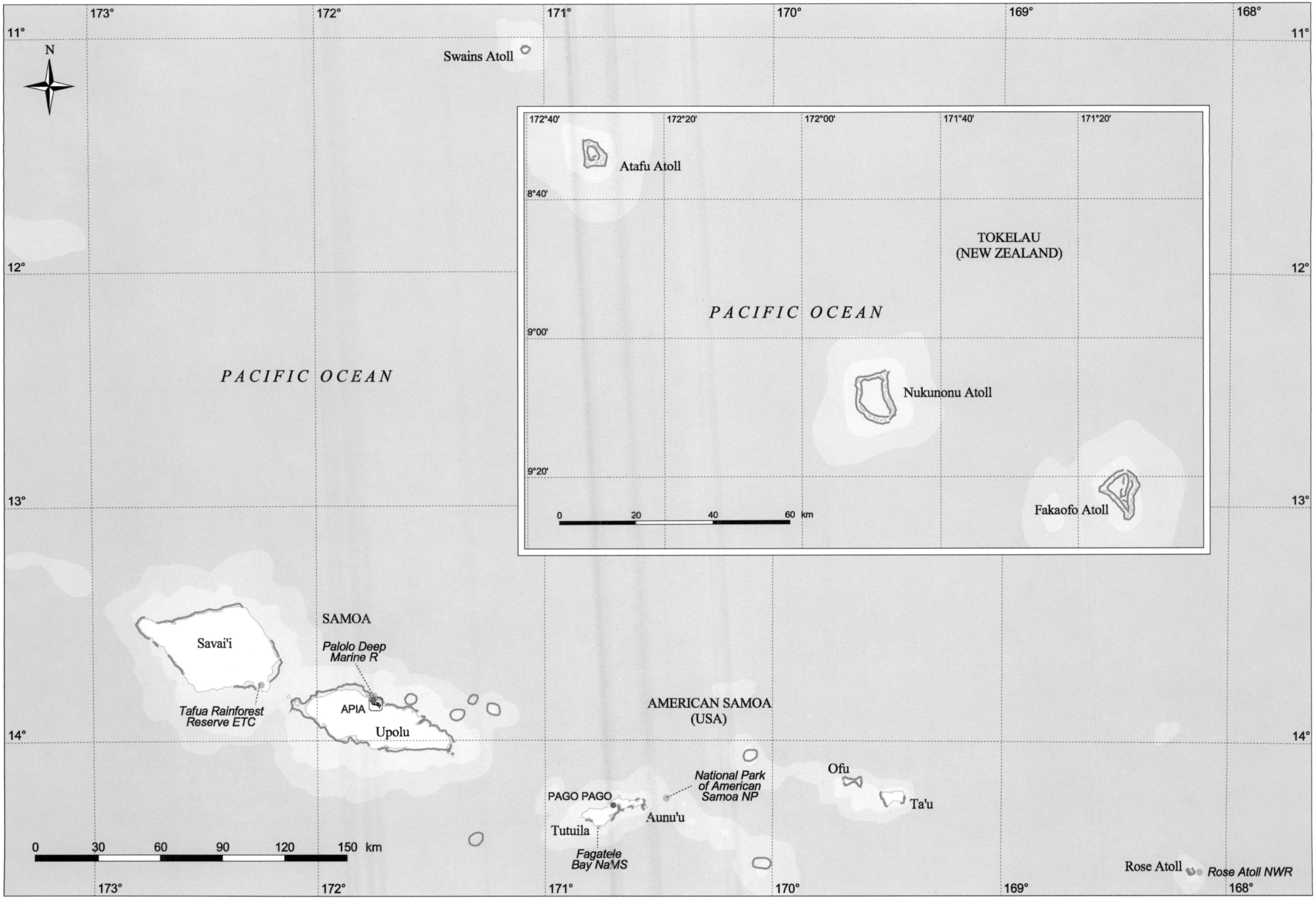

MAP 14b

173° 172° 171° 170° 169° 168°

11° 12° 13° 14°

N

Swains Atoll

PACIFIC OCEAN

172°40' 172°20' 172°00' 171°40' 171°20'

8°40' 9°00' 9°20'

Atafu Atoll

TOKELAU
(NEW ZEALAND)

PACIFIC OCEAN

Nukunonu Atoll

Fakaofo Atoll

0 20 40 60 km

SAMOA

Savai'i

Palolo Deep Marine R

APIA

Upolu

Tafua Rainforest Reserve ETC

AMERICAN SAMOA
(USA)

National Park of American Samoa NP

PAGO PAGO

Aunu'u

Tutuila

Fagatele Bay NaMS

Ofu

Ta'u

Rose Atoll

Rose Atoll NWR

0 30 60 90 120 150 km

Protected areas with coral reefs

Site name	Designation	Abbreviation	IUCN cat.	Size (km²)	Year
American Samoa					
Fagatele Bay	National Marine Sanctuary	NaMS	IV	0.64	1986
National Park of American Samoa	National Park	NP	II	37.25	1988
Rose Atoll	National Wildlife Refuge	NWR	Ia	6.53	1973
Samoa					
Palolo Deep Marine	Reserve	R	IV	0.22	1979
Tafua Rainforest Reserve	Other area	ETC	IV	60.00	1990

brackish atoll lagoon. There is a small population of some 50 people on the island.

The vast majority of the rapidly growing population of American Samoa lives on the southern shores of Tutuila. Although fisheries are very important, changes in the economy have meant that there is less reliance on subsistence fishing than in the past. About 150 tons of reef fish and invertebrates were taken by subsistence and small-scale artisanal fisheries in 1994, and overfishing has been shown to occur on Tutuila Island. There are further problems arising from land-derived sediments and pollutants. Although there is sewage treatment in the main population centers there are still some nutrient inputs from sewage in these areas and elsewhere. There are two tuna canneries which used to add considerable amounts of nutrients to Pago Pago harbor. These inputs have now been substantially reduced with the construction of a treated waste disposal pipe further offshore, and the dumping of high nutrient waste at some 8 kilometers distance. The coastline on Tutuila has been heavily impacted by road building and construction, and nesting turtles have largely stopped using the area.

A number of reefs and related habitats have been declared as protected areas.

	Tokelau	Samoa	American Samoa
General Data			
Population (thousands)	2	179	65
GDP (million US$)	na	90	na
Land area (km²)	20	2 803	187
Marine area (thousand km²)	290	120	390
Per capita fish consumption (kg/year)	129	32	na
Status and Threats			
Reefs at risk (%)	0	95	42
Recorded coral diseases	0	0	0
Biodiversity			
Reef area (km²)	<50	490	220
Coral diversity	na /210	na / 211	150 /212
Mangrove area (km²)	na	7	57
No. of mangrove species	na	3	3
No. of seagrass species	na	3	na

Tonga and Niue

MAP 14c

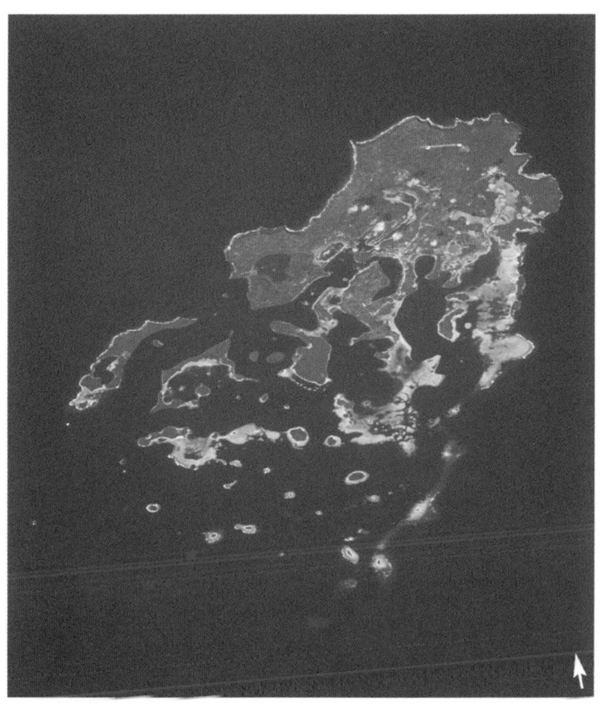

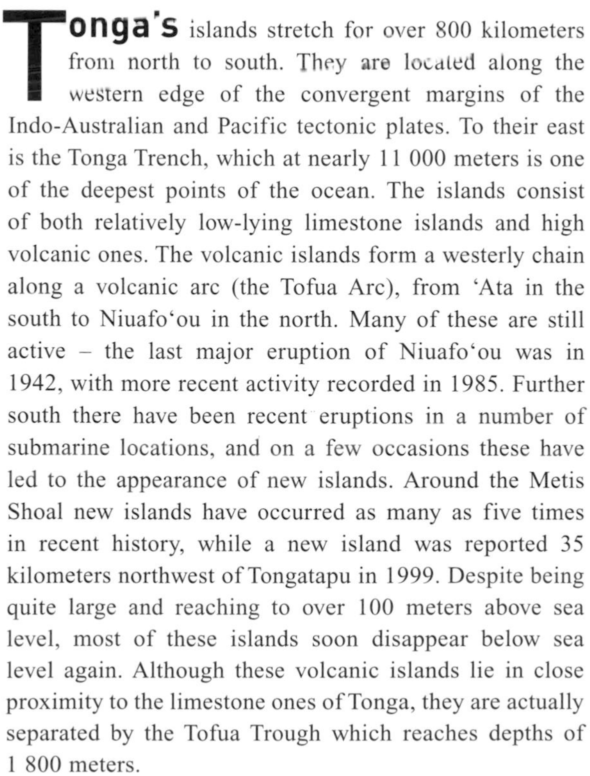

Tonga's islands stretch for over 800 kilometers from north to south. They are located along the western edge of the convergent margins of the Indo-Australian and Pacific tectonic plates. To their east is the Tonga Trench, which at nearly 11 000 meters is one of the deepest points of the ocean. The islands consist of both relatively low-lying limestone islands and high volcanic ones. The volcanic islands form a westerly chain along a volcanic arc (the Tofua Arc), from 'Ata in the south to Niuafo'ou in the north. Many of these are still active – the last major eruption of Niuafo'ou was in 1942, with more recent activity recorded in 1985. Further south there have been recent eruptions in a number of submarine locations, and on a few occasions these have led to the appearance of new islands. Around the Metis Shoal new islands have occurred as many as five times in recent history, while a new island was reported 35 kilometers northwest of Tongatapu in 1999. Despite being quite large and reaching to over 100 meters above sea level, most of these islands soon disappear below sea level again. Although these volcanic islands lie in close proximity to the limestone ones of Tonga, they are actually separated by the Tofua Trough which reaches depths of 1 800 meters.

The majority of Tonga's islands lie in the eastern arc. These are largely of reef origin, although there are also significant ash deposits, up to 5 meters thick on Tongatapu and 13 meters thick on Kotu. They fall into three main groups. The Tongatapu group in the south is dominated by the large uplifted islands of Tongatapu and 'Eua. The central Ha'apai group is a complex spread of reefs and low-lying islands. The Vava'u group in the north is dominated by the main island of Vava'u, but also includes a wide spread of islands and reefs on its southern side. Tonga also lays claim to the Minerva Reefs which lie to the southwest of 'Ata and south of Fiji's Lau Islands, although Fiji disputes this claim. In addition to these areas, there are reports of highly remote reefs to the south and east of Tonga, including Albert Meyer Reef about 300 kilometers east of Tongatapu and Gleaner Reef about 175 kilometers southeast of 'Eua.

Tectonic activity in the country is considerable. Many islands and shoals are too active at the present time for reef development and, in many of these areas, conditions of the substrate and even conditions of ocean chemistry may be unsuitable for the settlement of significant coral communities. Uplift and subsidence are relatively regular occurrences in many areas, particularly in the volcanic

Left: A marbled grouper Epinephelus polyphekadion. *Large groupers are common on less heavily fished reefs. Right: The Vava'u Islands and reefs in northern Tonga (STS068-252-50, 1994).*

MAP 14c

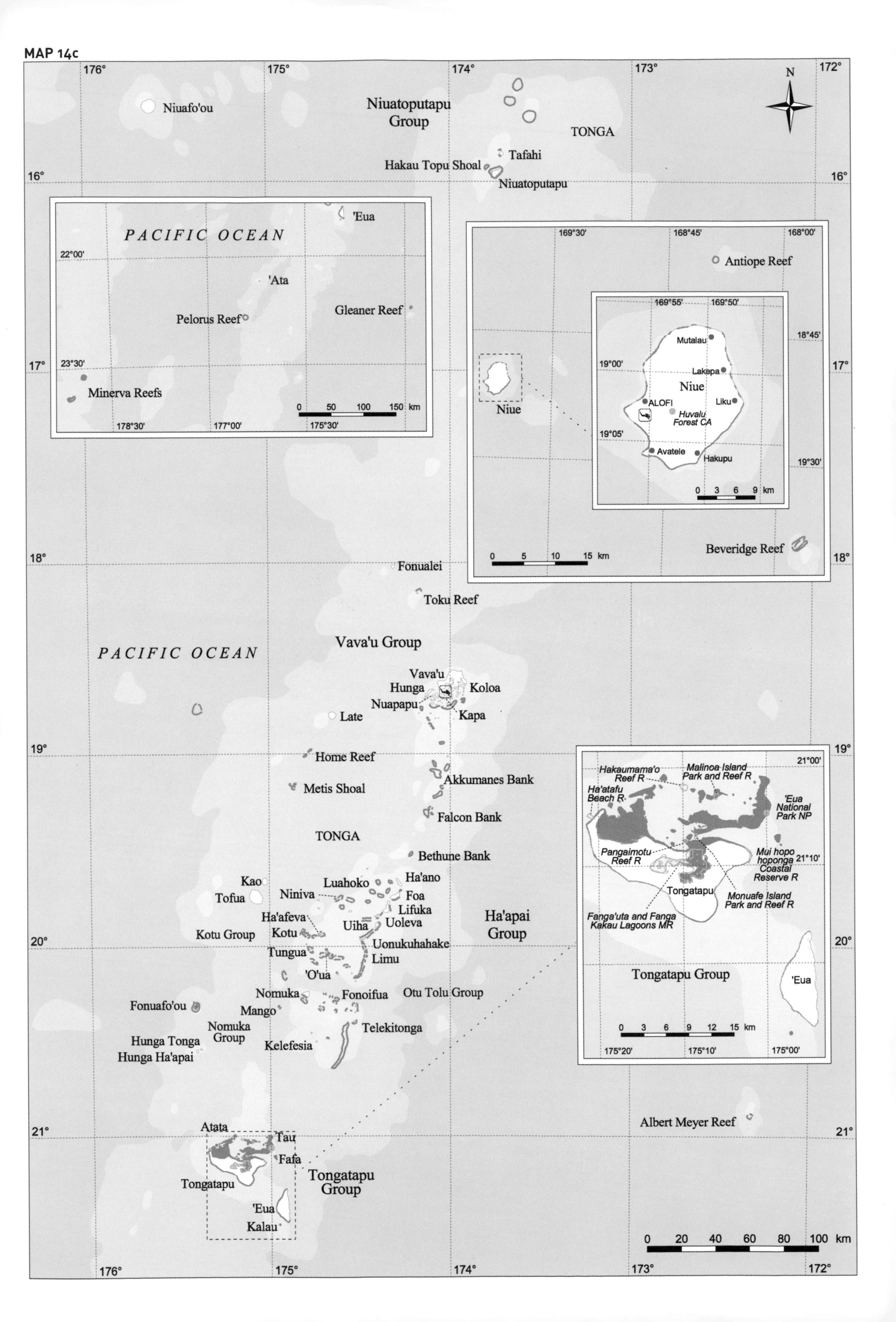

Niuafo'ou
Niuatoputapu Group
TONGA
Tafahi
Hakau Topu Shoal
Niuatoputapu
PACIFIC OCEAN
'Eua
'Ata
Pelorus Reef
Gleaner Reef
Minerva Reefs
Antiope Reef
Mutalau
Lakepa
Niue
ALOFI
Liku
Huvalu Forest CA
Avatele
Hakupu
Beveridge Reef
Fonualei
Toku Reef
Vava'u Group
PACIFIC OCEAN
Vava'u
Hunga
Koloa
Nuapapu
Kapa
Late
Home Reef
Metis Shoal
Akkumanes Bank
Falcon Bank
TONGA
Bethune Bank
Kao
Tofua
Luahoko
Ha'ano
Niniva
Foa
Lifuka
Ha'afeva
Uiha
Uoleva
Kotu Group
Kotu
Uonukuhahake
Tungua
Limu
'O'ua
Ha'apai Group
Nomuka
Fonoifua
Otu Tolu Group
Fonuafo'ou
Mango
Nomuka Group
Telekitonga
Hunga Tonga
Hunga Ha'apai
Kelefesia
Hakaumama'o Reef R
Malinoa Island Park and Reef R
Ha'atafu Beach R
'Eua National Park NP
Pangaimotu Reef R
Mui hopo hoponga Coastal Reserve R
Tongatapu
Monuafe Island Park and Reef R
Fanga'uta and Fanga Kakau Lagoons MR
Tongatapu Group
'Eua
Albert Meyer Reef
Atata
Tau
Fafa
Tongatapu
Tongatapu Group
'Eua
Kalau

islands. By contrast, reefs are widespread and well developed among the eastern islands. There are fringing reefs surrounding most coasts, while platform and barrier-type structures are also located in most of the main island groups. The most extensive areas of reef are in the Hapa'ai group. The dominant winds are southeast trade winds, particularly from March to October. From November to March there is a cyclone season, although trade winds still predominate much of the time.

Little information is available describing the biodiversity of Tongan reefs. Some 192 species of scleractinian corals have been recorded at 11 reefs around Tongatapu. Other studies have revealed 229 reef fish from 39 families, 55 bivalves, 83 gastropods and 13 holothurians. In all cases these are likely to be considerable underestimates. Coral cover on the reefs is likely to be highly variable, with records of only 2 percent in Monuafe but up to 50 percent at Hakaumama'o Reef. Although large numbers of crown-of-thorns starfish have been reported in many areas, major outbreaks have not been recorded. Cyclones cause periodic damage to reefs, with the latest ones reported in 1995, 1997, 1999 and 2000. Although largely unaffected by the 1998 bleaching, the Tongan reefs appeared to have suffered a major bleaching event similar to that reported in Fiji in 2000.

Artisanal fisheries are an important activity in the country. Studies have shown that almost 70 percent of the artisanal catch is made up of reef fish (notably emperors and mullet). Turtle eggs and meat are still eaten in many areas. There are also several important commercial fisheries, notably an aquarium trade dealing in fish, coral and live rock (small pieces of reef rock housing numerous species) together with limited numbers of invertebrates. Holothurians have also been widely taken, particularly the sandfish *Holothuria scabra*. Overfishing is a problem in areas of high population density, notably around Tongatapu. Target species for export have also been overfished throughout the country. Two giant clam species, *Tridacna gigas* and *Hippopus hippopus*, were thought to have become locally extinct, but were re-introduced in 1990 and 1991. A number of giant clam nurseries have been established to be managed at the community level. A complete ban on sea cucumber export was imposed for ten years in 1997 as a response to chronic overharvesting. These problems of overfishing have been exacerbated by a lack of local ownership of reef resources, enabling commercial collectors to harvest even close to local communities. In addition to overfishing, certain

Tonga

General Data	
Population (thousands)	102
GDP (million US$)	149
Land area (km²)	697
Marine area (thousand km²)	700
Per capita fish consumption (kg/year)	35
Status and Threats	
Reefs at risk (%)	46
Recorded coral diseases	0
Biodiversity	
Reef area (km²)	1 500
Coral diversity	na / 218
Mangrove area (km²)	10
No. of mangrove species	8
No. of seagrass species	na

Niue

General Data	
Population (thousands)	2
GDP (million US$)	na
Land area (km²)	228
Marine area (thousand km²)	390
Per capita fish consumption (kg/year)	62
Status and Threats	
Reefs at risk (%)	43
Recorded coral diseases	0
Biodiversity	
Reef area (km²)	170
Coral diversity	na / 189
Mangrove area (km²)	na
No. of mangrove species	1
No. of seagrass species	na

The lined butterflyfish Chaetodon lineolatus *is found from the Red Sea to the Pitcairn Islands.*

Protected areas with coral reefs

Site name	Designation	Abbreviation	IUCN cat.	Size (km²)	Year
Tonga					
'Eua National Park	National Park	NP	II	4.50	1992
Fanga'uta and Fanga Kakau Lagoons	Marine Reserve	MR	VI	28.35	1974
Ha'amonga Trilithon	Park	P	Unassigned	0.19	1972
Ha'atafu Beach	Reserve	R	IV	0.08	1979
Hakaumama'o Reef	Reserve	R	IV	2.60	1979
Malinoa Island Park and Reef	Reserve	R	IV	0.73	1979
Monuafe Island Park and Reef	Reserve	R	IV	0.33	1979
Mui hopo hoponga Coastal Reserve	Reserve	R	V	na	1972
Pangaimotu Reef	Reserve	R	IV	0.49	1979
Niue					
Beveridge Reef	Other Area	ETC	Unassigned	na	na
Huvalu Forest	Conservation Area	CA	VI	54.00	na

relatively destructive fishing practices appear to be degrading reefs, notably through smashing reefs to chase fish into nets, but also trampling of reef flats and the use of poisons such as bleach and pesticides.

Eutrophication is a problem in Tongatapu and Vava'u, arising from both untreated sewage and fertilizer runoff. This has been blamed for an increase in seagrass and mangroves and a decline in corals in Fanga'uta Lagoon, Tongatapu. Further problems have been noted resulting from the building of causeways in Ha'apai and Vava'u, while quarrying, construction and sand mining create problems in some areas.

Tourism is particularly important for Tonga, with over 30 000 visitors in 1999. A number of protected areas have been established, largely focussed around Tongatapu. Such protection measures are difficult to enforce without community involvement and there are now moves to establish more community-based management of marine resources. The whole of Ha'apai has been declared a Conservation Area following a recommendation by the South Pacific Regional Environment Programme (SPREP). Three proposed reserves were surveyed in the Vava'u group in 1997.

Niue

Niue consists of a single uplifted coral atoll, oval in shape and reaching a maximum height of about 70 meters above sea level. It is actually one of the largest carbonate islands in the Pacific. The island is almost surrounded by a narrow platform cut into the former reef structure and forming a modern reef flat, becoming discontinuous in the south and east. Few details are known about the diversity of the reefs, although 243 marine fish have been recorded and there are reported to be over 43 coral genera. Tropical Cyclone Ofa struck the island in 1990 and was reported to have caused considerable damage to the reefs, particularly on the western coast. Over 200 kilometers southeast of Niue there is a substantial seamount capped by a significant atoll-like structure: Beveridge Reef. Although there is no vegetated land, there is a significant and apparently permanent sand cay on the northern mouth of the lagoon channel, and there may be other cays. Coral cover is reported to be high, and the fish populations diverse and unfished.

Niue is an internally self-governing state which exists in free association with New Zealand. All Niueans have New Zealand citizenship, and in fact the majority live in New Zealand. The resident population live almost entirely along the coastal terrace which is typically about 500 meters wide. Fishing is an important activity, although mostly focussed on offshore pelagic species. There is no export fishery. There is a limited amount of tourism to the island and some diving. A substantial proportion of the eastern coastline, including the offshore reef, is included in the Huvalu Forest Conservation Area. Although the marine components of this site are not well known they are not considered to be heavily exploited. Beveridge Reef has also been declared as protected, although its legal status is unclear and there is no active management.

Cook Islands

MAPS 14d and e

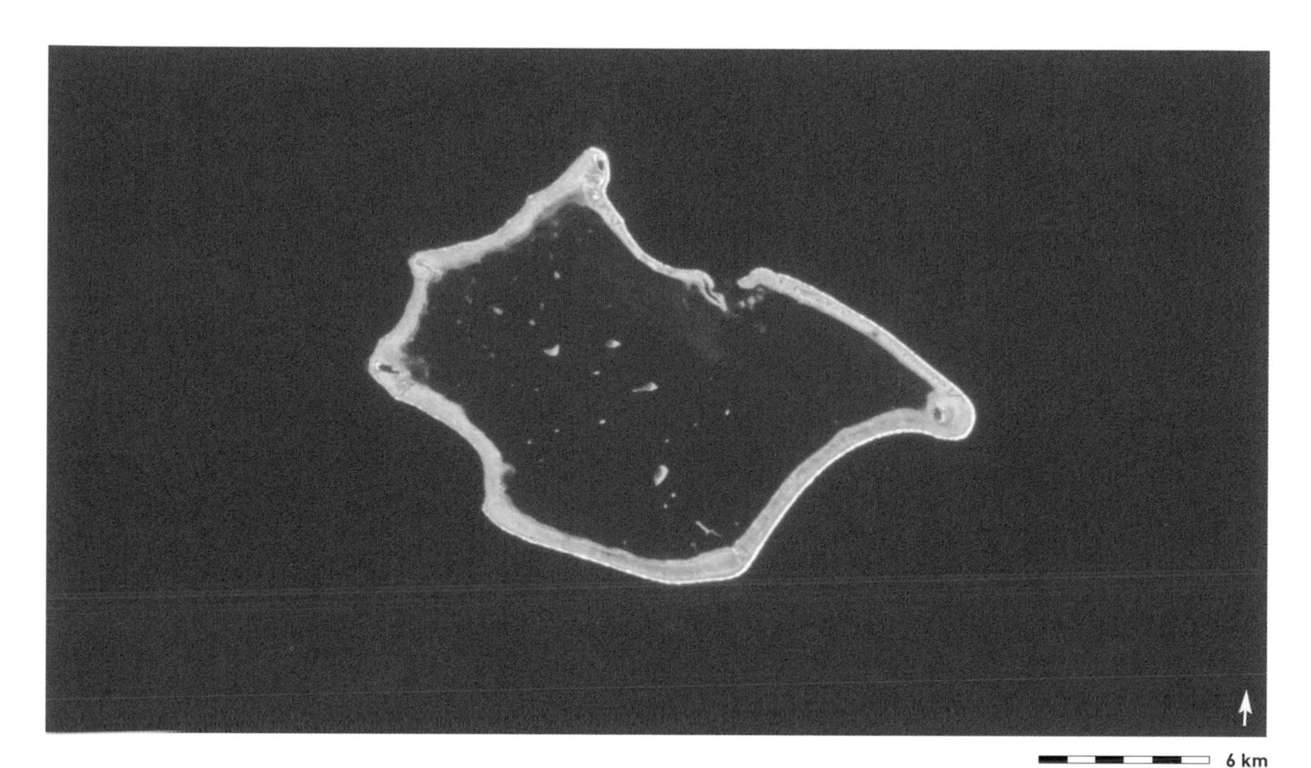

The Cook Islands are a group of 15 islands, or island groups, scattered across a wide expanse of ocean. Descriptions typically divide the islands geographically into a northern and a southern group. The northern Cook Islands are a group of five atolls and a platform island. With the exception of Penrhyn they are located on the Manihiki Plateau. It seems likely that they formed when this was a shallow volcanic feature and the atolls have grown up as the plateau itself subsided. Penrhyn, by contrast, is an isolated feature rising from the deep ocean. In addition to these islands, Tema Reef is a subsurface platform reef located between Pukapuka and Nassau Island. Flying Venus Reef is another platform structure lying on the same seamount as Penrhyn, but separated by a deep (more than 500 meters) channel.

The southern Cook Islands show a wide range of oceanic island types. They follow two parallel chains running from the northwest to the southeast, and they continue as the Austral Islands in French Polynesia, terminating in the volcanically active Macdonald Seamount, probably indicating some form of hotspot origin to the islands. Palmerston and Manuae are true atolls and Takutea is a platform island (possibly left behind after the partial collapse of an atoll structure). Aitutaki is a near-atoll with one large and two small volcanic islands located in the lagoon. Four islands, Mitiaro, Atiu, Mauke and Mangaia, have been described as *makatea* (fossil reef) islands. They have a volcanic center with a carbonate rim of reef origin, now uplifted to some height above sea level. Finally, the main island of Rarotonga is a high volcanic island (652 meters) with substantial ridges and steeply incised valleys. Fringing reefs occur around all of the *makatea* islands and around Rarotonga in the southern group, although these can be quite narrow. Reef flats are typically very level with a hard platform, while the crests are dominated by coralline algae, and spur and groove formations become characteristic offshore. The lagoons are highly varied among the atolls. Manuae, in the southern Cooks, has a very sandy shallow lagoon. Rakahanga's lagoon is virtually enclosed by islands and shows very little active coral growth, whereas most of the other northern islands have relatively deep lagoons (average depths of over 10 meters) with varying degrees of reef development. Winslow Reef, in the southern Cooks, is a shallow platform reef lying nearly 150 kilometers northeast of Rarotonga and lacking any above-water structures. Due to their southerly location, all the Cook Islands lie in the path of hurricanes and storms, which typically occur between January and March.

Suwarrow Atoll, an isolated atoll in the northern Cook Islands (STS055-97-58, 1993).

MAP 14d

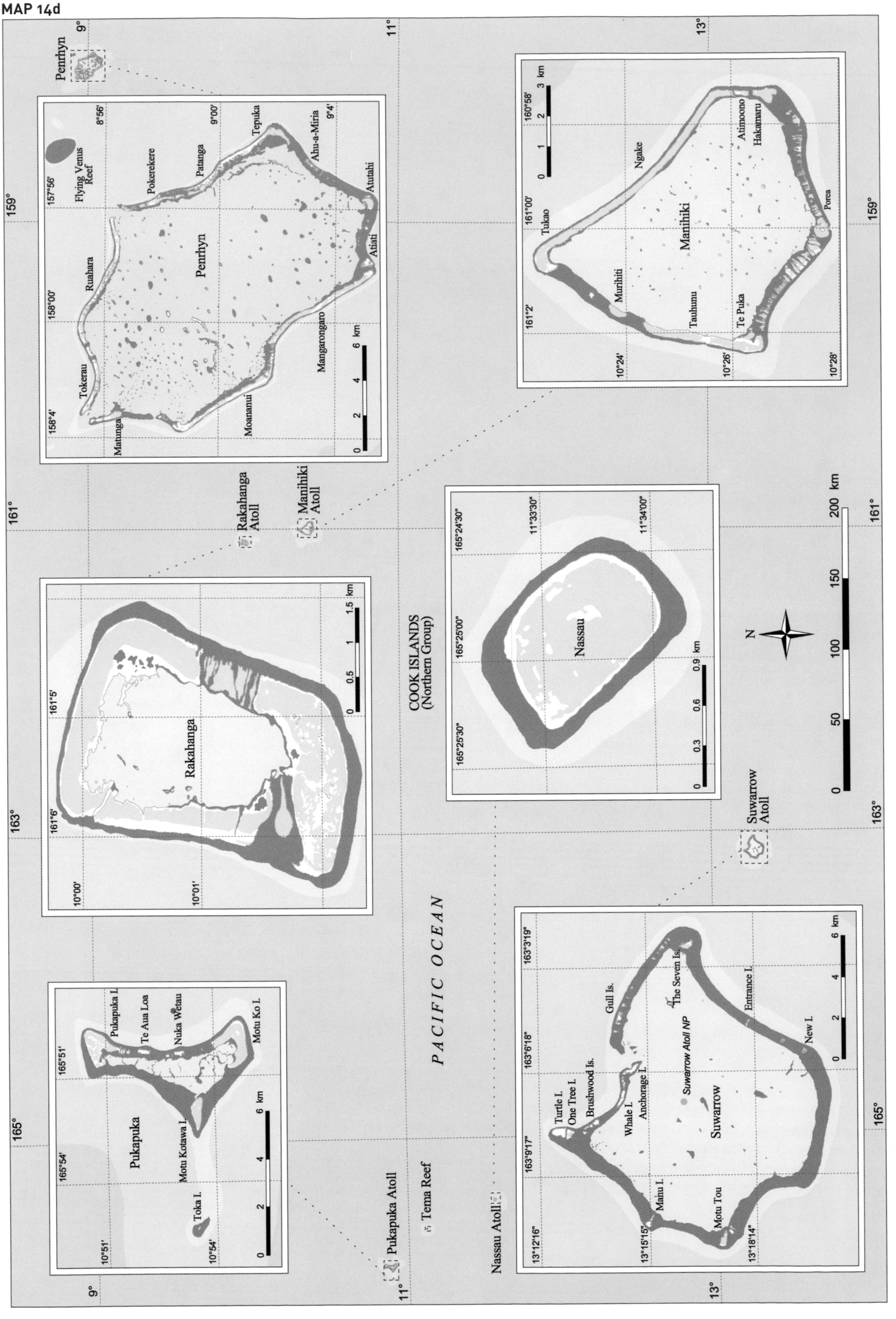

COOK ISLANDS
(Northern Group)

PACIFIC OCEAN

Pukapuka Atoll
Tema Reef
Nassau Atoll
Rakahanga Atoll
Manihiki Atoll
Penrhyn
Suwarrow Atoll

Pukapuka
Pukapuka I.
Te Aua Loa
Nuka Wetau
Motu Kotawa I.
Toka I.
Motu Ko I.

Rakahanga

Penrhyn
Tokerau
Ruahara
Flying Venus Reef
Matunga
Pokerekere
Patanga
Moananui
Tepuka
Mangarongaro
Ahu-a-Miria
Atiati
Atutahi

Nassau

Suwarrow
Turtle I.
One Tree I.
Brushwood Is.
Gull Is.
Whale I.
Anchorage I.
Manu I.
The Seven Is.
Suwarrow Atoll NP
Motu Tou
Entrance I.
New I.

Manihiki
Tukao
Murihiti
Ngake
Tauhunu
Te Puka
Atimoono
Hakamaru
Porea

0 50 100 150 200 km

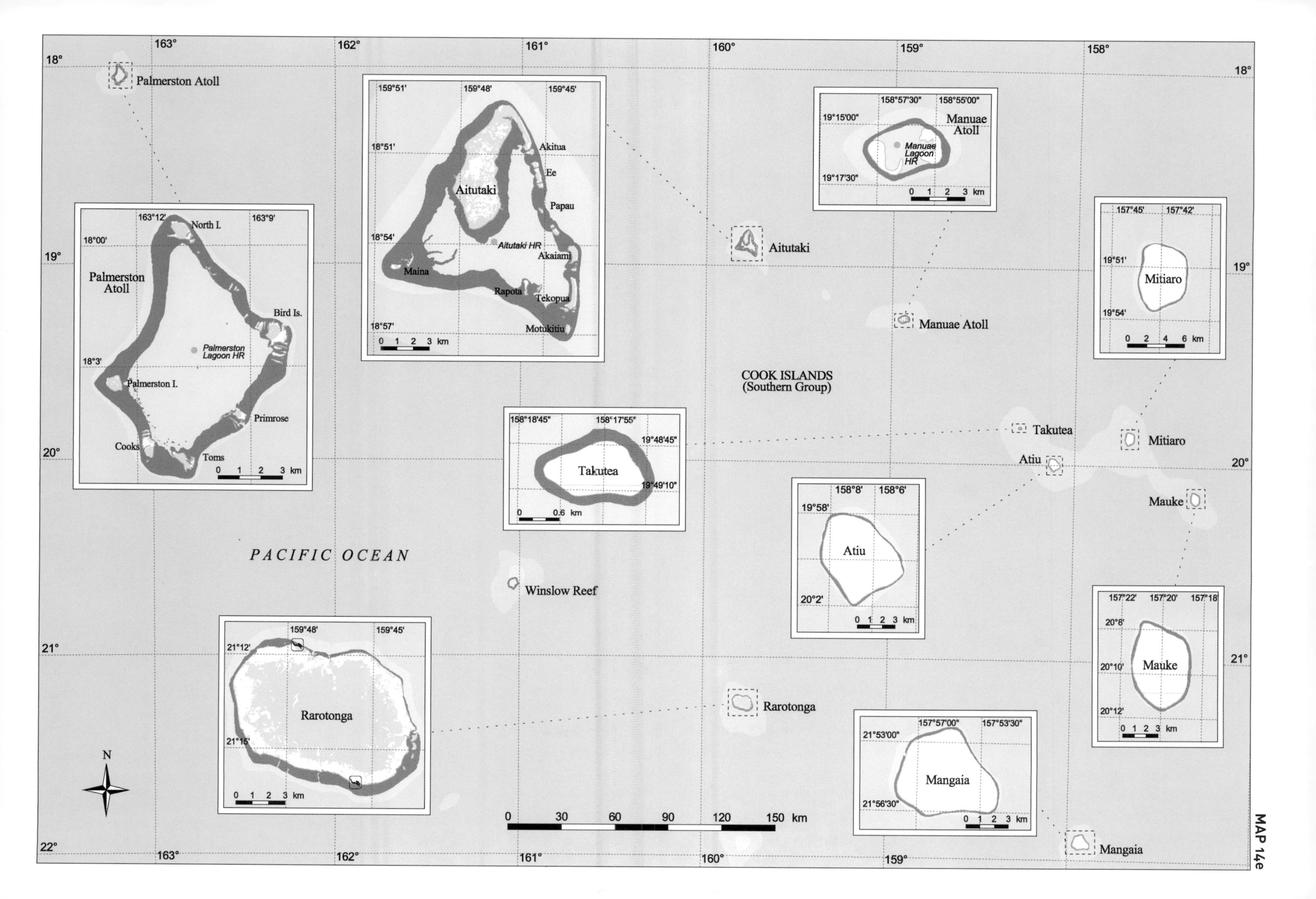

Palmerston Atoll
Aitutaki
Manuae Atoll
COOK ISLANDS
(Southern Group)
Takutea
Mitiaro
Atiu
Mauke
PACIFIC OCEAN
Winslow Reef
Rarotonga
Mangaia
Palmerston Atoll
North I.
Bird Is.
Palmerston Lagoon HR
Palmerston I.
Primrose
Cooks
Toms
Aitutaki
Akitua
Ee
Papau
Aitutaki HR
Akaiami
Maina
Rapota
Tekopua
Motukitiu
Manuae Atoll
Manuae Lagoon HR
Takutea
Atiu
Mitiaro
Mauke
Rarotonga
Mangaia
N
0 30 60 90 120 150 km

Protected areas with coral reefs

Site name	Designation	Abbreviation	IUCN cat.	Size (km²)	Year
Cook Islands					
Aitutaki	Hunting Reserve	HR	na	na	1981
Manuae Lagoon	Hunting Reserve	HR	na	na	na
Palmerston Lagoon	Hunting Reserve	HR	na	na	na
Suwarrow Atoll	National Park	NP	IV	1.60	1978

The biodiversity of Cook Island reefs has not received a great deal of attention, however it is clear that they lie at some distance from the mega-diverse areas of the Western Pacific. A natural heritage project is developing a database of species, and currently includes 578 fish species (of which 491 are shallow water or benthic species), 116 stony corals (excluding solitary species), 390 molluscs, 100 crustaceans and 50 echinoderms (including 20 sea cucumbers). It has been suggested that the diversity and abundance of species is greatest around the high (volcanic) islands and lowest around the uplifted (*makatea*) islands. The Cook Islands lie east of the natural distribution of mangroves and none are recorded.

Politically the Cook Islands are an internally self-governing state in free association with New Zealand. Like Niueans, Cook Islanders have New Zealand citizenship, and a large majority live in New Zealand. The remaining population is relatively small, probably far smaller than pre-European levels, with about half of the people living on Rarotonga. Utilization of, and reliance on, coral reef resources is considerable. In 1996 about 70 percent of all households undertook at least some form of subsistence fishing, including both gleaning from reef flats and boat-based fishing. Marine resources are also extensively used in exports, with large black pearl farms in Manihiki Atoll providing the main source of export income. There are also minor export industries associated with trochus and ornamental fish. Tourism is an important industry for the Cook Islands, which receive about 100 000 visitors per year. Snorkelling and diving are popular activities.

There are some human impacts on the reefs, particularly associated with urban and agricultural development. Sedimentation, chemical and nutrient pollution are all potential threats. Surveys of fringing reefs in the vicinity of the township area of Rarotonga in 1999 suggested that benthic cover of algal turf had increased by 20 percent to 90 percent, and that there had been decreases in diversity in some fish families over the previous five years. Such observations are probably linked to increased stress from urban development. Foreshore development has been poorly planned in a number of areas and is causing coastal erosion. There were reports of a crown-of-thorns starfish outbreak in 1998 on Rarotonga and Aitutaki. A mass bleaching event was also recorded in March 2000 with bleaching observed in up to 80 percent of corals. Sea-level rise associated with climate change could present significant problems for a number of the low-lying islands and associated reefs.

Although a number of protected areas are listed it is the low population densities and the remote location of many reefs that offer the greatest protection. In 1998 five coastal areas in Rarotonga were designated as temporary non-harvesting zones using a traditional (*Ra'ui*) system – these are not legally gazetted but are supported by the local communities. The benefits of these areas are already being observed and the system is being refined, with the addition of a permanent reserve area. The successful revival of this traditional system may lead to similar implementation in other areas.

Cook Islands

General Data	
Population (thousands)	20
GDP (million US$)	75
Land area (km²)	232
Marine area (thousand km²)	1 830
Per capita fish consumption (kg/year)	68
Status and Threats	
Reefs at risk (%)	57
Recorded coral diseases	0
Biodiversity	
Reef area (km²)	1 120
Coral diversity	51 / 172
Mangrove area (km²)	0
No. of mangrove species	0
No. of seagrass species	na

French Polynesia, the Pitcairn Islands and Clipperton Atoll

MAPS 14f, g and h

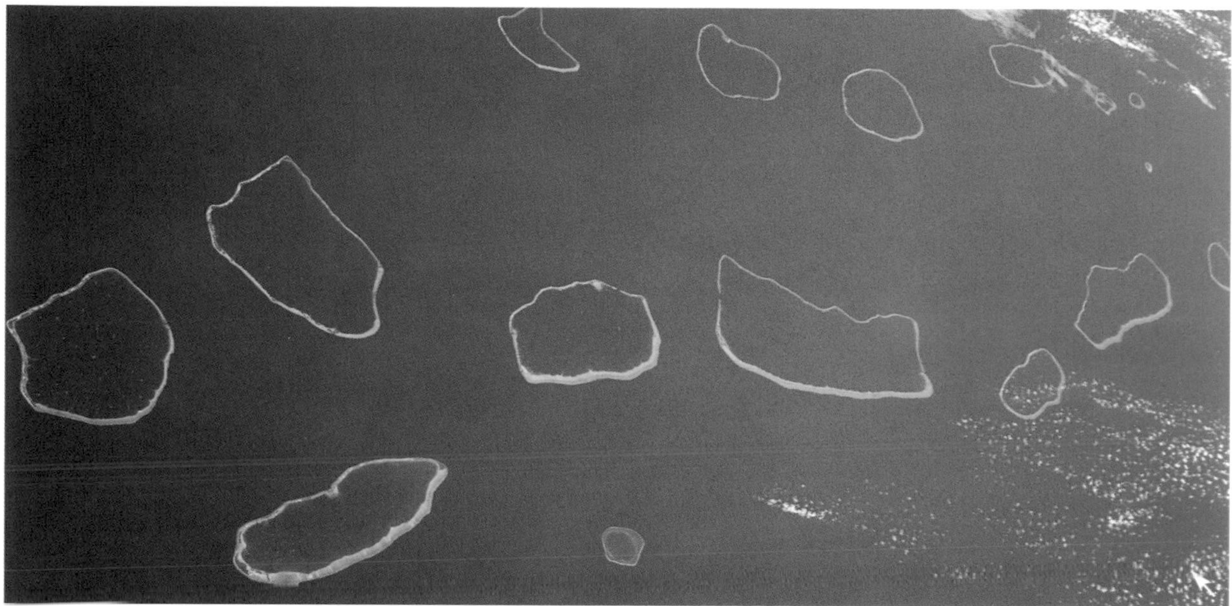

Broad view of the western Tuamotus, including five of the larger atolls: Arutua, Apataki, Kaukura, Toau and Fakarava (STS055-73-J, 1993).

French Polynesia represents one of the largest territories in the Pacific and incorporates some 6 000 square kilometers of coral reefs. It is divided into five distinct archipelagos, each following a chain oriented from northwest to southeast. Four of these archipelagos trace the movement of the Earth's crust over volcanic hotspots and their structures are younger in the southeast, where there are a number of high islands. The Tuamotu Archipelago in the center of the territory also originated from volcanic activity, but these volcanoes are associated with a shallow plateau on the edge of the spreading East Pacific Ridge.

The Marquesas, a group of high volcanic islands together with a number of smaller islets and shallow banks, form the most northerly archipelago. They lie in the path of the westward flowing South Equatorial Current and are climatically quite distinct from the other islands, with very low rainfall showing a peak in June. Their northerly latitude places them relatively close to the equator so they are rarely impacted by cyclones. Despite this, reef development is poor. There are short stretches of fringing reefs and many less clearly defined structures. These reefs are very young, and their diversity is low.

The Tuamotus make up the largest and geologically the oldest archipelago. They consist of low coralline atolls, with the exception of Makatea, which is raised, reaching some 113 meters above sea level. These atolls include some of the largest in the Pacific – Fakarava at 1 400 square kilometers and Rangiroa at nearly 1 800 square kilometers, with some 240 islands along its atoll rim. Taiaro Atoll near the center of the group is slightly uplifted, with a completely enclosed lagoon. The closure of this lagoon is thought to have been relatively recent, and there are some surviving reef communities despite their apparent isolation and the raised salinities.

The Society Islands are among the best known in the region. Mehetia in the southeast is an active volcano with only sparse coral development along its coastline. Tahiti is the largest island in the country. Like the neighboring Moorea it is a high volcanic island with vertiginous slopes. The coastlines of both islands have discontinuous fringing reefs and are surrounded by offshore barrier reefs, each broken by numerous passes. To the northwest many of the islands have similar structures with high central islands and barrier reefs; however, Maupiti is a near-atoll, and the remaining structures to the northwest are true atolls.

The Gambier Islands lie at the southeastern end of the Tuamotus, and are sometimes considered as part of this

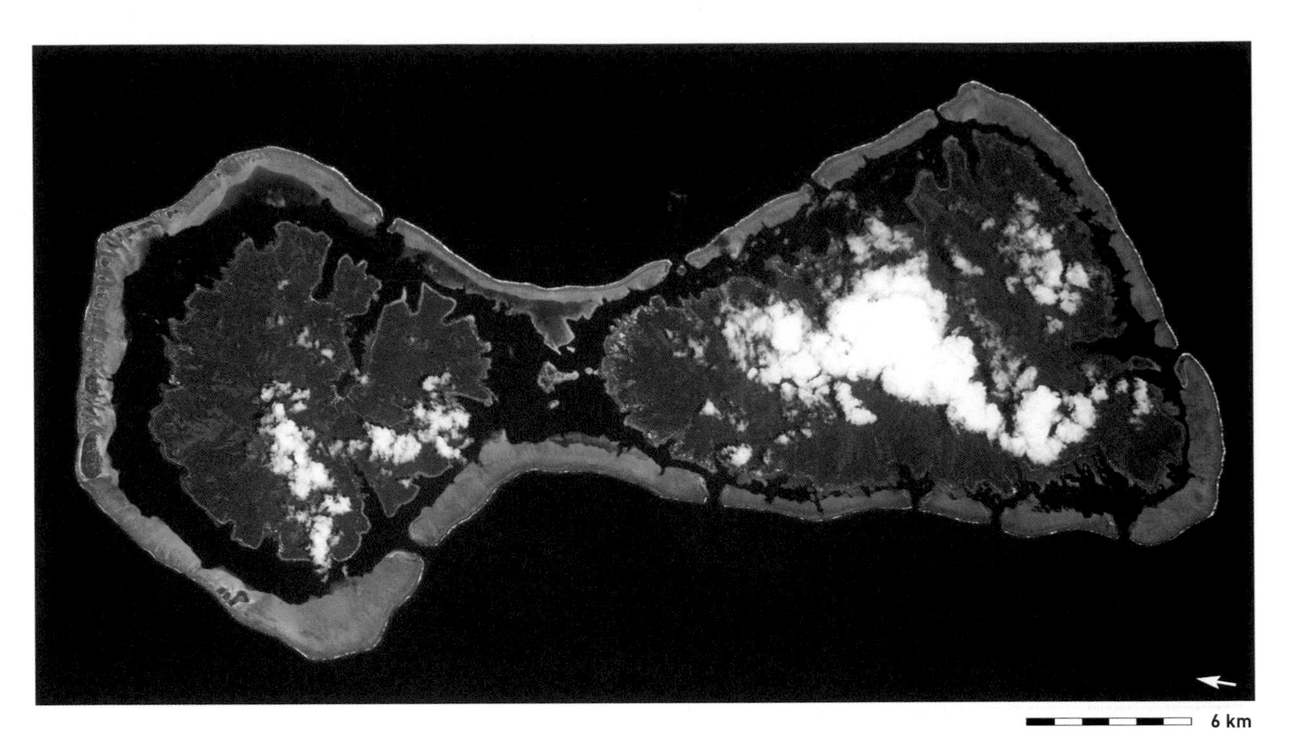

larger group. They represent the eastern extent of French Polynesia. The main islands are a cluster of four large volcanic islands (Mangareva, Taravai, Aukena and Akamaru) surrounded by a single barrier reef. These and other smaller islands in the lagoon also show some fringing reef development. The small atoll of Temoe is sometimes considered part of the Gambier group.

The Austral Islands lie to the southwest and include the rock pinnacles of Marotiri (Bass Islands) and the high volcanic island of Rapa in the far southeast. These are the southernmost islands of the region, with relatively low sea temperatures. There are no fringing reefs although there are significant coral communities. *Porites* and *Pachyseris* corals are absent, and algal cover is high. The remaining islands in the Austral group lie much further north and have well developed fringing and barrier reefs.

The reefs of French Polynesia include some of the best studied in the Pacific, but there are nonetheless some 50 islands or atolls which have never been visited by scientists. Information is particularly limited describing the Gambier, Marquesas and Austral archipelagos. Located in the most eastern region of the Indo-Pacific, diversity is generally relatively low, particularly on a unit-area basis. The vast total reef area and the considerable diversity of reef types and physico-climatic conditions mean that such low diversity is somewhat hidden in the species totals for the country. Some 168 coral species, about 800 reef fish, 30 echinoderms, 346 species of algae and 1 159 molluscs have been recorded. There appear to be some general patterns in the dominant species among different reef types. The lagoons of high volcanic islands are dominated by *Porites*, *Acropora*, *Psammocora* and *Synaraea*, while in atoll lagoons only *Porites* and *Acropora* dominate, and in the near-closed lagoons only *Acropora*. Outer slopes largely harbor *Pocillopora*, *Acropora* and *Porites*, with coral cover ranging from 40 to 60 percent at a depth of 15 meters. Even the deeper reef slopes have high coral cover, with over 90 percent cover in some atolls, down to 90 meters. There are some distinctive features about the communities in each of the archipelagos and, although general rates of endemism are low, there are a number of unique species recorded from the Marquesas and Gambier archipelagos.

Cyclones have regularly affected the reefs of the Tuamotus and Society Islands. A number of reefs were impacted by crown-of-thorns starfish in the 1970s and 1980s, exacerbated by cyclone damage in some areas. Bleaching at moderate levels was reported from various locations in the 1980s, and a more severe bleaching event in 1991 led to mortality of 20 percent of the live corals. Further bleaching recorded in 1998 was relatively patchy but led to significant mortalities in some areas.

The majority of islands and reefs in French Polynesia are remote from significant human populations and remain largely unimpacted by human activities. Fishing is critical for all of the populated islands, but particularly for those more remote from urban and tourism developments. In the late 1990s it was estimated that about 4 000-4 500 tons of lagoon fish were landed annually, about 3 500 of which were at subsistence level. In a few places there may be overfishing. One of the major industries in the country is black pearl culture, which employs around 5 000 people at 600 farms on some 50 islands. This industry supplies about 98 percent of the world market in black pearls and earns about US$130 million per annum. The oysters are

A barrier reef with broad reef flats has developed around Uturoa, Society Islands. Some of the channels through the barrier reef can be seen to correspond to inlets and river mouths of the adjacent island, a common occurrence on some barrier reefs (STS068-258-45, 1994).

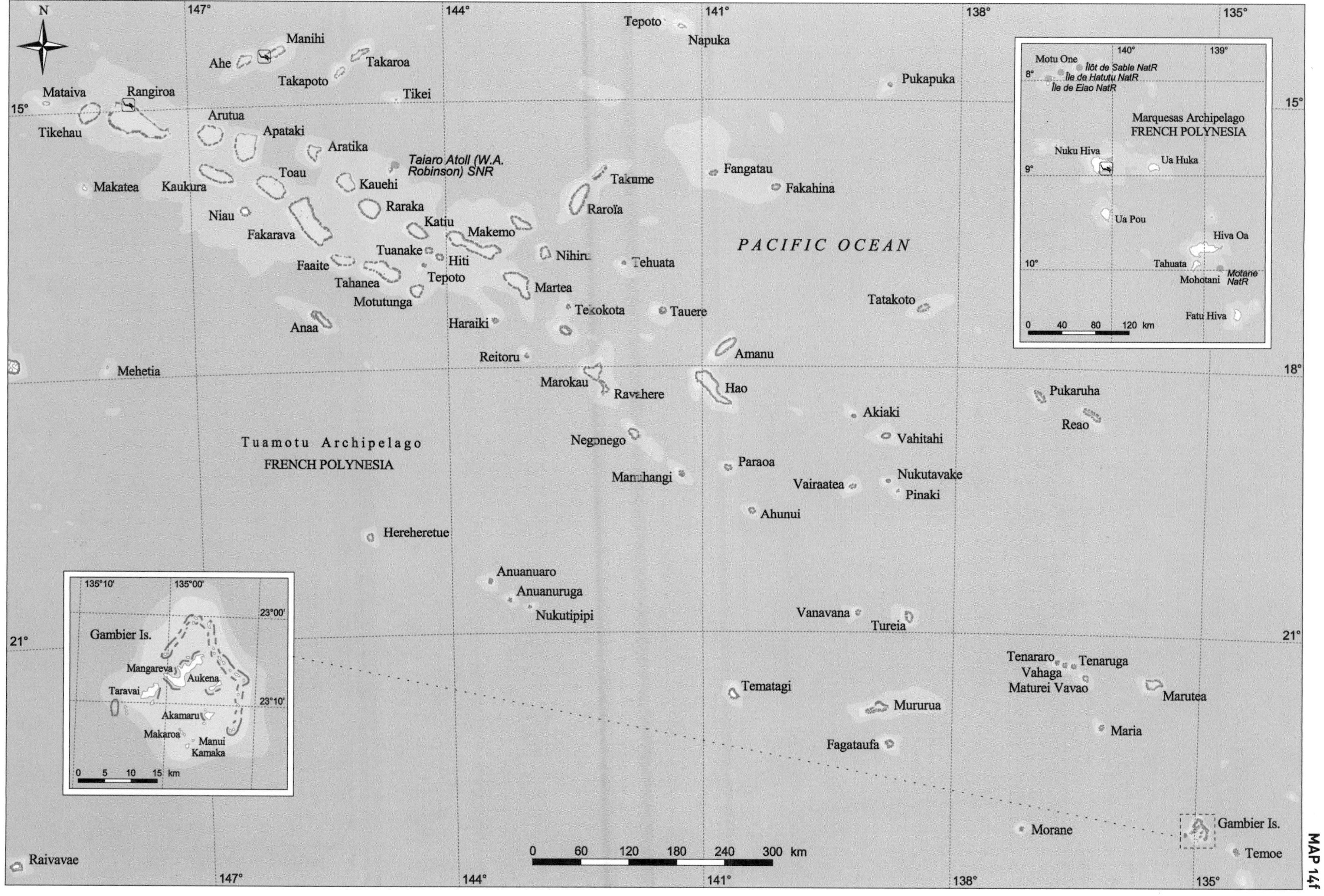

PACIFIC OCEAN
Tuamotu Archipelago
FRENCH POLYNESIA
Mataiva
Rangiroa
Tikehau
Ahe
Manihi
Takaroa
Takapoto
Tikei
Arutua
Apataki
Aratika
Taiaro Atoll (W.A. Robinson) SNR
Makatea
Kaukura
Toau
Kauehi
Raraka
Niau
Fakarava
Katiu
Makemo
Tuanake
Hiti
Tepoto
Faaite
Tahanea
Motutunga
Anaa
Haraiki
Reitoru
Mehetia
Nihiru
Tehuata
Martea
Takume
Raroïa
Tepoto
Napuka
Pukapuka
Fangatau
Fakahina
Tatakoto
Tauere
Amanu
Hao
Marokau
Negonego
Paraoa
Ahunui
Vairaatea
Akiaki
Vahitahi
Nukutavake
Pinaki
Pukaruha
Reao
Hereheretue
Anuanuaro
Anuanuruga
Nukutipipi
Vanavana
Tureia
Tematagi
Mururua
Fagataufa
Tenararo
Tenaruga
Vahaga
Maturei Vavao
Marutea
Maria
Morane
Gambier Is.
Temoe
Raivavae
15°
18°
21°
147°
144°
141°
138°
135°
0
60
120
180
240
300
km
N
Marquesas Archipelago
FRENCH POLYNESIA
Motu One
Îlôt de Sable NatR
Île de Hatutu NatR
Île de Eiao NatR
Nuku Hiva
Ua Huka
Ua Pou
Hiva Oa
Tahuata
Mohotani
Motane NatR
Fatu Hiva
140°
139°
8°
9°
10°
0
40
80
120
km
Gambier Is.
Mangareva
Aukena
Taravai
Akamaru
Makaroa
Manui
Kamaka
135°10'
135°00'
23°00'
23°10'
0
5
10
15
km

MAP 14g

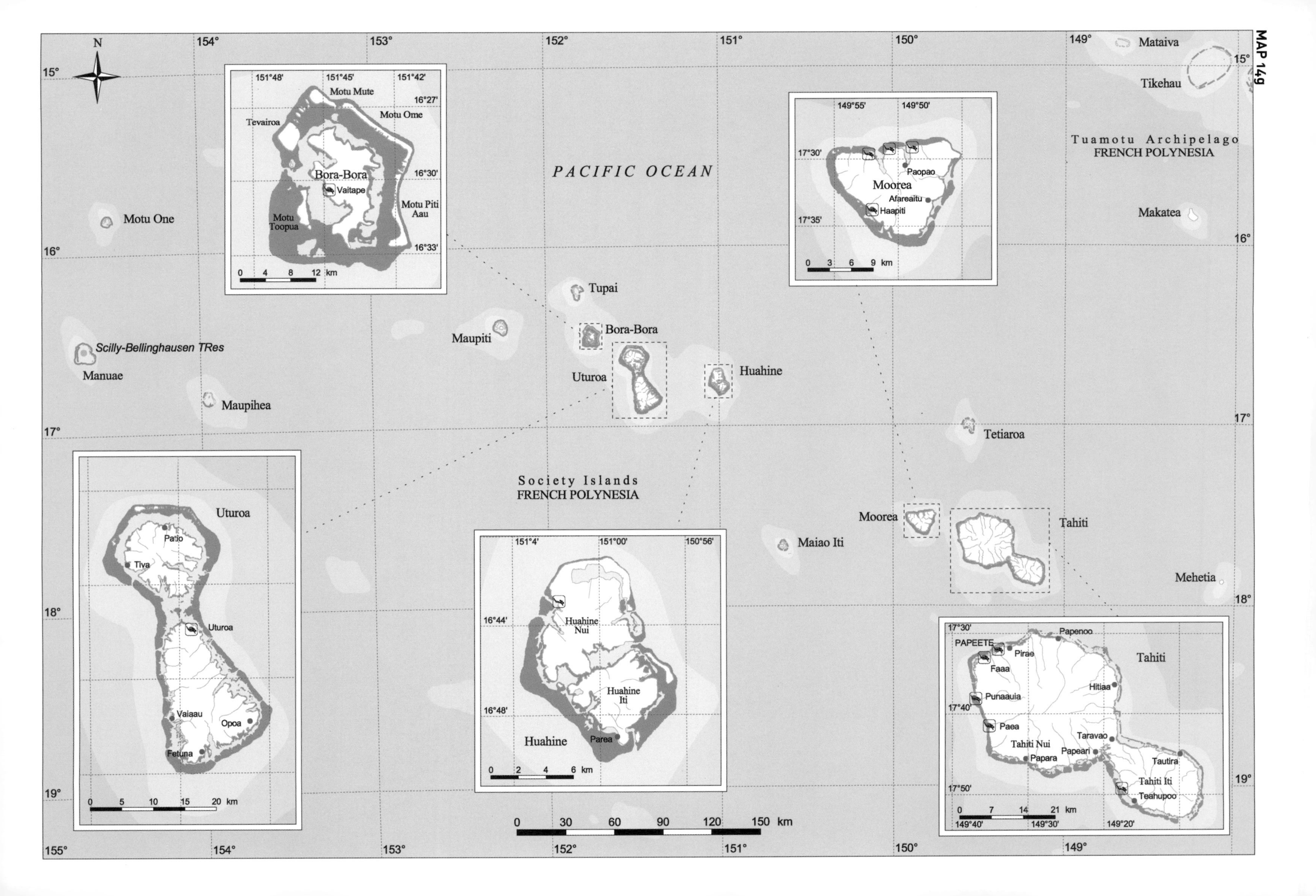
N
PACIFIC OCEAN
Tuamotu Archipelago
FRENCH POLYNESIA
Society Islands
FRENCH POLYNESIA
Mataiva
Tikehau
Makatea
Motu One
Scilly-Bellinghausen TRes
Manuae
Maupihea
Maupiti
Tupai
Bora-Bora
Uturoa
Huahine
Tetiaroa
Moorea
Tahiti
Maiao Iti
Mehetia
Bora-Bora
Motu Mute
Motu Ome
Tevairoa
Vaitape
Motu Piti Aau
Motu Toopua
Moorea
Paopao
Afareaitu
Haapiti
Uturoa
Patio
Tiva
Uturoa
Vaiaau
Opoa
Fetuna
Huahine Nui
Huahine Iti
Huahine
Parea
Tahiti
Papenoo
PAPEETE
Pirae
Faaa
Hitiaa
Punaauia
Paea
Taravao
Tahiti Nui
Papeari
Papara
Tautira
Tahiti Iti
Teahupoo
0 30 60 90 120 150 km

suspended in lagoons, and while there is no detailed knowledge of the impact they may have on planktonic systems or the predation of reef larvae, no major detrimental effects have been observed to date. There is also some prawn aquaculture for local markets. Trochus and green snails have been introduced and are now utilized for food and in local handicrafts.

Tourism is another major industry on some islands, with 164 000 visitors in 1996. All hotel developments are coastal, and a number of hotels extend over the reef flat on jetties or pontoons. While hotels are required to treat their wastewater there may be some introduction of nutrient-rich waters into the lagoon or via the groundwater. Coastal development more generally has led to considerable modifications of the coastline in Tahiti and Moorea. Sewage pollution is also a problem close to urban areas in these two islands, while they also have high levels of sediment runoff, possibly exacerbated by pesticides and fertilizers, which may have localized impacts. In Tahiti it has been estimated that 20 percent of reefs in urban areas have been destroyed, while 75 percent of the fringing reefs in Bora-Bora, which is one of the most popular tourist destinations, have been moderately to severely disturbed. Overall however, the total extent of these impacts is low or very low compared to the total area of coral reefs in French Polynesia.

The remote atolls of Mururoa and Fangataufa were used for nuclear tests, including atmospheric tests from 1966 to 1974, and for subsequent underground tests until 1996. The effects of these are not well known, although it is likely that the atmospheric tests in particular would have caused considerable damage to some areas. Only three protected areas currently incorporate coral reefs, making up a tiny proportion of the total reef area in the country. There are ongoing efforts to increase the total area protected, and to develop community-based management systems for a number of atoll lagoons.

Pitcairn

The easternmost islands of the Indo-Pacific region consist of the small group of the Pitcairn Islands. Pitcairn itself is a relatively recent volcanic island reaching some 347 meters above sea level. Henderson Island is a raised atoll which reaches 34 meters above sea level, while there are two small atolls, Oeno, with only one small island, and Ducie with one main island and three smaller islets. Coral cover is reported to be very high, typically 80-90 percent on Ducie at 10-30 meters, but slightly lower (40-70 percent) around Oeno, dominated by *Acropora* and *Montipora*. The fringing reefs around Henderson have lower coral cover, typically 10-30 percent on the fore reef slope, dominated by *Pocillopora*. There are no reefs around Pitcairn.

In terms of biodiversity these reefs have low species numbers, as might be expected from their easterly location and low latitude. Ducie is both the most easterly atoll of the Indo-Pacific and the most southerly atoll in the world. Pitcairn is the only inhabited island, with a low population. The islanders make very occasional visits to Oeno atoll for fishing but do not rely on these visits. The reefs and islands are largely protected by their isolation, whilst Henderson Island is now a World Heritage Site.

Extensive colonies of Porites arnaudi *on Clipperton Atoll. This species is probably restricted to the reefs of the Eastern Pacific, and is one of only 18 scleractinian corals recorded on the atoll (photo: JEN Veron).*

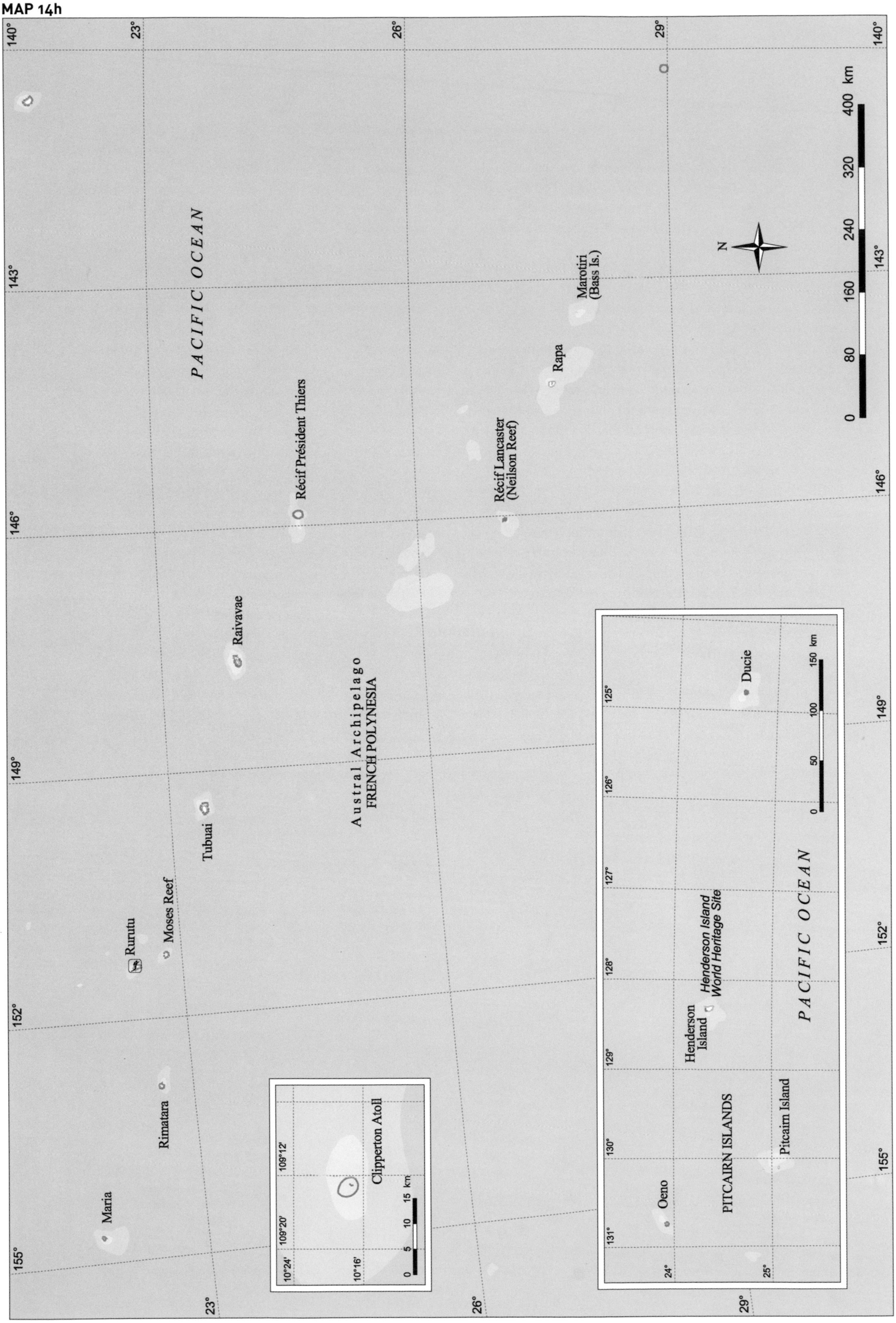

Maria
Rururu
Rimatara
Moses Reef
Tubuai
Raivavae
PACIFIC OCEAN
Récif Président Thiers
Austral Archipelago
FRENCH POLYNESIA
Récif Lancaster
(Neilson Reef)
Rapa
Marotiri
(Bass Is.)
N
0 80 160 240 320 400 km
155° 152° 149° 146° 143° 140°
23° 26° 29°

Clipperton Atoll
10°24' 10°16' 109°20' 109°12'
0 5 10 15 km

Oeno
PITCAIRN ISLANDS
Pitcairn Island
Henderson Island
Henderson Island World Heritage Site
Ducie
PACIFIC OCEAN
131° 130° 129° 128° 127° 126° 125°
24° 25°
0 50 100 150 km

Protected areas with coral reefs

Site name	Designation	Abbreviation	IUCN cat.	Size (km²)	Year
French Polynesia					
Taiaro Atoll (WA Robinson)	Strict Nature Reserve	SNR	IV	11.88	1973
Scilly (Manuae)	Territorial Reserve	TRes	IV	113.00	1992
Bellinghausen (Motu One)	Territorial Reserve	TRes	IV	12.40	1992
ATOLL DE TAIARO	UNESCO BIOSPHERE RESERVE			**20.00**	**1977**
Pitcairn Islands					
HENDERSON ISLAND	WORLD HERITAGE SITE			**37.00**	**1988**

Clipperton Atoll

Clipperton Atoll is located in the Eastern Pacific, about 1 100 kilometers southwest of the Mexican part of the American mainland. Although a considerable distance from French Polynesia it is administered by these islands. It is a roughly circular atoll about 4 kilometers across, with an island completely encircling its lagoon. There is a 50-200 meter wide reef flat terminating in a spur and groove system. The reef slope is relatively gentle with high coral cover (33-83 percent) in many areas. The deeper slope has more sand and rubble.

This atoll is extremely important in biogeographic terms. Its easterly location actually places it in a completely different biogeographic region from the other Pacific islands – the Tropical Eastern Pacific, with close affinities to the reefs and coral communities of the western coast of the Americas (see Chapter 5). It is in fact the best developed reef and the only atoll in the Tropical Eastern Pacific. Biodiversity is very low, with 18 recorded scleractinian coral species and 115 fishes, including 98 nearshore or demersal species. Nine of these fish are endemic to the island, including some of the most abundant species. The biogeographic affinities of these species include elements of true Indo-Pacific and Tropical Eastern Pacific species. The enclosed lagoon contains only brackish water and is devoid of corals or fish, dominated by algal growth.

The island is uninhabited and rarely visited. There is no legal protection of the natural resources. Sharks, originally reported to be abundant, are now scarce and it has been suggested that a group of Mexican fishing vessels may have intensively fished this population in 1993.

French Polynesia

GENERAL DATA	
Population (thousands)	249
GDP (million US$)	3 109
Land area (km²)	3 024
Marine area (thousand km²)	5 030
Per capita fish consumption (kg/year)	64
STATUS AND THREATS	
Reefs at risk (%)	29
Recorded coral diseases	0
BIODIVERSITY	
Reef area (km²)	6 000
Coral diversity	174 / 168
Mangrove area (km²)	0
No. of mangrove species	0
No. of seagrass species	2

Pitcairn Islands

GENERAL DATA	
Population (individuals)	c.50
GDP (million US$)	na
Land area (km²)	53
Marine area (thousand km²)	800
Per capita fish consumption (kg/year)	na
STATUS AND THREATS	
Reefs at risk (%)	0
Recorded coral diseases	0
BIODIVERSITY	
Reef area (km²)	<100
Coral diversity	60 / 42
Mangrove area (km²)	0
No. of mangrove species	0
No. of seagrass species	na

Hawai'i and the US minor outlying islands

MAPS 14i and j

The Hawaiian Islands are the most isolated archipelago in the world, lying considerably north and east of the vast majority of Pacific islands. They were formed as the Pacific tectonic plate moved northwest over a stationary hotspot beneath the ocean floor. The youngest island in the group is thus the island of Hawai'i in the southeast, which shows near continuous volcanic activity. Together with seven other main islands, it forms a distinct block of high volcanic islands making up the majority of the land area of the archipelago. Moving northwest, the older volcanic islands have largely subsided and there is a long chain of islands and reefs at or near sea level leading to Kure Atoll in the northwest. These latter reefs lie some distance north of the tropics. Further north and west of these is the sequence of Emperor Seamounts extending for thousands of kilometers towards the Kamchatka coastline of Siberia. These represent former reef-capped volcanic islands. Their northward migration took them to latitudes where reef growth was insufficiently rapid to keep up with crustal subsidence, and then out of the regions of hermatypic coral growth completely.

The Hawaiian Islands lie in the path of the westerly flowing North Pacific Equatorial Current. The surrounding pelagic waters are renowned for their low productivity and low nutrient content. Away from the high islands this is a significant factor for coastal ecosystem productivity. From March to October they are swept by the northeast trade winds which generate cool air temperatures and a pronounced windward to leeward gradient with greater wave energies, but also higher rainfall and runoff on eastern shores. There is a pronounced seasonality with distinctly sub-tropical temperatures during winter months, particularly in the northwestern islands. The islands are occasionally swept by tropical cyclones.

Fringing reefs are by no means continuous among the high islands of the southeast, though they are well developed in a number of places, particularly on leeward (southern and southwestern) shores. Along the coast of Hawai'i Island there are no true fringing reefs, but well developed submerged reefs occur along the western Kona coast. Recent underwater lava flows provide a substrate for looking at colonization by reef organisms in the south and east. Fringing reefs are better developed in a few locations along the western shores of Maui, the southern shores of Molokai and the northeastern shores of Lanai. Oahu has a number of well developed reefs, including a well studied fringing reef in Hanauma Bay and one of the only barrier reefs in Kanehoe Bay, which protects a large number of patch reefs and a coastal fringing reef. Fringing reefs are

Left: Honolulu, the largest city in the Pacific, with reef structures in the fore. Right: Ohau, Hawai'i (STS065-96-7, 1994).

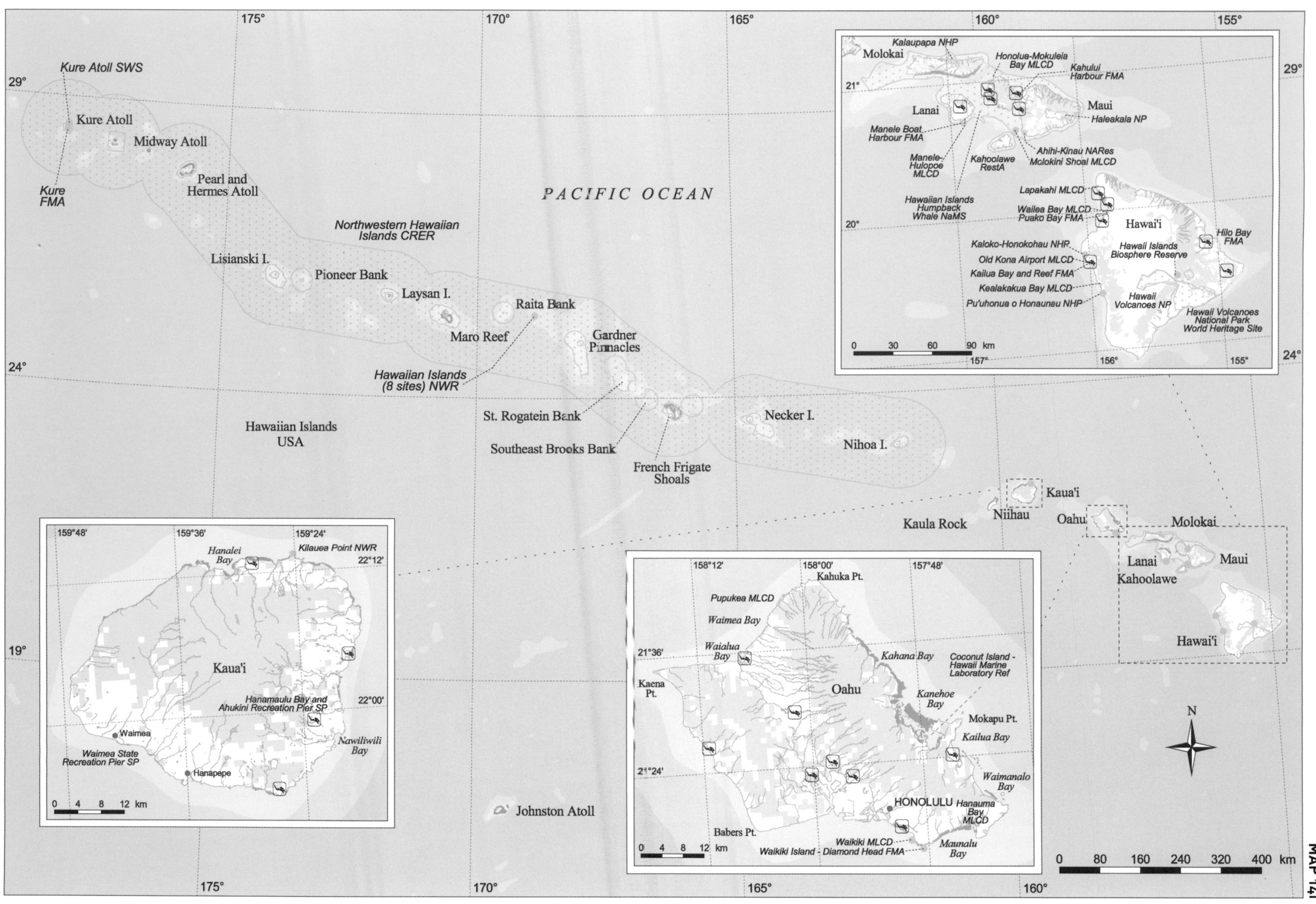

PACIFIC OCEAN
Kure Atoll SWS
Kure Atoll
Kure FMA
Midway Atoll
Pearl and Hermes Atoll
Northwestern Hawaiian Islands CRER
Lisianski I.
Pioneer Bank
Laysan I.
Raita Bank
Maro Reef
Gardner Pinnacles
Hawaiian Islands (8 sites) NWR
St. Rogatein Bank
Southeast Brooks Bank
French Frigate Shoals
Necker I.
Nihoa I.
Hawaiian Islands USA
Kaula Rock
Niihau
Kaua'i
Oahu
Molokai
Lanai
Kahoolawe
Maui
Hawai'i
Johnston Atoll
N
0 80 160 240 320 400 km
175°
170°
165°
160°
155°
29°
24°
19°
Molokai
Kalaupapa NHP
Honolua-Mokuleia Bay MLCD
Kahului Harbour FMA
Maui
Haleakala NP
Lanai
Manele Boat Harbour FMA
Manele-Hulopoe MLCD
Kahoolawe RestA
Ahihi-Kinau NARes
Molokini Shoal MLCD
Hawaiian Islands Humpback Whale NaMS
Lapakahi MLCD
Wailea Bay MLCD
Puako Bay FMA
Hawai'i
Hilo Bay FMA
Kaloko-Honokohau NHP
Hawaii Islands Biosphere Reserve
Old Kona Airport MLCD
Kailua Bay and Reef FMA
Kealakakua Bay MLCD
Pu'uhonua o Honaunau NHP
Hawaii Volcanoes NP
Hawaii Volcanoes National Park World Heritage Site
0 30 60 90 km
157° 156° 155°
21° 20°
159°48' 159°36' 159°24'
Hanalei Bay
Kilauea Point NWR
22°12'
Kaua'i
Hanamaulu Bay and Ahukini Recreation Pier SP
22°00'
Waimea
Waimea State Recreation Pier SP
Hanapepe
Nawiliwili Bay
0 4 8 12 km
158°12' 158°00' 157°48'
Kahuka Pt.
Pupukea MLCD
Waimea Bay
Waialua Bay
21°36'
Kahana Bay
Coconut Island - Hawaii Marine Laboratory Ref
Kaena Pt.
Oahu
Kanehoe Bay
Mokapu Pt.
Kailua Bay
21°24'
Waimanalo Bay
HONOLULU
Hanauma Bay MLCD
Babers Pt.
Waikiki MLCD
Maunalu Bay
Waikiki Island - Diamond Head FMA
0 4 8 12 km

MAP 14j

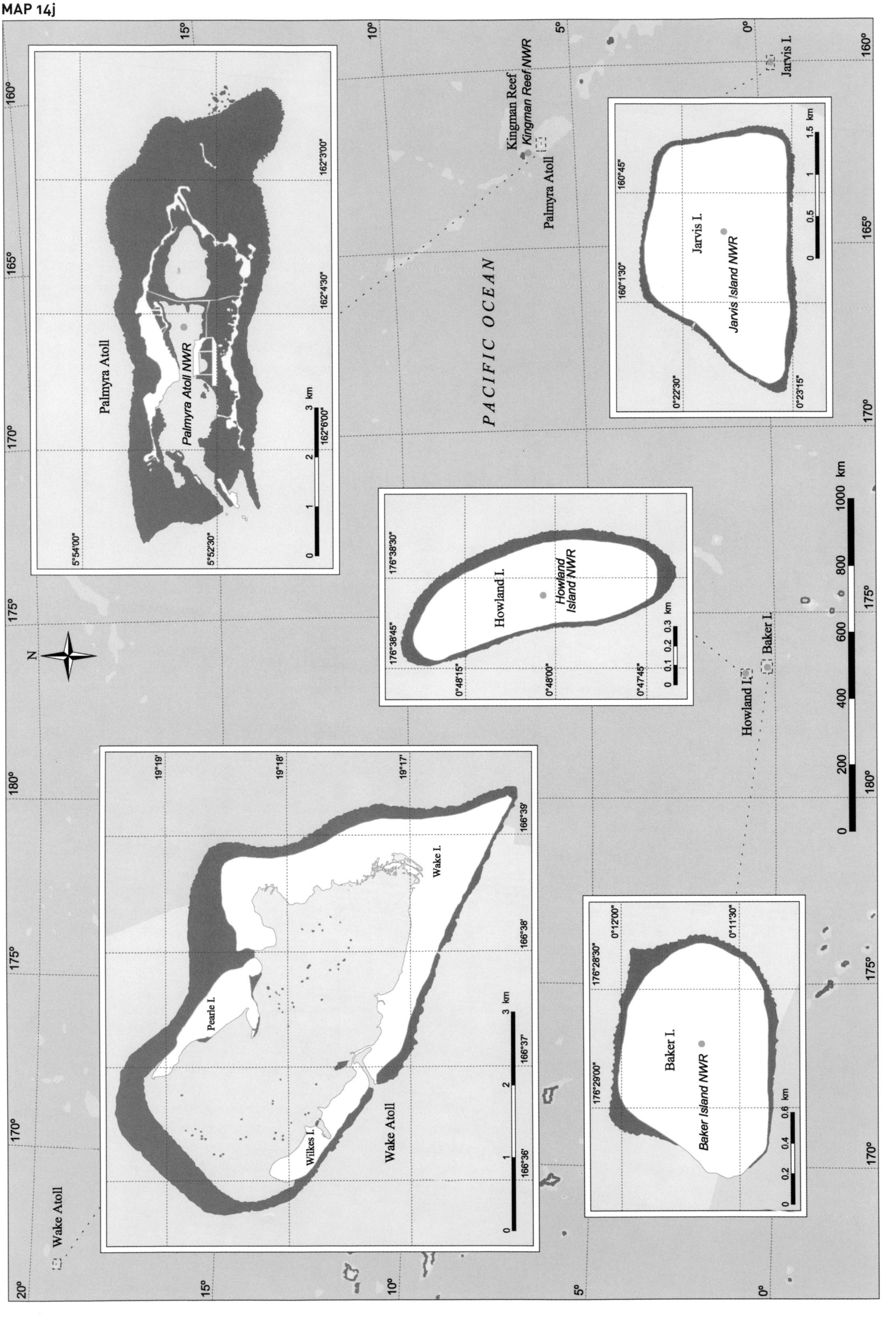

Wake Atoll
Pearle I.
Wilkes I.
Wake Atoll
Wake I.
19°19'
19°18'
19°17'
166°36'
166°37'
166°38'
166°39'
0 1 2 3 km
Palmyra Atoll
Palmyra Atoll NWR
5°54'00"
5°52'30"
162°6'00"
162°4'30"
162°3'00"
0 1 2 3 km
Howland I.
Howland Island NWR
176°38'45"
176°38'30"
0°48'15"
0°48'00"
0°47'45"
0 0.1 0.2 0.3 km
Baker I.
Baker Island NWR
176°29'00"
176°28'30"
0°12'00"
0°11'30"
0 0.2 0.4 0.6 km
Jarvis I.
Jarvis Island NWR
160°1'30"
160°45"
0°22'30"
0°23'15"
0 0.5 1 1.5 km
PACIFIC OCEAN
Kingman Reef
Kingman Reef NWR
Palmyra Atoll
Howland I.
Baker I.
Jarvis I.
N
0 200 400 600 800 1000 km
20°
15°
10°
5°
0°
170°
175°
180°
175°
170°
165°
160°

found around much of the coastline of Kaua'i but remain poorly developed around Niihau. There is a submerged barrier reef off the western coast of Kaua'i.

The majority of Hawai'i's reefs are located northwest of the main Hawaiian Islands. The first of the northwestern islands still have small basaltic elements. Nihoa and Necker are very isolated basalt islands, and there is little information about the marine communities nearby. The French Frigate Shoals are a near-atoll with just a small basaltic pinnacle at the western edge. The Gardner Pinnacles represent the last basaltic outcrop, and consist of three tiny, steep-sided rocks lying on a common platform with coral and sand at its base. Laysan and Lisianski Islands are both coral islands with submerged reefs around their edges, while west of these are three true atolls: Pearl and Hermes, Midway and Kure Atolls. In addition there are several reefs with no associated islands. These include the large Maro Reef northwest of the Gardner Pinnacles, a complex of shallow reticulated reef systems which might be considered an atoll or platform-type structure.

The location of the Hawaiian Islands, in considerable isolation and on the northern edge of the tropics, has meant that they are not highly diverse. Their isolation has been emphasized by the dominant oceanic currents which reduce the likelihood of pelagic larvae being carried to the archipelago. Over millions of years one consequence of this has been the considerable opportunities for the development of new species. Some 52 species of stony coral, 500 of nearshore fish, 1 000 of marine molluscs and 450 of marine algae have been recorded, of which typically about 25 percent of each of these and other groups are endemic to the Hawaiian Islands and a few nearby reefs. This is the highest level of endemism for any coral reef area in the world. One particular feature of the coral communities is the relative paucity of acroporid corals, the major reef builders in most parts of the Pacific. Although there are species of *Acropora* these are rarely dominant, the major reef-building species being *Porites* and *Pocillopora*. Another feature of the Hawaiian reefs has been the adaptation of at least some species to the relatively cool conditions, possibly indicating a genetic variance from the same species found in other locations.

There are no native mangroves in the Hawaiian Islands, although a few species have been introduced in modern times and two have become established and spread to several of the high islands. There is only one seagrass species *Halophila hawaiiana,* which is endemic to Hawai'i, and seagrass beds are unusual. Millions of seabirds nest in the northwest Hawaiian Islands, and a number of migratory shorebirds rest or overwinter here. These are among the largest and most important seabird colonies in the Pacific Ocean. Five species of marine turtle have been seen in Hawaiian waters, and the French Frigate Shoals are one of the largest remaining nesting grounds for the green turtle in the Pacific. These same islands are also of critical importance for the Hawaiian monk seal *Monachus shauinslandi*. The closest relations of this seal are the critically endangered Mediterranean monk seal and the extinct Caribbean monk seal. Although there remain about 1 500 individuals and they are highly protected, numbers are still declining. Several other marine mammals occur, including a large and growing population of humpback whales, *Megaptera novaeangliae*, which spend part of their migratory cycle in the islands between November and May during which time they mate and give birth.

The population of native Hawaiian people was much diminished by the arrival and colonization by Europeans and others, and it remains low to this day. Immigration and expansion of newly arrived populations, however, means that the total population of the islands is probably four to five times that of the original population. These people are all located on the main islands, with about 75 percent on Oahu. Honolulu is the largest city in the entire insular Pacific. Hawai'i is also one of the world's top tourism destinations, receiving about 6.5 million visitors annually.

Only a very few people now live in a manner which could be described as subsistence, or depend heavily on their own fish catches. There are, however, many who fish to supplement their diet, for recreational purposes or at small-scale commercial levels. All of these operate with modern equipment, including gill nets, spears, trolling, surround nets and traps. In addition to these, recreational and commercial collection of fish for aquaria is widespread and poorly regulated. The nearshore fish populations are reported to be depleted around nearly all the main islands. Larger-scale commercial fisheries primarily focus on pelagic species, but also operate on a number of reef species. A spiny lobster fishery in the northwestern islands is reported to have become severely overfished and may be closed. More recently a major shark fin fishery has developed which is extremely controversial, not least

An active hotspot. Lava flows are continually adding new land to the coast of Hawai'i.

Protected areas with coral reefs

Site name	Designation	Abbreviation	IUCN cat.	Size (km²)	Year
Hawai'i					
Coconut Island – Hawai'i Marine Laboratory	Refuge	Ref	IV	na	na
Hanauma Bay	Marine Life Conservation District	MLCD	IV	0.41	1967
Hawaiian Islands (8 sites)	National Wildlife Refuge	NWR	Ia	1 029.60	1945
Hawaiian Islands Humpback Whale	National Marine Sanctuary	NaMS	IV	3 548.13	1997
Kahoolawe	Restricted Area	RestA	Unassigned	na	na
Kealakakua Bay	Marine Life Conservation District	MLCD	IV	1.28	1969
Kure	Fisheries Management Area	FMA	Unassigned	na	na
Kure Atoll	State Wildlife Sanctuary	SWS	Ia	0.96	1981
Molokini Shoal	Marine Life Conservation District	MLCD	IV	na	1981
Northwestern Hawaiian Islands	Coral Reef Ecosystem Reserve	CRER	VI	341 362.00	2000
Puako Bay	Fisheries Management Area	FMA	Unassigned	na	na
Waikiki	Marine Life Conservation District	MLCD	IV	0.30	na
Waikiki Island – Diamond Head	Fisheries Management Area	FMA	Unassigned	na	na
Hawai'i Islands Biosphere Reserve	**UNESCO Biosphere Reserve**			**995.45**	**1980**
US minor outlying islands					
Baker Island	National Wildlife Refuge	NWR	Ia	128.43	1974
Howland Island	National Wildlife Refuge	NWR	Ia	131.73	1974
Jarvis Island	National Wildlife Refuge	NWR	Ia	151.83	1974
Johnston Island	National Wildlife Refuge	NWR	Unassigned	129.95	1926
Kingman Reef	National Wildlife Refuge	NWR	II	1 958.99	2001
Midway Atoll	National Wildlife Refuge	NWR	II	1 208.36	1988
Palmyra Atoll	National Wildlife Refuge	NWR	II	2 086.69	2001

because of the waste and perceived cruelty of returning living sharks to the water with no fins.

European settlement led to radical changes in the terrestrial environment, with the degradation or clearance of large tracts of native forest by settlers and the feral animals (goats and deer in particular) that accompanied them. These changes have also led to considerable sedimentation in nearshore environments and may have killed many reef communities. Hawai'i is also one of the few places where significant numbers of exotic species have been introduced to the coral reef environment, including marine algae and fish. There is evidence that these are spreading and may be displacing resident reef species in some areas.

With urbanization there have been problems of sewage discharge in a number of areas. Efforts to reduce this have led to better treatment in some places, but also to more remote discharge of sewage into deeper water. The overall effects of the latter approach remain controversial, but are still likely to be impacting reefs in some areas. Kanehoe Bay on Oahu has been the site of extensive monitoring over 30 years, including studies before, during and after the discharge of sewage into the bay. This site provides one of the best examples of coral reef restoration in the Pacific. Coastal erosion is a natural occurrence in the Hawaiian Islands, but urbanization has led to a large number of coastal developments now being threatened by such erosion. Efforts to prevent it are extremely costly and also disturb and interrupt natural sediment flows, leading to the damage or loss of offshore reefs. Construction in nearshore marine areas has also been extensive, including harbor and runway developments and road construction.

Tourism is the major industry in Hawai'i, heavily focussed on locations on the larger islands. Coastal tourism developments, including golf courses, may be adding to the stresses caused by wider coastal development, effluent discharge and the physical disturbance of nearshore communities. Diving and snorkelling are popular, although limited in some areas by rough seas and currents. Hana'uma Bay, the most popular snorkelling destination on Oahu, receives up to 10 000 visitors per day.

In the northwestern islands there are no permanent residents, although there are workers on Midway, French Frigate Shoals and, seasonally, on Laysan. Tourism to the outer islands is highly restricted, although ecotourism is developing, notably on Midway Atoll.

Efforts to regulate and protect coral reefs in Hawai'i include a large number of marine protected areas. Federal legislation covers most of the more remote islands and their surrounding reefs as national wildlife refuge. In 2000 all of the other reefs and shallow banks which lay outside these declared refuges, together with a very large area of surrounding seas, were declared a Coral Reef Ecosystem Reserve. The combined legislation for these areas has created a contiguous marine protected area, second in size only to the Great Barrier Reef in Australia. Among the main islands the Hawaiian Humpback Whale National Marine Sanctuary covers significant areas of offshore waters. Closer to centers of human population there are several sites protected at the state level. The degree of protection afforded by these varies considerably.

US minor outlying islands

In addition to Hawai'i, the USA holds jurisdiction over a number of other territories in the Pacific. These include American Samoa, Guam and the Northern Mariana Islands (see separate accounts), but also a number of remote atolls and reefs in the Central Pacific. Lacking any native population these islands have primarily been seen as important for defense, and they are administered at the federal level. Their importance in terms of biodiversity is now increasingly recognized.

Baker and Howland Islands are both low coral islands with surrounding fringing reefs, geologically associated with the Phoenix Islands of Kiribati to the south. Jarvis Island is similarly small with fringing reefs, an outlier to the west of the northern Line Islands in Kiribati. All three of these islands were exploited for guano in the 18th century. Johnston Atoll is one of the most remote reefs in the Pacific, lying midway between Hawai'i and the islands and reefs of Kiribati. Although it bears strong affinities to the Hawaiian fauna it may also provide a stepping stone connecting these faunas to those of the rest of the Pacific.

Wake Atoll is a small atoll lying to the north of the Marshall Islands, with three islands dominating most of the rim. Biodiversity is relatively low, which may be a product of the shallow lagoon combined with the isolation and the relatively northerly latitude.

From a geological perspective, Palmyra Atoll and Kingman Reef make up the northern end of the Line Islands of Kiribati. The 50 or so islets of Palmyra have been under private ownership since the 1920s, but The Nature Conservancy, a large US-based non-governmental organization, bought the entire atoll in early 2001, and will maintain it for conservation. Kingman Reef to the north of Palmyra is an atoll with a slightly submerged western edge and a deep lagoon, with no true islands but some exposed rocks (including at high tide).

Hawai'i, USA

General Data	
Population (thousands)	2 020
GDP* (million US$)	6 392 711
Land area (km²)	16 759
Marine area (thousand km²)	na
Per capita fish consumption* (kg/year)	21
Status and Threats	
Reefs at risk (%)	57
Recorded coral diseases	3
Biodiversity	
Reef area (km²)	1 180
Coral diversity	na / 49
Mangrove area (km²)	na
No. of mangrove species	na
No. of seagrass species	4

* National statistics

Johnston Island, USA

General Data	
Population	0
GDP (million US$)	0
Land area (km²)	3
Marine area (thousand km²)	444
Per capita fish consumption (kg/year)	0
Status and Threats	
Reefs at risk (%)	67
Recorded coral diseases	0
Biodiversity	
Reef area (km²)	220
Coral diversity	na / na
Mangrove area (km²)	0
No. of mangrove species	0
No. of seagrass species	na

Selected bibliography

TUVALU, WALLIS AND FUTUNA

Gabrié C (2000). *State of Coral Reefs in French Overseas Départements and Territories.* Ministry of Spatial Planning and Environment and State Secretariat for Overseas Affairs, Paris, France.

Sauni S (2000). The status of the coral reefs of Tuvalu. In: Salvat B, Wilkinson C, South GR (eds). Proceedings of the International Coral Reef Initiative Regional Symposium, Noumea, 22-24 May 2000.

South GR, Skelton PA (2000). Status of coral reefs in the southwest Pacific: Fiji, Nauru, New Caledonia, Samoa, Solomon Islands, Tuvalu and Vanuatu. In: Wilkinson CR (ed). *Status of Coral Reefs of the World: 2000.* Australian Institute of Marine Science, Cape Ferguson, Australia.

TOKELAU, SAMOA AND AMERICAN SAMOA

Craig P (in press). Status of coral reefs in 2000: American Samoa. In: *Status and Trends of US Coral Reefs 2000.* NOAA report to the US Coral Reef Task Force.

Craig P, Saucerman S, Wiegman S (2000). Central South Pacific Ocean (American Samoa). In: Sheppard C (ed). *Seas at the Millennium: An Environmental Evaluation.* Elsevier Science Ltd, Oxford, UK.

Skelton PA, Bell LJ, Mulipola A, Trevor A (2000). The status of the coral reefs and marine resources of Samoa. In: Salvat B, Wilkinson C, South GR (eds). Proceedings of the International Coral Reef Initiative Regional Symposium, Noumea, 22-24 May 2000.

South GR, Skelton PA (2000). Status of coral reefs in the southwest Pacific: Fiji, Nauru, New Caledonia, Samoa, Solomon Islands, Tuvalu and Vanuatu. In Wilkinson CR (ed). *Status of Coral Reefs of the World: 2000.* Australian Institute of Marine Science, Cape Ferguson, Australia.

Zann L (1994). The status of coral reefs in south western Pacific islands. *Mar Poll Bul* 29: 52-61.

Zann LP, Vuki V (2000). The South Western Pacific Islands Region. In: Sheppard C (ed). *Seas at the Millennium: An Environmental Evaluation.* Elsevier Science Ltd, Oxford, UK.

TONGA AND NIUE

Furness LJ (1997). Hydrogeology of Carbonate Islands of the Kingdom of Tonga. In: Vacher HL, Quinn T (eds). *Developments in Sedimentology, 54: Geology and Hydrology of Carbonate Islands.* Elsevier Science BV, Amsterdam, Netherlands.

Grano S (ed) (1993). *The Kingdom of Tonga: Action Strategy for Managing the Environment.* South Pacific Regional Environmental Programme, Apia, Western Samoa.

Lovell ER, Palaki A (2000). National coral reef status report for Tonga. In: Salvat B, Wilkinson C, South GR (eds). Proceedings of the International Coral Reef Initiative Regional Symposium, Noumea, 22-24 May 2000.

Mees CC (1997). Multispecies responses to fishing at Indian Ocean and Tongan offshore reefs. *Proc 8th Int Coral Reef Symp* 2: 2039-2044.

Wheeler C, Aharon P (1997). Geology and hydrogeology of Niue. In: Vacher HL, Quinn T (eds). *Developments in Sedimentology, 54: Geology and Hydrology of Carbonate Islands.* Elsevier Science BV, Amsterdam, Netherlands.

Zann L (1994). The status of coral reefs in south western Pacific islands. *Mar Poll Bul* 29: 52-61.

Zann LP, Vuki V (2000). The south western Pacific Islands region. In: Sheppard C (ed). *Seas at the Millennium: An Environmental Evaluation.* Elsevier Science Ltd, Oxford, UK.

COOK ISLANDS

Hein JR, Gray SC, Richmond BM (1997). Geology and hydrogeology of the Cook Islands. In: Vacher HL, Quinn T (eds). *Developments in Sedimentology, 54: Geology and Hydrology of Carbonate Islands.* Elsevier Science BV, Amsterdam, Netherlands.

Henson B (ed) (1993). *Cook Islands National Environmental Management Strategy.* South Pacific Regional Environmental Programme, Apia, Western Samoa.

Ponia B (2000). Coral reefs of the Cook Islands: national status report. In: Salvat B, Wilkinson C, South GR (eds). Proceedings of the International Coral Reef Initiative Regional Symposium, Noumea, 22-24 May 2000.

FRENCH POLYNESIA AND THE PITCAIRN ISLANDS

Adjeroud M, Salvat B (1996). Spatial patterns in biodiversity of a fringing reef community along Opunohu Bay, Mororea, French Polynesia. *Bull Mar Sci* 59(1): 175-187.

Allen GR, Robertson DR (1997). An annotated checklist of the fishes of Clipperton Atoll, tropical eastern Pacific. *Revistas de Biologia Tropical* 45(2): 813-844.

Benton TG, Spencer T (eds) (1995). *The Pitcairn Islands: Biogeography, Ecology and Prehistory.* Academic Press Ltd, London, UK.

Blake SG, Pandolfi JM (1997). Geology of selected islands of the Pitcairn Group, Southern Polynesia. In: Vacher HL, Quinn T (eds). *Developments in Sedimentology, 54: Geology and Hydrology of Carbonate Islands.* Elsevier Science BV, Amsterdam, Netherlands.

Buigues DC (1997). Geology and hydrogeology of Mururoa and Fangataufa, French Polynesia. In: Vacher HL, Quinn T (eds). *Developments in Sedimentology, 54: Geology and Hydrology of Carbonate Islands.* Elsevier Science BV, Amsterdam, Netherlands.

Carricart-Ganivet JP, Reyes-Bonilla H (1999). New and previous records of scleractinian corals from Clipperton Atoll, Eastern Pacific. *Pacific Science* 53: 370-375.

Gabrié C (2000). *State of Coral Reefs in French Overseas Départements and Territories.* Ministry of Spatial Planning and Environment and State Secretariat for Overseas Affairs, Paris, France.

Galzin R, Planes S, Dufour V, Salvat B (1994). Variation in diversity of coral reef fish between French Polynesian atolls. *Coral Reefs* 13: 175-180.

Hutchings P, Payri C, Gabrié C (1994). The current status of coral reef management in French Polynesia. *Mar Poll Bul* 29: 26-33.

Irving RA (1995). Near-shore bathymetry and reef biotopes of Henderson Island, Pitcairn Group. *Biol J Linn Soc* 56: 13-42.

Montaggioni LF, Camoin GF (1997). Geology of Makatea Island, Tuamotu Archipelago, French Polynesia. In: Vacher HL, Quinn T (eds). *Developments in Sedimentology, 54: Geology* *and Hydrology of Carbonate Islands.* Elsevier Science BV, Amsterdam, Netherlands.

Randall JE (1999). Report on fish collections from the Pitcairn Islands. *Atoll Res Bull* 461: 1-36.

Rougerie F, Fichez R, Déjardin P (1997). Geomorphology and hydrogeology of selected islands of French Polynesia: Tikehau (atoll) and Tahiti (barrier reef). In: Vacher HL, Quinn T (eds). *Developments in Sedimentology, 54: Geology and Hydrology of Carbonate Islands.* Elsevier Science BV, Amsterdam, Netherlands.

Salvat B, Hutchings P, Aubanel A, Tatarata M, Dauphin C (2000). The status of the coral reefs and marine resources of French Polynesia. In: Salvat B, Wilkinson C, South GR (eds). Proceedings of the International Coral Reef Initiative Regional Symposium, Noumea, 22-24 May 2000.

HAWAI'I AND THE US MINOR OUTLYING ISLANDS

Brainard R, Maragos J, DeMartini V, Wass R, Parrish F, Boland V, Newbold R (2000). A joint NOAA/USFWS coral reef assessment of the US Line and Phoenix Islands. In: Hopley D, Hopley PM, Tamelander J and Done T (eds). *Proc 9th Int Coral Reef Symp Abstracts*: 221.

DeMartini EE, Parrish FA, Parrish JD (1996). Interdecadal change in reef fish populations at French Frigate Shoals and Midway Atoll, Northwestern Hawaiian Islands: statistical power in retrospect. *Bull Mar Sci* 58(3): 804-825.

Friedlander AM, Parrish JD (1998). Habitat characteristics affecting fish assemblages on a Hawaiian coral reef. *J Exp Mar Biol Ecol* 224: 1-30.

Grigg RW (1988). Paleoceanography of coral reefs in the Hawaiian-Emperor Chain. *Science* 240: 1737-1743.

Jokiel PL, Coles SL (1990). Response of Hawaiian and other Indo-Pacific reef corals to elevated temperature. *Coral Reefs* 8: 155-162.

Maragos JE (2000). Hawaiian Islands (USA). In: Sheppard C (ed). *Seas at the Millennium: An Environmental Evaluation.* Elsevier Science Ltd, Oxford, UK.

Witte Maheney C, Witte Maheney A (2000). *Diving and Snorkelling Hawaii.* Lonely Planet Publications, Melbourne, Australia.

Map sources

Map 14a

Coral reef data have been taken as arcs from Petroconsultants SA (1990)*.

Map 14b

For all countries, coral reef data have been taken as arcs from Petroconsultants SA (1990)*.

Map 14c

In Tonga, for the main islands of 'Eua and Tongatapu coastline, coral reefs and mangroves were taken from DOS (1971, 1975). Information on these maps is based on aerial photographs taken in 1968 and field-checked in 1972. For Niue, coastline and coral reefs were taken from DLS (1985), based on 1965 photography. Remaining coral reef areas of Tonga, plus Beveridge Reef have been taken from Petroconsultants SA (1990)*.

DLS (1985). *Map of Niue,* 1:50 000. Universal Transverse Mercator. Department of Lands and Survey, New Zealand.

DOS (1971). *Tongatapu Island, Kingdom of Tonga.* Series X773 (DOS 6005) Sheet TONGATAPU, Edition 1-DOS 1971 (reprinted 1976). Directorate of Overseas Surveys, UK and Ministry of Lands and Survey, Tonga.

DOS (1975). *Kingdom of Tonga: Tongatapu Group – 'Eua.* 1:25 000. Series X872 (DOS 337) Sheet 23, Edition 1. Directorate of Overseas Surveys, UK and Ministry of Lands and Survey, Tonga.

Maps 14d and 14e

Coral reefs and coastline were taken from high resolution maps which were available for a number of islands (DLS, 1980s). These were based on photogrammetry from 1973-75 RNZAF photography. For Rarotonga further high resolution data were taken from Utanga and Lewis (1981). All remaining islands are based on lower resolution data from the sources listed.

DLS (1980 series). *Aitutaki* (1983). *Manihiki* (1986). *Mitiaro* (1983). *Palmerston* (1984). *Pukapuka and Nassau* (1986). *Rakahanga* (1989). *Suwarrow* (1986). 1:25 000. Department of Lands and Survey, New Zealand.

Lewis KB, Rongo TT, Utanga AT (1982). *Penrhyn* (includes Flying Venus Reef). 1:200 000. New Zealand Oceanographic Institute Chart, Island Series.

Lewis KB, Gilmore IP, Utanga AT (1982). *Pukapuka and Nassau* (includes Tema Reef). 1:200 000. New Zealand Oceanographic Institute Chart, Island Series.

Summerhayes CP (1968). *Manuae* (includes Eclipse Seamount). 1:200 000. New Zealand Oceanographic Institute Chart, Island Series.

Summerhayes CP, Kibblewhite AC (1966). *Aitutaki* (includes Eclipse Seamount). 1:200 000. New Zealand Oceanographic Institute Chart, Island Series.

Summerhayes CP, Kibblewhite AC (1968a). *Atiu* (includes Mitiaro and Takutea). 1:200 000. New Zealand Oceanographic Institute Chart, Island Series.

Summerhayes CP, Kibblewhite AC (1968b). *Mangaia.* 1:200 000. New Zealand Oceanographic Institute Chart, Island Series.

Summerhayes CP, Kibblewhite AC (1969). *Mauke* (includes Mitiaro). 1:200 000. New Zealand Oceanographic Institute Chart, Island Series.

Tupa V, Eade JV (1987). *Suwarrow.* 1:200 000. New Zealand Oceanographic Institute Chart, Island Series.

Utanga AT, Lewis KB (1981). *Rarotonga Nearshore Bathymetry,* 1:20 000. New Zealand Oceanographic Institute Chart, Misc. Series No. 56 (being also CCOP/SOPAC Misc. Series Chart. 1:20 000). Published by New Zealand Oceanographic Institute and Committee for Co-ordination of Joint Prospecting for Mineral Resources in South Pacific Offshore Areas.

Maps 14f, 14g and 14h

For most of this region, coral reef data have been taken as arcs from Petroconsultants SA (1990)*. For the Society Islands, reefs and coastline have been prepared from IGN (1988).

IGN (1988). *Archipel de la Société.* Map no. 513, 1:100 000, Edition 4. Institut Géographique National, Paris, France.

Maps 14i and 14j

In Hawai'i, coastline and coral reef data for Niihau, Kaua'i, Oahu, Molokai, Lanai, Maui, Kahoolawe and Hawai'i are taken from USFWS (1978). These maps were subsequently digitized by the USFWS. For the remaining islands, coral reef data have been taken as arcs from Petroconsultants SA (1990)*.

In the US outlying islands, coral reefs and coastline have

been taken from NOAA (1986, 1990, 1991) for Jarvis, Baker, Howland, Palmyra and Wake islands. For Johnson Atoll and Kingman Reef, coral reef data have been taken as arcs from Petroconsultants SA (1990).

NOAA (1986). Islands in the Pacific Ocean – Jarvis, Baker and Howland Islands. *NOAA (NOS) Chart 83116*. 1:15 000. 1978, revised 1986. Silver Spring, USA.

NOAA (1990). Wake Island. *NOAA (NOS) Chart 81664*. 1:15 000. Silver Spring, USA.

NOAA (1991). Palmyra Atoll – Approaches to Palmyra Atoll. *NOAA (NOS) Chart 83157*. 1:47 750. Silver Spring, USA.

USFWS (1978). *National Wetlands Inventory Maps*. 1:24 000. United States Fish and Wildlife Service, St. Petersburg, USA.

* See Technical notes, page 401

Technical notes

PART I

The initial sections of this book present a global overview of the world of coral reefs from a biological and a human perspective. They provide a holistic overview of reef ecology, human uses and threats, and also the techniques of reef mapping.

PARTS II-IV

For the purposes of this work, a number of geographic sub-divisions have been used. Firstly the world is divided into three broad realms: the Atlantic and Eastern Pacific; the wider Indian Ocean and Southeast Asia; and the Pacific (Parts II, III and IV respectively). These are then divided into regional chapters, each of which is sub-divided into sections, linked to particular maps. These sections do not strictly follow political boundaries – many do deal with individual countries or territories, but others contain parts of countries, or several countries combined.

Each section covers a range of issues, commencing with information about the physical geography of each region or country under consideration, followed by information on the reefs, both in terms of structure and biodiversity. Even where little information is available, every effort has been made to mention all the major reef features in each country, including remote and largely unknown formations. Information about the human utilization of the reefs follows, with information on their protection where applicable. These texts are followed by detailed data tables and references (see below).

Where possible, statistics describing biodiversity, fisheries and other information have been placed in the text.These are intended to complement the figures provided in the data tables. They are taken from various sources and hence are of limited value for country comparisons. For example, in providing biodiversity statistics, many sources supply information on coral diversity, but without a full definition. Rather than ignoring such statistics, the figure is reproduced using the wording of the original source ("corals", "hermatypic corals", "zooxanthellate corals", "hard corals", "scleractinian corals"). Similar problems arise in numbers of fish ("reef fish" or all fish species), and in fisheries statistics, which often do not separate coastal from pelagic fisheries.

THE MAPS

Two main styles of map are provided. At the start of each chapter are low resolution regional maps which provide a context for the more detailed maps that follow. The regional maps show coral reefs against a background of shaded bathymetry and shaded relief on the land areas, which have been generated using advanced GIS techniques on digital elevation model (DEM) data (from CRSSA, 1996 and USGS, 1996). They have been simply annotated to show the location of individual countries and a small number of major oceanographic features if these are not shown on the high resolution maps.

The high resolution maps are linked to the sections within each chapter. All have a common key, as provided on page 12. Apart from coral reefs, all maps also show major towns, rivers and the distribution of forests. Offshore, simple bathymetry is also provided, along with the distribution of mangrove forests along the coastlines. Every effort has been made to ensure that locations mentioned in the text are included on the maps. The source information for coral reef data on each map is provided at the end of each chapter. It is highly important to read this information in order to assess the accuracy, resolution and age of the information. Two standard sources have been used in a number of countries which are not detailed in every chapter, but are provided below: Petroconsultants SA (1990) and UNEP/IUCN (1988a, b).

Inland forest cover information is based on UNEP-WCMC data holdings which, like the coral reef cover, are derived from multiple sources. While this forest layer is probably the most detailed and accurate global dataset available it is not complete for all geographic areas. The definition of forest includes areas where tree canopy is greater than 10 percent. Where forest cover is absent for a particular country or island this may be due to a lack of data rather than an absence of forest cover.

Mangrove forest information is part of the UNEP-WCMC global forest map (above). This layer was largely developed for the *World Mangrove Atlas* (Spalding et al, 1997). Although this remains probably the most accurate global coverage available, it may be incomplete or outdated for a number of countries.

A major effort has been made to establish the location of dive centers. For the purposes of this work, these have been defined as any center that provides certified training. In this atlas a new dataset has been prepared of over 2 000 dive centers in coral reef regions.

Protected area information is taken directly from the UNEP-WCMC protected areas database. Within this database all marine protected areas are highlighted and all these sites, where their location is known, are marked on the maps. The type or designation of protection is given as an abbreviation, while for those protected areas containing coral reefs, the designations are provided in full in the data tables. A marine protected area in this context is defined by the IUCN World Commission on Protected Areas and includes all legally gazetted sites which incorporate at least some intertidal or subtidal areas. It does not therefore include private protected areas which do not enjoy any legal status, nor does it include sites which are proposed, but are not yet legally established.

CRSSA (1996). GlobalARC GIS Database '96. Center for Remote Sensing and Spatial Analysis, Cook College, Rutgers University, New Brunswick, NJ, USA.

Petroconsultants SA (1990). MUNDOCART/CD. Version 2.0. 1:1 000 000 world map prepared from the Operational Navigational Charts of the United States Defense Mapping Agency. Petroconsultants (CES) Ltd, London, UK.

Spalding MD, Blasco F, Field CD (1997). *World Mangrove Atlas.* The International Society for Mangrove Ecosystems, Okinawa, Japan.

UNEP/IUCN (1988a). *Coral Reefs of the World. Volume 1: Atlantic and Eastern Pacific.* UNEP Regional Seas Directories and Bibliographies. UNEP and IUCN, Nairobi, Kenya, Gland, Switzerland and Cambridge, UK.

UNEP/IUCN (1988b). *Coral Reefs of the World. Volume 2: Indian Ocean.* UNEP Regional Seas Directories and Bibliographies. UNEP and IUCN, Nairobi, Kenya, Gland, Switzerland and Cambridge, UK.

USGS (1996). GTOPO30. US Geological Survey EROS Data Center, Sioux Falls, SD, USA.

GENERAL DATA

Country areas: areas are from the World Resources Institute 1996-97 dataset.

Population: Population figures are estimates for the year 2000, taken from the US Census Bureau. These figures are estimated from the latest available national statistics.

GDP: GDP figures are taken from United Nations statistics and are based on 1996 data.

Marine area: Areas are draft estimates of the area of ocean up to the 200 nautical mile limit or equivalent EEZ boundary. They are provided to give an approximation of the marine waters, and the potential area of influence, of particular countries. In most cases these areas have been obtained by adding the EEZ area to the territorial sea area (both figures being obtained from WRI, 2000). It should be noted that, in a few cases this leads to overestimation of the total marine areas, as some countries claim a larger territorial sea than the standard 12 nautical miles. These figures have no political basis and do not imply any sovereignty. WRI does not provide data for all nations. For the remainder, figures have been obtained from a variety of sources, notably the South Pacific Island Web Atlas developed at the University of the South Pacific in Fiji (www.usp.ac.fj/~gisunit/pacatlas/atlas.htm). Figures are rounded to the nearest 1 000 square kilometers (or 100 square kilometers for countries with areas of less than 5 000 square kilometers).

Per capita fish consumption: Data are largely taken from FAOSTAT, the database operated by the United Nations Food and Agriculture Organization (http://apps.fao.org/). Some, particularly for the Pacific islands, are largely based on Gillett (1997) and are estimated averages for the 1990s. Others come from the World Resources Institute (WRI, 2000). All are based on all fish and seafood.

STATUS AND THREATS

Reefs at risk: Using the original threat coverage generated at the World Resources Institute (Bryant et al, 1998) new figures were generated, using the improved global reef map presented in this volume. The single figures presented are the percentage of each country's reefs which fall into either the medium or high level of threat. These threats, as explained in Chapter 2, were based on fishing, coastal pollution, marine pollution and sedimentation. It is important to note that these measures were produced at coarse resolutions, and are broadly indicative rather than highly accurate. They are also a measure of potential threat rather than actual reef state. In a number of countries threatened reefs remain in good condition. Threats can also be diminished or removed by active management interventions, which were not included in the model.

Coral diseases: Data are likely to be conservative. For each country the total number of recorded types of coral disease is listed. These figures are derived from a UNEP-WCMC database (www.unep-wcmc.org/marine/coraldis/index.htm), which has been developed from more than 150 published or authoritative sources. These cover some 29 different diseases, although there are likely to be problems over identification, and the true identification of some of these remains controversial. In many cases the total number of recorded diseases is little more than a measure of research effort, however general patterns can be drawn by looking at figures from a number of countries across a region or regions.

BIODIVERSITY

Please note that, where appropriate or available, biodiversity figures taken from national sources are provided in the text (see note above). These statistics are meant to complement the statistics found in the tables as described below. In these tables, the stastistics are based on standardized procedures and definitions, and hence are intended to be directly comparable between countries, although in some cases they are unlikely to include the latest information from all sources.

Reef area: Reef areas have been calculated from the UNEP-WCMC maps. In order to avoid the problems associated with scale and resolution, data were gridded to a 1 square kilometer grid prior to calculation (see Spalding and Grenfell, 1997). Although true error terms cannot be calculated caution should be applied in the use of these statistics for detailed analysis. Figures have been rounded to the nearest 100 square kilometers, while for those countries with small areas of coral reefs the terms <100, <50 and <10 km^2 have been used in order to provide an approximate estimation.

Coral diversity: Two global sources were available for the calculation of coral diversity, and their presentation together gives some idea of the probable range of estimates available for any country.

First figure: UNEP-WCMC maintains, on behalf of the CITES Secretariat, a database of all scleractinian corals. This dataset is based on around 1 000 published sources which provide known records from particular countries. All zooxanthellate Scleractinia (as defined by Veron, 2000) within this database were marked, and used in the generation of a list of recorded species by country. Taxonomic unreliability and problems of synonymy exist with this dataset, however the simple species totals generated from such data are only partially influenced by these errors and are still broadly accurate. These figures remain unreliable for smaller countries where there are few readily available published species inventories, and they have been omitted where they are clearly very inadequate. For larger countries, they may be accurate measures of described species, but in many cases such figures are as much a measure of research effort as they are true measures of diversity.

Second figure: These represent expected numbers of species by country and are based on the electronic (GIS) database used to generate the species distribution maps in Veron (2000). This database recognizes biogeographic regions rather than political boundaries. Note that, for localities where the geographic area is small (like Hawai'i and Singapore), species numbers are from original records (Veron, pers. com.) rather than from the database, and that for countries where a political boundary crosses a single geographic region (like Mozambique and Tanzania), only a combined total number of species is given. These figures are largely based on interpolations from distribution ranges, and as such they incorporate species which are predicted, but may not occur, in some countries. These are thus maximum figures, and may be exaggerated in a number of countries.

Mangrove area: Data are largely derived from the *World Mangrove Atlas* (Spalding et al, 1997).

No. of mangrove species: WCMC data, largely taken from the *World Mangrove Atlas*, with updates.

No. of seagrass species: These figures are likely to be conservative. They have been derived from a database under development at UNEP-WCMC. Records of species distribution have been incorporated into this database from over 60 sources, primarily published books and papers. In a number of cases, reviewers have pointed out that true totals are likely to be higher. In cases where a clearly documented figure was provided, the data were amended accordingly.

Bryant D, Burke L, McManus J, Spalding M (1998). *Reefs at Risk: A Map-Based Indicator of Threats to the World's Coral Reefs.* World Resources Institute, International Center for Living Aquatic Resources Management, World Conservation Monitoring Centre and United Nations Environment Programme, Washington DC, USA.
Gillett (1997). *The Importance of Tuna to Pacific Island Countries.* Forum Fisheries Agency Report 97/15. Honiara, Solomon Islands.
Spalding MD, Grenfell AM (1997). New estimates of global and regional coral reef areas. *Coral Reefs* 16: 225-230.
Spalding MD, Blasco F, Field CD (1997). *World Mangrove Atlas.* The International Society for Mangrove Ecosystems, Okinawa, Japan.
Veron JEN (2000). *Corals of the World.* 3 vols. Australian Institute of Marine Science, Townsville, Australia.
WRI (2000). *World Resources 2000-2001: People and Ecosystems: The Fraying Web of Life.* World Resources Institute, Washington DC, USA.

PROTECTED AREAS IN THE DATA TABLES

Within the UNEP-WCMC protected areas database, sites known to include coral reefs are further annotated and it is these sites only which have been incorporated into the data tables. Alongside the site name and designation a designation abbreviation is provided (this abbreviation is used on the maps). The IUCN management category is also provided, giving an indication of the legal regime protecting the site. This does not always equate with management effectiveness.

The following provides a short summary of the IUCN management categories, while more detailed information can be found at:
http://www.unep-wcmc.org/protected_areas/categories/index.html:

Ia: Strict Nature Reserve: protected area managed mainly for science
Ib: Wilderness Area: protected area managed mainly for wilderness protection
II: National Park: protected area managed mainly for ecosystem protection and recreation
III: Natural Monument: protected area managed mainly for conservation of specific natural features
IV: Habitat/Species Management Area: protected area managed mainly for conservation through management intervention
V: Protected Landscape/Seascape: protected area managed mainly for landscape/seascape conservation and for recreation
VI: Managed Resource Protected Area: protected area managed mainly for the sustainable use of natural ecosystems.

BIBLIOGRAPHY

Every chapter concludes with a bibliography which includes many of the sources used in the compilation of the texts, as well as additional reference materials which may be seen as further reading. In addition to these sources the authors relied heavily on gray literature and the Web for recent material.

Every effort has been made to ensure accuracy throughout this book, however in any compendium of this size there are likely to be oversights. The authors would welcome feedback (e-mail: information@unep-wcmc.org), and replacement data will be posted on our web-page as it comes to light (www.unep-wcmc.org/marine/coralatlas/index.html).

Index

A

B

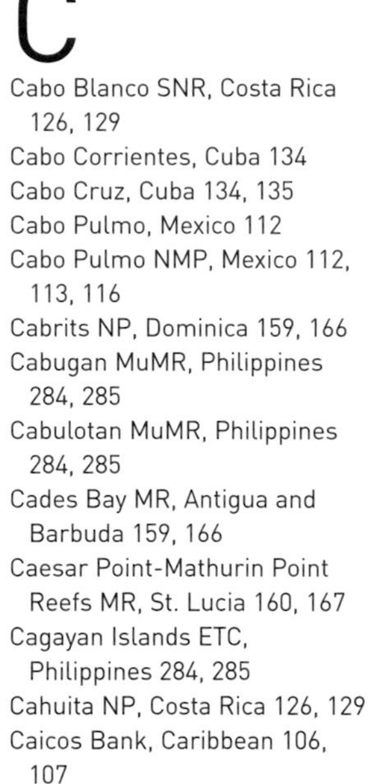

D

G

H

K

L

M

N

O

R

S

T